FLUORESCENCE LIFETIME SPECTROSCOPY AND IMAGING

FLUORESCENCE LIFETIME SPECTROSCOPY AND IMAGING

Principles and Applications in Biomedical Diagnostics

Edited by
Laura Marcu
Paul M. W. French
Daniel S. Elson

CRC Press
Taylor & Francis Group
Boca Raton London New York

CRC Press is an imprint of the
Taylor & Francis Group, an **informa** business

Cover Image. Reproduced from (Patalay 2012) PLoS ONE 7(9): e43460. doi:10.1371/journal.pone.0043460. See Chapter 16 - Oncology Applications: Skin Cancer by Rakesh Patalay, Paul French and Christopher Dunsby.

CRC Press
Taylor & Francis Group
6000 Broken Sound Parkway NW, Suite 300
Boca Raton, FL 33487-2742

First issued in paperback 2020

ISBN 13: 978-0-367-57610-3 (pbk)
ISBN 13: 978-1-4398-6167-7 (hbk)

Library of Congress Cataloging-in-Publication Data

Fluorescence lifetime spectroscopy and imaging : principles and applications in biomedical diagnostics /
   editors, Laura Marcu, Paul M.W. French, Daniel S. Elson.
      p. ; cm.
   Includes bibliographical references and index.
   ISBN 978-1-4398-6167-7 (hardcover : alk. paper)
      I. Marcu, Laura, editor of compilation. II. French, Paul M. W., editor of compilation. III. Elson, Daniel S.,
   editor of compilation.
      [DNLM: 1. Optical Imaging. 2. Spectrometry, Fluorescence. 3. Fluorescence. 4. Microscopy,
   Fluorescence. 5. Molecular Imaging. 6. Time Factors. WN 195]

   RC78.7.F5
   616.07'54--dc23   2014002766

Visit the Taylor & Francis Web site at
http://www.taylorandfrancis.com

and the CRC Press Web site at
http://www.crcpress.com

Dedicated to Professor Robert M. Clegg

Contents

Preface

Fluorescence provides a powerful means of achieving molecular contrast for a wide range of biological and medical applications. While fluorescence imaging of appropriately labeled molecules is well established for *ex vivo* samples, for example, for immunohistochemistry, and for imaging biological processes *in vivo*, particularly using genetically expressed fluorescent protein labels in live cell cultures or in disease models, label-free fluorescence imaging is less well established. However, autofluorescence from naturally occurring (endogenous) fluorophores in biological tissues can provide label-free contrast between different states of tissue for medical research and clinical applications. Targeted (exogenous) probes can also be applied that localize in specific cellular locations and provide extrinsic image contrast or that bind to specific receptors on the cell membrane. Quantitative fluorescence imaging in tissue is challenging, however, because the strong heterogeneity of structure and composition of biological tissue leads to significant optical scattering of light at visible wavelengths, and practical measurements are confounded by limited knowledge of its interaction with the imaging radiation. This book particularly concerns the application of time-resolved (lifetime) measurements to improve the quantitative molecular assessment of biological tissues using fluorescence, which can be usefully applied to both endogenous and exogenous fluorophores in biological tissue.

Over the past two decades, there has been an increasing appreciation of the significant value that lifetime-based techniques can add to biomedical studies and applications of fluorescence. This has been reflected in an increasing volume of journal articles, conference presentations, and research funding. Fluorescence lifetime measurements provide a means to contrast different fluorophores or states of fluorophores in tissues and thus can improve the specificity of fluorescence measurements. Because fluorescence lifetime measurements are inherently ratiometric (relying only on the assumption that the sample does not change during the fluorescence decay profile, which is typically on a nanosecond or picosecond timescale), they are independent of fluorophore concentration, excitation or detection efficiency, or attenuation of the fluorescence signal by the instrumentation or the sample. This means that fluorescence lifetime measurements are robust in heterogeneous samples such as biological tissue and can provide more robustly quantitative information than can be obtained from intensity measurements or even spectral ratiometric measurements (which depend on the specific properties of the instrument and the sample). These advantages have long been recognized, but the complexity and limited availability of the instrumentation to measure or image fluorescence signals on sub-nanosecond timescales—particularly for fluorescence lifetime imaging (FLIM)—formerly presented significant barriers to wide deployment. These issues have now been widely addressed as a result of tremendous progress in optical and electronics technologies, including ultrafast light sources and detectors, high-speed signal processing electronics, and the broad spectral coverage of solid-state and semiconductor lasers. These advances in hardware have been complemented by new approaches and tools for analysis of fluorescence lifetime data that now permit distributions of complex decay profiles to be imaged with modest numbers of detected photons compatible with live cell studies and *in vivo* applications.

This book aims to capture the current perspectives on tissue fluorescence lifetime metrology and imaging and their applications to biomedical investigations of intact tissues and medical diagnosis within the broader context of fluorescence spectroscopy and the important clinical applications. It is intended to be broadly accessible, introducing the fundamentals of fluorescence spectroscopy and imaging and describing the state-of-the-art techniques and their current applications, with a view to supporting the wider deployment of fluorescence lifetime technology in tissue diagnostics for both research and clinical application purposes. Although it has been particularly targeted at senior undergraduate and graduate students, we hope it will be useful to established researchers in biomedical optics, molecular imaging, and biomedical engineering and that it will be of interest to industry professionals in bioscience and medicine.

The book is organized in five major parts. Part I (Chapters 1–3) aims to provide an overview of the development of the field, the basic photophysics of fluorescence, and the origin of autofluorescence contrast in biological tissues. Part II (Chapters 4–9) reviews the instrumentation used for fluorescence lifetime measurements of biological tissues, including single-channel (point) spectroscopy utilizing ultrafast sampling (Chapter 4) and time-correlated single-photon counting (TCSPC) techniques (Chapter 5), FLIM microscopy including single-photon and multiphoton excitation (Chapter 6), and frequency-domain FLIM (Chapter 7) and time-domain FLIM based on time-gated imaging (Chapter 8) or multicolor TCSPC (Chapter 9). Part III (Chapters 10–13) addresses the methods used for the analysis of fluorescence lifetime data, including the measurement of complex fluorescence decay profiles. This begins with the graphical phasor approach that particularly suits frequency-domain or periodically sampled time domain data (Chapter 10) and then addresses time-domain analysis (Chapter 11) and global fitting of frequency-domain data (Chapter 12), followed by the determination of fluorescence lifetimes in turbid media (Chapter 13). Part IV (Chapters 14–21) describes the applications of label-free autofluorescence lifetime spectroscopy and imaging to a broad range of diseases and conditions in biological tissues. Chapters 14–18 focus on oncological applications, beginning with an overview of optical diagnosis of cancer (Chapter 14), followed by specific applications to diagnosis of brain cancer (Chapter 15), skin cancer and dermatologic diseases (Chapter 16), gastrointestinal (GI) cancer (Chapter 17), and head and neck cancer (Chapter 18). Part IV further reviews the application of autofluorescence lifetime spectroscopy and FLIM to atherosclerotic cardiovascular disease (Chapter 19), ophthalmologic diseases (Chapter 20), and engineered tissue (Chapter 21). Lastly, Part V (Chapters 22–24) covers the application of fluorescence lifetime techniques to diagnosis and monitoring of therapies using exogenous fluorescent molecular probes with discussions of tomographic FLIM of small animals (Chapter 22), FLIM of photosensitizers in tissue to diagnose disease (Chapter 23), and chemical sensing using FLIM of probes reporting specific ions in tissue (Chapter 24).

During the writing of this book, the fluorescence lifetime community lost one of its key pioneers, and we would like to dedicate this book to the memory of Robert M. Clegg, whose groundbreaking work on frequency-domain FLIM technology, analysis, and applications inspired us and many others in the field. His work continues to help experts and novices through his seminal papers, and we are grateful that he contributed Chapter 7 to this volume.

Laura Marcu
Paul M. W. French
Daniel S. Elson

Editors

Prof. Laura Marcu received her diploma in mechanical engineering—fine mechanics from the Polytechnic Institute of Bucharest in 1984, and her MS and PhD degrees (for work on time-resolved fluorescence of atherosclerotic lesions) in biomedical engineering from the University of Southern California in 1995 and 1998, respectively. From 2001 to 2005, she served as the director of Biophotonics Research and Technology Development–Department of Surgery at the Cedars-Sinai Medical Center in Los Angeles. While at Cedars-Sinai, she held a joint appointment as a research associate professor of Biomedical Engineering and Electrical Engineering—Electrophysics with the Viterbi School of Engineering, University of Southern California. She joined the University of California, Davis, in 2006, where she is currently a professor of biomedical engineering and neurological surgery. Her research group works on the development of fluorescence lifetime spectroscopy and imaging techniques and integration of these techniques in multimodal tissue diagnostic platforms (with ultrasound backscatter microscopy and photoacoustics). Applications of these systems include clinical studies for the characterization and diagnosis of athererosclerotic plaques at high risk, intraoperative delineation of brain tumors, intraoperative diagnosis of head and neck tumors, and evaluation of bioengineered tissue. She is a fellow of SPIE, OSA, BMES, and AIMBE.

Prof. Paul M. W. French received his BSc degree in physics in 1983 and PhD degree (for work on femtosecond dye lasers) in 1987 from Imperial College London. In 1988, he was a visiting professor at the University of New Mexico working on femtosecond dye lasers, and in 1989, he was awarded a Royal Society University Research Fellowship at Imperial, where he joined the academic staff in 1994. From 1990 to 1991, he worked on ultrafast all-optical switching in optical fibers at AT&T Bell Laboratories, Holmdel, NJ. He is currently a professor of physics at Imperial College London and has served as head of the Photonics Group. His research has evolved from ultrafast dye and solid-state laser physics to biomedical optics. Today, his group develops and applies multidimensional fluorescence imaging technology for molecular cell biology, drug discovery and clinical diagnosis with a strong emphasis on fluorescence lifetime imaging (FLIM) implemented with microscopy, high content analysis, endoscopy, and tomography.

Dr. Daniel S. Elson was awarded an MPhys degree in physics in 1999 and a PhD in 2003 ("Ultrafast Lasers Applied to Fluorescence Lifetime Imaging") from Imperial College London. He is currently a reader in the Department of Surgery and Cancer and Hamlyn Centre for Robotic Surgery. His research interests are based around the development and application of photonics technology to medical imaging, multispectral imaging, endoscopy, and fluorescence lifetime imaging (FLIM). This has included developing imaging catheters for FLIM and a multispectral polarization-sensitive laparoscope. Recent projects have involved the development of illumination and vision systems for endoscopy combining miniature light sources such as light emitting diodes (LEDs) and laser diodes with computer vision techniques for image mosaicing, and stereo detection. These devices are finding application in minimally invasive procedures and in the development of new robotic assisted surgery systems and new surgical approaches.

Contributors

Simon R. Arridge
Department of Computer Science
University College London
London, United Kingdom

Wolfgang Becker
Becker & Hickl GmbH
Berlin, Germany

Bernard Y. Binder
Department of Biomedical Engineering
University of California
Davis, California

Christoph Biskup
Biomolecular Photonics Group
Jena University Hospital
Jena, Germany

Pramod V. Butte
Department of Neurosurgery
Cedars-Sinai Medical Center
Los Angeles, California

Heejin Choi
Department of Mechanical Engineering
Massachusetts Institute of Technology
Cambridge, Massachusetts

Dusan Chorvat
Department of Biophotonics
International Laser Centre
Bratislava, Slovak Republic

Alzbeta Chorvatova
Department of Biophotonics
International Laser Centre
Bratislava, Slovak Republic

Robert M. Clegg*
Center of Biophysics and Computational Biology
University of Illinois at Urbana–Champaign
Urbana, Illinois

Sergio Coda
Department of Medicine
and
Department of Physics
Imperial College London
and
Department of Gastroenterology and
    Endoscopy Unit
Imperial College Healthcare NHS Trust
London, United Kingdom

Teresa M. Correia
Department of Computer Science
University College London
London, United Kingdom

Rinaldo Cubeddu
Dipartimento di Fisica
Politecnico di Milano
and
Istituto di Fotonica e Nanotecnologie
Consiglio Nazionale delle Ricerche
Milan, Italy

Michelle A. Digman
Laboratory for Fluorescence Dynamics
Department of Biomedical Engineering
University of California
Irvine, California

Christopher Dunsby
Department of Physics
and
Department of Experimental Medicine
Imperial College London
London, United Kingdom

John Paul Eichorst
Center of Biophysics and Computational Biology
University of Illinois at Urbana–Champaign
Urbana, Illinois

Daniel S. Elson
Department of Surgery and Cancer and Hamlyn
    Centre for Robotic Surgery
Imperial College London
London, United Kingdom

* deceased

D. Gregory Farwell
Department of Otolaryngology—Head and
    Neck Surgery
University of California Davis
Sacramento, California

Paul M. W. French
Department of Physics
Imperial College London
London, United Kingdom

Thomas Gensch
ICS-4 (Cellular Biophysics)
Institute of Complex Systems
Forschungszentrum Jülich
Jülich, Germany

Enrico Gratton
Laboratory for Fluorescence Dynamics
Department of Biomedical Engineering
University of California
Irvine, California

Hernán E. Grecco
Laboratory of Quantum Electronics
University of Buenos Aires
Buenos Aires, Argentina

Anand T. N. Kumar
Athinoula A. Martinos Center for Biomedical
    Imaging
Massachusetts General Hospital
Harvard Medical School
Charlestown, Massachusetts

J. Kent Leach
Department of Biomedical Engineering
University of California
Davis, California

Jing Liu
Department of Biomedical Engineering
University of California
Davis, California

Adam N. Mamelak
Department of Neurosurgery
Cedars-Sinai Medical Center
Los Angeles, California

Laura Marcu
Department of Biomedical Engineering and
    Neurological Surgery
University of California
Davis, California

James McGinty
Department of Physics
Imperial College London
London, United Kingdom

Rakesh Patalay
Department of Medicine
and
Department of Physics
Imperial College London
and
Department of Gastroenterology and Endoscopy Unit
Imperial College Healthcare NHS Trust
London, United Kingdom

Jennifer E. Phipps
Department of Medicine
The University of Texas Health Science Center at
    San Antonio
San Antonio, Texas

Christopher J. Rowlands
Department of Biomedical Engineering
Massachusetts Institute of Technology
Cambridge, Massachusetts

Dietrich Schweitzer
Experimental Ophthalmology
Department of Ophthalmology
University of Jena
Jena, Germany

Vijay R. Singh
Singapore–MIT Alliance for Research and
    Technology Center
Massachusetts Institute of Technology
Cambridge, Massachusetts

Peter T. C. So
Department of Mechanical Engineering
Department of Biomedical Engineering
and
Singapore–MIT Alliance for Research and
    Technology Center
Massachusetts Institute of Technology
Cambridge, Massachusetts

Vadim Y. Soloviev
Department of Computer Science
University College London
London, United Kingdom

Klaus Suhling
Department of Physics
King's College London
London, United Kingdom

Yang Sun
Department of Biomedical Engineering
University of California
Davis, California

Paola Taroni
Dipartimento di Fisica
Politecnico di Milano
Milan, Italy

Kai wen Teng
Center of Biophysics and Computational Biology
University of Illinois at Urbana–Champaign
Urbana, Illinois

Gianluca Valentini
Dipartimento di Fisica
Politecnico di Milano
Milan, Italy

Peter J. Verveer
Max Planck Institute of Molecular Physiology
Dortmund, Germany

Diego R. Yankelevich
Department of Electrical and Computer Engineering
and
Department of Biomedical Engineering
University of California
Davis, California

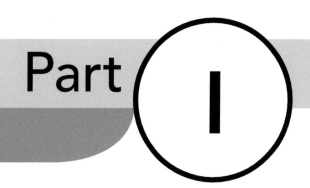

Part I

Overview of fluorescence measurements and concepts

1 Overview of fluorescence lifetime imaging and metrology

Daniel S. Elson, Laura Marcu, and Paul M. W. French

Contents

1.1 INTRODUCTION

This chapter aims to present an overview of fluorescence lifetime imaging (FLIM) and metrology in the context of their biomedical applications, introducing the main approaches that are discussed in detail in subsequent chapters of this book. Before discussing fluorescence lifetime measurements, however, it is important to understand the phenomenon of fluorescence, of which a brief discussion is provided here, and the reader is directed to the classic textbook by Lakowicz (1999) for further details.

Fluorescence is defined as the light emitted by an atom or molecule after a finite duration (typically $<10^{-8}$ s) following the absorption of photons (Lakowicz 1999). The emitted light results from the transition of the excited species from its first excited electronic singlet level to its ground electronic level. Fluorescence provides a powerful means of achieving optical molecular contrast for single-point (cuvette-based or fiber-optic probe-based) measurements in instruments such as fluorometers, cytometers, and cell sorters and for fluorescence imaging in microscopes, endoscopes, multiwell plate readers, and tomographic instruments. Fluorescence can provide information about the presence and distribution of specific molecules (fluorophores) and can also provide a *sensing* function since the fluorescence signal can be highly sensitive to the local environment of the molecules. For applications in cell biology, specialized fluorescent molecules are typically used as "labels" to tag specific molecules of interest. The advances in genetically expressed fluorophores such as green fluorescent protein (GFP) (Tsien 1998; Zimmer 2002) have enabled fundamental biological processes such as signaling to be directly visualized in live cells and organisms. For clinical applications, it is highly desirable to realize molecular contrast without using labels, for which it is possible to exploit the fluorescence properties of endogenous molecules to provide molecular contrast. Today, there is also increasing interest in utilizing readouts of endogenous fluorophores such as nicotinamide adenine dinucleotide (NADH) and flavin adenne dinucleotide (FAD) for cell biology applications and, conversely, significant activity developing and exploring exogenous labels for clinical applications to detect diseased tissue, for example, using photosensitizers for photodynamic or fluorescence detection (PDD/FD) (Ackroyd et al. 2001) or nanoparticles that bind to specific biological targets (Lapotko et al. 2006).

From a historical perspective, there are written reports on photoluminescence—and the phenomenon of fluorescence in particular—dating back to the 16th century, when Bernardino de Sahagún (1560) a missionary priest and ethnographer in colonial New Spain (now Mexico), and Nicolas Monardes (1565), a Spanish physician and botanist described, independently, the blue colored fluorescence of the water infusion of the wood from Mexico used to treat kidney and urinary diseases. Monardes's initial observations on this wood (Figure 1.1), subsequently named *lignum nephriticum* (kidney wood), inspired other early scientists to study its properties. Among those is the Flemish botanist Charles de L'Ecluse (1526–1609), the German Jesuit priest Athanasius Kircher (1601–1680), the Anglo-Irish scientist Robert Boyle (1627–1691), and the renowned Sir Issac Newton (1642–1727) (Valeur and Berberan-Santos 2011). Interestingly, the origin of the intense blue fluorescence of the infusion of *lignum nephriticum* was only recently identified (2009) as *matlaline* (after *matlali*—the Aztec word for blue) (Acuña et al. 2009).

More rigorous studies of the origin and physics of fluorescence and other photoluminescence phenomenon were reported much later in the 19th century and were stimulated by the discovery of new fluorescent materials (Berezin and Achilefu 2010; Valeur and Berberan-Santos 2011). At the forefront of this work were Edward D. Clarke (professor of mineralogy at the University of Cambridge) and Rene-Just Hauy (French mineralogist), who both reported (1819 and 1822, respectively) the optical properties of some crystals of fluorite. Moreover, the Scottish physicist, David Brewster, reported in 1833 the red fluorescence of chlorophyll, and the English polymath, John Herschel, described in 1845 the fluorescence of an acid solution of quinine (Valeur and Berberan-Santos 2011). Nevertheless, the most notable contribution to the earlier understanding of the physics behind photoluminescence was George Gabriel Stokes, a physicist and professor of mathematics at Cambridge (Figure 1.2). In his paper entitled "On the Refrangibility of Light," published in 1852 (Stokes 1852), Stokes presented detailed studies on samples of organic and inorganic fluorescent materials and noted that the fluorescence occurs at longer wavelengths than the excitation light, a phenomenon now known as "Stokes' shift." He is also credited with the introduction of the term *fluorescence* and with the transition from "observation" of fluorescence emission to actual "measurements" of such emission (Valeur and Berberan-Santos 2011).

The characteristics of fluorescence emission have been shown to give specific molecular information about a sample, and fluorescence measurements have become almost ubiquitous in many branches of science—particularly for biological and biochemical studies. Fluorescence can be analyzed with respect to the fluorescence intensity, excitation and emission spectra, quantum efficiency, polarization, and fluorescence lifetime. These parameters can depend on the properties of the fluorophore itself and/or on its local environment, including molecular interactions. A fluorophore's intrinsic optical properties include

Overview of fluorescence measurements and concepts

(a)  (b)

Figure 1.1 (a) Portrait of Spanish physician and botanist Nicolas Monardes (1493–1588). From the front page of the book *Dos libros, el uno que trata de todas las cosas que se traen de nuestras Indias Occidentales, que sirven al uso de la medicina, y el otro que trata de la piedra bezaar, y de la yerva escuerçonera.* Seville, Spain, 1569. (b) Absorption and fluorescence colors for infusions of *lignum nephriticum*. Cup of *lignum nephriticum* and flask containing its fluorescent solution. (From Valeur, B., and Berberan-Santos, M.N., *J. Chem. Educ.*, 88, 2011.)

(a)    (b)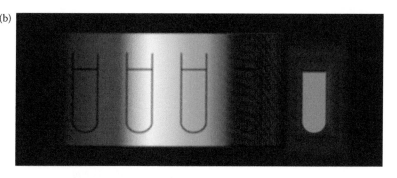

Figure 1.2 (a) Portrait of mathematician and physicist George Gabriel Stokes (1819 –1903). (b) The principle of Stokes's experiment showing that a solution of quinine irradiated with ultraviolet (UV) radiation emits blue light, but no effect is observed when it is placed in the visible part of the solar spectrum. (Reproduced from Valeur, B., and Berberan-Santos, M.N., *J. Chem. Educ.*, 88, 2011.)

its quantum efficiency, its temporal fluorescence decay profile, and its excitation and emission spectral profiles as well as its response to polarized light. These may be used to discriminate between different fluorophore species, or variations in their values may be quantified to sense changes in the local fluorophore environment modified through their molecular electronic configuration or de-excitation pathways.

Today, the most widespread application of fluorescence is probably microscopy of biological samples, where fluorescence imaging is predominantly used to obtain information about localization—with intensity images providing maps of fluorophore distribution and, therefore, of the distribution of labeled proteins. This can be extended by labeling different molecular species with fluorophores exhibiting distinct spectral properties, and so fluorescence imaging can also elucidate colocalization of proteins to within the spatial resolution of the imaging system. This can provide evidence for molecular interactions, although the diffraction-limited resolution of optical instruments normally limits this colocalization to >~200 nm. The recent development of super-resolved microscopy has pushed the achievable resolution to tens of nanometers, for example, in the work of Betzig et al. (2006), Gustafsson (2000), Klar et al. (2000), and Rust et al. (2006), although it is still practically challenging to achieve sub-50 nm resolution in live cells. Stronger evidence for interaction can be obtained using Förster resonant energy transfer (FRET) readouts that sense when two fluorescent labels with appropriate spectral properties are in very close proximity, that is, sufficiently close for energy to be transferred from one excited fluorophore (the "donor") to the other (the "acceptor") via a dipole–dipole interaction that only occurs with appreciable efficiency over distances of less than ~10 nm (Clegg et al. 2003; Förster 1948). FRET may be detected through changes in the "donor" or acceptor fluorescence and is an example of how a change in the local molecular environment can modify the emission characteristics of fluorophores. Other environmental changes that can be "sensed" by fluorophores include physical factors such as the local temperature, viscosity, refractive index or electric field, and chemical factors such as pH or the presence of species to which fluorophores may bind such as calcium, oxygen, and organic molecules, for example, molecules involved in cell signaling.

The quantum efficiency, η, of the fluorescence process is defined as the ratio of the number of fluorescent photons emitted compared to the number of excitation photons absorbed. Once a fluorophore has been optically excited to an upper-state energy level, it may decay back to the ground state either radiatively, with a rate constant, Γ, or nonradiatively, with a rate constant, k, as represented in Figure 1.3. The quantum efficiency is equal to the radiative decay rate, Γ, divided by the total (radiative + nonradiative) decay rate, that is, $\eta = \Gamma/\Gamma + k$, with Γ being sensitive to factors that change or modify the electronic properties of the fluorophore, for example, chemical binding, while k is sensitive to other changes in the molecular environment such as temperature or viscosity. Quantifying changes in the fluorophore environment through determination of the quantum efficiency, however, requires knowledge of the photon excitation and detection efficiencies, as well as the fluorophore concentration. Furthermore, quantitative measurements of intensity can be compromised by optical scattering, internal reabsorption of fluorescence (inner filter effect), and background fluorescence from other fluorophores present in a sample. Quantitative imaging and metrology based on quantum efficiency is therefore challenging, particularly for imaging

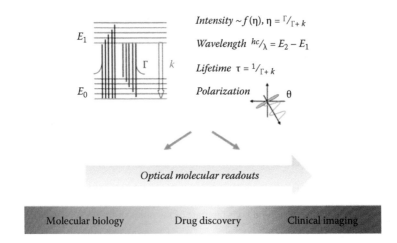

$Intensity \sim f(\eta), \eta = {}^{\Gamma}/_{\Gamma+k}$

$Wavelength \ {}^{hc}/_{\lambda} = E_2 - E_1$

$Lifetime \ \tau = {}^{1}/_{\Gamma+k}$

$Polarization$

Optical molecular readouts

Molecular biology   Drug discovery   Clinical imaging

Figure 1.3 Overview of multidimensional fluorescence imaging and metrology.

biological tissue. More robust measurements can be made using spectroscopic *ratiometric* techniques where spectroscopic parameters are sensitive to variations in the local fluorophore environment. For example, in the spectral domain, one can assume unknown quantities such as excitation and detection efficiency, fluorophore concentration, and signal attenuation to be approximately the same in two or more carefully chosen spectral windows such that they may be effectively "cancelled out" in a ratiometric measurement. This approach is used, for example, with excitation and emission (Grynkiewicz et al. 1985) ratiometric calcium-sensing dyes, and the approach has been demonstrated to provide useful contrast between, for example, malignant and normal tissue (Andersson-Engels et al. 1990a).

Fluorescence lifetime measurement is also a ratiometric technique in that it compares the fluorescence signal at different delays after excitation under the assumption that that the various unknown quantities do not change significantly during the fluorescence decay time (typically nanoseconds). The fluorescence lifetime refers to the average time the atom or molecule remains in its excited state before returning to its ground state and, like the quantum efficiency, is also a function of the radiative and nonradiative decay rates, that is, $\tau = 1/\Gamma + k$, and so can provide quantitative readouts of factors that impact the fluorescence process. A further ratiometric fluorescence parameter is polarization, which may be probed using polarized excitation sources and detectors to provide information concerning molecular orientation. Further, time-resolved polarization measurements can determine the rotational correlation time—or molecular tumbling time—which can be used to report ligand binding or local solvent properties.

The concept of luminescence lifetime was established with the first measurement, performed by a contemporary of Stokes, the French physicist Eduard Becquerel (Figure 1.4), who measured the decay times (lifetimes) of various uranyl salts and reported this in *La Lumiere—Ses Causes et ses Effets,* published in 1858 (Becquerel 1867). He is credited with building the first instrument (i.e., phosphoroscope) capable of measuring time intervals as short as 10^{-4} s and pioneering the work in time-resolved photoluminescence (Berezin and Achilefu 2010; Valeur and Berberan-Santos 2011). The instrument used the sun as a light source and consisted of two rotating disks with four windows positioned in such a way that they did not line up. The sample was placed between the two disks (Figure 1.4). Interestingly, Becquerel used a sum of exponential decays as a function of time to describe the phosphorescence temporal profile. While revolutionary and an excellent instrument capable of measuring long-lived phosphorescence, the pulsed light generated by the rotating disks in the phosphoroscope were not short enough to measure the much faster fluorescence decay. A faster pulsed light source was needed.

A solution came from the study of Kerr cells (1875) by the Scottish physicist John Kerr (1875). By using a Kerr cell with synchronized electrodes and a light source, Abraham and Lemoine (1899) demonstrated a new principle of measuring ultrafast intervals of time nearly 100,000 times faster when compared to phosphoroscope (Abraham and Lemoine 1900; Lemoine 1899). Their work not only established the foundation for studying short decays but also introduced the concepts of light modulation and phase

Figure 1.4 Portrait of French physicist Alexandre Edmond Becquerel (1820–1891) and the phosphoroscope apparatus invented in 1859 and described in *La Lumiere—Ses Causes et ses Effets*, Paris, 1868 (Becquerel 1867). This instrument was able to resolve time intervals as short as 8×10^{-4} s.

approaches used in current fluorescence lifetime instruments. The use of Kerr cells for fluorescence lifetime measurements, however, was implemented only later on by Robert W. Wood of John Hopkins University. In his paper entitled "The Time Interval Between Absorption and Emission of Light in Fluorescence," published in 1921 (Wood 1921), Wood demonstrated the ability to measure time intervals as short as 10^{-7} ns by using the method of Abraham and Lemoine. Shortly after, Philip F. Gottling (1923) and Jesse W. Beams (1926) expanded on Wood's instrument to measure time intervals on the order of 10^{-8} s.

The first nanosecond lifetime measurements were performed in the 1920s by the Argentinian physicist Ramon Enrique Gaviola, who established the theoretical and experimental basis for modern fluorescence lifetime spectroscopy, which is relevant for nowadays' biochemistry and biology (Figure 1.5). Gaviola had bridged between principles of Becquerel's phosphoroscope and the Abraham-Lemoire experiments and built a new instrument that allowed for measurements of nanosecond lifetimes. He called this instrument a "fluorometer." This new design was reported in 1926 in his seminal paper "Die Abklingungszeiten der Fluoreszenz von Farbstofflösungen" (Gaviola 1926). Gaviola was also the first to resolve and report the nanosecond lifetimes of fluorescein and Rhodamine B and to demonstrate that lifetimes depend on molecular species and that the lifetimes depend on the environment. The mathematical theory of the phase fluorometer, however, was only later reported by Von F. Duschinsky (1933). The first implementation of a pulsed method to measure fluorescence lifetimes in the time domain is attributed to Seymour Steven Brody (1957), who reported an instrument that utilized a hydrogen flash lamp and a high-speed oscilloscope to measure the fluorescence lifetimes in the order of microseconds.

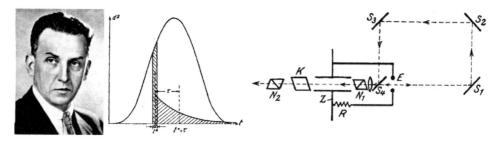

Figure 1.5 Portrait of Argentinian physicist Ramón Enrique Gaviola (1900–1989). Decay depicted as an exponential following excitation. From Gaviola's seminal paper "Die Abklingungszeiten der Fluoreszenz von Farbstofflösungen" published in 1926 (Gaviola 1926). This instrument is based on the Kerr cell modulation method reported by (Abraham and Lemoine 1899) (i.e., Figure 1.1, schematic at right) to generate the light pulse used for sample excitation.

Many subsequent refinements of instrumentation for time-resolved fluorescence spectroscopy have been developed, some of which are reviewed in the work of Lakowicz (1999), O'Connor and Phillips (1984), and Ross and Jameson (2008). These fluorometers can also include fluorescence excitation-, emission-, and polarization-based measurement capabilities, and such instrumentation has been widely applied to cuvette-based photochemical and photobiological studies, as exemplified by the work of Gregorio Weber and the many researchers who trained in his laboratory (Jameson 2001). Here, we are particularly interested in the application of time-resolved fluorometry of biological tissue, as discussed in Chapters 4 and 5.

While time-resolved and multidimensional fluorometry has been well established for several decades, it has been less common to exploit multidimensional spectroscopic information available from fluorescence in imaging applications. This was partly due to instrumentation limitations and partly due to the challenges associated with characterizing fluorescence signals available from typically heterogeneous and often sparsely labeled biological samples. In recent years, however, there have been tremendous changes in imaging tools and technology that are available to biologists and medical scientists. Key drivers have been the widespread adoption of confocal microscopy and digital image processing as standard tools together with dramatic advances in laser and photonics technology, including robust tunable and ultrafast excitation sources, relatively low-cost high-speed detection electronics, and high-performance imaging detectors. Technological advances have facilitated the development of powerful new techniques in biophotonics, such as multiphoton microscopy (Denk et al. 1990), which became widely deployed following its implementation with the conveniently tunable femtosecond Ti:sapphire lasers (Denk and Svoboda 1997). Such instruments confer the ability to excite most of the commonly used fluorophores with a single excitation laser—with computer-controlled operation to permit automated excitation spectroscopy (Dickinson et al. 2003). The proliferation of multiphoton microscopes, in turn, stimulated the uptake of FLIM (Cubeddu et al. 2002) as a relatively straightforward and inexpensive "add-on" that provided significant new spectroscopic functionality—taking advantage of the ultrafast excitation lasers and requiring only appropriate detectors and external electronic components to implement FLIM.

In parallel, the development and availability of genetically expressed fluorophores continues to provide new opportunities to study biological processes in live cells and organisms with highly specific labeling. For cell biology, there has been a tremendous uptake in the application of FRET, for example, to map protein interactions, such as receptor phosphorylation, ligand binding, protein translocation and aggregation, and sensing of signaling molecules. In some situations, it is possible to read out the distance between "FRETing" fluorophores—effectively implementing a so-called "spectroscopic ruler" (Dosremedios and Moens 1995; Stryer 1978) and obtaining information on molecular structure, for example, in the work of Chatterjee et al. (2012), and dynamics, for example, in the work of Kalinin et al. (2010). FRET is thus a powerful tool to study molecular cell biology, particularly when combined with genetically expressed fluorophores. However, quantifying the resonant energy transfer is often not straightforward when implementing FRET using intensity-based imaging—particularly in biological tissue—and increasingly fluorescence lifetime readouts and other techniques are being applied to FRET experiments; see, for example, the work of Bastiaens and Squire (1999), Jares-Erijman and Jovin (2003), Suhling et al. (2005).

For medical applications, the prospect of using autofluorescence to provide label-free readouts of molecular changes associated with the early manifestations of diseases such as cancer is particularly exciting. For example, accurate and early detection of cancer allows earlier treatment and significantly improves prognosis (American Cancer Society 2006). While exophytic tumors can often be visualized, the cellular and tissue perturbations of the peripheral components of neoplasia may not be apparent by direct inspection under visible light and are often beyond the discrimination of conventional minimally invasive diagnostic imaging techniques. In general, label-free measurement and imaging modalities are preferable for clinical applications, particularly for diagnosis and screening, as they avoid the need for administration of an exogenous agent, with associated considerations of toxicity and pharmacokinetics. A number of label-free modalities based on the interaction of light with tissue have been proposed to improve the detection of malignant change. These include fluorescence, elastic scattering, Raman, infrared absorption, and diffuse reflectance spectroscopies (Bigio and Mourant 1997; Richards-Kortum and Sevick-Muraca 1996; Sokolov et al. 2002; Wagnieres et al. 1998). To date, many reported systems have been limited to point measurements, whereby only a very small area of tissue is interrogated at a time, for example, via a fiber-optic contact probe. This enables the acquisition of biochemical information but provides no spatial or morphological information about the tissue. Autofluorescence can provide label-free "molecular" contrast for single-point measurements but can also be utilized as an imaging technique, allowing the rapid and relatively noninvasive collection of spatially resolved information from areas of tissue up to tens of centimeters in diameter. The actual autofluorescence signal excited in biological tissue will depend on the concentration and the distribution of the fluorophores present, on the presence of chromophores (principally hemoglobin) that absorb excitation and fluorescence light, and on the degree of light scattering that occurs within the tissue (Richards-Kortum and Sevick-Muraca 1996). Autofluorescence signals therefore reflect the biochemical and structural composition of the tissue and consequently are altered when tissue composition is changed by disease states such as atherosclerosis, cancer, and osteoarthritis.

While autofluorescence can be measured or imaged using conventional intensity measurements, it is challenging to make sufficiently quantitative measurements for diagnostic applications since the autofluorescence intensity signal may be affected by fluorophore concentration, variations in temporal and spatial properties of the excitation flux, the angle of the excitation light, the detection efficiency, attenuation by light absorption and scattering within the tissue, and spatial variations in the tissue microenvironment altering local quenching of fluorescence. Spectrally resolved imaging of autofluorescence has received significant attention, for example, from, Richards-Kortum and Sevick-Muraca (1996), and the use of two or more spectral emission windows can be used to improve quantitation through ratiometric imaging, but the heterogeneity in the distribution of tissue fluorophores and their broad, heavily overlapping, emission spectra can limit the discrimination achievable. This can be addressed using spectral unmixing approaches, including linear unmixing, for example, in the work of Chorvat et al. (2005), and multivariate statistical analysis, for example, in the work of Ramanujam et al. (1996), which are being developed to identify signatures for diagnostic applications, including in combination with tissue reflectance data (Weber et al. 2008). The complexity and heterogeneity of biological tissue, however, still presents significant challenges for this approach, and fluorescence imaging tools in clinical use have been hampered by low specificity and a high rate of false-positive findings (Bard et al. 2005; Beamis et al. 2004; Ohkawa et al. 2004).

Increasingly, there is interest in exploiting fluorescence lifetime contrast to analyze tissue autofluorescence signals since fluorescence lifetime measurements are inherently ratiometric and therefore largely unaffected by many factors that can compromise autofluorescence intensity measurements. Early time-resolved spectroscopy experiments (Andersson-Engels et al. 1990b; Glanzmann et al. 1996; Park et al. 1989; Wagnieres et al. 1995) made use of dye lasers and nitrogen lasers to distinguish cancerous tissues or atherosclerotic plaques. The principal endogenous tissue fluorophores are reviewed in Chapter 3 and include collagen and elastin cross-links, reduced NADH, oxidized flavins (FAD and flavin mononucleotide [FMN]), lipofuscin, keratin, and porphyrins. These exhibit characteristic lifetimes ranging from hundreds to thousands of picoseconds (Elson et al. 2004; Richards-Kortum and Sevick-Muraca 1996) that enable spectrally overlapping fluorophores to be distinguished. A recent review article presents a summary of time-resolved measurements in clinical applications (Marcu 2012). Furthermore, the sensitivity of fluorescence lifetime to changes in the local tissue microenvironment (e.g., pH, $[O_2]$, $[Ca^{2+}]$) (Lakowicz 1999) can provide readouts of biochemical changes indicating the onset or progression of disease. These endogenous fluorophores have excitation maxima in the ultraviolet A or blue (325–450 nm) spectral regions and emit Stokes-shifted fluorescence in the near-UV to the visible (390–520 nm) region of the spectrum.

The increasing availability of suitable excitation sources, particularly the mode-locked Ti:sapphire laser, gain-switched picosecond diode lasers, and mode-locked fiber laser-based sources that now complement the nitrogen, dye, and solid-state lasers previously applied to time-resolved fluorescence spectroscopy of biological tissues have prompted an increasing interest in studying and exploiting tissue autofluorescence lifetime, both *ex vivo* and *in vivo*. In particular, multiphoton excitation using femtosecond Ti:sapphire lasers has stimulated intravital microscopy, enabling optical sectioning at depths of hundreds of microns in turbid media such as biological tissue, and, compared to wide-field or confocal microscopy, confers several other advantages including reduced phototoxicity compared to UV or blue excitation. The increase in imaging depth is due partly to the reduced scattering experienced by the longer wavelength excitation radiation and partly to the nonlinear excitation permitting all of the fluorescence emitted from the focal volume to be collected without degrading the image—and also providing optical sectioning without

Overview of fluorescence measurements and concepts

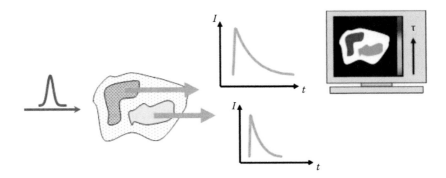

Figure 1.6 Schematic of fluorescence lifetime measurement and imaging.

confocal detection. The confinement of nonlinear excitation to the focal plane also reduces photobleaching through the sample, although the high intensities associated with multiphoton excitation can lead to enhanced nonlinear photobleaching in the focal volume. The ultrafast lasers used for multiphoton excitation are inherently convenient for FLIM, and the development and proliferation of multiphoton microscopes has been one of the drivers of FLIM studies of autofluorescence. Single-photon fluorescence lifetime measurements and imaging can also be conveniently implemented at a range of discrete spectral wavelengths using relatively low-cost semiconductor diode laser and LED technology. Gain-switched picosecond diode lasers have been applied to time-domain FLIM in scanning (Bohmer et al. 2001) and wide-field (Elson et al. 2002) FLIM systems, although their relatively low average output power (typically <1 mW) and sparse spectral coverage limit their range of applications. It is possible to reach higher average powers using sinusoidal modulation for frequency-domain FLIM (Booth and Wilson 2004; Dong et al. 2001), and for wide-field FLIM where their low spatial coherence is an advantage LEDs are potentially useful (van Geest and Stoop 2003), offering extended spectral coverage that extends to the deep ultraviolet (e.g., 280 nm) (McGuinness et al. 2004). An LED excitation source has also recently been used to realize frequency-domain FLIM in a scanning microscope (Herman et al. 2001).

This books aims to review the state of the art in this rapidly evolving field and, in the next section, aims to provide an introductory overview of the different technological approaches to measuring and mapping fluorescence lifetime (Figure 1.6).

1.2 TECHNIQUES FOR FLIM AND METROLOGY

The fluorescence lifetime is the average time the molecule spends in an upper energy level before returning to the ground state. For a large ensemble of identical molecules (or a large number of excitations of the same molecule), the time-resolved fluorescence intensity profile (detected photon histogram) following instantaneous excitation will exhibit a monoexponential decay that may be described as follows:

$$I(t) = I_0 e^{-\frac{t}{\tau}} + const. \tag{1.1}$$

where I_0 is the intensity of the fluorescence immediately after excitation (at time zero) and the constant term represents any background signal. In practice, the presence of multiple fluorophore species—or multiple states of a fluorophore species arising from interactions with the local environment—often result in more complex fluorescence decay profiles. This is the case for autofluorescence of biological tissue. Such complex fluorescence decays are commonly modeled by an N-component multiexponential decay model:

$$I(t) = \sum_{i=1}^{N} C_i e^{-\frac{t}{\tau_i}} + const. \tag{1.2}$$

where each pre-exponential amplitude is represented by the value C_i. Alternative approaches to model complex fluorescence decay profiles include fitting to a stretched exponential model that corresponds to a continuous lifetime distribution (Lee et al. 2001) or to a power law decay (Wlodarczyk and Kierdaszuk 2003) or performing a Laguerre polynomial expansion to generate empirical contrast (Jo et al. 2004) as discussed in Chapter 11.

In practice, experimental fluorescence decay profiles are recorded and fitted to model decay profiles according to prior knowledge of the sample—and hence the expected fluorescence decay profiles—or to the simplest model that provides an adequate fit to the experimental data. For complex fluorescence decay profiles, this is often undertaken with nonlinear iterative fitting algorithms such as the nonlinear Levenberg–Marquardt algorithm. Caution should be exercised when fitting to complex decay models, however, as increasing the complexity of the fitting model will tend to improve the goodness of fit (i.e., reduce the χ^2) but will also introduce errors if there is an insufficient signal-to-noise ratio to justify the number of parameters fitted. In general, as more complexity is used to describe a fluorescence decay profile, more detected photons are required to be able to measure it with sufficiently accuracy (Grinwald 1974; James and Ware 1985). Thus, fitting fluorescence decay profiles to complex models requires increased data acquisition times, which is often undesirable in terms of temporal resolution of dynamics or with respect to considerations of photobleaching or photodamage resulting from extended exposures to excitation radiation. Long acquisition times also require the sample to be stationary, which can be impractical, for example, for clinical applications. For many imaging applications, therefore, it is preferable to approximate complex fluorescence decay profiles with a single exponential decay model. The resulting average fluorescence lifetime can still provide useful contrast since it will usually reflect changes in the decay times or relative contributions of different lifetime components, but of course, the interpretation of a change in the average lifetime of a complex fluorescence decay profile can be subject to ambiguity.

Where quantitative analysis of complex fluorescence signals is required, as may be the case when analyzing tissue autofluorescence, it is often sensible to sacrifice image information and utilize all detected photons in a single-point measurement. If multicomponent FLIM of fluorophores exhibiting complex fluorescence is necessary, it is possible to reduce the number of photons required to be detected using a priori knowledge or assumptions, for example, about the magnitude of one or more component lifetimes, that is, τ_i, or about their relative contributions, that is, the C_i values of Equation 1.2. It can also be useful to assume that some parameters, for example, τ_i, are global, that is, they take the same value in each pixel of the image. Such global analysis (Verveer et al. 2000) is often used for the application of FLIM to FRET and is discussed in Chapter 12. Alternatively, fluorescence lifetime data can be represented graphically without any fitting using polar plots (Jameson et al. 1984; Redford and Clegg 2005) or phasor analysis (Digman et al. 2008; Hanley and Clayton 2005), as is discussed in Chapters 7 and 10.

In general, fluorescence lifetime measurement and FLIM techniques are categorized as time-domain or frequency-domain techniques, according to whether the instrumentation measures the fluorescence signal as a function of time delay following pulsed excitation or whether the lifetime information is derived from comparisons between a sinusoidally modulated excitation signal and the resulting sinusoidally modulated fluorescence signal. In principle, frequency and time-domain approaches can provide equivalent information, but specific implementations present different trade-offs with respect to cost and complexity, performance, and acquisition time, and the most appropriate method should be selected according to the target application. Historically, frequency-domain methods were initially developed with simpler electronic instrumentation and excitation source requirements, but time-domain techniques have benefitted from the tremendous advances in microelectronics and ultrafast laser technology, and today, the different approaches present a similar cost and complexity to most users. A further categorization of fluorescence lifetime measurement techniques can be made according to whether they are sampling techniques, which use gated detection to determine the relative timing of the fluorescence compared to the excitation signal, or photon counting techniques, which assign detected photons to different time bins. In general, wide-field FLIM is usually implemented with gated imaging detectors that sample the fluorescence signal, while photon counting techniques have been widely applied to single-point lifetime measurements and laser scanning microscopes. With respect to the photon counting, there is an important distinction between single-photon counting techniques, such as time-correlated single-photon counting (TCSPC), and techniques based on time binning (of photoelectrons) or high-speed analogue-to-digital converters that can detect more than one photon per excitation pulse. For laser scanning FLIM, single-point fluorescence lifetime measurement techniques can be readily implemented in laser scanning microscopes with electronic acquisition systems assigning detected photons to their respective image pixels. Some references (Cubeddu et al. 2002; Esposito et al. 2007; Gadella 2009; Lakowicz 1999) provide extensive reviews of the various techniques for fluorescence lifetime measurement, and there is further discussion in Chapters 4–12 of this book. In this chapter, we aim to briefly summarize the main time- and frequency-domain methods that are applied to single- and scanning-point fluorescence lifetime measurements and to wide-field FLIM.

1.2.1 SINGLE-POINT AND LASER SCANNING MEASUREMENTS OF FLUORESCENCE LIFETIME

1.2.1.1 Time-domain measurements

Single-point time-domain measurement of fluorescence decay profiles has been implemented using a wide range of instrumentation and has evolved from time-resolved fluorometry using fast flash lamp based measurements, having benefitted from the development of ultrafast photonics technology. Ultrashort pulsed excitation is, today, typically provided by ultrashort-pulse nitrogen lasers, mode-locked solid-state lasers, or gain-switched semiconductor diode lasers. For detection, fast (gigahertz bandwidth) sampling oscilloscopes, streak cameras, and transient digitizers have been used to directly measure the fluorescence decay profiles, as depicted in Figure 1.7a and discussed in Chapter 4. This approach has been complemented by photon counting techniques that build up histograms of the decay profiles, as indicated in Figure 1.7b. The current most widely used FLIM detection technique is probably TCSPC (O'Connor and Phillips 1984), which was demonstrated in the first laser scanning FLIM microscope (Bugiel et al. 1989) and is discussed in Chapters 5 and 9. The development of convenient and relatively low-cost TCSPC electronics, for example, in the work of Becker et al. (2004) and Zhang et al. (1999), for confocal and two-photon fluorescence scanning microscopes has greatly increased the uptake of FLIM generally, as well as impacting the development of single-point lifetime fluorometers.

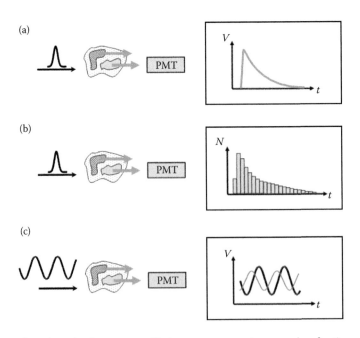

Figure 1.7 Schematic of single-point fluorescence lifetime measurement approaches for time-resolved fluorometry and laser scanning FLIM: (a) analogue measurement of decay profile, for example, using an oscilloscope, streak camera, or transient digitizer; (b) time-correlated single-photon counting (TCSPC); (c) frequency-domain measurement of change in phase and modulation depth of fluorescence with respect to excitation signal.

The principle of TCSPC is that, at low fluorescence fluxes, a histogram of photon arrival times can be built up by recording a series of voltage signals that depend on the arrival time of individual detected photons relative to the excitation pulse. By also recording spatial information from the scanning electronics of a confocal/multiphoton microscope, fluorescence lifetime images may be acquired. This is a straightforward way to implement FLIM on a scanning microscope since it only requires adding electronic components after the detector and may be "bolted on" to almost any system. TCSPC is widely accepted as one of the most accurate methods of lifetime determination due to its shot noise-limited detection, high photon economy, low temporal jitter, high temporal precision (offering large number of bins in the photon arrival time histogram), and high dynamic range (typically millions of photons can be recorded without saturation). Its main perceived drawback is a relatively low acquisition rate, owing to the requirement to operate at sufficiently low incident fluorescence intensity levels to ensure single-photon detection at a rate limited by the "dead time" between measurement events, which is associated with the electronic circuitry that determines the photon arrival times. However, for modern TCSPC instrumentation, this limitation of the electronic circuitry is usually less significant than problems caused by "classical" photon pileup, which limits all single-photon counting techniques. Pulse pileup refers to the issue of more than one photon arriving in a single-photon detection period, which results in apparently shorter lifetimes being "measured." This is avoided by decreasing the excitation power such that the excitation rate is much lower than the pulse repetition rate, which typically limits the maximum detection count rate to approximately 5% of the repetition rate of the laser.

An alternative single-point photon-counting time-domain technique is based on temporal photon binning, for which the photoelectrons arising from the detected photons are accumulated in a number of different time bins and a histogram is built up accordingly (Buurman et al. 1992). This is not a single-photon counting method and does not have the same dead-time and pulse-pileup limitations as TCSPC, so it may be used with higher photon fluxes to provide higher imaging rates when implemented on a scanning fluorescence microscope. To date, however, it has not been commercially implemented with the same precision as TCSPC and provides FLIM at faster rates but with lower lifetime precision (Gerritsen et al. 2002).

1.2.1.2 Frequency-domain measurements

Frequency-domain fluorescence lifetime measurements concern the demodulation of a fluorescence signal excited with modulated light. Typically, this approach utilizes sinusoidally modulated excitation sources that produce a fluorescence signal that is also sinusoidally modulated but with a different modulation depth and with a phase delay relative to the excitation signal, as illustrated in Figure 1.7c and discussed further in Chapter 7. The fluorescence lifetime can be determined from measurements of the relative modulation, m, and the phase delay, ϕ (Lakowicz 1999). These experimental parameters can be conveniently measured using appropriate electronic circuits for synchronous detection, such as a lock-in amplifier. Recent advances in electronics have resulted in relatively low-cost synchronous detection (Booth and Wilson 2004) and fast digitization circuitry (Colyer et al. 2008). For an ideal single exponential decay profile, the lifetime can be calculated from m or ϕ using the following equations:

$$\tau_{\varphi} = \frac{1}{\omega}\tan\phi \tag{1.3}$$

$$\tau_m = \frac{1}{\omega}\left(\frac{1}{m^2} - 1\right)^{\frac{1}{2}} \tag{1.4}$$

If the fluorescence excitation is not purely sinusoidal, the frequency-domain approach is still applicable—with the lifetime being readily calculated from the fundamental Fourier component of a modulated fluorescence signal that may be excited, for example, by a mode-locked laser (So et al. 1995).

In fact, the implementation of frequency-domain FLIM with periodically pulsed excitation provides an improved accuracy in lifetime determination compared to a purely sinusoidal excitation (Philip and Carlsson 2003). If the fluorescence does not manifest a single exponential decay, however, the lifetime values calculated from Equations 1.3 and 1.4 will not be equal, and it is necessary to repeat the measurement at multiple harmonics of the excitation modulation frequency to build up a more complete (multicomponent) description of the complex fluorescence profile. This may be obtained by fitting the measurements to a set of dispersion relationships, for example, in the work on Gratton and Limkeman (1983) and Lakowicz and Maliwal (1985). The increased acquisition and data processing time associated with multiple-frequency measurements of complex fluorescence signals can be undesirable. For applications requiring speed, it is possible to measure at only one modulation frequency and obtain an "average" fluorescence lifetime by calculating the mean of the lifetime values from the modulation depth and phase delay measurements.

The ability to use sinusoidally modulated diode lasers or LEDs as excitation sources makes frequency-domain measurements attractive for low-cost applications, for which frequencies of less than 100 MHz are sufficient to provide lifetime resolutions of hundreds of picoseconds—provided that the measurement exhibits sufficient signal-to-noise ratio. To achieve higher lifetime resolution requires higher frequency modulation, and the technology necessary for gigahertz measurements introduces comparable complexity and expense to that associated with time-domain measurements.

1.2.1.3 Laser scanning FLIM microscopy

Laser scanning microscopes essentially make sequential single-point measurements and are widely used for biological imaging because their implementation as confocal or multiphoton microscopes provide improved contrast and optical sectioning compared to wide-field microscopy. The approaches summarized in Figure 1.7 may all be used to realize FLIM in laser scanning microscopes. Having first been implemented using TCSPC (Bugiel et al. 1989), scanning FLIM microscopy was subsequently demonstrated using photon-binning (Buurman et al. 1992) and frequency-domain techniques, for example, in the work of Carlsson and Liljeborg (1998), but TCSPC remains the most widely implemented technique for laser scanning FLIM. When imaging typical biological samples, TCSPC typically requires tens of seconds to acquire sufficient photons for single-photon excited FLIM and longer for multiphoton excitation, which can limit some of its applications. Although recent technological advances have led to reductions in detector dead time and increased the maximum detection count rates of TCSPC, in practice, the maximum imaging rate is usually limited by the onset of significant photobleaching and/or phototoxicity. The photon time-binning and frequency-domain approaches are not limited to single-photon detection and may can provide faster imaging of bright samples compared to TCSPC, but photobleaching and/or phototoxicity considerations also limit the maximum practical imaging rates.

For frequency-domain laser scanning FLIM, one can use a sinusoidally modulated excitation laser and apply synchronous detection, for example, using a "lock-in" amplifier to determine the phase difference and change in modulation depth between the excitation signal and the resulting sinusoidally modulated fluorescence signal. One can also take advantage of pulsed excitation sources, for example, in two-photon microscopes, and exploit the harmonic content of the resulting fluorescence (So et al. 1995). As discussed above, this frequency-domain approach can be implemented using relatively low-cost electronic circuitry that is not limited by the dead-time or maximum count rates associated with single-photon counting detection and so can provide high-speed FLIM (Booth and Wilson 2004; Colyer et al. 2008).

For all laser scanning microscopes, the sequential pixel acquisition means that increasing the imaging speed requires a concomitant increase in excitation intensity, which can be undesirable due to photobleaching and phototoxicity considerations. This is a particular issue for FLIM, for which more photons need to be detected per pixel compared to intensity imaging (Kollner and Wolfrum 1992; Gerritsen et al. 2002). For example, it is estimated that detection of a few hundred photons is necessary to determine the lifetime (to ~10% accuracy) using a single exponential fluorescence decay model, while a double exponential fit requires ~10^4 detected photons. For multiphoton FLIM microscopy, for which excitation is relatively inefficient and photobleaching scales nonlinearly with intensity (Patterson and Piston 2000), these considerations can result in FLIM acquisition times of many minutes for biological samples. One way to significantly increase the imaging speed of multiphoton microscopy is to use multiple excitation beams in parallel (Bewersdorf et al. 1998; Fittinghoff et al. 2000; Kim et al. 2007), and this approach has been applied to TCSPC FLIM using 16 parallel excitation and detection channels (Kumar et al. 2007). In general, parallel pixel excitation and detection is a useful approach to increase the practical imaging speed of all laser scanning microscopes. This can be extended to optically sectioned line-scanning microscopy using a rapidly scanned multiple-beam array to produce a line of fluorescence emission that is relayed to the input slit of a streak camera (Krishnan et al. 2003). FLIM images have been acquired in less than 1 s using this approach, which was limited by the readout rate of the streak camera system. Alternatively, multiple scanning beam excitation can be applied with wide-field time-gated detection, as has been implemented with multibeam multiphoton microscopes (Benninger et al. 2005; Leveque-Fort et al. 2004; Straub and Hell 1998) and with single-photon excitation in spinning Nipkow disk microscopes (Grant et al. 2005, 2007; van Munster et al. 2007).

For the techniques indicated in Figure 1.7, the temporal instrument response function depends on the response time of the detector as well as the temporal jitter in the electronic circuitry. Photon counting photomultipliers typically exhibit a response time of ~200 ps, although faster multichannel plate (MCP) devices can have response times of a few tens of picoseconds. One significant exception to the above observation is the pump-probe approach, where a second (probe) beam, which is delayed with respect to the excitation (pump) beam, interrogates the upper-state population. A particularly elegant implementation of this technique for scanning fluorescence microscopy also provides optical sectioning in a manner analogous to two-photon microscopy (Dong et al. 1995).

1.2.2 WIDE-FIELD FLIM TECHNIQUES

The parallel pixel acquisition of wide-field imaging techniques can support FLIM imaging rates of tens to hundreds of hertz, for example, in the work of Requejo-Isidro et al. (2004) and Agronskaia et al. (2003), although the maximum acquisition speed is still, of course, limited by the number of photons/pixel available from the (biological) sample. Wide-field FLIM is most commonly implemented using modulated image intensifiers with frequency- or time-domain approaches, as represented in Figure 1.8. The frequency-domain approach was established first around 1990, for example, in the work of Morgan

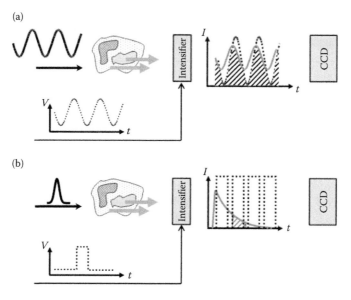

Figure 1.8 Wide-field FLIM techniques: (a) time-gated FLIM using a periodic pulsed excitation source and a periodic ultrashort time gate applied to an image intensifier read out using a CCD camera; (b) frequency-domain FLIM using a sinusoidally modulated excitation source with a sinusoidal modulation of the same period applied to an image intensifier read out using a CCD camera.

et al. (1990) and Gratton et al. (1990), utilizing frequency-modulated laser excitation and implementing frequency-modulated gain with a microchannel plate (MCP) optical image intensifier to analyze the resulting fluorescence by acquiring a series of gated "intensified" images acquired at different relative phases between the MCP modulation and the excitation signal. Originally, the optical output image from the intensifier was read out using a linear photodiode array, but this was rapidly superseded by charged-coupled device (CCD) camera technology, as represented in Figure 1.8a (Gadella et al. 1993; Lakowicz and Berndt 1991). It is possible to implement wide-field FLIM at a single modulation frequency using only three phase measurements (Lakowicz et al. 1992) to calculate the fluorescence lifetime map, although it is common to use eight or more phase-resolved images to improve the accuracy. This can be necessary if the system exhibits unwanted nonlinear behavior that produces modulated signals at harmonic frequencies.

The development of wide-field frequency-domain FLIM was complemented by the demonstration of a streak camera-based approach (Minami and Hirayama 1990) and by the application of short-pulse gated MCP image intensifiers coupled to CCD cameras for time-domain wide-field FLIM in an approach described as time-gated imaging, for example, in the work of Oida et al. (1993), Scully et al. (1996), and Dowling et al. (1998), depicted in Figure 1.8b. By gating the MCP image intensifier gain for short (picosecond to nanosecond) periods after excitation, the fluorescence decay can be sampled such that the fluorescence lifetime distribution can be calculated. Initially, the shortest MCP gate widths that could be applied were over 5 ns, but the technology quickly developed to provide subnanosecond gating times (Wang et al. 1991), and then, the use of a wire mesh proximity-coupled to the MCP photocathode led to devices with sub-100 ps resolution (Hares 1987).

While both frequency- and time-domain FLIM offer parallel pixel acquisition compared to scanning microscopy techniques, they both suffer from inherently reduced photon economy owing to the time-varying gain applied to the MCP image intensifier. This can be particularly significant in the time domain if very short (~100 ps) time gates are applied to sample the fluorescence decay profile. In many situations, however, such short time gates are not necessary since the fast-rising and -falling edges of the time gate provide the required time resolution, and so the gate can remain "open" for times comparable to the fluorescence decay time. Thus, the photon economy of time-gated imaging can approach that of wide-field-frequency FLIM. For monoexponential fluorescence decay profiles, the minimum resolvable lifetime difference depends on the signal-to-noise ratio and temporal jitter rather than the sampling gate width, and nanosecond gate widths can provide sub-100 ps lifetime discrimination (Munro et al. 2005). See Chapter 8 for further discussion of time-gated FLIM.

1.3 OUTLOOK

The future of FLIM and multidimensional flourescence imaging looks set to see increasing applications in fluorescence microscopy, including for FRET and other approaches to studying protein interactions, as well as in emerging clinical and preclinical applications. The functionality of FLIM instrumentation will continue to develop, and the technological implementations will improve in terms of higher performance, lower cost, and more compact and ergonomic instrumentation. Advances in imaging speed are likely to be driven by the development of detector technologies such as modulated CMOS, for example, in the work of Esposito et al. 2005, and the application of multipoint imaging in laser scanning systems, for example, extending the concept of Kumar et al. (2007). Spatial resolution may be realized below the diffraction limit using super-resolved microscopy techniques such as stimulated emission depletion microscopy (Auksorius et al. 2008). Lower costs may be achieved through the use of cheaper excitation sources such as modulated LEDs, diode lasers, and fiber laser-based sources, and higher functionality will continue to be realized through multidimensional fluoresce imaging approaches combining lifetime measurements with spectral resolution, as discussed in Chapters 4, 5, 8 and 9, and with polarization resolution. Some of the most exciting and challenging prospects for FLIM in biomedicine are associated with the translation to *in vivo* clinical and preclinical imaging, where the ability to study molecular function and tissue structure *in situ* may provide valuable insights into the mechanisms underlying disease. This goal can be approached by developing FLIM endoscopes and tomographic imaging systems as well as intravital microscopes.

REFERENCES

Abraham, H. & Lemoine, J. 1899. Disparition instantanée du phénomène de Kerr. *Comptes Rendus Hebdomadaines des Seances de Academie des Sciences. Sciences Naturelles*, 129, 206–208.

Abraham, H. & Lemoine, J. 1900. Nouvelle méthode de mesure des durées infinitesimals. Analyse de la disparition des phénomènes électro-optiques. *Ann Chim*, 20, 264–287.

Ackroyd, R., Kelty, C., Brown, N. & Reed, M. 2001. The history of photodetection and photodynamic therapy. *Photochemistry and Photobiology*, 74, 656–669.

Acuña, A. U., Amat-Guerri, F., Morcillo, P., Liras, M. & Rodriguez, B. 2009. Structure and formation of the fluorescent compound of lignum nephriticum. *Organic Letters*, 11, 3020–3023.

Agronskaia, A. V., Tertoolen, L. & Gerritsen, H. C. 2003. High frame rate fluorescence lifetime imaging. *Journal of Physics D-Applied Physics*, 36, 1655–1662.

American Cancer Society. 2006. *Cancer Facts and Figures 2006*. Atlanta, GA: American Cancer Society.

Andersson-Engels, S., Johansson, J., Stenram, U., Svanberg, K. & Svanberg, S. 1990a. Malignant tumour and atherosclerotic plaque diagnosis using laser induced fluorescence. *IEEE Journal of Quantum Electronics*, 26, 2207–2217.

Andersson-Engels, S., Johansson, J. & Svanberg, S. 1990b. The use of time-resolved fluorescence for diagnosis of atherosclerotic plaque and malignant-tumors. *Spectrochimica Acta Part A-Molecular and Biomolecular Spectroscopy*, 46, 1203–1210.

Auksorius, E., Boruah, B. R., Dunsby, C., Lanigan, P. M. P., Kennedy, G., Neil, M. A. A. & French, P. M. W. 2008. Stimulated emission depletion microscopy with a supercontinuum source and fluorescence lifetime imaging. *Optics Letters*, 33, 113–115.

Bard, M. P. L., Amelink, A., Skurichina, M., Den Bakker, M., Burgers, S. A., van Meerbeeck, J. P., Duin, R. P. W., Aerts, J., Hoogsteden, H. C. & Sterenborg, H. 2005. Improving the specificity of fluorescence bronchoscopy for the analysis of neoplastic lesions of the bronchial tree by combination with optical spectroscopy: Preliminary communication. *Lung Cancer*, 47, 41–47.

Bastiaens, P. I. H. & Squire, A. 1999. Fluorescence lifetime imaging microscopy: Spatial resolution of biochemical processes in the cell. *Trends in Cell Biology*, 9, 48–52.

Beamis, J. F., Ernst, A., Simoff, M., Yung, R. & Mathur, P. 2004. A multicenter study comparing autofluorescence bronchoscopy to white light bronchoscopy using a non-laser light stimulation system. *Chest*, 125, 148S–149S.

Beams, J. W. 1926. A method of obtaining light flashes of uniform intensity and short duration. *Journal of the Optical Society of America*, 13, 597–599.

Becker, W., Bergmann, A., Hink, M. A., Konig, K., Benndorf, K. & Biskup, C. 2004. Fluorescence lifetime imaging by time-correlated single-photon counting. *Microscopy Research and Technique*, 63, 58–66.

Becquerel, E. 1867. *La Lumiere. Ses Causes et ses Effects*. Paris: Firmin Didot, 1.

Benninger, R. K. P., Hofmann, O., McGinty, J., Requejo-Isidro, J., Munro, I., Neil, M. A. A., Demello, A. J. & French, P. M. W. 2005. Time-resolved fluorescence imaging of solvent interactions in microfluidic devices. *Optics Express*, 13, 6275–6285.

Berezin, M. Y. & Achilefu, S. 2010. Fluorescence lifetime measurements and biological imaging. *Chemical Reviews*, 110, 2641–2684.

Betzig, E., Patterson, G. H., Sougrat, R., Lindwasser, O. W., Olenych, S., Bonifacino, J. S., Davidson, M. W., Lippincott-Schwartz, J. & Hess, H. F. 2006. Imaging intracellular fluorescent proteins at nanometer resolution. *Science*, 313, 1642–1645.

Bewersdorf, J., Pick, R. & Hell, S. W. 1998. Multifocal multiphoton microscopy. *Optics Letters*, 23, 655–657.

Bigio, I. J. & Mourant, J. R. 1997. Ultraviolet and visible spectroscopies for tissue diagnostics: Fluorescence spectroscopy and elastic-scattering spectroscopy. *Physics in Medicine and Biology*, 42, 803–814.

Bohmer, M., Pampaloni, F., Wahl, M., Rahn, H. J., Erdmann, R. & Enderlein, J. 2001. Time-resolved confocal scanning device for ultrasensitive fluorescence detection. *Review of Scientific Instruments*, 72, 4145–4152.

Booth, M. J. & Wilson, T. 2004. Low-cost, frequency-domain, fluorescence lifetime confocal microscopy. *Journal of Microscopy-Oxford*, 214, 36–42.

Brody, S. S. 1957. Instrument to measure fluorescence lifetimes in the millimicrosecond region. *Review of Scientific Instruments*, 28, 1021.

Bugiel, I., Konig, K. & Wabnitz, H. 1989. Investigation of cells by fluorescence laser scanning microscopy with subnanosecond time resolution. *Lasers in the Life Sciences*, 3, 47–53.

Buurman, E. P., Sanders, R., Draaijer, A., Gerritsen, H. C., Vanveen, J. J. F., Houpt, P. M. & Levine, Y. K. 1992. Fluorescence lifetime imaging using a confocal laser scanning microscope. *Scanning*, 14, 155–159.

Carlsson, K. & Liljeborg, A. 1998. Simultaneous confocal lifetime imaging of multiple fluorophores using the intensity-modulated multiple-wavelength scanning (IMS) technique. *Journal of Microscopy*, 191, 119–127.

Chatterjee, S., Lee, J. B., Valappil, N. V., Luo, D. & Menon, V. M. 2012. Probing Y-shaped DNA structure with time-resolved FRET. *Nanoscale*, 4, 1568–1571.

Chorvat, D., Kirchnerova, J., Cagalinec, M., Smolka, J., Mateasik, A. & Chorvatova, A. 2005. Spectral unmixing of flavin autofluorescence components in cardiac myocytes. *Biophysical Journal*, 89, L55–L57.

Clegg, R. M., Holub, O. & Gohlke, C. 2003. Fluorescence Lifetime-resolved imaging: Measuring lifetimes in an image. Why do it? How to do it. How to interpret it. *In:* Marriott, G. and Parker, I. (eds) *Biophotonics Part A, Methods in Enzymology, vol. 360.* San Diego, CA: Academic Press.

Colyer, R. A., Lee, C. & Gratton, E. 2008. A novel fluorescence lifetime imaging system that optimizes photon efficiency. *Microscopy Research and Technique,* 71, 201–213.

Cubeddu, R., Comelli, D., D'Andrea, C., Taroni, P. & Valentini, G. 2002. Time-resolved fluorescence imaging in biology and medicine. *Journal of Physics D-Applied Physics,* 35, R61–R76.

Denk, W., Strickler, J. H. & Webb, W. W. 1990. Two-photon laser scanning fluorescence microscopy. *Science,* 248, 73–76.

Denk, W. & Svoboda, K. 1997. Photon upmanship: Why multiphoton imaging is more than a gimmick. *Neuron,* 18, 351–357.

Dickinson, M. E., Simbuerger, E., Zimmermann, B., Waters, C. W. & Fraser, S. E. 2003. Multiphoton excitation spectra in biological samples. *Journal of Biomedical Optics,* 8, 329–338.

Digman, M. A., Caiolfa, V. R., Zamai, M. & Gratton, E. 2008. The phasor approach to fluorescence lifetime imaging analysis. *Biophysical Journal,* 94, L14–L16.

Dong, C. Y., Buehler, C., So, P. T. C., French, T. & Gratton, E. 2001. Implementation of intensity-modulated laser diodes in time-resolved, pump-probe fluorescence microscopy. *Applied Optics,* 40, 1109–1115.

Dong, C. Y., So, P. T. C., French, T. & Gratton, E. 1995. Fluorescence lifetime imaging by asynchronous pump-probe microscopy. *Biophysical Journal,* 69, 2234–2242.

Dosremedios, C. G. & Moens, P. D. J. 1995. Fluorescence resonance energy-transfer spectroscopy is a reliable ruler for measuring structural-changes in proteins—Dispelling the problem of the unknown orientation factor. *Journal of Structural Biology,* 115, 175–185.

Dowling, K., Dayel, M. J., Lever, M. J., French, P. M. W., Hares, J. D. & Dymoke-Bradshaw, A. K. L. 1998. Fluorescence lifetime imaging with picosecond resolution for biomedical applications. *Optics Letters,* 23, 810–812.

Duschinsky, V. F. 1933. Eine allgemeine Theorie der zur Messung sehr kurzer Leuchtdauern dienenden Versuchsanordnungen (Fluorometer). *Zeitschrift für Physik,* 81, 23–42.

Elson, D., Requejo-Isidro, J., Munro, I., Reavell, F., Siegel, J., Suhling, K., Tadrous, P., Benninger, R., Lanigan, P., McGinty, J., Talbot, C., Treanor, B., Webb, S., Sandison, A., Wallace, A., Davis, D., Lever, J., Neil, M., Phillips, D., Stamp, G. & French, P. 2004. Time-domain fluorescence lifetime imaging applied to biological tissue. *Photochemical & Photobiological Sciences,* 3, 795–801.

Elson, D. S., Siegel, J., Webb, S. E. D., Leveque-Fort, S., Lever, M. J., French, P. M. W., Lauritsen, K., Wahl, M. & Erdmann, R. 2002. Fluorescence lifetime system for microscopy and multiwell plate imaging with a blue picosecond diode laser. *Optics Letters,* 27, 1409–1411.

Esposito, A., Gerritsen, H. C. & Wouters, F. S. 2007. Optimizing frequency-domain fluorescence lifetime sensing for high-throughput applications: Photon economy and acquisition speed. *Journal of the Optical Society of America A-Optics Image Science and Vision,* 24, 3261–3273.

Esposito, A., Oggier, T., Gerritsen, H. C., Lustenberger, F. & Wouters, F. S. 2005. All-solid-state lock-in imaging for wide-field fluorescence lifetime sensing. *Optics Express,* 13, 9812–9821.

Fittinghoff, D. N., Wiseman, P. W. & Squier, J. A. 2000. Widefield multiphoton and temporally decorrelated multifocal multiphoton microscopy. *Optics Express,* 7, 273–279.

Förster, T. 1948. Zwischenmolekulare Energiewanderung und Fluoreszenz. *Annals der Physik,* 2, 55–75.

Gadella, T. W. J. (ed.) 2009. *FRET and FLIM Techniques.* Amsterdam: Elsevier.

Gadella, T. W. J., Jovin, T. M. & Clegg, R. M. 1993. Fluorescence lifetime imaging microscopy (FLIM)—Spatial resolution of structures on the nanosecond timescale. *Biophysical Chemistry,* 48, 221–239.

Gaviola, E. 1926. Die Abklingungszeiten der Fluoreszenz von Farbstofflösungen. *Annals der Physik,* 386, 681–710.

Gerritsen, H. C., Asselbergs, M. A. H., Agronskaia, A. V. & van Sark, W. G. J. H. M. 2002. Fluorescence lifetime imaging in scanning microscopes: Acquisition speed, photon economy and lifetime resolution. *Journal of Microscopy,* 206, 218–224.

Glanzmann, T., Ballini, J. P., Jichlinski, P., Vandenbergh, H. & Wagnieres, G. 1996. Tissue characterization by time-resolved fluorescence spectroscopy of endogenous and exogenous fluorochromes: Apparatus design and preliminary results.

Gottling, P. F. 1923. The determination of the time between excitation and emission for certain fluorescent solids. *Physical Review,* 22, 566–573.

Grant, D. M., Elson, D. S., Schimpf, D., Dunsby, C., Requejo-Isidro, J., Auksorius, E., Munro, I., Neil, M. A. A., French, P. M. W., Nye, E., Stamp, G. & Courtney, P. 2005. Optically sectioned fluorescence lifetime imaging using a Nipkow disk microscope and a tunable ultrafast continuum excitation source. *Optics Letters,* 30, 3353–3355.

Grant, D. M., McGinty, J., McGhee, E. J., Bunney, T. D., Owen, D. M., Talbot, C. B., Zhang, W., Kumar, S., Munro, I., Lanigan, P. M. P., Kennedy, G. T., Dunsby, C., Magee, A. I., Courtney, P., Katan, M., Neil, M. A. A. & French, P. M. W. 2007. High speed optically sectioned fluorescence lifetime imaging permits study of live cell signaling events. *Optics Express*, 15, 15656–15673.

Gratton, E., Feddersen, B. & van de Ven, M. 1990. Parallel acquisition of fluorescence decay using array detectors. Time-resolved laser spectroscopy in biochemistry II. *SPIE*, 1204, 21–25.

Gratton, E. & Limkeman, M. 1983. A continuously variable frequency cross-correlation phase fluorometer with picosecond resolution. *Biophysical Journal,* 44, 315–324.

Grinwald, A. 1974. On the analysis of fluorescence decay kinetics by the method of least-squares. *Analytical Biochemistry,* 59, 583–593.

Grynkiewicz, G., Poenie, M. & Tsien, R. Y. 1985. A new generation of Ca2+ indicators with greatly improved fluorescence properties. *Journal of Biological Chemistry,* 260, 3440–3450.

Gustafsson, M. G. L. 2000. Surpassing the lateral resolution limit by a factor of two using structured illumination microscopy. *Journal of Microscopy-Oxford,* 198, 82–87.

Hanley, Q. S. & Clayton, A. H. A. 2005. AB-plot assisted determination of fluorophore mixtures in a fluorescence lifetime microscope using spectra or quenchers. *Journal of Microscopy-Oxford,* 218, 62–67.

Hares, J. D. 1987. Advances in sub-nanosecond shutter tube technology and applications in plasma physics. *In:* Richardson, M. C. (ed.) *X Rays from Laser Plasmas.* San Diego, CA: Academic Press, *Proceedings of SPIE,* 165–170.

Herman, P., Maliwal, B. P., Lin, H. J. & Lakowicz, J. R. 2001. Frequency-domain fluorescence microscopy with the LED as a light source. *Journal of Microscopy-Oxford,* 203, 176–181.

James, D. R. & Ware, W. R. 1985. A fallacy in the interpretation of fluorescence decay parameters. *Chemical Physics Letters,* 120, 455–459.

Jameson, D. M. 2001. The seminal contributions of Gregorio Weber to modern fluorescence spectroscopy. *New Trends in Fluorescence Spectroscopy,* 1, 35–58.

Jameson, D. M., Gratton, E. & Hall, R. D. 1984. The measurement and analysis of heterogeneous emissions by multifrequency phase and modulation fluorometry. *Applied Spectroscopy Reviews,* 20, 55–106.

Jares-Erijman, E. A. & Jovin, T. M. 2003. FRET imaging. *Nature Biotechnology,* 21, 1387–1395.

Jo, J. A., Fang, Q. Y., Papaioannou, T. & Marcu, L. 2004. Fast model-free deconvolution of fluorescence decay for analysis of biological systems. *Journal of Biomedical Optics,* 9, 743–752.

Kalinin, S., Valeri, A., Antonik, M., Felekyan, S. & Seidel, C. A. M. 2010. Detection of structural dynamics by FRET: A photon distribution and fluorescence lifetime analysis of systems with multiple states. *Journal of Physical Chemistry B,* 114, 7983–7995.

Kerr, J. 1875. A new relation between electricity and light: Dielectrified media birefringent. *Philosophical Magazine,* 50, 337–348.

Kim, K. H., Buehler, C., Bahlmann, K., Ragan, T., Lee, W.-C. A., Nedivi, E., Heffer, E. L., Fantini, S. & So, P. T. C. 2007. Multifocal multiphoton microscopy based on multianode photomultiplier tubes. *Optics Express,* 15, 11658–11678.

Klar, T. A., Jakobs, S., Dyba, M., Egner, A. & Hell, S. W. 2000. Fluorescence microscopy with diffraction resolution barrier broken by stimulated emission. *Proceedings of the National Academy of Sciences of the United States of America,* 97, 8206–8210.

Kollner, M. & Wolfrum, J. 1992. How many photons are necessary for fluorescence-lifetime measurements. *Chemical Physics Letters,* 200, 199–204.

Krishnan, R. V., Saitoh, H., Terada, H., Centonze, V. E. & Herman, B. 2003. Development of a multiphoton fluorescence lifetime imaging microscopy system using a streak camera. *Review of Scientific Instruments,* 74, 2714–2721.

Kumar, S., Dunsby, C., de Beule, P. A. A., Owen, D. M., Anand, U., Lanigan, P. M. P., Benninger, R. K. P., Davis, D. M., Neil, M. A. A., Anand, P., Benham, C., Naylor, A. & French, P. M. W. 2007. Multifocal multiphoton excitation and time correlated single photon counting detection for 3-D fluorescence lifetime imaging. *Optics Express,* 15, 12548–12561.

Lakowicz, J. R. 1999. *Principles of Fluorescence Spectroscopy*, 2nd edition. New York: Kluwer Academic/Plenum Publishers.

Lakowicz, J. R. & Berndt, K. W. 1991. Lifetime-selective fluorescence imaging using an rf phase-sensitive camera. *Review of Scientific Instruments,* 62, 1727–1734.

Lakowicz, J. R. & Maliwal, B. P. 1985. Construction and performance of a variable-frequency phase-modulation fluorometer. *Biophysical Chemistry,* 21, 61–78.

Lakowicz, J. R., Szmacinski, H., Nowaczyk, K., Berndt, K. W. & Johnson, M. 1992. Fluorescence lifetime imaging. *Analytical Biochemistry,* 202, 316–330.

Overview of fluorescence measurements and concepts

Lapotko, D., Lukianova, E., Potapnev, M., Aleinikova, O. & Oraevsky, A. 2006. Method of laser activated nano-thermolysis for elimination of tumor cells. *Cancer Letters,* 239, 36–45.

Lee, K. C. B., Siegel, J., Webb, S. E. D., Leveque-Fort, S., Cole, M. J., Jones, R., Dowling, K., Lever, M. J. & French, P. M. W. 2001. Application of the stretched exponential function to fluorescence lifetime imaging. *Biophysical Journal,* 81, 1265–1274.

Leveque-Fort, S., Fontaine-Aupart, M. P., Roger, G. & Georges, P. 2004. Fluorescence-lifetime imaging with a multifocal two-photon microscope. *Optics Letters,* 29, 2884–2886.

Marcu, L. 2012. Fluorescence lifetime techniques in medical applications. *Annals of Biomedical Engineering,* 40, 304–331.

McGuinness, C. D., Sagoo, K., McLoskey, D. & Birch, D. J. S. 2004. A new sub-nanosecond LED at 280 nm: Application to protein fluorescence. *Measurement Science & Technology,* 15, L19–L22.

Minami, T. & Hirayama, S. 1990. High-quality fluorescence decay curves and lifetime imaging using an elliptic scan streak camera. *Journal of Photochemistry and Photobiology A-Chemistry,* 53, 11–21.

Monardes, N. 1565. *Historia medicinal de las cosas que se traen de nuestras Indias Occidentales,* Seville: Hernando Diaz.

Morgan, C. G., Mitchell, A. C. & Murray, J. G. 1990. Nanosecond time-resolved fluorescence microscopy—Principles and practice. *Transactions of the Royal Microscopical Society: New Series,* 1, 463–466.

Munro, I., McGinty, J., Galletly, N., Requejo-Isidro, J., Lanigan, P. M. P., Elson, D. S., Dunsby, C., Neil, M. A. A., Lever, M. J., Stamp, G. W. H. & French, P. M. W. 2005. Towards the clinical application of time-domain fluorescence lifetime imaging. *Journal of Biomedical Optics,* 10, 051403.

O'Connor, D. V. & Phillips, D. 1984. *Time-Correlated Single-Photon Counting.* New York: Academic Press.

Ohkawa, A., Miwa, H., Namihisa, A., Kobayashi, O., Nakaniwa, N., Ohkusa, T., Ogihara, T. & Sato, N. 2004. Diagnostic performance of light-induced fluorescence endoscopy for gastric neoplasms. *Endoscopy,* 36, 515–521.

Oida, T., Sako, Y. & Kusumi, A. 1993. Fluorescence lifetime imaging microscopy (flimscopy)—Methodology development and application to studies of endosome fusion in single cells. *Biophysical Journal,* 64, 676–685.

Park, Y. D., An, K., Dasari, R. R. & Feld, M. S. 1989. UV time-resolved fluorescence of human aorta. *Lasers in Surgery and Medicine,* 3–3.

Patterson, G. H. & Piston, D. W. 2000. Photobleaching in two-photon excitation microscopy. *Biophysical Journal,* 78, 2159–2162.

Philip, J. & Carlsson, K. 2003. Theoretical investigation of the signal-to-noise ratio in fluorescence lifetime imaging. *Journal of the Optical Society of America A-Optics Image Science and Vision,* 20, 368–379.

Ramanujam, N., Mitchell, M. F., Mahadevanjansen, A., Thomsen, S. L., Staerkel, G., Malpica, A., Wright, T., Atkinson, N. & Richards-Kortum, R. 1996. Cervical precancer detection using a multivariate statistical algorithm based on laser-induced fluorescence spectra at multiple excitation wavelengths. *Photochemistry and Photobiology,* 64, 720–735.

Redford, G. I. & Clegg, R. M. 2005. Polar plot representation for frequency-domain analysis of fluorescence lifetimes. *Journal of Fluorescence,* 15, 805–815.

Requejo-Isidro, J., McGinty, J., Munro, I., Elson, D. S., Galletly, N., Lever, M. J., Neil, M. A. A., Stamp, G. W. H., French, P. M. W., Kellet, P. A., Hares, J. D. & Dymoke-Bradshaw, A. K. L. 2004. High-speed wide-field time-gated endoscopic fluorescence lifetime imaging. *Optics Letters,* 29, 2249–2251.

Richards-Kortum, R. & Sevick-Muraca, E. 1996. Quantitative optical spectroscopy for tissue diagnosis. *Annual Review of Physical Chemistry,* 47, 555–606.

Ross, J. A. & Jameson, D. M. 2008. Time-resolved methods in biophysics. 8. Frequency domain fluorometry: Applications to intrinsic protein fluorescence. *Photochemical & Photobiological Sciences,* 7, 1301–1312.

Rust, M. J., Bates, M. & Zhuang, X. 2006. Sub-diffraction-limit imaging by stochastic optical reconstruction microscopy (STORM). *Nature Methods,* 3, 793–795.

Sahagún, B. D. 1560. Florentine Codex: General History of the Things of New Spain.

Scully, A. D., Macrobert, A. J., Botchway, S., O'Neill, P., Parker, A. W., Ostler, R. B. & Phillips, D. 1996. Development of a laser-based fluorescence microscope with subnanosecond time resolution. *Journal of Fluorescence,* 6, 119–125.

So, P. T. C., French, T., Yu, W. M., Berland, K. M., Dong, C. Y. & Gratton, E. 1995. Time-resolved fluorescence microscopy using two-photon excitation. *Bioimaging,* 3, 49–63.

Sokolov, K., Follen, M. & Richards-Kortum, R. 2002. Optical spectroscopy for detection of neoplasia. *Current Opinion in Chemical Biology,* 6, 651–658.

Stokes, G. G. 1852. On the change of refrangibility of light. *Philosophical Transactions of the Royal Society of London,* 142, 463–562.

Straub, M. & Hell, S. W. 1998. Fluorescence lifetime three-dimensional microscopy with picosecond precision using a multifocal multiphoton microscope. *Applied Physics Letters,* 73, 1769–1771.

Stryer, L. 1978. Fluorescence energy transfer as a spectroscopic ruler. *Annual Review of Biochemistry,* 47, 819–846.

Suhling, K., French, P. M. W. & Phillips, D. 2005. Time-resolved fluorescence microscopy. *Photochemical & Photobiological Sciences,* 4, 13–22.

Tsien, R. Y. 1998. The green fluorescent protein. *Annual Review of Biochemistry, 67,* 509–544.

Valeur, B. & Berberan-Santos, M. N. 2011. A brief history of fluorescence and phosphorescence before the emergence of quantum theory. *Journal of Chemical Education,* 88, 731–738.

van Geest, L. K. & Stoop, K. W. J. 2003. FLIM on a wide field fluorescence microscope. *Letters in Peptide Science,* 10, 501–510.

van Munster, E. B., Goedhart, J., Kremers, G. J., Manders, E. M. M. & Gadella, T. W. J., Jr. 2007. Combination of a spinning disc confocal unit with frequency-domain fluorescence lifetime imaging microscopy. *Cytometry Part A,* 71A, 207–214.

Verveer, P. J., Squire, A. & Bastiaens, P. I. H. 2000. Global analysis of fluorescence lifetime imaging microscopy data. *Biophysical Journal,* 78, 2127–2137.

Wagnieres, G. A., Mizeret, J. C., Studzinski, A. & van den Bergh, H. 1995. Endoscopic frequency-domain fluorescence lifetime imaging for clinical cancer photodetection: Apparatus design. *Proceedings of the SPIE,* 2392, 42–54.

Wagnieres, G. A., Star, W. M. & Wilson, B. C. 1998. In vivo fluorescence spectroscopy and imaging for oncological applications. *Photochemistry and Photobiology,* 68, 603–632.

Wang, X. F., Uchida, T., Coleman, D. M. & Minami, S. 1991. A two-dimensional fluorescence lifetime imaging system using a gated image intensifier. *Applied Spectroscopy,* 45, 360–366.

Weber, C. R., Schwarz, R. A., Atkinson, E. N., Cox, D. D., Macaulay, C., Follen, M. & Richards-Kortum, R. 2008. Model-based analysis of reflectance and fluorescence spectra for in vivo detection of cervical dysplasia and cancer. *Journal of Biomedical Optics,* 13, 064016.

Wlodarczyk, J. & Kierdaszuk, B. 2003. Interpretation of fluorescence decays using a power-like model. *Biophysical Journal,* 85, 589–598.

Wood, R. W. 1921. The time interval between absorption and emission of light in fluorescence. *Proceedings of the Royal Society of London Series A,* 99, 362–371.

Zhang, Y. L., Soper, S. A., Middendorf, L. R., Wurm, J. A., Erdmann, R. & Wahl, M. 1999. Simple near-infrared time-correlated single photon counting instrument with a pulsed diode laser and avalanche photodiode for time-resolved measurements in scanning applications. *Applied Spectroscopy,* 53, 497–504.

Zimmer, M. 2002. Green fluorescent protein (GFP): Applications, structure, and related photophysical behavior. *Chemical Reviews,* 102, 759–781.

Overview of fluorescence measurements and concepts

2 Photophysics of fluorescence

Klaus Suhling

Contents

2.1 INTRODUCTION: A BRIEF HISTORY OF FLUORESCENCE, MICROSCOPY, AND ITS CONTEXT

2.1.1 FLUORESCENCE

Fluorescence is a phenomenon that has existed for millions of years in the form of bioluminescence of fireflies, glow worms and bioluminescent sea creatures, and the northern lights. A review of references to luminescence in ancient manuscripts in various cultures all over the world through the ages can be found in Harvey's (1957) detailed and comprehensive book, *A History of Luminescence from the Earliest Times until 1900*, as well as other historical studies on this subject (Lee 2008; Valeur and Berberan-Santos 2011) (Figure 2.1). The oldest account of what we now know to be fluorescence is said to date from 1565 when Nicolas Monardes observed and gave an account of a bluish glow in a cup made from wood used for medicinal purposes, lignum nephriticum (Acuña and Amat-Guerri 2008; Acuña et al. 2009). Observations with similar "glowing" features were made in 1646 by Athanasius Kircher and also later in the 18th and 19th centuries by others, for example, Boyle, Newton, and Brewster.

However, the understanding and explanation of this phenomenon of "cold light," and in particular, to distinguish it from incandescence or scattered light, took a long time. In 1845, Herschel described the phenomenon of fluorescence of quinine, an anti-malaria agent, and distinguished it from scattered light. In 1852, Stokes reinvestigated this phenomenon and finally explained that the emitted light was of a longer wavelength than the absorbed light (Stokes 1852)—an effect now known as the Stokes' shift. Above

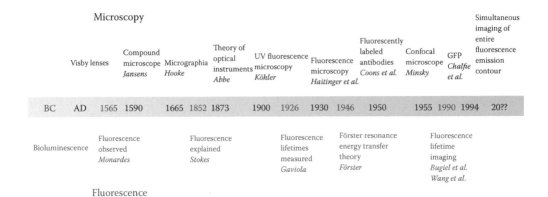

Figure 2.1 Brief history of microscopy and fluorescence. The combination of lasers, powerful computers, and cameras and the genetically encoded green fluorescent protein (GFP) have led to revolutionary advances in life sciences over the last two decades. The development of the microscope was essential for discovering and studying the cell, the basic and universal building blocks of all living organisms.

all, Stokes coined the term *fluorescence* (Stokes 1852, 1853). Despite this breakthrough, some confusion remained, but it eventually faded away like fluorescence itself (Malley 1991).

After some theoretical considerations regarding fluorescence lifetimes (Stern and Volmer 1919), the first reports on measuring nanosecond fluorescence lifetimes experimentally appeared in the mid-1920s (Gaviola 1926). The Perrin father and son team studied polarization-resolved fluorescence, also in the 1920s, and 1946 marks the landmark paper by Förster (1946), "Energy Migration and Fluorescence," on dipole–dipole resonance energy transfer, an important topic attracting much subsequent attention (Clegg 2006). This effect is now called Förster resonance energy transfer (FRET) in his honor.

Today, commercial instruments are available to study steady-state and time-resolved fluorescence (Udenfriend 1995), and the field of fluorescence spectroscopy is less about studying the basic phenomenon itself and its quantum mechanical explanation—although this is of course very important for computational modeling of, for example, excitation and emission or photoisomerization (Muguruza Gonzalez et al. 2009; Nieber and Doltsinis 2008)—but more about exploiting it as a tool: by employing fluorescent molecules as a probe to study their environment, particularly in life sciences (Birch 2011). Intrinsic or endogenous fluorescence of cells and tissues is also increasingly studied, often with a view to use it for clinical diagnostics, for example, by employing endoscopes. This is an emerging theme that is discussed in more detail in Chapter 6.

2.1.2 FLUORESCENCE IMAGING

Lenses were known to the Vikings (Schmidt et al. 1999), but the development of optical microscopy did not begin until the end of the 16th century when the Dutch spectacle makers Zaccharias Janssen and his son Hans noticed that two lenses in a tube allowed the magnification of small objects (Rost 1995). Later, Leuwenhoek used single lens microscopes and began to systematically study the microscopic world. In the 17th century, Robert Hooke started using compound microscopes to study small objects (while around the same time, Galileo began using telescopes to study large and distant objects). Hooke's "Micrographia or some physiological descriptions of minute bodies made by magnifying glasses with observations and inquiries thereupon," published in 1665, contained many images of small insects, fleas, spiders, seeds, plants, etc., appears to have inspired many others and was the scientific bestseller of its day (Figure 2.1). Indeed, to describe the compartmental structures he observed in cork, Hooke coined the term "cell." The development of the microscope was, of course, essential for discovering and studying the basic and universal building block of all living organisms, the cell (Harris 1999).

In the early 1900s, fluorescence microscopy developed from August Köhler's UV microscopy, when UV-excited autofluorescence was observed. However, it was not until the 1930s that Max Haitinger and others began systematically staining samples with fluorescent dyes, initially to make only weakly autofluorescent biological samples visible. In the early 1950s, Albert Coons developed the technique of labeling antibodies with fluorescent dyes (Rost 1995). This technique was extended to genetically encoded fluorescent labels, that is, GFP, in the 1990s, which revolutionized fluorescence imaging (Chalfie et al. 1994), eventually recognized by the award of the Nobel Prize in Chemistry in 2008 to Shimomura, Chalfie, and Tsien (Pieribone and Gruber 2005). Nature.com's "Nature Milestones in Light Microscopy" website provides a detailed historical overview of the development of optical microscopy.*

2.2 FLUORESCENCE AS A TOOL IN SENSING, IMAGING, AND MICROSCOPY

The great advantage of fluorescence as a tool is that, being an optical phenomenon, in most cases, it involves non-ionizing radiation, is nondestructive and minimally invasive, and can therefore be applied to living cells and tissues. This is a feature it shares with nonfluorescence-based, label-free (and thus truly noninvasive) optical microscopy techniques such as, for example, coherent anti-Stokes Raman scattering (CARS) and second harmonic generation (SHG) and third harmonic generation (THG) imaging (Michalet et al. 2003). These techniques are discussed in more detail in the context of nonlinear microscopy described in Chapter 6.

Compared to bioluminescence or chemiluminescence techniques, an advantage of using fluorescence is that a single fluorophore can be excited repeatedly (around 10^5 times and emitting as many fluorescence photons), unlike bioluminescence or chemiluminescence, which irreversibly produces only a single photon per chemical reaction of interacting molecules. Compared to radioactive labeling techniques, fluorescence methods are easier to manage, and the samples are less cumbersome to dispose of than those involving radioactive methods.

Although conventional fluorescence microscopy has a modest spatial resolution compared with electron or x-ray microscopy, compatibility with living systems is decisive for studying not only structure and morphology but also dynamics and function (Bastiaens and Pepperkok 2000; Harris 1999). Indeed, improving the spatial resolution in optical microscopy has recently become a field of great activity, which has spawned several promising methods, some of which have already been commercialized (Cox et al. 2012; Galbraith and Galbraith 2011; Heilemann 2010; Heintzmann and Ficz 2006; Hirvonen and Smith 2011; Schermelleh et al. 2010).

Conventional fluorescence microscopy relies on contrast according to the fluorescence intensity, with the fluorescence detection sensitivity extending down to the single molecule level (Prummer et al. 2004; Tinnefeld and Sauer 2005). Due to the Stokes shift of fluorescence emission, the exciting light can be eliminated from the image so that only fluorescence on a dark background is detected, leading to a high contrast (Michalet et al. 2003). Two- and three-dimensional fluorescence images to locate, for example, labeled proteins can be recorded using wide-field or one- or two-photon confocal scanning techniques. Confocal microscopy, the basic principle of which is a 1955 invention by Minsky (1988), is based on a spot of focused light scanned across the sample and detecting the fluorescence through a pinhole. This allows out-of-focus fluorescence emanating from above and below the focal plane to be eliminated from the image. This optical sectioning feature is also achieved by the more recent technique of multiphoton excitation microscopy (Denk et al. 1990; König 2000), as described in Chapter 6. The invention of laser and computers with image processing capacity and the demonstration of the advantages of this type of microscopy for cell biology (White et al. 1987) have led to a widespread use of confocal and multiphoton excitation microscopy over the last decade or so (Amos and White 2003). Time-lapse imaging allows the temporal evolution of the system to be studied, and fluorescence bleaching techniques can provide information about the diffusion of the fluorescent probe (Kang et al. 2009; Lippincott-Schwartz et al. 2001; Luby-Phelps et al. 1986). Imaging of FRET between a fluorescent donor and acceptor enables the determination of protein interaction and conformational changes well below the optical diffraction limit (Jares-Erijman and Jovin 2003).

The key purpose of fluorescence microscopy is the localization of fluorescent dyes, nanoparticles, or proteins in the specimen, which can be achieved by mapping the fluorescence intensity. However, exploiting more of the multidimensional fluorescence signal, for example, lifetime, spectrum, and

---

* http://www.nature.com/milestones/milelight/index.html.

Overview of fluorescence measurements and concepts

Table 2.1 **Some features and parameters that can be sensed with fluorescence techniques**

FEATURE TO BE SENSED	FLUORESCENCE PARAMETER YIELDING INFORMATION	FLUOROPHORE	REFERENCE	REMARKS
Hetero-FRET	Lifetime, spectrum, polarization	Many, provided there is spectral overlap, fluorescent proteins for polarization	Festy et al. 2007; Jares-Erijman and Jovin 2003; Mattheyses et al. 2004; Peter and Ameer-Beg 2004; Suhling et al. 2005	Conformational changes and interaction with other molecules
Homo-FRET	Polarization	Many, provided they have a small Stokes shift	Bader et al. 2007; Bader et al. 2009; Roberti et al. 2011; Thaler et al. 2009; Varma and Mayor 1998; Vishwasrao et al. 2012; Yeow and Clayton 2007	The only[a] way to detect homo-FRET is by polarization (when the fluorescent lifetimes of donor and acceptor are the same, see discussion in Suhling et al. 2014). For this approach fluorescent proteins are best, due to their large rotational correlation time
O_2	Lifetime	Ru	Gerritsen et al. 1997; Hosny et al. 2012	Long lifetimes
Viscosity	Polarization	Many, provided the rotational correlation time is no longer than 10 times the fluorescence lifetime	Suhling et al. 2004	
Viscosity	Lifetime of fluorescent molecular rotors	e.g., BODIPY-C_{12}, DCVJ, CCVJ, Thioflavin T, DASMPI, Cy dyes	Ghiggino et al. 2007; Haidekker et al. 2002, 2006; Hungerford et al. 2009; Kuimova et al. 2008, 2009; Levitt et al. 2009; Rumble et al. 2012	
Polarity	Spectrum	Nile red, lauradan, prodan, di-4-ANEPPDHQ	Giordano et al. 2012; Owen et al. 2006	

(continued)

Table 2.1 **(Continued) Some features and parameters that can be sensed with fluorescence techniques**

FEATURE TO BE SENSED	FLUORESCENCE PARAMETER YIELDING INFORMATION	FLUOROPHORE	REFERENCE	REMARKS
pH		BCECF	Hanson et al. 2002; Lin et al. 2003; Masters et al. 1997; Ogikubo et al. 2011	
Ca^{2+}	Intensity, lifetime	Calmodulin sensing GFP, Quin-2	Herman et al. 1997; Lakowicz et al. 1994; Miyawaki et al. 1997; Sanders et al. 1994;	
Cl^-	Lifetime	6-Methoxy-quinolyl acetoethyl ester (MQAE)	Kaneko et al. 2004	
Cu^{2+}	Lifetime	GFP-FRET sensor	Hötzer et al. 2011, 2012	
Na^+	Lifetime	CFP-YFP FRET, sodium green	Biskup et al. 2004; Szmacinski and Lakowicz 1997	
K^+	Lifetime	CFP-YFP FRET	Biskup et al. 2004	
Mg^{2+}	Lifetime	Mag-quin-2, magnesium green orange	Szmacinski and Lakowicz 1996	
Refractive index	Lifetime	GFP, quantum dots, nanodiamonds, fluorophores, which are not sensitive to anything else	Ma et al. 2011; Van Manen et al. 2008; Pliss et al. 2012; Suhling et al. 2002; Tisler et al. 2009; Wuister et al. 2004	Default effect due to Strickler Berg equation
Extrinsic fluorophore concentration	Lifetime		Bisby et al. 2012; Gadella et al. 1994, 1995	
Intrinsic fluorophore concentration	Lifetime	NADH, FAD	Cubeddu et al. 1999; König et al. 1999; Lakowicz et al. 1992; Schweitzer et al. 2004, 2007; Skala et al. 2007; Yu and Heikal 2009; Zhang et al. 2002	
Glucose	Lifetime	Ruthenium-malachite green FRET, glucose/galactose binding protein (GBP) labeled Badan	Saxl et al. 2009, 2011; Tolosa et al. 1997	

Note: The examples of fluorophores and the references are by no means exhaustive.
[a] In the special case of the cerulean fluorescent protein (CFP), the fluorescence lifetime has been reported to change due to homo-FRET (Koushik and Vogel 2008).

polarization (Bright 1995), can provide not only localization but also information about the biophysical environment of the probe.

Spectral imaging is most frequently applied to map the position of different fluorophores emitting in different wavelength regions, for example, for colocalization studies. Moreover, dyes like Nile red, laurdan, prodan, and di-4-ANEPPDHQ exhibit a spectral shift according to the polarity of their environment (see Table 2.1). This could be an indicator of, for example, water penetration into a membrane, as water has a high dielectric constant, 80, which causes a red shift of the probe's spectrum. Generally, the polarity of the probe's environment shifts the emission spectrum. In combination with imaging, this allows polarity mapping. Polarity is a measure of the dielectric constant and is, for example, important for solubilization. The dielectric constant also plays a role in shielding charged molecules, since the electrostatic attraction or repulsion is given by the Coulomb force, which is inversely proportional to the dielectric constant.

Fluorescence lifetime imaging microscopy (FLIM) is a powerful method to monitor the local environment of a molecular probe independent of the fluorescence intensity or local probe concentration. The fluorescence lifetime provides an absolute measurement, which, compared to fluorescence intensity-based imaging, is also less susceptible to artifacts arising from scattered light, photobleaching, nonuniform illumination of the sample, or light path length. Measurements can be made *in situ*, thus allowing access to biological function within a true physiological context without compromising the cell by biochemical assays (Becker 2012; Borst and Visser 2010; Festy et al. 2007; Peter and Ameer-Beg 2004; Suhling et al. 2005). Note that photophysical effects that are designed to occur in a sample, for example, FRET, can be observed by FLIM or intensity-based fluorescence imaging. FLIM is particularly good at detecting them, because it can be difficult to disentangle, interpret, or indeed quantify factors affecting the fluorescence intensity.

The most frequent use of fluorescence lifetime measurements is to detect FRET to identify protein interactions or conformational changes of proteins (Duncan 2006; Jares-Erijman and Jovin 2003; Peter and Ameer-Beg 2004; Wallrabe and Periasamy 2005). However, the fluorescence lifetime has also been used to probe the local environment of fluorophores, detecting or mapping, for example, ion concentrations such as Ca^{2+} (Herman et al. 1997; Lakowicz et al. 1994; Sanders et al. 1994), Cl^- (Kaneko et al. 2004), Cu^{2+} (Hötzer et al. 2011, 2012), Na^+ (Biskup et al. 2004), K^+ (Biskup et al. 2004), and Mg^{2+} (Szmacinski and Lakowicz 1996) in cells (see Table 2.1).

In addition, oxygen concentrations can be sensed using long-lived ruthenium-based sensors (Estrada et al. 2008; Finikova et al. 2008; Gerritsen et al. 1997; Hosny et al. 2012). Lifetime measurements are particularly advantageous, since intensity-based fluorescence imaging of oxygen in cells would require a calibration of the intensity of the probe unquenched by oxygen as well as knowing its concentration in the cell. This is not practically possible.

Other examples of FLIM are mapping the pH in single cells (Carlsson et al. 2000; Lin et al. 2003; Sanders et al. 1995) and skin (Behne et al. 2002; Hanson et al. 2002). Here, the pH sensor 2,7-bis-(2-carboxyethyl)-5-(and-6) carboxyfluorescein (BCECF) was used to image pH in the skin stratum corneum. The authors used two-photon excitation FLIM to nondestructively obtain pH maps at various depths, which is difficult to achieve by non-optical methods. Intensity-based fluorescence imaging of the pH probe could not have been used for their study as the observation of a variation in fluorescence intensity could be ascribed to either a change in pH or a variation of the local probe concentration.

Moreover, fluorescent molecular rotors have recently been used to image microviscosity, particularly that of a biological environment (Haidekker and Theodorakis 2007, 2010; Haidekker et al. 2001, 2010; Kuimova 2012a, b; Luby-Phelps et al. 1993; Uzhinov et al. 2011; Wandelt et al. 2005). Fluorescent molecular rotors are distinctive fluorophores whose quantum yield and fluorescence lifetime depend on the viscosity according to a model proposed by Förster and Hoffmann (1971) or later in a more general form by Loutfy (1986). Their radiative de-excitation pathway competes with intramolecular twisting in the excited state, which can lead to nonradiative deactivation of the excited state. The rate constant for the latter pathways decreases in viscous media, such that the fluorescence lifetime and quantum yield are high in viscous microenvironments and low in nonviscous microenvironments. Fluorescent molecular rotors have been used to measure the microviscosity in polymers (Loutfy 1986), sol-gels (Hungerford et al. 2009; Rei et al. 2008), micelles (Law 1981), ionic liquids (Gutkowski et al. 2006; Lu et al. 2003; Paul and Samanta

2008), blood plasma (Haidekker et al. 2002), liposomes (Kung and Reed 1986; Nipper et al. 2010), and biological structures such as tubulin (Kung and Reed 1989) and living cells (Haidekker et al. 2001; Kuimova et al. 2008; Levitt et al. 2009; Luby-Phelps et al. 1993; Peng et al. 2011; Wandelt et al. 2005). They can be accomplished either by ratiometric measurements (Ghiggino et al. 2007; Haidekker et al. 2006; Kuimova et al. 2009; Nipper et al. 2010; Peng et al. 2011; Wandelt et al. 2003, 2005) or by lifetime measurements (Kuimova et al. 2008; Levitt et al. 2009; Peng et al. 2011).

In combination with FLIM, molecular rotors can be used to map microviscosity, particularly in a biological environment. This approach does not require conjugation of the molecular rotor to another fluorescence label. It also decouples the influence of the viscosity on the fluorescence intensity from that of the rotor concentration and also allows detection of heterogeneous rotor environments via multiexponential fluorescence decays. In addition, spectral sensitivity variations of instruments measuring the calibration curves and performing the imaging do not bias quantitative measurements.

The use of FLIM in the cases mentioned above is more robust and reliable than fluorescence intensity-based imaging methods, since FLIM is unaffected by variations of illumination intensity, variations of fluorophore concentration, or photobleaching—provided the probes do not aggregate, and the photoproducts do not fluoresce.

In addition, FLIM applications in diverse areas such as forensic science (Bird et al. 2007), combustion research (Ni and Melton 1996), luminescence mapping in diamond (Liaugaudas et al. 2009, 2012), microfluidic systems (Benninger et al. 2005; Elder et al. 2006), art conservation (Comelli et al. 2004), and lipid order problems in physical chemistry (Togashi et al. 2005) have also been reported. Moreover, efforts are underway to use FLIM of autofluorescence, possibly combined with endoscopy, for clinical diagnostics (Fruhwirth et al. 2010; Requejo-Isidro et al. 2004; Skala et al. 2007; Yu and Heikal 2009). Indeed, FLIM of autofluorescence has been used to provide intrinsic contrast in unstained tissue (Cubeddu et al. 1999; Dowling et al. 1998; Elson et al. 2004; Siegel et al. 2003; Tadrous et al. 2003), the retina (Schweitzer et al. 2004, 2007), and teeth (Birmingham 1997; König et al. 1999), as reviewed previously (Elson et al. 2004; Urayama and Mycek 2003). The combination of multiphoton excitation for deep, sectioned, tissue imaging with FLIM yields contrast not available with fluorescence intensity-based imaging.

Fluorescence has also been used to monitor the uptake of antibody fragments labeled with photosensitizers (Kuimova et al. 2007; Stamati et al. 2010), and lifetime studies have been employed to study aggregation of sensitizers in photodynamic therapy (Connelly et al. 2001; Kinzler et al. 2007; Kress et al. 2003; Scully et al. 1997).

A summary of the features and parameters that can be sensed with fluorescence is presented in Table 2.1.

The versatility of fluorescence microscopy can be extended even further to obtain more information about the environment of the fluorescent probe by imaging the fluorescence lifetime, spectrum, and polarization (Levitt et al. 2009). To appreciate the benefits of this approach, some background and context of the phenomenon of fluorescence are essential. Indeed, fluorescence as an optical phenomenon can also be discussed in the context of interaction of light and matter and wave-particle duality.

The question what light is has occupied many minds throughout the history of mankind (Roychoudhuri et al. 2008). On the one hand, it can be described as a wave. This idea goes back to Huygens' idea of point sources being the sources of spherical waves, and Thomas Young's convincing interpretation of interference of light waves in a double slit experiment has been a milestone in the development of optics. In 1860, while at King's College London, James Clerk Maxwell unified electricity and magnetism, two hitherto separate fields, into a theory of electromagnetism (Tolstoy 1981). This approach allowed him to calculate the speed of propagation of electromagnetic waves, which turned out to be the speed of light, already known from Ole Rømer's astronomical observations of the moons of Jupiter in 1676 (Bobis and Lequeux 2008) and Fizeau's terrestrial measurements in 1849. In Maxwell's model, light is an electromagnetic wave with wavelengths of roughly 400–800 nm, that is, an octave in acoustic terms. The wave theory of light was extremely successful and allowed the explanation of phenomena such as the interference and diffraction of light, polarization, the dipole radiation characteristics of its emission, and the Doppler effect (Jenkins and White 1950).

However, it was discovered that there are experiments and phenomena that cannot be explained by assuming that light is a wave, for example, the blackbody spectrum (explained by Max Planck in 1900), the photoelectric effect (explained by Albert Einstein in 1905), and the Compton effect, scattering of light by electrons. The interpretation of these phenomena requires the particle nature of light, that is, the assumption that light energy is quantized into discrete energy packets known as photons (Roychoudhuri et al. 2008). A photon is defined as the smallest energy unit E that light can have, and it is quantitatively given by

$$E = h\nu \tag{2.1}$$

where h is Planck's constant, and ν is the frequency of light, that is, its speed c_0 divided by its wavelength λ, $\nu = c_0/\lambda$.

Technically, a photon is a quantized field (Roychoudhuri et al. 2008). So it appears that while it is not easy to answer the question what light really is in simple terms, there are two complementary models to describe its behavior. As a rough hand-waving guide, the wave model is predominantly used for the propagation of light, whereas the photon model is used for the interaction of light with matter (Jenkins and White 1950).

2.3 FLUORESCENT COMPOUNDS

Some minerals fluoresce, and naturally occurring fluorescent dyes have been known for a long time. The first synthetic dye was mauve, synthesized by Perkin in Manchester in 1856. It had a low quantum yield, but shortly afterward, the much brighter dye fluorescein was first synthesized by von Baeyer in 1871. This work was closely linked to color chemistry, that is, the research into dyes for staining fabrics (Zollinger 2003). Often, these dyes were not fluorescent, but they did absorb light, and were of major interest for the textile industry—not only in the West but also in China for staining silk, for example.

Today, fluorescence sensing and microscopy can be performed by labeling a sample with fluorescent dyes, quantum dots (Resch-Genger et al. 2008), or other nanoparticles (Green et al. 2009; Howes et al. 2010) (even nanodiamonds have been used; Faklaris et al. 2009; Neugart et al. 2007) or genetically encoded fluorescent proteins, and also by imaging autofluorescence, that is, endogenous fluorescence from tryptophan, collagen, elastin, FAD, flavins, NADH, or, in the case of plants, chlorophyll (Elson et al. 2004; Urayama and Mycek 2003).

2.4 THEORETICAL BACKGROUND

2.4.1 ABSORPTION OF LIGHT—THE LAMBERT–BEER LAW

The absorption of light passing through a sample is governed by the Lambert–Beer law. It states that the intensity of light I after passing through the sample is reduced compared to the incident intensity I_0 according to (Lakowicz 2006; Valeur 2002)

$$I = I_0 \, 10^{-\varepsilon[c]d} \tag{2.2}$$

where ε is the extinction coefficient, $[c]$ is the concentration of the sample, and d is the pathlength, as illustrated in Figure 2.2. The extinction coefficient is usually quoted at the absorption peak and has units of mol l^{-1} cm^{-1}, so that the product $\varepsilon[c]d$ is dimensionless, as required for Equation 2.2.

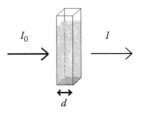

Figure 2.2 Illustration of the absorption of light when passing through a cuvette according to the Lambert–Beer law. I_0 is the incident light intensity, and I is the light intensity after passing through a sample of thickness d.

The logarithmic ratio of the incident intensity to the intensity after traversing the sample is known as the absorbance A:

$$\log \frac{I_0}{I} = \varepsilon[c]d = A \tag{2.3}$$

The absorbance is thus a quantitative but unitless measure, and it is independent of illumination intensity. An absorption spectrum yields information about the excited state, and it is not necessary for the molecules to fluoresce in order to obtain an absorption spectrum.

The Lambert–Beer law is effectively a measure of how much light is removed out of the beam path by absorption. Note that scattering also removes light out of the beam path, and a similar concept can be used to account for this, the turbidity. In contrast to light scattering, which is measured directly, the turbidity measures the light that is not scattered, so the difference of incident light intensity and light intensity after the sample is the scattered light into all angles. In the case of Rayleigh scattering, which is proportional to λ^{-4}, a double logarithmic plot should yield a straight line with gradient of -4. If absorbers are also present, their spectra should sit on top of this line. This can be helpful for absorption studies of dyes in small vesicles (Castanho et al. 1997).

The integral over the extinction coefficient yields the oscillator strength f (Huber and Sandeman 1986). It is defined as the fraction of a classical harmonic oscillator participating in a given transition. It can be expressed as

$$f = \frac{4mc_0^2\varepsilon_0 \ln 10}{e^2 N_A n} \int_{v_1}^{v_2} \varepsilon(\tilde{v})\,d\tilde{v} \tag{2.4}$$

where m is the mass of an electron, c_0 is the speed of light in vacuum, ε_0 is the permittivity of free space, m is the charge of an electron, N_A is Avogadro's number, and $\tilde{v}$ is the wavenumber, that is, $1/\lambda$, where λ is the wavelength. ε is the extinction coefficient.

The oscillator strength is a dimensionless quantity with $0 < f < 1$, and $\frac{c_0 \ln 10}{h N_A n} \int_{v_1}^{v_2} \varepsilon(\tilde{v})\,d\tilde{v}$ is known as the Einstein B-coefficient B_{km}, with h as Planck's constant, which can thus be calculated from the absorption spectrum. The order of magnitude of the oscillator strength gives an indication of the nature of the absorption transition:

- $f \approx 10^{-7}$ ($\varepsilon \approx 0.1$ mol l^{-1} cm^{-1}) spin forbidden and Laporte forbidden (within the same atomic shell, e.g., d–d transitions; Laporte 1924), for example, $[Mn(H_2O)_6]^{2+}$ ions.
- $f \approx 10^{-5}$ ($\varepsilon \approx 10$ mol l^{-1} cm^{-1}) spin-allowed but Laporte forbidden, for example, transition metal ions, $[Ni(H_2O)_6]^{2+}$, $[Co(H_2O)_6]^{2+}$, $[Cu(H_2O)_6]^{2+}$.
- $f \approx 10^{-3}$ ($\varepsilon \approx 1000$ mol l^{-1} cm^{-1}) spin-forbidden and Laporte forbidden, but mixing of orbitals occurs.
- $f \approx 10^{-1}$ ($\varepsilon \approx 100{,}000$ mol l^{-1} cm^{-1}) spin-allowed and Laporte allowed, for example, charge transfer transitions.

Note that care should be taken when the oscillator strength is expressed as a product of a numerical constant and the integral of the absorption spectrum. Cantor and Schimmel (1980) and Birks (1970) quote $f = 4.315 \times 10^{-9} \int_{v_1}^{v_2} \varepsilon(\tilde{v})\,d\tilde{v}$, whereas Straughan and Walker's (1976) value for f is $f = 4.33 \times 10^{-13} \int_{v_1}^{v_2} \varepsilon(\tilde{v})\,d\tilde{v}$. The difference of four orders of magnitude stems from the use of different units for the extinction coefficient ε. It is usually expressed in mol l^{-1} cm^{-1}, whereas Straughan and Walker use the unusual units of mol l^{-1} m^{-1}.

It should be noted that the fact that a sample absorbs light does not mean that it will also fluoresce. For example, some metal ions like $[Ni(H_2O)_6]^{2+}$, $[Co(H_2O)_6]^{2+}$, and $[Cu(H_2O)_6]^{2+}$ have an absorption spectrum, but they do not fluoresce. Technically, their quantum yield is zero. Interestingly, in FRET studies, this phenomenon has been exploited by designing "dark absorbers," which absorb but do not emit (Ganesan et al. 2006).

Absorption measurements are not as sensitive as fluorescence measurements, as discussed in Wolf (2003). They can also be difficult to interpret quantitatively when combined with imaging, because of generally unknown pathlength and also because of light-scattering effects.

2.4.2 WHAT IS FLUORESCENCE?

Upon excitation into an excited state, a fluorescent molecule—a fluorophore—can return to its ground state either radiatively by emitting a fluorescence photon,

$$A^* = A + h\nu$$

where A^* is an excited fluorophore, and A is a ground state fluorophore
or nonradiatively, for example, by dissipating the excited state energy as heat (Lakowicz 2006; Valeur 2002)

$$A^* = A + \text{heat}$$

This depends on the de-excitation pathways available. Fluorescence is the radiative deactivation of the lowest vibrational energy level of the first electronically excited singlet state S_1 back to the ground state. Radiative transitions between states of different multiplicities are called phosphorescence. As they require a spin flip, they are quantum-mechanically forbidden, and hence, emission rates are low and the decay times long, from microsecond to seconds. As triplet states are energetically below the corresponding singlet states, the phosphorescence spectrum lies at longer wavelengths than the fluorescence spectrum. Also, an excited fluorophore in its triplet state is paramagnetic, whereas a fluorophore excited into its singlet state is not. The absorption and emission processes are illustrated by an energy level diagram after Jablonski (1935), as shown in Figure 2.2.

Each electronic energy level consists of vibrational and rotational levels. The energy difference between adjacent rotational levels is only 10^{-2} to 10^{-3} eV, so they cannot be resolved in liquids at room temperature. However, the rotational levels lead to the characteristic broadening of the bands. By contrast, the energy difference between electronic levels and vibrational levels is several electron volts and about 0.1 eV, respectively, and those are the transitions that are seen and are of interest for fluorescence spectroscopy.

The population distribution of these energy levels is given by a Boltzmann distribution. In equilibrium, the number of molecules in state u (n_u) relative to the number in state l (n_l) with an energy difference $E_u - E_l = \Delta E$ is

$$n_u = n_l \frac{g_u}{g_l} e^{-\Delta E / kT} \tag{2.5}$$

where g_u and g_l are the degeneracies of the levels ($g_u = g_l = 1$ for singlet states). At room temperature ($T = 293$ K), $kT = 0.0252$ eV, and an absorption wavelength of 500 nm $\Delta E = 2.46$ eV, and thus, $e^{-98} = 2.8 \times 10^{-43}$. The number of molecules in the upper state n_u is 2.8×10^{-43} of the number of molecules in the lower state, n_l, that is, only a truly negligible fraction of the molecules are electronically excited. In fact, the Boltzmann distribution (Equation 2.5) states that at room temperature, most molecules are in the lowest vibrational level of the electronic ground state.

The fluorescence lifetime τ is the average time a fluorophore remains in the electronically excited state S_1 after excitation, and τ is defined as the inverse of the sum of the rate parameters for all depopulation processes:

$$\tau = \frac{1}{k_r + k_{nr}} \tag{2.6}$$

where k_r is the radiative rate constant, and the nonradiative rate constant k_{nr} is the sum of the rate constant for internal conversion, k_{ic}, and the rate constant for intersystem crossing to the triplet state, k_{isc}, so that $k_{nr} = k_{ic} + k_{isc}$ (Figure 2.3). The fluorescence emission always occurs from the lowest vibrational level of S_1, a rule known as Kasha's rule (Kasha 1950).

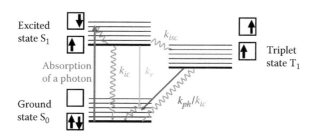

Figure 2.3 Schematic energy level diagram of a fluorescent molecule (Jablonski diagram). It depicts the ground state S_0, the first electronically excited singlet state S_1 (both with antiparallel electron spins), the triplet state T_1 (parallel electron spins), and the transitions between them. The thin lines represent vibrational energy levels. Nonradiative relaxation to S_0 can occur via internal conversion k_{ic} and intersystem crossing (from singlet to triplet) k_{isc}. k_r is the radiative rate constant, and k_{ph} is the rate constant for phosphorescence.

$\tau_0 = k_r^{-1}$ is the natural or radiative lifetime, which is related to the fluorescence lifetime τ via the fluorescence quantum yield ϕ:

$$\phi = \frac{\tau}{\tau_0} = \frac{k_r}{k_r + k_{nr}} = \frac{1}{1 + \frac{k_{nr}}{k_r}}$$

(2.7)

The fluorescence quantum yield can be thought of as the ratio of the number of fluorescence photons emitted to the number of photons absorbed (regardless of their energy) and is always less than 1. And due to $\phi\tau_0 = \tau$, τ_0 can be thought of as the longest lifetime the fluorophore can have.

The radiative rate constant k_r is related to the absorption and fluorescence spectra and is a function of the refractive index of the medium surrounding the fluorophore:

$$k_r = 2.88 \times 10^{-9}\, n^2\, \frac{\int F(\tilde{v})\,d\tilde{v}}{\int F(\tilde{v})\tilde{v}^{-3}\,d\tilde{v}} \int \frac{\varepsilon(\tilde{v})}{\tilde{v}}\,d\tilde{v}$$

(2.8)

where n is the refractive index, F is the fluorescence emission, ε is the extinction coefficient, and $\tilde{v}$ is the wavenumber. This equation is known as the Strickler–Berg equation (Strickler and Berg 1962). A recent more accurate treatment taking into account the transition dipole moment, an intrinsic property of the molecule, has been devised by Toptygin (2003) in an excellent review of the subject.

Essentially, the Strickler–Berg equation is a version of the Einstein coefficients for absorption and spontaneous and stimulated emission (Einstein 1916, 1917) but adapted for molecules with broad absorption and emission spectra, rather than atomic line spectra.

The wavelength of the absorption and emission spectra can yield information of the transition dipole moment of the fluorophore, and in particular, the interaction of the fluorophore with the surrounding solvent molecules. The excited state energy decreases with increasing polarity. The Lippert equation allows calculation of the transition dipole moment from measurements of the absorption and fluorescence peaks in solvents of varying refractive index and dielectric constant. It is based on general solvent effects and does not take into account chemical properties of the fluorophore or solvent molecules, for example, hydrogen bonding, but nonetheless, it is a useful general model.

Lippert plots of the Stokes shift $\tilde{v}_{abs} - \tilde{v}_{em}$ versus the orientation polarizability Δf can be created according to

$$\tilde{v}_{abs} - \tilde{v}_{em} = \frac{2}{hc_0}(\mu_e - \mu_g)^2 a^{-3} \Delta f + \text{constant}$$

(2.9)

where μ_e and μ_g are excited and ground state dipole moments, respectively, a is the cavity radius, h is Planck's constant, c_0 is the speed of light in vacuum, and

$$\Delta f = f(\varepsilon) - f(n^2) = \frac{\varepsilon - 1}{2\varepsilon + 1} - \frac{n^2 - 1}{2n^2 + 1} \tag{2.10}$$

is the orientation polarizability. $f(\varepsilon)$ is the low-frequency polarizability, which depends on the electronic and molecular redistribution of the solvent molecules upon excitation. $f(n)$ is the high-frequency polarizability, which depends on the rapid motion of electrons in the solvent molecules. n and ε have opposite effects on the Stokes shift, an increase in n will increase $f(n)$ and thus reduce the Stokes shift, whereas an increase in ε and thus $f(\varepsilon)$ will increase the Stokes shift. μ_e and μ_g are excited and ground state dipole moments, which are measured in Debye (D). 4.8 D corresponds to 1 Å change in μ_g. The reorientation of the solvent molecules around an excited fluorophore takes place within 100 ps or less and can be studied with time-resolved emission spectroscopy. The absorption spectra are less sensitive to the solvent polarity than the emission spectra due to the fast absorption process, 10^{-15} s—the excited state energy decreases only after the absorption has occurred (Lakowicz 2006; Valeur 2002).

2.4.3 DECAY OF EXCITED STATE

After excitation, N fluorophores will leave the excited state S_1 according to the following rate equation:

$$dN = (k_r + k_{nr})\, N(t)dt \tag{2.11}$$

where t is the time. Integration, using Equation 2.1, and taking into account that the fluorescence intensity $F(t)$ is proportional to the excited state population $N(t)$ yields

$$F(t) = F_0 e^{-t/\tau} \tag{2.12}$$

where F_0 represents the fluorescence intensity at $t = 0$. The decay of the fluorescence intensity thus follows an exponential decay law (Istratov and Vyvenko 1999), schematically shown in Figure 2.4. Note that on a logarithmic fluorescence intensity scale, a monoexponential decay conveniently appears as a straight line. This way of plotting the data thus aids simple visual inspection of the decay behavior.

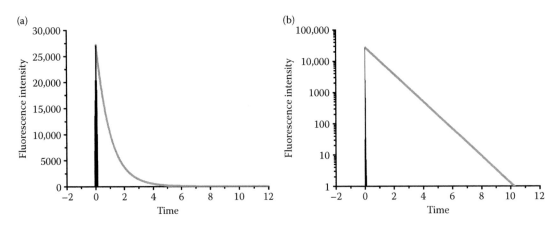

Figure 2.4 Schematic fluorescence decay following a monoexponential decay law. The instrumental response is shown in black, the decay in gray. (a) On a linear fluorescence intensity scale. (b) On a logarithmic intensity scale, a monoexponential decay conveniently appears as a straight line. Simple visual inspection thus clearly shows a monoexponential decay over five orders of magnitude.

2.4.4 POLARIZATION-RESOLVED LIFETIME MEASUREMENTS—TIME-RESOLVED FLUORESCENCE ANISOTROPY

Fluorescence anisotropy involves measuring fluorescence at polarizations parallel and perpendicular to that of the exciting light. Steady state, that is, non-time-resolved anisotropy imaging, can be used to detect energy migration or homo-FRET (resonance energy transfer between the same type of fluorophore), since it leads to a depolarization of the emitted fluorescence (Lidke et al. 2003; Squire et al. 2004; Vogel et al. 2010). The technique has been used to study the proximity of isoforms of the GPI-anchored folate receptor bound to a fluorescent analog of folic acid to detect lipid rafts (Varma and Mayor 1998).

Upon excitation with linearly polarized light, Brownian rotational diffusion of the fluorophore in its excited state or homo-FRET results in a depolarization of the fluorescence emission. Time-resolved fluorescence anisotropy imaging measures fluorescence decays at polarizations parallel and perpendicular to that of the exciting light (Figure 2.5). The time-resolved fluorescence anisotropy $r(t)$ can be defined as

$$r(t) = \frac{I_{\parallel}(t) - GI_{\perp}(t)}{I_{\parallel}(t) + 2GI_{\perp}(t)} \tag{2.13}$$

where $I_{\parallel}(t)$ and $I_{\perp}(t)$ are the fluorescence intensity decays parallel and perpendicular to the polarization of the exciting light, respectively (Lakowicz 2006; Valeur 2002). G accounts for different transmission and detection efficiencies of the imaging system at parallel and perpendicular polarization, and, if necessary, an appropriate background has to be subtracted (Suhling et al. 2004). For a spherical molecule, $r(t)$ decays as a single exponential and is related to the rotational correlation time θ according to

$$r(t) = (r_0 - r_{\infty}) \exp\left(-\frac{t}{\theta}\right) + r_{\infty} \tag{2.14}$$

where r_0 is the initial anisotropy, and r_{∞} accounts for a restricted rotational mobility.

The initial anisotropy r_0 of a fluorophore is given by (Gryczynski et al. 1995)

$$r_0 = \frac{2k}{2k+3}\left(\frac{3\overline{\cos}^2\alpha - 1}{2}\right) \tag{2.15}$$

where α is the angle between the absorption and emission dipole moments of the molecule, and $\overline{\cos}^2\alpha$ is the average of $\cos^2\alpha$ (Lakowicz 2006; Valeur 2002). k is the number of photons creating the excited state. The absorption probability is proportional to $\cos^2\alpha$ and for single-photon excitation ($k = 1$), and molecules

(a)   (b)

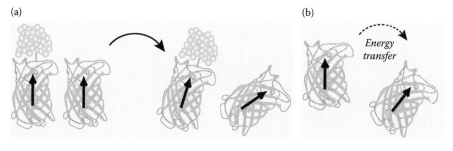

Energy transfer

Figure 2.5 Schematic representation of the fluorescence depolarization mechanisms. (a) Brownian rotational motion. This is affected by the viscosity of its surroundings, or by binding or conformational changes, and is characterized by the rotational correlation time θ. A fast rotational motion leads to a rapid depolarization. (b) Homo-FRET, resonance energy transfer between the same type of fluorophore. The anisotropy can be calculated from the difference between the polarization-resolved fluorescence decays $I_{\parallel}$ and $I_{\perp}$.

Overview of fluorescence measurements and concepts

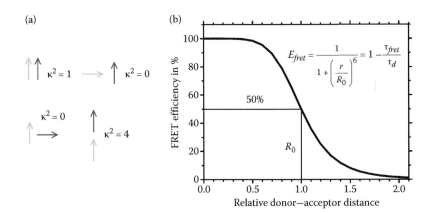

Figure 2.6 (a) Dipole orientation in FRET. The transition dipole moment of the donor is shown in green, and that of the acceptor in red. The corresponding κ^2 values are also indicated. (b) FRET efficiency as a function of relative donor–acceptor distance.

with parallel absorption and emission dipole moments $\alpha = 0$, the maximum value for the initial anisotropy is 0.4. However, due to the nature of the photoselection for absorption and emission transition dipoles, multiphoton excitation provides a greater dynamic range for anisotropy measurements than single-photon excitation (Birch 2001).

For a spherical molecule in an isotropic medium, θ is directly proportional to the viscosity η of the solvent and the volume V of the rotating molecule:

$$\theta = \frac{\eta V}{kT} \tag{2.16}$$

where k is the Boltzmann constant and T is the absolute temperature. Thus, imaging θ with time-resolved fluorescence anisotropy imaging (TR-FAIM) can measure the rotational mobility of a fluorophore in its environment (Figure 2.5a) and has been used to image the viscosity in living cells (Botchway et al. 2011; Suhling et al. 2004). Moreover, as the rotational diffusion can be slowed down by binding, TR-FAIM has the potential to visualize the binding of ligands and receptors in the cell. In addition, a hindered rotation of the fluorophore due to geometrical restrictions, for example, in the cell membrane, can also be detected with this method. Homo-FRET (Figure 2.5b) has been used to elucidate oligomerization states of proteins (Bader et al. 2007, 2009) or structural changes in proteins (Roberti et al. 2011; Thaler et al. 2009).

Note that the steady-state anisotropy is calculated from the fluorescence intensities, that is, integrated fluorescence decays, from Equation 2.13 and obeys the Perrin equation

$$r = \frac{r_0 - r_\infty}{1 + \dfrac{\tau}{\theta}} + r_\infty \tag{2.17}$$

where τ is the fluorescence lifetime.

2.4.5 FLUORESCENCE QUENCHING

The previous definition of k_{nr} in Equations 2.6 and 2.7 assumes no quenching processes. If there is quenching of the excited fluorophore by another molecule, Q, then this represents another pathway for the fluorophore to return into its ground state with a rate constant k_Q. So in addition to the terms in the denominator, there will appear the term $k_Q[Q]$ so that $k_{nr} = k_{ic} + k_{isc} + k_Q[Q]$. There are two types of quenching: static and dynamic.

In static quenching, ground state complex formation between the fluorophore and the quencher disables the former from absorbing a photon, and thus from fluorescing. In other words, a certain fraction of the fluorophores escapes observation, and only the fluorescence of the unperturbed species can be seen (Lakowicz 2006).

In dynamic quenching, the fluorophore does absorb a photon, but during the lifetime of the excited state, the quencher de-excites the fluorophore. Excited state reactions such as excimer and exciplex formation, energy transfer, electron transfer, and molecular collisions are all dynamic quenching mechanisms. Dynamic quenching thus reduces the fluorescence quantum yield ϕ and fluorescence lifetime τ (Lakowicz 2006).

Quenching studies can be very useful: In diffusion-controlled quenching processes, for example, diffusion rates of quenchers and fluorophores can be worked out, as the quenching strongly depends on their diffusion. The accessibility of fluorophores for the quencher can also be revealed, for example, if a fluorophore is bound to a protein or a membrane that is impermeable to the quencher, then neither static nor collisional quenching can occur.

Qualitatively, diffusion-controlled quenching is described by the Stern–Volmer equation (Stern and Volmer 1919):

$$\frac{F_0}{F} = \frac{\tau}{\tau_q} = 1 + K[c] \tag{2.18}$$

where F_0 and F are the fluorescence intensities in the absence and presence of the quencher, τ and τ_q are the fluorescence lifetimes in the absence and presence of the quencher, K is the quenching constant and $[c]$ is the quencher concentration. So for diffusion-controlled quenching, a plot of F_0/F versus $[c]$ yields a straight line with gradient K and intercept 1—which is expected if a single class of fluorophores is present, and all fluorophores are accessible to the quencher (Birks 1970; Lakowicz 2006; Valeur 2002).

If the quenching is occurs via collisional encounters, then $K = K_D = k_q \tau_q$ where k_q is the bimolecular quenching constant.

How does one distinguish between static and collisional quenching? If the quenching is static, then
- F_0/F should decrease with increasing temperature, and consequently K_d, the gradient of the Stern–Volmer plot, should decrease (as long as the quantum yield is constant), reflecting a decreased stability of the quencher/fluorophore complex.
- As the emitting fluorophore is unperturbed, there is no change in the fluorescence lifetime, i.e. $\tau/\tau_q = 1$.
- The absorption spectrum is likely to exhibit changes caused by ground state complexation.
  On the other hand, if the quenching occurs by collision, then
- F_0/F and K_d should increase with increasing temperature as long as the quantum yield is constant, because the solvent viscosity η decreases, and consequently, the diffusion rates increase, resulting in more collisional encounters between the fluorophore and the quencher. k_q is proportional to T/η.
- As quenching is an additional rate process which depopulates the excited state without fluorescence emission, a decrease in lifetime occurs. $F_0/F = \tau/\tau_q$.
- No changes in the absorption spectrum occur.

It is also possible to observe both static and collisional quenching for the same quencher/fluorophore system, that is, complex formation and collision. Then the fractional fluorescence remaining is given by the product of the fraction not complexed and the fraction not quenched by collisional encounters (Lakowicz 2006):

$$\frac{F_0}{F} = (1 + K_s[c])(1 + K_d[c]) \tag{2.19}$$

This modified form of the Stern–Volmer equation has a quadratic term in $[c]$ that causes a characteristic upward curvature, sometimes referred to as a positive deviation.

If the Stern–Volmer plots show a downward curvature, then this is an indication that not all of the fluorophores are accessible to the quencher as is the case in, for example, micelles or lipid bilayers. Consider two ensembles of fluorophores: (a) one is accessible to the quencher and (b) the other is not—it is buried. So the total fluorescence is (Lakowicz 2006)

$$F_0 = F^{(a)} + F^{(b)} \tag{2.20}$$

On adding quencher, only the accessible fraction is quenched:

$$F_0 = \frac{F^{(a)}}{(1 + K_s[c])} + F^{(b)} \tag{2.21}$$

Combination of Equations 2.20 and 2.21 yields

$$\frac{F_0}{F_0 - F} = \frac{F_0}{\Delta F} = \frac{1}{f_a K[c]} + \frac{1}{f_a}$$

where $f_a = F_0/\left(F_0^{(a)} + F_0^{(b)}\right)$ is the fraction of initial fluorescence, which is accessible to the quencher. Again, a straight line plot of F_0/F versus $1/[c]$ yields a straight line with intercept $1/f$ and gradient $1/(fK)$ (Lakowicz 2006).

2.4.6 FÖRSTER RESONANCE ENERGY TRANSFER

FRET is a bimolecular fluorescence quenching process where the excited state energy of a donor fluorophore is nonradiatively transferred to an acceptor molecule by a dipole–dipole coupling process (Förster 1946, 1948; Jares-Erijman and Jovin 2003; Lakowicz 2006; Valeur 2002). The FRET efficiency varies with the inverse sixth power of the distance between donor and acceptor and is usually negligible beyond 10 nm, as shown in Figure 2.6a. FRET can therefore be used as a "spectroscopic ruler" to probe intermolecular distances on the scale of the dimensions of the proteins themselves (Stryer and Haugland 1967). This is a significant advantage over colocalization of two fluorophores, which is limited to the optical resolution limit ($\approx$200 nm laterally, $\approx$500 nm axially). For FRET to occur, the donor emission spectrum and the acceptor absorption spectrum must overlap, and the transition dipole moments of the donor and acceptor must be oriented favorably (the orientation factor κ^2, a function of the angles between the transition dipole moments of the donor and acceptor and their line of separation, must not be zero, $\kappa^2 \neq 0$, as shown in Figure 2.6b.) (Jares-Erijman and Jovin 2003; Lakowicz 2006; Valeur 2002). If the fluorophore's Brownian rotation is slow compared to the energy transfer rate, then "static averaging" of κ^2 in three dimensions yields 0.476, whereas when the donor and the acceptor can have all orientations during the transfer time, "dynamic averaging" yields $\kappa^2 = 2/3$.

FRET can take place between dye pairs, fluorescent protein pairs, or quantum dots and dyes, where quantum dots are usually the donor, because of their broad absorption spectrum. The largest FRET distances can be obtained with quantum dots and lanthanides (Charbonnière and Hildebrandt 2008).

2.5 OUTLOOK

Fluorescence is widely used in many fields, from physical chemistry to clinical and pharmaceutical applications, drug discovery, forensic science, fluorescence microscopy in cell biology, and the life sciences in general, but also in environmental monitoring and art conservation. There are no signs that its role is going to be diminished in the future. New developments in the future will probably focus on new probes, particularly in the infrared (Gerega et al. 2011; Yazdanfar et al. 2010) and also multifunctional probes, which will allow multimodal imaging, for example, fluorescence and MRI.

Moreover, recent camera development has increased the detection sensitivity to near 100% quantum efficiency, and photocathode-based single point detectors are also now very efficient with around 50%

quantum efficiency (Becker et al. 2011) as discussed in Chapter 6. Any further developments in this field could focus on the acquisition speed, photon counting imaging with a 40 kHz camera frame rate has been demonstrated (Sergent et al. 2010), and SPAD array development has also come of age to be promising for applications in fluorescence microscopy (Li et al. 2011). In the longer term, photon energy-resolving detectors (Fraser et al. 2003, 2006) and superconducting nanowires (Stevens et al. 2006) may be employed for fluorescence spectroscopy and microscopy.

The combination of fluorescence enhancement by metallic surfaces via plasmonic effects also offers interesting possibilities and has already been applied in cell imaging (Le Moal et al. 2007) to map receptor internalization (Cade et al. 2010).

Technological advances in excitation sources, detectors, probes, detection methods, data processing, and analysis will ensure that fluorescence as a tool will continue to flourish as a powerful, sensitive, and versatile optical spectroscopic technique. Indeed, tissue autofluorescence may one day be used as a routine diagnostic tool for clinical applications.

REFERENCES

Acuña, A. U. and Amat-Guerri, F. (2008). "Early history of solution fluorescence: The lignum nephriticum of Nicolás Monardes." *Springer Series in Fluorescence* **4**: 3–20.

Acuña, A. U., Amat-Guerri, F., Morcillo, P., Liras, M. and Rodriguez, B. (2009). "Structure and formation of the fluorescent compound of lignum nephriticum." *Organic Letters* **11**(14): 3020–3023.

Amos, W. B. and White, J. G. (2003). "How the confocal laser scanning microscope entered biological research." *Biology of the Cell* **95**(6): 335–342.

Bader, A. N., Hofman, E. G., van Bergen en Henegouwen, P. M. P. and Gerritsen, H. C. (2007). "Imaging of protein cluster sizes by means of confocal time-gated fluorescence anisotropy microscopy." *Optics Express* **15**(11): 6934–6945.

Bader, A. N., Hofman, E. G., Voortman, J., van Bergen en Henegouwen, P. M. P. and Gerritsen, H. C. (2009). "Homo-FRET imaging enables quantification of protein cluster sizes with subcellular resolution." *Biophysical Journal* **97**(9): 2613–2622.

Bastiaens, P. I. H. and Pepperkok, R. (2000). "Observing proteins in their natural habitat: The living cell." *Trends in Biochemical Sciences* **25**: 631–637.

Becker, W. (2012). "Fluorescence lifetime imaging—Techniques and applications." *Journal of Microscopy* **247**(2): 119–136.

Becker, W., Su, B., Holub, O. and Weisshart, K. (2011). "FLIM and FCS detection in laser-scanning microscopes: Increased efficiency by GaAsP hybrid detectors." *Microscopy Research and Technique* **74**(9): 804–811.

Behne, M. J., Meyer, J. W., Hanson, K. M., Barry, N. P., Murata, S., Crumrine, D., Clegg, R. W., Gratton, E., Holleran, W. M. and Elias, P. M. (2002). "NHE1 regulates the stratum corneum permeability barrier homeostasis. Microenvironment acidification assessed with fluorescence lifetime imaging." *Journal of Biological Chemistry* **277**(49): 47399–47406.

Benninger, R. K. P., Hofmann, O., McGinty, J., Requejo-Isidro, J., Munro, I., Neil, M. A. A., deMello, A. J. and French, P. M. W. (2005). "Time-resolved fluorescence imaging of solvent interactions in microfluidic devices." *Optics Express* **13**(16): 6275–6285.

Birch, D. J. S. (2001). "Multiphoton excited fluorescence spectroscopy of biomolecular systems." *Spectrochimica Acta Part A-Molecular and Biomolecular Spectroscopy* **57**(11): 2313–2336.

Birch, D. J. S. (2011). "Fluorescence detections and directions." *Measurement Science & Technology* **22**(5): 052002.

Bird, D. K., Agg, K. M., Barnett, N. W. and Smith, T. A. (2007). "Time-resolved fluorescence microscopy of gunshot residue: An application to forensic science." *Journal of Microscopy-Oxford* **226**(1): 18–25.

Birks, J. B. (1970). *Photophysics of Aromatic Molecules*. Chichester: Wiley-Interscience.

Birmingham, J. J. (1997). "Frequency-domain lifetime imaging methods at Unilever Research." *Journal of Fluorescence* **7**(1): 45–54.

Bisby, R. H., Botchway, S. W., Hadfield, J. A., McGown, A. T., Parker, A. W. and Scherer, K. M. (2012). "Fluorescence lifetime imaging of E-combretastatin uptake and distribution in live mammalian cells." *European Journal of Cancer* **48**(12): 1896–1903.

Biskup, C., Kelbauskas, L., Zimmer, T., Benndorf, K., Bergmann, A., Becker, W., Ruppersberg, J. P., Stockklausner, C. and Klocker, N. (2004). "Interaction of PSD-95 with potassium channels visualized by fluorescence lifetime-based resonance energy transfer imaging." *Journal of Biomedical Optics* **9**(4): 753–759.

Biskup, C., Zimmer, T. and Benndorf, K. (2004). "FRET between cardiac Na$^+$ channel subunits measured with a confocal microscope and a streak camera." *Nature Biotechnology* **22**(2): 220–224.

Overview of fluorescence measurements and concepts

Bobis, L. and Lequeux, J. (2008). "Cassini, Rømer and the velocity of light." *Journal of Astronomical History and Heritage* **11**(2): 97–105.

Borst, J. W. and Visser, A. J. W. G. (2010). "Fluorescence lifetime imaging microscopy in life sciences." *Measurement Science & Technology* **21**(10): 102002.

Botchway, S. W., Lewis, A. M. and Stubbs, C. D. (2011). "Development of fluorophore dynamics imaging as a probe for lipid domains in model vesicles and cell membranes." *European Biophysics Journal with Biophysics Letters* **40**(2): 131–141.

Bright, F. V. (1995). "Modern molecular fluorescence spectroscopy." *Applied Spectroscopy* **49**(1): A14–A19.

Cade, N. I., Fruhwirth, G., Archibald, S. J., Ng, T. and Richards, D. (2010). "A cellular screening assay using analysis of metal-modified fluorescence lifetime." *Biophysical Journal* **98**(11): 2752–2757.

Cantor, C. R. and Schimmel, P. R. (1980). *Biophysical Chemistry, Part 2: Techniques for the Study of Biological Structure and Function.* Oxford: W. H. Freeman and Company (Oxford).

Carlsson, K., Liljeborg, A., Andersson, R. M. and Brismar, H. (2000). "Confocal pH imaging of microscopic specimens using fluorescence lifetimes and phase fluorometry: Influence of parameter choice on system performance." *Journal of Microscopy* **199**(2): 106–114.

Castanho, M. A. R. B., Santos, N. C. and Loura, L. M. S. (1997). "Separating the turbidity spectra of vesicles from the absorption spectra of membrane probes and other chromophores." *European Biophysics Journal with Biophysics Letters* **26**(3): 253–259.

Chalfie, M., Tu, Y., Euskirchen, G., Ward, W. W. and Prasher, D. C. (1994). "Green fluorescent protein as a marker for gene expression." *Science* **263**: 802–805.

Charbonnière, L. J. and Hildebrandt, N. (2008). "Lanthanide complexes and quantum dots: A bright wedding for resonance energy transfer." *European Journal of Inorganic Chemistry* **2008**(21): 3241–3251.

Clegg, R. M. (2006). The history of FRET: From conception through the labors of birth. In *Reviews in Fluorescence*, C. D. Geddes and J. R. Lakowicz, eds. New York: Springer Science and Business Inc, 1–45.

Comelli, D., D'Andrea, C., Valentini, G., Cubeddu, R., Colombo, C. and Toniolo, L. (2004). "Fluorescence lifetime imaging and spectroscopy as tools for nondestructive analysis of works of art." *Applied Optics* **43**(10): 2175–2183.

Connelly, J. P., Botchway, S. W., Kunz, L., Pattison, D., Parker, A. W. and MacRobert, A. J. (2001). "Time-resolved fluorescence imaging of photosensitiser distributions in mammalian cells using a picosecond laser line-scanning microscope." *Journal of Photochemistry and Photobiology A Chemistry* **142**(2–3): 169–175.

Cox, S., Rosten, E., Monypenny, J., Jovanovic-Talisman, T., Burnette, D. T., Lippincott-Schwartz, J., Jones, G. E. and Heintzmann, R. (2012). "Bayesian localization microscopy reveals nanoscale podosome dynamics." *Nature Methods* **9**(2): 195–200.

Cubeddu, R., Pifferi, A., Taroni, P., Torricelli, A., Valentini, G., Rinaldi, F. and Sorbellini, E. (1999). "Fluorescence lifetime imaging: An application to the detection of skin tumors." *IEEE Journal of Selected Topics in Quantum Electronics* **5**(4): 923–929.

Denk, W., Strickler, J. H. and Webb, W. W. (1990). "Two-photon laser scanning fluorescence microscopy." *Science* **248**: 73–76.

Dowling, K., Dayel, M. J., Lever, M. J., French, P. M. W., Hares, J. D. and Dymoke-Bradshaw, A. K. L. (1998). "Fluorescence lifetime imaging with picosecond resolution for biomedical applications." *Optics Letters* **23**(10): 810–812.

Duncan, R. R. (2006). "Fluorescence lifetime imaging microscopy (FLIM) to quantify protein-protein interactions inside cells." *Biochemical Society Transactions* **34**(5): 679–682.

Einstein, A. (1916). "Strahlungsemission und -absorption nach der Quantentheorie." *Berichte der Deutschen Physikalischen Gesellschaft* **13–14**: 3128–3323.

Einstein, A. (1917). "Zur Quantentherie der Strahlung." *Physikalische Zeitschrift* **18**: 121–128.

Elder, A. D., Matthews, S. M., Swartling, J., Yunus, K., Frank, J. H., Brennan, C. M., Fisher, A. C. and Kaminski, C. F. (2006). "The application of frequency-domain Fluorescence Lifetime Imaging Microscopy as a quantitative analytical tool for microfluidic devices." *Optics Express* **14**(12): 5456–5467.

Elson, D. S., Requejo-Isidro, J., Munro, I., Reavell, F., Siegel, J., Suhling, K., Tadrous, P. J., Benninger, R., Lanigan, P. M. P., McGinty, J., Talbot, C., Treanor, B., Webb, S., Sandison, A., Wallace, A., Davis, D. M., Lever, J., Neil, M. A. A., Phillips, D., Stamp, G. W. and French, P. M. W. (2004). "Time-domain fluorescence lifetime imaging applied to biological tissue." *Photochemical & Photobiological Sciences* **3**: 795–801.

Estrada, A. D., Ponticorvo, A., Ford, T. N. and Dunn, A. K. (2008). "Microvascular oxygen quantification using two-photon microscopy." *Optics Letters* **33**(10): 1038–1040.

Faklaris, O., Joshi, V., Irinopoulou, T., Tauc, P., Sennour, M., Girard, H., Gesset, C., Arnault, J. C., Thorel, A., Boudou, J. P., Curmi, P. A. and Treussart, F. (2009). "Photoluminescent diamond nanoparticles for cell labeling: Study of the uptake mechanism in mammalian cells." *ACS Nano* **3**(12): 3955–3962.

Festy, F., Ameer-Beg, S. M., Ng, T. and Suhling, K. (2007). "Imaging proteins in vivo using fluorescence lifetime microscopy." *Molecular Biosystems* **3**(6): 381–391.

Finikova, O. S., Lebedev, A. Y., Aprelev, A., Troxler, T., Gao, F., Garnacho, C., Muro, S., Hochstrasser, R. M. and Vinogradov, S. A. (2008). "Oxygen microscopy by two-photon-excited phosphorescence." *Chemphyschem* **9**(12): 1673–1679.

Förster, T. (1946). "Energiewanderung und Fluoreszenz." *Naturwissenschaften* **33**(6): 166–175, translated into English by Klaus Suhling, *Journal of Biomedical Optics* **117**(161): 011002, 012012.

Förster, T. (1948). "Zwischenmolekulare Energiewanderung und Fluoreszenz." *Annalen der Physik* **2**(6): 55–75, translated into English by Robert Knox, in *Biological Physics*, E. V. Mielczarek, E. Greenbaum and R. S Knox, eds. New York: Americal Institute of Physics, 148–160.

Förster, T. and Hoffmann, G. (1971). "Die Viskositätsabhängigkeit der Fluoreszenzquantenausbeuten einiger Farbstoffsysteme." *Zeitschrift für Physikalische Chemie Neue Folge* **75**: 63–76.

Fraser, G. W., Heslop-Harrison, J. S., Schwarzacher, T., Holland, A. D., Verhoeve, P. and Peacock, A. (2003). "Detection of multiple fluorescent labels using superconducting tunnel junction detectors." *Review of Scientific Instruments* **74**(9): 4140–4144.

Fraser, G. W., Heslop-Harrison, J. S., Schwarzacher, T., Verhoeve, P., Peacock, A. and Smith, S. J. (2006). "Optical fluorescence of biological samples using STJs." *Nuclear Instruments & Methods in Physics Research Section A-Accelerators Spectrometers Detectors and Associated Equipment* **559**(2): 782–784.

Fruhwirth, G. O., Ameer-Beg, S., Cook, R., Watson, T., Ng, T. and Festy, F. (2010). "Fluorescence lifetime endoscopy using TCSPC for the measurement of FRET in live cells." *Optics Express* **18**(11): 11148–11158.

Gadella, B. M., Gadella, T. W. J., Colenbrander, B. and Van Golde, L. M. G. (1994). "Visualization and quantification of glycolipid polarity dynamics in the plasma membrane of the mammalian spermatozoon." *Journal of Cell Science* **107**(8): 2151–2163.

Gadella, B. M., Lopez-Cardozo, M., Van Golde, L. M. G. and Colenbrander, B. (1995). "Glycolipid migration from the apical to the equatorial subdomains of the sperm head plasma membrane precedes the acrosome reaction. Evidence for a primary capacitation event in boar spermatozoa." *Journal of Cell Science* **108**(3): 935–945.

Galbraith, C. G. and Galbraith, J. A. (2011). "Super-resolution microscopy at a glance." *Journal of Cell Science* **124**(10): 1607–1611.

Ganesan, S., Ameer-Beg, S. M., Ng, T. T. C., Vojnovic, B. and Wouters, F. S. (2006). "A dark yellow fluorescent protein (YFP)-based Resonance Energy-Accepting Chromoprotein (REACh) for Förster resonance energy transfer with GFP." *Proceedings of the National Academy of Sciences of the United States of America* **103**(11): 4089–4094.

Gaviola, E. (1926). "Die Abklingungszeiten der Fluoreszenz von Farbstofflösungen." *Annalen der Physik* **386**(23): 681–710.

Gerega, A., Zolek, N., Soltysinski, T., Milej, D., Sawosz, P., Toczylowska, B. and Liebert, A. (2011). "Wavelength-resolved measurements of fluorescence lifetime of indocyanine green." *Journal of Biomedical Optics* **16**(6): 067010.

Gerritsen, H. C., Sanders, R., Draaijer, A., Ince, C. and Levine, Y. K. (1997). "Fluorescence lifetime imaging of oxygen in living cells." *Journal of Fluorescence* **7**(1): 11–16.

Ghiggino, K. P., Hutchison, J. A., Langford, S. J., Latter, M. J., Lee, M. A. P., Lowenstern, P. R., Scholes, C., Takezaki, M. and Wilman, B. E. (2007). "Porphyrin-based molecular rotors as fluorescent probes of nanoscale environments." *Advanced Functional Materials* **17**(5): 805–813.

Giordano, L., Shvadchak, V. V., Fauerbach, J. A., Jares-Erijman, E. A. and Jovin, T. M. (2012). "Highly solvatochromic 7-Aryl-3-hydroxychromones." *Journal of Physical Chemistry Letters* **3**: 1011–1016.

Green, M., Howes, P., Berry, C., Argyros, O. and Thanou, M. (2009). "Simple conjugated polymer nanoparticles as biological labels." *Proceedings of the Royal Society A-Mathematical Physical and Engineering Sciences* **465**(2109): 2751–2759.

Gryczynski, I., Malak, H. and Lakowicz, J. R. (1995). "3-photon induced fluorescence of 2,5-diphenyloxazole with a femtosecond Ti-Sapphire laser." *Chemical Physics Letters* **245**(1): 30–35.

Gutkowski, K. I., Japas, M. L. and Aramendia, P. F. (2006). "Fluorescence of dicyanovinyl julolidine in a room-temperature ionic liquid." *Chemical Physics Letters* **426**(4–6): 329–333.

Haidekker, M., Brady, T. P., Lichlyter, D. and Theodorakis, E. A. (2006). "A ratiometric fluorescent viscosity sensor." *Journal of the American Chemical Society* **128**: 398–399.

Haidekker, M. A., Ling, T., Anglo, M., Stevens, H. Y., Frangos, J. A. and Theodorakis, E. A. (2001). "New fluorescent probes for the measurement of cell membrane viscosity." *Chemistry and Biology* **8**(2): 123–131.

Haidekker, M. A., Nipper, M., Mustafic, A., Lichlyter, D., Dakanali, M. and Theodorakis, E. A. (2010). Dyes with segmental mobility: Molecular rotors. In *Advanced Fluorescence Reporters in Chemistry and Biology I. Fundamentals and Molecular Design*, volume 8, A. P. Demchenko, ed. Berlin Heidelberg: Springer, 267–308.

Haidekker, M. A. and Theodorakis, E. A. (2007). "Molecular rotors–fluorescent biosensors for viscosity and flow." *Organic and Biomolecular Chemistry* **5**(11): 1669–1678.

Haidekker, M. A. and Theodorakis, E. A. (2010). "Environment-sensitive behavior of fluorescent molecular rotors." *Journal of Biological Engineering* **4**: 11.

Haidekker, M. A., Tsai, A. G., Brady, T., Stevens, H. Y., Frangos, J. A., Theodorakis, E. and Intaglietta, M. (2002). "A novel approach to blood plasma viscosity measurement using fluorescent molecular rotors." *American Journal of Physiology-Heart and Circulatory Physiology* **282**(5): H1609–H1614.

Hanson, K. M., Behne, M. J., Barry, N. P., Mauro, T. M., Gratton, E. and Clegg, R. M. (2002). "Two-photon fluorescence lifetime imaging of the skin stratum corneum pH gradient." *Biophysical Journal* **83**(3): 1682–1690.

Harris, H. (1999). *The Birth of the Cell*. New Haven, CT: Yale University Press.

Harvey, E. N. (1957). *A History of Luminescence from the Earliest Times Until 1900*. Philadelphia, PA: American Philosophical Society.

Heilemann, M. (2010). "Fluorescence microscopy beyond the diffraction limit." *Journal of Biotechnology* **149**(4): 243–251.

Heintzmann, R. and Ficz, G. (2006). "Breaking the resolution limit in light microscopy." *Briefings in Functional Genomics and Proteomics* **5**: 289–301.

Herman, B., Wodnicki, P., Kwon, S., Periasamy, A., Gordon, G. W., Mahajan, N. and Xue Feng, W. (1997). "Recent developments in monitoring calcium and protein interactions in cells using fluorescence lifetime microscopy." *Journal of Fluorescence* **7**(1): 85–92.

Hirvonen, L. M. and Smith, T. A. (2011). "Imaging on the nanoscale: Super-resolution fluorescence microscopy." *Australian Journal of Chemistry* **64**(1): 41–45.

Hosny, N. A., Lee, D. A. and Knight, M. M. (2012). "Single photon counting fluorescence lifetime detection of pericellular oxygen concentrations." *Journal of Biomedical Optics* **17**: 016007.

Hötzer, B., Ivanov, R., Altmeier, S., Kappl, R. and Jung, G. (2011). "Determination of copper(II) ion concentration by lifetime measurements of green fluorescent protein." *Journal of Fluorescence* **21**(6): 2143–2153.

Hötzer, B., Ivanov, R., Brumbarova, T., Bauer, P. and Jung, G. (2012). "Visualization of Cu²⁺ uptake and release in plant cells by fluorescence lifetime imaging microscopy." *FEBS Journal* **279**(3): 410–419.

Howes, P., Green, M., Levitt, J., Suhling, K. and Hughes, M. (2010). "Phospholipid encapsulated semiconducting polymer nanoparticles: Their use in cell imaging and protein attachment." *Journal of the American Chemical Society* **132**(11): 3989–3996.

Huber, M. C. E. and Sandeman, R. J. (1986). "The measurement of oscillator strength." *Reports of Progress in Physics* **49**: 397–490.

Hungerford, G., Allison, A., McLoskey, D., Kuimova, M. K., Yahioglu, G. and Suhling, K. (2009). "Monitoring sol-to-gel transitions via fluorescence lifetime determination using viscosity sensitive fluorescent probes." *Journal of Physical Chemistry B* **113**(35): 12067–12074.

Istratov, A. A. and Vyvenko, O. F. (1999). "Exponential analysis in physical phenomena." *Review of Scientific Instruments* **70**(2): 1233–1257.

Jablonski, A. (1935). "Über den Mechanismus der Photolumineszenz von Farbstoffphosphoren." *Zeitschrift für Physik* **94**: 38–46.

Jares-Erijman, E. A. and Jovin, T. M. (2003). "FRET imaging." *Nature Biotechnology* **21**(11): 1387–1396.

Jenkins, F. A. and White, H. E. (1950). *Fundamentals of Optics*, 4th edition. New York: McGraw Hill.

Kaneko, H., Putzier, I., Frings, S., Kaupp, U. B. and Gensch, T. (2004). "Chloride accumulation in mammalian olfactory sensory neurons." *Journal of Neuroscience* **24**(36): 7931–7938.

Kang, M., Day, C. A., Drake, K., Kenworthy, A. K. and DiBenedetto, E. (2009). "A generalization of theory for two-dimensional fluorescence recovery after photobleaching applicable to confocal laser scanning microscopes." *Biophysical Journal* **97**(5): 1501–1511.

Kasha, M. (1950). "Characterization of electronic transitions in complex molecules." *Discussions of the Faraday Society* **9**: 14–19.

Kinzler, I., Haseroth, E., Hauser, C. and Rück, A. (2007). "Role of mitochondria in cell death induced by Photofrin (R)-PDT and ursodeoxycholic acid by means of SLIM." *Photochemical & Photobiological Sciences* **6**(12): 1332–1340.

König, K. (2000). "Multiphoton microscopy in life sciences." *Journal of Microscopy* **200**(2): 83–104.

König, K., Schneckenburger, H. and Hibst, R. (1999). "Time-gated in vivo autofluorescence imaging of dental caries." *Cellular and Molecular Biology* **45**(2): 233–239.

Koushik, S. V. and Vogel, S. S. (2008). "Energy migration alters the fluorescence lifetime of Cerulean: Implications for fluorescence lifetime imaging Forster resonance energy transfer measurements." *Journal of Biomedical Optics* **13**(4): 031204.

Kress, M., Meier, T., Steiner, R., Dolp, F., Erdmann, R., Ortmann, U. and Rück, A. (2003). "Time-resolved microspectrofluorometry and fluorescence lifetime imaging of photosensitizers using picosecond pulsed diode lasers in laser scanning microscopes." *Journal of Biomedical Optics* **8**(1): 26–32.

Kuimova, M. K. (2012a). "Mapping viscosity in cells using molecular rotors." *Physical Chemistry Chemical Physics* **14**(37): 12671–12686.

Kuimova, M. K. (2012b). "Molecular rotors image intracellular viscosity." *Chimia* **66**(4): 159–165.

Kuimova, M. K., Bhatti, M., Deonarain, M., Yahioglu, G., Levitt, J. A., Stamati, I., Suhling, K. and Phillips, D. (2007). "Fluorescence characterisation of multiply-loaded anti-HER2 single chain Fv conjugates with photosensitizers suitable for photodynamic therapy." *Photochemical and Photobiological Sciences* **6**: 933–939.

Kuimova, M. K., Botchway, S. W., Parker, A. W., Balaz, M., Collins, H. A., Anderson, H. L., Suhling, K. and Ogilby, P. R. (2009). "Imaging intracellular viscosity of a single cell during photoinduced cell death." *Nature Chemistry* **1**: 69–73.

Kuimova, M. K., Yahioglu, G., Levitt, J. A. and Suhling, K. (2008). "Molecular rotor measures viscosity of live cells via fluorescence lifetime imaging." *Journal of the American Chemical Society* **130**(21): 6672–6673.

Kung, C. E. and Reed, J. K. (1986). "Microviscosity measurements of phospholipid bilayers using fluorescent dyes that undergo torsional relaxation." *Biochemistry* **25**: 6114–6121.

Kung, C. E. and Reed, J. K. (1989). "Fluorescent molecular rotors—A new class of probes for tubulin structure and assembly." *Biochemistry* **28**(16): 6678–6686.

Lakowicz, J. R. (2006). *Principles of Fluorescence Spectroscopy*. New York: Springer.

Lakowicz, J. R., Szmacinski, H., Nowaczyk, K. and Johnson, M. L. (1992). "Fluorescence lifetime imaging of free and protein-bound NADH." *Proceedings of the National Academy of Sciences of the United States of America* **89**(4): 1271–1275.

Lakowicz, J. R., Szmacinski, H., Nowaczyk, K. and Lederer, W. J. (1994). "Fluorescence lifetime imaging of intracellular calcium in COS cells using Quin-2." *Cell Calcium* **15**(1): 7–27.

Laporte, O. (1924). "Die Struktur des Eisenspektrums." *Zeitschrift für Physik* **23**: 135–175.

Law, K. Y. (1981). "Fluorescence probe for micro-environments—A new probe for micelle solvent parameters and premicellar aggregates." *Photochemistry and Photobiology* **33**(6): 799–806.

Le Moal, E., Fort, E., Lévêque-Fort, S., Cordelieres, F. P., Fontaine-Aupart, M. P. and Ricolleau, C. (2007). "Enhanced fluorescence cell imaging with metal-coated slides." *Biophysical Journal* **92**(6): 2150–2161.

Lee, J. (2008). "Bioluminescence: The first 3000 years (Review)." *Journal of the Siberian Federal University. Biology* **3**: 194–205.

Levitt, J. A., Kuimova, M. K., Yahioglu, G., Chung, P. H., Suhling, K. and Phillips, D. (2009). "Membrane-bound molecular rotors measure viscosity in live cells via fluorescence lifetime imaging." *Journal of Physical Chemistry C* **113**(27): 11634–11642.

Levitt, J. A., Matthews, D. R., Ameer-Beg, S. M. and Suhling, K. (2009). "Fluorescence lifetime and polarization-resolved imaging in cell biology." *Current Opinion in Biotechnology* **20**: 28–36.

Li, D. D. U., Arlt, J., Tyndall, D., Walker, R., Richardson, J., Stoppa, D., Charbon, E. and Henderson, R. K. (2011). "Video-rate fluorescence lifetime imaging camera with CMOS single-photon avalanche diode arrays and high-speed imaging algorithm." *Journal of Biomedical Optics* **16**(9): 096012.

Liaugaudas, G., Collins, A. T., Suhling, K., Davies, G. and Heintzmann, R. (2009). "Luminescence-lifetime mapping in diamond." *Journal of Physics: Condensed Matter* **21**: 364210.

Liaugaudas, G., Davies, G., Suhling, K., Khan, R. U. A. and Evans, D. J. F. (2012). "Luminescence lifetimes of neutral nitrogen-vacancy centres in synthetic diamond containing nitrogen." *Journal of Physics-Condensed Matter* **24**(43): 435503.

Lidke, D. S., Nagy, P., Barisas, B. G., Heintzmann, R., Post, J. N., Lidke, K. A., Clayton, A. H. A., Arndt-Jovin, D. J. and Jovin, T. M. (2003). "Imaging molecular interactions in cells by dynamic and static fluorescence anisotropy (rFLIM and emFRET)." *Biochemical Society Transactions* **31**(5): 1020–1027.

Lin, H. J., Herman, P. and Lakowicz, J. R. (2003). "Fluorescence lifetime-resolved pH imaging of living cells." *Cytometry* **52A**: 77–89.

Lippincott-Schwartz, J., Snapp, E. and Kenworthy, A. (2001). "Studying protein dynamics in living cells." *Nature Reviews Molecular Cell Biology* **2**(6): 444–456.

Loutfy, R. O. (1986). "Fluorescence probes for polymer free-volume." *Pure and Applied Chemistry* **58**(9): 1239–1248.

Lu, J., Liotta, C. L. and Eckert, C. A. (2003). "Spectroscopically probing microscopic solvent properties of room-temperature ionic liquids with the addition of carbon dioxide." *Journal of Physical Chemistry A* **107**(19): 3995–4000.

Luby-Phelps, K., Mujumdar, S., Mujumdar, R., Ernst, L. A., Galbraith, W. and Waggoner, A. S. (1993). "A novel fluorescence ratiometric method confirms the low solvent viscosity of the cytoplasma." *Biophysical Journal* **65**(1): 236–242.

Luby-Phelps, K., Taylor, D. L. and Lanni, F. (1986). "Probing the structure of cytoplasm." *Journal of Cell Biology* **102**(6): 2015–2022.

Ma, Y. J., Rajendran, P., Blum, C., Cesa, Y., Gartmann, N., Bruhwiler, D. and Subramaniam, V. (2011). "Microspectroscopic analysis of green fluorescent proteins infiltrated into mesoporous silica nanochannels." *Journal of Colloid and Interface Science* **356**(1): 123–130.

Malley, M. (1991). "A heated controversy on cold light." *Archive for History of Exact Sciences* **42**(2): 173–186.

Masters, B. R., So, P. T. C. and Gratton, E. (1997). "Multiphoton excitation fluorescence microscopy and spectroscopy of in vivo human skin." *Biophysical Journal* **72**: 2405–2412.

Mattheyses, A. L., Hoppe, A. D. and Axelrod, D. (2004). "Polarized fluorescence resonance energy transfer microscopy." *Biophysical Journal* **87**(4): 2787–2797.

Michalet, X., Kapanidis, A. N., Laurence, T., Pinaud, F., Doose, S., Pflughoefft, M. and Weiss, S. (2003). "The power and prospects of fluorescence microscopies and spectroscopies." *Annual Review of Biophysics and Biomolecular Structure* **32**: 161–182.

Minsky, M. (1988). "Memoir on inventing the confocal scanning microscope." *Scanning* **10**: 128–138.

Miyawaki, A., Llopis, J., Helm, R., McCaffery, J. M., Adams, J. A., Ikura, M. and Tsien, R. Y. (1997). "Fluorescent indicators for Ca^{2+} based on green fluorescent proteins and calmodulin." *Nature* **388**(6645): 882–887.

Muguruza Gonzalez, E., Guidoni, L. and Molteni, C. (2009). "Chemical and protein shifts in the absorption spectrum of the photoactive yellow protein: A time-dependent density functional theory/molecular mechanics study." *Physical Chemistry Chemical Physics* **11**: 4556.

Neugart, F., Zappe, A., Jelezko, F., Tietz, C., Boudou, J. P., Krueger, A. and Wrachtrup, J. (2007). "Dynamics of diamond nanoparticles in solution and cells." *Nano Letters* **7**(12): 3588–3591.

Ni, T. and Melton, L. A. (1996). "Two-dimensional gas-phase temperature measurements using fluorescence lifetime imaging." *Applied Spectroscopy* **50**(9): 1112–1116.

Nieber, H. and Doltsinis, N. L. (2008). "Elucidating ultrafast nonradiative decay of photoexcited uracil in aqueous solution by ab initio molecular dynamics." *Chemical Physics* **347**(1–3): 405–412.

Nipper, M. E., Dakanali, M., Theodorakis, E. A. and Haidekker, M. A. (2010). "Detection of liposome membrane viscosity perturbations with ratiometric molecular rotors." *Biochimie* **93**: 988–994.

Ogikubo, S., Nakabayashi, T., Adachi, T., Islam, M. S., Yoshizawa, T., Kinjo, M. and Ohta, N. (2011). "Intracellular pH sensing using autofluorescence lifetime microscopy." *Journal of Physical Chemistry B* **115**(34): 10385–10390.

Owen, D. M., Lanigan, P. M. P., Dunsby, C., Munro, I., Grant, D., Neil, M. A. A., French, P. M. W. and Magee, A. I. (2006). "Fluorescence lifetime imaging provides enhanced contrast when imaging the phase-sensitive dye di-4-ANEPPDHQ in model membranes and live cells." *Biophysical Journal* **90**: L80–L82.

Paul, A. and Samanta, A. (2008). "Free volume dependence of the internal rotation of a molecular rotor probe in room temperature ionic liquids." *Journal of Physical Chemistry B* **112**(51): 16626–16632.

Peng, X., Yang, Z., Wang, J., Fan, J., He, Y., Song, F., Wang, B., Sun, S., Qu, J., Qi, J. and Yan, M. (2011). "Fluorescence ratiometry and fluorescence lifetime imaging: Using a single molecular sensor for dual mode imaging of cellular viscosity." *Journal of the American Chemical Society* **133**: 6626–6635.

Peter, M. and Ameer-Beg, S. M. (2004). "Imaging molecular interactions by multiphoton FLIM." *Biology of the Cell* **96**(3): 231–236.

Pieribone, V. and Gruber, D. F. (2005). *Aglow in the Dark: The Revolutionary Science of Biofluorescence*. Cambridge, MA: The Belknap Press of Harvard University Press.

Pliss, A., Zhao, L. L., Ohulchanskyy, T. Y., Qu, J. L. and Prasad, P. N. (2012). "Fluorescence lifetime of fluorescent proteins as an intracellular environment probe sensing the cell cycle progression." *ACS Chemical Biology* **7**(8): 1385–1392.

Prummer, M., Sick, B., Renn, A. and Wild, U. P. (2004). "Multiparameter microscopy and spectroscopy for single-molecule analytics." *Analytical Chemistry* **76**(6): 1633–1640.

Rei, A., Hungerford, G. and Ferreira, M. I. C. (2008). "Probing local effects in silica sol-gel media by fluorescence spectroscopy of p-DASPMI." *Journal of Physical Chemistry B* **112**(29): 8832–8839.

Requejo-Isidro, J., McGinty, J., Munro, I., Elson, D. S., Galletly, N. P., Lever, M. J., Neil, M. A. A., Stamp, G. W. H., French, P. M. W. and Kellett, P. A. (2004). "High-speed wide-field time-gated endoscopic fluorescence-lifetime imaging." *Optics Letters* **29**: 2249–2251.

Resch-Genger, U., Grabolle, M., Cavaliere-Jaricot, S., Nitschke, R. and Nann, T. (2008). "Quantum dots versus organic dyes as fluorescent labels." *Nature Methods* **5**(9): 763–775.

Roberti, M. J., Jovin, T. M. and Jares-Erijman, E. (2011). "Confocal fluorescence anisotropy and FRAP imaging of alpha-synuclein amyloid aggregates in living cells." *Plos One* **6**(8): e23338.

Rost, F. W. D. (1995). The history of fluorescence microscopy. *Fluorescence Microscopy*. Cambridge: Cambridge University Press, 183–195.

Roychoudhuri, C., Kracklauer, A. F. and Creath, K. (2008). *The Nature of Light: What is a Photon?* Boca Raton, FL: CRC Press, Taylor & Francis Group.

Rumble, C., Rich, K., He, G. and Maroncelli, M. (2012). "CCVJ is not a simple rotor probe." *Journal of Physical Chemistry A* **116**(44): 10786–10792.

Sanders, R., Draaijer, A., Gerritsen, H. C., Houpt, P. M. and Levine, Y. K. (1995). "Quantitative pH imaging in cells using confocal fluorescence lifetime imaging microscopy." *Analytical Biochemistry* **227**: 302–308.

Sanders, R., Gerritsen, H. C., Draaijer, A., Houpt, P. M. and Levine, Y. K. (1994). "Fluorescence lifetime imaging of free calcium in single cells." *Bioimaging* **2**: 131–138.

Saxl, T., Khan, F., Ferla, M., Birch, D. and Pickup, J. (2011). "A fluorescence lifetime-based fibre-optic glucose sensor using glucose/galactose-binding protein." *Analyst* **136**(5): 968–972.

Saxl, T., Khan, F., Matthews, D. R., Zhi, Z. L., Rolinski, O., Ameer-Beg, S. and Pickup, J. (2009). "Fluorescence lifetime spectroscopy and imaging of nano-engineered glucose sensor microcapsules based on glucose/galactose-binding protein." *Biosensors & Bioelectronics* **24**(11): 3229–3234.

Schermelleh, L., Heintzmann, R. and Leonhardt, H. (2010). "A guide to super-resolution fluorescence microscopy." *Journal of Cell Biology* **190**(2): 165–175.

Schmidt, O., Wilms, K. H. and Lingelbach, B. (1999). "The Visby lenses." *Optometry and Vision Science* **76**(9): 624–630.

Schweitzer, D., Hammer, M., Schweitzer, F., Anders, R., Doebbecke, T., Schenke, S., Gaillard, E. R. and Gaillard, E. R. (2004). "In vivo measurement of time-resolved autofluorescence at the human fundus." *Journal of Biomedical Optics* **9**(6): 1214–1222.

Schweitzer, D., Schenke, S., Hammer, M., Schweitzer, F., Jentsch, S., Birckner, E., Becker, W. and Bergmann, A. (2007). "Towards metabolic mapping of the human retina." *Microscopy Research and Technique* **70**(5): 410–419.

Scully, A. D., Ostler, R. B., Phillips, D., O'Neill, P., Parker, A. W. and MacRobert, A. J. (1997). "Application of fluorescence lifetime imaging microscopy to the investigation of intracellular PDT mechanisms." *Bioimaging* **5**: 9–18.

Sergent, N., Levitt, J. A., Green, M. and Suhling, K. (2010). "Rapid wide-field photon counting imaging with microsecond time resolution." *Optics Express* **18**(24): 25292–25298.

Siegel, J., Elson, D. S., Webb, S. E. D., Lee, K. C. B., Vlandas, A., Gambaruto, G. L., Lévêque-Fort, S., Lever, M. J., Tadrous, P. J. and Stamp, G. W. H. (2003). "Studying biological tissue with fluorescence lifetime imaging: Microscopy, endoscopy, and complex decay profiles." *Applied Optics* **42**(16): 2995–3004.

Skala, M. C., Riching, K. M., Gendron-Fitzpatrick, A., Eickhoff, J., Eliceiri, K. W., White, J. G. and Ramanujam, N. (2007). "In vivo multiphoton microscopy of NADH and FAD redox states, fluorescence lifetimes, and cellular morphology in precancerous epithelia." *Proceedings of the National Academy of Sciences of the United States of America* **104**(49): 19494–19499.

Squire, A., Verveer, P. J., Rocks, O. and Bastiaens, P. I. (2004). "Red-edge anisotropy microscopy enables dynamic imaging of homo-FRET between green fluorescent proteins in cells." *Journal of Structural Biology* **147**(1): 62–69.

Stamati, I., Kuimova, M. K., Lion, M., Yahioglu, G., Phillips, D. and Deonarain, M. P. (2010). "Novel photosensitisers derived from pyropheophorbide-a: Uptake by cells and photodynamic efficiency in vitro." *Photochemical & Photobiological Sciences* **9**(7): 1033–1041.

Stern, O. and Volmer, M. (1919). "Über die Abklingungszeit der Fluoreszenz." *Physikalische Zeitschrift* **20**: 183–188.

Stevens, M. J., Hadfield, R. H., Schwall, R. E., Nam, S. W., Mirin, R. P. and Gupta, J. A. (2006). "Fast lifetime measurements of infrared emitters using a low-jitter superconducting single-photon detector." *Applied Physics Letters* **89**: 031109.

Stokes, G. G. (1852). "On the change of refrangibility of light." *Philosophical Transactions of the Royal Society of London* **142**: 463–562.

Stokes, G. G. (1853). "On the change of refrangibility of light II." *Philosophical Transactions of the Royal Society of London* **143**: 385–396.

Straughan, B. P. and Walker, S. (1976). *Spectroscopy*, volume 3. London: Chapman and Hall.

Strickler, S. J. and Berg, R. A. (1962). "Relationship between absorption intensity and fluorescence lifetime of molecules." *Journal of Chemical Physics* **37**(4): 814–820.

Stryer, L. and Haugland, R. P. (1967). "Energy transfer: A spectroscopic ruler." *Proceedings of the National Academy of Sciences of the United States of America* **58**: 719–726.

Suhling, K., French, P. M. W. and Phillips, D. (2005). "Time-resolved fluorescence microscopy." *Photochemical and Photobiological Sciences* **4**: 13–22.

Suhling, K., Levitt, J., and Chung, P.-H. (2014). "Time-resolved fluorescence anisotropy imaging." In *Fluorescence Spectroscopy and Microscopy: Methods and Protocols, Methods in Molecular Biology*, vol. 1076, Y. Engelborghs and A. J. W. G. Visser (eds.). New York: Springer, pp. 503–519.

Suhling, K., Siegel, J., Lanigan, P. M. P., Lévêque-Fort, S., Webb, S. E. D., Phillips, D., Davis, D. M. and French, P. M. W. (2004). "Time-resolved fluorescence anisotropy imaging applied to live cells." *Optics Letters* **29**(6): 584–586.

Suhling, K., Siegel, J., Phillips, D., French, P. M. W., Lévêque-Fort, S., Webb, S. E. D. and Davis, D. M. (2002). "Imaging the environment of green fluorescent protein." *Biophysical Journal* **83**(6): 3589–3595.

Szmacinski, H. and Lakowicz, J. R. (1996). "Fluorescence lifetime characterization of magnesium probes: Improvement of Mg^{2+} dynamic range and sensitivity using phase-modulation fluorometry." *Journal of Fluorescence* **6**: 83–85.

Szmacinski, H. and Lakowicz, J. R. (1997). "Sodium Green as a potential probe for intracellular sodium imaging based on fluorescence lifetime." *Analytical Biochemistry* **250**(2): 131–138.

Tadrous, P. J., Siegel, J., French, P. M. W., Shousha, S., Lalani, E. N. and Stamp, G. W. (2003). "Fluorescence lifetime imaging of unstained tissues: Early results in human breast cancer." *Journal of Pathology* **199**(3): 309–317.

Thaler, C., Koushik, S. V., Puhl, H. L., Blank, P. S. and Vogel, S. S. (2009). "Structural rearrangement of CaMKII alpha catalytic domains encodes activation." *Proceedings of the National Academy of Sciences of the United States of America* **106**(15): 6369–6374.

Tinnefeld, P. and Sauer, M. (2005). "Branching out of single-molecule fluorescence spectroscopy: Challenges for chemistry and influence on biology." *Angewandte Chemie-International Edition* **44**(18): 2642–2671.

Tisler, J., Balasubramanian, G., Naydenov, B., Kolesov, R., Grotz, B., Reuter, R., Boudou, J. P., Curmi, P. A., Sennour, M., Thorel, A., Borsch, M., Aulenbacher, K., Erdmann, R., Hemmer, P. R., Jelezko, F. and Wrachtrup, J. (2009). "Fluorescence and spin properties of defects in single digit nanodiamonds." *ACS Nano* **3**(7): 1959–1965.

Togashi, D. M., Romao, R. I. S., da Silva, A. M. G., Sobral, A. J. F. N. and Costa, S. M. B. (2005). "Self-organization of a sulfonamido-porphyrin in Langmuir monolayers and Langmuir-Blodgett films." *Physical Chemistry Chemical Physics* **7**: 3875–3884.

Tolosa, L., Szmacinski, H., Rao, G. and Lakowicz, J. R. (1997). "Lifetime-based sensing of glucose using energy transfer with a long lifetime donor." *Analytical Biochemistry* **250**(1): 102–108.

Tolstoy, I. (1981). *James Clerk Maxwell*. Chicago: University of Chicago Press.

Toptygin, D. (2003). "Effects of the solvent refractive index and its dispersion on the radiative decay rate and extinction coefficient of a fluorescent solute." *Journal of Fluorescence* **13**(3): 201–219.

Udenfriend, S. (1995). "Development of the spectrophotofluorometer and its commercialization." *Protein Science* **4**(3): 542–551.

Urayama, P. and Mycek, M.-A. (2003). Fluorescence lifetime imaging microscopy of endogenous biological fluorescence. In *Handbook of Biomedical Fluorescence*, M.-A. Mycek and B. W. Pogue, eds. New York, Marcel Dekker.

Uzhinov, B. M., Ivanov, V. L. and Melnikov, M. Y. (2011). "Molecular rotors as luminescence sensors of local viscosity and viscous flow in solutions and organized systems." *Russian Chemical Reviews* **80**(12): 1179–1190.

Valeur, B. (2002). *Molecular Fluorescence*. Weinheim: Wiley-VCH.

Valeur, B. and Berberan-Santos, M. (2011). "A brief history of fluorescence and phosphorescence before the emergence of quantum theory." *Journal of Chemical Education* **88**: 731–738.

Van Manen, H. J., Verkuijlen, P., Wittendorp, P., Subramaniam, V., van den Berg, T. K., Roos, D. and Otto, C. (2008). "Refractive index sensing of green fluorescent proteins in living cells using fluorescence lifetime imaging microscopy." *Biophysical Journal* **94**: L67–L69.

Varma, R. and Mayor, S. (1998). "GPI-anchored proteins are organized in submicron domains at the cell surface." *Nature* **394**: 798–801.

Vishwasrao, H. D., Trifilieff, P. and Kandel, E. R. (2012). "In vivo imaging of the actin polymerization state with two-photon fluorescence anisotropy." *Biophysical Journal* **102**(5): 1204–1214.

Vogel, S. S., Thaler, C., Blank, P. S. and Koushik, S. V. (2010). Time-resolved fluorescence anisotropy. In *FLIM Microscopy in Biology and Medicine*, A. Periasamy and R. M. Clegg, eds. Boca Raton, FL: Chapman & Hall, Taylor & Francis Group: 245–288.

Wallrabe, H. and Periasamy, A. (2005). "Imaging protein molecules using FRET and FLIM microscopy." *Current Opinion in Biotechnology* **16**(1): 19–27.

Wandelt, B., Cywinski, P., Darling, G. D. and Stranix, B. R. (2005). "Single cell measurement of micro-viscosity by ratio imaging of fluorescence of styrylpyridinium probe." *Biosensors & Bioelectronics* **20**(9): 1728–1736.

Wandelt, B., Mielniczak, A., Turkewitsch, P., Darling, G. D. and Stranix, B. R. (2003). "Substituted 4-[4-(dimethylamino)styryl]pyridinium salt as a fluorescent probe for cell microviscosity." *Biosensors & Bioelectronics* **18**(4): 465–471.

White, J. G., Amos, W. B. and Fordham, M. (1987). "An evaluation of confocal versus conventional imaging of biological structures by fluorescence light microscopy." *Journal of Cell Biology* **105**: 41–48.

Wolf, D. E. (2003). Fundamentals of fluorescence and fluorescence microscopy. In *Methods in Cell Biology*, volume 72, G. Sluder and D. E. Wolf, eds. Amsterdam: Elsevier, 157–184.

Wuister, S. F., de Mello Donega, C. and Meijerink, A. (2004). "Local-field effects on the spontaneous emission rate of CdTe and CdSe quantum dots in dielectric media." *Journal of Chemical Physics* **121**(9): 4310–4315.

Yazdanfar, S., Joo, C., Zhan, C., Berezin, M. Y., Akers, W. J. and Achilefu, S. (2010). "Multiphoton microscopy with near infrared contrast agents." *Journal of Biomedical Optics* **15**(3): 030505.

Yeow, E. K. L. and Clayton, A. H. A. (2007). "Enumeration of oligomerization states of membrane proteins in living cells by homo-FRET spectroscopy and microscopy: Theory and application." *Biophysical Journal* **92**(9): 3098–3104.

Yu, Q. R. and Heikal, A. A. (2009). "Two-photon autofluorescence dynamics imaging reveals sensitivity of intracellular NADH concentration and conformation to cell physiology at the single-cell level." *Journal of Photochemistry and Photobiology B-Biology* **95**(1): 46–57.

Zhang, Q., Piston, D. W. and Goodman, R. H. (2002). "Regulation of corepressor function by nuclear NADH." *Science* **295**(5561): 1895–1897.

Zollinger, H. (2003). *Color Chemistry: Syntheses, Properties, and Applications of Organic Dyes and Pigments*. Zurich: Helvetica Chimica Acta.

3 Tissue fluorophores and their spectroscopic characteristics

Alzbeta Chorvatova and Dusan Chorvat

Contents

3.1 INTRODUCTION

Noninvasive monitoring of processes underlying structural and functional changes of living systems in their real environmental conditions is a prerequisite for understanding their alterations in pathological situations. Internal and external fluorophores provide an effective mean for noninvasive investigation of living cells and tissues using various methods/techniques of optical diagnostics, such as microscopy or spectroscopy. Spectral characteristics of each endogenous fluorophore are unique and can therefore provide a base for their identification and separation in complex biological samples, when an appropriate detection technique is used. Recent advances in technology, particularly combination of fluorescence spectroscopy and imaging with time-resolved detection, provide great potential for gathering detailed information on biochemical, functional, and structural changes in biomolecular complexes directly in living cells and tissues. This potential is reflected in better diagnostics capabilities to distinguish, for example, between

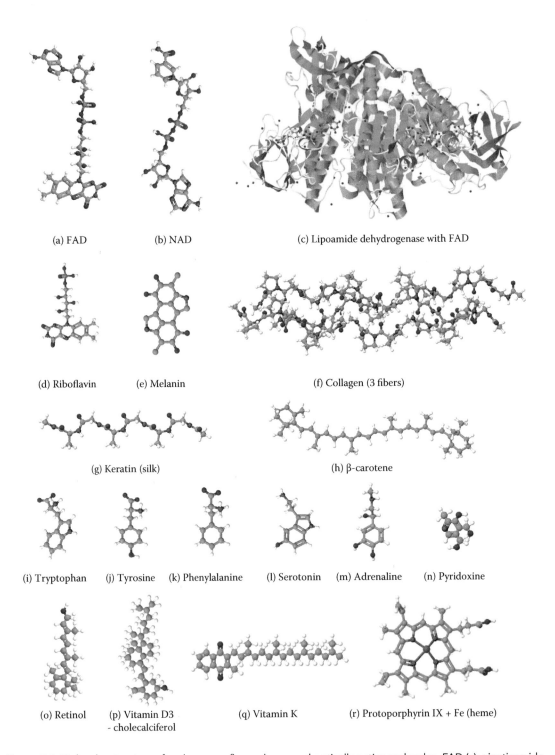

(a) FAD   (b) NAD   (c) Lipoamide dehydrogenase with FAD

(d) Riboflavin   (e) Melanin   (f) Collagen (3 fibers)

(g) Keratin (silk)   (h) β-carotene

(i) Tryptophan   (j) Tyrosine   (k) Phenylalanine   (l) Serotonin   (m) Adrenaline   (n) Pyridoxine

(o) Retinol   (p) Vitamin D3   (q) Vitamin K   (r) Protoporphyrin IX + Fe (heme)
              - cholecalciferol

Figure 3.1 Molecular structure of endogenous fluorophores and optically active molecules. FAD (a); nicotinamide adenine dinucleotide—NAD (b); example of a flavoprotein–lipoamide dehydrogenase with FAD cofactors (c); riboflavin—vitamin B2 (d); generic melanin (e); collagen—visualization of three interconnected fibers (f); example of keratin—silk (g); β-carotene (h); amino acid residues: Trp, tyrosine, and phenylalanine (i–k); hormones: serotonin (l) and adrenaline (m); vitamins: pyridoxine—vitamin B6 (n), retinol—vitamin A (o), cholecalciferol—vitamin D3 (p), vitamin K (q), complex of PPIX with iron atom (heme; r). Visualizations of molecules were created using open-source Java-based viewer *Jmol* (www.jmol.org) using molecular models from *Protein Databank* (www.pdb.org), *Drug Bank* (www.drugbank.ca), and *Klotho* database of molecular structures (www.biocheminfo.org).

normal and diseased tissues. A number of combined experimental techniques, based on various hybrids between time-resolved, spectroscopic, and imaging approaches, have recently been developed to explore the use of fluorescence in cells and tissues for biomedical research investigations.

Natural tissue autofluorescence (AF) originates primarily from specific endogenous fluorophores located in mitochondria (NAD(P)H and flavin coenzymes, porphyrins, and lipopigments), structural proteins in extracellular matrix (ECM; collagen and elastin), cell lysosomes (lipofuscins), and various aromatic amino acids appearing in many proteins within the living body. Amino acid tryptophan (Trp) with two absorption maxima at 220 and 287 nm, both emitting at 350 nm, is one of the most abundant protein fluorophores but, due to lack of specificity, is seldom used for biomedical diagnostics purpose. The major endogenous fluorophores excited above 300 nm are structural proteins such as collagen, elastin, or keratin; metabolites; and enzyme cofactors, namely, cellular flavins and reduced nicotinamide adenine dinucleotide (phosphate), also known as NAD(P)H. Reduced NAD(P)H is one of the most studied endogenous fluorophores, with excitation maximum at 350–380 nm and emission near 440 nm. Oxidized flavins are, on the other hand, excited near 450 nm with emission maximum near 515 nm. Other endogenous fluorescence compounds include lipofuscins, age-related lipophilic materials with wide excitation and emission bands spanning over the visible spectral range. Important sources of endogenous fluorescence also include pigments (such as melanin or bilirubin, porphyrins, as well as different vitamins and pterins). Table 3.1 summarizes excitation, emission, and lifetime characteristics of endogenous fluorophores found in biological systems, while Table 3.2 enumerates fluorescence parameters of known endogenous fluorophores in different cells and tissues. Molecular structure of the representative fluorophores and optically active molecules found in tissues and discussed further in this chapter is shown in Figure 3.1, while their excitation and emission spectra are illustrated in Figure 3.2.

In this chapter, we characterize excitation, emission, and fluorescence lifetime properties of the most important endogenous fluorophores found in mammalian cells and tissues that can serve for human biomedical diagnostics. We also give a brief overview of individual tissues and modifications of their endogenous fluorophores in pathophysiological situations. Finally, we succinctly mention the properties and applications of fluorophores synthesized externally or administered into tissues by genetic manipulation, with the aim of biomedical diagnostics and treatment of diseases.

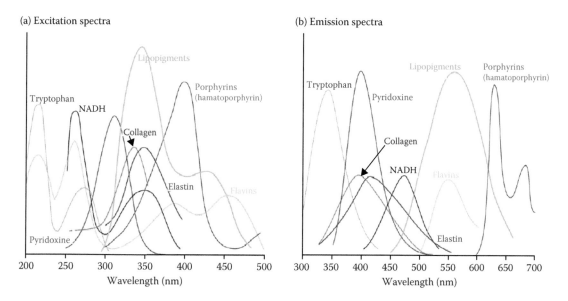

Figure 3.2 Spectroscopy characteristics of biological fluorophores. Excitation (a) and emission (b) spectra of the main types of biological fluorophores in tissue. (Adapted from Wagnieres G. A. et al., *Photochem. Photobiol.* 68, 605, 1998.)

Overview of fluorescence measurements and concepts

Table 3.1 **Excitation, emission, and lifetime characteristics of known endogenous fluorophores**

TISSUE FLUOROPHORE	EXCITATION (nm)	EXCITATION MAXIMA (nm)	EMISSION (nm)	EMISSION MAXIMA (nm)	FLUORESCENCE LIFETIME (ns)	REFERENCES
Enzymes and Coenzymes						
NADH and NADPH	330–450	290, 340–351	350–450	440, 450–460	0.3–0.65, 1.9–2.3 in tissues	Konig & Riemann 2003; Ramanujam 2000; Richards-Kortum & Sevick-Muraca 1996; Schweitzer et al. 2007
	355				0.4, 0.78 in phosphate buffered saline pH 7.2, 20°C	De Beule et al. 2007
Free NADH	300–380		450–500	460	0.3	Konig 2008
				470	$\langle\tau\rangle$ of 0.64–0.74 in Mops buffer pH 7 at 24°C $\langle\tau\rangle$ of 0.43 in aqueous solution Tris pH 7.4 at 23°C	Lakowicz et al. 1992; Wakita et al. 1995
Enzyme-bound NADH	300–380		450–500	440	2.0–2.3	Konig 2008
				470	$\langle\tau\rangle$ of 2.0 in Mops buffer pH 7 at 24°C with malate dehydrogenase	Lakowicz et al. 1992
FAD, flavins	370–450	450	490–530	515, 535	0.15–0.33 0.5–0.95 4.5–5.7 in tissues	Konig & Riemann 2003; Ramanujam 2000; Richards-Kortum & Sevick-Muraca 1996; Schweitzer et al. 2007
	440				$\langle\tau\rangle$ of 2.5 2.8 in phosphate buffered saline pH 7.2, 20°C	De Beule et al. 2007; Visser 1984
FAD	420–500 420–500 (bound)		520–570 520–270 (bound, weak)		2.91 <0.01 (bound, weak)	Berezin & Achilefu 2010

	Excitation	Emission	Lifetime / $\langle\tau\rangle$	Reference
FAD in LipDH	450	550	$\langle\tau\rangle$ of 2.0	Bastiaens et al. 1992
FAD in GR	450	557.9	$\langle\tau\rangle$ of 0.027 in potassium phosphate buffer at 293 K, pH 7.6	van den Berg et al. 2004
FMN	444	558	$\langle\tau\rangle$ of 4.7 0.12 1.41 5.27 $\langle\tau\rangle$ of 0.30 at 5°C in solution (Na phosphate)	Leenders et al. 1993; Visser 1984
	420–500	520–570	4.27–4.67	Berezin & Achilefu 2010
Riboflavin (vitamin B2)	420–500	520–570	4–12	Berezin & Achilefu 2010
Vitamins				
Vitamin A	340	490		Billinton & Knight 2001
	327	510		Ramanujam 2000
Retinol palmitate	325–330	505	2.17	Singh & Das 1998
Vitamin B6 compounds		380–395 (pyridoxinic compounds)		
Pyridoxine	332, 340	400	$\langle\tau\rangle$ of 0.78 in water, pH 7	Bueno et al. 2010; Ramanujam 2000; Richards-Kortum & Sevick-Muraca 1996
Pyridoxamine	335	400	$\langle\tau\rangle$ of 1.60 in water, pH 7	Bueno et al. 2010; Ramanujam 2000; Richards-Kortum & Sevick-Muraca 1996
Pyridoxal	330; 290–320	320, 390; 380–385	$\langle\tau\rangle$ of 0.60 in water, pH 7	Bueno et al. 2010; Ramanujam 2000; Richards-Kortum & Sevick-Muraca 1996

(continued)

Table 3.1 **(Continued) Excitation, emission, and lifetime characteristics of known endogenous fluorophores**

TISSUE FLUOROPHORE	EXCITATION (nm)	EXCITATION MAXIMA (nm)	EMISSION (nm)	EMISSION MAXIMA (nm)	FLUORESCENCE LIFETIME (ns)	REFERENCES
Vitamins						
4-Pyridoxic acid		315		425	$\langle\tau\rangle$ of 8.4 in water, pH 7	Bueno et al. 2010; Richards-Kortum & Sevick-Muraca 1996
Vitamin B12		275		305		Ramanujam 2000
Pterins (xantoperins)						
Biopterins	350			441 455	9.1 in acid 7.6 in basic	Lorente et al. 2004
Neopterins	350			440 454	8.9 in acid 7.4 in basic	Lorente et al. 2004
6-carboxypterins (used for folic acid assay)	350			439 451	5.8 in acid 4.1 in basic	Lorente et al. 2004
Vitamin K		335		480		Ramanujam 2000
Vitamin D		390		470–480		Billinton & Knight 2001; Ramanujam 2000
Vitamin C		350		430		Billinton & Knight 2001
β-Carotene	535		660–1250		0.010–0.015, 0.2–1 ns synthetic	Bachilo & Gillbro 1995
Carotenoids from yellow atherosclerotic plaques		325		538		Billinton & Knight 2001

Structural Proteins

Collagen	280–350		370–440		≤5.3	Berezin & Achilefu 2010
		280, 265, 325–330, 360–450		310, 385, 390–400, 405–530	Powder	Ramanujam 2000; Richards-Kortum & Sevick-Muraca 1996
	300–340	275	420–460	300	0.2–0.4 0.4–2.5 in skin	Konig & Riemann 2003; Konig 2008
					2 8.9 powder	Konig et al. 1994
Collagen 1–4	446		510–700		0.47–0.74 2.8–4.04 in endogenous fundus substances	Schweitzer et al. 2007
Elastin	300–370		420–460		≤2.3	Berezin & Achilefu 2010
	446	290, 325	510–700	340, 400		Ramanujam 2000; Richards-Kortum & Sevick-Muraca 1996
					0.38 3.59 in endogenous fundus substances	Schweitzer et al. 2007
					2 6.7 in powder	Konig et al. 1994
	300–340		420–460		0.2–0.4 0.4–2.5 in skin	Konig & Riemann 2003
Keratin	720–930		700 short pass	70 short pass	1.4 in blond human hair	Ehlers et al. 2007

(continued)

Overview of fluorescence measurements and concepts

Overview of fluorescence measurements and concepts

Table 3.1 (Continued) Excitation, emission, and lifetime characteristics of known endogenous fluorophores

TISSUE FLUOROPHORE	EXCITATION (nm)	EXCITATION MAXIMA (nm)	EMISSION (nm)	EMISSION MAXIMA (nm)	FLUORESCENCE LIFETIME (ns)	REFERENCES
Lipids						
Lipofuscins		340–395		430–460, 540	1.34	Berezin & Achilefu 2010; Ramanujam 2000; Richards-Kortum & Sevick-Muraca 1996
	446		510–700		0.39 ns / 2.24 ns in endogenous fundus substances	Schweitzer et al. 2007
Lipofuscin A2E	446		510–700		0.17 / 1.12 in endogenous fundus substances	Schweitzer et al. 2007
Ceroid/lipofuscin		340–395		430–460, 540–640		Ramanujam 2000; Richards-Kortum & Sevick-Muraca 1996
AGE (advanced glycation end products)	446		510–700		0.865 / 4.17 in endogenous fundus substances	Schweitzer et al. 2007
Phospholipids		436		540, 560		Ramanujam 2000
DPH (1,6-diphenyl-1,3,5-hexatriene)	360		>480		0.81 / 7.46 in ethanol at 20°C / 13 in cyclohexane	Parasassi et al. 1991
Amino Acids						
Tryptophan	250–310	280	350–355	350	3.03 / 0.2–1.0 / 1.4–3.2 / 4.4–6.6 in solution (phosphate buffer) at 5°C	Berezin & Achilefu 2010; Pan & Barkley 2004; Ramanujam 2000; Richards-Kortum & Sevick-Muraca 1996

Substance	Excitation (nm)	Emission (nm)	Fluorescence lifetime / value	References
L-Trp	275	350	1.05 2.91 3.04 in water at 25°C	Wlodarczyk & Kierdaszuk 2003
Tyrosine	250–290		2.5	Berezin & Achilefu 2010
	275	300		Ramanujam 2000; Richards-Kortum & Sevick-Muraca 1996
L-Tyr	275	320	3.21 in water	Wlodarczyk & Kierdaszuk 2003
Phenylalanine	250–270	280	7.5	Berezin & Achilefu 2010
	258, 260		7.4	Duneau et al. 1998; Richards-Kortum & Sevick-Muraca 1996
Serotonin	340		3.8 ns in solutions; 2.5 ns in cell cytosol	Botchway et al. 2008
Adrenalin	330, 420	525		Billinton & Knight 2001
	280		0.87 in buffer 2.11 in ethanol 0.9 2.2 in β-cyclodextrine	Polewski 2003
Noradrenalin	325	515		Billinton & Knight 2001
Pigments				
Melanin	300–800	440, 520, 575	0.1 1.9 8	Berezin & Achilefu 2010
			1.2 (artificial DOPA-melanin)	Ehlers et al. 2007
	400	520		De Beule et al. 2007

(continued)

Overview of fluorescence measurements and concepts

Table 3.1 (Continued) **Excitation, emission, and lifetime characteristics of known endogenous fluorophores**

TISSUE FLUOROPHORE	EXCITATION (nm)	EXCITATION MAXIMA (nm)	EMISSION (nm)	EMISSION MAXIMA (nm)	FLUORESCENCE LIFETIME (ns)	REFERENCES
Pigments						
	446		510–700		0.28 2.4 in endogenous fundus substances	Schweitzer et al. 2007
	UV/visible		440, 520, 575		0.2 1.9 7.9 in skin	Konig & Riemann 2003
Eumelanin	355		520		0.058 0.51 2.9 7	Liu & Simon 2003
z,z-bilirubin IXalpha	350–520	453	480–650	518–521	14 ps in chloroform at 20°C	Plavskii et al. 2008
Porphyrins		400–450		630, 690		Ramanujam 2000
Protoporphyrin PpIX	400–450		635, 710		10–12 monomers 2–3 dimers	Berezin & Achilefu 2010; Konig 2008
	410		633		1.76 8.14 in normal oral mucosa (human patients)	Chen et al. 2005
	425				15 in normal urothelium (human bladder)	Glanzmann et al. 1996
	398		LP590		3.6 7.4 in rat epithelial cells in culture	Kress et al. 2003

Note: <τ>: mean lifetime; 2p: 2-photon excitation.

Table 3.2 Fluorescence parameters of endogenous fluorophores in different tissues

CELL TYPE/TISSUE	FLUOROPHORE	EXCITATION (nm)	EMISSION (nm)	FLUORESCENCE LIFETIME (ns)	CHANGES IN PATHOPHYSIOLOGICAL CONDITIONS	REFERENCES
Brain						
Brain (hippocampal tissue)	NAD(P)H	740 (2p)	350–550	0.155 0.599 2.154 6.040 0.948 ns (<τ>)	↓ <τ> to 0.780 in hypoxia	Vishwasrao et al. 2005
Normal mouse brain	Unspecified tissue AF	750 (2p)	Short pass	1.4	↑ <τ> in tumor cells to 1.6 ns	Leppert et al. 2006
Skin						
Skin	NAD(P)H	340	450–470	0.3 2 (bound)		Konig & Riemann 2003
	FAD	370, 450	530	5.2 1 (bound)		Konig & Riemann 2003
	Lipofuscins	UV/visible	570–590	Multiexponential		Konig & Riemann 2003
	Collagen	300–340	420–460	0.2–0.4 0.4–2.5		Konig & Riemann 2003
	Elastin	300–340	420–460	0.2–0.4 0.4–2.5		Konig & Riemann 2003
	Melanin	UV/visible	440, 520, 575	0.2 1.9 7.9		Konig & Riemann 2003
	Melanin	(300–800)				Birch et al. 2005

(*continued*)

Table 3.2 (Continued) Fluorescence parameters of endogenous fluorophores in different tissues

CELL TYPE/TISSUE	FLUOROPHORE	EXCITATION (nm)	EMISSION (nm)	FLUORESCENCE LIFETIME (ns)	CHANGES IN PATHOPHYSIOLOGICAL CONDITIONS	REFERENCES
Skin						
Uninvolved skin (human patients)	Unspecified tissue AF	355	375 (also 455)	1.551 1.617 1.567 (<τ>)	↓ <τ> in basal cell carcinomas 1.398 ns 1.617 ns 1.417 ns (<τ>)	Galletly et al. 2008
Human skin (left palm)	Unspecified tissue AF	375	440–500 (442 measured)	0.5 2.74 9.35	↑ τ in diabetic subjects to 0.504 2.82 9.79	Blackwell et al. 2008
Hair						
Black human hair	Eumelanin	720–930	700 short pass	0.03		Ehlers et al. 2007
Red and blond human hair	Pheomelanin	720–930	700 short pass	0.34		Ehlers et al. 2007
Blond human hair	Keratin	720–930	700 short pass	1.4		Ehlers et al. 2007
Blond human hair	Unspecified tissue AF	720–930	700 short pass	0.4 2.2	↓ τ in black human hair to 0.2 1.3	Ehlers et al. 2007
Heart and cardiovascular						
Heart (human cardiac myocytes)	NAD(P)H	375	397 long pass	0.62 2.33 13.64		Cheng et al. 2007
Heart (human cardiac myocytes)	NAD(P)H	375	397 long pass	1st component: 0.32 1.61 2nd component: 0.42 by linear unmixing	↑ τ of the 1st component to 0.44 and 2.18 in patients with rejection of transplanted hearts	Chorvat et al. 2010

Sample	Fluorophore	Excitation	Emission	Lifetime	Notes	Reference
Heart (rat cardiac myocytes)	NAD(P)H	375	397 long pass	1st component: 0.55 ± 0.01; 2nd component: 1.94 ± 0.01 by linear unmixing		Chorvatova et al. 2013b
Heart (rat left ventricular cardiac myocytes)	FAD	438	470 long pass	0.15–0.20 0.46–0.94 1.48–3.67		Chorvat & Chorvatova 2006
Normal human aortic samples	Unspecified tissue AF	337	370–510	2.4	↑ τ in advanced atherosclerotic lesions to 3.9	Maarek et al. 2000a
Intact cardiac mitochondria	NAD(P)H	335	440	0.4 1.9 5.7		Blinova et al. 2005
Eye						
Eye: human fundus	Unspecified tissue AF	446	510–700	0.32–0.38 2–5	↑ τ1 in age-related macular degeneration: 0.52–0.60 ns	Schweitzer et al. 2004
Eye: human retina	Unspecified tissue AF	446	510–700	0.26 2.79		Schweitzer et al. 2007
Eye: human retinal pigment epithelium (RPE)	Unspecified tissue AF	446	510–700	0.21 1.80		Schweitzer et al. 2007
Eye: human choroids sclera	Unspecified tissue AF	446	510–700	0.50 3.40		Schweitzer et al. 2007
Eye: human cornea	Unspecified tissue AF	446	510–700	0.57 3.57		Schweitzer et al. 2007
Eye: human lens	Unspecified tissue AF	446	510–700	0.49 3.60		Schweitzer et al. 2007
Endogenous fundus substances	NAD(P)H	446	510–700	0.387 3.65		Schweitzer et al. 2007
	FAD	446	510–700	0.33 2.81		Schweitzer et al. 2007

(continued)

Overview of fluorescence measurements and concepts

Table 3.2 (Continued) Fluorescence parameters of endogenous fluorophores in different tissues

CELL TYPE/TISSUE	FLUOROPHORE	EXCITATION (nm)	EMISSION (nm)	FLUORESCENCE LIFETIME (ns)	CHANGES IN PATHOPHYSIOLOGICAL CONDITIONS	REFERENCES
Eye						
	Lipofuscins	446	510–700	0.39 2.24		Schweitzer et al. 2007
	Lipofuscin A2E	446	510–700	0.17 1.12		Schweitzer et al. 2007
	AGE (advanced glycation end products)	446	510–700	0.865 4.17		Schweitzer et al. 2007
	Collagen 1	446	510–700	0.67 4.04		Schweitzer et al. 2007
	Collagen 2	446	510–700	0.47 3.15		Schweitzer et al. 2007
	Collagen 3	446	510–700	0.345 2.80		Schweitzer et al. 2007
	Collagen 4	446	510–700	0.74 3.67		Schweitzer et al. 2007

	Elastin	446	510–700	0.38 3.59		Schweitzer et al. 2007
	Melanin	446	510–700	0.28 2.4		Schweitzer et al. 2007
Organotypic perfusion culture of the porcine ocular fundus	Lipofuscin granules	450–490	>530	0.71 3.23		Hammer et al. 2008
Porcine ocular fundus	A2E, A2PE, A2PE-H2	450–490		Not quantified		Hammer et al. 2008
Skin Constructs and Stem Cells						
Skin constructs (dermal fibroblasts)	NAD(P)H	765 (2p)		0.438 2.286		Niesner et al. 2004
Human mesenchymal stem cells	NAD(P)H	740 (2p)	450 ± 40 band pass	1.011 (<τ>)	↑ <τ> during differentiation to 1.165 ns at day 21	Guo et al. 2008
Others						
Rat liver mitochondria	NAD(P)H	362	450	2.8 (<τ>)		Wakita et al. 1995
Isolated hepatocytes	NAD(P)H	362	450	3.39–5.0 (<τ>)		Wakita et al. 1995

Note: Symbols are equivalent to the ones used in Table 3.1. ↑: increase, ↓: decrease.

Overview of fluorescence measurements and concepts

3.2 SPECTROSCOPIC AND FLUORESCENCE LIFETIME PROPERTIES OF ENDOGENOUS FLUOROPHORES

3.2.1 METABOLITES, ENZYMES, AND COENZYMES

NAD(P)H, FADH$_2$, oxygen, and adenosine triphosphate (ATP) are key biomolecules consumed or generated in the metabolic process of oxidative phosphorylation (Cortassa et al. 2003; Stanley et al. 1997). This cycle is the end point of ATP generation from energy sources, namely, glucose and fatty acids that are initially metabolized in the cytoplasm. NADH feeds electrons into the respiratory chain at complex I, while succinate enters via FADH$_2$ at complex II (Williamson 1979). Resulting electrons, travelling down the respiratory chain by sequential redox reactions at complexes I–IV, thus provide large amounts of free energy for ATP synthesis at complex V. The blue AF has long been correlated with metabolic changes and ascribed to mitochondrial NAD(P)H (Chance & Thorell 1959; Chance et al. 1962, 1979; Chance 1976; Estabrook 1962; Griffiths et al. 1998; Huang et al. 2002; Nakayama et al. 2002; Schneckenburger & Konig 1992), while yellow/green fluorescence has been assigned to oxidized flavoproteins in respiratory complexes (Chance et al. 1979; Eng et al. 1989; Huang et al. 2002; Nakayama et al. 2002; Romashko et al. 1998; Schneckenburger & Konig 1992), after excitation by ultraviolet or visible light, respectively (see Table 3.1 and Figure 3.2 for details). Fluorescence lifetimes of NAD(P)H and flavins are highly sensitive to their microenvironment, particularly the viscosity and polarity. NAD(P)H is fluorescent in its reduced form and loses fluorescence under oxidation. In contrast, flavins are fluorescent in their oxidized form (FAD$^+$) and lose fluorescence when reduced (to FADH$_2$). For these reasons, NADH and flavins have been widely exploited to evaluate oxidative metabolic state in cells and tissues (Chance & Jobsis 1959). *NAD(P)H is the major source of UV-excited fluorescence in cells and tissues, with two molecules, NADH and NADPH, being its main contributors.* NADH and NADPH play a significant role not only in cell metabolism, cell signaling, and antioxidant processes but also in aging and apoptosis.

The main biological function of *NADH* is to act as electron carriers in oxidative phosphorylation, and these molecules are therefore abundantly present in the mitochondria. Time-resolved characteristics of free NADH fluorescence have been relatively well studied *in vitro* (Gafni & Brand 1976; Konig et al. 1997; Lakowicz et al. 1992), where a two-exponential model was most usually applied for decay fitting. It is now generally accepted that the short lifetime (0.4 ns) of endogenous NAD(P)H fluorescence is used as an indicator of "free" NAD(P)H molecules in cells and tissues. Quantum yield of free NADH is lower than that of riboflavin (Zipfel et al. 2003), but, upon binding, the quantum yield of the NADH generally increases fourfold (Lakowicz 2006), which is interpreted as binding of the NADH in an elongated fashion. Protein-bound NADH was described to have longer lifetimes (~2 ns) (Jameson et al. 1989; Lakowicz et al. 1992); thus, in tissues, fluorescence lifetime of 1.6–2.3 ns was attributed to "bound" NAD(P)H molecules (Bird et al. 2005; Chorvat et al. 2010; Konig et al. 1996; Niesner et al. 2004; Provenzano et al. 2008; Schneckenburger et al. 2004). It should be noted, however, that in some circumstances, such as in viscous environment (glycerol), the decay kinetics of NADH should be described by biexponential kinetics (Gafni & Brand 1976). This phenomenon indicates that simple separation of the short and long lifetime pools is an oversimplification of the real de-excitation kinetics of NAD(P)H and should be thoroughly scrutinized.

NADPH is another intrinsically fluorescing molecule found in cells and tissues. Unlike NADH acting as an electron carrier, the NADPH is essential for antioxidant processes. Unfortunately, at the moment, there are no known spectroscopical means to distinguish between fluorescence from NADH and NADPH. In healthy cells, the concentration of reduced NADPH is five times lower than that of NADH (Klaidman et al. 1995). In addition, in mitochondria, the contribution of NADPH is considered to represent a small fraction of the NAD(P)H fluorescence due to enhanced relative quantum yield of NADH by a factor of 1.25 to 2.5 (Avi-Dor et al. 1962). For these reasons, the contribution of NADPH to the overall tissue AF is often considered negligible. However, NADPH is a cofactor for enzymes implicated in antioxidant processes put in place to counteract the generation of reactive oxygen species by oxidative respiration and/ or oxidative stress (Benderdour et al. 2004). For example, NADPH is an important cofactor for glutathione reductase reactions, facilitating glutathione recycling by converting its oxidized form to reduced glutathione. Oxidative stress can also modulate cellular NADPH content through the release of peroxides and various by-products that have been shown to decrease the activity of several enzymes, such as NADP-isocitrate dehydrogenase (Chorvatova et al. 2013a). Its contribution, therefore, needs to be taken into account in pathological situations, where a significant rise in oxidative stress can occur, such as cancer.

Flavins (isoalloxazines) are ubiquitous coenzymatic redox carriers in the metabolism of mammals. Best-known flavins are the molecules of flavin adenine dinucleotide (FAD) and flavin mononucleotide (FMN), both derivates of riboflavin (vitamin B2). Most flavins found in biological systems are bound to proteins, where they act as cofactors, which are referred to as flavoproteins or flavoenzymes. These oxidoreductases require FAD as a prosthetic group for electron transport. Riboflavin and FMN are the most intensely fluorescent flavins, while FAD is about 10 times less fluorescent (Billinton & Knight 2001). Flavin fluorescence has absorption maximum around 450 nm and maximal fluorescence emission around 520–530 nm. The fluorescence quantum yield and lifetime of riboflavin depend on the solvent, pH, and presence of amino acids in the system (Drossler et al. 2003). Fluorescence lifetime of FAD^+ was described to be reduced from 2.8 ns to under 0.4 ns upon protein binding (Bastiaens et al. 1992; van den Berg et al. 1998, 2001, 2002; Yang et al. 2003). Such dramatic reduction of the lifetime was proposed to be induced by the presence of nearby aromatic residues in protein binding group, including tyrosine and Trp (Berezin & Achilefu 2010).

In cell research, the ratio between the relative concentration of free NAD(P)H and of NAD(P)H involved in metabolic processes, in other words, the free/bound component amplitude ratio, was proposed to correspond to $NADH/NAD^+$ reduction/oxidation pair (Bird et al. 2005; Niesner et al. 2004). For these reasons, time-resolved NAD(P)H fluorescence is widely studied for estimation of metabolic oxidative state and thus for metabolic redox imaging. In addition, Chance and Nielands (1952) observed a spectral shift in the NADH spectrum upon binding to lactate dehydrogenation, implicating possible structural alterations in the coenzyme following binding. This feature can also be used for spectral separation of time-resolved fluorescence of the two molecular forms (Chorvat et al. 2010; Chorvatova et al. 2013b).

Fluorescence anisotropy is an additional parameter to spectral and lifetime characteristics proposed to improve discrimination between NADH in bound and free states (Vishwasrao et al. 2005). A large increase in fluorescence anisotropy decay time following NADH binding to proteins (Hones et al. 1986) was proposed to reflect large difference in the size of the free NADH molecule vs. its binding to enzymes. This feature, which allows discriminating the bound and free states of NADH in the living mitochondria with higher specificity, is described in more detail in Chapter 2.

The original investigations of Chance et al. (1962) and Barlow and Chance (1976) demonstrated that differences in partial oxygen pressure can be determined by measuring alterations in AF of redox pairs of coenzymes (oxidized and reduced NAD–NADH and oxidized and reduced $FAD–FADH_2$). Consequently, NADH/FAD ratio is another approach used for the redox state estimation, based on the assumption that, *in vitro*, protein-bound state of the NADH has a longer lifetime (Jameson et al. 1989; Lakowicz et al. 1992) while that of FAD is characterized by a shorter lifetime (Nakashima et al. 1980). This property, possible result of dynamic quenching by adenine moiety (Lakowicz 1999; Maeda-Yorita & Aki 1984), is applied as a discrimination method of metabolic state changes. This approach has also its advantages in regard to differences in the excitation/emission characteristics of the two molecules. Such ratio was applied, for example, to distinguish cancer cells from healthy ones (Skala et al. 2007a, b), as discussed in Chapter 14.

3.2.2 VITAMINS

Vitamins are organic compounds, obtained from diet, which serve as vital nutrients. Classified by their biological and chemical activity, most vitamins exhibit endogenous fluorescence (listed in Table 3.1) due to their aromatic ring structures with delocalized electrons: *vitamin A* is fluorescent, as both retinol (vitamin A alcohol) and retinol palmitate (vitamin A palmitate) are fluorescent compounds (Singh & Das 1998). Retinol palmitate is excitable at 325–330 nm, with emission at 505–510 nm. Relatively long lifetime of 2.17 ns was identified for this molecule. *Vitamin B2*, also called riboflavin, is also a well-known fluorescent compound, as described in Section 3.2.1. *B3 vitamins* have different forms, including nicotinamide (niacinamid) and niacin, which can be converted *in vivo* to NADH and NADP. Niacin is thus a precursor to NAD+/NADH and NADP+/NADPH, which plays essential metabolic role in living cells, as described in Section 3.2.1.

Endogenous fluorescence was also noted for *vitamin B6* pyridoxinic compounds (see Table 3.1 and Figure 3.2 for details) with excitation/emission maximum near 330/420 nm (Richards-Kortum & Sevick-Muraca 1996) and lifetime ranging from 0.6 to 8.4 ns (Bueno et al. 2010). Pyridoxinic compounds are constituents of vitamin B6, a group composed of three natural compounds: pyridoxine, pyridoxamine, and pyridoxal (Bueno et al. 2010). They are required by many of the enzymes involved in amino acid metabolism (Wilson & Davis 1983). A crucial stage in the metabolism is the conversion of these compounds to the active form, which is mediated by a single kinase enzyme in the brain, liver, and erythrocytes (McCormick 2001; Snell 1990). The vitamin B6 group in excess is degraded in the liver to 4-pyridoxic acid and eliminated by urinary secretion. All of these compounds are fluorescent: the UVA-induced phototoxicity has been reported for all vitamin B6 compounds (Maeda et al. 2000). As a result, they are considered as important endogenous skin photosensitizers, with relevance to human skin *in vivo*, as human skin contains various B6 forms (Coburn et al. 2003). Spectral properties of pyridoxinic compounds are very sensitive to the environment, such as the pH. Pyridoxamine and 4-pyridoxic acid are the most efficient chromophores with higher quantum yield and longer lifetime, and these photophysic properties make them very attractive to be used for production of fluorescent biocompatible probes. Quantum yields for pyridoxal, pyridoxamine, and pyridoxamine-5-phosphate at 25°C in neutral aqueous solution were evaluated to 0.048, 0.11, and 0.14, respectively (Chen 1965). Pyridoxinic compounds were also shown to interact with DNA bases and amino acids through an electron transfer process, and this photoinduced electron transfer reaction was proposed to provide mechanism for initiation of DNA damage (Bueno et al. 2010). Several diseases are associated with vitamin B6 metabolism defects, including cirrhosis, renal disease, and cancer.

Folic acid (also called *vitamin B9*) or pteroyl-L-glutamic acid plays an important role in the biosynthesis of nucleic acids and is an effective prenatal vitamin, which decreases occurrence of spina bifida. Derived from pterins are heterocyclic compounds, which are widespread in biological systems and participate in relevant biological functions. Some pterin derivates (such as xantoperin and leucopterin) are present in butterflies as natural pigments. Others, such as neopterins, are synthesized in active macrophages, and their high levels were related to activity in cell-mediated immune system, particularly during infections caused by viruses and intracellular bacteria and parasites (Lorente et al. 2004). Pterins are involved in different photobiological processes; for example, they act as sensitizers in photochemical reactions inducing DNA damage and are able to produce singlet oxygen. Pterin fluorescence is observed upon excitation at 350 nm and monitored at 450 nm. Steady-state and time-resolved studies of pterins were evaluated in aqueous solution. Fluorescence lifetimes of biopterin fluorescence were showed dependent on the phosphate concentration (Lorente et al. 2004). The fluorescence of pterins has been used for analytical purposes. Some essays use the emission of 6-carboxypterin for analyzing the concentration of folic acid (Herbert and Bertino 1967), as this molecule is composed of a pterin moiety, p-aminobenzoic acid, and glutamic acid (Ouellette et al. 2002). The quantum yield of folic acid is, however, very low compared to those of pterins (Swarna et al. 2012; Thomas et al. 2002).

Vitamin B12 is another fluorescent compound, originally identified in liver and liver extract that can be found in living tissues. Biological active form of the fully oxidized folate, or folic acid, is serving as an essential cofactor in methylation reactions, including the vitamin B12-dependent formation of methionine from homocystein (see review by Birn 2006). The renal uptake of both folate and vitamin B12 involves glomerular filtration followed by tubular reabsorption and is affected in renal disease.

Endogenous fluorescence of *other vitamins* is less understood, although, as listed in Table 3.1, fluorescence of vitamins D and C, or the ascorbic acid (Billinton & Knight 2001), was reported together with that of vitamin K (Ramanujam 2000). The latter was proposed to be applicable for the quantitative antioxidant assay in a synthesized form, where it exhibited weak fluorescence around 440 nm in ethanol (Ohara et al. 2010). In addition, carotenoids, fluorescing at 538 nm following 325 nm excitation, were proposed responsible for the fluorescence of yellow plaques (Richards-Kortum & Sevick-Muraca 1996).

3.2.3 STRUCTURAL PROTEINS

Structural proteins are responsible for the constitution and the shape of cells in organs. Collagen and elastin are two fluorophores found in the ECM, which are often modified in pathological conditions (Richards-Kortum & Sevick-Muraca 1996). In tissues, the ECM often contributes to the AF emission more than the cellular component because collagen and elastin have, among the endogenous fluorophores, a relatively high quantum yield.

Collagen is a major component of the ECM, as well as many tissue-engineered models, with the main function of holding different organs in place and/or holding cells together in organs. The ECM provides a structural lattice for cells in the tissue, thus facilitating cellular communication. It is a key component during processes such as angiogenesis and neoplasia. Collagens are found in several tissues, including skin, eye, and cardiovascular system. Collagen fibrils and elastin are stabilized by covalent cross-links, which are essential for their mechanical stability (for both collagen and elastin) using a unique mechanism based on aldehyde formation (reviewed in Eyre et al. 1984). Reduced collagen cross-linking, as recorded in photodamaged skin, results in a posttranscriptionally diminished collagen synthesis and, consequently, increased second harmonic generation (SHG) signal and a decreased fluorescence lifetime *in vitro* (Lutz et al. 2012). Cross-linking of collagen is also important for diagnostics and treatment of patients with corneal diseases (Ashwin & McDonnell 2010) or aging of human skeletal muscle (Haus et al. 2007). Importantly, separate from the enzymatically derived, mature cross-links are an additional form of biochemical linkage found on connective tissues known as advanced glycation end products (AGE) (Goldin et al. 2006; Haus et al. 2007), which also exhibit endogenous fluorescence (see Section 3.2.4 for more details). Accumulation of these compounds is the result of prolonged exposure to monosaccharides, in which a spontaneous nonenzymatic bound is formed between a reducing sugar and a protein residue. Collagen-linked fluorescence is sensitive to aging that causes predictable alterations in both epidermal and dermal fluorescence (Kollias et al. 1998) and can be used as a quantitative marker of aging and photoaging, including biological age (Odetti et al. 1992).

Collagen endogenous fluorescence arises from lysyl and its enzymatic cross-links, but additional fluorescence can be observed from nonenzyme-dependent cross-linking such as glycation (Kirkpatrick et al. 2006) Collagens have excitation/emission maxima at 325/400 nm (Table 3.1 and Figure 3.2). The blue emission of collagen was attributed to tyrosine and dityrosine, while the yellow-green fluorescence was assigned to pentosidine (Sell et al. 1991; Berezin & Achilefu 2010). Collagen has relatively long lifetime from around 2 ns in solutions to 8.9 ns in powder (Konig et al. 1994). A double exponential decay of collagen was reported in skin and eye tissues (Konig & Riemann 2003; Schweitzer et al. 2007).

Type I collagen, one of the main components of the interstitial matrix, exhibits both optical scattering and endogenous fluorescence, properties that can be exploited for noninvasive estimation of changes in the ECM (Kirkpatrick et al. 2006). Consequently, in addition to fluorescence techniques, SHG imaging is a well-suited microscopy technique to investigate collagen–fiber orientation and structure of connective tissues thanks to an important nonlinear (second-order) susceptibility of collagen (Brown et al. 2003; Cicchi et al. 2009, 2013; Lin et al. 2006; Provenzano et al. 2008).

Elastin is another intrinsically fluorescing protein in connective tissues, which allows tissues such as blood vessels, lungs, skin, stomach, etc. to change shape undergoing substantial deformation over the life span of the organism (Berezin & Achilefu 2010). Elastin is also fluorescent at 400 nm following excitation at 325 nm, which complicates its spectroscopic separation from collagen. Elastin fluorescence was showed to be due to a tricarboxylic, triamino pyridinium derivate, which is very similar to the fluorophore in collagen (Deyl et al. 1980; Thornhill 1975). Importantly, in human skin, ECM components like elastin fibers can be distinguished from collagen (Konig et al. 2005) by spectrally resolved lifetime detection, because collagen has much longer lifetime than elastin (Table 3.1 and Figure 3.2). In addition to skin, elastin and different types of collagens were also found in the eye (Schweitzer et al. 2007), and in tumors, collagen and elastin are attenuated, leading to better isolation of cancer tissue (Provenzano et al. 2006).

Keratin is an endogenously fluorescing substance found in the hair and skin; it was shown to have fluorescent lifetime of 1.4 ns (Ehlers et al. 2007). Identification of endogenous melanin and keratin by time-resolved spectroscopy has the potential to gather information on intrahair dye accumulation, or to monitor intratissue diffusion of pharmaceutical and cosmetic components along hair shafts (Ehlers et al. 2007). In addition, keratinization of epithelial tissues, particularly oral, but also esophageal and cervical, was described (Wu & Qu 2006).

3.2.4 LIPIDS

Lipids are one of the cornerstones of living organism, and to the great extent, they are not fluorescent. However, in tissues, fluorescent pigments associated with aging and various pathological processes such as atherosclerosis and retinal degeneration have been identified (Richards-Kortum & Sevick-Muraca 1996). Linked to lipid oxidation, these pigments have been named lipopigments, from which two are closely related: lipofuscin (dark fat) and ceroid (waxy), which are undigested components that accumulate over a human life. Lipopigments exhibit endogenous fluorescence with broad emission spectra (Figure 3.2).

Lipofuscins, yellow/brown AF pigments also known as age pigments, mostly visible in skin as the "age" or "liver" pigment, are mixture of compounds derived from lipids and post-translational protein modifications, which are stored in lysosomal cell compartments (Sulzer et al. 2008). Widely distributed throughout the animal kingdom, lipofuscin is most highly expressed in postmitotic cells and considered the most consistent cellular morphological change of the normal aging process. Lipofuscin is a ubiquitous material, highly insoluble and reactive, and present in granules, with a characteristic UV-excitable fluorescence with broad emission spectra and excitation/emission maxima of 360–420/540–650 nm. Lipofuscin pigmented granules are 1–5 µm in diameter in long-lived cells, such as neurones and cardiac myocytes. They are important sources of AF in tissues, particularly in the eye (Delori et al. 1995, 2001; Schweitzer et al. 2007), where they manifest changes in pathological conditions, such as age-related degeneration (Delori et al. 2001; Schweitzer et al. 2004, 2007). Lipofuscin consists of 30%–58% protein, primarily malonaldehyde, and 19%–51% lipid-like material, possibly oxidation products of polyunsaturated fatty acids (Jolly et al. 2002), highly enriched in metals such as iron. Heterogeneity in lipofuscin emission maxima was found in different cells and related, for example, to the presence of retinol in the retinal pigment epithelia; indeed, in the eye, this age-related lipophilic material is accounted in part by A2E, the product of hydrolysis of vitamin A aldehyde and phosphatidylethanolamine (Marmorstein et al. 2002), which was proposed to be a major fluorophore in the eye (Eldred & Lasky 1993; Schweitzer et al. 2007; Hammer et al. 2008). In endogenous fundus substances, fluorescence decay of lipofuscins was described to have a double exponential decay with 0.39 and 2.24 ns (Schweitzer et al. 2007; Hammer et al. 2008). Spectroscopic and morphological studies were also performed in human retinal lipofuscin granules (Haralampus-Grynaviski et al. 2003). In neurons, lipofuscin is considered to result from incomplete digestion of mitochondrial products. Its principal components were suggested to be derived from reactions of a highly reactive lipid derivate, 4-hydroxy-2-nonenal (HNE) (Tsai et al. 1998). As a result, peroxide and iron-mediated oxidation was proposed to favor lipofuscin production, whereas iron chelators and antioxidants block the pigment synthesis (Brunk & Terman 2002).

Ceroids are similar structures to lipofuscins, possibly early stages of lipofuscins (Sohal 1984). However, as opposed to age-related lipofuscin, ceroid pigment in tissues is considered to be closely associated with disease, such as vitamin E deficiency. In the brain, ceroid is particularly linked to neuronal ceroid lipofuscinoses, a common childhood disease. Ceroid diseases are strongly related to mitochondria, as the major protein component of ceroid has been identified to be a F0 portion of mitochondrial ATP synthase (Hall et al. 1991).

Another lipidic source of endogenous fluorescence in tissues are *advanced glycation end-products (AGE)*, final products of complex chemical reactions between sugars, proteins, lipids, and nucleic acids (Singh et al. 2001), which accumulate naturally during certain conditions, including aging and diabetes. AGE-related AF has been described in the skin (Konig & Riemann 2003; Na et al. 2001) and eye (Schweitzer et al. 2007). AGE formation often induces structural changes in proteins, such as collagen (see Section 3.2.3). AGE are oxidized after contact with aldose sugars and may fluoresce, produce ROS, bind to specific cell surface receptors, or form cross-links. The fluorescence of some AGE, for example, pentosidine, has been well characterized (Sell et al. 1992). Excited around 450 nm with emission between 510 and 700 nm, AGE were described to have fluorescence lifetime of 0.865 and 4.17 ns in endogenous fundus substances (Hammer et al. 2008; Schweitzer et al. 2007). AGE have been linked to disease and aging process and have been demonstrated to be strongly correlating with plasma glucose concentration and chronological age. Consequently, they contribute to pathophysiology of vascular disease in diabetes (Goldin et al. 2006).

AGE of hemoglobin (Hb-AGE) are formed as a result of slow, spontaneous, and nonenzymatic glycation reactions. They possess a characteristic AF at 308/345 nm. Despite the presence of heme as a quenching molecule, they were described to have AF with quantum yield of 0.19 (Vigneshwaran et al. 2005). As glucose reacts nonenzymatically with the protein amino groups to initiate a post-translational modification process known as glycation, Hb-AGE had been suggested to provide a better index for diabetes.

3.2.5 AMINO ACIDS

Amino acids are crucial organic components, structural units forming proteins implicated in major biological roles in neurotransmission, nutrition, and biosynthesis. The most pronounced fluorescent amino acids are phenylalanine, tyrosine, and tryptophan (Table 3.1), with fluorescence excitation below 300 nm and fluorescence emission below 450 nm. Only amino acids with aromatic ring exhibit significant AF, and majority of amino acids useful for fluorescence imaging purposes are not present in a free form but are bound in proteins. Quantum yields of tyrosine and Trp were estimated at 0.13 and 0.14 in water at neutral pH, while that of phenylalanine was evaluated only to 0.02 (Lakowicz 2006). In the tissues, however, both tyrosine and phenylalanine feature quite low emission intensity compared to that of tryptophan, which fluoresces with the highest quantum yield and, accordingly, high intensity. The decay of protein fluorescence is usually described by multiexponential model (Alcala et al. 1987). The fluorescence intensity decay kinetics of tryptophan and tyrosine residues has been widely used to obtain information about structure and dynamics of proteins and peptides, and their complexes with ligands (Wlodarczyk & Kierdaszuk 2003). Events such as association of proteins (enzymes) with ligands (substrates, inhibitors) are usually accompanied by shift in excitation and emission spectra, as well as by changes in fluorescence quantum yields and decay kinetics (Cherednikova et al. 2001).

Tryptophan (Trp) is an essential amino acid in the human diet, the building blocks of proteins and precursor to a number of biologically active compounds, particularly the neurotransmitter serotonin, which, in turn, is converted into melatonin. In this way, Trp is playing a key role in balancing mood and sleep patterns and/or in the vitamin B3 (niacin) deficiency, which leads to pellagra. Trp residues were proposed as fluorescent reporters that are able to provide information about the dynamic nature of proteins (Alcala et al. 1987). While many proteins contain only one Trp residue, it is not uncommon to observe proteins that include multiple Trp residues within their amino acid sequence (Pokalsky et al. 1995). This feature was proposed to be used in the study of protein phosphorylation and dephosphorylation–cellular mechanisms that regulate cell growth, proliferation, and transformation. The environmental sensitivity of Trp fluorescence is relatively well understood in model systems, where the dominant role of peptide bond quenching was demonstrated in Trp fluorescence (Pan & Barkley 2004). However, in cells and tissues and *in vivo*, the use of Trp fluorescence for diagnostic applications is highly compromised due to significant phototoxicity of radiation below 300 nm and also due to its high absorption in most optical systems and strong background fluorescence. Nevertheless, with the development of novel imaging systems, especially the two photon excited time-resolved spectroscopy, Trp AF is becoming more and more interesting in the study of cells and their compartments. This is particularly related to the high sensitivity of the Trp fluorescence lifetime to its environment. Time-resolved fluorescence images of Trp emission can be used for differentiation of different organelles in cells. Trp fluorescence was also recorded in cancerous tissues: increase in the fluorescence of Trp residues (emission peak at 350 nm, excitation under 300 nm) was noted in bladder tumors (Zheng et al. 2003), as well as in nonmetastatic malignant cells from different species (Pradhan et al. 1995). In the latter study, lifetimes for Trp ranged between 2.5 and 3.7 ns.

Endogenous fluorescence of *serotonin*, a monoamine (5-hydroxytryptamine) biochemically derived from Trp, can be evaluated by two- and three-photon microscopy following 340 nm excitation. Fluorescence lifetimes of serotonin were found decreased in the cell cytosol when compared to solutions from 2.5 to 3.8 ns (Botchway et al. 2008).

Time-resolved fluorescence properties of *phenylalanine* residues were described in different molecular environments, with fluorescence lifetime values varying from 1.7 ns in epidermal growth factor receptor transmembrane sequence to 7.4 ns in water (Duneau et al. 1998). However, despite nanosecond lifetimes, low quantum yield of the phenylalanine fluorescence (around 0.024) is responsible for its little contribution to the overall AF, and this molecule is much less used for biomedical studies than Trp.

Tyrosine is implicated in the production of neurotransmitters. Tyrosine derivates, namely, adrenaline and noradrenalin, also exhibit endogenous fluorescence, which is significantly red-shifted from tyrosine (Billinton & Knight 2001). Adrenalin, excited at 330 and 420 nm, is emitting at 525 nm, while noradrenalin, excitable at 325 and 395 nm, has maximum emission at 515 nm. Fluorescence lifetimes of adrenaline were showed to vary from 0.8 ns in the pH 6 buffer up to 1.95–2.2 ns in methanol, ethanol, or propanol, while exhibiting a double exponential decay in environments such as cyclohexane or β-cyclodextrin (Polewski 2003).

3.2.6 PIGMENTS

Melanin, or black skin pigment, is responsible for the coloration of eye, hair, and skin. This substance is classically labeled eumelanin (or true melanin) if dark brown or black, or pheomelanin (dusky melanin) if reddish, because of its high sulfur content (Sulzer et al. 2008). Melanin is the pigment produced by L-DOPA substance within melanin granules-specialized lysosomes (Orlow 1995), which are secreted from the melanocytes and then endocytosed by the pigmented cells. It is composed of a complex mixture of largely unknown biopolymers derived from tyrosine (Berezin & Achilefu 2010). Melanin is the dominant endogenous pigment and the body's natural sunscreen, with its major function being to protect the body from exposure to UV radiation with absorption ranging from near-infrared at 800 nm to UV below 300 nm (Birch et al. 2005). Consequently, it represents a major obstacle to optically detect and spectroscopically isolate other metabolites in the whole body, especially the skin. There are four main categories of melanins—eumelanin, pheomelanin, allomelanin, and neuromelanin (Wakamatsu et al. 2003)—but only eumelanin and pheomelanin are present in human skin. Melanin is also found in other tissues, particularly the eye (Schweitzer et al. 2007), where eumelanin is the predominant pigment (Wielgus & Sarna 2005). Although eumelanin has strong broadband emission, quantum yield for eumelanin is low (Nighswander-Rempel et al. 2005). Sensing fluorescence lifetime of the melanin represents an opportunity to better isolate signal from this molecule from other fluorophores in tissues. The emission decay of melanin is complex with the lifetime ranging from picoseconds to 8 ns with average lifetime of 1.2 ns (Ehlers et al. 2007; Forest & Simon 1998). Eumelanin in black hair and pheomelanin in red and blond human hair were proposed to account for much shorter fluorescence lifetime in black vs. blond hair ($\tau 1/\tau 2$ of 0.03/0.8 ns vs. 0.34/2.3 ns) (Ehlers et al. 2007). In addition to hair analysis, melanin fluorescence lifetime was also used for diagnostics and treatment of skin cancer (Teuchner et al. 1999), as the selective excitation of melanin facilitated by fluorescence lifetime imaging microscopy (FLIM) can also serve to discriminate melanin-filled melanomas (see Chapter 14 for more details).

Neuromelanin is another form of melanin found in neuronal tissues, particularly in the *substantia nigra* or black substance (Sulzer et al. 2000). It is a brown/black pigment, which aggregates to about 30 nm diameter spheres with a pheomelanin core and eumelanin surface in 3:1 eumelanin/pheomelanin ratio. The eumelanin is formed by dopamine oxidation. Neuromelanin accumulates in *substantia nigra* throughout the lifetime, suggesting its poor degradation by neurons. Neuromelanin levels have been reported decreased in patients with Parkinson's disease (Kastner et al. 1992), possibly as a consequence of increased oxidative processes. Hearing loss has also been associated with the loss of neuromelanin (Goding 2007).

Mucin is a glycoprotein found in most epithelial cells (Cheng et al. 2009a). It is another pigment that was described to be a source of small tissue AF when excited with UV/VIS light, with maximum emission recorded following excitation at 436 nm for goblet cell mucin (Castillo et al. 1986).

Yellow pigment *bilirubin* (Plavskii et al. 2008) is an important pigment playing a role in phototherapy for hyperbilirubinemia (jaundice) in newborn infants. This disease is caused by excess accumulation of bilirubin in the blood and subcutaneous layer of the infant, because of its hyperproduction and/or the slow rate of excretion. Excited at 350–520 nm, with the emission at 480–650 nm, bilirubin has a relatively short excited state lifetime of around 14 ps. Because of the high affinity of bilirubin for membranes and low fluorescence quantum yield of bilirubin in water, membrane-bound bilirubin is considered to be fluorescent (Zucker et al. 2001).

3.2.7 PORPHYRINS

Porphyrins are substances widely distributed in living systems. The most distinguished members of the porphyrin family are bacteriochlorophylls in phototrophic bacteria, chlorophylls in plants, and heme in mammalian red blood cells. Porphyrins are dominant autofluorescing substances in the red spectral region, as the majority of porphyrins emit strongly around 600–800 nm with the exception of heme, whose fluorescence is quenched by the coordinated iron (Berezin & Achilefu 2010). The presence of endogenous fluorescence from porphyrins in tissues and in blood has been linked with serious pathologies, such as cancer (Berezin & Achilefu 2010), as porphyrins were described to be the source of red fluorescence noted in necrotic tumors (Richards-Kortum & Sevick-Muraca 1996) and distinct red fluorescence spots can also be observed in oral squamous cell carcinoma. In addition, blood analysis of patients presenting gastric cancer, breast cancer, or Hodgkin's lymphoma presented enhanced AF around 630 nm because of high concentration of porphyrins and its derivatives, namely, protoporphyrin IX (PPIX) (Berezin & Achilefu 2010). During biosynthesis of PPIX, σ-aminolevulinic acid (ALA), a nonphotoactivable precursor, produces fluorescent PPIX via the cellular heme synthesis pathway (Fukuda et al. 2005; Ajioka et al. 2006). Ferrochelatase then catalyzes the insertion of iron into PPIX to form heme, which again is nonfluorescent. PPIX itself is a strongly fluorescent compound, excitable at 400–450 nm, with emission in red spectral region between 635 and 710 nm (Figure 3.2) (Kress et al. 2003). Different fluorescence lifetimes were described for PPIX monomers (10–15 ns) vs. dimers (2–3 ns) (Konig 2008). In healthy cells, rapid conversion of PPIX to heme is responsible for relatively low PPIX fluorescence. However, in pathological situations such as cancer, excess of ALA results in accumulation of significant amounts of PPIX in cells, which are fluorescently detectable, allowing exploiting PPIX as a fluorescence marker for cancer (see Chapter 23 for more details).

Mitochondria were proposed to be the primary cellular sites of PPIX (Ji et al. 2006). Many environmental factors, such as glucose, or oxygenation levels, temperature, pH, etc., affect PPIX fluorescence due to the dependence of its production on sequential enzymatic pathways. Selective accumulation of PPIX in cancer cells and tissues also provides possibility to use these molecules as sensitizers for photodynamic therapy and diagnostics (PDT and PDD; see Chapters 14 and 23 for more details).

3.3 TISSUE FLUOROPHORES

Tissue observation using light is among the most common imaging practices in medicine and biomedical research, ranging from visual observation of patients to advanced nonlinear microscopy techniques. Many diseases are associated with alterations in tissue and cell structure and/or function, reflected in changes of their optical properties. In addition to investigation of tissue reflectance and absorption properties, it is possible to utilize AF of cells, which can be observed without addition of any fluorescent markers that may interfere with natural cell functioning. Spectroscopy of endogenous fluorophores has been thus widely used to monitor the physiological state of tissues. Differences in tissue spectral and lifetime characteristics have been identified, for example, between malignant and normal tissues (see Chapter 14 for more details) and in other disease processes, but also during aging and photoaging. An example of fluorescence lifetime images recorded in various cells and tissues is shown in Figure 3.3. The figure clearly shows the variability of AF lifetime patterns recorded in different tissues (a, b), in comparison to two-photon/SHG signal recorded from collagen fibers in muscle tissue (c) and/or the signal derived from fluorescence protein expressed in cultured cells (d).

It should be noted that study of AF in biological tissues is challenging because of low intrinsic levels of their fluorescence intensity, the overlapping excitation and emission spectra of many endogenous fluorophores, and strong scattering and heterogeneity of the biological tissues. For these reasons, measurement of fluorescent intensity alone is commonly not sufficient to gather reliable quantitative data. Study of additional parameters, such as excitation, emission spectra, and fluorescence lifetime, is often necessary. Time-resolved fluorescence spectroscopy (TRFS) is a powerful method to quantitatively study endogenous fluorescence species in cells and tissues, and it has been used to identify different

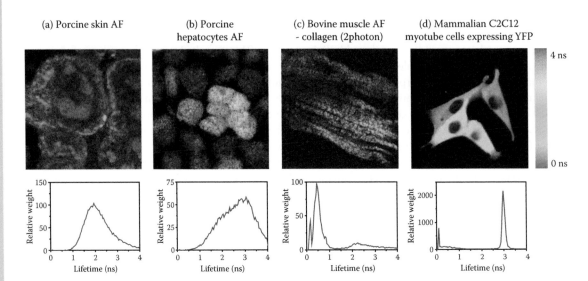

Figure 3.3 Fluorescence lifetime imaging of cells and tissues. Images of time-resolved endogenous fluorescence of cells recorded *in vitro* using a laser scanning microscope with pulsed laser excitation. Upper row: fluorescence lifetime (FLIM) images; bottom row: fluorescence lifetime distributions corresponding to the images in the upper row. Samples: porcine skin (a), porcine hepatocytes (b), bovine muscle–collagen (c); and mammalian C2C12 cultured myotube cells expressing YFP (d). Fluorescence was excited by 473 nm line of picosecond diode laser (a, b, d) or by femtosecond NIR laser at 1038 nm (c). Emission was recorded with 490 nm LP filter using an LSM 510 META NLO confocal microscope (Zeiss, Germany) fiber-optically coupled to TCSPC instrumentation (PML-Spec and SPC-830, Becker-Hickl, Germany).

tissue types such as colon, lung, cervix, and skin. As a next step in this approach, spectrally resolved fluorescence lifetime analysis of endogenous fluorescence facilitates even better recognition, discrimination, and diagnosis of biologically active tissues and their pathological modifications. This includes structural discrimination of anatomical features of cellular morphology or changes of metabolic state in various pathological conditions. Here, we give a brief overview of endogenous fluorophores found in specific tissues, in which spectral and lifetime characteristics are used for cell and tissue biomedical diagnostics.

3.3.1 BRAIN

The brain is one of the most fascinating and complex organs. It contains numerous optically active substances, including blood, cytochromes, neuromelanin, and lipofuscin. The spectral profile of blood correlates with oxygenation levels, whereas melanin characterizes certain structures and lipofuscin is an indicator of tissue damage (Terman & Brunk 2006). In pathological situations, such as hypoxia in the brain, alterations in mitochondrial energy metabolism associated with mitochondrial dysfunction often occur (Dhalla et al. 1993). Fluorescence lifetimes have been reported to help in the analysis of hypoxic conditions in the brain, namely, by investigation of endogenous NAD(P)H fluorescence (Table 3.2) (Vishwasrao et al. 2005). In hypoxia, due to osmotic swelling of the mitochondria, intracellular viscosity was proposed to be reduced, and these conformational and environmental changes effectively decreased average tissue fluorescence lifetimes.

Multiphoton excitation and four-dimensional microscopy were used to generate fluorescence lifetime maps of the murine brain anatomy, experimental glioma tissue, as well as biopsy specimens of human glial tumors (Kantelhardt et al. 2007). In this work, distinct excitation profiles and lifetimes of endogenous fluorophores served to identify specific brain regions. Multiphoton excited fluorescence of endogenous fluorophores also allows structural imaging of tumor and central nervous system histoarchitecture at a subcellular level, as fluorescent lifetimes permit discrimination of glioma cells from normal brain (Leppert et al. 2006). These studies demonstrated that tumor can be distinguished from normal brain on the basis of fluorescence intensity and fluorescence lifetime profiles, and consequently, noninvasive optical tissue analysis has the potential to be applied in the tumor surgery. Compact optical fiber-based apparatus for *in situ* time-resolved laser-induced fluorescence spectroscopy of biological systems was now constructed to identify fluorescence decay lifetimes and emission spectra of endogenous fluorophores, namely, elastin, collagen, NADH, and FADs (Butte et al. 2003; Fang et al. 2004). Measurement of endogenous fluorescence has displayed the diagnostics potential for diagnosing atherosclerotic plaques and tumors and/or for intraoperative diagnosis of primary brain tumors (Maarek et al. 2000a, b; Marcu et al. 2001a; Richards-Kortum & Sevick-Muraca 1996), and FLIM was also applied to discriminate advanced injuries based on increased tissue fluorescence lifetime in this condition (see Chapter 15 for more details).

3.3.2 SKIN

Human skin exhibits weak AF based on naturally occurring fluorophores (Table 3.2), namely, NAD(P)H and flavins (Konig & Riemann 2003; Masters et al. 1997), lipofuscins, collagen, melanin, elastin, and keratin (Konig & Riemann 2003). These compounds are found in specific skin layers and often exhibit changes in fluorescence characteristics with modifications of skin functionality. In such specific skin regions, FLIM is an effective means to distinguish different fluorophores, namely, elastic fibers from collagen (Konig et al. 2005) and/or for melanin identification (Teuchner et al. 1999). Fluorescence lifetimes are particularly useful in clinical dermatology for the evaluation of skin condition and monitoring of cutaneous response to noninvasive therapies (Astner et al. 2008), cancer detection and diagnosis (De Beule et al. 2007), and human skin aging (Koehler et al. 2008), and/or for study of penetration and localization of drugs within the skin (Teichmann et al. 2007). Lately, time-resolved AF measurements also served to test glycation of human skin for screening of type 2 diabetes (Blackwell et al. 2008). Furthermore, fluorescence lifetime spectroscopy also proved to be useful for the analysis of components of human hair (Ehlers et al. 2007) (see Chapter 16 for more details).

3.3.3 CARDIOVASCULAR TISSUE

In the heart, the endogenous fluorescence of NAD(P)H (Barlow & Chance 1976; Blinova et al. 2005; Eng et al. 1989; Griffiths et al. 1998; Huang et al. 2002) and flavins (Chorvat et al. 2005; Eng et al. 1989; Huang et al. 2002; Koke et al. 1981; Romashko et al. 1998) has long been used for noninvasive fluorescence probing of the metabolic state. Two lifetime pools were proposed to correspond to "free" and "protein-bound" NAD(P)H in porcine heart mitochondria (Table 3.2) (Blinova et al. 2005). In living cardiac cells, spectra and lifetime values were identified for NAD(P)H in rats (Chorvatova et al. 2012, 2013a, b) as well as in humans (Cheng et al. 2007, 2009b; Chorvat et al. 2010). In addition, endogenous flavin and NAD(P)H fluorescence was studied in rats both in steady-state conditions (Chorvat & Chorvatova 2006) and during fast processes of cellular contraction (Chorvat et al. 2008; Chorvatova et al. 2011). In human cardiac cells, spectral unmixing revealed two NAD(P)H components, first with two fluorescence lifetimes of 0.32 and 1.61 ns, and second with monoexponential decay of 0.42 ns (Chorvat et al. 2010). In this study, mild rejection of transplanted hearts in patients affected primarily the first component, which lifetimes were slowed to 0.44 and 2.18 ns (Table 3.2), resulting in higher NAD(P)H fluorescence intensities. In the cardiovascular system, FLIM was also used for characterization and diagnostics of atherosclerotic plaques (Marcu 2010), as described in details in Chapter 19.

In cardiovascular tissue, SHG was applied to study the collagen fibers of the ECM, a connective tissue in the heart and vessels (Schenke-Layland et al. 2006, 2007). In unstained myocytes, the SHG signal from myosin filaments, which constitute muscle sarcomeres of the cardiac cell contractile apparatus, was recorded at 415–450 nm following excitation at 830–900 nm (Plotnikov et al. 2006; Wallace et al. 2008) and applied for evaluation of cell volume and/or sarcomere length (Schenke-Layland et al. 2007). In addition, elastin fluorescence was recorded in arterial (Maarek et al. 2000b; Marcu et al. 2001b) and aortic (Liu et al. 2010) tissues.

3.3.4 OPHTHALMOLOGICAL APPLICATIONS

AF is one of the most versatile noninvasive tools for mapping the metabolic state in the eye (reviewed in Schweitzer 2010). Spectrally resolved lifetime analysis of endogenous fluorophores has facilitated metabolic mapping of the human retina *in vivo* (Table 3.2) (Schweitzer et al. 2007). Even with limited excitation

Overview of fluorescence measurements and concepts

and emission ranges, fluorescence lifetimes allow separating individual fluorophores that can be identified in distinct tissue slices or specific cell types. Importantly, specific excitation/emission spectra and lifetime properties of the intrinsic fluorophores in the eye also allow identifying functional modifications—the first signs of a starting pathological process. Mapping of human retina in healthy subjects and patients revealed the possibility of using lifetime spectroscopy to facilitate clinical diagnosis, as differences in the distribution of fluorescence lifetimes were observed between age-related macular degeneration patients when compared to healthy subjects (Schweitzer et al. 2004). Studies of the eyes showed that aging and age-related macular degeneration were associated with increase in the fluorescence lifetimes in fundus AF (Schweitzer et al. 2004), and this process was proposed as a key feature of aging (Delori et al. 2001; Eldred & Lasky 1993; Schweitzer et al. 2004). The principal fluorophore implicated in these changes was lipofuscin (Delori et al. 1995, 2001; Hammer et al. 2008; Schweitzer et al. 2007), particularly A2E (Eldred & Lasky 1993; Hammer et al. 2008), a common morphological result of the aging process, which is manifested as a heterogeneous complex of fluorescent, lipid-protein aggregates found in the cytoplasm of postmitotic cells. In the retinal pigment epithelium (RPE) of the human eye, the formation of LF was attributed to the accumulation of indigestible end products from the phagocytosis of the photoreceptor outer segment (see Chapter 20 for more details).

3.3.5 STEM CELLS AND TISSUE CONSTRUCTS

During stem cell differentiation, the relative increase in bound NAD(P)H was observed together with an increase in the ratio of bound/free NAD(P)H (Konig et al. 2011). An overall decrease in fluorescence intensity was noted during differentiation, indicating drop of reduced form of NAD(P)H, compared to the oxidized one, pointing to an increase in oxygen consumption. Metabolic functions were reported modified during differentiation of human stem cells, as mean fluorescence lifetime of NAD(P)H was found increased (Guo et al. 2008). Consequently, NADH lifetimes were proposed to serve as an optical biomarker for noninvasive selection of stem cells from differentiated progenies, with the decrease in $NAD(P)H/NAD^+$ associated with an increased metabolism (Guo et al. 2008). Flavin fluorescence was also showed decreased during adipogenic differentiation—both NAD(P)H and flavin fluorescence dropped, but the NAD(P)H/flavin ratio increased. In addition, bright fluorescent granules were observed in older stem cells, indicating accumulation of lipopigments in these cells. Progress of ECM formation is evaluated in human stem cells by advanced optical methods, namely, using SHG by monitoring the state of the matrix collagen (Lee et al. 2006; Dumas et al. 2010). This method was also used for comparison of the collagen deposition between the stimulated human mesenchymal stem cells and controls (Hronik-Tupaj et al. 2011), or to monitor the effect of mechanical stress on mesenchymal stem cell collagen production (Chen et al. 2008). A bimodal technique integrating TRFS and ultrasound backscatter microscopy (UBM) was applied for nondestructive detection of changes in the biochemical, structural, and mechanical properties of self-assembled engineered articular cartilage constructs (Sun et al. 2012). The technique was capable of nondestructively evaluating the composition of the ECM and the microstructure of engineered tissue, demonstrating great potential as an alternative to traditional destructive assays. SHG combined with fluorescence lifetime microscopy is also applied to monitor structural information on the state of cell collagen in human mesenchymal stem cells (see Chapter 21 for more details).

3.3.6 OTHER TISSUES AND CELL SYSTEMS

In gastrointestinal tract, endogenous fluorophores were studied, primarily in relation to cancer diagnostics, in pancreas by spectroscopy, and in esophagus and colon by time-resolved spectroscopy, as described in Chapters 14 and 17. NAD(P)H and flavins were the main fluorophores found in these epithelial tissues. Importantly, strong fluorescence of keratin, observed in the top layer of the keratinized squamous epithelium (Wu & Qu 2006), displays excitation/emission properties similar to collagen and creates interference in the assessment of endogenous signals elicited by NADH/FAD fluorescence in epithelium and collagen fluorescence in stroma. The degree of keratinization identified in oral but also esophageal and cervical epithelial tissues (Wu & Qu 2006) was reflected in signals emitted at 420–435 nm following excitation at 355 nm, which are most likely a mixture of keratin and NADH fluorescence. At 405 nm excitation, NADH, keratin, and FAD fluorescence overlap in these tissues, while a peak at 680 nm was attributed to porphyrin derivates (Wu & Qu 2006). Using the time-resolved fluorescence measurement, it is thus possible to resolve the signal from these molecules in specific sublayers, as keratin is mostly pronounced in the keratinized epithelium, NADH in the normal one, and collagen in the stroma.

Other applications of multispectral FLIM include the study of articular cartilage and different stages of arthritis disease (Talbot et al. 2005) and/or noninvasive metabolic monitoring of human diabetes, after observing glucose-dependent changes in NAD(P)H-related fluorescence lifetimes of adipocytes and fibroblasts *in vivo* (Evans et al. 2005). Cervix, bladder, and urinary tract tissue endogenous fluorescence was mainly examined in relation to cancer. In the immune system, white cell eosinophils were described to exhibit intense fluorescence following excitation at 370–450 nm and emission around 520 nm derived from eosinophilic granules, which decreased significantly in patients with eosinophilia (Richards-Kortum & Sevick-Muraca 1996; Weil & Chused 1981).

3.4 EXOGENOUS MOLECULAR PROBES FOR CELL AND TISSUE DIAGNOSTICS

One of the most promising applications of fluorescence lifetime imaging is in the field of functional imaging using fluorescent dyes. Various molecular probes that report on the biological status or biochemical function have been demonstrated and became very important in contemporary medical and pathological diagnostics. Imaging based on fluorescence lifetimes of molecular probes has numerous advantages, particularly independence on intensity, caused by inhomogeneous probe distribution. The fluorescence lifetime of many probes is sensitive to environmental conditions such as pH, polarity, the presence of chemical species allowing development of molecular probes for sensing protein binding, and solvent polarity. In this section, we specifically focus on probes that are currently used in cell and tissue investigation in animals and humans, namely, fluorescent proteins, near-infrared (NIR) probes, and other dyes including PDT agents and phospholipid markers.

To overcome limitations in specificity of AF signal, a wide spectrum of fluorophores has been specifically engineered for investigation of structure and function of living cells. Cells typically interpret their biochemical environment by transformation of external information, initiated by ligand binding to cell-surface receptors, into intracellular signals and conformational changes of macromolecules such as proteins and lipids, accompanied by changes in concentration of small metabolites (e.g., cAMP) or ions (e.g., calcium). In FLIM, the decay kinetics of the excited state of fluorophores, characterized by fluorescence lifetime, is sensitive to excited-state reactions such as FRET (typically used to detect macromolecular protein–protein associations within living cells), independently on the fluorophore concentration or light-path length (Bastiaens & Squire 1999). Fluorescence lifetime acquisition is also rapid enough (in order of seconds) to make measurements from live cells and has therefore been exploited to resolve physiological parameters such as pH, calcium and sodium concentrations, or molecular associations. With the aim of application directly in living cells or tissues, a number of probes and labels have been recently synthesized for investigation of cell morphology (actin/myosin analogues, membrane probes, nuclear probes), physiology (ion concentration, membrane potential,...), etc. A specific example of such probes represents the NIR dyes, in which photons are the only ones capable of penetrating deep tissues and are therefore particularly useful for the whole animal studies (Berezin et al. 2009; Berezin & Achilefu 2010; Wessels et al. 2007, 2010).

Despite tremendous effort toward noninvasiveness, most of the externally administered fluorescent molecules had been reported to significantly alter the cell homeostasis and usually show a number of subtle to major physiological alterations. The reactive potential of most fluorophores cannot be handled by the living cell by prolonged periods of time, thus hampering their possible direct applications in clinical diagnostics except as a part of biochemical kits for histology or cytometry or their eventual use for light-induced therapy. One of the important exceptions to this behavior is the family of fluorescent proteins, owing ability to specifically label biomolecules during the time of their synthesis in living cells. Extensive application of fluorescent proteins in biochemistry and developmental biology had a crucial supporting role for parallel development of many experimental techniques aiming to refine the available information gathered from detected optical signal, such as multispectral imaging or fluorescence lifetime imaging. As a consequence, the newly engineered approaches and devices proved to be extremely capable of gaining a deep insight into the rather overlooked AF signal, bringing it back on the stage of viable tools of biomedical diagnostics.

3.4.1 FLUORESCENT PROTEINS

Genetically engineered fluorescent protein reporters such as GFP and its spectral derivates have been used for years in living animals for cell tracking and investigation of molecular biology. GFP from the jellyfish *Aequorea victoria* have attracted tremendous interest thanks to their ability to clone and heterologously express GFP genes in a diverse range of cells and organisms with many favorable properties, such as high stability, minimal toxicity, noninvasive detection, and the ability to generate highly visible fluorophores *in vivo* in the absence of external cofactors (Billinton & Knight 2001). Consequently, FP became a truly versatile marker for visualizing physiological processes, monitoring intracellular protein localization, etc. (see example at Figure 3.3d) and thus qualify as major biological reporters for gene expression (Chalfie et al. 1994). Their discoverers, Osamu Shimomura, Martin Chalfie, and Roger Y. Tsien, were awarded a Nobel Prize in Chemistry in 2008; interested readers should consult a comprehensive book on GFP edited by Chalfie and Kain (2006), where detailed information on the history, principles, technical issues, and applications of GFP and its variants can be found.

One of the most promising and scientifically interesting applications of FPs is to monitor the Föster resonance energy transfer (FRET) (Selvin 2000; Takanishi et al. 2006). FRET is a nonradiative transfer between two fluorescent proteins that translates energy from the donor—an excited fluorophore—to an acceptor molecule. Such energy transfer results in changes in the fluorophore optical properties, namely, fluorescence intensity, spectral shape, and importantly, fluorescence lifetimes. FRET is frequently employed as an imaging method to enhance knowledge on the molecular structure of cellular proteins (Lauterborn et al. 2003; Periasamy & Diaspro 2003; Lakowicz 2006; Verbiest et al. 2009; Sun et al. 2011; Periasamy 2001). Based on visualization of fluorescence, which lights up when a resonance transfer occurs between two very close atoms of specific proteins, researchers were capable of establishing the ultrastructure of a great number of proteins or protein machines (Periasamy et al. 2008). FRET was used to visualize biological processes such as protein–protein interactions, protein trafficking, or sensing of the molecular environment (Voss et al. 2005). Constructs featuring FRET between two fluorescent proteins also provide a variety of possible biomedical assays; for example, in the immune system, an enterovirus 71 infection was detected using FLIM-monitored FRET between fluorescent proteins at a plasmid construct (Ghukasyan et al. 2007).

3.4.2 NIR PROBES

Most NIR dyes engineered for fluorescent lifetime contrast are more and more being applied in medicine (see reviews by Berezin et al. 2009 and Berezin & Achilefu 2010) in combination with other fluorescence technologies (reviewed by Wessels et al. 2007, 2010). Biological applications of exogenous contrast agents, detected in the NIR wavelengths by FLIM, allow improving deep tissue imaging and thus monitoring small changes in physiologic functions, cell trafficking, controlled drug release, and cancer biology *in vivo* (Akers et al. 2007, 2010). This approach has several applications, including assessing renal function or measuring the rate of metabolism of biodegradable nanoparticles. Most NIR dyes (used as contrast agents; Akers et al. 2010; Berezin et al. 2009) have the capability of polarity sensing; their distribution in tumor-bearing rodents has been recovered by lifetime gating. In diseases such as cancer and inflammation, such probes accumulate in pathological tissues in response to physiological alterations such as enhanced blood flow and vascular permeability. While conventional molecular probes often lack specificity in pathology, contrast agents accumulate in diseased tissues upon interaction with disease-specific conditions or molecular events, thus enhancing contrast. For example, noninvasive fluorescence lifetime imaging of tumors has been achieved with receptor-targeted NIR fluorescent probes (Bloch et al. 2005). In addition, monitoring fluorescence lifetime of encapsulated NIR dyes (Almutairi et al. 2008) is employed to identify the biodegradability of the drug polymer carrier *in vivo* or the delivery of the therapeutic vehicle toward the target disease (Tarte & Klein 1999).

Time-resolved measurements of NIR dyes were used in combination with genetically engineered probes for quantitative *in vivo* imaging of the lung (Ma et al. 2009). Dynamic noninvasive monitoring of renal function *in vivo* by lifetime imaging of a NIR fluorescent probe LS-288 was used in mice to investigate excess of serum protein in urine (proteinuria) in the aim to better understand the protein-losing nephropathy due to diabetes mellitus and other diseases (Goiffon et al. 2009). Engineered NIR fluorescent probes that exhibit changes in time-resolved fluorescence characteristics after binding to amyloid-beta peptide, the main constituent of amyloid plaques in the brains of patients with Alzheimer's disease, have been synthesized (Chia et al. 2008). *In vivo* fluorescence tomography of multiple fluorophores is also using evaluation of the lifetime component for quantitative separation of the fluorophores (Raymond et al. 2010). NIR femtosecond excitation is used as an innovative invasiveless technique to facilitate monitoring of cell/scaffold combination of tissue repair (Dumas et al. 2010).

Indocyanine green (ICG) analogs are important NIR dyes, which can be attached to biomolecules, such as short peptides at designed positions, to create optical bioprobes particularly useful in cancer diagnostics (Achilefu et al. 2000, 2005; Bloch et al. 2005; Carter et al. 2005; Frangioni 2003; Zhang et al. 2005, 2010). ICG is thus capable of providing exogenous contrast for rapid NIR fluorescence imaging (Kwon & Sevick-Muraca 2011), and consequently, dynamic NIR fluorescence imaging with injection of ICG intravenously to patients is now proposed to provide a method for diagnostic motility testing, for intestinal motility disorder or dysfunction, and for potential evaluation of therapeutic agents. ICG can also be used to monitor molecular events involved in normal and pathological processes (Jose et al. 2011), or assist medical procedures, such as cardiac surgery (Soltesz et al. 2007), or noninvasive human brain imaging (Liebert et al. 2006).

In vivo and *ex vivo* canine mammary gland tissue was imaged by NIR with lifetime sensitive detection and localization of exogenous fluorescent contrast agents, confirming the ability of this technique to detect spontaneous mammary tumors and regional lymph nodes (Reynolds et al. 1999). NIR fluorescence lymph imaging was also demonstrated as a new method to sensitively image lymph vasculature and to quantitatively assess the lymph function in a swine model (Sharma et al. 2007), suggesting possible clinical translation of the enhanced optical imaging for quantifying lymph function in lymphatic disease. See Chapter 22 for detailed information on the use of exogenous contrast agents for clinical FLIM.

3.4.3 OTHER FLUORESCENT PROBES

As indicated above, in pathological situations such as cancer, excess of ALA results in accumulation of significant amounts of PPIX in cells, which are fluorescently detectable, allowing exploiting PPIX as a fluorescence marker for cancer. For these reasons, porphyrins are widely used as PDT photosensitizers. On the basis of the improved characterization of specific biological mechanisms involved in PDT, molecular imaging brought new opportunities for *in vivo* monitoring of tumor responses to PDT in real time (reviewed by Celli et al. 2010). Fluorescence lifetime imaging is one of the new techniques employed to help the evaluation of the PDT action. Time-resolved techniques are, for example, used to monitor fluorescence lifetimes of the photosensitizer used in the PDT to ensure unambiguous detection, for example, in FRET studies. More information on photosensitizers, photodynamic detection (PDD), and clinical PDT studies can be found in Chapters 14 and 23.

Fluorescence investigation of *phospholipids* is possible thanks to existence of membrane-specific probes, namely, that of 1,6-diphenyl-1,3,5-hexatriene or DPH (Fiorini et al. 1988; Parasassi et al. 1991). Time-resolved fluorescence of DPH was studied as the most popular hydrophilic probes used to investigate structural and dynamical properties of synthetic and natural membrane systems. Excitable at 360 nm, with emission above 480 nm, DPH was described to have monoexponential fluorescence decay of 13 ns in cyclohexane, while multiple lifetimes were described in other solvents, such as ethanol, or in phospholipid vesicles (Fiorini et al. 1988; Parasassi et al. 1991). In addition, low-density lipoprotein (LDL) can be characterized by Apo-B100 fluorescence (Zorn et al. 2001).

Other probes such as organelle, membrane, or ion-sensitive probes, as well as probes sensing the physicochemical environment, glucose, oxygen, reactive nitrogen or reactive oxygen species, and/or metal-based nanoparticle probes and quantum dots, are not evaluated in this chapter.

Overview of fluorescence measurements and concepts

3.5 CONCLUSION

Fluorescence lifetime spectroscopy and imaging of tissues play an important role in biomedical diagnostics. To assess and properly understand the data from complex patterns of tissue AF, there is a need for insight knowledge of its molecular origins. In this chapter, we characterized the most important fluorophores responsible for the endogenous fluorescence of mammalian tissues, as well as chosen external fluorophores. In details, we described fluorescence properties and biological function of various metabolites, enzymes and coenzymes, structural proteins, vitamins, lipid substances, amino acids, pigments, and porphyrins. Based on this information, spectrally and time-resolved detection of endogenous fluorescence can be used to discriminate specific cell types in tissues and/or to study functional changes in cells and tissues, such as metabolic oxidative state. As a result, new clinical applications are being developed based on correlations between fluorescence lifetimes and/or spectra of endogenous fluorophores and biological pathologies.

ACKNOWLEDGMENTS

We would like to acknowledge Dr. B. Motro and Prof. S. Michaeli, Bar-Ilan University, Israel, for providing the mammalian C2C12 myotube cells with GBP-FluoCFP12-based expression vectors. We would also like to acknowledge support from Integrated Initiative of European Laser Infrastructures LaserLab Europe III (EC's FP7 under grant agreement n° 284464) and *P.Cezanne* project (grant agreement IP031867 under 6FP of EC), the EC Structural funds within the frames of the project NanoNet 2 (ITMS code 26240120018), the research grant agency of the Ministry of Education, Science, Research and Sport of the Slovak Republic VEGA No. 1/0296/11, and the Slovak Research and Development Agency, APVV-0242-11.

REFERENCES

Achilefu S., Bloch S., Markiewicz M. A., Zhong T., Ye Y., Dorshow R. B., Chance B., & Liang K. (2005). Synergistic effects of light-emitting probes and peptides for targeting and monitoring integrin expression. *Proc Natl Acad Sci U S A* 102, 7976–7981.

Achilefu S., Dorshow R. B., Bugaj J. E., & Rajagopalan R. (2000). Novel receptor-targeted fluorescent contrast agents for in vivo tumor imaging. *Invest Radiol* 35, 479–485.

Ajioka R. S., Phillips J. D., & Kushner J. P. (2006). Biosynthesis of heme in mammals. *Biochim Biophys Acta* 1763, 723–736.

Akers W. J., Berezin M. Y., Lee H., Guo K., Almutairi A., Frechet J. M. J., Fischer G. M., Daltrozzo E., & Achilefu S. (2010). Biological applications of fluorescence lifetime imaging beyond microscopy. *Proc. SPIE* 7576, pp. 7576-1–7576-9. SPIE-Int Soc Optical Engineering, Bellingham, WA.

Akers W., Lesage F., Holten D., & Achilefu S. (2007). In vivo resolution of multiexponential decays of multiple near-infrared molecular probes by fluorescence lifetime-gated whole-body time-resolved diffuse optical imaging. *Mol Imaging* 6, 237–246.

Alcala J. R., Gratton E., & Prendergast F. G. (1987). Fluorescence lifetime distributions in proteins. *Biophys J* 51, 597–604.

Almutairi A., Akers W. J., Berezin M. Y., Achilefu S., & Frechet J. M. J. (2008). Monitoring the biodegradation of dendritic near-infrared nanoprobes by in vivo fluorescence imaging. *Mol Pharm* 5, 1103–1110.

Ashwin P. T., & McDonnell P. J. (2010). Collagen cross-linkage: A comprehensive review and directions for future research. *Br J Ophthalmol* 94, 965–970.

Astner S., Dieterle S., Otberg N., Rowert-Huber H. J., Stockfleth E., & Lademann J. (2008). Clinical applicability of in vivo fluorescence confocal microscopy for noninvasive diagnosis and therapeutic monitoring of nonmelanoma skin cancer. *J Biomed Opt* 13, 014003-1–014003-12.

Avi-Dor Y., Olson J. M., Doherty M. D., & Kaplan N. O. (1962). Fluorescence of pyridine nucleotides in mitochondria. *J Biol Chem* 237, 2377–2383.

Bachilo S. M., & Gillbro T. (1995). Beta-carotene S1 fluorescence. *Proc. SPIE* 2370, pp. 719–723.

Barlow C. H., & Chance B. (1976). Ischemic areas in perfused rat hearts: Measurement by NADH fluorescence photography. *Science* 193, 909–910.

Bastiaens P. I. H., & Squire A. (1999). Fluorescence lifetime imaging microscopy: Spatial resolution of biochemical processes in the cell. *Trends Cell Biol* 9, 48–52.

Bastiaens P. I. H., Van Hoek A., Benen J. A. E., Brochon J. C., & Visser A. J. W. G. (1992). Conformational dynamics and intersubunit energy-transfer in wild-type and mutant lipoamide dehydrogenase from Azotobacter-vinelandii—A multidimensional time-resolved polarized fluorescence study. *Biophys J* 63, 839–853.

Benderdour M., Charron G., Comte B., Ayoub R., Beaudry D., Foisy S., Deblois D., & Des Rosiers C. (2004). Decreased cardiac mitochondrial NADP+-isocitrate dehydrogenase activity and expression: A marker of oxidative stress in hypertrophy development. *Am J Physiol Heart Circ Physiol* 287, H2122–H2131.

Berezin M. Y., & Achilefu S. (2010). Fluorescence lifetime measurements and biological imaging. *Chem Rev* 110, 2641–2684.

Berezin M. Y., Lee H., Akers W., Guo K., Goiffon R. J., Almutairi A., Frechet J. M., & Achilefu S. (2009). Engineering NIR dyes for fluorescent lifetime contrast. *Conf Proc IEEE Eng Med Biol Soc* 2009, 114–117.

Billinton N., & Knight A. W. (2001). Seeing the wood through the trees: A review of techniques for distinguishing green fluorescent protein from endogenous autofluorescence. *Anal Biochem* 291, 175–197.

Birch D. J., Ganesan A. N., & Karolin J. (2005). Metabolic sensing using fluorescence. *Synth Met* 155, 410–413.

Bird D. K., Yan L., Vrotsos K. M., Eliceiri K. W., Vaughan E. M., Keely P. J., White J. G., & Ramanujam N. (2005). Metabolic mapping of MCF10A human breast cells via multiphoton fluorescence lifetime imaging of the coenzyme NADH. *Cancer Res* 65, 8766–8773.

Birn H. (2006). The kidney in vitamin B-12 and folate homeostasis: Characterization of receptors for tubular uptake of vitamins and carrier proteins. *Am J Physiol Renal Physiol* 291, F22–F36.

Blackwell J., Katika K. M., Pilon L., Dipple K. M., Levin S. R., & Nouvong A. (2008). In vivo time-resolved autofluorescence measurements to test for glycation of human skin. *J Biomed Opt* 13, 014004-1–014004-15.

Blinova K., Carroll S., Bose S., Smirnov A. V., Harvey J. J., Knutson J. R., & Balaban R. S. (2005). Distribution of mitochondrial NADH fluorescence lifetimes: Steady-state kinetics of matrix NADH interactions. *Biochemistry* 44, 2585–2594.

Bloch S., Lesage F., McIntosh L., Gandjbakhche A., Liang K. X., & Achilefu S. (2005). Whole-body fluorescence lifetime imaging of a tumor-targeted near-infrared molecular probe in mice. *J Biomed Opt* 10, 054003-1–054003-8.

Botchway S. W., Parker A. W., Bisby R. H., & Crisostomo A. G. (2008). Real-time cellular uptake of serotonin using fluorescence lifetime imaging with two-photon excitation. *Microsc Res Tech* 71, 267–273.

Brown E., McKee T., diTomaso E., Pluen A., Seed B., Boucher Y., & Jain R. K. (2003). Dynamic imaging of collagen and its modulation in tumors in vivo using second-harmonic generation. *Nat Med* 9, 796–800.

Brunk U. T., & Terman A. (2002). Lipofuscin: Mechanisms of age-related accumulation and influence on cell function. *Free Radical Biol Med* 33, 611–619.

Bueno C., Pavez P., Salazar R., & Encinas M. V. (2010). Photophysics and photochemical studies of the vitamin B6 group and related derivatives. *Photochem Photobiol* 86, 39–46.

Butte P. V., Vishwanath K., Pikul B., Mycek M. A., & Marcu L. (2003). Effects of tissue optical properties on time-resolved fluorescence measurements from brain tumors: An experimental and computational study. *Opt Tomogr Spectrosc Tissue V* 4955, 600–608.

Carter S. G., Birkedal V., Wang C. S., Coldren L. A., Maslov A. V., Citrin D. S., & Sherwin M. S. (2005). Quantum coherence in an optical modulator. *Science* 310, 651–653.

Castillo E. J., Koenig J. L., Anderson J. M., & Jentoft N. (1986). Protein adsorption on soft contact-lenses. 3. Mucin. *Biomaterials* 7, 9–16.

Celli J. P., Spring B. Q., Rizvi I., Evans C. L., Samkoe K. S., Verma S., Pogue B. W., & Hasan T. (2010). Imaging and photodynamic therapy: Mechanisms, monitoring, and optimization. *Chem Rev* 110, 2795–2838.

Chalfie M., & Kain S. R. (2006). *Green Fluorescent Protein: Properties, Applications, and Protocols*, 2nd ed. John Wiley & Sons, New York.

Chalfie M., Tu Y., Euskirchen G., Ward W. W., & Prasher D. C. (1994). Green fluorescent protein as a marker for gene expression. *Science* 263, 802–805.

Chance B. (1976). Pyridine nucleotide as an indicator of the oxygen requirements for energy-linked functions of mitochondria. *Circ Res* 38, 131–138.

Chance B., Cohen P., Jobsis F., & Schoener B. (1962). Intracellular oxidation-reduction states in vivo. *Science* 137, 499–508.

Chance B., & Jobsis F. (1959). Changes in fluorescence in a frog sartorius muscle following a twitch. *Nature* 185, 195–196.

Chance B., & Neilands J. B. (1952). Studies on lactic dehydrogenase of heart. II. A compound of lactic dehydrogenase and reduced pyridine nucleotide. *J Biol Chem* 199, 383–387.

Chance B., Schoener B., Oshino R., Itshak F., & Nakase Y. (1979). Oxidation-reduction ratio studies of mitochondria in freeze-trapped samples. NADH and flavoprotein fluorescence signals. *J Biol Chem* 254, 4764–4771.

Chance B., & Thorell B. (1959). Fluorescence measurements of mitochondrial pyridine nucleotide in aerobiosis and anaerobiosis. *Nature* 184, 931–934.

Chen H. M., Chiang C. P., You C., Hsiao T. C., & Wang C. Y. (2005). Time-resolved autofluorescence spectroscopy for classifying normal and premalignant oral tissues. *Lasers Surg Med* 37, 37–45.

Chen R. F. (1965). Fluorescence quantum yield measurements: Vitamin B6 compounds. *Science* 150, 1593–1595.

Chen W. L., Chang C. C., Chiou L. L., Li T. H., Liu Y., & Dong C. Y. (2008). Monitoring the effect of mechanical stress on mesenchymal stem cell collagen production by multiphoton microscopy. *Proc SPIE* 6858, 685806-1–685806-8.

Cheng A. K. H., Su H. P., Wang A., & Yu H. Z. (2009a). Aptamer-based detection of epithelial tumor marker mucin 1 with quantum dot-based fluorescence readout. *Anal Chem* 81, 6130–6139.

Cheng Y., Dahdah N., Porier N., Miro J., Chorvat D., Jr., & Chorvatova A. (2007). Spectrally and time-resolved study of NADH autofluorescence in cardiac myocytes from human biopsies. *Proc SPIE* 6771, 677104-1–677104-13.

Cheng Y., Mateasik A., Dahdah N., Poirier N., Miro J., Chorvat D., Jr., & Chorvatova A. (2009b). Analysis of NAD(P)H fluorescence components in cardiac myocytes from human biopsies: A new tool to improve diagnostics of rejection of transplanted patients. *Proc SPIE* 7183, 718319-1–718319-8.

Cherednikova E. Y., Chikishev A. Y., Dementieva E. I., Kossobokova O. V., & Ugarova N. N. (2001). Quenching of tryptophan fluorescence of firefly luciferase by substrates. *J Photochem Photobiol B-Biol* 60, 7–11.

Chia T. H., Williamson A., Spencer D. D., & Levene M. J. (2008). Multiphoton fluorescence lifetime imaging of intrinsic fluorescence in human and rat brain tissue reveals spatially distinct NADH binding. *Opt Express* 16, 4237–4249.

Chorvat D., Jr., Abdulla S., Elzwiei F., Mateasik A., & Chorvatova A. (2008). Screening of cardiomyocyte fluorescence during cell contraction by multi-dimensional TCSPC. *Proc SPIE* 6860, 686029-1–686029-12.

Chorvat D., Jr., & Chorvatova A. (2006). Spectrally resolved time-correlated single photon counting: A novel approach for characterization of endogenous fluorescence in isolated cardiac myocytes. *Eur Biophys J* 36, 73–83.

Chorvat D., Jr., Kirchnerova J., Cagalinec M., Smolka J., Mateasik A., & Chorvatova A. (2005). Spectral unmixing of flavin autofluorescence components in cardiac myocytes. *Biophys J* 89, L55–L57.

Chorvat D., Jr., Mateasik A., Cheng Y., Poirier N., Miro J., Dahdah N. S., & Chorvatova A. (2010). Rejection of transplanted hearts in patients evaluated by the component analysis of multi-wavelength NAD(P)H fluorescence lifetime spectroscopy. *J Biophotonics* 3, 646–652.

Chorvatova A., Aneba S., Mateasik A., Chorvat D. Jr, and Comte B. (2013a). Time-resolved fluorescence spectroscopy investigation of the effect of 4-hydroxynonenal on endogenous NAD(P)H in living cardiac myocytes. *J Biomed Opt* 18, 67009-1–67009-11.

Chorvatova A., Mateasik A., and Chorvat D., Jr. (2013b). Spectral decomposition of NAD(P)H fluorescence components recorded by multi-wavelength fluorescence lifetime spectroscopy in living cardiac cells. *Laser Physics Letters* 10, 125703-1–125703-10.

Chorvatova A., Elzwiei F., Mateasik A., & Chorvat D., Jr. (2012). Effect of ouabain on metabolic oxidative state in living cardiomyocytes evaluated by time-resolved spectroscopy of endogenous NAD(P)H fluorescence. *J Biomed Opt* 17, 101505-1–101505-7.

Chorvatova A., Mateasik A., & Chorvat D., Jr. (2011). Laser-induced photobleaching of NAD(P)H fluorescence components in cardiac cells resolved by linear unmixing of TCSPC signals. *Proc SPIE* 7903, 790326-1–790326-9.

Cicchi R., Crisci A., Nesi G., Cosci A., Giancane S., Carini M., & Pavone F. S. (2009). Multispectral multiphoton lifetime analysis of human bladder tissue. *Proc. SPIE* 7161, pp. 716116-1–716116-9. SPIE-Int Soc Optical Engineering, Bellingham, WA.

Cicchi R., Volger N., Kapsokalyvas D., Dietzek B., Popp J., & Pavone F. S. (2013). From molecular structure to tissue architecture: Collagen organization probed by SHG microscopy. *J Biophotonics* 6, 129–142.

Coburn S. P., Slominski A., Mahuren J. D., Wortsman J., Hessle L., & Millan J. L. (2003). Cutaneous metabolism of vitamin B-6. *J Invest Dermatol* 120, 292–300.

Cortassa S., Aon M. A., Marban E., Winslow R. L., & O'Rourke B. (2003). An integrated model of cardiac mitochondrial energy metabolism and calcium dynamics. *Biophys J* 84, 2734–2755.

De Beule P. A., Dunsby C., Galletly N. P., Stamp G. W., Chu A. C., Anand U., Anand P., Benham C. D., Naylor A., & French P. M. (2007). A hyperspectral fluorescence lifetime probe for skin cancer diagnosis. *Rev Sci Instrum* 78, 123101.

Delori F. C., Dorey C. K., Staurenghi G., Arend O., Goger D. G., & Weiter J. J. (1995). In vivo fluorescence of the ocular fundus exhibits retinal pigment epithelium lipofuscin characteristics. *Invest Ophthalmol Vis Sci* 36, 718–729.

Delori F. C., Goger D. G., & Dorey C. K. (2001). Age-related accumulation and spatial distribution of lipofuscin in RPE of normal subjects. *Invest Ophthalmol Vis Sci* 42, 1855–1866.

Deyl Z., Macek K., Adam M., & Vancikova O. (1980). Studies on the chemical nature of elastin fluorescence. *Biochim Biophys Acta* 625, 248–254.

Dhalla N. S., Afzal N., Beamish R. E., Naimark B., Takeda N., & Nagano M. (1993). Pathophysiology of cardiac dysfunction in congestive heart failure. *Can J Cardiol* 9, 873–887.

Drossler P., Holzer W., Penzkofer A., & Hegemann P. (2003). Fluorescence quenching of riboflavin in aqueous solution by methionin and cystein. *Chem Phys* 286, 409–420.

Dumas D., Henrionnet C., Hupont S., Werkmeister E., Stoltz J. F., Pinzano A., & Gillet P. (2010). Innovative TCSPC-SHG microscopy imaging to monitor matrix collagen neo-synthetized in bioscaffolds. *Biomed Mater Eng* 20, 183–188.

Duneau J. P., Garnier N., Cremel G., Nullans G., Hubert P., Genest D., Vincent M., Gallay J., & Genest M. (1998). Time resolved fluorescence properties of phenylalanine in different environments. Comparison with molecular dynamics simulation. *Biophys Chem* 73, 109–119.

Ehlers A., Riemann I., Stark M., & Konig K. (2007). Multiphoton fluorescence lifetime imaging of human hair. *Microsc Res Tech* 70, 154–161.

Eldred G. E., & Lasky M. R. (1993). Retinal age pigments generated by self-assembling lysosomotropic detergents. *Nature* 361, 724–726.

Eng J., Lynch R. M., & Balaban R. S. (1989). Nicotinamide adenine dinucleotide fluorescence spectroscopy and imaging of isolated cardiac myocytes. *Biophys J* 55, 621–630.

Estabrook R. W. (1962). Fluorometric measurement of reduced pyridine nucleotide in cellular and subcellular particles. *Anal Biochem* 4, 231–245.

Evans N. D., Gnudi L., Rolinski O. J., Birch D. J., & Pickup J. C. (2005). Glucose-dependent changes in NAD(P)H-related fluorescence lifetime of adipocytes and fibroblasts in vitro: Potential for non-invasive glucose sensing in diabetes mellitus. *J Photochem Photobiol B* 80, 122–129.

Eyre D. R., Paz M. A., & Gallop P. M. (1984). Cross-linking in collagen and elastin. *Annu Rev Biochem* 53, 717–748.

Fang Q., Papaioannou T., Jo J. A., Vaitha R., Shastry K., & Marcu L. (2004). Time-domain laser-induced fluorescence spectroscopy apparatus for clinical diagnostics. *Rev Sci Instrum* 75, 151–162.

Fiorini R., Curatola G., Bertoli E., & Gratton E. (1988). Dph fluorescence lifetime distributions in cholesterol-egg lecithin multilamellar liposomes. *Biophys J* 53, A491.

Forest S.E., & Simon J. D. (1998). Wavelength-dependent photoacoustic calorimetry study of melanin. *Photochem Photobiol* 68, 296–298.

Frangioni J. V. (2003). In vivo near-infrared fluorescence imaging. *Curr Opin Chem Biol* 7, 626–634.

Fukuda H., Casas A., & Batlle A. (2005). Aminolevulinic acid: From its unique biological function to its star role in photodynamic therapy. *Int J Biochem Cell Biol* 37, 272–276.

Gafni A., & Brand L. (1976). Fluorescence decay studies of reduced nicotinamide adenine dinucleotide in solution and bound to liver alcohol dehydrogenase. *Biochemistry* 15, 3165–3171.

Galletly N. P., McGinty J., Dunsby C., Teixeira F., Requejo-Isidro J., Munro I., Elson D. S., Neil M. A., Chu A. C., French P. M., & Stamp G. W. (2008). Fluorescence lifetime imaging distinguishes basal cell carcinoma from surrounding uninvolved skin. *Br J Dermatol* 159, 152–161.

Ghukasyan V., Hsu Y. Y., Kung S. H., & Kao F. J. (2007). Application of fluorescence resonance energy transfer resolved by fluorescence lifetime imaging microscopy for the detection of enterovirus 71 infection in cells. *J Biomed Opt* 12, 024016-1–024016-8.

Glanzmann T. M., Ballini J. P., Jichlinski P., van den Bergh H., & Wagnieres G. A. (1996). Tissue characteristics by time-resolved fluorescence spectroscopy of endogenous and exogenous fluorophores. *Proc SPIE* 2926, 41–50.

Goding C. R. (2007). Melanocytes: The new Black. *Int J Biochem Cell Biol* 39, 275–279.

Goiffon R. J., Akers W. J., Berezin M. Y., Lee H., & Achilefu S. (2009). Dynamic noninvasive monitoring of renal function in vivo by fluorescence lifetime imaging. *J Biomed Opt* 14, 020501-1–020501-3.

Goldin A., Beckman J. A., Schmidt A. M., & Creager M. A. (2006). Advanced glycation end products—Sparking the development of diabetic vascular injury. *Circulation* 114, 597–605.

Griffiths E. J., Lin H., & Suleiman M. S. (1998). NADH fluorescence in isolated guinea-pig and rat cardiomyocytes exposed to low or high stimulation rates and effect of metabolic inhibition with cyanide. *Biochem Pharmacol* 56, 173–179.

Guo H. W., Chen C. T., Wei Y. H., Lee O.K., Gukassyan V., Kao F. J., & Wang H. W. (2008). Reduced nicotinamide adenine dinucleotide fluorescence lifetime separates human mesenchymal stem cells from differentiated progenies. *J Biomed Opt* 13, 050505.

Hall N. A., Lake B. D., Dewji N. N., & Patrick A. D. (1991). Lysosomal storage of subunit-c of mitochondrial ATP synthase in Batten's disease (ceroid-lipofuscinosis). *Biochem J* 275, 269–272.

Hammer M., Richter S., Kobuch K., Mata N., & Schweitzer D. (2008). Intrinsic tissue fluorescence in an organotypic perfusion culture of the porcine ocular fundus exposed to blue light and free radicals. *Graefes Arch Clin Exp Ophthalmol* 246, 979–988.

Haralampus-Grynaviski N. M., Lamb L. E., Clancy C. M. R., Skumatz C., Burke J. M., Sarna T., & Simon J. D. (2003). Spectroscopic and morphological studies of human retinal lipofuscin granules. *Proc Natl Acad Sci U S A* 100, 3179–3184.

Haus J. M., Carrithers J. A., Trappe S.W., & Trappe T. A. (2007). Collagen, cross-linking, and advanced glycation end products in aging human skeletal muscle. *J Appl Physiol* 103, 2068–2076.

Overview of fluorescence measurements and concepts

Herbert V., & Bertino J. R. (1967). Folic acid. In *The Vitamins*, eds. P. György and W.N. Pearson. Academic Press, New York.

Hones G., Hones J., & Hauser M. (1986). Studies of enzyme-ligand complexes using dynamic fluorescence anisotropy. I. The substrate-binding site of malate dehydrogenase. *Biol Chem Hoppe Seyler* 367, 95–102.

Hronik-Tupaj M., Rice W. L., Cronin-Golomb M., Kaplan D. L., & Georgakoudi I. (2011). Osteoblastic differentiation and stress response of human mesenchymal stem cells exposed to alternating current electric fields. *Biomed Eng Online* 10, 9.

Huang S., Heikal A. A., & Webb W. W. (2002). Two-photon fluorescence spectroscopy and microscopy of NAD(P)H and flavoprotein. *Biophys J* 82, 2811–2825.

Jameson D. M., Thomas V., & Zhou D. M. (1989). Time-resolved fluorescence studies on NADH bound to mitochondrial malate dehydrogenase. *Biochim Biophys Acta* 994, 187–190.

Ji Z., Yang G., Vasovic V., Cunderlikova B., Suo Z., Nesland J. M., & Peng Q. (2006). Subcellular localization pattern of protoporphyrin IX is an important determinant for its photodynamic efficiency of human carcinoma and normal cell lines. *J Photochem Photobiol B* 84, 213–220.

Jolly R. D., Palmer D. N., & Dalefield R. R (2002). The analytical approach to the nature of lipofuscin (age pigment). *Arch Gerontol Geriatr* 34, 205–217.

Jose I., Deodhar K. D., Desai U. B., & Bhattacharjee S. (2011). Early detection of breast cancer: Synthesis and characterization of novel target specific NIR-fluorescent estrogen conjugate for molecular optical imaging. *J Fluoresc* 21, 1171–1177.

Kantelhardt S. R., Leppert J., Krajewski J., Petkus N., Reusche E., Tronnier V. M., Huttmann G., & Giese A. (2007). Imaging of brain and brain tumor specimens by time-resolved multiphoton excitation microscopy ex vivo. *Neuro Oncol* 9, 103–112.

Kastner A., Hirsch E. C., Lejeune O., Javoyagid F., Rascol O., & Agid Y. (1992). Is the vulnerability of neurons in the substantia-nigra of patients with Parkinson's-disease related to their neuromelanin content. *J Neurochem* 59, 1080–1089.

Kirkpatrick N. D., Hoying J. B., Botting S. K., Weiss J. A., & Utzinger U. (2006). In vitro model for endogenous optical signatures of collagen. *J Biomed Opt* 11, 054021-1–054021-8.

Klaidman L. K., Leung A. C., & Adams J. D., Jr. (1995). High-performance liquid chromatography analysis of oxidized and reduced pyridine dinucleotides in specific brain regions. *Anal Biochem* 228, 312–317.

Koehler M. J., Hahn S., Preller A., Elsner P., Ziemer M., Bauer A., Konig K., Buckle R., Fluhr J. W., & Kaatz M. (2008). Morphological skin ageing criteria by multiphoton laser scanning tomography: Non-invasive in vivo scoring of the dermal fibre network. *Exp Dermatol* 17, 519–523.

Koke J. R., Wylie W., & Wills M. (1981). Sensitivity of flavoprotein fluorescence to oxidative state in single isolated heart cells. *Cytobios* 32, 139–145.

Kollias N., Gillies R., Moran M., Kochevar I. E., & Anderson R. R. (1998). Endogenous skin fluorescence includes bands that may serve as quantitative markers of aging and photoaging. *J Invest Dermatol* 111, 776–780.

Konig K. (2008). Clinical multiphoton tomography. *J Biophotonics* 1, 13–23.

Konig K., Berns M. W., & Tromberg B. J. (1997). Time-resolved and steady-state fluorescence measurements of beta-nicotinamide adenine dinucleotide-alcohol dehydrogenase complex during UVA exposure. *J Photochem Photobiol B* 37, 91–95.

Konig K., & Riemann I. (2003). High-resolution multiphoton tomography of human skin with subcellular spatial resolution and picosecond time resolution. *J Biomed Opt* 8, 432–439.

Konig K., Schenke-Layland K., Riemann I., & Stock U. A. (2005). Multiphoton autofluorescence imaging of intratissue elastic fibers. *Biomaterials* 26, 495–500.

Konig K., Schneckenburger H., Hemmer J., Tromberg B., & Steiner R. (1994). In-vivo fluorescence detection and imaging of porphyrin-producing bacteria in the human skin and in the oral cavity for diagnosis of acne-vulgaris, caries, and squamous-cell carcinoma. *Proc. SPIE* 2135, 129–138.

Konig K., So P. T., Mantulin W. W., Tromberg B. J., & Gratton E. (1996). Two-photon excited lifetime imaging of autofluorescence in cells during UVA and NIR photostress. *J Microsc* 183, 197–204.

Konig K., Uchugonova A., & Gorjup E. (2011). Multiphoton fluorescence lifetime imaging of 3D-stem cell spheroids during differentiation. *Microsc Res Tech* 74, 9–17.

Kress M., Meier T., Steiner R., Dolp F., Erdmann R., Ortmann U., & Ruck A. (2003). Time-resolved microspectrofluorometry and fluorescence lifetime imaging of photosensitizers using picosecond pulsed diode lasers in laser scanning microscopes. *J Biomed Opt* 8, 26–32.

Kwon S., & Sevick-Muraca E. M. (2011). Non-invasive, dynamic imaging of murine intestinal motility. *Neurogastroenterol Motil* 23, 881-e344.

Lakowicz J. R. (1999). *Principles of Fluorescence Spectroscopy Introducing the Phase-Modulation Methods*, 3rd ed. Springer, New York.

Lakowicz J. R. (2006). *Principles of Fluorescence Spectroscopy*, 3rd ed. Springer, New York.

Lakowicz J. R., Szmacinski H., Nowaczyk K., & Johnson M. L. (1992). Fluorescence lifetime imaging of free and protein-bound NADH. *Proc Natl Acad Sci U S A* 89, 1271–1275.

Lauterborn W., Kurz T., & Wiesenfeldt M. (2003). *Coherent Optics: Fundamentals and Applications.* Springer-Verlag, Berlin Heidelberg.

Lee H. S., Teng S. W., Chen H. C., Lo W., Sun Y., Lin T. Y., Chiou L. L., Jiang C. C., & Dong C. Y. (2006). Imaging human bone marrow stem cell morphogenesis in polyglycolic acid scaffold by multiphoton microscopy. *Tissue Eng* 12, 2835–2841.

Leenders R., Van Hoek A., Vaniersel M., Veeger C., & Visser A. J. W. G. (1993). Flavin dynamics in oxidized clostridium-beijerinckii flavodoxin as assessed by time-resolved polarized fluorescence. *Eur J Biochem* 218, 977–984.

Leppert J., Krajewski J., Kantelhardt S. R., Schlaffer S., Petkus N., Reusche E., Huttmann G., & Giese A. (2006). Multiphoton excitation of autofluorescence for microscopy of glioma tissue. *Neurosurgery* 58, 759–767.

Liebert A., Wabnitz H., Obrig H., Erdmann R., Moller M., Macdonald R., Rinneberg H., Villringer A., & Steinbrink J. (2006). Non-invasive detection of fluorescence from exogenous chromophores in the adult human brain. *Neuroimage* 31, 600–608.

Lin S. J., Jee S. H., Kuo C. J., Wu R. J., Lin W. C., Chen J. S., Liao Y. H., Hsu C. J., Tsai T. F., Chen Y. F., & Dong C. Y. (2006). Discrimination of basal cell carcinoma from normal dermal stroma by quantitative multiphoton imaging. *Opt Lett* 31, 2756–2758.

Liu C. H., Wang W. B., Kartazaev V., Savag H., & Alfano R. R. (2010). Changes of collagen, elastin and tryptophan contents in laser welded porcine aorta tissues studied using fluorescence spectroscopy. *Proc. SPIE* 7561, pp. 756115-1–756115-8. SPIE-Int Soc Optical Engineering, Bellingham, WA.

Liu Y., & Simon J. D. (2003). Isolation and biophysical studies of natural eumelanins: Applications of imaging technologies and ultrafast spectroscopy. *Pigment Cell Res* 16, 606–618.

Lorente C., Capparelli A. L., Thomas A. H., Braun A. M., & Oliveros E. (2004). Quenching of the fluorescence of pterin derivatives by anions. *Photochem Photobiol Sci* 3, 167–173.

Lutz V., Sattler M., Gallinat S., Wenck H., Poertner R., & Fischer F. (2012). Impact of collagen crosslinking on the second harmonic generation signal and the fluorescence lifetime of collagen autofluorescence. *Skin Res Technol* 18, 168–179.

Ma G. B., Jean-Jacques M., Melanson-Drapeau L., & Khayat M. (2009). Quantitative in-vivo imaging of the lung using time-domain fluorescence measurements. *Proc. SPIE* 7191, pp. 71910C-1–71910C-8. SPIE-Int Soc Optical Engineering, Bellingham, WA.

Maarek J. M., Marcu L., Fishbein M. C., & Grundfest W. S. (2000a). Time-resolved fluorescence of human aortic wall: Use for improved identification of atherosclerotic lesions. *Lasers Surg Med* 27, 241–254.

Maarek J. M., Marcu L., Snyder W. J., & Grundfest W. S. (2000b). Time-resolved fluorescence spectra of arterial fluorescent compounds: Reconstruction with the Laguerre expansion technique. *Photochem Photobiol* 71, 178–187.

Maeda T., Taguchi H., Minami H., Sato K., Shiga T., Kosaka H., & Yoshikawa K. (2000). Vitamin B6 phototoxicity induced by UVA radiation. *Arch Dermatol Res* 292, 562–567.

Maeda-Yorita K., & Aki K. (1984). Effect of nicotinamide adenine dinucleotide on the oxidation-reduction potentials of lipoamide dehydrogenase from pig heart. *J Biochem (Tokyo)* 96, 683–690.

Marcu L. (2010). Fluorescence lifetime in cardiovascular diagnostics. *J Biomed Opt* 15, 011106-1–011106-10.

Marcu L., Fishbein M. C., Maarek J. M., & Grundfest W. S. (2001a). Discrimination of human coronary artery atherosclerotic lipid-rich lesions by time-resolved laser-induced fluorescence spectroscopy. *Arterioscler Thromb Vasc Biol* 21, 1244–1250.

Marcu L., Grundfest W. S., & Maarek J. M. I. (2001b). Arterial fluorescent components involved in atherosclerotic plaque instability: Differentiation by time-resolved fluorescence spectroscopy. *Proc. SPIE* 4244, pp. 428–433. SPIE-Int Soc Optical Engineering, Bellingham, WA.

Marmorstein A. D., Marmorstein L. Y., Sakaguchi H., & Hollyfield J. G. (2002). Spectral profiling of autofluorescence associated with lipofuscin, Bruch's membrane, and sub-RPE deposits in normal and AMD eyes. *Invest Ophthalmol Vis Sci* 43, 2435–2441.

Masters B. R., So P. T., & Gratton E. (1997). Multiphoton excitation fluorescence microscopy and spectroscopy of in vivo human skin. *Biophys J* 72, 2405–2412.

McCormick D. B. (2001). Vitamin B-6. In *Present Knowledge in Nutrition*, eds. A. Bownam and R.M. Rusell. ILSI Press, Washington, DC, pp. 207–213.

Na R., Stender I. M., Henriksen M., & Wulf H. C. (2001). Autofluorescence of human skin is age-related after correction for skin pigmentation and redness. *J Invest Dermatol* 116, 536–540.

Nakashima N., Yoshihara K., Tanaka F., & Yagi K. (1980). Picosecond fluorescence lifetime of the coenzyme of D-amino acid oxidase. *J Biol Chem* 255, 5261–5263.

Nakayama S., Sakuyama T., Mitaku S., & Ohta Y. (2002). Fluorescence imaging of metabolic responses in single mitochondria. *Biochem Biophys Res Commun* 290, 23–28.

Overview of fluorescence measurements and concepts

Niesner R., Peker B., Schlusche P., & Gericke K. H. (2004). Noniterative biexponential fluorescence lifetime imaging in the investigation of cellular metabolism by means of NAD(P)H autofluorescence. *Chemphyschem* 5, 1141–1149.

Nighswander-Rempel S. P., Riesz J., Gilmore J., & Meredith P. (2005). A quantum yield map for synthetic eumelanin. *J Chem Phys* 123, 194901-1–194901-6.

Odetti P. R., Borgoglio A., & Rolandi R. (1992). Age-related increase of collagen fluorescence in human subcutaneous tissue. *Metab-Clin Exp* 41, 655–658.

Ohara K., Mitsumori R., Takebe M., Kuzuhara D., Yamada H., & Nagaoka S. (2010). Vitamin K analogue as a new fluorescence probe for quantitative antioxidant assay. *J Photochem Photobiol A-Chem* 215, 52–58.

Orlow S. J. (1995). Melanosomes are specialized members of the lysosomal lineage of organelles. *J Invest Dermatol* 105, 3–7.

Ouellette M., Drummelsmith J., El Fadili A., Kundig C., Richard D., & Roy G. (2002). Pterin transport and metabolism in Leishmania and related trypanosomatid parasites. *Int J Parasitol* 32, 385–398.

Pan C. P., & Barkley M. D. (2004). Conformational effects on tryptophan fluorescence in cyclic hexapeptides. *Biophys J* 86, 3828–3835.

Parasassi T., De Stasio G., Rusch R. M., & Gratton E. (1991). A photophysical model for diphenylhexatriene fluorescence decay in solvents and in phospholipid-vesicles. *Biophys J* 59, 466–475.

Periasamy A. (2001). Fluorescence resonance energy transfer microscopy: A mini review. *J Biomed Opt* 6, 287–291.

Periasamy A., & Diaspro A. (2003). Multiphoton microscopy. *J Biomed Opt* 8, 327–328.

Periasamy A., Wallrabe H., Chen Y., & Barroso M. (2008). Chapter 22: Quantitation of protein-protein interactions: Confocal FRET microscopy. *Methods Cell Biol* 89, 569–598.

Plavskii V. Y., Mostovnikov V. A., Tret'yakova A. I., & Mostovnikova G. R. (2008). Sensitizing effect of Z,Z-Bilirubin Ix Alpha and its photoproducts on enzymes in model solutions. *J Appl Spectrosc* 75, 407–419.

Plotnikov S. V., Millard A. C., Campagnola P. J., & Mohler W. A. (2006). Characterization of the myosin-based source for second-harmonic generation from muscle sarcomeres. *Biophys J* 90, 693–703.

Pokalsky C., Wick P., Harms E., Lytle F. E., & Vanetten R. L. (1995). Fluorescence resolution of the intrinsic tryptophan residues of bovine protein tyrosyl phosphatase. *J Biol Chem* 270, 3809–3815.

Polewski K. (2003). Spectral properties of adrenaline in micellar environment. *Physiol Chem Phys Med NMR* 35, 13–25.

Pradhan A., Pal P., Durocher G., Villeneuve L., Balassy A., Babai F., Gaboury L., & Blanchard L. (1995). Steady state and time-resolved fluorescence properties of metastatic and non-metastatic malignant cells from different species. *J Photochem Photobiol B* 31, 101–112.

Provenzano P. P., Eliceiri K. W., Campbell J. M., Inman D. R., White J. G., & Keely P. J. (2006). Collagen reorganization at the tumor-stromal interface facilitates local invasion. *BMC Med* 4, 38.

Provenzano P. P., Eliceiri K. W., & Keely P. J. (2008). Multiphoton microscopy and fluorescence lifetime imaging microscopy (FLIM) to monitor metastasis and the tumor microenvironment. *Clin Exp Metastasis* 26(4), 357–370.

Ramanujam N. (2000). Fluorescence spectroscopy of neoplastic and non-neoplastic tissues. *Neoplasia* 2, 89–117.

Raymond S. B., Boas D. A., Bacskai B. J., & Kumar A. T. N. (2010). Lifetime-based tomographic multiplexing. *J Biomed Opt* 15, 046011-1–046011-9.

Reynolds J. S., Troy T. L., Mayer R. H., Thompson A. B., Waters D. J., Cornell K. K., Snyder P. W., & Sevick-Muraca E. M. (1999). Imaging of spontaneous canine mammary tumors using fluorescent contrast agents. *Photochem Photobiol* 70, 87–94.

Richards-Kortum R., & Sevick-Muraca E. (1996). Quantitative optical spectroscopy for tissue diagnosis. *Annu Rev Phys Chem* 47, 555–606.

Romashko D. N., Marban E., & O'Rourke B. (1998). Subcellular metabolic transients and mitochondrial redox waves in heart cells. *Proc Natl Acad Sci U S A* 95, 1618–1623.

Schenke-Layland K., Riemann I., Damour O., Stock U. A., & Konig K. (2006). Two-photon microscopes and in vivo multiphoton tomographs—Powerful diagnostic tools for tissue engineering and drug delivery. *Adv Drug Deliv Rev* 58, 878–896.

Schenke-Layland K., Xie J., Heydarkhan-Hagvall S., Hamm-Alvarez S. F., Stock U. A., Brockbank K. G., & Maclellan W. R. (2007). Optimized preservation of extracellular matrix in cardiac tissues: Implications for long-term graft durability. *Ann Thorac Surg* 83, 1641–1650.

Schneckenburger H., & Konig K. (1992). Fluorescence decay kinetics and imaging of NAD(P)H and flavins as metabolic indicators. *Opt Eng* 31, 1447–1451.

Schneckenburger H., Wagner M., Weber P., Strauss W. S., & Sailer R. (2004). Autofluorescence lifetime imaging of cultivated cells using a UV picosecond laser diode. *J Fluoresc* 14, 649–654.

Schweitzer D. (2010). Metabolic mapping. In *Medical Retina-Focus on Retinal Imaging*, eds. G.K. Kriegelstein and R.N. Weinreb. Springer-Verlag, Berlin, Heidelberg, pp. 107–123.

Schweitzer D., Hammer M., Schweitzer F., Anders R., Doebbecke T., Schenke S., Gaillard E. R., & Gaillard E. R. (2004). In vivo measurement of time-resolved autofluorescence at the human fundus. *J Biomed Opt* 9, 1214–1222.

Schweitzer D., Schenke S., Hammer M., Schweitzer F., Jentsch S., Birckner E., Becker W., & Bergmann A. (2007). Towards metabolic mapping of the human retina. *Microsc Res Tech* 70, 410–419.

Sell D. R., Nagaraj R. H., Grandhee S. K., Odetti P., Lapolla A., Fogarty J., & Monnier V. M. (1991). Pentosidine—A molecular marker for the cumulative damage to proteins in diabetes, aging, and uremia. *Diabetes Metab Rev* 7, 239–251.

Sell D. R., Lapolla A., Odetti P., Fogarty J., & Monier V. M. (1992). Pentosidine formation in skin correlates with severity of complications in individuals with long-standing IDDM. *Diabetes* 41(10), 1286–1292.

Selvin P. R. (2000). The renaissance of fluorescence resonance energy transfer. *Nat Struct Biol* 7, 730–734.

Sharma R., Wang W., Rasmussen J. C., Joshi A., Houston J. P., Adams K. E., Cameron A., Ke S., Kwon S., Mawad M. E., & Sevick-Muraca E. M. (2007). Quantitative imaging of lymph function. *Am J Physiol Heart Circ Physiol* 292, H3109–H3118.

Singh A. K., & Das J. (1998). Liposome encapsulated vitamin A compounds exhibit greater stability and diminished toxicity. *Biophys Chem* 73, 155–162.

Singh R., Barden A., Mori T., & Beilin L. (2001). Advanced glycation end-products: A review. *Diabetologia* 44, 129–146.

Skala M. C., Riching K. M., Bird D. K., Gendron-Fitzpatrick A., Eickhoff J., Eliceiri K. W., Keely P. J., & Ramanujam N. (2007a). In vivo multiphoton fluorescence lifetime imaging of protein-bound and free nicotinamide adenine dinucleotide in normal and precancerous epithelia. *J Biomed Opt* 12, 024014.

Skala M. C., Riching K. M., Gendron-Fitzpatrick A., Eickhoff J., Eliceiri K. W., White J. G., & Ramanujam N. (2007b). In vivo multiphoton microscopy of NADH and FAD redox states, fluorescence lifetimes, and cellular morphology in precancerous epithelia. *Proc Natl Acad Sci U S A* 104, 19494–19499.

Snell E. E. (1990). Vitamin-B6 and decarboxylation of histidine. *Ann N Y Acad Sci* 585, 1–12.

Sohal R. S. (1984). Assay of lipofuscin ceroid pigment in vivo during aging. *Methods Enzymol* 105, 484–487.

Soltesz E. G., Laurence R. G., De Grand A. M., Cohn L. H., Mihaljevic T., & Frangioni J. V. (2007). Image-guided quantification of cardioplegia delivery during cardiac surgery. *Heart Surg Forum* 10, E381–E386.

Stanley W. C., Lopaschuk G. D., Hall J. L., & McCormack J. G. (1997). Regulation of myocardial carbohydrate metabolism under normal and ischaemic conditions. Potential for pharmacological interventions. *Cardiovasc Res* 33, 243–257.

Sulzer D., Bogulavsky J., Larsen K. E., Behr G., Karatekin E., Kleinman M. H., Turro N., Krantz D., Edwards R. H., Greene L. A., & Zecca L. (2000). Neuromelanin biosynthesis is driven by excess cytosolic catecholamines not accumulated by synaptic vesicles. *Proc Natl Acad Sci U S A* 97, 11869–11874.

Sulzer D., Mosharov E., Talloczy Z., Zucca F. A., Simon J. D., & Zecca L. (2008). Neuronal pigmented autophagic vacuoles: Lipofuscin, neuromelanin, and ceroid as macroautophagic responses during aging and disease. *J Neurochem* 106, 24–36.

Sun Y., Responte D., Xie H., Liu J., Fatakdawala H., Hu J., Athanasiou K. A., & Marcu L. (2012). Nondestructive evaluation of tissue engineered articular cartilage using time-resolved fluorescence spectroscopy and ultrasound backscatter microscopy. *Tissue Eng Part C Methods* 18, 215–226.

Sun Y., Wallrabe H., Seo S. A., & Periasamy A. (2011). FRET microscopy in 2010: The legacy of Theodor Forster on the 100th anniversary of his birth. *Chemphyschem* 12, 462–474.

Swarna S., Lorente C., Thomas A. H., & Martin C. B. (2012). Rate constants of quenching of the fluorescence of pterins by the iodide anion in aqueous solution. *Chem Phys Lett* 542, 62–65.

Takanishi C. L., Bykova E. A., Cheng W., & Zheng J. (2006). GFP-based FRET analysis in live cells. *Brain Res* 1091, 132–139.

Talbot C. B., Benninger R. K. P., De Beule P. A., Requejo-Isidro J., Elson D. S., Dunsby C., Munro I., Neil M. A., Sandison A., Sofat N., Nagase H., French P. M. W., & Lever M. J. (2005). Application of hyperspectral fluorescence lifetime imaging to tissue autofluorescence: Arthritis. *Proc SPIE* 5862, 58620T-1–58620T-6.

Tarte K., & Klein B. (1999). Dendritic cell-based vaccine: A promising approach for cancer immunotherapy. *Leukemia* 13, 653–663.

Teichmann A., Heuschkel S., Jacobi U, Presse G., Neubert R. H., Sterry W., & Lademann J. (2007). Comparison of stratum corneum penetration and localization of a lipophilic model drug applied in an o/w microemulsion and an amphiphilic cream. *Eur J Pharm Biopharm* 67, 699–706.

Terman A., & Brunk U. T. (2006). Oxidative stress, accumulation of biological 'garbage', and aging. *Antioxid Redox Signal* 8, 197–204.

Teuchner K., Freyer W., Leupold D., Volkmer A., Birch D. J., Altmeyer P., Stucker M., & Hoffmann K. (1999). Femtosecond two-photon excited fluorescence of melanin. *Photochem Photobiol* 70, 146–151.

Thomas A. H., Lorente C., Capparelli A. L., Pokhrel M. R., Braun A. M., & Oliveros E. (2002). Fluorescence of pterin, 6-formylpterin, 6-carboxypterin and folic acid in aqueous solution: pH effects. *Photochem Photobiol Sci* 1, 421–426.

Thornhill D. P. (1975). Separation of a series of chromophores and fluorophores present in elastin. *Biochem J* 147, 215–219.

Tsai L., Szweda P. A., Vinogradova O., & Szweda L. I. (1998). Structural characterization and immunochemical detection of a fluorophore derived from 4-hydroxy-2-nonenal and lysine. *Proc Natl Acad Sci U S A* 95, 7975–7980.

van den Berg P. A., van Hoek A., & Visser A. J. (2004). Evidence for a novel mechanism of time-resolved flavin fluorescence depolarization in glutathione reductase. *Biophys J* 87, 2577–2586.

van den Berg P. A., van Hoek A., Walentas C. D., Perham R. N., & Visser A. J. (1998). Flavin fluorescence dynamics and photoinduced electron transfer in Escherichia coli glutathione reductase. *Biophys J* 74, 2046–2058.

van den Berg P. A., Widengren J., Hink M. A., Rigler R., & Visser A. J. (2001). Fluorescence correlation spectroscopy of flavins and flavoenzymes: Photochemical and photophysical aspects. *Spectrochim Acta A Mol Biomol Spectrosc* 57, 2135–2144.

van den Berg P. A. W., Feenstra K. A., Mark A. E., Berendsen H. J. C., & Visser A. J. W. G. (2002). Dynamic conformations of flavin adenine dinucleotide: Simulated molecular dynamics of the flavin cofactor related to the time-resolved fluorescence characteristics. *J Phys Chem B* 106, 8858–8869.

Verbiest T., Clays K., & Rodriguez V. (2009). *Second-Order Nonlinear Optical Characterization Techniques: An Introduction*. CRC Press, Boca-Raton, FL, pp. 1–192.

Vigneshwaran N., Bijukumar G., Karmakar N., Anand S., & Misra A. (2005). Autofluorescence characterization of advanced glycation end products of hemoglobin. *Spectrochim Acta A Mol Biomol Spectrosc* 61, 163–170.

Vishwasrao H. D., Heikal A. A., Kasischke K. A., & Webb W. W. (2005). Conformational dependence of intracellular NADH on metabolic state revealed by associated fluorescence anisotropy. *J Biol Chem* 280, 25119–25126.

Visser A. J. (1984). Kinetics of stacking interactions in flavin adenine dinucleotide from time-resolved flavin fluorescence. *Photochem Photobiol* 40, 703–706.

Voss T. C., Demarco I. A., & Day R. N. (2005). Quantitative imaging of protein interactions in the cell nucleus. *Biotechniques* 38, 413–424.

Wagnieres G. A., Star W. M., & Wilson B. C. (1998). In vivo fluorescence spectroscopy and imaging for oncological applications. *Photochem Photobiol* 68(5), 603–632.

Wakamatsu K., Fujikawa K., Zucca F. A., Zecca L., & Ito S. (2003). The structure of neuromelanin as studied by chemical degradative methods. *J Neurochem* 86, 1015–1023.

Wakita M., Nishimura G., & Tamura M. (1995). Some characteristics of the fluorescence lifetime of reduced pyridine nucleotides in isolated mitochondria, isolated hepatocytes, and perfused rat liver in situ. *J Biochem (Tokyo)* 118, 1151–1160.

Wallace S. J., Morrison J. L., Botting K. J., & Kee T. W. (2008). Second-harmonic generation and two-photon-excited autofluorescence microscopy of cardiomyocytes: Quantification of cell volume and myosin filaments. *J Biomed Opt* 13, 064018-1–064018-5.

Weil G. J., & Chused T. M. (1981). Eosinophil autofluorescence and its use in isolation and analysis of human eosinophils using flow microfluorometry. *Blood* 57, 1099–1104.

Wessels J. T., Busse A. C., Mahrt J., Dullin C., Grabbe E., & Mueller G. A. (2007). In vivo imaging in experimental preclinical tumor research—A review. *Cytometry A* 71A, 542–549.

Wessels J. T., Yamauchi K., Hoffman R. M., & Wouters F. S. (2010). Advances in cellular, subcellular, and nanoscale imaging in vitro and in vivo. *Cytometry A* 77, 667–676.

Wielgus A. R., & Sarna T. (2005). Melanin in human irides of different color and age of donors. *Pigment Cell Res* 18, 454–464.

Williamson J. R. (1979). Mitochondrial function in the heart. *Annu Rev Physiol* 41, 485–506.

Wilson R. G., & Davis R. E. (1983). Clinical chemistry of vitamin B6. *Adv Clin Chem* 23, 1–68.

Wlodarczyk J., & Kierdaszuk B. (2003). Interpretation of fluorescence decays using a power-like model. *Biophys J* 85, 589–598.

Wu Y. C., & Qu J. N. Y. (2006). Autofluorescence spectroscopy of epithelial tissues. *J Biomed Opt* 11, 054023-1–054023-11.

Yang H., Luo G. B., Karnchanaphanurach P., Louie T. M., Rech I., Cova S., Xun L. Y., & Xie X. S. (2003). Protein conformational dynamics probed by single-molecule electron transfer. *Science* 302, 262–266.

Zhang Z., Kao J., D'Avignon A., & Achilefu S. (2010). Understanding dichromic fluorescence manifested in certain ICG analogs. *Pure Appl Chem* 82, 307–311.

Zhang Z., Liang K., Bloch S., Berezin M., & Achilefu S. (2005). Monomolecular multimodal fluorescence-radioisotope imaging agents. *Bioconjug Chem* 16, 1232–1239.

Zheng W., Lau W., Cheng C., Soo K. C., & Olivo M. (2003). Optimal excitation-emission wavelengths for autofluorescence diagnosis of bladder tumors. *Int J Cancer* 104, 477–481.

Zipfel W. R., Williams R. M., Christie R., Nikitin A. Y., Hyman B. T., & Webb W. W. (2003). Live tissue intrinsic emission microscopy using multiphoton-excited native fluorescence and second harmonic generation. *Proc Natl Acad Sci U S A* 100, 7075–7080.

Zorn U., Haug C., Celik E., Wennauer R., Schmid-Kotsas A., Bachem M. G., & Grunert A. (2001). Characterization of modified low density lipoprotein subfractions by capillary isotachophoresis. *Electrophoresis* 22, 1143–1149.

Zucker S. D., Goessling W., Bootle E. J., & Sterritt C. (2001). Localization of bilirubin in phospholipid bilayers by parallax analysis of fluorescence quenching. *J Lipid Res* 42, 1377–1388.

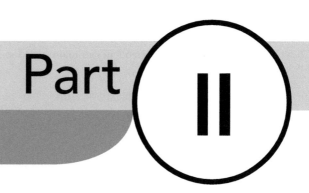

Part **II**

Principles of fluorescence lifetime instrumentation

4 Pulse sampling technique

Diego R. Yankelevich, Daniel S. Elson, and Laura Marcu

Contents

4.1 PRINCIPLE OF DIRECTLY RECORDING FLUORESCENCE DECAYS

The acquisition of time-resolved data using the pulse sampling technique is perhaps the most intuitive of the data acquisition methods described in this book. A schematic behind the pulse sample principle is shown in Figure 4.1. The sample is excited by a short pulse of light, and the emitted fluorescence is detected by a photodetector and electronics that have a response that is fast enough to record the temporal decay of the fluorescence directly. Typically, these instruments have a response time of 100 ps to a few nanoseconds, and therefore deconvolution, of the instrument response function is often required to extract the intrinsic fluorescence decay. The simplicity of the technology and acquisition method make these instruments well suited to clinical use, and there have been numerous reports of applications *in vivo* that will be reviewed in this chapter.

The temporal response of pulse sampling methods is determined by the laser pulse shape, the response of the photodetector, the response of the digitizer electronics, and the processing methods. Due to the requirements for a fast detector and digitization electronics these instruments are typically nonimaging single-point detectors, although scanning instruments have been reported for scanning of arterial samples (Bec et al. 2012; Sun et al. 2011; Xie et al. 2012) and oral carcinoma in animal models (Sun et al. 2011). The single-point technique is also suited to single-point spectroscopy measurements using fiber probes, which are often compatible with the working channel of endoscope systems, and therefore, the measurements can be applied during endoscopic investigations, for instance, in the colon (Mycek et al. 1998) or esophagus (Pfefer et al. 2003b) and other organs.

This chapter presents an overview of the instrumentation used during the development of the pulse sampling technique, the advantages and limitations of this technique, and examples of applications of this technique. Section 4.2 reviews the requirements for the excitation laser light sources (e.g., pulse energy and laser types). Section 4.3 examines configurations for light delivery to and from the tissue. Section 4.4 examines specific aspects pertaining to the detection of the fluorescence signal, including a review of the spectral and temporal resolution for a variety of detector technologies. The method used to deconvolve the

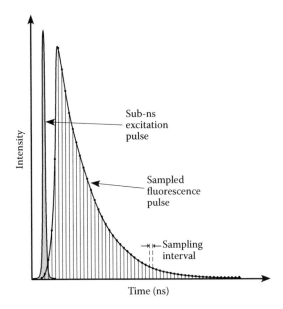

Figure 4.1 Schematic of the principle of pulse sampling performed by a high-sampling-rate digitizer. The fluorophore is excited by a subnanosecond laser pulse, and the fluorescence pulse transient is sampled by a gigasample/s digitizer with large analog bandwidth.

broadening of fluorescence pulses, as well as the origins of fluorescence pulse dispersion and electrical pulse broadening, are presented in Section 4.5. Section 4.6 is an overview of methodologies used for the measurement of fluorescence lifetimes as well as their applicability for the *in vivo* characterization of tissue biochemical features. Section 4.7 presents newly developed configurations for the multispectral time-resolved fluorescence technique. A summary and conclusion is given in Section 4.8.

4.2 FLUORESCENCE LASER EXCITATION SOURCES

There are a number of requirements for the excitation pulses in the pulse sampling method so that the signal levels are high enough to allow direct digitization of the fluorescence decay with sufficient temporal resolution. For instance, nanojoule pulse energies and subnanosecond pulse widths are desirable to resolve the fluorescent lifetimes of tissue fluorophores that are as short as a few 100 ps. In addition, the pulse wavelength must be in the near ultraviolet (UV) to fall within the absorption band of most tissue fluorophores. Lastly, the pulse repetition rate must be preferably in the kilohertz regime or higher to operate with short averaging time. There are multiple laser sources available that can provide these specifications and therefore yield fluorescence signals above the noise floor of the fluorescence pulse detection system. Pulsed laser systems used for pulse sampling in the past and present span a wide range of technologies. These includes electric discharge lasers, dye lasers, semiconductor lasers, rare-earth-doped glass fiber lasers, and diode-pumped crystalline lasers, often combined with nonlinear optical crystals to shorten the emitted wavelength by a factor that can range from 2 to 4. Examples from these laser technologies will be presented below.

Many early pulse sampling instruments used a nitrogen laser as the fluorescence excitation source, since they provide a high-enough pulse energy to yield sufficient detected fluorescence photons for every excitation. The molecular nitrogen laser (Lengyel 1971) operates at a wavelength of 337.1 nm and, as its name indicates, uses nitrogen gas as active medium at pressures between 100 and 10^5 Pa. Today, most commercial designs use flowing gas, but sealed tube systems do exist, although these operate at low repetition rates. The nitrogen laser is a three-level system, but since the upper laser level of nitrogen is directly pumped, population inversion can take place. Pumping is normally provided by direct electron impact where the gain medium is usually pumped by a transverse electrical discharge. Laser pulses with energy range from microjoules to millijoules, and a peak power in the range of kilowatts to more than

3 MW can be achieved. The main disadvantages are the long pulse lengths—between a few hundred picoseconds and a maximum of approximately 30 ns depending on the gas pressure—and the low repetition rate, although this is not necessarily a problem for point spectroscopy measurements unless significant averaging over multiple pulses is required to improve the signal-to-noise ratio. Nitrogen lasers only offer limited control of pulse energy, have poor beam quality, need regeneration of the laser discharge electrodes after a few million pulses, and require a bulky gas supply that can compromise portability. However, the nitrogen gas supply is economical, and modern systems have a lower complexity compared with other subnanosecond pulsed nitrogen laser systems used in the past (Mycek et al. 1998).

The pulse length of early pulse sampling systems using nitrogen lasers was greater than 1 ns, which placed a limit on the minimum fluorescence lifetime that could be measured, and simple processing or deconvolution methods were required to determine the fluorescence lifetimes (Marcu et al. 1999). For instance, one useful and simple processing measure used was the difference in time from the point when the signal rises above 0.2 to the point where it falls below 0.2 of the peak value (Mycek et al. 1998). There are also variations in the pulse energy and arrival time when using a nitrogen laser (Pitts and Mycek 2001). Intensity variations may be detected and corrected for by using a separate photodetector (Marcu et al. 1999), and the detection system can also be gated using this signal (Pitts and Mycek 2001). A further disadvantage for this type of laser is the electronic noise that results from the discharge of high voltages required to obtain laser emission from N_2 molecules, and care must be taken to protect the detectors from this interference by using electromagnetic screening such as Faraday cages (Pitts and Mycek 2001).

Nitrogen lasers can also be used to pump dye lasers to allow the light to be downconverted into wavelengths across the visible and near-infrared (Pitts and Mycek 2001) so that other fluorophores may be excited. More generally, the dye laser plays an important role as a tunable laser pulse source covering the spectral range from 365 to 1000 nm (Dienes and Yankelevich 1995). The active medium is an organic fluorescent dye dissolved in common solvents that is optically pumped either by a Xenon flashlamp or, most commonly, by a laser with a wavelength matched to the absorption band of the dye. Changing the wavelength of operation of the laser is as simple as replacing the laser dye, which, in many cases, does not require replacement of the pump laser, and then tuning the wavelength within the fluorescence absorption spectrum of the dye. Dye lasers are capable of operation from continuous wave to femtosecond pulse duration (Duarte and Hillman 1990; Ippen et al. 1972), and with pulse energies in the millijoule to microjoule range. Owing to its simple construction, the dye laser by itself is an inexpensive source of tunable laser pulses provided that a pump laser is already available. Contrary to solid-state or crystalline laser media, the laser dye is robust, self-regenerating, and inexpensive. Their main drawback is the need to replace the laser dye every few months and, in the event that a pump laser is not available, high cost.

More recently, fiber lasers have begun to be applied with the pulse sampling technique. These lasers have been used for decades in the field of optical communications and have a proven high reliability. However, more recently, ytterbium-doped glass fiber lasers are finding a role in biomedical applications (Duarte 2009; Jackson and Lauto 2002). In this laser, the excitation energy to achieve population inversion is provided by powerful laser diodes, operating at 950 nm, coupled into the core of a fiber doped with the gain media. The gain medium, instead of being distributed along a few centimeters, for example, 1 cm in a dye cuvette, is distributed over the length of a fiber that is many meters long, resulting in a highly efficient laser. The laser beam wavelength is typically in the 1.03 to 1.09 μm range; therefore, to make them applicable to spectroscopy of biological tissue, it is necessary to use third harmonic generation with nonlinear crystals, such as borate-based crystals, to generate near-UV pulses (Boyd 1992). This type of laser is capable of generating pulses with picosecond to microsecond duration and power in the milliwatt to kilowatt range. The emitted beam has very good quality, and since it is generated in a fiber, it can easily be delivered to the sample.

4.3 LIGHT DELIVERY AND COLLECTION OPTICS

In the pulse sampling technique, the laser light is coupled into an optical fiber, and a quartz (UV-grade silica) core is often chosen since the low absorption and fluorescence result in very small background signals (Maarek et al. 2000b; Mycek et al. 1998). With long pulses from nitrogen lasers, the temporal dispersion of the optical fiber is not necessarily a major consideration, and large step-index core diameters greater than 500 μm are commonly used. Fiber-optic probes used in fluorescence spectroscopy are characterized mainly by the number, core size, and numerical aperture (NA) of source and collection/detector fibers and whether they are single-fiber or bifurcated, as illustrated in Figure 4.2. The emitted fluorescence may be collected via the same optical fiber (Figure 4.2a) or via a single or multiple emission fibers (Figure 4.2b). The general consideration when designing such fiber probes is to optimize the overlap between the illuminated region of the tissue and the collection area. In the single-fiber case, this is automatically achieved, whereas for multiple fibers, a transparent quartz spacer may be attached to the end of the probe (Utzinger and Richards-Kortum 2003), or a tip may be used that ensures a certain length of free space, to allow the effective illumination and collection areas to overlap. Since the collection solid angle subtended by the fiber at the surface is monotonically reduced, the collected signal decays with increasing probe-to-target distance (Papaioannou et al. 2004), with smaller probes decaying much more rapidly. In a bifurcated probe, the maximum light collection occurs several millimeters away from the target. The reflected signal as a function of probe-to-target distance resembles a positively skewed bell-shaped curve with a sharp increase for small probe-to-target distance and slower falloff past the curve's maximum. These and other general considerations are described in Chapter 6 of this book and in a review article (Utzinger and Richards-Kortum 2003). Further details of the light delivery optics are provided in this section.

Several theoretical and experimental investigations of fiber-optic probe design (Cooney et al. 1996a,b; Marcu et al. 1999; Papaioannou et al. 2004; Shim et al. 1999; Zhu and Yappert 1992a,b) have involved clear solutions or other weakly absorbing nonturbid media and have shown that the aforementioned characteristics affect the probe's performance with respect to light collection efficiency and volume sampling. Tissue examination, however, involves additional factors such as (1) the tissue's optical absorption (μ_a) and scattering (μ_s) coefficients; (2) heterogeneities and intervening blood flow; and (3) the illumination collection geometry. These factors can significantly affect the delivery of light, coupling efficiency, and sampling volume of the collected photons, which, in turn, can influence the signal-to-noise ratio (SNR) and determine the volume of interrogated tissue. Therefore, overall optimization of probe design and usage requires understanding of the interplay among probe design, tissue characteristics, and excitation collection geometry.

In a clinical environment, the influence of probe-to-sample distance on collection of light stems from the need to accommodate various practical limitations such as the presence of blood or other biological fluids between the probe and the tissue, which could contaminate the probe and/or induce undesirable modulation of the spectroscopic signal. Thus, spacers or other protective shields may be incorporated into the distal part of the probe, increasing the probe-to-tissue distance and enabling a larger interrogation area. The combination of the overall excitation and collection geometry along with the tissue's optical properties can then influence the SNR, thus compromising the quality of the data. Additionally, changes in the penetration depth of the collected photons may also ensue (Pfefer et al. 2001, 2002, 2003a). Accurate knowledge of this effect is necessary for correlation of the spectroscopic

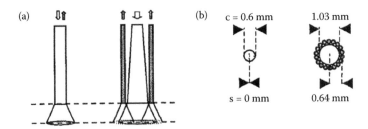

Figure 4.2 (a) Side and (b) cross-sectional views of two commonly used fiber-optic probes. Probe 1 is a single-fiber configuration with a common illumination and collection channel. Probe 2 is bifurcated, with a central illumination channel and a peripheral collection ring. In this particular implementation, the illumination channel of probe 2 is a tapered fiber. Typical diameters of the illumination cores and the source-to-detector fiber distances (center to center) are indicated as c and s, respectively. (From Papaioannou, T. et al., *Appl. Opt.*, 43, 2004.)

signatures with the corresponding local tissue pathology. Therefore, the diagnostic value of such techniques could be compromised unless further information on the effect of probe-to-tissue distance on the interrogated tissue volume is available.

4.4 FLUORESCENCE EMISSION DETECTION

The detection of fluorescence photons during pulse sampling is mainly performed with photomultiplier tubes (PMTs) and avalanche photodiodes. These convert the time-varying optical signal into an electronic pulse, and both have a high temporal resolution, as described in Chapter 6. Microchannel plate PMT tubes are usually used since they have very large gain and can be gated to avoid the detection of background light reaching the detector between laser pulses, and the signals are amplified before digitization (Marcu et al. 1999). Both of these methods are briefly described in this section.

The major difference between the pulse sampling technique and time-correlated single photon counting (TCSPC) is in the sampling of the fluorescence signal by the detector. As described in Chapters 5 and 9, TCSPC records single-photon events and gradually builds a histogram depicting the probability of photon detection as a function of time. Using multidetector schemes, TCSPC has been extended to the simultaneous detection at several emission spectral channels where histograms at all wavelengths are built simultaneously (Becker et al. 2005; Ruck et al. 2007). However, this scheme can be very costly if subnanosecond detectors are used, and it requires multiple high-voltage sources that considerably complicate the system. Also, to avoid photon pileup, the count rate should be limited to 1%–5% of the excitation rate (Erdmann 2005), resulting in histogram acquisition times that may be seconds long. By comparison, in the pulse sampling method, fluorescence detection can be performed with a single detector with spectral responsivity in the near UV and visible, subnanosecond instrument response function and gain in the 10^4–10^5 range. This section contains a brief description of commonly used detectors that meet these requirements.

The PMT consist of a photocathode and a series of electrodes, called dynodes, which are biased with a voltage divider resistor ladder at incremental potentials with respect to the cathode. The last electrode is the anode, which collects the electrons emitted by the dynodes. In order to avoid electronic collisions between electrons and gas molecules, the electrode assembly has to be contained within a vacuum. Electrons emitted from the photocathode are accelerated toward the first dynode arriving with kinetic energy, resulting in secondary electron emission. This process repeats until the initial number of electrons emitted by the photocathode is amplified by a factor given by the average secondary emission raised to a power set by the number of dynodes. This amplification has typical values of 10^6 (Donati 2000). The spectral response of multichannel plate (MCP) PMT tubes spans 160–850 nm interval, and depending on the material used for the photocathode, it can be extended to ~1700 nm (Hamamatsu Corporation [a]). A more detailed description can be found in Chapter 6.

An alternative PMT configuration, the MCP PMT, is a design where the dynodes have been substituted by a MCP. An MCP PMT, shown in Figure 4.3, consists of a circular array of capillary tubes 6–30 μm in diameter, with an inner wall surface treated to have optimum electron secondary emission, forming a continuous dynode. By applying a potential V_{AL} to one of the metal-coated edges, an axial electric field accelerates the emitted electrons, providing kinetic energy for additional secondary electrons. The distance between the metalized ends is only a few millimeters, resulting in a very compact configuration.

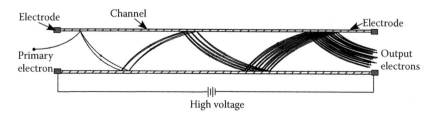

Figure 4.3 Schematic diagram of a multichannel plate PMT.

Principles of fluorescence lifetime instrumentation

In a PMT, the electron emission angle, photoelectron initial velocity, and secondary electron distributions will result in time broadening during the multiplication process. In contrast, in an MCP PMT, the accelerating field is applied at a small angle, whereby the above-mentioned distributions can be neglected. Furthermore, the very small channel diameter results in very short secondary electron transit time, allowing a major improvement in transit time spread, making the MCP one of the fastest available detectors. It is not uncommon for an MCP PMT to have impulse response functions of less than 100 ps (Hamamatsu Corporation). However, one of the disadvantages of MCPs is the gain saturation found with increasing supply voltage due to the decrease in the number of multiplications, exponentially related to the gain, as the accelerating field increases (Donati 2000). In addition, due to their labor-intensive fabrication process, MCPs are expensive detectors.

If the fluorescence signal is confined to a very short amount of time, gated detection can be a good choice for signal detection, whereby the detector is only sensitive to arriving photons during the fluorescence decay and not during the long inactive time between excitation pulses. Note that this type of PMT gating is different from the gated detection method of Chapter 8, where the gate width is short enough to sample the fluorescence decay itself. A typical pulse sampling experiment is performed with a pulsed laser excitation where the signal lasts for only a few nanoseconds, at a repetition rate in hertz to kilohertz, resulting in a signal duty cycle of a few percent.

A gating function can be built into the PMT tube design by adding a gate mesh positioned in close proximity to the photocathode. If the mesh is biased at a reverse potential with respect to the photocathode, the detector will be off, and vice versa (Hamamatsu Corporation [b]). Nonetheless, gating may have drawbacks. In order to maximize the gate response time and avoid gate pulse oscillations, the output impedance of the pulse generator and gate input stage are matched (usually to 50 Ω). Gated PMT tubes typically require gate pulses with 20–40 V amplitude, which dissipate considerable power in the gate input impedance. However, the power dissipation capabilities of the gate-matching impedance are limited, and therefore, the gating pulse duty cycle has to be restricted to a few percent (Hamamatsu Corporation [c]). This can severely restrain the data acquisition frequency, especially if a laser with a high pulse repetition rate is used. For example, if a 1 MHz laser is used, and the duty cycle is limited to 1%, the pulse generator driving the gate must produce a 10 ns pulse with 20–40 V amplitude, which is not a trivial requirement. In many cases, the pulse width for large voltage amplitudes is in microseconds, translating to kilohertz data acquisition frequencies that waste laser pulses. In addition, the gating pulse induces oscillations in the baseline and increases the price of the detector. Therefore, if gating is not essential, nongated MCPs should be used that can take advantage of the large repetition rate.

An alternative to photon tube detectors is semiconductor pn-junction detectors, where the absorption of photons generates electron–hole pairs. In an avalanche photodetector, the electron–hole pair is accelerated through the depletion region, where they generate extra carriers that contribute to a current. In an avalanche process, the accelerated electrons collide with valance-lattice electrons and generate additional free electron–hole pairs. These pairs are also accelerated and generate additional pairs, resulting in a chain-reaction-type process (Yariv and Yeh 2007). As a result of the avalanche process, the ensuing current is much greater than the current that would be produced by a photoionization process. Unfortunately, since the avalanche current multiplication is a random process, the SNR may be degraded for large current magnification factors when compared to a photodiode. However, if the current multiplication factor is optimized, considerable improvements in detector sensitivity can be expected with similar SNR to a photodiode (Yariv and Yeh 2007). In order to extend the spectral response of avalanche photodiodes to cover fluorescence wavelengths of organic fluorophores, the device is designed to collect photons from the p-doped layer (Hamamatsu Corporation [d]). A gain of 70 can be obtained for a spectral range of 200 to 700 nm, and using a small effective area device, cutoff frequencies as high as 250 MHz can be expected (Hamamatsu Corporation [e]). Since this gain is orders of magnitude smaller than in PMTs, they are used with high-gain large-bandwidth amplifiers. Avalanche photodetectors are affordable, and for optimal operation, they require a 150–200 V DC power supply and, in most cases, an amplification stage prior to signal sampling. See Chapter 6 for further details.

The time-varying electrical signal generated by the detector is then digitized for computational analysis, which is achieved with either a dedicated digitizer or a fast digital oscilloscope. The acquisition is usually controlled by a personal computer so that the fluorescence decays can be further preprocessed, averaged, and analyzed to extract characteristic temporal features such as the fluorescence lifetime(s) (e.g., using least-squares fitting [Pitts and Mycek 2001]) and Laguerre parameters (Jo et al. 2006, 2007; Liu et al. 2012), as described in Chapter 11. Techniques such as linear discriminant analysis can then be used to allow classification of the data into different disease groups (Pfefer et al. 2003b; Phipps et al. 2012).

4.5 INSTRUMENT RESPONSE FUNCTION AND DECONVOLUTION

A measured time-resolved signal can be determined by considering it as a convolution of the fluorescence impulse response function (fIRF) and the instrument response function (Liu et al. 2012; Marcu 2012). To estimate the fIRF of a sample, the instrument response must be deconvolved from the measured fluorescence pulse. The measurement of the system instrument response is performed by exciting a fluorophore with as short a lifetime as possible or by detecting the reflected excitation light for a short excitation pulse (delta function). The broadening of the fluorescence pulse is due to both electronic and optical effects. These effects are described in the following paragraphs.

Electronic pulse broadening is mainly due to the finite bandwidth of the amplifiers used prior to signal digitalization. If large-bandwidth amplifiers designed for high-frequency communications are used, care must be taken so that they have adequate low-frequency cutoff. Pulse broadening can also originate in the analog bandwidth of the digitizer input, which amplifies and filters the signal to optimize the digitization. To acquire the signal with minimal distortion, it is recommended that the digitizer's analog bandwidth be in the order of 3–5 times the highest frequency component of interest in the measured signal (National Instruments 2006). A second digitizer parameter that is not related to the analog bandwidth is the sampling rate, which is the speed at which the digitizer's analog-to-digital converter acquires the input signal into the computer memory. This rate is set by the sample clock that directs the instrument to convert an analog voltage to a digital value. Based on the Nyquist theorem (Oppenheim and Schafer 2010), a signal must be sampled at a rate greater than twice the highest frequency component of the signal to accurately reconstruct the waveform. If the data are sampled at a rate that is too low, a false lower-frequency component, known as an alias, will appear in sampled data. Finally, as was mentioned in Section 4.4, in a PMT, the electron emission angle, photoelectron initial velocity, and secondary electron distributions will result in the broadening of the electronic florescence pulse during the multiplication process; however, this effect can be neglected if an MCP detector is used.

Optical pulse broadening is due to the dispersion in the optical fibers of pulses with subnanosecond duration. This dispersion may become a significant factor in the overall instrument response function. The fluorescence pulses traveling in a multimode optical fiber are subject to broadening due to intermodal dispersion from the difference in group velocities between the modes traveling in the fiber. In a step-index multimode fiber, where a very large number of modes propagate (typically thousands), the broadening is given by the temporal difference between the fundamental mode and the highest order mode given by (Keiser 1983)

$$\tau_{mod} = \frac{n_1 L \Delta}{c_1},$$
(4.1)

where $\Delta = (n_i - n_2)/n_1$; n_1 and n_2 are the index of refraction of the fiber core and cladding, respectively; L is the fiber length; and c_1 is the speed of light in fused silica. In addition, the limited bandwidth of the combined MCP detector/amplifier also contributes to fluorescence pulse broadening, resulting in a response time given by

$$\tau = \sqrt{\tau_{mod}^2 + \tau_{det}^2},$$
(4.2)

where τ_{det} is the detector broadening.

In Figure 4.4, the dispersion introduced by multimode fiber of different lengths was measured after excitation with visible 30 ps fluorescence pulses from 2-(p-dimethylaminosotyryl)pyridylmethyl iodide (2-DASPI) (Kim et al. 2000), detected by an MCP PMT with 95 ps instrument response time and 1.5 GHz amplifier. An alternative fluorophore that has been used for this purpose is Rose Bengal dissolved in deionized water with lifetime of 90 ps (Pitts and Mycek 2001). The measurement was made for the multichannel device described in Section 4.7, which incorporates four channels with fibers of different lengths. The multimodal dispersion results in a ~54 ps pulse broadening per meter of fiber, which is very similar to 56 ps per meter obtained from Equation 4.1 for $\Delta = 0.0115$ and $n_1 = 1.4366$ ($\lambda = 400$ nm). The deconvolution is performed with a model-free method, introduced in Chapter 11, which relies on the expansion of the fIRF as a Laguerre basis of functions.

Theoretical calculations from Equation 4.2 for the 1, 10, 19, and 28 m fibers result in a multimodal dispersion of 692, 888, 1235, and 1713 ps, respectively. The small discrepancies between these results and the experimental example shown in Figure 4.5 may originate from the uncertainty of how the modes are actually populated in a multimode fiber. Given that the multimodal pulse dispersion for the 1 m fiber pulse dispersion is less than 100 ps, the experimental dispersion is dominated by the bandwidth-limiting effects of the MCP PMT and amplifier used prior to digital conversion. For the longer fibers, pulse dispersion is dominated by multimodal dispersion. The secondary peak observed for the 1 m trace, but masked by dispersion in the traces from longer fibers, is attributed to impedance mismatch effects.

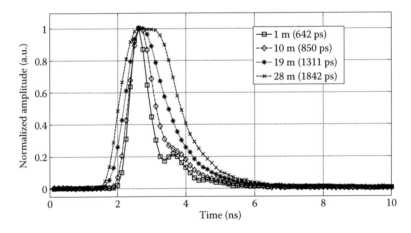

Figure 4.4 Fluorescent pulses from 2-DASPI, excited by 355 nm, 30 ps, 6 nJ pulses, after propagation through 1, 10, 19, and 28 m multimode fibers.

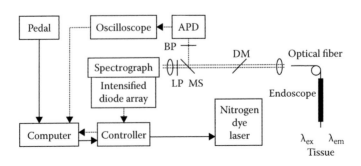

Figure 4.5 System for simultaneously recording time-resolved fluorescence and the fluorescence spectrum. Abbreviations: APD, avalanche photodiode; BP, band-pass filter; DM, dichroic mirror; LP, long-pass filter; MS, microscope slide. (From Pfefer, T.J. et al., *Lasers Surg. Med.*, 32, 2003.)

4.6 EXAMPLES OF PULSE SAMPLING IN TISSUE SPECTROSCOPY, PRECLINICAL AND CLINICAL APPLICATIONS

By dispersing the fluorescence using a monochromator or other spectral filtering system, the time-resolved measurements can be further spectrally resolved. For instance, Marcu et al. (1999) connected the output ends of a series of fibers to a 5 nm band-pass scanning monochromator equipped with a PMT detector, which allowed spectrally resolved temporal decays to be recorded. These fluorescence wavelengths were measured across the UV and visible spectrum; for instance, Maarek et al. (2000a) recorded 29 individual time-resolved spectra between 370 and 510 nm with this instrument. The temporal response of the system could be found by tuning the monochromator close to the wavelength of the laser. This provided a 2-D array of spectral lifetime data for analysis using various discrimination techniques (Maarek et al. 2000b). The most intuitive analysis involves the calculation of the fluorescence lifetime(s) at each emission wavelength and using knowledge of the spectral lifetime signatures of different fluorophores to match the results to the underlying tissue biochemistry. Other methods of analysis such as the Laguerre deconvolution technique can also be used, either by calculating the parameters at each wavelength or by averaging the data over a spectral region (Maarek et al. 2000b).

If a spectrograph is used instead of a monochromator together with a linear detector array, then multiple time-resolved measurements can be recorded simultaneously (see Figure 4.5), and spectral lifetime plots can be generated (see Figure 4.6). It was suggested that detection with separate time and spectral resolution could also increase the information content from the sample (Mycek et al. 2000). In further work, the spectrograph was fitted with an intensified CCD to record fluorescence spectra with 3 nm resolution, with intensifier gating to reject background light (Pitts and Mycek 2001). A similar system used an intensified diode array to record the spectrum that was dispersed by a spectrograph (Pfefer et al. 2003b).

The identification of adenomatous polyps in the colon is one suggested application area of the pulse sampling method, where the higher sensitivity of time-resolved over steady-state fluorescence could allow polyps to be classified *in vivo* rather than requiring excision and biopsy (Mycek et al. 1998). This was the first endoscopic application of the pulse sampling method *in vivo*. Pfefer (2003b) saw potential for the detection of high-grade dysplasia in Barrett's esophagus. Arterial fluorescence has also been extensively studied, particularly for the characterization of atherosclerotic plaque (Maarek et al. 2000a,b).

The combination of time-resolved fluorescence and spectrally resolved fluorescence has been tested in Barrett's esophagus surveillance, and the effect of the excitation wavelength on the detection of high-grade dysplasia on Barrett's esophagus was evaluated (Pfefer et al. 2003b). To accomplish this, laser-induced fluorescence was measured in patients with Barrett's esophagus. The measured data were subject to statistical analysis and classified with different algorithms. The excitation source for the fluorescence spectroscopy system consisted of a 337 nitrogen laser pumping a dye laser module emitting pulses at 400 nm coupled into a 600 μm fused silica fiber. The delivery of pulses to the esophageal wall required the insertion of the fiber into the accessory port of an endoscope. The fluorescence pulses emitted by the esophagus were collected by the excitation fiber, and, after removal of the excitation pulses from both the nitrogen and dye lasers using long-pass filters, a portion was guided using a glass plate through a 550 ± 20 nm band-pass filter onto an avalanche photodiode. To ensure that the photodiode was kept in its linear regime, neutral density filters were used before digitizing the signal with an oscilloscope. The remaining portion of the spectrum was dispersed by a monochromator and detected by a 1 μs-gate-width intensified photodiode and a multichannel analyzer. The monochromator grating had a grating blazed at 450 nm with 150 lines/mm, and its slits were set at 25 μm. The experimental apparatus was controlled with a personal computer. Background and residual excitation peak subtraction was performed by subtracting a scan measured with the excitation fiber pointing to free space. In addition, to adjust for the changes in the system efficiency due to aging of the laser and optics, a spectrum from the laser dye DCM, prepared with an exact concentration, was also recorded before the clinical trial. The instrument spectral response was taken into account by dividing the background-subtracted data by the instrument's spectral response curve measured with a calibrated 1 kW quartz halogen lamp.

Principles of fluorescence lifetime instrumentation

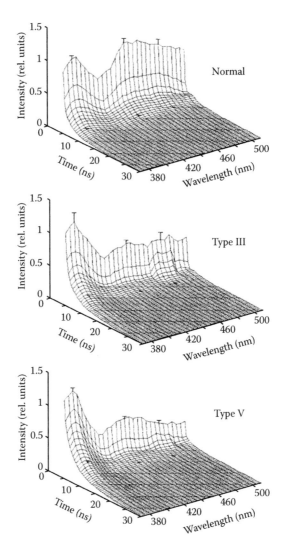

Figure 4.6 Example time-resolved fluorescence spectroscopy data comparing type III and type V collagen with normal aortic wall. (From Maarek, J.M.I. et al., *Lasers Surg. Med.*, 27, 2000.)

The ability of time-resolved fluorescence spectroscopy to characterize *in vivo* the biochemical features of clinically relevant primary brain tumors has been demonstrated (Butte et al. 2010). The experimental apparatus to measure fluorescence temporal features at spectral emission bands centered at 390, 440, and 460 nm is shown in Figure 4.7. Briefly, it consisted of a pulsed nitrogen laser with 337 nm wavelength, 700 ps full width half maximum (FWHM) pulse width as the excitation source. The laser pulses were delivered and collected with a custom-made sterilizable bifurcated fiber probe. A monochromator with F-number 4.4 and a grating with 600 lines per mm blazed at 450 nm were used to disperse the fluorescence pulses, which, in turn, were detected with a gated MCP PMT with 180 ps rise time. The electrical pulses derived from the MCP PMT were amplified with a fast 1.5 GHz amplifier and digitized with a 5 gigasample/s oscilloscope. The instrument was mobile, as it was contained in a standard endoscopic cart.

The study showed that normal cortex, normal white matter, and low-grade glioma tissue are well characterized by distinct fluorescence emission features. It demonstrated that normal cortex and normal white matter show fluorescence peak emissions at 390 nm with a lifetime of 1.8 ± 0.3 ns and at 460 nm with a lifetime of 0.8 ± 0.1 ns. In low-grade glioma, the 390 nm emission peak is absent, and it is reduced in high-grade glioma with a lifetime of 1.7 ± 0.4 ns. The emission characteristics at 460 nm in all tissues

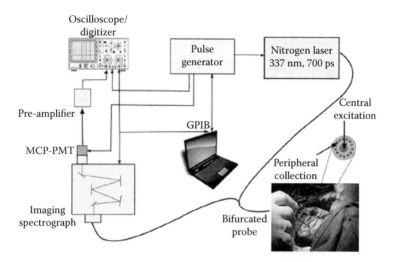

Figure 4.7 Schematic of the pulse sampling time-resolved laser-induced fluorescence spectroscopy apparatus, including a fast digitizer and gated detection used for the *in vivo* evaluation of brain tumor time-resolved fluorescence emission. The photo shows the tip of the optical probe used on the brain tissue cortex of a patient undergoing brain tumor surgery. (Adapted from Butte, P.V. et al., *J. Biomed. Opt.*, 15, 2010.)

correlated with the fluorescence peak in the 440 to 460 nm range, lifetime of 0.8 to 1.0 ns, characteristic of nicotinamide adenine dinucleotide. These findings demonstrate the potential of using time-resolved laser-induced fluorescence spectroscopy as a tool for enhanced delineation of brain tumors during surgery. In addition, this study evaluated similarities and differences between time-resolved laser-induced fluorescence spectroscopy signatures of brain tumors obtained *in vivo* and those previously reported in *ex vivo* brain tumor specimens.

Further application areas are discussed in more detail in the other chapters of this book, particularly Chapters 14–19.

4.7 NEWER DEVELOPMENTS: FAST MULTISPECTRAL SAMPLING, SCANNING POINT SPECTROSCOPY

Time-resolved and spectrally resolved fluorescence spectroscopy has proven to be a reliable tissue characterization technique capable of analyzing multiple parameters including fluorescence intensity, spectrum, and lifetime (Fang et al. 2004; Pfefer et al. 2003b; Sun et al. 2008). However, the wavelength-scanning data acquisition required by this technique is time consuming and can limit its applicability. Newly developed configurations for simultaneous acquisition of fluorescence decays in multiple spectral bands have offered improvements over previous systems (Sun et al. 2008) and are reviewed in this section.

Initial prototypes successfully demonstrated that three to four spectrally resolved fluorescence pulse transients (decays) induced by a single laser pulse excitation event can be can be acquired simultaneously (Sun et al. 2008). The exposure to a reduced number of optical pulses, and hence, low average optical power delivered to tissue, can minimize the effect of photobleaching on the measurements. An initial prototype, shown in Figure 4.8, demonstrated the ability to analyze the biochemical features of arterial tissues (Sun et al. 2011) and to retrieve robust time-resolved data during tissue scanning. The design consisted of a combination of dichroic filters and different lengths of optical fiber. In this setup, the laser delivery and fluorescence collection was executed through the same optical fiber, which provided a self-aligned fluorescence excitation and collection probe with a high optical efficiency.

The initial system was designed with an ability to resolve three key autofluorescent spectral bands from tissue that are sensitive to collagen, NADH, elastin, lipopigment, and flavin emissions. This also enabled the design of intravascular catheters of relatively small diameter. Secondly, the pulsed nitrogen laser ($\lambda = 377$ nm) used in this prototype offered microjoule pulse energies (at the fiber output), at a repetition rate of 50 Hz and

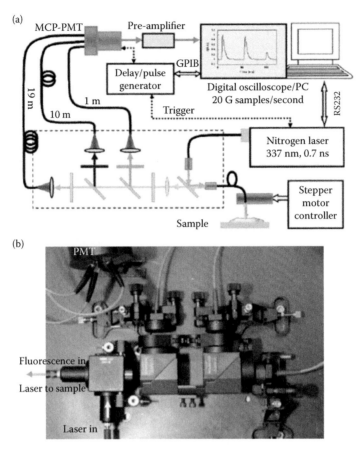

Figure 4.8 Time- and wavelength-resolved spectroscopic system with a single optical fiber as a probe for excitation delivery and fluorescence collection. (a) System diagram including optical design and electronic configuration. (b) Photograph of the portable optical setup. (Adapted from Sun, Y.H. et al., *Opt. Express*, 19, 2011.)

a 700 ps pulse width. This laser was compact enough to allow the implementation of the system in a mobile cart appropriate for *in vivo* experiments in the large animal surgery lab. A line scanning function and the incorporation of intravascular ultrasound (IVUS) imaging allowed for *in vivo* multimodal dynamic tissue characterization. Results from the scanning-mode experiments are depicted in Figure 4.9a and b for intensity and average lifetime at different positions along the artery wall. It was observed that the fluorescence intensity fluctuated significantly when the catheter was moved along the lumen (Figure 4.9a). This effect is most likely a result of changes in the light excitation-emission geometry due to fluctuations of the distance between the probe and sidewall, and changes in the lumen diameter. In addition, the excitation and fluorescent pulse intensity may be attenuated by diffused blood between the fiber-optic tip and arterial wall.

The operation of the system was tested in an intravascular setting in conjunction with IVUS and under blood flow conditions by performing catheter pullback experiments in healthy arterial walls of juvenile Yorkshire pigs. The catheter consisted of a commercial 3 Fr IVUS catheter and a customized side-view optical fiber with a 400 μm core and 780 μm outer diameter fitted in the distal end with a 45° polished metalized mirror termination for beam deflection at 90° to the fiber axis.

The results indicate that while the intensity measurements showed large fluctuations due to a range of factors that affect the light intensity, the fluorescence lifetime values provided a reliable means for the characterization of tissue composition independent of light intensity.

An improved configuration has been implemented with enhanced portability, speed, and the ability to resolve a fourth autofluorescent spectral band matched to the emission of porphyrins (Bec et al. 2013). By taking advantage of a more compact, robust, low-maintenance, and stable diode-pumped mode-locked

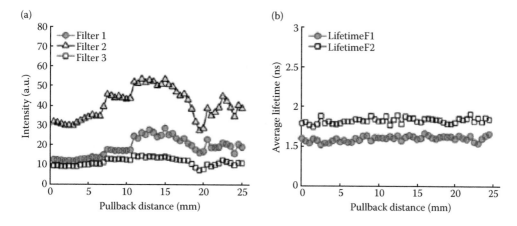

Figure 4.9 *In vivo* test of simultaneous time- and wavelength-resolved fluorescence spectroscopy (STWRFS) in pig femoral artery from a pullback measurement when the catheter was moved along the axial direction for 25 mm. (a) Fluorescence intensity for corresponding subbands for F1(390/40), F2(452/45), and F3(542/50) nm. (b) Measured lifetimes. (Adapted from Sun, Y.H. et al., *Opt. Express*, 19, 2011.)

fiber laser instead of the electrical discharge gas laser, together with a fast data storage scheme, an increased scanning speed was achieved. The portability of the system was also enhanced by using a compact digitizer instead of a high-speed digital oscilloscope. The excitation source was a compact and robust diode-pumped mode-locked ytterbium fiber laser that emits pulses at 355 nm, 30 ps pulse width, 60 nJ energy per pulse at a 1 MHz repetition rate (Bec et al. 2013). Even though it has two orders of magnitude lower energy per pulse than the nitrogen laser, its higher repetition rate results in a data acquisition rate increase of at least two orders of magnitude, which enables fast signal averaging for the rapid acquisition of fluorescent lifetime images. The two orders of magnitude shorter excitation pulse, width is insignificant when compared with the lifetime of biological molecules or the temporal instrument response, enabling the assumption of quasi delta function excitation for a simplified instrument response deconvolution. The 355 nm laser wavelength is well matched to the absorption spectra of collagen, NADH, flavins, and porphyrins, as shown in Chapter 3. This configuration has been recently integrated to a multimodal catheter system consisting of a 40 MHz IVUS and fluorescence lifetime imaging using fast helical motion scanning able to acquire *in vivo* in pulsatile blood flow the autofluorescence emission of arterial vessels.

4.8 SUMMARY AND CONCLUSIONS

The point sampling technique relies on relatively simple instrumentation, is capable of fast collection of fluorescence decays, and can be easily integrated with scanning configurations, making it well suited for clinical use. Reports of numerous *in vivo* applications have been reviewed in this chapter, including *in vivo* endoscopic applications and brain tumor biochemical characterization.

The interplay between the laser pulse, detector, and digitizer parameters, in addition to the data processing algorithms, determines the temporal resolution of the pulse sampling technique. For the most part, the laser pulse requirements are UV emission wavelength, subnanosecond pulse duration, nJ pulse energy, low jitter, and pulse-to-pulse stability. Multiple laser technologies are capable of delivering these parameters, although, due to its low cost and microjoule energy pulses, the nitrogen laser has been the most prevalent. At present, the ytterbium-doped fiber laser is gaining rapid acceptance because it can deliver the above requirements with the additional advantage of having variable pulse repletion rate in the kilohertz to megahertz range and a very robust and reliable construction.

Fluorescence spectroscopy fiber-optic probes can be differentiated by their core diameter, NA, and number of excitation and collection fibers. They may have either a single or bifurcated geometry, and for the case of the single core fiber, there is an inherent overlap between the excitation and collection volumes, with maximum collection when the probe and sample are in contact. In contrast, the bifurcated probe has maximum collection at a sample-to-fiber distance of a few millimeters, which may be maintained through the use of an optimized fused silica spacer. For both types of fibers, the emission of background fluorescence must be minimized; therefore, the glass used for their construction is UV-grade fused silica. In addition to the fiber design parameters, the tissue optical properties also affect the performance of the probe, and therefore, these play a role during probe design.

The large gain and subnanosecond response time of the MCP PMT optical detector make it the detector of choice for point sampling measurements. In applications where the IRF can be increased to a few nanoseconds, the more economical dynode-based PMT can be used. An even more economical alternative to photon tube detectors is semiconductor pn-junction avalanche detectors. These detectors have IRFs comparable to the PMT and can be constructed to have UV sensitivity, but the low gain requires use in conjunction with an amplifier.

The finite bandwidth of the electronic components used for signal detection and amplification and the multimodal optical dispersion of the fluorescence pulses traveling in the optical fiber will result in the broadening of detected fluorescence pulses. This broadening can be compensated by deconvolving the instrument response function with a model-free method based on Laguerre functions. Time-resolved fluorescence measurements collected using the pulse sampling technique can be spectrally dispersed by using a monochromator or spectral filtering elements. The combined time and spectrally resolved fluorescence has been used for the *in vivo* identification of adenomatous colon polyps as well as the *in vivo* characterization of the biochemical features of clinically relevant primary brain tumors and for demonstrating the potential of using time-resolved laser-induced fluorescence spectroscopy as a tool for enhanced delineation of brain tumors during surgery. This technique also emerges as a promising solution in intravascular diagnosis of atherosclerotic plaques.

Newly developed configurations for the simultaneous acquisition of fluorescence decays in three to four spectral bands have been successfully demonstrated. The reduced number of laser pulses minimizes potential photobleaching effects, and the spectral bands can be tuned to the emission of NADH, elastin, lipoligaments, flavins, and porphyrins. Compact configurations have been integrated with commercial IVUS for *in vivo* intravascular pullback experiments in healthy pigs, demonstrating a reliable method for the characterization of tissue composition.

REFERENCES

Bec, J., Ma, D. M., Yankelevich, D. R., Liu, J., Ferrier, W. T., Southard, J. & Marcu, L. 2013. Multispectral fluorescence lifetime imaging system for intravascular diagnostics with ultrasound guidance: In vivo validation in swine arteries. *Journal of Biophotonics,* doi: 10.1002/jbio.201200220.

Bec, J., Xie, H., Yankelevich, D. R., Zhou, F., Sun, Y., Ghata, N., Aldredge, R. & Marcu, L. 2012. Design, construction, and validation of a rotary multifunctional intravascular diagnostic catheter combining multispectral fluorescence lifetime imaging and intravascular ultrasound. *Journal of Biomedical Optics,* 17, 106012-1–106012-10.

Becker, W., Castleman, A. W., Toennies, J. P. & Zinth, W. 2005. *Advanced Time-Correlated Single Photon Counting Techniques.* Berlin, Heidelberg: Springer Verlag.

Boyd, R. W. 1992. *Nonlinear Optics.* Boston, MA: Academic Press.

Butte, P. V., Fang, Q. Y., Jo, J. A., Yong, W. H., Pikul, B. K., Black, K. L. & Marcu, L. 2010. Intraoperative delineation of primary brain tumors using time-resolved fluorescence spectroscopy. *Journal of Biomedical Optics,* 15, 027008-1–027008-8.

Cooney, T. F., Skinner, H. T. & Angel, S. M. 1996a. Comparative study of some fiber-optic remote raman probe designs. Part I: Model for liquids and transparent solids. *Applied Spectroscopy,* 50, 836–848.

Cooney, T. F., Skinner, H. T. & Angel, S. M. 1996b. Comparative study of some fiber-optic remote raman probe designs. Part II: Tests of single-fiber, lensed, and flat- and bevel-tip multi-fiber probes. *Applied Spectroscopy,* 50, 849–860.

Dienes, A. & Yankelevich, D. R. 1995. Continuous wave dye lasers. In: Dunning, F. B. & Hulet, R. G. (eds) *Atomic, Molecular and Optical Physics.* San Diego, CA: Academic Press.

Donati, S. 2000. *Photodetectors: Devices, Circuits, and Applications.* Upper Saddle River, NJ: Prentice Hall.

Duarte, F. J. 2009. *Tunable Laser Applications.* Boca Raton, FL: CRC/Taylor & Francis.

Duarte, F. J. & Hillman, L. W. 1990. *Dye Laser Principles with Applications.* Boston, MA: Academic Press.

Erdmann, R. 2005. *Time Correlated Single-Photon Counting and Fluorescence Spectroscopy.* Weinheim: Wiley-VCH.

Fang, Q. Y., Papaionnou, T., Jo, J. A., Vaitha, R., Shastry, K. & Marcu, L. 2004. Time-domain laser-induced fluorescence spectroscopy apparatus for clinical diagnostics. *Review of Scientific Instruments,* 75, 151–162.

Hamamatsu Corporation (a), Electron Tube Division. Photomultiplier Tubes, Basics and Applications.

Hamamatsu Corporation (b), Electron Tube Division. Technical Specifications, R3809U-69 Microchannel Plate Photomultiplier Tube.

Hamamatsu Corporation (c), Electron Tube Division. Technical Specifications, R5916U Microchannel Plate Photomultiplier Tube.

Hamamatsu Corporation (d), Solid State Division. Characteristics and use of Si APD (Avalanche Photodiode). *Technical Information SD-28* [Online].

Hamamatsu Corporation (e), Solid State Division. S5343 Short Wavelength APD technical specifications.

Ippen, E. P., Shank, C. V. & Dienes, A. 1972. Passive mode locking of the cw dye laser. *Applied Physics Letters,* 21, 348–350.

Jackson, S. D. & Lauto, A. 2002. Diode-pumped fiber lasers: A new clinical tool? *Lasers in Surgery and Medicine,* 30, 184–190.

Jo, J. A., Fang, Q., Papaioannou, T., Baker, J. D., Dorafshar, A. H., Reil, T., Qiao, J. H., Fishbein, M. C., Freischlag, J. A. & Marcu, L. 2006. Laguerre-based method for analysis of time-resolved fluorescence data: Application to in-vivo characterization and diagnosis of atherosclerotic lesions. *Journal of Biomedical Optics,* 11, 021004-1-021004-13.

Jo, J. A., Marcu, L., Fang, Q., Papaioannou, T., Qiao, J. H., Fishbein, M. C., Beseth, B., Dorafshar, A. H., Reil, T., Baker, D. & Freischlag, J. 2007. New methods for time-resolved fluorescence spectroscopy data analysis based on the Laguerre expansion technique—Applications in tissue diagnosis. *Methods of Information in Medicine,* 46, 206–211.

Keiser, G. 1983. *Optical Fiber Communications.* New York: McGraw-Hill.

Kim, J., Lee, M., Yang, J. H. & Choy, J. H. 2000. Photophysical properties of hemicyanine dyes intercalated in Na-fluorine mica. *Journal of Physical Chemistry A,* 104, 1388–1392.

Lengyel, B. A. 1971. *Lasers.* New York: Wiley-Interscience.

Liu, J., Sun, Y., Qi, J. Y. & Marcu, L. 2012. A novel method for fast and robust estimation of fluorescence decay dynamics using constrained least-squares deconvolution with Laguerre expansion. *Physics in Medicine and Biology,* 57, 843–865.

Maarek, J.-M. I., Marcu, L., Snyder, W. J. & Grundfest, W. S. 2000a. Time-resolved fluorescence spectra of arterial fluorescent compounds: Reconstruction with the Laguerre expansion technique. *Photochemistry and Photobiology,* 71, 178–187.

Maarek, J. M. I., Marcu, L., Fishbein, M. C. & Grundfest, W. S. 2000b. Time-resolved fluorescence of human aortic wall: Use for improved identification of atherosclerotic lesions. *Lasers in Surgery and Medicine,* 27, 241–254.

Marcu, L. 2012. Fluorescence lifetime techniques in medical applications. *Annals of Biomedical Engineering,* 40, 304–331.

Marcu, L., Grundfest, W. S. & Maarek, J.-M. I. 1999. Photobleaching of arterial fluorescent compounds: Characterization of elastin, collagen and cholesterol time-resolved spectra during prolonged ultraviolet irradiation. *Photochemistry and Photobiology,* 69, 713–721.

Mycek, M. A., Schomaker, K. T. & Nishioka, N. S. 1998. Colonic polyp differentiation using time-resolved autofluorescence spectroscopy. *Gastrointestinal Endoscopy,* 48, 390–394.

Mycek, M. A., Vishwanath, K., Schomaker, K. T. & Nishioka, N. S. 2000. Fluorescence spectroscopy for in vivo discrimination of pre-malignant colonic lesions. *Biomedical Topical Meetings, Trends in Optics and Photonics. Optical Society of America Technical Digest,* 11–13.

National Instruments 2006. Bandwidth, Sample Rate, and Nyquist Theorem. Trends in Optics and Photonics: Optical Society of America.

Oppenheim, A. V. & Schafer, R. W. 2010. *Discrete-Time Signal Processing.* Upper Saddle River, NJ: Pearson.

Papaioannou, T., Preyer, N. W., Fang, Q. Y., Brightwell, A., Carnohan, M., Cottone, G., Ross, R., Jones, L. R. & Marcu, L. 2004. Effects of fiber-optic probe design and probe-to-target distance on diffuse reflectance measurements of turbid media: An experimental and computational study at 337 nm. *Applied Optics,* 43, 2846–2860.

Pfefer, T. J., Matchette, L. S., Ross, A. M. & Ediger, M. N. 2003a. Selective detection of fluorophore layers in turbid media: The role of fiber-optic probe design. *Optics Letters,* 28, 120–122.

Pfefer, T. J., Paithankar, D. Y., Poneros, J. M., Schomaker, K. T. & Nishioka, N. S. 2003b. Temporally and spectrally resolved fluorescence spectroscopy for the detection of high grade dysplasia in Barrett's esophagus. *Lasers in Surgery and Medicine,* 32, 10–16.

Pfefer, T. J., Schomaker, K. T., Ediger, M. N. & Nishioka, N. S. 2001. Light propagation in tissue during fluorescence spectroscopy with single-fiber probes. *IEEE Journal of Selected Topics in Quantum Electronics,* 7, 1004–1012.

Pfefer, T. J., Schomaker, K. T., Ediger, M. N. & Nishioka, N. S. 2002. Multiple-fiber probe design for fluorescence spectroscopy in tissue. *Applied Optics,* 41, 4712–4721.

Phipps, J. E., Sun, Y. H., Fishbein, M. C. & Marcu, L. 2012. A fluorescence lifetime imaging classification method to investigate the collagen to lipid ratio in fibrous caps of atherosclerotic plaque. *Lasers in Surgery and Medicine,* 44, 564–571.

Principles of fluorescence lifetime instrumentation

Pitts, J. D. & Mycek, M. A. 2001. Design and development of a rapid acquisition laser-based fluorometer with simultaneous spectral and temporal resolution. *Review of Scientific Instruments,* 72, 3061–3072.

Ruck, A., Hulshoff, C., Kinzler, I., Becker, W. & Steiner, R. 2007. SLIM: A new method for molecular imaging. *Microscopy Research and Technique,* 70, 485–492.

Shim, M. G., Wilson, B. C., Marple, E. & Wach, M. 1999. Study of fiber-optic probes for *in vivo* medical raman spectroscopy. *Applied Spectroscopy,* 53, 619–627.

Sun, Y., Liu, R., Elson, D. S., Hollars, C. W., Jo, J. A., Park, J., Sun, Y. & Marcu, L. 2008. Simultaneous time- and wavelength-resolved fluorescence spectroscopy for near real-time tissue diagnosis. *Optics Letters,* 33, 630–632.

Sun, Y. H., Sun, Y., Stephens, D., Xie, H. T., Phipps, J., Saroufem, R., Southard, J., Elson, D. S. & Marcu, L. 2011. Dynamic tissue analysis using time- and wavelength-resolved fluorescence spectroscopy for atherosclerosis diagnosis. *Optics Express,* 19, 3890–3901.

Utzinger, U. & Richards-Kortum, R. R. 2003. Fiber optic probes for biomedical optical spectroscopy. *Journal of Biomedical Optics,* 8, 121–147.

Xie, H. T., Bec, J., Liu, J., Sun, Y., Lam, M., Yankelevich, D. R. & Marcu, L. 2012. Multispectral scanning time-resolved fluorescence spectroscopy (TRFS) technique for intravascular diagnosis. *Biomedical Optics Express,* 3, 1521–1533.

Yariv, A. & Yeh, P. 2007. *Photonics: Optical Electronics in Modern Communications.* New York: Oxford University Press.

Zhu, Z. Y. & Yappert, M. C. 1992a. Determination of effective depth and equivalent pathlength for a single-fiber fluorometric sensor. *Applied Spectroscopy,* 46, 912–918.

Zhu, Z. Y. & Yappert, M. C. 1992b. Determination of the effective depth for double-fiber fluorometric sensors. *Applied Spectroscopy,* 46, 919–924.

Single-point probes for lifetime spectroscopy: Time-correlated single-photon counting technique

Christopher Dunsby and Paul M. W. French

Contents

5.1 INTRODUCTION

Naturally occurring autofluorescence from biological tissue can be exploited for basic studies of physiology and pathology and may provide label-free diagnostic tools. It is therefore useful to develop instrumentation that can be used to characterize tissue autofluorescence *in situ*. While spectroscopic imaging modalities such as fluorescence lifetime imaging (FLIM) and multispectral imaging are clearly desirable to correlate biochemical information with morphology and facilitate direct comparisons with histopathology, the amount of useful spectroscopic information available can be limited by the finite numbers of detected photons associated with clinically practical data acquisition times and by difficulties associated with engineering appropriately compact and sterilizable clinical endoscopes. Fiber-optic-based probes delivering single-point measurements of autofluorescence can be made much more compact and at lower cost compared to imaging instruments and are simpler to engineer. They also provide more opportunities for more sophisticated multidimensional analysis of autofluorescence signals because all the detected photons contribute to a single (spatial) detection channel and measurement times of a few seconds, which are practical for *in vivo* applications and can provide sufficient photons to permit, for example, spectrally resolved analysis of multiexponential fluorescence decay profiles. Furthermore, the compact nature of nonimaging fiber-optic probes means that they can be readily applied via the biopsy port of a standard clinical endoscope with minimal perturbation to existing clinical procedures.

Tissue autofluorescence emanates from a broad range of endogenous fluorophores whose fluorescence emission spans the ultraviolet (UV) and visible spectrum. Radiation below 300 nm can excite the fluorescent amino acids phenylalanine, tyrosine, and tryptophan, producing fluorescence emission mainly below 450 nm. It is difficult to study such fluorophores, however, particularly in tissue and *in vivo*, because of the high absorption of radiation at wavelengths below 300 nm in most optical systems and because of the background fluorescence that such radiation can excite in optical components, including optical fibers. The phototoxicity of radiation below 300 nm wavelength is also a significant issue, so longer wavelength excitation is most interesting for *in vivo* applications. As discussed in Chapter 2 of this book, the major endogenous intracellular fluorophores excited above 300 nm are cellular flavins (e.g., flavin adenine dinucleotide [FAD]) and reduced nicotinamide adenine dinucleotide (phosphate), known as NAD(P)H. Other important sources of autofluorescence include keratin, melanin, and porphyrins, as well as cross-linkages between the structural proteins collagen and elastin found in the tissue extracellular matrix. Because a single-channel fiber-optic-based probe integrates the autofluorescence from a range of spatial locations, the detected signal will comprise contributions from endogenous fluorophores distributed throughout the measurement volume in the tissue, so the autofluorescence measurements will be influenced by the structure of the tissue, the local optical (scattering and absorption) properties, and the geometry of the optical fiber probe.

Autofluorescence studies on biological tissue are challenging because of the overlapping excitation and emission spectra of these endogenous fluorophores, their intrinsically low fluorescence intensity levels, and the strong scattering and heterogeneity of biological tissue. It is therefore desirable to resolve autofluorescence with respect to multiple spectroscopic parameters in order to distinguish different tissue types and pathologies. To date, the main spectroscopic parameters studied have been the autofluorescence emission spectrum and lifetime (Richards-Kortum and Sevick-Muraca 1996), with lifetime measurements being predominantly undertaken in the time domain. Fluorescence lifetime measurements can be made in a single spectral channel, for example, as applied to the *in vivo* study of colonic polyps (Mycek et al. 1998); fluorescence lifetime and emission spectrum can be measured separately, for example, in the work of Pitts and Mycek (2001); or spectrally resolved fluorescence decay profiles can be acquired, for example, by using a spectrograph and streak camera as applied *in vivo* to the bladder, bronchi, esophagus, and oral cavity (Glanzmann et al. 1999) or by using a scanning monochromator combined with single-channel lifetime measurements, as has been applied *ex vivo* to brain tissue (Marcu et al. 2004) and *in vivo* in rabbit models (Marcu et al. 2005) and in humans (Butte et al. 2010; Marcu et al. 2009). An alternative approach is to use temporal multiplexing to simultaneously measure fluorescence decay profiles in multiple spectral windows (Sun et al. 2008). With the exception of the streak-camera based system (Glanzmann et al. 1999), these instruments have utilized fast digitizers to measure the temporal characteristics of autofluorescence, with the requirement for ultrashort excitation pulses being initially met by nitrogen lasers or nitrogen laser–pumped dye lasers that could provide the pulsed UV/blue radiation required to excite the most common endogenous fluorophores. Nitrogen lasers are relatively compact and low cost but deliver pulses of several nanoseconds' duration at repetition rates in the range of tens of hertz, which can impact the time resolution, measurement speed, and signal-to-noise ratio compared to the performance achievable with continuous-wave mode-locked lasers. Nevertheless, this approach has proved practically useful for research and clinical measurements. Recently, the fast digitizer approach to multispectral lifetime measurement has been implemented with a rapidly Q-switched microchip laser providing nanosecond pulses at 3 kHz to improve data acquisition times (Lloyd et al. 2010). More details of the fast digitizer-based approach and its clinical applications are given in Chapter 4 of this book.

At Imperial College London, our approach has been to develop single-point autofluorescence probes utilizing time-correlated single-photon counting (TCSPC) (Becker 2005) combined with high-repetition-rate (>~40 MHz) mode-locked or gain-switched laser excitation sources, as is commonly used to implement FLIM in laser scanning microscopes. TCSPC entails exciting the sample with ultrashort pulses and measuring the relative arrival times of single fluorescence photons at the detector with respect to the excitation pulses. This requires that the excitation power is sufficiently low to ensure that less than one photon per excitation pulse is detected. Thus, the arrival times of individual detected photons relative to the train of excitation pulses can be recorded to build up a histogram of photon arrival times. In laser scanning microscopes, the spatial information from the scanning electronics is also recorded such that the photon arrival times can be allocated to their respective image pixels to form fluorescence lifetime images. This is straightforward to implement in most multiphoton microscopes since it requires only additional electronic components (after the detector), and TCSPC-based FLIM is discussed in further detail in Chapter 9.

There are several different approaches to implementing TCSPC electronics, and Figure 5.1 illustrates a commonly used method. TCSPC (Becker 2005) entails detecting fluorescence photons, for example, using a high gain photomultiplier, and photons from the excitation pulses, for example, using a photodetector

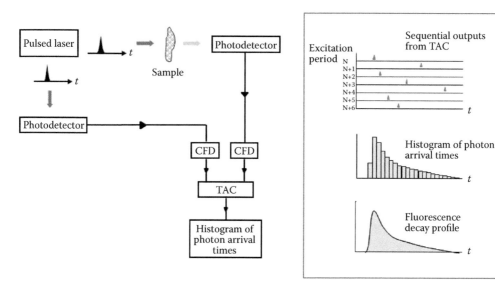

Figure 5.1 Schematic representation of TCSPC measurement of fluorescence decay profiles implemented using constant fraction discriminators (CFDs) and a time-to-amplitude converter (TAC).

such as a photodiode to measure the signal from a small fraction of the excitation beam picked off by a beam splitter. If available, a direct electrical synchronization signal from the excitation laser may be used in place of the photodiode signal. These two electrical signals are then passed to constant fraction discriminators (CFDs) to minimize triggering jitter. The detection of a fluorescence photon then initiates a time-to-amplitude converter (TAC), which outputs a voltage signal that increases with time. This voltage ramp is then stopped by arrival of the next excitation pulse, thereby providing a voltage signal that is proportional to the photon arrival time. A histogram of fluorescence photon arrival times is built up in computer memory from each photon arrival event as it is processed. The envelope of this histogram provides the fluorescence decay profile.

TCSPC offers shot noise-limited detection, high photon economy, low temporal jitter, high temporal precision (with the photon arrival time histogram resolved into thousands of time bins), and high dynamic range (typically millions of photons can be recorded without saturation), making it widely accepted as one of the most accurate methods of lifetime determination. Its main perceived drawback is the limited data acquisition rate, owing to the requirement to operate at sufficiently low incident fluorescence intensity levels to ensure single photon detection. The maximum detection rate is partly limited by the "dead time" between measurement events, which is imposed by the TAC and CFD circuitry that determine the photon arrival times (Becker 2005). However, for modern TCSPC instrumentation, this limitation of the electronic circuitry is usually less significant than problems caused by "classical" photon pileup, which limits all single photon counting techniques. Photon pileup refers to the issue of more than one photon arriving in a single photon detection period, which results in the apparent measured lifetimes being shorter than the real fluorescence lifetimes. This can be avoided by decreasing the excitation power such that the excitation rate is much lower than the pulse repetition rate—a practical limit is to maintain the maximum detection count rate below approximately 5% of the laser repetition rate (Becker 2005).

TCSPC measurements of tissue autofluorescence typically require less than 1 s per measurement for excitation average power levels of the order of 10–1000 µW. When combined with multichannel detectors, parallel TCSPC measurements can provide multispectral measurements of complex fluorescence decay profiles, as discussed in Chapter 9, and are applied to microspectrofluorometry of cells (Becker et al. 2007; Bird et al. 2004; Chorvat and Chorvatova 2006; Tramier et al. 2004).

We have implemented single-point TCSPC with a range of picosecond and femtosecond excitation sources, including relatively low-cost and compact picosecond diode lasers (De Beule et al. 2007; Thompson et al. 2012), mode-locked fiber lasers (De Beule et al. 2007), fiber laser–pumped

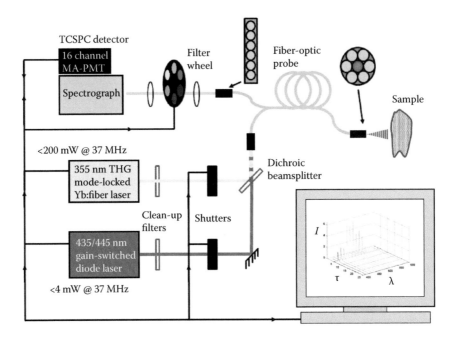

Figure 5.2 Schematic of dual-wavelength excitation autofluorescence hyperspectral lifetime probe.

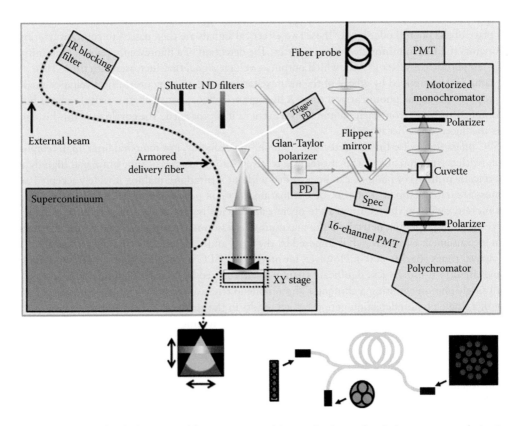

Figure 5.3 Schematic of multidimensional fluorometer resolving excitation and emission spectra, polarization, and fluorescence lifetime with measurements applied to samples in a cuvette, in a multiwell plate array, and via a fiber-optic probe (inset, bottom right). Translation of V-shaped aperture to control center wavelength and spectral width of excitation radiation is indicated (inset, bottom left), with arrows showing direction of translation. (Adapted from Manning, H.B. et al., *J. Biophotonics*, 1, 2008.)

supercontinuum sources (Manning et al. 2008), and mode-locked solid-state lasers (Manning et al. 2008, 2013), for fluorescence lifetime measurements. The short pulse durations and high repetition rates of such sources provide high temporal resolution with low jitter, and data acquisition times of a few seconds provide sufficient detected photons to permit analysis of complex fluorescence decay profiles. Fiber-optic probe-based fluorescence decays can also be recorded using pulsed LED excitation and with fluorescence emission detected using a scanning monochromator and TCSPC, as has been used for studies of skin in healthy and diabetic patients by Blackwell et al. (2008).

Figure 5.2 represents the first TCSPC instrument that we developed for tissue autofluorescence measurements (designated AFHL1), which was applied to investigate the potential to discriminate between healthy and cancerous tissue. This "hyperspectral fluorescence lifetime probe" integrated a custom-built fiber-optic bundle with a spectrograph and multianode photomultiplier (MA-PMT) to facilitate simultaneous *in situ* measurements of spectrally resolved fluorescence decay profiles for UV and blue excitation wavelengths that are intended to excite NADH and flavoproteins respectively, as well as collagen and elastin. Figure 5.3 shows a schematic of a laboratory-based instrument that we apply to a wide range of multidimensional fluorescence studies in cuvette for solution phase measurements and via an optical fiber probe for *ex vivo* tissue studies.

5.2 INSTRUMENTATION

5.2.1 DUAL EXCITATION CHANNEL AUTOFLUORESCENCE HYPERSPECTRAL LIFETIME PROBE

5.2.1.1 Experimental configuration (autofluorescence hyperspectral lifetime)

The multidimensional fluorometer (MDF) depicted in Figure 5.2 employs ultrafast excitation lasers operating at 37.1 MHz together with a multichannel TCSPC detection system to provide parallel fluorescence lifetime measurement in 16 spectrally resolved channels. The whole instrument was mounted on a 60 × 60 cm breadboard, and it was designed to maximize the throughput of the limited blue excitation light, to ensure that reflected excitation light could not reach the detector, and to prevent any scattered excitation light reaching the sample after the main excitation pulse. Laser excitation at 355 and 440 nm was realized using two compact, pulsed laser sources. The first was an ultrafast frequency-tripled Yb:glass fiber laser system (Fianium UVPower355, Fianium Ltd, UK) generating up to 200 mW average output power and light pulses at 355 nm of 10 ps duration with a 37.1 MHz repetition rate. The second laser source was a pulsed diode laser emitting pulses of 50–150 ps duration (LDH-P-C-440B, PicoQuant GmbH, Germany) with a user-adjustable repetition rate up to 80 MHz. During our experiments, a diode laser with an average output power of 1 mW at 435 nm was replaced by a higher-power version that provided 3.7 mW average output power at 445 nm.

In our first experimental configuration, we implemented interleaved excitation using a chopper wheel to alternately select the UV (355 nm) or blue (435 nm) excitation light. Data acquisition was controlled using the "SPCM" software package (SPCM, Becker-Hickl GmbH, Germany). The 435 nm diode laser radiation was "cleaned up" using a filter with a 1.7 nm wide band-pass region centered at 441.6 nm (MaxLine helium cadmium laser line filter, Semrock, USA) that was angle-tuned to downshift the central band-pass wavelength 435 nm. An infrared stopping filter was deployed in front of the UV fiber laser to block any residual fundamental radiation. Any stray or reflected excitation radiation was blocked at the detector using a UV long-pass filter and a blue notch filter (Stopline, Semrock). In a second configuration, we incorporated a higher-power blue laser diode operating at 445 nm and implemented computer-controlled mechanical shutters (VS14, Uniblitz, USA) and an emission filter wheel (Lambda 10-3, Sutter, USA) to alternately select the UV and blue excitation light and prevent reflected excitation light reaching the detector. For this setup, the acquisition was controlled by a custom-written LabVIEW program (LabVIEW, National Instruments, UK). For both configurations, the excitation laser beams were combined using a dichroic mirror and subsequently coupled via a 10× microscope objective into the central 200 μm diameter multimode fiber of a custom-built fiber probe (custom built, FiberTech Optica, Canada). In this fiber probe, the central excitation fiber was surrounded by six identical peripheral fibers to collect the emitted fluorescence. For the fluorescence detection arm, the six fluorescence collection fibers were arranged in a line to achieve efficient coupling into the input slit of a grating spectrometer attached to am MA-PMT detector (PML-SPEC, Becker-Hickl GmbH). This unit covered a total spectral observation band between 390 and 600 nm and provided the fluorescence temporal decay profile in 16 spectral channels. Multichannel TCSPC was achieved using commercially available routing electronics and a TCSPC card (SPC-730, Becker-Hickl GmbH).

5.2.1.2 System calibration

Reference solutions of NADH and FAD in phosphate-buffered saline (PBS) at pH 7.2 and 20°C were used to evaluate the performance of the experimental setups. Figure 5.4 shows the uncorrected spectrotemporal profiles acquired for each of these fluorophores. The observed fluorescence lifetimes for the spectrally averaged decay profile are consistent with values in the literature: NADH exhibits a complex decay profile that can be reasonably approximated to a biexponential decay with lifetime components of 400 and 780 ps with relative contributions of 71% and 29% respectively, while a single exponential decay lifetime of 2.8 ns was observed for FAD. The fluorescence emission spectrum obtained by temporal integration of the photon counts, however, did not correspond to reference spectra acquired with a spectrofluorimeter (RF-5301PC, Shimadzu, Japan). The discrepancies are due to the combined effects of spectral instrument response functions (IRFs) of the spectrograph and the nonuniform channel sensitivity of MA-PMT, as well as some contributions from cross talk.

In order to correct for the effects mentioned above, the spectral location of each PMT element was determined using a white light source projected through a calibrated monochromator. The overall spectral

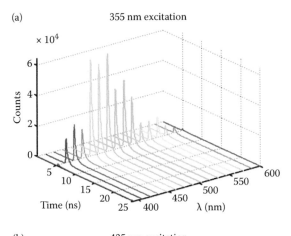

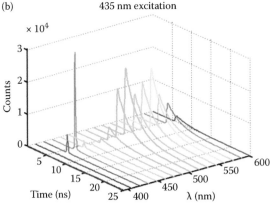

Figure 5.4 Uncorrected hyperspectral fluorescence decay profiles for (a) 20 μM NADH and (b) FAD dissolved in PBS, excited at 355 and 435 nm, respectively. Some reflected excitation light is observed in the 440 nm channel in (b). (Adapted from De Beule, P.A. et al., *Rev. Sci. Instrum.*, 78, 2007.)

IRF was then determined using a radiometric calibration standard (LS-1-CAL, Ocean Optics, USA). Apart from a small ~10 nm discrepancy in absolute wavelength, the corrected fluorescence emission profiles obtained agreed well with literature values and time-integrated measurements obtained using a conventional spectrofluorimeter. The peak observed at 440 nm for the FAD curve, which is apparent in Figure 5.2a and b, is due to residual leakage of reflected blue excitation light. The measured spectral resolution was 14 nm.

In order to accurately fit nonsingle exponential decay profiles to cell or tissue autofluorescence data, we also measured the temporal IRF of the instrument. However, this is complicated by the channel-dependent transit time for the MA-PMT, which results in a different IRF for each spectral channel. The detector elements each have a similar temporal IRF profile (Becker 2006) but with a slightly different temporal offset. The IRF for each PMT channel was therefore measured using 2-(p-dimethylaminostyryl) pyridylmethyl iodide (DASPI) dissolved in ethanol, a fluorophore with a short fluorescence lifetime (<50 ps) (Taylor et al. 1980) that is significantly less than the PMT response time. A FWHM of 170 and 325 ps was measured for the 355 and 435 nm excitation, respectively. The increased width of the IRF for the blue laser diode is caused by the longer pulse duration of this source.

5.2.2 MDF

To expand the excitation capabilities and experimental versatility of our hyperspectral fluorescence lifetime probe, we refined the instrument shown in Figure 5.2 and developed a MDF that incorporates a fiber laser-pumped supercontinuum source (Fianium SC400-4-PP) with a prism spectrometer to provide tunable excitation from the blue (~400 nm) to near infrared (above 1 μm). A "V-shaped" aperture in front of the mirror that returns the radiation dispersed by the prism can be translated to adjust both the center wavelength and the spectral width of the selected radiation as indicated in Figure 5.3 (inset, bottom left). The instrument, which is depicted in Figure 5.3, also includes polarizers in the excitation and emission path to permit time-resolved analysis of fluorescence anisotropy (Manning et al. 2008) and can be configured for solution phase measurements in cuvette or for measurements via an external fiber-optic probe. This probe, which is similar to that described above but with 3 excitation fibers and 16 collection fibers (custom built, FiberTech Optica), can be applied to remote samples including biological tissue. When using the fiber-optic probe, the excitation light is reflected from the normal light path by a flipper mirror to couple into a 600 μm core fiber that is butted against three 200 μm excitation fibers in a triangular formation in the first leg of the probe. These three fibers then deliver the excitation light to the sample at which fluorescence is collected by sixteen 200 μm core fibers in a circular pattern as shown in the diagram (bottom right of Figure 5.3). The proximal ends of the detection fibers are arranged in a line that is mounted vertically such that the light is reflected from a 45° mirror and imaged (through the same collection optics as for cuvette measurements) onto the input slit of either the monochromator or polychromator for spectrally resolved detection. This instrument also has the capability to use external excitation lasers such as a mode-locked UV laser (Vanguard 350-HM355, Spectra-Physics, UK) providing 12 ps pulses at 355 nm. The sample emission can be analyzed by the 16-channel spectrally resolved TCSPC system described above or be continuously spectrally resolved using a motorized monochromator (CM110, CVI Inc, UK) and single-channel TCSPC system (PMC-100 photomultiplier and SPC-730 TCSPC electronics, Becker-Hickl GmbH). The MA-PMT was calibrated in a similar manner as for the dual-channel instrument discussed above.

5.3 APPLICATIONS OF TCSPC METROLOGY TO AUTOFLUORESCENCE

5.3.1 APPLICATION TO EX VIVO SKIN LESIONS

Label-free *in situ* noninvasive diagnosis of skin cancer could be of significant clinical benefit since conventional diagnosis, based on biopsy followed by histopathological examination, is time consuming and invasive. Accordingly, we applied the autofluorescence hyperspectral lifetime probe (AFHL) depicted in Figure 5.2 to the first combined study of fluorescence emission spectra and fluorescence decay profiles for skin lesion diagnosis (Thompson et al. 2012), noting that autofluorescence had previously been shown to be useful in distinguishing skin tumors from healthy skin using spectrum alone (Brancaleon et al. 2001; Lohmann et al. 1991; Panjehpour et al. 2002; Sterenborg et al. 1994) and also to distinguish cancer from normal tissue via lifetime measurements (Pradhan et al. 1995; Skala et al. 2007).

This preliminary study of 25 *ex vivo* skin samples included 16 unpigmented lesions, 4 seborrheic keratosis (benign), 10 basal cell carcinomas (BCCs), and 2 squamous cell carcinomas (SCCs), and 9 pigmented lesions, 6 (benign) nevi and 3 melanomas. Our results indicated that the AHFL1 instrument could help distinguish uninvolved skin and BCC lesions, melanoma, and nevi and potentially could contrast SCC and BCC from seborrheic keratosis lesions, although the sample numbers were too small in this study to draw any significant conclusions.

Figure 5.5 shows data obtained with AFHL probe using the 1 mW blue diode emitting at 435 nm setup, which were, to the best of our knowledge, the first reported fluorescence lifetime measurements of skin lesion autofluorescence for 435 nm fluorescence excitation. Autofluorescence from both lesions and from regions of skin surrounding the lesion, designated as "uninvolved" skin, was sampled using the AFHL probe as indicated in Figure 5.5a, except for punch biopsy samples where only the diseased area could be studied due to the limited sample size.

All measurements were made within 2 h of excision of skin tissue to obtain similar fluorescence measurements to observations of *in vivo* tissue (Palmer et al. 2002) and avoid an artificial change in NADH concentration (Gupta et al. 1997). After excision, skin samples were placed immediately into chilled tissue culture medium without phenol red (Dulbeccos's Modified Eagle Medium, Invitrogen, UK), and samples were washed twice in chilled Hank's Buffered Salt Solution (HBSS; Invitrogen) prior to the measurements. The autofluorescence was collected through a glass-bottomed culture dish in a setup designed to maintain equal probe sample distances for the uninvolved skin and lesion, respectively.

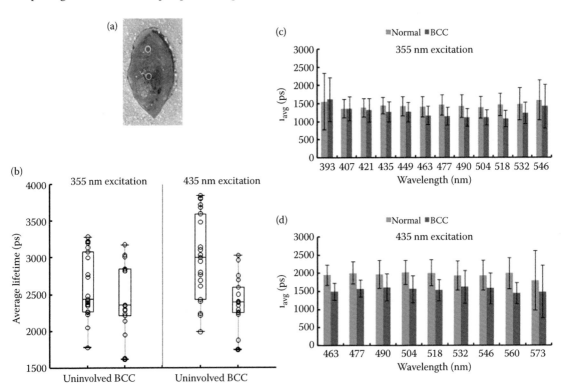

Figure 5.5 (a) Photograph of skin sample with measurement sites indicated. (b) Distribution of average spectrally integrated autofluorescence lifetimes of BCC lesions and surrounding uninvolved skin for excitation at 355 nm and 435 nm (box plots indicate the minimum, first quartile, median, third quartile, and maximum lifetimes). (c and d) Spectrally resolved average autofluorescence fluorescence lifetimes of BCC (red) and surrounding uninvolved skin (blue) for excitation at (c) 355 and (d) 435 nm. Error bars indicate standard deviation. (Adapted from De Beule, P.A. et al., *Rev. Sci. Instrum.*, 78, 2007.)

Following the probe data acquisition, the samples were fixed in 10% formaldehyde for subsequent classification by conventional histopathology.

To analyze the hyperspectral lifetime data, we initially considered the spectrally integrated fluorescence lifetime data shown in Figure 5.5b, obtained by summing fluorescence decay profiles across the spectral bins and fitting to a double exponential decay model using SPCImage (Becker-Hickl GmbH). This decay model provided an improved fit compared to a single exponential decay model with low χ^2 values (average of 2.7) when calculating the average fluorescence lifetime. A lower value for the mean spectrally integrated average fluorescence lifetime was observed for BCC for both UV and blue excitation, but the difference is only statistically significant for the data obtained with 435 nm excitation, for which a Wilcoxon signed-rank hypothesis test of the difference in average fluorescence lifetime between uninvolved skin and BCC cancer yielded a P value of .03. Figure 5.5c and d shows the spectrally resolved average autofluorescence for normal (uninvolved) skin and BCC. The difference in the 435 nm-excited autofluorescence lifetime is apparent across the spectral range, but for 355 nm-excited autofluorescence, there is only a difference apparent for emission above 421 nm. This could result from changes in contributions from the different tissue matrix components or from a decrease in the fluorescence of lifetime of the NAD(P)H fluorescence—noting that the majority of oral epithelium autofluorescence excited at 349 nm and emitted below ~425 nm was previously reported (Wu et al. 2004) to be from keratin and collagen, while NAD(P)H is expected to be the major fluorophore contributing to autofluorescence emission above ~425 nm.

Our study (De Beule et al. 2007) also suggested that it may also be possible to discriminate BCC and SCC from seborrheic keratosis lesions via the 355 nm-excited autofluorescence emission spectrum, since the latter was slightly blue-shifted (potentially attributable to higher concentrations of keratin) and further contrast may be available from the spectrally resolved average fluorescence lifetime data, but a more extended study is necessary to evaluate this diagnostic potential. Similarly, melanoma and nevi appeared to be potentially separable via the 435 nm-excited autofluorescence for which melanoma exhibited a red-shifted emission spectrum and a lower average fluorescence lifetime compared to nevi, but insufficient samples were measured to realize a statistically significant conclusion.

Since human skin is a highly layered structure, the detected autofluorescence signal is a mixture of emission from the various endogenous fluorophores found in the different layers. The most abundant cells in the outermost stratum of the epidermis are the keratinocytes, which are rich in keratin (Wu et al. 2004) that has an emission maximum at ~400 nm for excitation at 355 nm. Cells in the lower sections of the epidermis have less keratin and exhibit higher relative contributions from NAD(P)H and flavins, while the layer beneath the epidermis (dermis) is constructed of a dense collagen network. The relative contributions from these different layers will depend on the structure of the tissue, for example, the thickness of the epidermis, as well as the geometry of the optical fiber probe and on the scattering and absorption properties of the skin. Thus, caution must be exercised when comparing single-point autofluorescence data between different lesions. Nevertheless, this preliminary study indicated some contrast between normal tissue and basal cell carcinoma and demonstrated that the implementation of the dual UV/blue excitation using compact pulsed lasers permits the clinical deployment of this instrument.

5.3.2 APPLICATION TO SKIN CANCER *IN VIVO*

Following the preliminary *ex vivo* study of the application of the autofluorescence hyperspectral lifetime probe (AFHL1) to skin cancer, the instrument was applied *in vivo* in a clinical study (Thompson et al. 2012) at the Lund University Hospital. Patients were assessed and recruited by the clinical team during outpatient clinics. The study was conducted with the approval of the local ethics committee and in accordance with the ethical principles of the Declaration of Helsinki. For this work, the AFHL1 probe was mainly employed to measure the autofluorescence lifetime, for which we employed the 3.7 mW pulsed laser diode emitting at 445 nm for blue excitation of autofluorescence with the same frequency-tripled Yb:glass fiber laser providing the excitation at 355 nm. For the *in vivo* skin measurements, the average power at the sample was limited to 10 μW of UV and 50 μW of blue light, and the irradiation time was 5 s, with each measurement being repeated three times at each sample location. A transparent spacer was used to maintain a constant distance from the tip of the fiber-optic probe to the sample (tissue) surface that resulted in illumination of a spot of ~0.4 mm diameter. The spacer was hollow and did not overlap the illumination beam path. Measurements were taken from two sites of the perilesional skin surrounding each lesion and between one and four sites in the lesion itself (depending on its size). A total of 27 lesions on 25 patients (all of skin phototypes I–III) were investigated prior to surgical excision of the measured regions. Following surgical excitation, the lesions were analyzed using conventional histopathology and classified as eight BCC, one SCC, two benign nevi, one dysplastic nevus, three malignant melanomas, three actinic keratosis, and four lesions in other categories.

For the analysis of these *in vivo* autofluorescence lifetime data, we first quantified the intrapatient variability by comparing the measurements of the two "normal" perilesional sites on each patient. Figure 5.6a and b shows the differences observed in the spectrally averaged mean lifetimes for UV and blue excitation, respectively. These differences are evenly distributed about zero, with mean values of –18 and 19 ps respectively. Lesion 6 shows the largest difference in mean lifetime between the two normal sites for both excitation wavelengths, and we attribute this to the presence of scar tissue in the first perilesional measurement site. To provide an estimate of the error on in the measured autofluorescence lifetime, we calculated the mean standard deviations of the normal tissue measurements, and these were found to be 75 ps with 355 nm excitation and 175 ps with 445 nm excitation.

Since BCC was the only type of lesion for which there were sufficient samples to permit statistically significant conclusions, we focused our analysis on tissue from patients diagnosed with this lesion. It was observed that neither the mean fluorescence lifetimes nor the lifetime contrast between normal and diseased tissue varied significantly with emission wavelength. Integrating across all spectral channels to calculate spectrally averaged lifetimes and averaging lifetimes across all BCC patients, the healthy and lesional lifetimes were respectively calculated to be 2770 ± 250 and 2880 ± 409 ps with UV excitation and 3130 ± 413 and 2240 ± 480 ps with blue excitation (where the values report the mean ± one standard

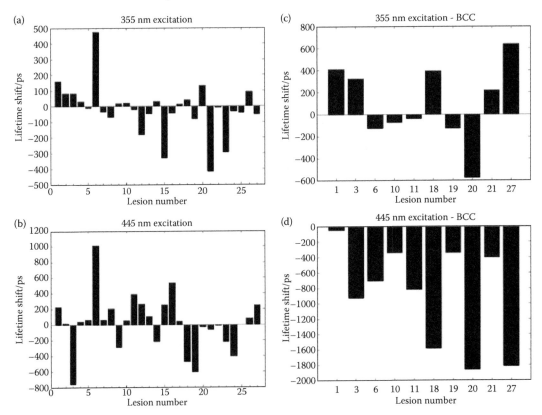

Figure 5.6 (a and b) Autofluorescence lifetime shifts between two measurements of normal (perilesional) tissue for (a) UV and (b) blue excitation. (c and d) The (spectrally integrated) lifetime shifts between lesion and normal tissue (i.e., $\Delta\tau = \tau_{lesion} - \tau_{normal}$) for (c) UV and (d) blue excitation for all BCC lesions. (Adapted from Thompson, A.J. et al., *J. Biophotonics*, 5, 2012.)

deviation). We did, however, observe significant interpatient variation in the measured autofluorescence lifetime, which could be due to interpatient differences in factors such as skin phototype, level of sun exposure, age, and the location of the measurement site on the body. We therefore chose to analyze the autofluorescence lifetime differences between the healthy and lesional tissue measured on each individual in order to circumvent the issue of interpatient variability and to calculate spectrally integrated mean fluorescence lifetimes for each measurement point to increase the accuracy of the lifetime determination.

Figure 5.6c and d shows the paired difference in spectrally integrated fluorescence lifetime between the healthy and diseased regions for each BCC lesion for excitation at 355 and 445 nm, respectively. While there is no clear trend in the lifetime shifts of BCCs observed with UV excitation (Figure 5.6c), it is apparent that for blue (445 nm) excitation, the autofluorescence lifetime of BCCs is consistently lower than that of the surrounding perilesional skin (Figure 5.6d). The mean lifetime decrease was calculated as 886 ps, and the lifetime shifts observed in all but one case (lesion 1) were also seen to exceed our (175 ps) value of the mean intrapatient variability. For the BCC patient data based on 445 nm excitation, a Wilcoxon signed-rank test on the autofluorescence lifetime shift yielded a statistically significant decrease in fluorescence lifetime (i.e., $\tau_{lesion} < \tau_{normal}$), with a P value of .002. We note that we did not observe any significant difference in the mean emission wavelength between healthy skin and BCCs for either excitation wavelength.

In general, the trends observed for BCCs were also seen to hold for all other lesion types, and Wilcoxon signed-rank tests performed on the lifetime shifts for all lesion types combined as an ensemble yielded a P value of 5.6×10^{-6}. Thus, our *in vivo* study suggests that there is a statistically significant difference between the autofluorescence lifetime of healthy tissue and skin cancer when exciting with radiation centered at 445 nm, although the absolute autofluorescence lifetime values can vary significantly from patient to patient. It was not possible to evaluate the sensitivity and specificity of the lifetime measurements as we calculated lifetime differences between normal and lesional skin within each patient. Future work is needed to assess the diagnostic accuracy and to ascertain whether autofluorescence lifetime can be used to differentiate between different lesion types and not just between healthy and diseased tissue.

5.3.3 APPLICATION TO *EX VIVO* CARTILAGE

Multidimensional autofluorescence measurements can be used to study and detect diseases through changes in metabolic pathways or in tissue matrix components. While the former are particularly important for cancer, changes in the latter—particularly collagen—may be useful to detect and monitor the progress of degenerative diseases, for example, affecting cartilage and articular tissue. We have applied the MDF with the fiber-optic probe shown in Figure 5.3 to a preliminary study investigating the potential of autofluorescence lifetime to report the degradation of cartilage that occurs in osteoarthritis.

Figure 5.7 shows data acquired from unstained human cartilage from freshly excised metatarsal heads, for which a full (16 spectral channel-resolved) lifetime data set was acquired in 2 s using 10 ps excitation pulses at 355 nm radiation from the frequency-tripled mode-locked Nd:vanadate laser (Vanguard 350-HM355, Spectra-Physics). For these experiments, the temporal response was calibrated using the fluorescence lifetime standard 1,4-bis(5-phenyloxazol-2-yl)benzene (POPOP) (Boens et al. 2007). Specimens of intact metatarsal heads were obtained with consent from patients undergoing lower-limb amputation surgery at Charing Cross Hospital, London. Ethical approval was granted from the Riverside Research Ethics Committee (RREC 2064, 1752). Metatarsal heads were placed immediately in transport medium and analyzed within 48 h. Prior to measurement, the tissue was washed thoroughly with PBS and placed on a petri dish; the probe was then introduced perpendicular to the cartilage surface. The tissue was frequently washed in PBS to maintain moisture.

The autofluorescence spectra acquired are consistent with the fluorescence expected from cartilage that is mainly attributed to collagen cross-links (Richards-Kortum and Sevick-Muraca 1996), for example, (hydroxyl) lysyl-pyrodinoline, vesperlysine, and pentosidine. The data fitted well to a double exponential decay profile with lifetime components of 0.53 ns and 3.61 ns. These values were observed consistently across the emission spectrum, but the relative contributions of these components varied as shown in Figure 5.7b. These preliminary data indicated the potential for future clinical instruments and motivated further study of the changes in the complex collagen fluorescence signal as tissue is degraded by

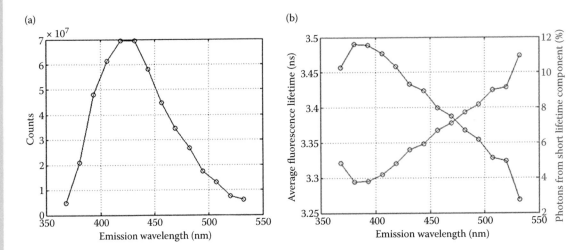

Figure 5.7 (a) Autofluorescence emission spectrum and (b) mean lifetime from a double exponential fit (left axis, black) and relative contribution from short lifetime component (right axis, red) of human cartilage. (Adapted from Manning, H.B. et al., *J. Biophotonics*, 1, 2008.)

disease. In an ongoing investigation (Manning et al. 2013), we have simulated the breakdown of collagen during osteoarthritis by digesting porcine and bovine cartilage samples with bacterial collagenase and have observed that, while the autofluorescence spectrum does not change during the degradation of the cartilage, the mean fluorescence lifetime appears to systematically decrease.

5.4 SUMMARY

We have presented a review of previously published work to show that fiber-optic probe-based instrumentation combining ultrafast excitation with spectrally resolved fluorescence lifetime measurements of autofluorescence can provide contrast between normal and diseased tissues. In particular, we have described two instruments utilizing mode-locked or gain-switched excitation lasers and TCSPC combined with spectrally resolved detection that provide sufficient signal-to-noise ratios for acquisition times of a few seconds to enable the analysis of complex autofluorescence signals fitted to multiexponential decay profiles. While interanimal/interpatient variability remains a significant issue for the clinical application of tissue autofluorescence lifetime-based readouts, we believe that the flexibility, convenience, and rapid data acquisition of single-point multidimensional measurements holds promise for future fundamental investigations of disease mechanisms and pathology—particularly for *in vivo* studies via fiber-optic probes. To this end, we are also investigating the potential for *in vivo* optical biopsy in the gastrointestinal (GI) tract; in the heart; and for ear, nose, and throat disease using tissue autofluorescence lifetime.

We believe that single-point probes utilizing TCSPC or other lifetime measurement approaches have many other potential applications for *in situ* measurements, for example, in bioreactors for tissue engineering, and for a wide range of *ex vivo* studies. For the latter, we have also implemented a motorized sample stage in the MDF shown in Figure 5.3 to enable automated measurements of multiwell plate sample arrays.

REFERENCES

Becker, W. 2005. *Advanced Time-Correlated Single Photon Counting Techniques*. Berlin, Heidelberg, New York: Springer.

Becker, W. 2006. *User Manual—PML-16-C, 16 Channel Detector Head for Time-Correlated Single Photon Counting.* Becker & Hickl GmbH, Berlin, Germany.

Becker, W., Bergmann, A. & Biskup, C. 2007. Multispectral fluorescence lifetime imaging by TCSPC. *Microscopy Research and Technique,* 70, 403–409.

Bird, D. K., Eliceiri, K. W., Fan, C. H. & White, J. G. 2004. Simultaneous two-photon spectral and lifetime fluorescence microscopy. *Applied Optics,* 43, 5173–5182.

Blackwell, J., Katika, K. M., Pilon, L., Dipple, K. M., Levin, S. R. & Nouvong, A. 2008. In vivo time-resolved autofluorescence measurements to test for glycation of human skin. *Journal of Biomedical Optics,* 13, 014004.

Boens, N., Qin, W. W., Basaric, N., Hofkens, J., Ameloot, M., Pouget, J., Lefevre, J. P., Valeur, B., Gratton, E., Vandeven, M., Silva, N. D., Engelborghs, Y., Willaert, K., Sillen, A., Rumbles, G., Phillips, D., Visser, A., van Hoek, A., Lakowicz, J. R., Malak, H., Gryczynski, I., Szabo, A. G., Krajcarski, D. T., Tamai, N. & Miura, A. 2007. Fluorescence lifetime standards for time and frequency domain fluorescence spectroscopy. *Analytical Chemistry,* 79, 2137–2149.

Brancaleon, L., Durkin, A. J., Tu, J. H., Menaker, G., Fallon, J. D. & Kollias, N. 2001. In vivo fluorescence spectroscopy of nonmelanoma skin cancer. *Photochemistry and Photobiology,* 73, 178–183.

Butte, P. V., Fang, Q. Y., Jo, J. A., Yong, W. H., Pikul, B. K., Black, K. L. & Marcu, L. 2010. Intraoperative delineation of primary brain tumors using time-resolved fluorescence spectroscopy. *Journal of Biomedical Optics,* 15(2), 027008.

Chorvat, D. & Chorvatova, A. 2006. Spectrally resolved time-correlated single photon counting: A novel approach for characterization of endogenous fluorescence in isolated cardiac myocytes. *European Biophysics Journal,* 36, 73–83.

De Beule, P. A. A., Dunsby, C., Galletly, N. P., Stamp, G. W., Chu, A. C., Anand, U., Anand, P., Benham, C. D., Naylor, A. & French, P. M. W. 2007. A hyperspectral fluorescence lifetime probe for skin cancer diagnosis. *Review of Scientific Instruments,* 78, 123101.

Glanzmann, T., Ballini, J.-P., Bergh, H. V. D. & Wagnieres, G. 1999. Time-resolved spectrofluorometer for clinical tissue characterization during endoscopy. *Review of Scientific Instruments,* 70, 4067–4077.

Gupta, P. K., Majumder, S. K. & Uppal, A. 1997. Breast cancer diagnosis using N-2 laser excited autofluorescence spectroscopy. *Lasers in Surgery and Medicine,* 21, 417–422.

Lloyd, W. R., Wilson, R. H., Chang C.-W., Gillispie, G. D. & Mycek, M.-A. 2010. Instrumentation to rapidly acquire fluorescence wavelength-time matrices of biological tissues. *Biomedical Optics Express,* 1, 574–586.

Lohmann, W., Nilles, M. & Bodeker, R. H. 1991. In situ differentiation between nevi and malignant melanomas by fluorescence measurements. *Naturwissenschaften,* 78, 456–457.

Manning, H. B., Kennedy, G. T., Owen, D. M., Grant, D. M., Magee, A. I., Neil, M. A. A., Itoh, Y., Dunsby, C. & French, P. M. W. 2008. A compact, multidimensional spectrofluorometer exploiting supercontinuum generation. *Journal of Biophotonics,* 1(6), 494–505.

Manning, H. B., Nickdel, M. B., Yamamoto, K., Lagarto, J. L., Kelly, D. J., Talbot, C. B., Kennedy, G., Dudhia, J., Lever, J., Dunsby, C., French, P. & Itoh, Y. 2013. Detection of cartilage matrix degradation by autofluorescence lifetime. *Matrix Biology,* 32, 32–38.

Marcu, L., Fang, Q., Jo, J. A., Papaioannou, T., Dorafshar, A., Reil, T., Qiao, J. H., Baker, J. D., Freischlag, J. A. & Fishbein, M. C. 2005. In vivo detection of macrophages in a rabbit atherosclerotic model by time-resolved laser-induced fluorescence spectroscopy. *Atherosclerosis,* 181, 295–303.

Marcu, L., Jo, J. A., Butte, P. V., Yong, W. H., Pikul, B. K., Black, K. L. & Thompson, R. C. 2004. Fluorescence lifetime spectroscopy of glioblastoma multiforme. *Photochemistry and Photobiology,* 80, 98–103.

Marcu, L., Jo, J. A., Fang, Q. Y., Papaioannou, T., Reil, T., Qiao, J. H., Baker, J. D., Freischlag, J. A. & Fishbein, M. C. 2009. Detection of rupture-prone atherosclerotic plaques by time-resolved laser-induced fluorescence spectroscopy. *Atherosclerosis,* 204, 156–164.

Mycek, M. A., Schomaker, K. T. & Nishioka, N. S. 1998. Colonic polyp differentiation using time-resolved autofluorescence spectroscopy. *Gastrointestinal Endoscopy,* 48, 390–394.

Palmer, G. M., Marshek, C. L., Vrotsos, K. M. & Ramanujam, N. 2002. Optimal methods for fluorescence and diffuse reflectance measurements of tissue biopsy samples. *Lasers in Surgery and Medicine,* 30, 191–200.

Panjehpour, M., Julius, C. E., Phan, M. N., Vo-Dinh, T. & Overholt, S. 2002. Laser-induced fluorescence spectroscopy for in vivo diagnosis of non-melanoma skin cancers. *Lasers in Surgery and Medicine,* 31, 367–373.

Pitts, J. D. & Mycek, M. A. 2001. Design and development of a rapid acquisition laser-based fluorometer with simultaneous spectral and temporal resolution. *Review of Scientific Instruments,* 72, 3061.

Pradhan, A., Pal, P., Durocher, G., Villeneuve, L., Balassy, A., Babai, F., Gaboury, L. & Blanchard, L. 1995. Steady state and time-resolved fluorescence properties of metastatic and non-metastatic malignant cells from different species. *Journal of Photochemistry and Photobiology B,* 31, 101–112.

Richards-Kortum, R. & Sevick-Muraca, E. 1996. Quantitative optical spectroscopy for tissue diagnosis. *Annual Review of Physical Chemistry,* 47, 555–606.

Skala, M. C., Riching, K. M., Bird, D. K., Gendron-Fitzpatrick, A., Eickhoff, J., Eliceiri, K. W., Keely, P. J. & Ramanujam, N. 2007. In vivo multiphoton fluorescence lifetime imaging of protein-bound and free NADH in normal and pe-cancerous epithelia. *Journal of Biomedical Optics,* 12(2), 024014.

Sterenborg, H., Motamedi, M., Wagner, R. F., Duvic, M., Thomsen, S. & Jacques, S. L. 1994. In-vivo fluorescence spectroscopy and imaging of human skin tumors. *Lasers in Medical Science,* 9, 191–201.

Sun, Y., Liu, R., Elson, D. S., Hollars, C. W., Jo, J. A., Park, J., Sun, Y. & Marcu, L. 2008. Simultaneous time- and wavelength-resolved fluorescence spectroscopy for near real-time tissue diagnosis. *Optics Letters,* 33, 630–632.

Taylor, J. R., Adams, M. C. & Sibbett, W. 1980. Investigation of viscosity dependent fluorescence lifetime using a synchronously operated picosecond streak camera. *Applied Physics,* 21, 13–17.

Principles of fluorescence lifetime instrumentation

Thompson, A. J., Coda, S., Sorensen, M. B., Kennedy, G., Patalay, R., Waitong-Bramming, U., De Beule, P. A. A., Neil, M. A. A., Andersson-Engels, S., Bendsoe, N., French, P. M. W., Svanberg, K. & Dunsby, C. 2012. In vivo measurements of diffuse reflectance and time-resolved autofluorescence emission spectra of basal cell carcinomas. *Journal of Biophotonics,* 5(3), 240–254.

Tramier, M., Kemnitz, K., Durieux, C. & Coppey-Moisan, M. 2004. Picosecond time-resolved microspectrofluorometry in live cells exemplified by complex fluorescence dynamics of popular probes ethidium and cyan fluorescent protein. *Journal of Microscopy,* 213, 110–118.

Wu, Y., Xi, P., Qu, J., Cheung, T.-H. & Yu, M.-Y. 2004. Depth-resolved fluorescence spectroscopy reveals layered structure of tissue. *Optics Express,* 12, 3218–3223.

Optical instrumentation design for fluorescence lifetime spectroscopy and imaging

Peter T. C. So, Heejin Choi, Christopher J. Rowlands, and Vijay R. Singh

Contents

6.1 OVERVIEW

The history of our understanding of molecular phosphorescence and fluorescence has been reviewed in Chapter 2 and several recent publications (Masters and So 2008; Valeur 2002, 2011). The phenomenon of phosphorescence has been noted since the Middle Ages; some minerals were observed to continuously emit "cold" light in the dark with no obvious source of energy. Therefore, the unusually long temporal delay between energy absorption and energy emission of these minerals, the phosphorescence lifetime, is partly responsible for sparking the fascination of researchers in the 19th century. Their interests and studies resulted in fundamental understanding of these phenomena and ushering in numerous applications using fluorescence and phosphorescence spectroscopy and imaging, especially in biology and medicine.

With the advent of quantum mechanics, the phenomena of fluorescence and phosphorescence are now fully understood. A molecule with an electronic energy level difference comparable to that of an optical photon can be excited by absorbing a photon quantum. After excitation, thermal relaxation occurs on the timescale of picoseconds, and a molecule rapidly relaxes to the lowest vibronic level in the excited electronic state. Fluorescence occurs via a dipole coupling between an excited singlet electronic state and the singlet ground state with a time constant on the order of picoseconds and nanoseconds. This time delay is called the fluorescence lifetime. Instead of fluorescence relaxation, the energy of an excited molecule may also be transferred to a triplet electronic state via a spin-orbit coupling process. The relaxation back to the singlet ground state is "forbidden" with a long time constant on the order of microseconds to milliseconds. These processes can be summarized by a Jablonski diagram (Figure 6.1).

This chapter focuses on the technology development required for precise measurement of these lifetimes and the applications of these measurements in a broad range of biological and medical problems. The most important challenge of these measurements lies in the development of measurement methodologies that can precisely quantify these time constants, especially for very short fluorescence lifetimes. Methodologies such as time-correlated single photon counting and frequency domain measurements are covered extensively in other chapters. In this chapter, we focus on common optical instrument considerations underlying all these different methodologies. We will cover the selection of optical and optoelectronic components, and special consideration will be given to optical path design for accurate lifetime measurement, such as the inclusion of a reference path and the concept of the magic angle. We will subsequently review the basic optical designs of different instruments where lifetime measurements are commonly performed in the laboratory and in the clinic. We divide these instruments into two main classes: nonimaging-based devices and imaging-based devices. Within the nonimaging-based devices, we cover the designs of different lifetime-resolved fluorometers used for laboratory bioanalytical applications. Current research focuses on the more challenging problem of developing imaging-based lifetime-resolved instruments. The class of imaging-based lifetime-resolved instruments can be categorized by their resolution: different optical microscopic methods for the study of cellular and subcellular systems (nanometer to micrometer scale), sheet light-based imaging methods for studying the developing embryo (micrometer to millimeter scale), and diffusive optical imaging method for studies on the clinical scale (millimeter to centimeter scale).

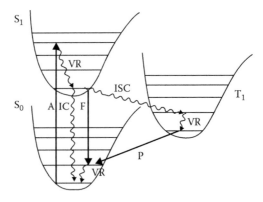

Figure 6.1 Jablonski diagram depicting electronic and vibronic transitions responsible for fluorescence and phosphorescence molecular emission. A, absorption; F, fluorescence; IC, internal conversion; ISC, intersystem crossing; P, phosphorescence; S_1, singlet excited state; S_0, singlet ground state; T_1, triplet excited state; VR, vibrational relaxation.

6.2 OVERVIEW OF OPTICAL INSTRUMENT DESIGN AND COMPONENT SELECTION FOR FLUORESCENCE LIFETIME MEASUREMENTS

While an optical instrument for lifetime-resolved measurement is not very different from any other biomedical optical device, the successful development of an optical instrument for fluorescence lifetime measurements requires keeping several additional design principles in mind. First, an efficient optical system that wastes no photons is important in general, but this is particularly important for fluorescence lifetime imaging. For example, tissue morphology may be quantified with sufficient signal-to-noise ratio (SNR) from a fluorescence image with a mean pixel intensity of tens of photons. However, it should be noted that several hundreds of photons are required to obtain the mean lifetime of a pixel. If there are two chemical species with distinct lifetimes within this pixel, the resolution of these two species on the basis of fluorescence lifetime will require at least several thousand photons (Gratton et al. 2003). Clearly, the optimization of excitation and detection efficiency in the optical system is much more important for successful lifetime measurement. Second, fluorescence lifetime measurement is model dependent. Often, a researcher does not know, *a priori*, the number of fluorescence species within a biological specimen and the magnitude of the lifetime differences of these components. Given this unknown, the lifetime resolution of an instrument required to study a particular biomedical problem may also not be known *a priori*. Since the design of an instrument with 10 ps lifetime resolution is more expensive and more challenging than constructing an instrument with 500 ps lifetime resolution, this concern is significant. Third, since the required fluorescence lifetime measurement resolution can be on the order of a few picoseconds, if the specimen is an extended object or if the detector has a large active surface area, we must ensure that the optical path differences between the different locations in the specimen and on the detector surface are minimized. Otherwise, lifetime measurement uncertainty may result from these optical path differences. Fourth, the accuracy of lifetime measurement can be highly sensitive to optical noise sources, including stray room light, residual excitation light, and fluorescent contaminants in the specimen. It should be noted that the influence of these noise sources on the final measurement may be somewhat complex, being dependent on both their intensities and lifetimes.

Taking into account these general considerations, the successful design of an optical instrument for fluorescence lifetime measurement requires careful selection of system components that may be grouped and categorized as light source, excitation and emission intermediate optics, detector, and specimen.

6.2.1 LIGHT SOURCES FOR FLUORESCENCE LIFETIME MEASUREMENT

Light sources for biophotonic measurements can, in general, be divided into two classes: non-laser light sources and laser light sources. Non-laser light sources were the most typical choices until a decade ago due to their low cost and ease of use. However, with the advent of a variety of lower cost and robust diode lasers, the use of these lasers as excitation light sources for lifetime-resolved measurements is becoming more common. Regardless of specific light source choice, it is important to note that every modality of fluorescence lifetime measurement requires measuring how the fluorescence signal tracks excitation intensity changes. Since fluorescence lifetime is on the order of picoseconds to nanoseconds, the excitation light source should have a temporal bandwidth up to at least tens to hundreds of megahertz. A general overview of common sources in these two categories will be described.

The two most important non-laser light sources are probably gas lamps and light-emitting diodes (LEDs). Lamp-based light sources, such as xenon or mercury arc lamps, are based on electrical discharge across a gaseous medium where gas molecules are excited, resulting in light generation based both on black body thermal emission and molecular electronic level transitions. Lamp systems typically generate constant intensity emission. In order to use these sources for lifetime measurement, additional optoelectronic components must be installed in the excitation light path to generate the necessary temporal intensity modulation. A lamp system may also be configured in a pulsed configuration based on rapid electrical discharge. While pulse lamps were standards for fluorescence lifetime measurement many decades ago, their usage is rare today since the nanosecond-scale pulse width achievable with lamps significantly limits fluorescence lifetime measurement resolution, providing significantly worse performance compared with the relatively low-cost picosecond laser sources available today. However, lamp-based sources do have the advantage of providing a very broad spectral range that can span easily from near-ultraviolet to the near-infrared spectral range, allowing the excitation of almost any fluorophore of interest. In many lamp-based sources, spectral brilliance (the amount of power within a given spectral band) is low compared with laser sources, and so a lamp-based source may not be suitable for applications where rapid data acquisition is required, such as in microscopic imaging.

Today, the lowest cost option for fluorescence lifetime measurement is the use of LEDs. With the recent improvements made in LED technology, their intensity is sufficient for many fluorescence spectroscopic measurement applications. Importantly, the LEDs' stronger emission in the blue and even ultraviolet range is becoming more available, allowing the excitation of a broader range of common fluorescent probes. It is also important to note that LEDs can be intensity modulated with simple electronic circuits allowing fluorescence lifetime measurement, but the maximum bandwidth is limited to tens of megahertz to about 100 MHz. LED sources have the advantages of being very low cost and easily modulated but are also limited by low spectral brilliance and relatively narrow excitation spectral range (about 10–20 nm) as compared with lamp sources.

Laser light sources are probably the most common sources used today for lifetime-resolved fluorescence spectroscopy. There are many laser sources available dependent on both performance requirements and budget. In general, lasers can be classified as continuous wave (constant intensity, CW) or pulsed lasers. CW lasers are not often used for fluorescence lifetime measurement today since pulsed lasers are now readily available at comparable cost. Nonetheless, CW lasers can be used by adding intensity-modulating optoelectronic components into the excitation light path and provide spectral brilliance greatly exceeding that achievable with lamp or LED sources, allowing fast data acquisition. There are also a large variety of pulsed laser sources based on different pulse generation mechanisms: Q-switching, gain switching, and mode locking. These pulsed laser sources may also be classified based on their use of different lasing media. The cost of laser sources also ranges widely, dependent on the required laser peak power, average power, and pulse width. In general, there are really only two main choices: lower cost, lower power, picosecond pulsed diode lasers, or higher cost, higher power, femtosecond solid-state lasers. With the recent commercialization of diode laser technology in the consumer market, more advanced picosecond pulsed diode laser technologies are also becoming more available. Today, many of these picosecond pulsed diode lasers are based on gain switch technology. For gain switching, the pump energy is modulated, resulting in a large number of excited electrons being injected into the active region of a diode laser at short intervals. This increase in carrier density results in photon generation above the lasing threshold. The lasing of the device continues until the store of excited electrons is depleted and the carrier density falls below lasing threshold. By appropriate engineering of the diode structure and pump modulation, diode lasers with pulse widths down to tens of picoseconds can be fabricated.

If lifetime measurement with resolution on the order of picoseconds is required, femtosecond laser sources may be used. Today, the most common femtosecond laser sources are based on mode-locked solid-state devices such as titanium–sapphire lasers. The basic idea of mode locking relies on the fact that the fluorescence of the laser medium such as the titanium–sapphire crystal contains a broad spectral content. These different spectral components have longitudinal modes that normally have no fixed phase relationship resulting in CW operation. However, if all these longitudinal modes can be synchronized with a fixed phase relationship, the different modes will add constructively at a specific time while destructively at other times. The resultant pulse width can be as short as tens of femtoseconds, dependent on the spectral content of the laser and the dispersion compensation mechanism that keeps the different spectral components within a short laser pulse in phase with each other. Since light traverses the laser cavity periodically, the cavity length of the laser determines the pulse repetition rate; this rate is typically on the order of 100 MHz to several gigahertzes. There are also many mechanisms to initiate mode locking in pulsed lasers, including active processes such as the use of an acousto-optical modulator (AOM), which acts as a weak periodic gate. The AOM attenuates laser modes that are not synchronized with this gate. Since the laser is an amplifier, it will settle into a state in which all the longitudinal modes are in synchrony with the gate and the laser will become phase locked.

There are also passive mechanisms, such as Kerr mode locking and the use of saturable absorbers, that are regularly used to achieve mode locking. For example, in Kerr mode locking, the higher peak intensity mode-locked mode exhibits self-focusing due to the Kerr effect and consequently has a narrower spatial transverse mode than the lower peak intensity CW mode. By putting a slit into the cavity, it is possible to favor the spatially narrower mode-locked modes resulting in femtosecond pulse generation.

It should be noted that measurement with a time resolution on the order of tens of picoseconds is typically suitable for the majority of biomedical applications, and a picosecond diode laser is therefore often sufficient. However, femtosecond lasers are now routinely used in nonlinear microscopy (to be discussed in Section 6.4.1.3) and provide a convenient light source for fluorescence lifetime measurement. While femtosecond lasers have the advantage of shorter pulse widths and a fairly good range of excitation wavelength tunability, picosecond diode lasers have the advantage of being lower cost and their repetition rates are often adjustable.

6.2.2 EXCITATION AND EMISSION OPTICAL PATH COMPONENTS

An important consideration of optical design for both the excitation and emission optical paths is to maximize the throughput of light. For the excitation light path, the maximization of light throughput ensures that the specimen can be efficiently excited without having to use higher power laser or lamp sources, which in turn reduces system cost. Equally importantly, the maximization of light transmission through the intended optical path ensures that stray light within the instrument is minimized, reducing potential sources for lifetime measurement error.

Since most fluorophores are excited in the UV, blue, or green excitation spectral regions, impurities in the optical components may produce unwanted fluorescence. An important consideration in the excitation light path of a fluorescence lifetime measurement system is to ensure that fluorescence generation from the optical component of the excitation light path is minimized. This can be partly accomplished by using high-quality glass optical components. For experiments in the deeper UV region, high-quality fused silica components may be needed to minimize unwanted fluorescence generation. Since fluorescence generation along the excitation path is sometimes inevitable, for high sensitivity measurements, it is often useful to insert a band-pass interference filter (the "exciter") right before the specimen to ensure that the spectrum of the excitation light incident upon the specimen is pure and predictable. This issue is particularly severe in fiber optic-based instruments such as in an endoscope, where the path length of the excitation light through the fiber is long.

An intensity modulation element is often needed in the excitation light path of a fluorescence lifetime measurement system. When lamps, LEDs, or CW laser light sources are used for fluorescence excitation, it is clear that an additional intensity modulation component is required to provide the required temporal bandwidth. For fast intensity modulation, up to hundreds of megahertz, there are several classes of devices that may be used: AOMs, electro-optical modulators, and Pockels cells. AOMs control the intensity of the light beam by deflecting the laser beam using a transiently formed Bragg grating inside a solid state crystal, such as quartz. The transiently formed Bragg grating is produced by driving the crystal acoustically, resulting in local material density variation and a corresponding variation in the index of refraction. This index variation forms the required grating. Electro-optical modulators can produce intensity variation by controlling the propagation speed of light through a nonlinear crystal by electrically controlling the crystal refractive index. By incorporating the nonlinear crystal in one arm of an interferometer, the output signal intensity can be modulated by controlling the phase relationship, and consequently the degree of constructive interference, of the light going through the two arms of the interferometer. Finally, another way to modulate the excitation light intensity is based on the Pockels effect, where the polarization of light going through a nonlinear optical (NLO) crystal can be varied based on the voltage across the crystal. By placing a polarizer after the Pockels cell, the intensity of transmitted light can be controlled depending on the degree of alignment between the polarization of light after the Pockels cell and the orientation of the polarizer. It should be noted that additional intensity modulation devices are also sometimes used to reduce the repetition rate of femtosecond lasers that have 100 MHz or gigahertz repetition rate, when studying molecules with very long fluorescence or phosphorescence lifetimes. Slower intensity modulation is needed to ensure that fluorophores with long lifetimes are not re-excited before a majority of the excited molecules have relaxed back to the ground state after the initial excitation.

The efficiency of the excitation light path is important, but one may argue that the efficiency of the emission light path is even more critical. In the excitation light path, inefficiency can be partly compensated by using high-cost and higher power light source; the emission light path inefficiency cannot be readily compensated without increasing photodamage to the biological specimen. Since the emission spectra of most typical fluorophores are often in the visible range, using optical components with antireflection coatings along the optical path is important. Since light loss at each surface of an uncoated

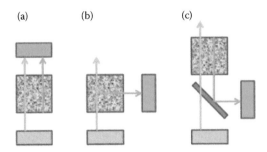

(a)   (b)   (c)

Figure 6.2 Common intermediate optics geometries: (a) trans-geometry, (b) T-geometry, and (c) epi-geometry.

optical component is about 5% as can be estimated from the Fresnel equations, significant loss can result in a typical system if the number of optical components in the emission path is not minimized and if uncoated components are used.

Besides maximizing the transmission of emission light through the optical path, there are two more important factors that should be optimized. First, since fluorescence cross sections are low, the excitation light may be over ten orders of magnitude higher in intensity than the fluorescence signal, depending on the quantity of fluorophores in the specimen. While barrier interference filters are very efficient, with typical optical densities of over six for excitation light (blocking excitation light relative to emission light by a factor of 10^6), more than one emission filter may be necessary to fully attenuate the excitation light. However, instead of relying entirely on the emission barrier filter, when it is possible, it is preferable to organize the geometry of the emission beam path that the excitation light does not directly enter the emission beam path. Since fluorescence emission is isotropic, there is significant flexibility in the geometric arrangement of the emission light path (Figure 6.2).

The worst choice is arranging the excitation and the emission light path in a "trans" geometry where the excitation light passes through the specimen and is incident directly onto the emission light path. Instead, many fluorescence detection systems are arranged in a "T" geometry, where the excitation light is transmitted along one direction while detection is arranged along an orthogonal direction. Another effective option is to organize the fluorescent detection in an "epi" geometry. A dichroic mirror is inserted in the excitation path that serves to separate the excitation light and the emission light based on their spectra. For example, the dichroic filter may transmit the light from the excitation path toward the specimen while the backward-going emission light is reflected by the dichroic mirror and enters into the emission light path. In general, these geometries will improve the signal-to-noise level of the detection system by two to four orders of magnitude prior to the emission barrier filter.

Besides minimizing excitation light contamination relative to fluorescence signal, another important function of the emission optical path components is the optimization of fluorescence signal collection. There is no simple optical arrangement that will always maximize signal collection as the instrument design must be application dependent. However, the basic principle is to maximize the solid angle subtended by the detector throughout the excitation region of the specimen. In general, it is easier to maximize the solid angle of collection if the excitation region is more confined, as in a point scanning fluorescence microscope. It is often more difficult to maximize the solid angle of collection if the excitation region is more extended, as in typical fluorometers that use a 1 cm cuvette. In the best case, if only a single emission light path is installed along any given direction, approximately 30% of the emitted fluorescence may be collected. High collection efficiency may be achieved by installing additional emission light collection paths along orthogonal directions limited by specimen accessibility and geometry.

6.2.3 DETECTORS FOR FLUORESCENCE LIFETIME MEASUREMENT

Photomultiplier tubes (PMTs), hybrid-photomultiplier (HPD) tubes, and intensified charge-coupled device (CCD)/complementary metal oxide semiconductor (CMOS) detectors are the most common devices used for lifetime-resolved imaging. A PMT is a photoelectric effect device with substantial internal gain (Hamamatsu 2006; Kaufman 2005). The cathode of the device is typically held at high negative voltage relative to the anode, which is set at ground. A series of electrodes, known as dynodes, are placed between the cathode and the anode at successively lower voltages set by a voltage divider chain (Figure 6.3).

As a result of the photoelectric effect, photoelectrons are generated at the cathode; these photoelectrons are subsequently accelerated from the cathode to the first dynode due to the voltage differential. The electrons gain kinetic energy during the transit, and the bombardment of the dynode by these high-energy electrons results in the production of multiple secondary electrons from each incident electron. The amplified electron burst is then accelerated again toward the next dynode. As a result of this multiplicative process, a photoelectron is amplified exponentially as a function of the number of the dynode stages. The amplification efficiency is a function of dynode material, geometry, and voltage differential. Typical PMTs have eight to ten dynode stages providing an electron gain factor of 10^6 to 10^8. The electron burst is subsequently incident upon the anode and is directed to the preamplifier electronics of the readout circuit, where it is converted to a current or voltage pulse. Due to the large gain possible, single photon detection can be readily accomplished with a PMT. The choice of photocathode materials is critical in PMT selection. Photocathodes composed of traditional bialkali materials have quantum efficiency (QE) of 20%–25% in the 300–500 nm wavelength region. Multialkali cathodes extend the usable range to about 700 nm but have QE of only a few percent at red wavelengths. An important recent development is the introduction of GaAsP cathode materials that have extended maximum QE to 40% in the blue-green and maintain sensitivity above 10% up to almost 700 nm. Another new class of PMTs features GaAs cathodes that have over 10% QE from 500 nm up to almost 850 nm. Other relevant parameters of the PMT include dark noise, pulse height distribution, electron transition time, and its spread. The PMT dark noise is a result of the thermally driven emission of electrons at the cathode and the dynodes. Due to the amplification nature of the PMT, electrons emitted at the cathode and the initial stages of the dynode chain will contribute to higher dark current at the anode. The dark current of a PMT is dependent on its cathode material and the design of its dynode chain. The dark current of typical PMTs at room temperature can vary from tens to thousands of electrons per second. As in any thermally driven process, reducing its absolute temperature can exponentially reduce the dark current rate of PMT. Since electron generation at each dynode is a stochastic process, the current burst produced from each photon at the anode can vary, resulting in multiplicative noise. This variability in electron pulse magnitude due to a single photon is quantified by the pulse height distribution of the PMT. Finally, the time required for the detection of a current burst after photon–electron conversion at the cathode is dependent on the geometry of the dynode chain design. The typical electron transit time is on the order of nanoseconds. Again, due to the stochastic nature of electron generation at each dynode, there is a substantial variation in the arrival time of the electrons generated from a single photon. This is characterized by the transit time spread, which is on the order of hundreds

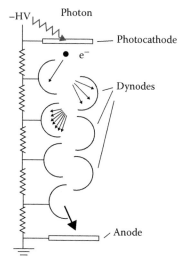

Figure 6.3 Typical structure of PMTs and the associated voltage divider for the dynode chain (HV indicates a negative high-voltage source).

of picoseconds to nanoseconds. The transit time spread sets a lower limit on the lifetime measurement resolution using PMTs.

HPD tubes (Figure 6.4a) (Cushman and Heering 2002; Cushman et al. 1997; Hayashida et al. 2005) are becoming more widely used when high-precision lifetime measurement is required, since these devices feature excellent pulse height distribution and tight electron transition time (Hayashida et al. 2005). HPDs merge PMT and avalanche photodiode (APD) technology (Figure 6.4b).

An APD is a type of photodiode. A photodiode is a photovoltaic solid-state device composed of p-type and n-type semiconductor layers in contact with each other. At the junction of the p-type and n-type semiconductors, electrons from the n-type region diffuse into the p-type region, and holes from the p-type region diffuse into the n-type region. This diffusion is stopped as an electrical potential develops due to carrier separation. At steady state, a central region, the depletion zone, is formed where there are no free charge carriers. A positive potential is present on the n-type side of the depletion zone, while a negative potential is present on the p-type side. When a photon is absorbed in the depletion zone, resulting in the promotion of an electron to the conduction band, a pair of free charge carriers is produced. The potential across the depletion zone causes the electron to drift toward the n-type region while the hole toward the p-type. If an electrical circuit is connected between the p- and n-type terminals, a current will flow resulting in the recombination of the electron and holes. The current produced in the circuit is proportional to the intensity of light impinged upon the device.

An APD modifies the standard diode design by adding internal gain (Cova et al. 1996; Dautet 1993). With high reverse bias voltage across its depletion region, the drift velocity of the carriers increases with increasing bias voltage. At a field strength of about 10^4 V/cm, drift velocity plateaus due to charge carrier collision with the crystal lattice. With further increase in field strength beyond 10^5 V/cm, secondary carriers are created upon collision. This process is called impact ionization. These secondary carriers further gain kinetic energy and again create more carriers in an avalanche fashion, providing internal gain for APDs.

For the HPD, photoelectrons are produced at the cathode in a manner similar to any PMT. High voltage accelerates these photoelectrons toward an APD. As the electron is deposited into the silicon substrate of the APD, the electron bombardment effect causes the generation of many secondary electrons

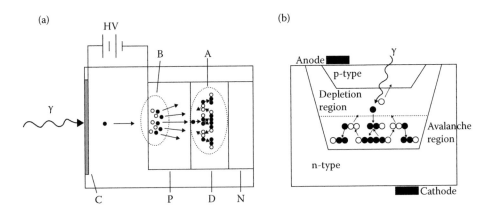

Figure 6.4 (a) Typical structure of a hybrid-PMT detector. The incident photon (γ) is converted into an electron (solid circle) at the photocathode (C). The electron is accelerated by a high voltage (HV) toward essentially an APD consisting of a pair of p-type (P) and n-type (N) semiconductors. Due to the high voltage, instead of a single electron (solid circle)–hole (empty circle) pair being generated in the depletion region (D), many electron–hole pairs are generated by the high kinetic energy electron in a process called bombardment gain (B). These electron–hole pairs undergo further amplification by the standard avalanche gain processes (A). The electron burst is then read out after this two-step amplification process. (b) A more detailed explanation of an APD. A depletion region is formed between a p-type and an n-type semiconductor. An incident photon (γ) creates an electron (solid circle)–hole (empty circle) at the depletion region. The presence of a high voltage creates an avalanche region where the electrons and holes created are accelerated and their collision creates more electron–hole pairs. The migration of these amplified electrons and holes can be detected as a current by completing a circuit via the cathode and the anode terminals.

with a gain on the order of 1000. These electrons are subsequently amplified in the APD by impact ionization in the depletion zone, producing another gain of 50 to 100, resulting in a total amplification factor on the order of 10^4 to 10^5.

PMTs and HPDs are very effective detectors for single incidence lifetime measurement. However, there are situations where it is advantageous to perform multiple lifetime measurements in parallel, such as mapping the lifetime distribution of fluorophores in an image or simultaneously measuring lifetimes across an emission spectrum. In this case, an intensified CCD (Inoué and Spring 1997) or an intensified CMOS (Janesick 2002) camera may be used. This type of device consists of a standard CCD/CMOS camera with a microchannel plate (MCP) coupled to its front end. MCPs are a highly conductive glass substrate with a dense array of micron-size holes coated with a secondary electron emitter. The cathode and anode surfaces of the MCP are held at high voltage similar to a PMT. The front end of the MCP has a photocathode composed of material similar to that in a PMT. Electrons are produced at the cathode due to the photoelectric effect. The high voltage accelerates the photoelectron into a nearby channel in the array. An electron avalanche is produced into the channel similarly to that in a PMT, resulting in an amplified electron burst exiting the glass substrate. The anode of the MCP is typically a phosphor screen that emits light in response to the incident electron burst. Since the electrons are spatially confined by the hole, the spatial location of the incident photon is preserved. MCPs have QE similar to a PMT but provide spatial resolution. Furthermore, since MCPs have a more compact design than the bulkier PMT, the transition time spread of the electron burst can be very short, allowing tens of picoseconds of time resolution using these devices.

The optical output of the MCP can be transferred to the CCD/CMOS by either direct or lens coupling. The readout noise is no longer a limiting factor in the sensitivity of an intensified CCD/CMOS due to the gain of the MCP. Instead, the dark current of the MCP influences the SNR, and high sensitivity devices require cooling of both the MCP and the CCD/CMOS camera. Furthermore, while the MCP glass channels are in a closely packed array, some spatial resolution degradation is often observed in intensified CCDs/CMOSs as compared with systems without an MCP.

Streak cameras are another class of photoemissive device that are extremely fast and can resolve photon arrival times differing by fractions of a picosecond. They are often used in high temporal resolution lifetime-resolved microscopes (Hamamatsu 2002; Krishnan et al. 2003) (Figure 6.5). After electron emission from the photocathode, the electrons are accelerated toward an MCP. In between the photocathode and the MCP is a pair of electrodes positioned transversely. The voltage across this pair of electrodes can be ramped up rapidly, called a swipe. The transverse displacement of the electron on

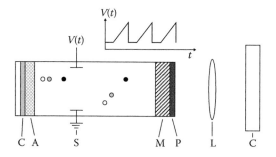

Figure 6.5 Principle of a streak camera. Photons arriving at the photocathode (C) are converted into electrons. The electrons are accelerated through a high-voltage region (A). Three electrons are depicted that are generated by photons arriving at successive times. The black circle denotes the electron generated by the earliest photon, the gray circle denotes the electron generated by the next photon, and the white circle denotes the electron generated by the last photon. These electrons go through a region in the camera where there is a scanning transverse voltage (S), which is a repetitive ramp in time. As a result of the linearly increasing transverse voltage, the electrons are deflected laterally depending on their arrival time. This process encodes the electron's timing information along the horizontal spatial dimension. The electrons then enter a microchannel plate (M) where they are amplified with spatial resolution. The amplified electron bursts are then incident upon a phosphor screen (P) and are converted back to light. A CCD camera (C) images the phosphor screen via a lens L.

the phosphor screen is then a function of the ramp voltage. Electrons that arrive at the start of the swipe are minimally deflected laterally, while electrons that arrive later have greater deflection. Therefore, the arrival time of the electron at the photocathode is encoded by the electron lateral position at the MCP. The electrons are amplified and converted back to photons by the output phosphor screen of the MCP. The output of the MCP can be imaged by a high-sensitivity CCD detector providing both intensity and timing information of the photons. Since the MCP and CCD are 2D devices, multiple optical channels can be positioned along the y-axis, allowing simultaneous measurement. An example is to spectrally resolve the emission light along the y-axis allowing simultaneous spectral and lifetime measurements. While the streak camera cathode has similar sensitivity to PMTs, the complex multiple stage design introduces noise for intensity measurement and degrades spatial resolution. The duty cycle of a streak camera suffers from long setup time between swipes that are typically over tens of nanoseconds.

6.2.4 SPECIMEN HANDLING CONSIDERATIONS

Since fluorescence lifetime-resolved measurements have found applications ranging from studying protein folding in solution to cancer diagnosis *in vivo*, the designs of the specimen stages in these instruments vary greatly. However, there are two common issues that should be noted regarding specimen handling. First, all the different lifetime measurement methods rely on quantifying the delay between the excitation light and the fluorescence emission. Since the inherent optical path delay and electronic propagation delay in the instrument cannot be precisely controlled, all lifetime-resolved instruments rely on performing a calibration measurement using a lifetime reference standard to "zero" out these delays. The reference standard may be the fluorescence generated from a dye with a known lifetime, or a scattering solution with a zero lifetime, or induced second harmonic generation (SHG) with a zero lifetime. In any case, depending on the timing stability of the instrument, such as the phase and pointing stability of the laser and the phase drift of the electronic circuitry, measurements of the reference sample are often intermixed with measurements of the specimen to ensure that the effect of any instrumental timing drift is minimized. Therefore, many lifetime measuring instruments incorporate an efficient method to quickly exchange the sample with the reference specimen. Second, it should be noted that the lifetime of many fluorophores is highly environment dependent. Factors such as temperature, pH, and other chemical environmental changes will change the measured lifetime. Therefore, an environmental control stage is often necessary.

6.3 LIFETIME-RESOLVED FLUORESCENCE SPECTROSCOPY WITH NONIMAGING DEVICES

The measurement of fluorescence lifetime is challenging as it is necessary to quantify timing differences of optical signals on the picosecond timescale. This precise quantification requires measurement with a sufficient SNR, a property that often improves with the number of photons acquired. Since the fluorescence photon generation rates of biological specimens are often limited, fluorescence lifetime measurement is, in general, easier with longer data acquisition time. While this book focuses more on lifetime-resolved measurement in images where the need for a high frame rate limits the data acquisition time per frame, we first discuss nonimaging lifetime-resolved imaging instruments that are often less technically challenging but remain one of the key techniques for studying protein structures and dynamics in solution.

6.3.1 LIFETIME-RESOLVED SPECTROPHOTOMETER

A lifetime-resolved spectrophotometer simultaneously quantifies the fluorescence emission spectrum and associates lifetime measurements at each spectral range of interest. The quantification of the emission spectrum can be based on band-pass devices and spectrally dispersive devices. Band-pass devices include band-pass filter wheels or electrically tunable filters. Mechanical filter wheels are the most common band-pass devices. A complete spectral data set is recorded by mechanically placing filters of different pass bands into the imaging light path. The spectral resolution and transmission efficiency is determined by the filter type used. Typical efficiencies of holographic band-pass filters are approximately 70%–80%. Since the filters are exchanged mechanically, data acquisition is slow and sequential in terms of wavelength scanning. Electrically tunable filters such as liquid crystal tunable filters (LCTFs) or acoustic-optical tunable filters

(AOTF) are alternative band-pass devices that provide fast (microseconds to milliseconds) and random switching through a sequence of wavelengths (Morris et al. 1994, 1996; Wachman et al. 1996, 1997). A typical LCTF employs a stack of polarizers and birefringent liquid crystal plates that act as voltage-tunable retarders. The transmission efficiency of LCTFs is approximately 40% across the VIS and near-IR spectral range but can drop to a few percent in the near-UV. AOTFs are solid-state electro-optical devices and apply radio frequency (RF) acoustic waves in dielectric materials to diffract one (or a few) specific wavelength of light. The performance characteristics of AOTFs are comparable to those of LCTFs, that is, AOTFs have transmission efficiencies of approximately 40% (VIS to near IR), pass bands of a few nanometers, and submillisecond switching speed.

A simple spectrophotometer may be based on these band-pass devices and coupled with the different lifetime measurement techniques described in the different chapters of this book. However, most modern spectrophotometers are based on spectrally dispersive devices. The main reason is that band-pass devices are wasteful of fluorescence photons, since out-of-spectral band photons are rejected. In contrast, spectrally dispersive devices separate the different spectral component spatially. In the configuration of a monochromator, photons from outside the spectral band of interest can be rejected. In the configuration of a polychromator, photons from many spectral bands of interest can be acquired simultaneously without loss using a parallelized lifetime-resolved detector, such as a lifetime-resolved image intensified camera.

Spectrally dispersive devices spread out the spectral components of the incident light spatially. Typical spectrally dispersive devices are prisms (refraction based) and gratings (diffraction based). The efficiency of these optical elements is quantified by their angular dispersion, characterizing the change in angular deviation of different colors of light as a function of wavelength. For a refractive prism, it is often positioned such that it is angled relative to incident light at Brewster's angle, in order to minimize incident light reflection loss. Angular dispersion is defined as the change of angle of deviation ε through the prism as a function of wavelength, λ (Born et al. 1999):

$$\frac{d\varepsilon}{d\lambda} \propto \frac{dn(\lambda)}{d(\lambda)} \tag{6.1}$$

where $n(\lambda)$ is the wavelength-dependent index of refraction of the prism. It should be noted that the angular dispersion of a prism depends on its material properties. Prisms fabricated from strongly dispersive materials are more efficient. Further, since the index of refraction is not a linear function of wavelength, angular dispersion is not constant over a broad wavelength range.

A grating is a diffractive element with an array of equally spaced, identical diffracting elements, where the distance between the elements (the groove spacing or pitch) is comparable to the wavelength of incident light. Angular dispersion of diffractive gratings is dependent on grating parameters (Born et al. 1999):

$$\frac{d\beta}{d\lambda} = Gm\sec(\beta) \tag{6.2}$$

where β is the angle of diffraction, G is the groove frequency of the grating, and m is the diffraction order (zero diffraction order corresponds to the undiffracted light and the first diffraction is often used). Unlike the prism, angular dispersion is constant over a broad wavelength range since it is independent of wavelength. Most importantly, gratings are often more efficient than prisms since the angular dispersion of most materials used for prism fabrication is limited. One drawback of using gratings is their higher light loss, since first-order diffraction even with proper grating design (blazing) results in only ~70%–80% diffraction efficiency. Nonetheless, most lifetime-resolved spectrophotometers are grating-based.

Two typical designs of spectrophotometers are shown in Figure 6.6. In both systems, intensity-modulated light sources are used. Broadband lamp sources coupled with intensity modulators provide a low-cost option and allow flexible selection of excitation wavelength when coupled with an excitation monochromator. More robust and higher intensity excitation may be provided by using several intensity-modulated diode lasers. The most flexible, highest intensity, but expensive, option is to use a femtosecond

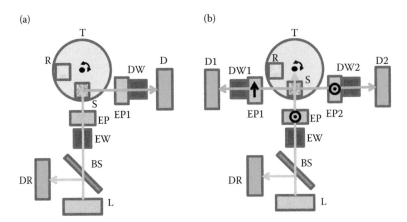

Figure 6.6 Typical optical designs of (a) a lifetime-resolved fluorometer and (b) a lifetime- and polarization-resolved fluorometer. BS, beam splitter; D, D1, D2, fluorescence detectors; DR, reference detector; DW1, DW2, detection wavelength selectors (monochromator or interference filter); EP, excitation polarizer; EP1, EP2, emission polarizers; EW, excitation wavelength; L, light source; R, lifetime reference sample; S, sample; T, turret. The reference detector is used to normalize light source intensity fluctuation. Bulls-eye patterns in the polarizers indicate polarization perpendicular to the plane of the drawing. Arrow patterns in the polarizers indicate polarization parallel to the plane of the drawing.

pulsed broadband light source (produced by white light continuum generation) coupled with an excitation monochromator. The excitation light typically excites a solution specimen held in a low fluorescent background cuvette made from a material such as quartz. Typically, excitation and emission light paths are often organized in a T-geometry to ensure low excitation light leakage into the detection channel. The cuvette is often placed in a rotary turret such that a reference compound may be placed beside the sample, allowing calibration to be readily and frequently performed. Since most lifetime-resolved spectrophotometers require seconds to minutes for data acquisition, the need for efficient fluorescent signal detection is less stringent than in some other cases. Nonetheless, a high-performance spectrophotometer should have photon collection optics with as high a numerical aperture (NA) as possible without compromising the ease of sample access. One form of the spectrophotometer uses PMTs as detectors. In this case, lifetime measurement is performed sequentially across the emission wavelengths by using a motorized monochromator in the emission path that varies its grating angle to selectively to diffract a different portion of the emission spectrum toward the PMT. Alternatively, spectral acquisition may be performed in parallel using a polychromator coupled in a lifetime-resolved imager, such as a modulated intensified camera.

For lifetime measurement, it is important to be aware that many of these detectors can suffer from artifacts. For example, many PMTs have a substantial color effect. The time response speed of a PMT may be dependent on the wavelength of the incident fluorescence signal. Photoelectrons ejected by incident photons may have different average kinetic energies depending on the color, that is, energy, of the fluorescence photons. This difference in kinetic energies can result in different electron transition time as a function of fluorescence spectrum. A color effect resulting in a nanosecond time shift across the visible spectrum is not unusual for some detectors. The color effect can be minimized by better choice of detector. It may also be minimized by calibrating against a fluorescent lifetime standard with known lifetime and similar spectrum as the sample; in this case, the common color effect is "subtracted out" in the calibration procedure. Another important artifact comes from the geometry of light incident upon the sample and the detector. As discussed earlier, if an extended region of the specimen is illuminated and if an extended area of the detector surface is used, the resolution of the measured lifetime may be limited by the temporal differences between these different optical paths. Therefore, for the best lifetime resolution, it is often better to ensure that only a small region of the cuvette is optically coupled to the detector and the light incident upon the detector is focused, limiting the variability of the optical path lengths. However, the improvement in lifetime resolution obtained by limiting the usable excitation volume in the specimen is traded off with reduced optical signal.

6.3.2 POLARIZATION AND LIFETIME-RESOLVED SPECTROPHOTOMETER

Lifetime-resolved polarization measurements are very useful in quantifying how biomolecules interact with their environment, including the quantification of protein residue flexibility, the diffusivity of protein in membrane bilayers, the binding of ligands to receptors, and the fluorescence energy transfer between two molecules of the same type. In the absence of motion, after a subset of randomly oriented molecules is excited by polarized light, their emission will remain mostly polarized. In the presence of motion, the degree of polarization of the emitted light will decrease with a time constant characterizing the motion of interest. Specifically, the time-resolved polarization $p(t)$ and anisotropy $r(t)$ is defined as

$$p(t) = \frac{I_\parallel(t) - I_\perp(t)}{I_\parallel(t) + I_\perp(t)} \quad r(t) = \frac{I_\parallel(t) - I_\perp(t)}{I_\parallel(t) + 2I_\perp(t)} \tag{6.3}$$

where $I_\parallel(t)$ is the time-dependent fluorescence intensity emitted at a polarization parallel to the excitation polarization, and $I_\perp(t)$ is the time-dependent fluorescence intensity emitted at a polarization perpendicular to the excitation polarization. The two quantities provide equivalent information and have a simple algebraic relationship. Clearly, in the absence of motion, polarization and anisotropy are high, while in the presence of fast motion, both quantities are low. In the simple case of isotropy rotational diffusion of a molecule, the time-dependent anisotropy has a simple form: $r(t) = r_0 e^{-t/\phi}$, where r_0 is the anisotropy immediately after excitation and ϕ is the rotational correlation time.

The measurement of time-resolved anisotropy requires quantifying the temporal evolution of the fluorescence intensities with polarizations parallel and perpendicular to the excitation polarization. This measurement can be accomplished by simple modification of a lifetime-resolved spectrophotometer (Figure 6.6b). In this design, a polarizer is placed at the excitation light path. For detection in a T-geometry, the most efficient approach is to implement two emission light paths, where one path detects the polarization component parallel to the excitation light, and a second path detects the perpendicular one. The simultaneous detection of both polarizations allows more precise measurement of anisotropy by rejecting common mode noise sources, such as the fluctuation of the excitation light source.

Finally, even when we are not interested in quantifying molecular rotational motion, an understanding of molecular rotation remains important for accurate lifetime measurement. It should be noted that the efficiency of gratings used in monochromators and polychromators is polarization dependent (which is also true for dichroic mirrors). Therefore, in the presence of molecular rotation, the measured fluorescence lifetime will be dependent on polarization artifacts of these gratings. In a T-geometry detection scheme, the excitation and emission polarizers can be set in a "magic angle" configuration where the measured fluorescence signal is always proportional to the total fluorescence intensity, $I_\parallel + 2I_\perp$, accounting for the one parallel polarization direction and the two perpendicular ones, irrespective of the sample polarization.

6.4 LIFETIME-RESOLVED FLUORESCENCE SPECTROSCOPY WITH IMAGING DEVICES

6.4.1 MICROSCOPES

The optical design of a microscope is often very simple. The main complexity lies in the microscope objective. The objective is required to focus light with minimal aberration and pulse dispersion, and to optimize the transmission of the emitted signal. Excellent microscope objectives are currently available commercially and can be readily incorporated into any microscope design. The rest of the microscope design is often satisfactorily modeled by the paraxial approximation and can be understood using simple ray tracing.

All the microscope designs in the following sections contain lens pairs with separation equal to the sum of their focal lengths (4-f geometry) and can be understood based on just these four ray-tracing rules (Figure 6.7).

Principles of fluorescence lifetime instrumentation

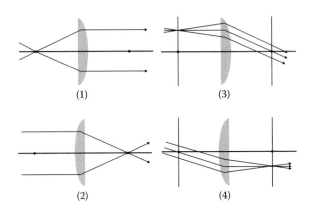

Figure 6.7 Basic ray tracing can be codified in four simple rules. These four rules can be summarized as follows: (1) A light ray emitted from the front focal point propagates parallel to the optical axis after the lens; (2) A light ray incident upon the lens parallel to the optical axis goes through the back focal point after the lens; (3) A divergent fan of light rays emitted from a point at the front focal plane becomes collimated after the lens. Coupled with rule (2), one can determine the propagation angle of the collimated light after the lens; and (4) Collimated light rays incident upon a lens are focused at the back focal plane. The focus position at the back focal plane can be determined with the aid of rule (1).

6.4.1.1 Wide-field fluorescence microscopy

A wide-field microscope is one of the most basic designs and consists of three elements: a microscope objective, a tube lens, and a spatially resolved detector such as a CMOS or CCD camera. Although a microscope objective consists of multiple lens elements, it is considered as an ideal singlet lens for simplicity in this section. Simple ray tracing of two objects at the specimen plane shows that they are imaged at the detector surface with a separation increased by a factor equal to the ratio of the focal length of the tube lens and that of the objective. This factor is called the magnification of the objective, and it achieves the goal of making small objects appear larger.

For practical imaging, most microscopes further require an illumination source, which can be arranged in either epi- or trans-illumination geometry (Figure 6.8). Most microscopes have wide-field white light illumination, which is extremely useful for the rapid inspection of the specimen. The white light

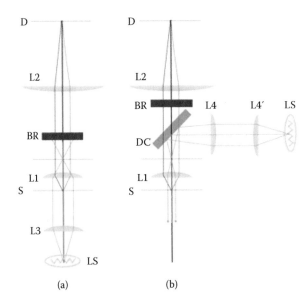

Figure 6.8 (a) Trans- and (b) epi-illumination microscope designs. BR, barrier filter; D, detector; DC, dichroic filter; L1, L2, L3, L4, L4′, lenses; LS, light source; S: specimen.

illuminator is often arranged in a trans-illumination geometry based on scattering, absorption, phase change, or polarization change as contrast mechanisms. Microscope illumination uses the Köhler design. The light source intensity distribution is typically inhomogeneous, such as that of the filament of a Xenon arc lamp. It is therefore undesirable to directly project an image of the light source directly onto the specimen plane as it will result in uneven illumination. Instead, the light emitted from each point of the light source is focused to the back focal plane of the objective using a pair of intermediate lenses in a 4-f geometry. The focused excitation beams from different points of the light source are then collimated and uniformly illuminate the whole specimen plane from different incidence angles.

Fluorescence wide-field illumination often uses mercury arc lamp or LED illumination. An excitation interference filter is often used to better define the excitation spectrum for these relatively broadband light sources. Similar to white light illumination, the fluorescence excitation light sources also use Köhler illumination geometry to ensure uniform illumination. However, for fluorescence imaging mode, epi-illumination is typically employed instead of trans-illumination (Figure 6.8). The importance of epi-illumination for low excitation light background fluorescence imaging has been discussed previously. For epi-illumination, a dichroic filter is needed to reflect the shorter wavelength excitation light toward the objective while allowing the longer wavelength emission light to pass toward the detector. Although the majority of the excitation light propagates away from the detector and most optical surfaces in the microscope are antireflection coated, typically the several percent of excitation light intensity that is reflected back towards the detector can swamp the fluorescence signal. As discussed previously, the fluorescent signal is weak and is low even when compared with this small fraction of the excitation light. Therefore, a barrier filter is still needed for further rejection of the excitation light. Barrier filters used in fluorescence microscopes can have over 4 to 6 optical density (OD) blocking power.

Today, microscope objectives are very well optimized, typically containing over ten individual optical elements, aiming to minimize all lower order monochromatic and chromatic aberrations. Furthermore, because of the number of optical elements used, it is critical that these lens be antireflection-coated to minimize light loss. Most objectives used today are antireflection-coated in the visible spectrum. Objectives can be further classified in many different ways. The most important parameter for a microscope objective is its NA. This specifies its ability (in terms of solid angle) to gather light and consequently its resolution. The NA of the objective is defined as $NA = n \sin \alpha$ where α is the half solid angle subtended by the objective lens at the focal point and n is the refractive index of the immersion medium. Therefore, the maximum NA is always equal to the index of refraction of the immersion medium. The immersion medium is the fluid coupling between the lens and the specimen; air, water, and oil are the typical media used. Today, the highest NA objective has a NA of 1.6 using a very high index oil. Another important parameter of the objective is its working distance, which quantifies how deep one may image into a specimen. In general, higher NA implies shorter working distance and *vice versa*. Note that the magnification of the microscope objective is not an inherent property of the lens itself because magnification is defined by the combination of both the objective and the tube lens. Therefore, the magnification of the objective lens is only correct as long as it is used with a specific matching tube lens (i.e., objectives should be mostly used with microscopes of the same manufacturer).

While the magnification is an important parameter of the microscope, resolution is more important. Magnification can be increased almost arbitrarily by increasing the ratio of the tube lens focal distance relative to that of the objective. However, even at very high magnification with a high NA objective, neighboring point emitters in a specimen plane cannot be distinguished if their separation is below approximately half the wavelength of light. This observation is often described as the Abbe limit and has its origin in the diffraction of light. Due to diffraction, even at high magnification, the image of a point emitter is always blurry on the detector (Figure 6.9).

This phenomenon can be readily understood if one accepts a basic result of Fourier optics stating that the electric field at back focal plane of a lens (under the far field Fraunhofer approximation) is the spatial Fourier transform of the field distribution at the front focal plane:

$$E_2'(x', y') = \Im_{2d}(E_1(x, y)) \tag{6.4}$$

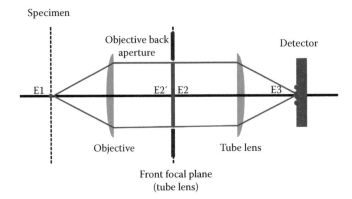

Figure 6.9 Optical modeling of a wide-field fluorescence microscope. A point source at the specimen plane with electric field E1 is imaged by the objective and becomes a plane wave, E2′, at the objective back aperture. The plane wave is truncated by the finite extent of the aperture with a field E2 propagating toward the tube lens. The tube lens subsequently projects the diffracted wavefront, E3, on the detector.

Therefore, the Fourier transform of an incoherent fluorescent point emitter, modeled as a delta function, $\delta(x, y)$, is a plane wave at the back focal plane of the objective. We have

$$E_2'(x', y') = \Im_{2d}(\delta(x, y)) \tag{6.5}$$

It is important to note that the objective has a finite-size back aperture; therefore, plane wave is truncated by the aperture stop and the actual field at the back aperture plane is actually

$$E_2(x', y') = E_2'(x', y')\mathrm{circ}(x', y', r_0) = \Im_{2d}(\delta(x, y))\mathrm{circ}(x', y', r_0) \tag{6.6}$$

where $\mathrm{circ}(x', y', r_0) = 1$ if $x'^2 + y'^2 \leq r_0^2$; else $\mathrm{circ}(x, y, r_0) = 0$.

Since the two lenses have a 4-f configuration, the back focal plane of the first lens is the front focal plane of the second lens. The electric field of the point source at the detector is then another Fourier transform of the field at the back focal plane of the first lens:

$$\begin{aligned} E_3(x', y'') &= \Im_{2d}\left[E_2(x', y')\right] \\ &= \Im_{2d}\left[\Im_{2d}(\delta(x, y))\right] \otimes \Im_{2d}\left(\mathrm{circ}(x', y', r_0)\right) = \delta(x'', y'') \otimes \Im_{2d}\left(\mathrm{circ}(x', y', r_0)\right) \\ &= \Im_{2d}\left(\mathrm{circ}(x', y', r_0)\right) \end{aligned} \tag{6.7}$$

We have used the fact that the convolution of delta function with another function returns that function itself. Therefore, the point source at the specimen is mapped to a point source at the detector convoluted with, that is, broadened by, the Fourier transform of the circ function. The intensity distribution at the detector for a fluorescent point source is the square of the field distribution, which has the form of an Airy function (see below). Extending into 3D, the field distributions at different axial depths close to the image plane can be found via Fresnel propagation and the result is (Born and Wolf 1987)

$$\mathrm{PSF}(u, v) \propto \left| \int e^{-\frac{iu}{2}\rho^2} J_0(v\rho)\rho \, d\rho \right|^2 \tag{6.8}$$

where $v = \dfrac{2\pi NA}{\lambda} r$, $u = \dfrac{2\pi NA^2}{\lambda} z$, NA is the numerical aperture of the objective, λ is the wavelength of light, r is the radial coordinate at the image space, z is the axial coordinate of the image space, J_0 indicates a Bessel function of the first kind of order 0, ρ is the radial coordinate, and u and v are the reduced dimensionless optical coordinates. Focusing on the image plane, the point spread function (PSF) has the form of an Airy disc as expected:

$$PSF(0, v) \propto \left[\frac{2 J_1(v)}{v} \right]^2 \qquad (6.9)$$

The size of the PSF is limited by the NA of the lens and the wavelength of light. The PSF is often characterized by its width. The width can be characterized by two parameters: (1) the position of the first zero and (2) the full width at half maximum (FWHM) of the central maximum. The first minimum position of the PSF can be found numerically. From the property of $J_1(v) = 0$ when $v = 3.83$, we find that the first minimum will occur at

$$\frac{2\pi}{\lambda} r NA = 3.83 \qquad (6.10)$$

where r is the radial distance from the focal point. We therefore have (where d is the diameter for the first zero)

$$d = 2r = 1.22 \frac{\lambda}{NA} \qquad (6.11)$$

Using the criterion first defined by Rayleigh stating that two objects are distinguishable if the maximum of one point source is located no closer than the first zero of the PSF of the second point source, therefore the resolution of a fluorescence wide-field microscope is

$$r_{res} \sim \frac{d}{2} = 0.6 \frac{\lambda}{NA} \qquad (6.12)$$

A quantity that is partly related to resolution is contrast. Specifically, the imaging of a grid (with light and dark stripes) with increasing spatial frequency will become increasingly blurred due to the convolution with the PSF and its contrast will decrease. The contrast of an image (with maximum and minimum intensity values of I_{max} and I_{min}, respectively) can be defined as

$$V = \frac{I_{max} - I_{min}}{I_{max} + I_{min}} \qquad (6.13)$$

In addition to morphological imaging capabilities, fluorescence microscopy by its nature has several advantages for biomedical applications. Most importantly, fluorescence imaging is specific to particular chemical and biological features. Biological structures can often be identified by their endogenous emission characteristics, but, more importantly, the abundance of contrast agents developed in the past several decades allows specific labeling of biological specimens. It is also very important that the Stokes shift between the excitation and emission wavelengths allows highly sensitive detection. The detection of single protein tagged by a single fluorophore is routine today. Importantly, the location of single fluorophore can be determined much below the diffraction limit.

Principles of fluorescence lifetime instrumentation

6.4.1.2 Fluorescence confocal microscopy

The main disadvantage of wide-field fluorescence microscopy is its lack of depth resolution, meaning that it cannot study 3D structures of many biological systems. There are now a variety of methods to extract depth information from biological specimens. The most common methods are confocal and NLO microscopy. This section focuses on confocal methods. Confocal microscopes (Masters 1996; Pawley 1995; Wilson and Sheppard 1984) may be based on reflected light (similar to white light modalities of wide-field microscopy) and fluorescence signals. The image contrast in reflected light confocal microscopes is generated by index of refraction mismatch in the specimen (Corcuff et al. 1993); the image contrast in two-photon microscopy is based on the nonuniform distribution of endogenous fluorophores. The concept of confocal detection may be considered to have started with Naora (1951) where conjugated pinholes were used in a microspectrophotometer to reject background for better measurement of DNA content in solution. Subsequently, the concept of the beam scanning microscope using a cathode ray tube was introduced by Roberts and Young (1952). In 1961, Minsky filed the first patent on confocal microscopy where conjugated pinholes were used to reject out-of-focal-plane background, while point scanning was used for mapping volumetric information (Minsky 1961). Due to a number of technological limitations, these first attempts at confocal microscopes were soon forgotten. With the advent of computer visualization and control technology, more mature scanner technology, and high-sensitivity detectors, commercial confocal microscopes were reintroduced by companies such as BioRad, Zeiss, and Nikon in the late 1980s. Confocal imaging is now a standard tool for the 3D study of cellular and tissue structures. Conceptually, this form of 3D microscopy works by placing a spatial filter (pinhole) in the confocal plane and a photodetector; only light emanating from the focal volume can efficiently pass through the pinhole in the confocal plane, in front of the photodetector. Out-of-focus illuminated objects form a defocused spot at the pinhole and consequently only contribute weakly to the signal. This optical sectioning effect allows imaging in three dimensions based on the confocal principle.

For a quantitative understanding of confocal microscopy, we will return to basic Fourier optics. The confocal microscope may be modeled as two lens pairs, a point light source, and a pinhole aperture in front of a detector (Figure 6.10).

Let us denote the optical transfer function of the first and second lens pairs as $h_1(u, v)$ and $h_2(u, v)$, respectively. Each lens pair can be exactly modeled as described in the Section 6.4.1.1; specifically, the square magnitude of $h_1(u, v)$ and $h_2(u, v)$ have the form of Equation 6.9. The intensity distribution resulting from imaging the point light source inside the specimen is described exactly by Equation 6.9. Since the field distribution at the specimen is $h_1(u, v)$ instead of a point source, the field distribution, $h(u, v)$, at the pinhole aperture plane is, in general, the convolution of the transfer functions of both lens pairs. To account for the presence of the specimen, we will treat two cases: scattered light confocal microscope (coherent case) and fluorescence confocal microscope (incoherent case).

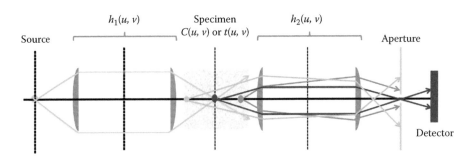

Figure 6.10 Design of a confocal microscope. Light from a point source (yellow) illuminates a double inverted cone region in the specimen. For a qualitative view of the confocal principal, ray tracing shows that fluorescence generated at the focal point (red) transmits through the pinhole aperture while fluorescence signal generated out of the focal plane (green and blue) is defocused at the pinhole aperture and is attenuated. For a quantitative description, the lens pair close to the light source and the aperture can be modeled as $h_1(u, v)$ and $h_2(u, v)$, respectively. For coherent and incoherent processes, the specimen can be modeled as a transmission function $t(u, v)$ or a fluorophore distribution $C(u, v)$.

For the coherent case, we can consider the transmission function $t(u, v)$ at the specimen plane. In this case, the light distribution at the focal volume is $h_1(u, v)\, t(u - u_s, v - v_s)$, where u_s, v_s are the positions of the specimen relative to the optical system (for simplicity, we assume that we scan the specimen instead of the beam). The intensity at the emission aperture plane will then depend on the specimen:

$$h(u, v, u_s, v_s) = h_1(u, v)\, t(u - u_s, v - v_s) \otimes h_2(u, v) \tag{6.14}$$

$$h(u,v,u_s,v_s) = \iint h_1(u',v')t(u' - u_s, v' - v_s)h_2(u - u', v - v')\, du'\, dv' \tag{6.15}$$

The field after an infinitely small pinhole

$$h(u_s, v_s) = h(0,0,u_s,v_s) = \iint h_1(u',v')t(u' - u_s, v' - v_s)h_2(u', v')\, du'\, dv' \tag{6.16}$$

$$h(u_s, v_s) = h_1(u_s, v_s)h_2(u_s, v_s) \otimes t(u_s, v_s) \tag{6.17}$$

We have used the fact that the transfer function is symmetric relative to the origin. The intensity transfer function of the system is

$$I(u_s, v_s) = |h_1(u_s, v_s)h_1(u_s, v_s) \otimes t(u_s, v_s)|^2 \tag{6.18}$$

In the above case, the transfer function is derived for the imaging of scattered or transmitted light where the light detected at the aperture plane is coherent with the signal at the specimen plane. For fluorescence confocal microscopy, the situation is actually simpler. The fluorescence signal generated at the focal point, I_s, after the first lens pair is

$$I_s(u, v, u_s, v_s) = |h_1(u, v)|^2\, C(u - u_s, v - v_s) \tag{6.19}$$

where $C(u, v)$ is the fluorophore distribution in the specimen.

The fluorescence intensity distribution at the aperture plane, $I(u,v)$, is then

$$I(u, v, u_s, v_s) = I_s(u, v, u_s, v_s) \otimes |h_2(u, v)|^2 \tag{6.20}$$

To calculate the PSF for fluorescence confocal microscopy, we set $C(u, v) = \delta(u, v)$ and evaluate the intensity after the pinhole at the origin of the aperture plane.

$$\begin{aligned} I(u,v,u_s,v_s) &= I(0,0,u_s,v_s) \\ &= \iint du'\, dv' |h_1(u',v')|^2\, \delta(u' - u_s, v' - v_s)|h_2(0 - u', 0 - v')|^2 \end{aligned} \tag{6.21}$$

$$I(u_s, v_s) = |h_1(u_s, v_s)|^2\, |h_2(u_s, v_s)|^2 \tag{6.22}$$

In the case that both lenses are identical, as in the epidetection geometry, $h_1(u, v) = h_2(u, v) = h(u, v)$,

$$I(u_s, v_s) = |h(u_s, v_s)|^4 \tag{6.23}$$

Note that the PSF for fluorescence confocal microscopy is essentially the square of the PSF of the wide-field fluorescence microscope. The intensity dependence of the excitation field at the focal point can be evaluated numerically (Figure 6.11).

Principles of fluorescence lifetime instrumentation

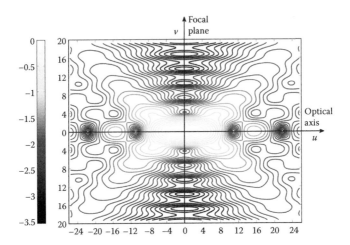

Figure 6.11 Intensity distribution at the vicinity of the focal point of a lens.

There are two characteristics of a confocal microscope that are of particular importance. Firstly, the confocal microscope provides modest improvement in the lateral resolution. Secondly, the confocal microscope has depth resolution. The resolution of a confocal microscope (or any microscope for that matter) is often quantified by its PSF (in positional space) and its modulation transfer function (MTF; in spatial frequency space). The PSF of the confocal signal is

$$\text{PSF}_{\text{conf}}(u,v) \propto \left| \int e^{-\frac{iu}{2}\rho^2} J_0(v\rho)\rho \, d\rho \right|^4 \tag{6.24}$$

Note that the incoherent confocal PSF is the square of that of one-photon wide-field system. Let us consider the PSF along the radial and axial directions. Along the radial direction at the center of the PSF (Figure 6.12),

$$\text{PSF}_{\text{conf}}(0,v) \propto \left[\frac{2J_1(v)}{v} \right]^4 \tag{6.25}$$

The radial PSF of the confocal system is the square of the wide-field system resulting in resolution improvement (about 40%) and side band suppression and can be observed graphically (Figure 6.13).

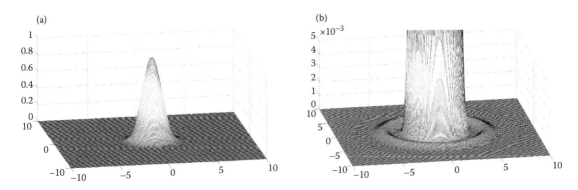

Figure 6.12 (a) Radial intensity distribution of confocal microscopy is a squared Airy function. (b) Expanded view of the lower intensity higher order rings.

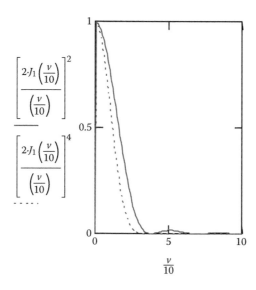

Figure 6.13 Comparing the radial intensity distributions of wide-field (red) and confocal (blue) microscopy. Note the narrowing of the FWHM and the suppression of higher order diffraction rings in the confocal case.

Along the axial direction at the center of the PSF (Figure 6.14),

$$
\mathrm{PSF}_{\mathrm{conf}}(u,0) \propto \left[\frac{\sin\left(\dfrac{u}{4}\right)}{\left(\dfrac{u}{4}\right)} \right]^{4}
\tag{6.26}
$$

This is the fourth power of a sinc function, which is another decaying oscillatory function (Figure 6.14).

The most important advantage of the confocal microscope is its depth resolution. The resolution of an optical system is sometimes better considered in the spatial frequency space. The Fourier transform of the PSF, the impulse response of the system, is called the optical transfer function.

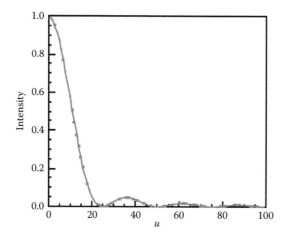

Figure 6.14 Axial intensity distribution of confocal microscopy is a squared sinc function.

Principles of fluorescence lifetime instrumentation

The wide-field optical transfer function can be expressed as

$$\text{OTF} = (\vec{k}) = \Im\left(\text{PSF}(\vec{r})\right) = M(\vec{k})e^{i\Phi(\vec{k})} \tag{6.27}$$

where $M(\vec{k})$ is called the MTF and $\Phi(\vec{k})$ is the phase transfer function where $\vec{k} = \dfrac{2\pi}{\vec{r}}$.

In 3D, the functional forms of the MTFs of the wide-field and confocal cases are rather complicated and are not too illuminating. Instead, it is more instructive to study the support of the MTF. The support is defined as the region where the MTF is nonzero. A comparison of normal and confocal MTF support is shown in Figure 6.15 (red: confocal; blue: normal).

Three features of the MTF support are of particular interest. First, the support of the MTF of the confocal is wider than that of a normal microscope. This implies that the confocal system can transfer an image with higher spatial frequency—better resolution. Second, the supports of the MTFs of both confocal and normal microscopes are wider along the radial direction. This indicates that microscopes tend to have better resolution in the radial direction than in the axial direction. Third, the MTF of the normal microscope has two missing cones—regions around the k_z axis where MTF has zero support. This implies that the normal microscope cannot transmit lower spatial frequencies along the axial direction. This is a result of normal microscopy lacking axial (depth) resolution. The MTF of confocal microscopy has no missing cone (Figure 6.15).

The effect of the missing cone can be seen by considering depth discrimination. For a uniform specimen, we can ask how much fluorescence is generated at each z-section above and below the focal plane, assuming that negligible amount of light is absorbed throughout (Figure 6.16a). In this case, the fluorescence signal is

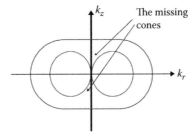

Figure 6.15 Frequency supports of the MTF for the wide-field (blue) and the confocal (red) systems, respectively. Note the missing cone in the wide-field case.

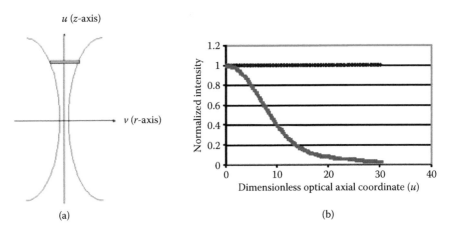

(a) (b)

Figure 6.16 (a) Profile of the radially integrated intensity function in the vicinity of the focal point of the microscope objective. (b) Radially integrated fluorescence intensity at each depth. Blue: wide field; red: confocal.

proportional to the excitation intensity, which is a quadratic function of the electric field. Without confocal detection, we have

$$I^{zsec}(u) = 2\pi \int \left| \int e^{-\frac{iu}{2}\rho^2} J_0(v\rho)\rho \, d\rho \right|^2 v \, dv \qquad (6.28)$$

The behavior of this integral can be evaluated numerically (Figure 6.16b). It may be surprising to see that the fluorescence emission from each z-plane is equal (blue line). This actually should be expected. The number of excitation photons at each z-plane is constant because of the conservation of energy. This implies that the fluorescence from each z-plane should be equal. For a normal fluorescence microscopy, each z-plane contributes equally and there is no "depth discrimination." In contrast, since the detected intensity goes as the fourth power of the electric field,

$$I^{zsec}_{conf}(u) = 2\pi \int \left| \int e^{-\frac{iu}{2}\rho^2} J_0(v\rho)\rho \, d\rho \right|^4 v \, dv \qquad (6.29)$$

The important difference after numerical integration is shown in Figure 6.16 (red line). The fluorescence intensity quickly drops away from the focal point. Therefore, confocal microscopy has depth discrimination.

The instrumentation design (Figure 6.17) of the confocal microscope has many similarities to the standard wide-field microscope with three major differences. Firstly, confocal microscopy almost always uses a laser as its light source. Extended light sources such as lamps cannot be focused down to a diffraction-limited spot, which affects final resolution. Secondly, confocal microscopy generates images at a single point at a time. Therefore, in order to map out the 3D fluorophore distribution, raster scanning of either the specimen or the light beam must be performed. While raster scanning of the specimen ensures minimal aberration degradation of the image, it is too slow and not very compatible with many biological preparations. Generally, modern confocal microscopes implement beam scanning, where the light is deflected by mirrors to focus at different lateral locations. Typical confocal microscopes use a galvanometer system consisting of two synchronized scanners that move in tandem to produce an x–y raster scanning pattern. The scanners typically run in servo mode. In servo mode, the scanners have a bandwidth up to about 1 kHz. In resonance mode, the scanners can reach about 10 kHz. Other high-speed mechanical scanners include rotation polygonal mirror systems that can generate a line rate up to 10–30 kHz. For nonmechanical scanners, acousto-optical scanners have also been used for beam diffraction. An acousto-optical system has the advantage of allowing the generation of an arbitrary scanning pattern. The axial

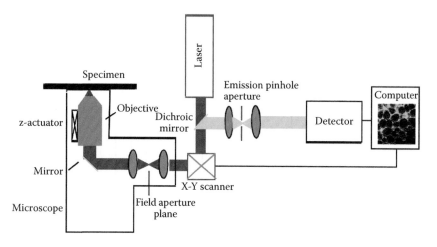

Figure 6.17 Design of a typical confocal microscope.

Principles of fluorescence lifetime instrumentation

position of the excitation light is typically accomplished by either piezo-driven or mechanically driven objective positioners since axial raster scanning often does not require high speed. Thirdly, with the use of a beam scanning approach, a major hurdle in the design of a beam scanning confocal microscope is the need to de-scan the emission light. Since the emission light originates from different locations in the specimen during raster scanning, the emission light travels back through the microscope along a different path. To achieve confocal imaging, we need to focus the emission light and pass it through a pinhole. It is clearly impractical to place the pinhole in the optical path where the beam moves as a function of mirror position. If we trace the emission light path of the microscope, it is clear that the beam is nonstationary up-stream from the servo mirror scanner. However, since the mirror movement is slow compared with the speed of light, the mirror remains at virtually the same angle when the emission light passes through as when the excitation light goes in. By symmetry, it is clear that the emission light will become stationary after it passes through the scanner a second time. By placing a beam splitter or a dichroic mirror in between the scanner and the light source, one can separate the excitation light from the emission. The emission light can then be focused down and filtered through a pinhole. Confocal detection is accomplished.

6.4.1.3 NLO microscopes

The extended discussion on confocal microscopy provides an excellent foundation for us to understand the operation of nonlinear microscopy. Today, there are many modalities of NLO microscopy based on physical processes such as two- and three-photon fluorescence, second- and third-harmonic generation, coherent anti-Stokes Raman, and stimulated Raman. The diagrammatic representation of these processes is shown (Figure 6.18).

Among all these nonlinear processes, nonlinear microscopy based on two-photon fluorescence is the most relevant in terms of lifetime spectroscopy. Denk et al. (1990) introduced two-photon fluorescence microscopy. Fluorophores can be excited by the simultaneous absorption of two photons, each having half the energy needed for the excitation transition. Since the two-photon excitation probability is significantly less than the one-photon probability, two-photon excitation occurs only at appreciable rates in regions of high temporal and spatial photon concentration. The high spatial concentration of photons can be achieved by focusing the laser beam with a high NA objective to a diffraction-limited spot. The high temporal concentration of photons is made possible by the availability of high peak power mode-locked lasers. In general, two-photon excitation allows 3D biological structures to be imaged with resolution comparable to confocal microscopes but with a number of significant advantages: (1) Conventional confocal techniques obtain 3D resolution by using a detection pinhole to reject out-of-focal-plane fluorescence. In contrast, two-photon excitation achieves a similar effect by limiting the excitation region to a submicron volume at the focal point. This capability of limiting the region of excitation instead of the region of detection is critical. Photodamage of biological specimens is restricted to the focal point. Since out-of-plane chromophores are not excited, they are not subject to photobleaching; (2) Two-photon excitation wavelengths are typically red-shifted to about twice the one-photon excitation wavelengths. The significantly lower absorption and scattering coefficients at these longer wavelengths ensure deeper tissue penetration; and (3) The wide separation between the excitation and emission spectra ensures that

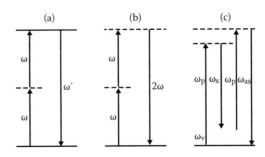

Figure 6.18 Jablonski diagrams for two-photon fluorescence (a), second harmonic generation (b), and coherent anti-Stokes Raman scattering (c).

the excitation light and any Raman scattering can be rejected without filtering out any of the fluorescence photons. This sensitivity enhancement improves the detection-signal-to-background ratio.

The image formation principle of two-photon fluorescence microscope is actually simpler than for confocal systems. The geometry of the optical system is identical to that of a confocal system. Let us denote the transfer function of the excitation lens pair (consisting of the excitation tube lens and the excitation objective) as $h_1(u, v)$ and the transfer function of the detection lens pair (consisting of the emission tube lens and the emission objective) as $h_2(u, v)$. Typically, a two-photon system is arranged in an epi-illumination geometry, in which case the excitation and emission objectives are the same and there is a dichroic mirror inserted to separate the excitation and emission optical paths. Since two-photon excitation is a second-order process, the fluorescence signal generated in the specimen is proportional to the square of the excitation transfer function. The intensity distribution $I(u, v)$ at the image plane is the convolution of the fluorescence distribution and the detection path intensity transfer function:

$$I(u, v, u_s, v_s) = I_s(u, v, u_s, v_s) \otimes \left| h_2(u, v) \right|^2 \tag{6.30}$$

$$I(u, v, u_s, v_s) = \left| h_1(u, v) \right|^4 C(u - u_s, v - v_s) \tag{6.31}$$

where $C(u, v)$ is the fluorophore distribution, and u_s and v_s are the raster scan coordinates. The fourth-power dependence on $h_1(u, v)$ originates from the quadratic two-photon process. In a typical two-photon microscope, a nonimaging detector such as a PMT is often used that integrates all the intensity at the image plane. To calculate the PSF of the fluorescence confocal microscopy, we set $C(u, v) = \delta(u, v)$.

$$
\begin{aligned}
\text{PSF}_{2P}(u_s, v_s) &= \iint du\, dv \iint du'\, dv' \left| h_1(u', v') \right|^4 \delta(u' - u_s, v' - v_s) \left| h_2(u - u', u - v') \right|^2 \\
&= \left| h_1(u_s, v_s) \right|^4 \iint du\, dv \left| h_2(u - u_s, u - v_s) \right|^2 \\
&= \left| h_1(u_s, v_s) \right|^4
\end{aligned}
\tag{6.32}
$$

Note that the final point spread of the two-photon fluorescence microscope is virtually identical in functional form to that of the confocal microscope. However, it should be noted that in this case, the PSF depends solely on the excitation optics but not the detection. It is also most important to note that the two-photon microscope has the same depth discrimination property as the confocal microscope. For a typical two-photon microscope, over 80% of the total fluorescence intensity comes from a 1 μm thick region about the focal point for objectives with NA of 1.25. Thus, 3D images can be constructed as in confocal microscopy, but without confocal pinholes. For a given fluorophore, since two-photon microscopes typically use excitation light at about twice the wavelength as that of confocal microscopes, the spatial resolution of two-photon microscopy is roughly half the one-photon confocal resolution. For a 1.25 NA objective using excitation wavelength of 960 nm, the typical PSF has FWHM of 0.3 μm in the radial direction and 0.9 μm in the axial direction. Two-photon excitation provides better suppression of higher order Airy rings.

The instrumentation design of the two-photon microscope is very similar to that of the confocal microscope but with several major differences. Firstly, while two-photon imaging has been shown to be possible with CW lasers and picosecond pulsed lasers, it is not practical to use in most systems. Instead, for efficient excitation of typical fluorophores, most two-photon microscopes require high peak power femtosecond lasers such as the titanium–sapphire systems.

Secondly, a two-photon system and a one-photon system use different dichroic mirrors. The dichroic mirror in a one-photon system typically reflects shorter wavelength excitation light and passes the longer wavelength emission light. The dichroic mirror in a two-photon system typically reflects longer wavelength excitation light and passes the shorter wavelength emission light. Furthermore, since the excitation and emission spectra of a two-photon system typically do not overlap (as they do in the one-photon case), more optimized dichroic mirrors and barrier filters can be used to improve emission signal detection while minimizing the amount of excitation light leaking through.

Thirdly, the most important difference between confocal and two-photon microscope design lies in the way a two-photon system obtains its depth discrimination ability entirely from the quadratic excitation process. There is no need for a confocal pinhole, allowing the use of a large area nonimaging detector to integrate all the fluorescence emission from the sample, including photons that have been scattered by the biological specimen. It has been shown that over 90% of emission photons are scattered at imaging depth beyond a couple of optical mean free path lengths. The ability to use these scattered emission photons accounts for much of the improved image penetration depth of two-photon microscopy.

Finally, excellent microscope objectives are made by major microscope manufacturers for linear optical applications in the visible range. These objectives have low intrinsic fluorescence and have high transmission efficiency throughout the near-ultraviolet and visible spectrum. Unfortunately, many commercial microscope objectives are not optimized for NLO microscopy; they do not have high transmission throughout the spectrum spanning the near-ultraviolet to the near-infrared wavelengths. The issue is particularly severe for applications such as third harmonic generation using excitation light at 1.2–1.3 μm. In addition, nonlinear microscopes require the transmission of laser light with minimal pulse dispersion, which reduces two-photon excitation efficiency. Femtosecond pulses are broadened as the blue and red edges of their spectra travel at different speed through dielectric media. However, due to the presence of multiple optical components in these microscope objectives, many microscope objectives have significant pulse dispersion as measured by the group delay dispersion (GDD). Microscope objectives often have GDD on the order of 3000 fs^2 (Wokosin 2005). There are microscope objectives with significantly higher GDD, which are less suitable for nonlinear microscopy. This level of GDD causes broadening of the 100 fs pulses by a factor of two or three.

6.4.2 ENDOMICROSCOPES

An endomicroscope operates on the same resolution and depth penetration regime as a typical laboratory microscope, but its miniaturization enables clinical applications. Endomicroscopes allow noninvasive examination of the gastrointestinal (GI) tract and the epithelial surface of other internal organs of a patient. Since many cancer types occur on the epithelial surface of interior body cavities, endomicroscopes are routinely used in clinical diagnosis and treatment today.

The importance of developing more effective endomicroscopes can be seen from the incidence rate of some of these diseases. For example, the incidences of colorectal, esophageal, and cervical cancers are over 12,000, 150,000, and 15,000 new cases per year in the United States, respectively. To further illustrate the use of this new technology for cancer detection, we will examine the situation with cervical cancer in greater depth (Agnantis et al. 2003; Apgar and Brotzman 2004; De Palo 2004; Dunleavey 2004; Gupta and Sodhani 2004; Lambrou and Twiggs 2003).

The onset of cervical cancers has been associated with a sexually transmitted human papillomavirus (HPV) infection, an infection that affects over 40 million people. Approximately 1 million new cases of HPV are detected each year. Some forms of HPV have been identified as the primary cause of cervical intraepithelial neoplasia. HPV is often detected either because of the observation of genital warts or from an abnormal Pap smear. The common Pap smear cannot localize the lesions. Patients with an abnormal Pap smear are often referred to colposcopy, a conventional microscopic imaging method of the cervix. Colposcopy has had false positive rates as high as 40%–50%. Furthermore, high-grade disease may be misclassified as low grade. Punch biopsy, another common technique, is subject to false negative results through incomplete sampling.

This is an area where endomicroscopy may be applied to identify the most critical biopsy sites and provide more informative diagnosis, since endomicroscopy can provide subcellular level images similar to histopathology. In addition to the initial diagnosis, endomicroscopy may also provide a benefit of long-term, periodic monitoring of premalignant conditions. For example, Barrett's esophagus is very common and is the result of acid reflux. Barrett's esophagus has a high chance of developing into cancer and must be carefully monitored clinically (Ban et al. 2004; Belo and Playford 2003; Falk 2002; Faruqi et al. 2004; Guelrud and Ehrlich 2004; Kara et al. 2004; Lambert 2004; Pacifico et al. 2003; Paulson and Reid 2004; Peters and Wang 2004; Sharma 2004).

Endomicroscopy may permit more accurate noninvasive monitoring of the potential progress of the disease at an organ site where frequent and high-coverage excisional biopsy is highly undesirable. In the case of cervical cancer, genital warts often involve surgical intervention, with all the associated risks thereof. Low-grade lesions may be topically treated with trichloroacetic acid or 20% podophyllin solution. More extensive cases are treated with cryosurgery, laser vaporization, or loop electrosurgical excision. Since early-stage genital warts often regress and excision procedures may compromise patient fertility, a conservative management scheme has often been suggested. The success of conservative management schemes depends on the existence of reliable clinical markers for cancer progression and imaging techniques to monitor them. Endomicroscopy may allow a more accurate, noninvasive determination of the propensity for Barrett's esophagus and cervical lesions to progress into a more invasive form.

6.4.2.1 Non-depth-resolved endoscopes

Conventional endoscopes operate in a wide-field imaging mode and most often at low resolution. Endoscopes come in different formats but typically consist of a light source for illumination, a lens for image formation, a fiber bundle for relaying the image to the image sensor in the proximal end, and an instrumentation channel for other treatments such as excisional biopsy. Currently, endoscopy is routinely used in medical procedures for the diagnosis and treatment of some diseases. For example, the bronchoscope, the colonoscope, and the hysteroscope are used for the inspection of the trachea and bronchi, the colon and large intestine, and the inside of the uterus, respectively (Golding et al. 2005). Recently, the capsule-type endoscope has been introduced that takes pictures as it travels through the GI tract by the bowel movement and sends the images wirelessly to the receiver (Qureshi 2004). Patients swallow the pill-type capsule, and it can acquire images in either from a head-on view (Eliakim 2006) or from a side view (Fisher and Hasler 2012). The advantage of this type of endoscope is that it can examine the whole GI tract, which is especially useful for imaging the small intestine where upper endoscopy and colonoscopy cannot easily reach. The capsule-type endoscope is also compatible with examination on an outpatient basis, providing patient convenience and minimizing medical cost. However, capsule-type devices typically do not provide as high-quality images as conventional endoscopy.

In addition to generating image contrast based on scattered light, other image contrast mechanisms such as tissue autofluorescence, spectroscopy, Raman scattering, and fluorescence lifetime can increase the endoscopic diagnostic accuracy (Dacosta et al. 2002). Richards-Kortum and coworkers have demonstrated that a high-resolution microendoscopy of intact oral mucosa has similar diagnostic accuracy to the histopathological diagnosis result (Muldoon et al. 2012). However, these techniques are still confined to the surface imaging of the tissues and subject to limited utility for other types of diseases, such as cancer occurring in the epithelial layer of tissue.

6.4.2.2 Depth-resolved endomicroscopes

Since wide-field endoscope imaging can only provide tissue surface information, it cannot replace traditional histopathological diagnosis where microscopic examination of histological tissue sections provides morphological information on individual cells and their organization within tissues. Therefore, wide-field endoscopy is typically used to identify suspicious tissue sites where an excisional biopsy will be taken so that a definitive clinical diagnosis based on histopathological analysis can be performed. However, these procedures are clearly invasive. More importantly, excisional biopsy is time consuming, involving tissue excision, fixation, and subsequent examination by trained pathologists; the long cycle time of this process often extends the time required to keep a patient in the operating room when excision biopsy is being used for surgical margin determination. With the advent of high-resolution 3D endoscopic imaging technologies with the potential for providing images comparable with excisional histopathology, there is a clear opportunity today to significantly extend the 3D, depth-resolved imaging capability of endoscopes and laparoscopes in clinical diagnosis. The three classes of high-resolution endomicroscopes that have been developed today are based on optical coherence tomography (OCT), confocal microscopy, and NLO microscopy.

OCT is based on the optical ranging of reflected light pulses from surfaces based on precise timing of femtosecond pulses interferometrically. Among the three 3D endomicroscope approaches, endoscopes based on OCT are the most advanced because the miniaturization of this technology is easier for two important reasons. Firstly, OCT systems inherently obtain depth information based on the timing of the low-coherence optical pulses such that no mechanical scanning in the depth direction is required. 2D sagittal or coronal sections of the tissue can be generated by only mechanically scanning the light ray along a single direction either radially or axially. Since scanning along only one direction is required, this greatly simplifies microscanner design. Secondly, OCT automatically provides depth-resolved information when the imaging optics have a large depth of field, that is, a relatively low NA. This longer depth of field, and hence longer working distance, significantly reduces the complexity involved in assembling the micro-optical components in the distal end of the endoscope at the expense of lower resolution. OCT endoscopes have entered into a number of clinical trials including the identification of Barrett's esophagus (Brand et al. 2000; Faruqi et al. 2004; Jacobson and Van Dam 2002; Li et al. 2000; Nishioka 2003; Pitris et al. 2000; Poneros 2004; Poneros and Nishioka 2003; Sharma 2004; Zuccaro et al. 2001) and colonic polyps (Asano and McLeod 2002; Brand et al. 2000; Nishioka 2003). OCT endoscope development is advanced enough that devices with an outer diameter less than 2 mm have been fabricated and have been applied to the evaluation of intracoronary stenting and the study of atherosclerotic plaques in patients (Bouma et al. 2003a, b; Chau et al. 2004; Fujimoto et al. 1999; Tearney et al. 2003; Yabushita et al. 2002). While OCT endoscopes are finding important clinical applications, the combination of using low NA optics, the relatively narrow bandwidth of light sources, and the inherently low contrast of the OCT signal in tissues limits the lateral and axial resolutions of these systems to about 5–7 μm (Adler et al. 2007). Therefore, OCT endoscopes often do not provide images with a resolution comparable to standard histopathology (Figure 6.19).

Today, endoscopes with subcellular resolution are based on confocal detection or NLO excitation principles. Unlike in an OCT system, confocal and NLO endomicroscopes obtain information from a single 3D-resolved location in the tissues. Scanning along two directions must be performed to generate a 2D image. The need to incorporate scanning along at least two directions presents a challenge in endoscope design. Nevertheless, significant progress has been made in confocal-based systems. Today, some confocal endomicroscopes are becoming commercially available, and preliminary clinical tests using these devices have started (Nguyen and Leong 2008). Three classes of confocal endomicroscopes have been developed.

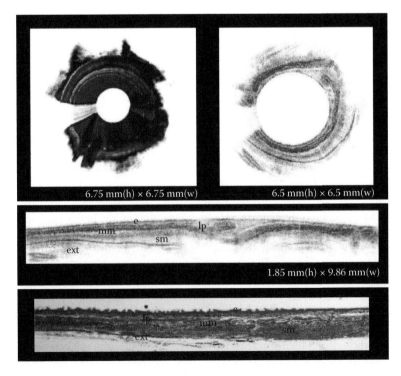

Figure 6.19 OCT endoscopic 3D rendering of *in vivo* rabbit esophagus. (From Su, J. P. et al., *Optics Express* 15, 10390–6, 2007.)

The most popular approach for confocal endomicroscopes utilizes high-density, flexible fiber optical bundles and performs scanning at the proximal end of the device (Liang et al. 2002; Rouse et al. 2004; Sakashita et al. 2003; Watson et al. 2002). The advantage of this class of endomicroscope is its simplicity. Since scanning is performed at the proximal end, outside the patient, the scanning mechanism does not require miniaturization, but the resolution of these systems is limited by the fiber bundle pitch size and the static pattern noise caused by imperfections in fiber bundle fabrication.

The second class of device incorporates microscanning devices that scan micro-optic components, such as the tip of a fiber optic, which are integrated in the distal end of the endomicroscope (Kiesslich et al. 2004). This class of device is very promising as high-resolution images have been demonstrated (Figure 6.20). However, the state-of-the art device is still fairly large with over 6 mm outer diameter and relatively modest resolution (0.7 mm lateral and 7 mm axial).

Significant progress was also made in a third class of devices that is based on microelectromechanical system (MEMS) mirrors or a MEMS mirror array integrated at the distal end of an endomicroscope (Dickensheets and Kino 1996; Himmer et al. 2001; Rector et al. 2003). While MEMS mirror-based confocal endoscopes have shown potential for clinical imaging at an early date and with good promise, there have been few significant clinical instrument breakthroughs along this direction recently due to the difficulty and cost of MEMS mirror fabrication.

Endomicroscopy based on NLO excitation provides complementary information to confocal systems. NLO endomicroscopes can better utilize molecular level contrast. Fluorescence and SHG are efficient contrast mechanisms allowing molecular imaging and metabolic imaging, although preliminary work is underway using gold nanoparticles as a contrast agent in reflected light confocal endomicroscopes (Sokolov et al. 2003). NLO microendoscopy has comparable excitation penetration as reflected light confocal endomicroscopes when operated in the infrared wavelengths. For fluorescence imaging, a confocal endomicroscope has shallower penetration depth due to the use of shorter excitation wavelengths. Furthermore, the use of a detection aperture in confocal microendoscopy results in significant rejection of scattered photons that can be retained in NLO endomicroscopes allowing deeper imaging with a higher SNR in principle. Finally, NLO endomicroscopy produces less tissue photodamage as compared to a fluorescence confocal endomicroscope, due to the inherent localization of the excitation volume. Given these potential advantages, a number of NLO endomicroscopes have been developed despite the greater technical challenges in constructing these systems.

Endoscope imaging started with preliminary work on human subjects using laboratory format microscopes (Masters et al. 1997). These instruments were ultimately unsuitable for clinical work. The most significant progress in clinical applications using laboratory-scale NLO microscopes is the work of Konig et al. (2009) on dermal lesion diagnosis. This group of investigators obtained, for the first time, regulatory approval to evaluate the clinical utility of NLO microscopy in the diagnosis of skin disorders with a substantial population of patients.

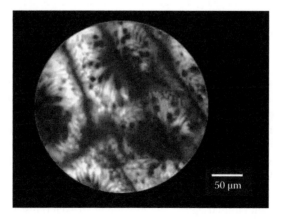

Figure 6.20 Confocal endoscopic images of healthy colonic crypts stained with fluorescein. (From Jean, F. et al., *Optics Express* 15, 4008–17, 2007.)

Principles of fluorescence lifetime instrumentation

The extension of these systems for clinical examination of internal organs faces three major technical challenges. Firstly, the efficiency of second-order NLO excitation is a linear function of laser pulse width. Due to effects such as group velocity dispersion and self-phase modulation that occur during transmission of high-intensity light through a typical silica core optical fiber, femtosecond pulses can be significantly broadened in a fiber endomicroscope, which greatly decreases the excitation efficiency of these systems. Secondly, due to pulse dispersion considerations, the successful confocal endomicroscope designs that utilize distal scanning through a fiber optics bundle are not feasible for NLO excitation. The results from initial experiments are not optimal (Gobel et al. 2004). Therefore, microscanning mechanisms must be packaged in the proximal end of the device, while the instrument outer diameter must be kept within a few millimeters. Thirdly, fluorescence and second harmonic signals are weak. Typically, each pixel of a video rate image has a maximum photon count between hundreds and thousands of photons when the sample is labeled with a good fluorophore such as Rhodamine. However, the photon count can be as few as tens to hundreds when imaging is based on tissue endogenous fluorophores, such as NAD(P)H. Therefore, unlike a reflected light confocal system where a strong optical signal can be obtained, an NLO endomicroscope must be designed to maximize light collection efficiency.

Significant progress has been made in overcoming the first challenge as photonic bandgap crystal technology is coming of age, allowing the creation of ultra-broadband, omnireflective mirrors and waveguides (Hart et al. 2002; Temelkuran et al. 2002). Photonic bandgap crystal technology is poised to produce novel optical switches and multiplexers (Attard 2003; Ibrahim et al. 2004; Soljacic et al. 2002). Most importantly for the future of endomicroscopy, photonic bandgap crystal technology allows the creation of novel waveguides for femtosecond pulses with controllable dispersion (McConnell and Riis 2004; Myaing et al. 2003).

Initial design of the nonlinear endomicroscope used the conventional bench top multiphoton microscope and scanning mechanism with a long gradient index lens (GRIN) for relaying the excitation light. Practically, the depth of the NLO microscopy is ultimately limited by the scattering of the excitation light by tissues. For example, the typical penetration depth is about 500 µm in brain tissue, which corresponds to about 2 to 3 scattering mean free path lengths (Theer and Denk 2006). Rigid-type endomicroscopes have been used to image deep in the tissues such as green fluorescent protein (GFP) expressing neurons in the hippocampus of the mice brain (Jung and Schnitzer 2003; Levene et al. 2004) and to image the long-term tumor progression in the mice colon (Kim et al. 2010) that is not easily accessible with the conventional technique. Initial clinical studies have been conducted by Konig et al. (2007) for the deep tissue skin imaging with SHG and autofluorescence such as contrast mechanism. While these devices allow cutting-edge study of neurobiology and tumor biology in small animals, they are not feasible for clinical human use other than the skin due to their length limitation and their rigidity.

Significant progress has been made in overcoming the need for multiple axis scanning mechanisms. Fiber optic-based nonlinear endomicroscopes started with miniature, handheld two-photon systems using Lissajous resonant scanning of a short-length single mode fiber was first developed by Helmchen et al. (2001). This system was used to study the vasculature structure and dendritic calcium dynamics of a freely moving rat. Schnitzer and coworkers developed a similar system but with much lighter weight that can be directly mounted on the mouse brain. The unique feature of this device is that the miniaturized objective lens at the tip of a long relay lens was inserted into the mouse brain and could image hippocampal vessels as well as vessels near the neocortical surface (Deisseroth et al. 2006; Flusberg et al. 2005). These kinds of head-mountable multiphoton endoscopes can provide a useful tool for the neuroscientist to study the intact living mammalian brain without constraining the motion of the mice.

The development of a more clinically applicable endomicroscope started with a number of parallel developments, including further miniaturization of the Helmchen design by resonantly scanning a dual-core photonic crystal fiber instead of single mode fiber for more compact design of the imaging probe by the groups of Gu and Li (Bao et al. 2008; Myaing et al. 2006; Wu et al. 2009a). The major advance here involves the use of a dual-core fiber with single mode, low dispersion transmission of excitation light and multimode, higher efficiency collection of the emitted signal. The addition of an axial translation capability with a shape memory alloy actuator in a recent design enabled 3D scanning NLO endomicroscopy (Wu et al. 2010). As opposed to Li's design that performs spiral scanning using a tube-type piezo actuator, the recent development

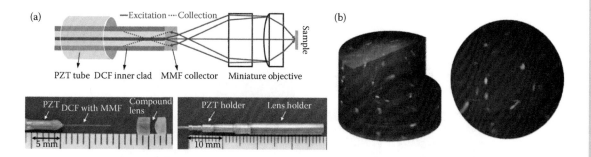

Figure 6.21 Fiber scanning nonlinear endomicroscope (a). (From Wu, Y. C. et al., *Optics Letters* 34, 953–5, 2009b.) 3D image of Pig cornea tissue (b). (From Wu, Y. C. et al., *Optics Express* 17, 7907–15, 2009a.)

by Rivera et al. (2011) performs raster scanning, which generates more uniform illumination pattern. A parallel development by Chen and coworkers utilizing 2D scanning MEMS mirrors has resulted in another class of NLO endomicroscopes (Jung et al. 2008; Tang et al. 2009). In addition to actuation of the scanning algorithm, distal scanning endoscopes are limited by a number of other optical constraints, such as the NA of miniaturized lenses. Recent advances in coupling a spherical lens to the top of a GRIN lens has first resulted in NLO endomicroscopes with objective NAs approaching 0.8 (Barretto et al. 2009).

For compactness and simplicity, most NLO endomicroscopes use the same optical fiber for excitation and detection. Since the excitation and emission wavelengths in nonlinear microscope imaging are widely separated, the endomicroscope optics optimized for excitation are, in general, very suboptimal for emission detection due to chromatic aberration. Using the single fiber for the excitation beam delivery and the emission beam collection results in significant loss in collection efficiency. The Li group has partially solved this problem by splicing the tip of the dual-core fiber with multimode fiber, which maintains the single mode excitation beam delivery and effectively increases the signal collection area (Figure 6.21). By further introducing an effective achromatic triplet as the objective, the chromatic misalignment problem is now largely overcome (Wu et al. 2009b). The main contrast mechanisms implemented in most NLO endomicroscopes are based on two-photon fluorescence and SHG. A system has been recently designed based on third harmonic generation and has been applied to study the dermal fibrosis and hyperkeratosis (Lee et al. 2009). In addition, an exciting new system has been developed by Hoy et al. (2008) that combines imaging and microsurgery capability in the same compact system. While normal nonlinear endomicroscopes produce a 3D image stack by scanning the focal spot in 3D, the nonlinear excitation based on the principle of spatiotemporal focusing pioneered by Silberberg and Xu groups greatly simplifies instrumentation design by producing a depth-resolved 2D section without using any scanning mechanisms (Oron et al. 2005; Zhu et al. 2005). Depth scanning can be accomplished by tuning the group velocity dispersion at the proximal end as proposed and demonstrated by Durst et al. (2006).

6.5 SHEET LIGHT (PROJECTION TOMOGRAPHY, SPIM) BASED IMAGER

Going beyond the realm of clinical medicine, 3D microscopic imaging through thick tissue in real time is one of the most efficient tools to directly visualize the mechanism of biological processes in different developmental or immunological contexts. Development of microscopic methods for 3D imaging of thick tissues with single cell resolution inside a live small animal is one of the major challenges. Additionally, high content imaging information of the volumetric tissue with cellular resolution is invaluable for the study of bimolecular signaling and disease mechanisms. 3D pattern mapping of gene expression and their superposition with anatomical information is obtained from the optical properties of tissues (Davidson and Baldock 2001). With great advances in the labeling technology, for example, genetically expressed fluorescence proteins, quantitative fluorescence microscopy methods potentially fulfill the major requirements of high content 3D tissue imaging. High-resolution imaging methods such as confocal and multiphoton microscopy provide optical sectioned imaging of specimens with relatively small dimensions

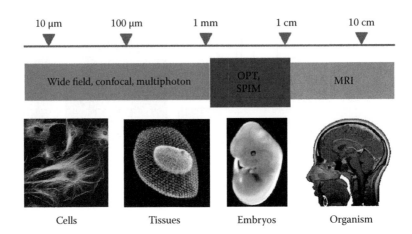

Figure 6.22 Imaging methods based on length scale of specimens.

(<1 mm). However, 3D imaging of specimens with dimensions of several millimeters is not practical with highly time-consuming confocal microscopy methods; additionally, there is significant limitation in the fluorescence mode as photobleaching occurs along the entire optical axis of excitation. Sheet light tomographic methods typically have slightly lower resolution than microscopic approaches but provide deeper imaging in relative transparent specimens, such as embryos. However, the most important advantage of sheet light imaging is its high throughput.

Tomography is a well-known technology used for the reconstruction of the 3D structure of the specimen by using a 2D image stack. In x-ray tomography (Momose et al. 1996), the specimen is illuminated from behind and rotated to make sure that the transmission from each angle is captured, which is used for the final 3D reconstruction. To overcome the challenges for optical 3D imaging of a specimen several millimeters in size (1–15 mm), other tomographic-based techniques, such as optical projection tomography (OPT) (Sharpe et al. 2002) and selective plane illumination microscopy (SPIM) (Huisken et al. 2004), have been developed. OPT is the optical equivalent of x-ray computed tomography, where a sequence of optically transmitted images are recorded through the specimen, which is rotated at various angles. The acquired projected images are later combined for the reconstruction of the 3D structure of the specimen. In addition to the transmission mode of operation, OPT can also be used for imaging fluorescence from selectively labeled tissues. OPT development leads its wide range of applications such as imaging of gene expression (Oldham et al. 2006), embryonic development (Sharpe et al. 2002), transgenic imaging (Caruana et al. 2006), and fluorescence immunohistochemistry (Delaurier et al. 2006) and in different tissue type and organism phenotyping (Arques et al. 2007; Lee et al. 2006). Most recently, OPT has been used to obtain molecular signaling information and extended to fluorescence lifetime imaging applications (McGinty et al. 2011). Figure 6.22 shows the currently available imaging methods based on the length scale of the specimens.

The major limitation of optical tomographic techniques is their applicability for mainly optically transparent specimens. However, the tissue optical properties strongly depend on the wavelength, which results to significant scattering while light passes through the biological tissues. Chemical clearing techniques are required for optical tomographic imaging applications in less transparent and thick tissue specimens. This, however, not only restricts it from *in vivo* imaging applications but also may have an untoward effect on the properties of genetically expressed fluorescence proteins (Sakhalkar et al. 2007).

6.5.1 OPTICAL PROJECTION TOMOGRAPHY

Figure 6.23 shows the optical schematic and optical light path of the OPT. Homogeneous brightness is used to illuminate the specimen, and the light can be absorbed either due to optical properties of the specimen or by contrast agents used by fluorescence probes. The specimen is rotated about a particular axis, and the imaging is performed from an axis perpendicular to the rotation axis. The images are captured

(a)

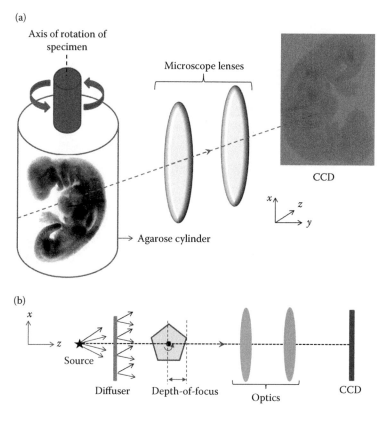

Axis of rotation of specimen

Microscope lenses

CCD

Agarose cylinder

(b)

x

z

Source

Diffuser

Depth-of-focus

Optics

CCD

Figure 6.23 (a) Optical schematic of OPT. The light is transmitted through the specimen, which is illuminated homogeneously. The specimen is rotating as shown in the figure, and images corresponding to multiple angles are recorded by the CCD. (b) Light path shows that the DOF of the imaging optics covers the front half of the specimen. (From Sharpe, J., *Annual Review of Biomedical Engineering* 6, 209–28, 2004.)

corresponding to the series of angular rotation (usually <1°) of the specimen. An absorption tomogram is recorded for the light influenced either by the specimen's optical properties or due to fluorescence probes, and thus its reconstruction provides the tomographic map of the corresponding absorption coefficient of the specimen. The reconstruction method commonly used in projection tomography is back-projection (Kak and Slaney 1999). In case of fluorescence imaging applications, emitted light from the contrast agents was detected by using a fluorescence filter in front of the CCD detector. For the applications in less transparent and/or thick tissue imaging applications of OPT, the common approach is to place the specimen in the index-matching liquid. It helps to further suppress the heterogeneities of the refractive index of the specimen.

For a given specimen, a series of projection images are recorded corresponding to the different orientations. For an *XZ* plane of the specimen, let θ be the angle for the projection and *r* the transverse coordinate. Suppose *I* is the integrated signal at the CCD plane and the recorded image is represented by $I_\theta(r)$. In order to perform the tomographic reconstruction, first a Fourier transform on each projection is performed in 1D. It can be represented by $\tilde{I}_\theta(w)$, where *w* is the spatial frequency coordinate. A filtering process is performed to compensate the unequal sample effect, followed by the inverse Fourier transform. The intensity reconstruction of the *XZ* plane is back-projected corresponding to the projection angles. The reconstructed intensity can be written as (McGinty et al. 2008)

$$I_R(x,z) = \int_0^\pi \left[\int_{-\infty}^\infty \tilde{I}_\theta(w)|w|e^{2\pi iwr}\,dw \right] d\theta \tag{6.33}$$

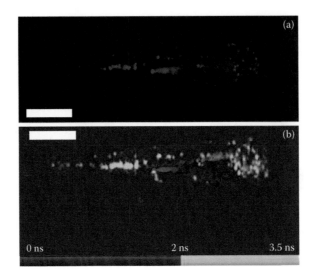

Figure 6.24 (a) Time-integrated 3D fluorescence intensity; (b) 3D fluorescence lifetime rending of a mouse embryo. (From McGinty, J. et al., *Biomedical Optics Express* 2, 1340–50, 2011.)

Here $|w|$ is the ramp filter. This reconstruction, which only holds for the object plane, is located at the plane of the axis of rotation.

The imaging lenses of the microscope, used to record the images in the OCT system, provide a limited depth-of-focus (DOF) range. This limited DOF usually does not comprise the entire specimen size (in axial direction). One way to address this issue is to record the tomogram at different depths, that is, for a particular z position, a series of projections are acquired at different orientations. The alternative way is to reduce the NA of the detection optics, which provides the larger DOF. This method partially solves the problem as the lower NA also reduces the lateral resolution of the system. To address the issue of limited DOF in OPT, the most effective solution is to position the focal plane of the microscope between the axis of rotation and the edge of the specimen closest to the objective lens as shown in Figure 6.23b (Sharpe 2004; Sharpe et al. 2002). In such a situation, each image recorded comprises the focused data from the front half of the specimen (close to the imaging lens side) and the out-of-focus data from the remaining half of the specimen. This method avoids the need to record the images at multiple depths, and the reconstruction provides the maximized focused information after using the modified back-projection algorithm. Application of OPT for 3D time-integrated fluorescence intensity and 3D fluorescence lifetime rending in a mouse embryo is shown in Figure 6.24.

6.5.2 SELECTIVE PLANE ILLUMINATION MICROSCOPY

In SPIM, the specimen is illuminated by a "sheet" of light, perpendicular to the imaging axis, defining the optical sectioning of this technique. In the ideal case, fluorescence signal should only be generated at the plane of excitation of the specimen, and the emission fluorescence can be collected at either side of the light "sheet." The imaging is performed using the conventional wide-field microscopy system that determines its lateral resolution. As the thickness of the "sheet" is tailored to be thin, it is possible to achieve excellent optical sectioning of the specimen and suppression of out-of-focus light. Compared to confocal microscopy, SPIM provides significantly less photobleaching because the excitation occurs only at one plane, at a time, of the specimen perpendicular to the imaging axis.

Figure 6.25 shows the experimental schematic of SPIM. Different types of lasers can be used to excite different biological fluorophores, depending on the applications. A laser beam is passed through a cylindrical lens to generate the laser "sheet" at the focal plane of the collection objective, which penetrates into the sample. Thus, only the fluorophores in the vicinity of the focal plane of the collection objective are excited. The collection objective is inserted into a tightly sealed liquid-filled chamber, and the sample is mounted into a water- or buffer-filled sample chamber. Precise alignment of the excitation beam and collection objective is critical. Collected fluorescence emission is passed through an emission filter for

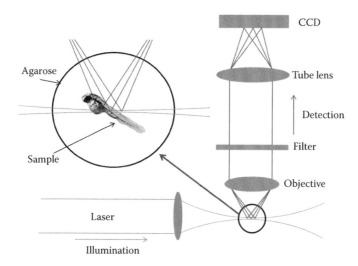

Figure 6.25 Optical system arrangement of SPIM. A cylindrical lens used to create a sheet of light that penetrates the sample. Inset shows the close look of the illumination and selection mechanism in SPIM. (From Huisken, J. et al., *Science* 305, 1007–9, 2004.)

fluorescence detection by using the camera, for example, an electron-multiplying CCD (EMCCD) or single-photon avalanche diode (SPAD) array.

SPIM was demonstrated for various biological applications such as optical sectioning deep inside live embryos (Huisken et al. 2004), 3D visualization of the neuronal network in the whole mouse brain (Dodt et al. 2007), single molecule imaging in cells (Tokunaga et al. 2008), and deep and fast live imaging using two-photon SPIM (Truong et al. 2011). Figure 6.26 demonstrates the long-term development imaging of zebrafish embryos for 3D cell tracking and migration.

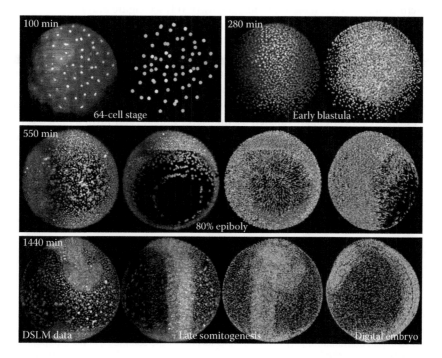

Figure 6.26 Long-term development imaging of zebrafish embryo (diameter ~700 μm) with SPIM allowing 3D tracking of cell division and migration over 24 h. (From Keller, P. J. et al., *Science* 322, 1065–9, 2008.)

6.6 DIFFUSIVE OPTICAL TOMOGRAPHY-BASED IMAGER

Endomicroscopy and sheet light microscopy focus on high-resolution imaging with a maximum penetration depth on the order of millimeters. If spatial resolution on the order of the millimeter scale is acceptable, then optical imaging on the centimeter scale can be achieved using diffusive light scattering. Interested readers are also directed to Chapter 22, in which tomographic FLIM of animals using exogenous probes is discussed.

The region of the electromagnetic spectrum between approximately 650 and 1300 nm is often referred to as the "therapeutic window." In this region, absorption by tissue chromophores is sufficiently low that absorption measurements can be carried out over several centimeters of tissue (Jobsis 1977). Absorption is not the only process undergone by photons, however; many tissues are highly scattering, with scattering coefficients on the order of 1–4 mm^{-1} (Simpson et al. 1998; Tromberg et al. 2000). It is scattering, which is the primary limitation on imaging depth and resolution for most tissues. Several approaches for overcoming scattering have been tried, including imaging using ballistic (i.e., only unscattered) photons (Liu et al. 1993), optical phase conjugation (Yaqoob et al. 2008), or wave-front shaping (Vellekoop and Mosk 2008).

Rather than trying to overcome scattering to achieve large imaging depths, it is possible to partially overcome the effects of scattering on conventional imaging by utilizing scattering itself. When a photon scatters multiple times in a sample, its direction of propagation becomes increasingly randomized. Its position over time therefore increasingly resembles a random walk. While the position at a given time of any single particle undergoing a random walk is difficult to predict, the law of large numbers permits the particle density as a function of position and time for a large number of particles to be very well characterized.

Diffuse optical tomography (DOT) experiments place multiple light sources and detectors around a scattering sample (Boas et al. 2001; Gibson et al. 2005; Ntziachristos et al. 2005). By modulating the light sources and measuring the light intensity at each detector, it is possible to determine the attenuation of light intensity for each source–detector pair. With a sufficiently large number of source–detector pairs, and a suitable model of the optical properties of the sample, it is possible to reconstruct images of features that are hidden from view, buried several scattering lengths below the surface. The use of fluorescence with DOT is often termed fluorescence molecular tomography (FMT; see Figure 6.27).

Fluorescence lifetime measurements extend DOT primarily by providing functional information regarding the nature of the buried feature (Das et al. 1997). Different fluorophores can be distinguished by their characteristic lifetimes, and the same fluorophore may have a different lifetime depending on how it is being bound, providing information on the local chemical environment around the fluorophore. Förster Resonance Energy Transfer (FRET) exploits this principle further. A donor and an acceptor chromophore are linked in some manner, and the lifetime of the donor chromophore gets considerably reduced when the donor–acceptor distance is below approximately 10 nm (Jares-Erijman and Jovin 2003), due to efficient energy transfer between the donor and the acceptor. This makes the FRET probe very sensitive to binding

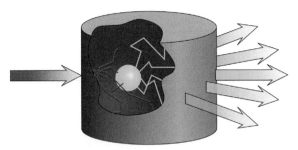

Figure 6.27 Illustration of FMT. Light illuminates one point in the sample, and some scattered photons are absorbed by the fluorescent feature. Fluorescence photons from this feature undergo multiple scattering events before exiting the sample, and the intensity profile on the surface of the sample is, in part, dependent on the location of the fluorophore and the internal structure of the sample. The experiment is repeated several times, illuminating at different locations, and the resulting intensity distributions analyzed to determine the location of the fluorescent feature.

state. In contrast, DOT measurements of absorbing features are comparatively nonspecific and do not provide the same amount of information as fluorescence.

Deducing the sample structure from the transmission between each source–detector pair is an inverse problem and consequently difficult to solve in the general case. There are many proposed methods for recovering the structure from the source–detector transmission data, as will be discussed later, but since any potential solution must be evaluated to ensure that it is consistent with the data, it is necessary to model the propagation of photons in the postulated structure.

Modeling the propagation of each photon by ray optics is avoided due to the number of scattering events that must be considered, so one approach is to use the *radiative transport equation* (RTE) (Boas et al. 2001). This equation assumes that all photons are independent of each other, that polarization effects are negligible, and that they can undergo propagation, absorption, or scattering. Scattering is assumed to be elastic, nonlinear effects are ignored, and the optical properties of the material are time-invariant.

The left-hand side of the RTE expresses how a beam of light passing through a small, homogeneous area changes. The right-hand side expresses how these changes occur. The terms on the right-hand side, in order, describe firstly the ordinary propagation of light through the homogeneous area; the negative sign demonstrates that the beam is passing through the area so radiance is being "lost." The second term describes the decrease in radiance due to both scattering and absorption. The third term is a correction term, accounting for the fact that some photons may be scattered back into the beam. The final term is the source term and describes radiance increase due to the light source:

$$
\frac{1}{v}\frac{\partial L(r,\hat{\Omega},t)}{\partial t} = -\hat{\Omega}\cdot\nabla L(r,\hat{\Omega},t) - \mu_t L(r,\hat{\Omega},t)
$$
$$
+ \mu_s \int_{4\pi} f(\hat{\Omega},\hat{\Omega}') L(r,\hat{\Omega}',t)\, d\hat{\Omega}' + Q(r,\hat{\Omega},t)
$$

(6.34)

where v is the speed of light in the medium, and $L(r,\hat{\Omega},t)$ describes the radiance (power per unit area or solid angle) as a function of position r, direction of propagation $\hat{\Omega}$, and time t. $\mu_t = \mu_a + \mu_s$ is the extinction coefficient and is equal to the sum of the absorption coefficient and the scattering coefficient. $f(\hat{\Omega},\hat{\Omega}')$ quantifies the probability that a beam of light propagating in direction $\hat{\Omega}'$ will scatter into the direction $\hat{\Omega}$. Finally, $Q(r,\hat{\Omega},t)$ describes the radiance of any light sources.

The radiation transport equation is very complex and, except in a few simple cases, impossible to solve analytically. Owing to the large number of terms, it is also computationally expensive to solve numerically. In situations where the scattering coefficient is much larger than the absorption coefficient, and the minimum optical thickness through which a photon can travel is much larger than a few scattering lengths, it may be simplified into the *photon diffusion equation*.

The diffusion equation assumes that all the photons are undergoing a random walk as they propagate; hence, there is no $\hat{\Omega}$ term. The $L(r,\hat{\Omega},t)$ term can therefore be replaced by a weighted sum of $\Phi(r,t)$, which is equivalent to $L(r,\hat{\Omega},t)$ integrated over all propagation directions, and a further term $J(r,t)$, which describes the photon flux. By making the assumption that $\partial J(r,t)/\partial t = 0$, the substitution $J(r,t) = -\nabla\Phi(r,t)/3(\mu_a + \mu_s')$ can be made, simplifying the equation further.

The description of the photon propagation is equal to a hypothetical perfectly isotropic source $S(r,t)$ and is a function of $D = 1/3(\mu_a + \mu_s')$, the diffusion constant, which is itself a function of $\mu_s' = (1-g)\mu_s$, the reduced scattering coefficient, where g is the scattering anisotropy:

$$
-\nabla\cdot D\nabla\Phi(r,t) + \mu_a\Phi(r,t) + \frac{1}{v}\frac{\partial\Phi(r,t)}{\partial t} = S(r,t)
$$

(6.35)

A rigorous justification for this transformation, including further details on precisely when these approximations are suitable, can be found in Arridge (1999).

Several research groups have constructed systems to reconstruct an image of the fluorescence intensity and lifetime of a highly scattering sample. Since the focus of this chapter is on instrumentation, the focus of this review will be on instruments that have actually been built and tested, with an emphasis on demonstrating how FMT has become more sophisticated over time. Since scattering and absorption in a sample can alter the measured fluorescence lifetime, it is important to consider these effects when performing an imaging study. The nature of these effects is beyond the scope of this chapter, but interested readers are directed to Chapter 22, which covers these effects in detail.

The Feld group first tried to overcome the scattering problem by time-gating and imaging using only the unscattered photons (Wu et al. 1995), but it would be two more years until Randall Barbour's group demonstrated that it was theoretically possible to recover lifetime data from a cylindrical tissue phantom containing a balloon of Rhodamine 6G. An argon ion laser was used to illuminate the sample from different angles, and several detectors surrounding the cylinder were used to reconstruct the fluorescence intensity distribution. A theoretical method for reconstructing the lifetime was also included but not experimentally demonstrated (Chang et al. 1997a, b).

Another interesting instrument where lifetime properties must be determined in a highly scattering medium (namely milk) was described by the group of Sergio Fantini (Cerussi et al. 1999). In this case, the change in lifetime of ethidium bromide was used as a measure for the concentration of cells present in milk, which is diagnostic for subclinical mastitis in cows. The instrument consisted of a modulated argon ion laser with the fluorescence signal detected using a PMT, and was successful in detecting cell concentration over a range of 10^3 to 10^6 cells per milliliter. Shives et al. (2002) showed in a similar application that FMT could be used to measure the oxygen concentration in a phantom, measured using 16 source fibers and 16 detector fibers with diode laser as the light source and a photomultiplier as the detector.

Graves et al. (2003) have constructed a number of FMT instruments, including a compression chamber for live small-animal imaging. Light from a laser diode is guided to the sample via an optical fiber and imaged using a 512 × 512 cooled CCD array. Similarly, the group of Eva M. Sevick-Muraca has constructed a breast-tissue phantom, which they have imaged using frequency-domain photon migration (Godavarty et al. 2003, 2005; Roy et al. 2007). The setup consists of a 783 nm diode laser modulated at 100 MHz illuminating the phantom. Fluorescent light from the phantom is filtered and either imaged directly with an intensified CCD (iCCD) or optical fibers capture light from the surface of the phantom and transfer them to an "interface plate" where they can be imaged by the iCCD. This iCCD is modulated at a frequency equal to the 100 MHz source plus a small phase delay (Darne et al. 2011). By capturing images with different phase delays, the amplitude and phase delay of the fluorescence signal can be obtained.

Researchers in Paul French's group, in collaboration with the Simon Arridge's group at the University College London, also demonstrated fluorescence lifetime measurements in a phantom, but captured images in the time domain, by time gating, rather than in the frequency domain (Soloviev et al. 2007b). A pulsed supercontinuum source was filtered to obtain a desired wavelength. A photodiode measured the modulation frequency, and a digital delay generator ensured that the iCCD was triggered with a defined pulse width and delay. Later developments demonstrated the measurement of fluorescent cells (Soloviev et al. 2007a) and FRET probes, both in a phantom (Soloviev et al. 2010b) and small animals (Soloviev et al. 2011). A similar system, using a laser diode instead of a supercontinuum source, was constructed by Rinaldo Cubeddu's group also in collaboration with Simon Arridge and co-workers, and was used to measure fluorescent inclusions in scattering phantoms (Brambilla et al. 2008; Soloviev et al. 2008, 2009, 2010a).

Researchers working with Joseph Culver also demonstrated imaging *in vivo*, with an instrument consisting of a Ti:Sapphire laser scanned using a galvanometer and an iCCD measuring transmission of light through a mouse. This enabled a very large number of source–detector pairs to be measured, but, like other CCD-based approaches, at the cost of not being able to image from all directions (Nothdurft et al. 2009). A demonstration, imaging the fluorescence lifetime of a tumor-targeting agent, is shown in Figure 6.28; accumulation of the fluorophore in the tumor and the liver is apparent, and the lifetime is consistent with previous results using the same dye in model systems. At about the same time, Kepshire et al. (2009)

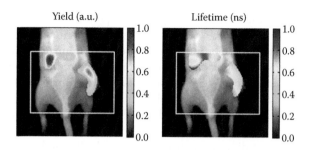

Figure 6.28 Fluorescence intensity and lifetime of the fluorophore cypate-cyclic[RGDfK] accumulating in the liver and implanted tumor of a mouse. Images are a 2D slice from a volumetric dataset; the slice is located 8 mm from the surface. (From Nothdurft, R. E. et al., *Journal of Biomedical Optics* 14, 024004, 2009.)

demonstrated that an autoexposure algorithm could be used to increase the SNR and significantly increase the dynamic range of a FMT instrument.

Gao et al. (2010) have constructed an instrument with a different design, based on time-correlated single-photon counting (TCSPC), using 16 optical fiber light sources and 16 fibers for detection. The instrument is partially parallelized, being capable of photon-counting from four fibers simultaneously, and while the experimental geometry is planar, unlike imaging using an iCCD, the fibers can be moved to other locations to capture light scattering from different directions.

Gultekin Gulsen's group has made more recent developments, combining a frequency-domain fluorescence tomography system with an MRI scanner (Lin et al. 2011). A modulated photomultiplier was used as the detector, and a modulated 785 nm laser diode was used as the light source. Optical fibers transferred light to and from the sample, and a mechanical translation stage controlled which fibers were in use at any particular time. A similar instrument was also proposed by Seetamraju et al. (2011).

Other recent developments include an analysis of the noise in frequency-domain FMT (Kang and Kupinski 2012) and a similar investigation of the uncertainty in measuring lifetime and concentration of a buried fluorophore (Reinbacher-Koestinger 2011). As a sign of the degree to which FMT is becoming an accepted imaging technique, Kumar (2010) published a protocol for the imaging of small animals with a time-domain system, permitting researchers who are not trained in FMT to achieve repeatable results. Another development to make FMT more readily accessible is Freiberger et al.'s (2011) development of a graphics processing unit (GPU)-based algorithm for significantly faster image reconstructions. Speed increases on the order of 15 times have been claimed, with no loss of accuracy.

A slightly different approach to reconstructing the position and shape of a buried fluorescent sample is to combine fluorescent and acoustic tomography (Kobayashi et al. 2006). Because acoustic waves have much longer scattering lengths compared to photons in turbid media, and because the acoustic wave can be used to modulate light at the focal point of the transducer, the optical properties of the fluorophore can be determined from the modulated scattered light, and the spatial resolution is determined by the focused acoustic beam.

To the authors' knowledge, there is one commercial FMT instrument available, the eXplore Optix from Advanced Research Technologies. The latest version at the time of writing, the Optix M3, steers one of up to four lasers onto the sample and collects the light scattered from a location 3 mm away, which serves to highlight deeper fluorescent features rather than those closer to the surface. Various characterizations of the system have been performed (De la Zerda et al. 2009; Keren et al. 2008; Ma et al. 2007). While this system does not have quite the versatility of a full FMT setup, it is similar to at least two other papers in the literature, which use a similar fixed distance between a source and a detector (Han and Hall 2008; Laidevant et al. 2007).

While the focus of this section is on instrumentation, it is necessary to discuss the more popular methods used to reconstruct the sample structure from the data. One highly cited paper on the topic of reconstructing images of fluorescent samples is by O'Leary et al. (1996), who described the use of diffuse photon-density waves to model the imaging of fluorescent entities in a highly scattering medium. Other early papers emphasized the finite element model aspect of the problem (Jiang 1998) and methods for attaining a solution (Paithankar et al. 1997). The general approach was further refined by Ntziachristos and Weissleder (2001), who introduced an algorithm that required no measurements prior to the introduction of the fluorophore. They proposed a normalized Born approximation to simplify the reconstruction using frequency-domain data and demonstrated that two fluorescent Cy5.5 samples could be resolved within an Intralipid bath. Milstein et al. (2003, 2004) cast the problem as a Bayesian optimization, claiming improved performance over that of Ntziachristos and Weissleder when incorporating data from more than one wavelength. Furthermore, Gao et al. (2006) extended the method into the time domain by performing a Laplace transform. A different approach was taken by Roy and Sevick-Muraca (1999a, b). They focused more on the optimization methods, finite element model, and boundary conditions in order to demonstrate robust reconstruction of the sample structure. Fedele et al. (2003) also attempted to reduce the calculational requirements for reconstructing an image, developing a vectorized implementation to take advantage of modern computer architectures. In contrast, Klose and Hielscher (2003) opted to solve the RTE rather than the photon diffusion equation, although no accommodation was made for the fluorescence lifetime. Graves et al. (2004) investigated the optimal size and distribution of sources and detectors for small animal imaging. For a 20 mm fluorophore, they found that source and detector fields of view should be around 30 mm, and for parallel-plate geometry, equal numbers of sources and detectors are optimal. An early development in the modeling of photon transport was made by Chang et al. (1997b), who not only used the RTE but also included a correction for fluorescence saturation. Their model was easily solved using iterative methods but crucially was a perturbation-based method, and therefore required knowledge of the homogeneous medium in which the fluorophores were embedded. Lin et al. (2007) later demonstrated that attempting to recover the shape and lifetime of a fluorescent feature without this *a priori* information significantly increased the error in the reconstruction. Kumar et al. (2006) provided a method for reconstructing the fluorescence lifetime distribution entirely in the time domain, without the need for transformation into the frequency domain. They demonstrated several advantages to this approach, such as the ability to localize each fluorescence lifetime component separately (Kumar et al. 2006).

Overall, the combination of DOT and fluorescence lifetime enables the acquisition of images that are difficult or impossible to obtain by any other method. Fluorescence lifetime provides chemospecific information for many important biological features and processes using either intrinsic autofluorescence or probes that are well understood from conventional fluorescence microscopy. The primary limitation in optical tissue imaging, that of low penetration, can be partially overcome using DOT, and with a sufficiently large number of sources and detectors, penetration depths on the order of centimeters can be achieved with spatial resolution on the order of millimeters or less. It is likely that, in future, this technique will see increased usage, particularly as instruments become easier to use, with large numbers of source and detector pairs all prealigned or under computer control.

6.7 CONCLUSION

This chapter covers a broad range of imaging and nonimaging instruments that are capable of performing fluorescence lifetime-resolved measurement in biological specimens ranging from protein solution, to cell cultures, to animal models, and finally, to patients. While these optical instruments have different degrees of complexity, a relatively thorough overview of typical optical components (e.g., light sources, light path parts, detectors) and optical principles (e.g., physics underlying different nonlinear processes, image formation theories) should enable the readers to understand the principles of their operation. Importantly, we hope that these basic overviews will enable readers to understand many other lifetime-resolved optical instruments that are not explicitly covered here.

ACKNOWLEDGMENTS

PS, HC, CJR, and VS acknowledge NIH NIBIB support from Laser Biomedical Research Center, NSF support from Emergent Behavior of Integrated Cellular System Center, support from Singapore-MIT Alliance for Research and Technology Center, and support from Singapore-MIT Alliance 2. CJR also acknowledges support from the Wellcome Trust.

REFERENCES

Adler, D. C., Chen, Y., Huber, R., Schmitt, J., Connolly, J. & Fujimoto, J. G., "Three-dimensional endomicroscopy using optical coherence tomography," *Nature Photonics* 1 (2007): 709–16.

Agnantis, N. J., Sotiriadis, A. & Paraskevaidis, E., "The current status of HPV DNA testing," *European Journal of Gynaecological Oncology* 24 (2003): 351–6.

Apgar, B. S. & Brotzman, G., "Management of cervical cytologic abnormalities," *American Family Physician* 70 (2004): 1905–16.

Arques, C. G., Doohan, R., Sharpe, J. & Torres, M., "Cell tracing reveals a dorsoventral lineage restriction plane in the mouse limb bud mesenchyme," *Development* 134 (2007): 3713–22.

Arridge, S. R., "Optical tomography in medical imaging," *Inverse Problems* 15 (1999): R41–93.

Asano, T. K. & Mcleod, R. S., "Dietary fibre for the prevention of colorectal adenomas and carcinomas," *Cochrane Database of Systematic Reviews* 1 (2002): CD003430.

Attard, A. E., "Optical bandwidth in coupling: The multicore photonic switch," *Applied Optics* 42 (2003): 2665–73.

Ban, S., Mino, M., Nishioka, N. S., Puricelli, W., Zukerberg, L. R., Shimizu, M. & Lauwers, G. Y., "Histopathologic aspects of photodynamic therapy for dysplasia and early adenocarcinoma arising in Barrett's esophagus," *The American Journal of Surgical Pathology* 28 (2004): 1466–73.

Bao, H. C., Allen, J., Pattie, R., Vance, R. & Gu, M., "Fast handheld two-photon fluorescence microendoscope with a 475 mu m X 475 mu m field of view for in vivo imaging," *Optics Letters* 33 (2008): 1333–5.

Barretto, R. P. J., Messerschmidt, B. & Schnitzer, M. J., "In vivo fluorescence imaging with high-resolution microlenses," *Nature Methods* 6 (2009): 511–12.

Belo, A. C. & Playford, R. J., "Surveillance for Barrett's oesophagus: Is there light at the end of the metaplastic tunnel?," *Surgeon* 1 (2003): 152–6.

Boas, D. A., Brooks, D. H., Miller, E. L., Dimarzio, C. A., Kilmer, M., Gaudette, R. J. & Zhang, Q., "Imaging the body with diffuse optical tomography," *IEEE Signal Processing Magazine* 18 (2001): 57–75.

Born, M. & Wolf, E., *Principles of Optics*. Oxford, Pergamon, 1987.

Born, M., Wolf, E. & Bhatia, A. B., *Principles of Optics: Electromagnetic Theory of Propagation, Interference and Diffraction of Light*. New York, Cambridge University Press, 1999.

Bouma, B. E., Tearney, G. J., Yabushita, H., Shishkov, M., Kauffman, C. R., DeJoseph Gauthier, D., MacNeill, B. D., Houser, S. L., Aretz, H. T., Halpern, E. F. & Jang, I. K., "Evaluation of intracoronary stenting by intravascular optical coherence tomography," *Heart* 89 (2003a): 317–20.

Bouma, J. L., Aronson, L. R., Keith, D. G. & Saunders, H. M., "Use of computed tomography renal angiography for screening feline renal transplant donors," *Veterinary Radiology & Ultrasound* 44 (2003b): 636–41.

Brambilla, M., Spinelli, L., Pifferi, A., Torricelli, A. & Cubeddu, R., "Time-resolved scanning system for double reflectance and transmittance fluorescence imaging of diffusive media," *Review of Scientific Instruments* 79 (2008): 013103.

Brand, S., Poneros, J. M., Bouma, B. E., Tearney, G. J., Compton, C. C. & Nishioka, N. S., "Optical coherence tomography in the gastrointestinal tract," *Endoscopy* 32 (2000): 796–803.

Caruana, G., Young, R. J. & Bertram, J. F., "Imaging the embryonic kidney," *Nephron. Experimental Nephrology* 103 (2006): e62–8.

Cerussi, A. E., Gratton, E. & Fantini, S., "Fluorescence lifetime spectroscopy in multiple-scattering environments: An application to biotechnology," *Proc. SPIE* 3600 (July 2, 1999), Biomedical Imaging: Reporters, Dyes, and Instrumentation, 171.

Chang, J., Graber, H. L. & Barbour, R. L., "Imaging of fluorescence in highly scattering media," *IEEE Transactions on Biomedical Engineering* 44 (1997a): 810–22.

Chang, J., Graber, H. L. & Barbour, R. L., "Luminescence optical tomography of dense scattering media," *Journal of the Optical Society of America A-Optics Image Science and Vision* 14 (1997b): 288–99.

Chau, A. H., Chan, R. C., Shishkov, M., MacNeill, B., Iftimia, N., Tearney, G. J., Kamm, R. D., Bouma, B. E. & Kaazempur-Mofrad, M. R., "Mechanical analysis of atherosclerotic plaques based on optical coherence tomography," *Annals of Biomedical Engineering* 32 (2004): 1494–503.

Corcuff, P., Bertrand, C. & Leveque, J. L., "Morphometry of human epidermis in vivo by real-time confocal microscopy," *Archives of Dermatological Research* 285 (1993): 475–81.

Cova, S., Ghioni, M., Lacaita, A., Samori, C. & Zappa, F., "Avalanche photodiodes and quenching circuits for single photon detection," *Applied Optics* 35 (1996): 1956–76.

Cushman, P. B. & Heering, A. H., "Problems and solutions in high-rate multichannel hybrid photodiode design: The CMS experience," *IEEE Transactions on Nuclear Science* 49 (2002): 963–70.

Cushman, P., Heering, A., Nelson, J., Timmermans, C., Dugad, S. R., Katta, S. & Tonwar, S., "Multi-pixel hybrid photodiode tubes for the CMS hadron calorimeter," *Nuclear Instruments and Methods* A387 (1997): 107–12.

Dacosta, R. S., Wilson, B. C. & Marcon, N. E., "New optical technologies for earlier endoscopic diagnosis of premalignant gastrointestinal lesions," *Journal of Gastroenterology and Hepatology* 17 (2002): S85–104.

Darne, C., Zhu, B., Lu, Y., Tan, I. C., Rasmussen, J. & Sevick-Muraca, E. M., "Radiofrequency circuit design and performance evaluation for small animal frequency-domain NIR fluorescence optical tomography," *Proc. SPIE* 7896 (February 17, 2011), Optical Tomography and Spectroscopy of Tissue IX, 789621.

Das, B. B., Liu, F. & Alfano, R. R., "Time-resolved fluorescence and photon migration studies in biomedical and model random media," *Reports on Progress in Physics* 60 (1997): 227–92.

Dautet, H. L., "Photon-counting techniques with silicon avalanche photodiode," *Applied Optics* 32 (1993): 3894–900.

Davidson, D. & Baldock, R., "Bioinformatics beyond sequence: Mapping gene function in the embryo," *Nature Reviews. Genetics* 2 (2001): 409–17.

De la Zerda, A., Bodapati, S., Teed, R., Schipper, M. L., Keren, S., Smith, B. R., Ng, J. S. & Gambhir, S. S., "A comparison between time domain and spectral imaging systems for imaging quantum dots in small living animals," *Molecular Imaging and Biology* 12 (2009): 500–8.

De Palo, G., "Cervical precancer and cancer, past, present and future," *European Journal of Gynaecological Oncology* 25 (2004): 269–78.

Deisseroth, K., Feng, G., Majewska, A. K., Miesenbock, G., Ting, A. & Schnitzer, M. J., "Next-generation optical technologies for illuminating genetically targeted brain circuits," *The Journal of Neuroscience* 26 (2006): 10380–6.

Delaurier, A., Schweitzer, R. & Logan, M., "Pitx1 determines the morphology of muscle, tendon, and bones of the hindlimb," *Developmental Biology* 299 (2006): 22–34.

Denk, W., Strickler, J. H. & Webb, W. W., "Two-photon laser scanning fluorescence microscopy," *Science* 248 (1990): 73–6.

Dickensheets, D. L. & Kino, G. S., "Micromachined scanning confocal optical microscope," *Optics Letters* 21 (1996): 764–6.

Dodt, H. U., Leischner, U., Schierloh, A., Jahrling, N., Mauch, C. P., Deininger, K., Deussing, J. M., Eder, M., Zieglgansberger, W. & Becker, K., "Ultramicroscopy: Three-dimensional visualization of neuronal networks in the whole mouse brain," *Nature Methods* 4 (2007): 331–6.

Dunleavey, R., "Incidence, pathophysiology and treatment of cervical cancer," *Nursing Times* 100 (2004): 38–41.

Durst, M. E., Zhu, G. H. & Xu, C., "Simultaneous spatial and temporal focusing for axial scanning," *Optics Express* 14 (2006): 12243–54.

Eliakim, A. R., "Video capsule endoscopy of the small bowel (PillCam SB)," *Current Opinion in Gastroenterology* 22 (2006): 124–7.

Falk, G. W., "Barrett's esophagus," *Gastroenterology* 122 (2002): 1569–91.

Faruqi, S. A., Arantes, V. & Bhutani, M. S., "Barrett's esophagus: Current and future role of endosonography and optical coherence tomography," *Diseases of the Esophagus* 17 (2004): 118–23.

Fedele, F., Laible, J. P. & Eppstein, M. J., "Coupled complex adjoint sensitivities for frequency-domain fluorescence tomography: Theory and vectorized implementation," *Journal of Computational Physics* 187 (2003): 597–619.

Fisher, L. R. & Hasler, W. L., "New vision in video capsule endoscopy: Current status and future directions," *Nature Reviews. Gastroenterology & Hepatology* 9 (2012): 392–405.

Flusberg, B. A., Lung, J. C., Cocker, E. D., Anderson, E. P. & Schnitzer, M. J., "In vivo brain imaging using a portable 3.9 gram two-photon fluorescence microendoscope," *Optics Letters* 30 (2005): 2272–4.

Freiberger, M., Egger, H., Liebmann, M. & Scharfetter, H., "High-performance image reconstruction in fluorescence tomography on desktop computers and graphics hardware," *Biomedical Optics Express* 2 (2011): 3207–22.

Fujimoto, J. G., Boppart, S. A., Tearney, G. J., Bouma, B. E., Pitris, C. & Brezinski, M. E., "High resolution in vivo intra-arterial imaging with optical coherence tomography," *Heart* 82 (1999): 128–33.

Gao, F., Li, J., Zhang, L., Poulet, P., Zhao, H. & Yamada, Y., "Simultaneous fluorescence yield and lifetime tomography from time-resolved transmittances of small-animal-sized phantom," *Applied Optics* 49 (2010): 3163–72.

Gao, F., Zhao, H., Tanikawa, Y. & Yamada, Y., "A linear, featured-data scheme for image reconstruction in time-domain fluorescence molecular tomography," *Optics Express* 14 (2006): 7109–24.

Gibson, A. P., Hebden, J. C. & Arridge, S. R., "Recent advances in diffuse optical imaging," *Physics in Medicine and Biology* 50 (2005): R1–43.

Gobel, W., Kerr, J. N., Nimmerjahn, A. & Helmchen, F., "Miniaturized two-photon microscope based on a flexible coherent fiber bundle and a gradient-index lens objective," *Optics Letters* 29 (2004): 2521–3.

Godavarty, A., Eppstein, M. J., Zhang, C., Theru, S., Thompson, A. B., Gurfinkel, M. & Sevick-Muraca, E. M., "Fluorescence-enhanced optical imaging in large tissue volumes using a gain-modulated ICCD camera," *Physics in Medicine and Biology* 48 (2003): 1701–20.

Godavarty, A., Sevick-Muraca, E. M. & Eppstein, M. J., "Three-dimensional fluorescence lifetime tomography," *Medical Physics* 32 (2005): 992–1000.

Golding, M. I., Doman, D. B. & Goldberg, H. J., "Take your Pill(Cam): It might save your life," *Gastrointestinal Endoscopy* 62 (2005): 196–8.

Gratton, E., Breusegem, S., Sutin, J., Ruan, Q. & Barry, N., "Fluorescence lifetime imaging for the two-photon microscope: Time-domain and frequency-domain methods," *Journal of Biomedical Optics* 8 (2003): 381–90.

Graves, E. E., Culver, J. P., Ripoll, J., Weissleder, R. & Ntziachristos, V., "Singular-value analysis and optimization of experimental parameters in fluorescence molecular tomography," *Journal of the Optical Society of America A-Optics Image Science and Vision* 21 (2004): 231–41.

Graves, E. E., Ripoll, J., Weissleder, R. & Ntziachristos, V., "A submillimeter resolution fluorescence molecular imaging system for small animal imaging," *Medical Physics* 30 (2003): 901–11.

Guelrud, M. & Ehrlich, E. E., "Enhanced magnification endoscopy in the upper gastrointestinal tract," *Gastrointestinal Endoscopy Clinics of North America* 14 (2004): 461–73, viii.

Gupta, S. & Sodhani, P., "Why is high grade squamous intraepithelial neoplasia under-diagnosed on cytology in a quarter of cases? Analysis of smear characteristics in discrepant cases," *Indian Journal of Cancer* 41 (2004): 104–8.

Hamamatsu 2002. *Guide to Streak Cameras.* Hamamatsu, Japan, Hamamatsu Inc.

Hamamatsu 2006. *Photomultiplier Tubes: Basics and Applications.* Hamamatsu, Japan, Hamamatsu, Inc.

Han, S. H. & Hall, D. J., "Estimating the depth and lifetime of a fluorescent inclusion in a turbid medium using a simple time-domain optical method," *Optics Letters* 33 (2008): 1035–7.

Hart, S. D., Maskaly, G. R., Temelkuran, B., Prideaux, P. H., Joannopoulos, J. D. & Fink, Y., "External reflection from omnidirectional dielectric mirror fibers," *Science* 296 (2002): 510–13.

Hayashida, M., Hose, J., Laatiaoui, M., Lorenz, E., Mirzoyan, R., Teshima, M., Fukasawa, A., Hotta, Y., Errando, M. & Martinez, M. Development of HPDs with an 18-mm-diameter GaAsP photo cathode for the MAGIC-II project. In: 29th International Cosmic Ray Conference, Pune, 2005.

Helmchen, F., Fee, M. S., Tank, D. W. & Denk, W., "A miniature head-mounted two-photon microscope: High-resolution brain imaging in freely moving animals," *Neuron* 31 (2001): 903–12.

Himmer, P. A., Dickensheets, D. L. & Friholm, R. A., "Micromachined silicon nitride deformable mirrors for focus control," *Optics Letters* 28 (2001): 1280–2.

Hoy, C. L., Durr, N. J., Chen, P. Y., Piyawattanametha, W., Ra, H., Solgaard, O. & Ben-Yakar, A., "Miniaturized probe for femtosecond laser microsurgery and two-photon imaging," *Optics Express* 16 (2008): 9996–10005.

Huisken, J., Swoger, J., Del Bene, F., Wittbrodt, J. & Stelzer, E. H., "Optical sectioning deep inside live embryos by selective plane illumination microscopy," *Science* 305 (2004): 1007–9.

Ibrahim, T. A., Amarnath, K., Kuo, L. C., Grover, R., Van, V. & Ho, P. T., "Photonic logic NOR gate based on two symmetric microring resonators," *Optics Letters* 29 (2004): 2779–81.

Inoué, S. & Spring, K. R., *Video Microscopy: The Fundamentals.* New York, Plenum Press, 1997.

Jacobson, B. C. & Van Dam, J., "Enhanced endoscopy in inflammatory bowel disease," *Gastrointestinal Endoscopy Clinics of North America* 12 (2002): 573–87.

Janesick, J., "Dueling detectors," *OE Magazine* 2 (2002): 30–3.

Jares-Erijman, E. A. & Jovin, T. M., "FRET imaging," *Nature Biotechnology* 21 (2003): 1387–95.

Jean, F., Bourg-Heckly, G. & Viellerobe, B., "Fibered confocal spectroscopy and multicolor imaging system for in vivo fluorescence analysis," *Optics Express* 15 (2007): 4008–17.

Jiang, H., "Frequency-domain fluorescent diffusion tomography: A finite-element-based algorithm and simulations," *Applied Optics* 37 (1998): 5337–43.

Jobsis, F. F., "Noninvasive, infrared monitoring of cerebral and myocardial oxygen sufficiency and circulatory parameters," *Science* 198 (1977): 1264–7.

Jung, J. C. & Schnitzer, M. J., "Multiphoton endoscopy," *Optics Letters* 28 (2003): 902–4.

Jung, W. Y., Tang, S., Mccormic, D. T., Xie, T. Q., Ahn, Y. C., Su, J. P., Tomov, I. V., Krasieva, T. B., Tromberg, B. J. & Chen, Z. P., "Miniaturized probe based on a microelectromechanical system mirror for multiphoton microscopy," *Optics Letters* 33 (2008): 1324–6.

Kak, A. C. & Slaney, M., *Principles of Computerized Tomographic Imaging.* New York, IEEE Press, 1999.

Kang, D. & Kupinski, M. A., "Noise characteristics of heterodyne/homodyne frequency-domain measurements," *Journal of Biomedical Optics* 17 (2012): 015002.

Kara, M., Dacosta, R. S., Wilson, B. C., Marcon, N. E. & Bergman, J., "Autofluorescence-based detection of early neoplasia in patients with Barrett's esophagus," *Digestive Diseases* 22 (2004): 134–41.

Kaufman, K., "Choosing your detector," *OE Magazine* (2005): 25–7.

Keller, P. J., Schmidt, A. D., Wittbrodt, J. & Stelzer, E. H., "Reconstruction of zebrafish early embryonic development by scanned light sheet microscopy," *Science* 322 (2008): 1065–9.

Kepshire, D. L., Dehghani, H., Leblond, F. & Pogue, B. W., "Automatic exposure control and estimation of effective system noise in diffuse fluorescence tomography," *Optics Express* 17 (2009): 23272–83.

Keren, S., Gheysens, O., Levin, C. S. & Gambhir, S. S., "A comparison between a time domain and continuous wave small animal optical imaging system," *IEEE Transactions on Medical Imaging* 27 (2008): 58–63.

Kiesslich, R., Burg, J., Vieth, M., Gnaendiger, J., Enders, M., Delaney, P., Polglase, A., McLaren, W., Janell, D., Thomas, S., Nafe, B., Galle, P. R. & Neurath, M. F., "Confocal laser endoscopy for diagnosing intraepithelial neoplasias and colorectal cancer in vivo," *Gastroenterology* 127 (2004): 706–13.

Kim, P., Chung, E., Yamashita, H., Hung, K. E., Mizoguchi, A., Kucherlapati, R., Fukumura, D., Jain, R. K. & Yun, S. H., "In vivo wide-area cellular imaging by side-view endomicroscopy," *Nature Methods* 7 (2010): 303–5.

Klose, A. D. & Hielscher, A. H., "Fluorescence tomography with simulated data based on the equation of radiative transfer," *Optics Letters* 28 (2003): 1019–21.

Kobayashi, M., Mizumoto, T., Shibuya, Y., Enomoto, M. & Takeda, M., "Fluorescence tomography in turbid media based on acousto-optic modulation imaging," *Applied Physics Letters* 89 (2006): 181102.

Konig, K., Ehlers, A., Riemann, I., Schenkl, S., Buckle, R. & Kaatz, M., "Clinical two-photon microendoscopy," *Microscopy Research and Technique* 70 (2007): 398–402.

Konig, K., Speicher, M., Buckle, R., Reckfort, J., McKenzie, G., Welzel, J., Köhler, M. J., Elsner, P. & Kaatz, M., "Clinical optical coherence tomography combined with multiphoton tomography of patients with skin diseases," *Journal of Biophotonics* 2 (2009): 389–97.

Krishnan, R. V., Masuda, A., Centonze, V. E. & Herman, B., "Quantitative imaging of protein-protein interactions by multiphoton fluorescence lifetime imaging microscopy using a streak camera," *Journal of Biomedical Optics* 8 (2003): 362–7.

Kumar, A. T., "Fluorescence lifetime-based optical molecular imaging," *Methods in Molecular Biology* 680 (2010): 165–80.

Kumar, A. T., Raymond, S. B., Boverman, G., Boas, D. A. & Bacskai, B. J., "Time resolved fluorescence tomography of turbid media based on lifetime contrast," *Optics Express* 14 (2006): 12255–70.

Laidevant, A., Da Silva, A., Berger, M., Boutet, J., Dinten, J. M. & Boccara, A. C., "Analytical method for localizing a fluorescent inclusion in a turbid medium," *Applied Optics* 46 (2007): 2131–7.

Lambert, R., "Diagnosis of esophagogastric tumors," *Endoscopy* 36 (2004): 110–19.

Lambrou, N. C. & Twiggs, L. B., "High-grade squamous intraepithelial lesions," *Cancer Journal* 9 (2003): 382–9.

Lee, J. H., Chen, S. Y., Yu, C. H., Chu, S. W., Wang, L. F., Sun, C. K. & Chiang, B. L., "Noninvasive in vitro and in vivo assessment of epidermal hyperkeratosis and dermal fibrosis in atopic dermatitis," *Journal of Biomedical Optics* 14 (2009).

Lee, K., Avondo, J., Morrison, H., Blot, L., Stark, M., Sharpe, J., Bangham, A. & Coen, E., "Visualizing plant development and gene expression in three dimensions using optical projection tomography," *Plant Cell* 18 (2006): 2145–56.

Levene, M. J., Dombeck, D. A., Kasischke, K. A., Molloy, R. P. & Webb, W. W., "In vivo multiphoton microscopy of deep brain tissue," *Journal of Neurophysiology* 91 (2004): 1908–12.

Li, X. D., Boppart, S. A., Van Dam, J., Mashimo, H., Mutinga, M., Drexler, W., Klein, M., Pitris, C., Krinsky, M. L., Brezinski, M. E. & Fujimoto, J. G., "Optical coherence tomography: Advanced technology for the endoscopic imaging of Barrett's esophagus," *Endoscopy* 32 (2000): 921–30.

Liang, C., Sung, K. B., Richards-Kortum, R. R. & Descour, M. R., "Design of a high-numerical-aperture miniature microscope objective for an endoscopic fiber confocal reflectance microscope," *Applied Optics* 41 (2002): 4603–10.

Lin, Y., Gao, H., Nalcioglu, O. & Gulsen, G., "Fluorescence diffuse optical tomography with functional and anatomical a priori information: Feasibility study," *Physics in Medicine and Biology* 52 (2007): 5569–85.

Lin, Y., Ghijsen, M. T., Gao, H., Liu, N., Nalcioglu, O. & Gulsen, G., "A photo-multiplier tube-based hybrid MRI and frequency domain fluorescence tomography system for small animal imaging," *Physics in Medicine and Biology* 56 (2011): 4731–47.

Liu, F., Yoo, K. M. & Alfano, R. R., "Ultrafast laser-pulse transmission and imaging through biological tissues," *Applied Optics* 32 (1993): 554–8.

Ma, G., Gallant, P. & McIntosh, L., "Sensitivity characterization of a time-domain fluorescence imager: eXplore Optix," *Applied Optics* 46 (2007): 1650–7.

Masters, B. R., *Selected Papers on Confocal Microscopy*. Bellingham, WA, SPIE, 1996.

Masters, B. R. & So, P. T. C. (eds.), *Handbook of Biomedical Non-Linear Optical Microscopy*. New York, Oxford University Press, 2008.

Masters, B. R., So, P. T. C. & Gratton, E., "Multiphoton excitation fluorescence microscopy and spectroscopy of in vivo human skin," *Biophysical Journal* 72 (1997): 2405–12.

McConnell, G. & Riis, E., "Two-photon laser scanning fluorescence microscopy using photonic crystal fiber," *Journal of Biomedical Optics* 9 (2004): 922–7.

McGinty, J., Tahir, K. B., Laine, R., Talbot, C. B., Dunsby, C., Neil, M. A., Quintana, L., Swoger, J., Sharpe, J. & French, P. M., "Fluorescence lifetime optical projection tomography," *Journal of Biophotonics* 1 (2008): 390–4.

McGinty, J., Taylor, H. B., Chen, L., Bugeon, L., Lamb, J. R., Dallman, M. J. & French, P. M. W., "In vivo fluorescence lifetime optical projection tomography," *Biomedical Optics Express* 2 (2011): 1340–50.

Milstein, A. B., Oh, S., Webb, K. J., Bouman, C. A., Zhang, Q., Boas, D. A. & Millane, R. P., "Fluorescence optical diffusion tomography," *Applied Optics* 42 (2003): 3081–94.

Milstein, A. B., Stott, J. J., Oh, S., Boas, D. A., Millane, R. P., Bouman, C. A. & Webb, K. J., "Fluorescence optical diffusion tomography using multiple-frequency data," *Journal of the Optical Society of America A-Optics Image Science and Vision* 21 (2004): 1035–49.

Minsky, M., Microscopy apparatus. US Patent 3,013,467, 1961.

Momose, A., Takeda, T., Itai, Y. & Hirano, K., "Phase-contrast X-ray computed tomography for observing biological soft tissues," *Nature Medicine* 2 (1996): 473–5.

Morris, H. R., Hoyt, C. C., Miller, P. & Treado, P. J., "Liquid crystal tunable filter Raman chemical imaging," *Applied Spectroscopy* 50 (1996): 805–11.

Morris, H. R., Hoyt, C. C. & Treado, P. J., "Imaging spectrometers for fluorescence and Raman microscopy— Acoustooptic and liquid-crystal tunable filters," *Applied Spectroscopy* 48 (1994): 857–66.

Muldoon, T. J., Roblyer, D., Williams, M. D., Stepanek, V. M. T., Richards-Kortum, R. & Gillenwater, A. M., "Noninvasive imaging of oral neoplasia with a high-resolution fiber-optic microendoscope," *Head and Neck-Journal for the Sciences and Specialties of the Head and Neck* 34 (2012): 305–12.

Myaing, M. T., MacDonald, D. J. & Li, X. D., "Fiber-optic scanning two-photon fluorescence endoscope," *Optics Letters* 31 (2006): 1076–8.

Myaing, M. T., Ye, J. Y., Norris, T. B., Thomas, T., Baker, J. R., Jr., Wadsworth, W. J., Bouwmans, G., Knight, J. C. & Russell, P. S., "Enhanced two-photon biosensing with double-clad photonic crystal fibers," *Optics Letters* 28 (2003): 1224–6.

Naora, H., "Microspectrophotometry and cytochemical analysis of nucleic acids," *Science* 114 (1951): 279–80.

Nguyen, N. Q. & Leong, R. W. L., "Current application of confocal endomicroscopy in gastrointestinal disorders," *Journal of Gastroenterology and Hepatology* 23 (2008): 1483–91.

Nishioka, N. S., "Optical biopsy using tissue spectroscopy and optical coherence tomography," *Canadian Journal of Gastroenterology* 17 (2003): 376–80.

Nothdurft, R. E., Patwardhan, S. V., Akers, W., Ye, Y., Achilefu, S. & Culver, J. P., "In vivo fluorescence lifetime tomography," *Journal of Biomedical Optics* 14 (2009): 024004.

Ntziachristos, V., Ripoll, J., Wang, L. V. & Weissleder, R., "Looking and listening to light: The evolution of whole-body photonic imaging," *Nature Biotechnology* 23 (2005): 313–20.

Ntziachristos, V. & Weissleder, R., "Experimental three-dimensional fluorescence reconstruction of diffuse media by use of a normalized Born approximation," *Optics Letters* 26 (2001): 893–5.

O'Leary, M. A., Boas, D. A., Li, X. D., Chance, B. & Yodh, A. G., "Fluorescence lifetime imaging in turbid media," *Optics Letters* 21 (1996): 158–60.

Oldham, M., Sakhalkar, H., Oliver, T., Wang, Y. M., Kirpatrick, J., Cao, Y., Badea, C., Johnson, G. A. & Dewhirst, M., "Three-dimensional imaging of xenograft tumors using optical computed and emission tomography," *Medical Physics* 33 (2006): 3193–202.

Oron, D., Tal, E. & Silberberg, Y., "Scanningless depth-resolved microscopy," *Optics Express* 13 (2005): 1468–76.

Pacifico, R. J., Wang, K. K., Wongkeesong, L. M., Buttar, N. S. & Lutzke, L. S., "Combined endoscopic mucosal resection and photodynamic therapy versus esophagectomy for management of early adenocarcinoma in Barrett's esophagus," *Clinical Gastroenterology and Hepatology* 1 (2003): 252–7.

Paithankar, D. Y., Chen, A. U., Pogue, B. W., Patterson, M. S. & Sevick-Muraca, E. M., "Imaging of fluorescent yield and lifetime from multiply scattered light reemitted from random media," *Applied Optics* 36 (1997): 2260–72.

Paulson, T. G. & Reid, B. J., "Focus on Barrett's esophagus and esophageal adenocarcinoma," *Cancer Cell* 6 (2004): 11–16.

Pawley, J. B. (ed.), *Handbook of Confocal Microscopy*. New York, Plenum, 1995.

Peters, J. H. & Wang, K. K., "How should Barrett's ulceration be treated?," *Surgical Endoscopy* 18 (2004): 338–44.

Pitris, C., Jesser, C., Boppart, S. A., Stamper, D., Brezinski, M. E. & Fujimoto, J. G., "Feasibility of optical coherence tomography for high-resolution imaging of human gastrointestinal tract malignancies," *Journal of Gastroenterology* 35 (2000): 87–92.

Poneros, J. M., "Diagnosis of Barrett's esophagus using optical coherence tomography," *Gastrointestinal Endoscopy Clinics of North America* 14 (2004): 573–88, x.

Poneros, J. M. & Nishioka, N. S., "Diagnosis of Barrett's esophagus using optical coherence tomography," *Gastrointestinal Endoscopy Clinics of North America* 13 (2003): 309–23.

Qureshi, W. A., "Current and future applications of the capsule camera," *Nature Reviews Drug Discovery* 3 (2004): 447–50.

Rector, D. M., Ranken, D. M. & George, J. S., "High-performance confocal system for microscopic or endoscopic applications," *Methods* 30 (2003): 16–27.

Reinbacher-Koestinger, A., "Uncertainty analysis for fluorescence tomography with Monte Carlo method," *Proc. SPIE* 8088 (June 14, 2011), Diffuse Optical Imaging III, 80881P.

Rivera, D. R., Brown, C. M., Ouzounov, D. G., Pavlova, I., Kobat, D., Webb, W. W. & Xu, C., "Compact and flexible raster scanning multiphoton endoscope capable of imaging unstained tissue," *Proceedings of the National Academy of Sciences of the United States of America* 108 (2011): 17598–603.

Roberts, F. & Young, J. Z., "The flying-spot microscope," *Proc. IEE—Part IIIA (1952): Television* 99, 20 (1952): 747–7.

Rouse, A. R., Kano, A., Udovich, J. A., Kroto, S. M. & Gmitro, A. F., "Design and demonstration of a miniature catheter for a confocal microendoscope," *Applied Optics* 43 (2004): 5763–71.

Roy, R., Godavarty, A. & Sevick-Muraca, E. M., "Fluorescence-enhanced three-dimensional lifetime imaging: A phantom study," *Physics in Medicine and Biology* 52 (2007): 4155–70.

Roy, R. & Sevick-Muraca, E., "Truncated Newton's optimization scheme for absorption and fluorescence optical tomography: Part I theory and formulation," *Optics Express* 4 (1999a): 353–71.

Roy, R. & Sevick-Muraca, E., "Truncated Newton's optimization scheme for absorption and fluorescence optical tomography: Part II Reconstruction from synthetic measurements," *Optics Express* 4 (1999b): 372–82.

Sakashita, M., Inoue, H., Kashida, H., Tanaka, J., Cho, J. Y., Satodate, H., Hidaka, E., Yoshida, T., Fukami, N., Tamegai, Y., Shiokawa, A. & Kudo, S., "Virtual histology of colorectal lesions using laser-scanning confocal microscopy," *Endoscopy* 35 (2003): 1033–8.

Sakhalkar, H. S., Dewhirst, M., Oliver, T., Cao, Y. & Oldham, M., "Functional imaging in bulk tissue specimens using optical emission tomography: Fluorescence preservation during optical clearing," *Physics in Medicine and Biology* 52 (2007): 2035–54.

Seetamraju, M., Zhang, X., Davis, S., Gurjar, R., Myers, R., Pogue, B. W. & Entine, G., "Concurrent magnetic resonance and diffuse luminescence imaging for hypoxic tumor characterization," *Proc. SPIE* 7892 (February 10, 2011), Multimodal Biomedical Imaging VI, 78920K.

Sharma, P., "Review article: Emerging techniques for screening and surveillance in Barrett's oesophagus," *Alimentary Pharmacology & Therapeutics* 20 Suppl 5 (2004): 63–70; discussion 95–6.

Sharpe, J., "Optical projection tomography," *Annual Review of Biomedical Engineering* 6 (2004): 209–28.

Sharpe, J., Ahlgren, U., Perry, P., Hill, B., Ross, A., Hecksher-Sorensen, J., Baldock, R. & Davidson, D., "Optical projection tomography as a tool for 3D microscopy and gene expression studies," *Science* 296 (2002): 541–5.

Shives, E., Xu, Y. & Jiang, H., "Fluorescence lifetime tomography of turbid media based on an oxygen-sensitive dye," *Optics Express* 10 (2002): 1557–62.

Simpson, C. R., Kohl, M., Essenpreis, M. & Cope, M., "Near-infrared optical properties of ex vivo human skin and subcutaneous tissues measured using the Monte Carlo inversion technique," *Physics in Medicine and Biology* 43 (1998): 2465–78.

Sokolov, K., Follen, M., Aaron, J., Pavlova, I., Malpica, A., Lotan, R. & Richards-Kortum, R., "Real-time vital optical imaging of precancer using anti-epidermal growth factor receptor antibodies conjugated to gold nanoparticles," *Cancer Research* 63 (2003): 1999–2004.

Soljacic, M., Ibanescu, M., Johnson, S. G., Fink, Y. & Joannopoulos, J. D., "Optimal bistable switching in nonlinear photonic crystals," *Physical Review. E, Statistical, Nonlinear, and Soft Matter Physics* 66 (2002): 055601.

Soloviev, V. Y., D'Andrea, C., Brambilla, M., Valentini, G., Schulz, R. B., Cubeddu, R. & Arridge, S. R., "Adjoint time domain method for fluorescent imaging in turbid media," *Applied Optics* 47 (2008): 2303–11.

Soloviev, V. Y., D'Andrea, C., Mohan, P. S., Valentini, G., Cubeddu, R. & Arridge, S. R., "Fluorescence lifetime optical tomography with Discontinuous Galerkin discretisation scheme," *Biomedical Optics Express* 1 (2010a): 998–1013.

Soloviev, V. Y., D'Andrea, C., Valentini, G., Cubeddu, R. & Arridge, S. R., "Combined reconstruction of fluorescent and optical parameters using time-resolved data," *Applied Optics* 48 (2009): 28–36.

Soloviev, V. Y., McGinty, J., Stuckey, D. W., Laine, R., Wylezinska-Arridge, M., Wells, D. J., Sardini, A., Hajnal, J. V., French, P. M. W. & Arridge, S. R., "Forster resonance energy transfer imaging in vivo with approximated radiative transfer equation," *Applied Optics* 50 (2011): 6583–90.

Soloviev, V. Y., McGinty, J., Tahir, K. B., Laine, R., Stuckey, D. W., Mohan, P. S., Hajnal, J. V., Sardini, A., French, P. M. W. & Arridge, S. R., "Tomographic imaging of flourescence resonance energy transfer in highly light scattering media," *Proc. SPIE* 7573 (February 26, 2010b), Biomedical Applications of Light Scattering IV, 75730G.

Soloviev, V. Y., McGinty, J., Tahir, K. B., Neil, M. A., Sardini, A., Hajnal, J. V., Arridge, S. R. & French, P. M., "Fluorescence lifetime tomography of live cells expressing enhanced green fluorescent protein embedded in a scattering medium exhibiting background autofluorescence," *Optics Letters* 32 (2007a): 2034–6.

Soloviev, V. Y., Tahir, K. B., McGinty, J., Elson, D. S., Neil, M. A., French, P. M. & Arridge, S. R., "Fluorescence lifetime imaging by using time-gated data acquisition," *Applied Optics* 46 (2007b): 7384–91.

Su, J. P., Zhang, J., Yu, L. F. & Chen, Z. P., "In vivo three-dimensional microelectromechanical endoscopic swept source optical coherence tomography," *Optics Express* 15 (2007): 10390–6.

Tang, S., Jung, W. G., McCormick, D., Xie, T. Q., Su, J. P., Ahn, Y. C., Tromberg, B. J. & Chen, Z. P., "Design and implementation of fiber-based multiphoton endoscopy with microelectromechanical systems scanning," *Journal of Biomedical Optics* 14 (2009): 034005.

Tearney, G. J., Yabushita, H., Houser, S. L., Aretz, H. T., Jang, I. K., Schlendorf, K. H., Kauffman, C. R., Shishkov, M., Halpern, E. F. & Bouma, B. E., "Quantification of macrophage content in atherosclerotic plaques by optical coherence tomography," *Circulation* 107 (2003): 113–19.

Temelkuran, B., Hart, S. D., Benoit, G., Joannopoulos, J. D. & Fink, Y., "Wavelength-scalable hollow optical fibres with large photonic bandgaps for CO2 laser transmission," *Nature* 420 (2002): 650–3.

Theer, P. & Denk, W., "On the fundamental imaging-depth limit in two-photon microscopy," *Journal of the Optical Society of America A-Optics Image Science and Vision* 23 (2006): 3139–49.

Tokunaga, M., Imamoto, N. & Sakata-Sogawa, K., "Highly inclined thin illumination enables clear single-molecule imaging in cells," *Nature Methods* 5 (2008): 159–61.

Tromberg, B. J., Shah, N., Lanning, R., Cerussi, A., Espinoza, J., Pham, T., Svaasand, L. & Butler, J., "Non-invasive in vivo characterization of breast tumors using photon migration spectroscopy," *Neoplasia* 2 (2000): 26–40.

Truong, T. V., Supatto, W., Koos, D. S., Choi, J. M. & Fraser, S. E., "Deep and fast live imaging with two-photon scanned light-sheet microscopy," *Nature Methods* 8 (2011): 757–60.

Valeur, B., "A brief history of fluorescence and phosphorescence before the emergence of quantum theory," *Journal of Chemical Education* 88 (2011): 731–8.

Valeur, B., *Molecular Fluorescence: Principles and Applications*. New York, Wiley-VCH, 2002.

Vellekoop, I. M. & Mosk, A. P., "Universal optimal transmission of light through disordered materials," *Physical Review Letters* 101 (2008): 120601.

Wachman, E. S., Niu, W. H. & Farkas, D. L., "Imaging acousto-optic tunable filter with 0.35-micrometer spatial resolution," *Applied Optics* 35 (1996): 5220–6.

Wachman, E. S., Niu, W. H. & Farkas, D. L., "AOTF microscope for imaging with increased speed and spectral versatility," *Biophysical Journal* 73 (1997): 1215–22.

Watson, T. F., Neil, M. A. A., Juskaitis, R., Cook, R. J. & Wilson, T., "Video-rate confocal endoscopy," *Journal of Microscopy-Oxford* 207 (2002): 37–42.

Wilson, T. & Sheppard, C. J. R., *Theory and Practice of Scanning Optical Microscopy*. New York, Academic Press, 1984.

Wokosin, D. L., "Pulse duration spectra and measurements for laser scanning microscope systems," *Proc. SPIE* 5700 (April 11, 2005), Multiphoton Microscopy in the Biomedical Sciences V, 1.

Wu, J., Wang, Y., Perelman, L., Itzkan, I., Dasari, R. R. & Feld, M. S., "Three-dimensional imaging of objects embedded in turbid media with fluorescence and Raman spectroscopy," *Applied Optics* 34 (1995): 3425–30.

Wu, Y. C., Leng, Y. X., Xi, J. F. & Li, X. D., "Scanning all-fiber-optic endomicroscopy system for 3D nonlinear optical imaging of biological tissues," *Optics Express* 17 (2009a): 7907–15.

Wu, Y. C., Xi, J. F., Cobb, M. J. & Li, X. D., "Scanning fiber-optic nonlinear endomicroscopy with miniature aspherical compound lens and multimode fiber collector," *Optics Letters* 34 (2009b): 953–5.

Wu, Y. C., Zhang, Y. Y., Xi, J. F., Li, M. J. & Li, X. D., "Fiber-optic nonlinear endomicroscopy with focus scanning by using shape memory alloy actuation," *Journal of Biomedical Optics* 15 (2010): 060506.

Yabushita, H., Bouma, B. E., Houser, S. L., Aretz, H. T., Jang, I. K., Schlendorf, K. H., Kauffman, C. R., Shishkov, M., Kang, D. H., Halpern, E. F. & Tearney, G. J., "Characterization of human atherosclerosis by optical coherence tomography," *Circulation* 106 (2002): 1640–5.

Yaqoob, Z., Psaltis, D., Feld, M. S. & Yang, C. H., "Optical phase conjugation for turbidity suppression in biological samples," *Nature Photonics* 2 (2008): 110–15.

Zhu, G. H., Van Howe, J., Durst, M., Zipfel, W. & Xu, C., "Simultaneous spatial and temporal focusing of femtosecond pulses," *Optics Express* 13 (2005): 2153–9.

Zuccaro, G., Gladkova, N., Vargo, J., Feldchtein, F., Zagaynova, E., Conwell, D., Falk, G., Goldblum, J., Dumot, J., Ponsky, J., Gelikonov, G., Davros, B., Donchenko, E. & Richter, J., "Optical coherence tomography of the esophagus and proximal stomach in health and disease," *American Journal of Gastroenterology* 96 (2001): 2633–9.

Principles of fluorescence lifetime instrumentation

7

Fluorescence lifetime imaging techniques: Frequency-domain FLIM

John Paul Eichorst, Kai wen Teng, and Robert M. Clegg

Contents

7.1 FUNDAMENTAL FLUORESCENCE RESPONSE

When a fluorescent probe is excited to a higher energy level, the excited molecule returns to the ground state in a statistical manner (with a certain probability per unit time). The fluorescence lifetime is a measure of the average time that the molecule remains in the excited state following excitation.

The rate by which a molecule returns to the ground state after being excited is a first-order kinetic process competitive with other pathways of de-excitation (Birks 1970; Lakowicz 2006; Noomnarm and Clegg 2009; Valeur 2002). In the case of spontaneous emission, soon after a molecule has been excited, it relaxes within picoseconds to the lowest vibrational level of the lowest excited singlet state (the S1 singlet state). During the thermal relaxation, some excitation energy is lost to the environment. From the S1 state, several pathways of de-excitation are available and simultaneously compete kinetically to return the molecule to the ground state. Each pathway, s, of de-excitation (e.g., fluorescence, dynamic quenching, nonradiative decay, Förster resonance energy transfer [FRET]) has an associated rate constant (k_s). The rate constant for each pathway is simply the probability per unit time that the excited molecule will return to the ground state through this pathway. The rate constant for the decay from the excited state (the inverse of the average measured lifetime) is the sum of the separate rate constants of all the pathways.

In general, the rate constant for the fluorescence pathway is invariant and time-independent. However, because the lifetime of an excited fluorophore is the inverse of the total rate of leaving the excited state, the measured lifetime will exhibit different values if the rate constant of a pathway other than fluorescence changes (e.g., a variable extent of dynamic quenching, or the presence of FRET). Therefore, measured fluorescence lifetimes are excellent evaluators of the molecular environment of the fluorophore. As a result, FLIM has become popular for investigating the molecular environment in cells and tissues.

A fluorescence lifetime can be most easily understood by considering the time course of the fluorescence emission following an excitation light pulse that is short compared to the lifetime (a delta function light pulse). In this type of experiment, the fluorescence emission is referred to as the fundamental fluorescence response $F_\delta(t)$. If we assume a single lifetime (τ) with the pre-exponential amplitude (a), the emission following the short pulse will decay exponentially with time.

$$F_\delta(t) = ae^{-\frac{t}{\tau}} \tag{7.1}$$

Any additional pathways of de-excitation present in the system will increase the probability per unit time that the molecule will leave the excited state, and thereby shorten the fluorescence lifetime (Figure 7.1).

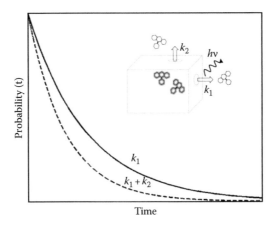

Figure 7.1 Probability that a molecule will remain in the excited state as a function of time: Following excitation by an extremely short pulse of light, the probability that a molecule will remain in the excited state decays exponentially (solid line) in time. As more de-excitation pathways become available, the molecule exits the excited state at a faster rate (dashed line), lowering the overall probability that the molecule will remain in the excited state.

For multiple individual lifetime components, the fundamental fluorescence response can be written as a sum of exponentials,

$$F_\delta(t) = \sum_{s=1}^{S} a_s e^{\frac{-t}{\tau_s}},$$ (7.2)

where (a_s) is the amplitude of the sth fluorescence component with lifetime (τ_s).

7.2 LIFETIMES IN THE FREQUENCY DOMAIN

7.2.1 THE CONVOLUTION INTEGRAL—PURE SINUSOIDAL EXCITATION; SINGLE OR MULTIPLE LIFETIME COMPONENTS

Lifetimes are measured in the frequency domain by detecting the fluorescence emission of a sample in response to a repetitively intensity-modulated excitation light. The repetition pattern of the intensity-modulated excitation light does not have to be purely sinusoidal, because any repetitive waveform can be decomposed into a sum of sinusoids by Fourier analysis (see Appendix 7.1). Each sinusoidal frequency component contributing to the repetitive excitation waveform is also present in the fluorescence signal. Each sinusoidal component can be analyzed independently as though the excitation light were modulated as a pure sinusoid at just one frequency (see Section 7.2.2). For this reason, in this section, the intensity of the modulated excitation light $E(t)$ is represented by a single harmonic oscillating at the repetition excitation radial frequency ($\omega_E = 2\pi/T$),

$$E(t) = E_o + E_{\omega_E} \cos(\omega_E t + \varphi_E).$$ (7.3)

(E_o) is the steady-state intensity, $\left(E_{\omega_E}\right)$ is the amplitude of the time-varying intensity, and (T) is the period of repetition.

The fluorescence emission of the sample emitted at time (t), $F(t)$, resulting from any waveform of excitation light, $E(t')$, is the convolution of the fundamental fluorescence $F_\delta(t)$ of Equation 7.2 with $E(t')$,

$$F(t) = \int_0^t E(t') \cdot F_\delta(t - t') \, dt'$$ (7.4)

Equations 7.2 and 7.3 can be substituted into Equation 7.4 to evaluate the convolution integral (Clegg 1996) for any number S of fluorescence species with different lifetimes.

$$F(t) = E_o \sum_{s=1}^{S} a_s \tau_s + E_{\omega_n} \sum_{s=1}^{S} a_s \tau_s M_s \cos(\omega_E t + \varphi_E - \varphi_{F,s})$$ (7.5)

Equation 7.5 shows that the contribution of each lifetime species oscillates at the same frequency (ω_E) as the intensity-modulated excitation (Figure 7.2).

Each term corresponding to a different lifetime component has a characteristic phase delay ($\varphi_{F,s}$) and modulation ratio (M_s).

$$M_s = \frac{1}{\sqrt{1 + (\omega_E \tau_s)^2}}$$ (7.6)

$$\varphi_{F,s} = \tan^{-1}(\omega_E \tau_s).$$ (7.7)

Principles of fluorescence lifetime instrumentation

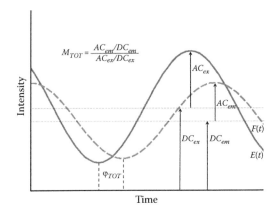

Figure 7.2 Detecting a sample's time-dependent fluorescence in the frequency domain: The intensity of both the emitted fluorescence (dashed line) and the exciting light oscillate in time. The emitted fluorescence is shifted in phase, and its amplitude of oscillation is dampened with respect to the excitation waveform (solid line). The ratio of amplitudes of oscillation ($[AC_{ex}]$ and $[AC_{em}]$) relative to their corresponding DC offsets ($[DC_{ex}]$ and $[DC_{em}]$) is measured to calculate the modulation ratio. Both the ratio of amplitudes $\left(M_{TOT} = \dfrac{AC_{em}/DC_{em}}{AC_{ex}/DC_{ex}} \right)$ and the shift in phase (φ_{TOT}) can be used to resolve the fluorescence lifetimes.

To obtain the overall measured phase delay (φ_{TOT}) and modulation ratio (M_{TOT}), $F(t)$ is first normalized to the steady-state (average) intensity (SS).

$$\frac{F(t)}{SS} = 1 + \frac{E_{\omega_n}}{E_o} \sum_{s=1}^{S} \alpha_s M_s \cos(\omega_E t + \varphi_E - \varphi_{F,s}) \tag{7.8}$$

The fractional intensity contribution of each component lifetime to the steady-state intensity is defined as $\alpha_s = a_s \tau_s \left/ \sum\limits_{s=1}^{S} a_s \tau_s \right.$. Because every term of the sum in Equation 7.8 varies at the same frequency (ω_E), the measured normalized fluorescence signal (Equation 7.8) can then be written as shown in Equation 7.9.

$$\frac{F(t)}{SS} = 1 + \frac{E_{\omega_E}}{E_o} M_{TOT} \cos(\omega_E t + \varphi_E - \varphi_{TOT}) \tag{7.9}$$

In Equation 7.9, we define the measured phase delay, (φ_{TOT}), and the measured modulation ratio, (M_{TOT}); the subscript "TOT" indicates that the measured phase delay and modulation ratio do not refer to any specific lifetime component if multiple lifetimes are present but refer to the total measured system. If there are multiple lifetimes, one sometimes still formally defines a (τ_{Mod}) and (τ_φ) as in Equations 7.10 and 7.11, but in this case, ($\tau_{Mod} > \tau_\varphi$), and neither ($\tau_{Mod}$) nor ($\tau_\varphi$) corresponds to a real lifetime.

$$M_{TOT} = \frac{1}{\sqrt{1 + (\omega_E \tau_{Mod})^2}} \tag{7.10}$$

$$\varphi_{TOT} = \tan^{-1}(\omega_E \tau_\varphi). \tag{7.11}$$

However, for "S" lifetime components, the correct relationship between (φ_{TOT}), (M_{TOT}), and the lifetimes is given by Equations 7.12 and 7.13.

$$\varphi_{TOT} = \tan^{-1}\left(\frac{\displaystyle\sum_s^S \frac{\alpha_s \omega_s \tau_s}{1+(\omega_E \tau_s)^2}}{\displaystyle\sum_s^S \frac{\alpha_s}{1+(\omega_E \tau_s)^2}}\right) \tag{7.12}$$

$$M_{TOT} = \sqrt{\left(\sum_s^S \frac{\alpha_s}{1+(\omega_E \tau_s)^2}\right)^2 + \left(\sum_s^S \frac{\alpha_s \omega_s \tau_s}{1+(\omega_E \tau_s)^2}\right)^2} \tag{7.13}$$

As discussed later, several methods are available to extract multiple component lifetimes from complex systems measured in the frequency domain.

7.2.2 THE FOURIER REPRESENTATION OF THE FLUORESCENCE RESPONSE FOR NONSINUSOIDAL EXCITATION

Very often, the modulated intensity of the excitation light is not a perfect sinusoid. However, any repetitive waveform can be represented by a Fourier series expansion as a sum of sine and cosine functions, as shown in Equation 7.14 (see also Appendix 7.1).

$$\begin{aligned}
Signal &= A_o + \sum_{n-1}^{\infty} A_n \cos\left(\frac{n2\pi}{T}t\right) + \sum_{n-1}^{\infty} B_n \sin\left(\frac{n2\pi}{T}t\right) \\
&= A_o + \sum_{n-1}^{\infty} A_n \cos(\omega_n t) + \sum_{n-1}^{\infty} B_n \sin(\omega_n t)
\end{aligned} \tag{7.14}$$

Such a Fourier series expansion is used to describe both the repetitive excitation waveform and the fluorescence response (of course, with different (A_n) and (B_n) values). The frequencies of the cosine and sine terms of the Fourier series, (ω_n), are integer multiples of the repetition frequency $\left(\omega_o = \frac{2\pi}{T}\right)$, where (T) is the period of repetition. Because each frequency component of the Fourier series expansion is orthogonal to all other terms, the analysis of the data can be carried out for each frequency component individually (see Figure 7.3 and Appendix 7.1).

The Fourier expansion coefficients, (A_n) and (B_n), which are the coefficients of the nth frequency component of the fluorescence response $F(t)$ (Bracewell 1978; Brigham 1988), are calculated from an integration over the period of repetition (T).

$$A_n = \frac{2}{T}\int_0^T F_\delta(t)\cos(\omega_n t)\,dt \tag{7.15}$$

$$B_n = \frac{2}{T}\int_0^T F_\delta(t)\sin(\omega_n t)\,dt \tag{7.16}$$

A similar Fourier series expansion is carried out for the excitation, where in the integrals of Equations 7.15 and 7.16, $F_\delta(t)$ is replaced by $E(t)$. Even in the case of multiple lifetimes, the excitation and fluorescence

Principles of fluorescence lifetime instrumentation

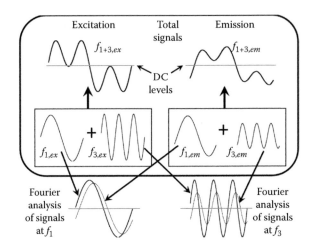

Figure 7.3 Simulating a Fourier decomposition of the excitation and fluorescence response for the case of a repeating excitation wave composed of two sinusoidal components $Excitation(t) = (DC_{ex}) + (AC_{ex})\sin\left(\frac{2\pi}{T}t\right) + (AC_{ex})\sin\left(3\frac{2\pi}{T}t\right)$. In this example, both frequency terms, (AC_{ex}), are equal, and the average steady-state excitation intensity (DC_{ex}) is arbitrary but constant. A single fluorescence lifetime is assumed and is set equal to $\tau_{fluroescence} = \frac{1}{2\left(\frac{2\pi}{T}\right)} = \frac{T}{4\pi}$. The upper part of the figure (top of the box) shows the waveforms for the excitation and the emission over one period of repetition. These are the signals that would be directly measured as a function of time. In the middle (bottom of the box) are the separate contributions of the two frequencies (f_1) and (f_3) for excitation and emission. At the bottom of the figure, the separate Fourier components are compared for the excitation and emission signals for the frequencies $f_1 = \frac{1}{T}$ and $f_3 = \frac{3}{T}$. Note the reduction in amplitude of oscillation of the nth frequency components of the emission waveform when compared to the excitation waveform, and the difference in phase between the emission and excitation waveforms at each of the frequencies $f_1 = \frac{1}{T}$ and $f_3 = \frac{3}{T}$. See Equations 7.10 and 7.11.

waveforms are composed of sinusoidal oscillations at each repetition frequency (ω_n). From the Fourier coefficients of each of the frequency components, the phase delay $\varphi_{n,TOT} = \tan^{-1}\left(\frac{B_n}{A_n}\right)$ and modulation ratio $M_{n,TOT} = \sqrt{A_n^2 + B_n^2}$ can be calculated. The phase and modulation ratio of the fluorescence signal for the nth frequency are relative to the phase and modulation depth of the excitation waveform at that frequency. Expressed in terms of the contributing lifetime components, for every frequency (ω_n), the phase delay and modulation ratio of multiple lifetime components is given by equations similar to Equations 7.12 and 7.13.

$$\varphi_{n,TOT} = \tan^{-1}\left(\frac{\sum_s^S \frac{\alpha_s \omega_s \tau_s}{1+(\omega_n \tau_s)^2}}{\sum_s^S \frac{\alpha_s}{1+(\omega_n \tau_s)^2}}\right) \tag{7.17}$$

$$M_{n,TOT} = \sqrt{\left(\sum_s^S \frac{\alpha_s}{1+(\omega_n \tau_s)^2}\right)^2 + \left(\sum_s^S \frac{\alpha_s \omega_s \tau_s}{1+(\omega_n \tau_s)^2}\right)^2} \tag{7.18}$$

The only difference between Equations 7.12 and 7.13 and Equations 7.17 and 7.18 is that in the latter case, the excitation is not a single pure sinusoid; each frequency (ω_n) refers to the nth coefficient of the Fourier series expansion of the signal. Also, most importantly, as said earlier, every frequency component can be analyzed independently; this is a significant advantage of analyzing the repetitive signal in the frequency domain.

Using a trigonometric identity, the nth term of the Fourier expansion of Equation 7.14 can be expressed in a form that provides a more intuitive understanding of the modulation and phase (similar to the expressions of Equations 7.5, 7.8, and 7.9) in terms of (A_n) and (B_n).

$$A_n \cos(\omega_n t) + B_n \sin(\omega_n t) = \sqrt{A_n^2 + B_n^2} \cos(\omega_n t - \varphi_n) = M_n \cos(\omega_n t - \varphi_n) \tag{7.19}$$

7.3 DETECTING AND CHARACTERIZING LIFETIMES IN AN IMAGE

7.3.1 MIXING TECHNIQUES—HOMODYNE AND HETERODYNE

7.3.1.1 Frequency mixing

Most fluorescent probes commonly used in lifetime imaging, including a majority of the fluorescent proteins, have lifetimes on the order of nanoseconds, requiring modulation frequencies, ($f_E = \omega_E/2\pi$), in the range of 10–200 MHz for their detection. The detection of signals directly at such high frequencies is accompanied by noise with a broad bandwidth. To reduce the noise bandwidth and avoid high frequencies during data acquisition, techniques known as heterodyning and homodyning are used, whereby two high-frequency (HF) signals are mixed (by multiplication) in order to produce a low (beat) frequency (as well as several other HF terms). The frequency of the term oscillating at the low (beat) frequency is the difference frequency of the original two fundamental components of the HF repetitive signals. In order to retain steady phase coherence, the two signals (two original HF signals) are also synchronously phase-locked (Piston et al. 1989).

To apply HF frequency mixing to the detection of lifetimes in the frequency domain, the gain of the detector is modulated by an oscillating voltage $G(t)$.

$$G(t) = G_o + 2G_\omega \cos(\omega_G t + \varphi_G) \tag{7.20}$$

Again, for the sake of discussion, we can assume a sinusoidal modulation because $G(t)$ can also be expanded in a Fourier series. The signal output by the detector is the product of the intensity-modulated fluorescence emission of the sample and the signal modulating the detector's gain. The terms resulting from this multiplication (e.g., of Equations 7.5 and 7.20) include a DC offset (DC offset is the non-oscillating portion of the sine wave), a term oscillating at the difference frequency of ($\omega_G - \omega_E$) and HF terms oscillating at frequencies of (ω_E), (ω_G), and ($\omega_G + \omega_E$), as shown in Equation 7.21.

$$[G(t) \cdot F(t)]_{LF} = G_o E_o \sum_s a_s \tau_s$$
$$+ 2G_\omega E_\omega \sum_s \frac{a_s \tau_s}{\sqrt{1 + (\omega_E \tau_s)^2}} \cos\left((\omega_G - \omega_E)t + \varphi_G - \varphi_E + \varphi_{F,s}\right) \tag{7.21}$$

The DC offset and the term oscillating at the difference frequency contain all the necessary information to determine the phase delay and modulation ratio of the sample. The three HF terms average to zero during the acquisition integration time.

7.3.1.2 Homodyne detection

For homodyne operation, $\omega_E = \omega_G$, so the time dependence of Equation 7.21 drops out (see $[G(t) \cdot F(t)]_{LF,Homo}$ of Appendix 7.2, ($\omega_G - \omega_E$) $t = 0$). The phase delay and modulation ratio of the sample are determined by

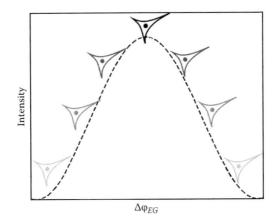

Figure 7.4 Demonstration of frequency mixing for detecting lifetimes in the frequency domain: In the homodyne detection of lifetimes, the measured fluorescence intensity is a function of the phase difference between the detector gain and the excitation, ($\Delta\varphi_{E-G} = \varphi_G - \varphi_E$). The phase of either the detector's gain modulation (φ_G) or excitation's intensity modulation (φ_E) is shifted over the whole period in order to extract the phase delay and modulation ratio of the sample.

acquiring a series of steady-state intensity images ($[G(t) \cdot F(t)]_{LF,Homo}$ at every pixel) at different values of ($\varphi_G - \varphi_E$) spanning a full period of oscillation (Figure 7.4). Either (φ_G) or (φ_E) can be independently varied when acquiring a set of phase images in homodyne operation (Clegg 1996; Clegg and Schneider 1996; Schneider and Clegg 1997).

7.3.1.3 Heterodyne detection

The heterodyne mode occurs when the frequencies of the two signals (excitation modulation and intensifier gain modulation) are not identical; the difference frequency is usually very small compared to the frequencies of the original HF signals. (The HF signals are typically in the range of 10–200 MHz.) The modulation frequency of the detector's gain and that of the excitation modulation differ typically by only 10^2 to 10^4 hertz (for the detection of nanosecond lifetimes). The difference frequency, $\Delta f = (\omega_G - \omega_E)/2\pi$, is sometimes referred to as the cross-correlation frequency. The amplitude and phase of $[G(t) \cdot F(t)]_{LF}$ can be determined efficiently by analog or digital filtering (to remove all frequency components other than those close to the difference frequency) and subsequent data analysis of the difference frequency signal. The modulation ratio and phase delay of the fluorescence sample are determined by comparing to the corresponding parameters of the excitation light (see Section 7.3.1.4).

Because the subsequent data acquisition is limited to low frequencies, heterodyning increases the signal-to-noise ratio and reliability of lifetime measurements by significantly reducing HF noise (Gadella et al. 1993; Gratton and Limkeman 1983; Spencer and Weber 1969). This is a common method for single-channel (cuvette) data acquisition in the frequency domain. However, for full-field imaging, heterodyning is difficult to apply because the entire image must be read out at rates that are high compared to the difference frequency.

A recent version of heterodyne-based FLIM detection exists that does not require direct modulation of the detector's gain, relying instead on detection with digital photon counting combined with heterodyning in the detection circuitry (Colyer et al. 2008).

7.3.1.4 Additional experimental parameters

When lifetimes are detected by heterodyning or homodyning, knowledge of the steady-state coefficients, amplitudes (modulation depths) of oscillation, and phase of both $G(t)$ and $E(t)$ are necessary to determine the phase delay and modulation ratio of the unknown fluorescence. These parameters $\left(\dfrac{E_\omega}{E_o}, \dfrac{G_\omega}{G_o}, \varphi_E, \varphi_G \right)$ are determined by carrying out a measurement with a fluorescent sample with a well-established single lifetime (Boens et al. 2007; Lakowicz 2006).

Other experimental parameters such as the apparent quantum yield of the sample, and in some cases, changes in the brightness of the excitation source and the gain of the detector are accounted for by normalizing the data to the measured DC offset. Very often, $[G(t) \cdot F(t)]_{LF}$ (Equation 7.21) is presented with a scaling factor for these sorts of experimental parameters (Schneider and Clegg 1997).

7.3.2 LIFETIME CHARACTERIZATION IN IMAGES

7.3.2.1 Multiple modulation frequencies

As discussed previously, apparent lifetimes derived from the phase delay and modulation ratio (Equations 7.10 and 7.11) acquired at a single modulation frequency from a sample with multiple lifetimes do not correspond to lifetimes of the separate components. Nevertheless, several methods exist to estimate component lifetimes, species fractions, and fractional contributions to the steady-state intensity (intensity fractions) (Gratton et al. 1984b; Jameson et al. 1984; Weber 1981). Also, if individual lifetimes from a multicomponent system are known, together with their species and intensity fractions, one can accurately map the type and relative concentration of different fluorescent probes in an image.

Multicomponent lifetimes and corresponding species and intensity fractions can be resolved into separate components by measuring the phase delay and modulation ratio at a set of modulation frequencies followed by iterative fitting based on Equations 7.17 and 7.18 (Gratton and Limkeman 1983; Gratton et al. 1984a, b). If only a few frequencies are available, an analysis following the Weber (1981) algorithm is sometimes useful. Although collecting data at multiple modulation frequencies is common and accurate in cuvette-based measurements, this technique is not as practical for imaging applications, because of the time required for the analysis of hundreds of thousands of pixels at every frequency. However, multifrequency lifetime imaging has been applied by Fourier analyzing the higher harmonics of a nonsinusoidal repetitive excitation (Schlachter et al. 2009; Squire et al. 2000); see Section 7.2.2.

7.3.2.2 Polar plots

Without fitting to any predetermined model of a lifetime distribution, the measured modulation ratio (M_{TOT}) and the phase delay (φ_{TOT}) can be displayed and analyzed on a polar plot (Equations 7.22 and 7.23 and Figure 7.5).

This involves transforming the primary measured data $(M_{TOT}$ and $\varphi_{TOT})$ to a functional space where the (M_s) and $(\varphi_{F,s})$ of the separate contributing lifetime components add like vectors (weighted by their intensity fractions, $[\alpha_s]$) to generate the coordinate of the measured polar plot point describing the corresponding multicomponent system (Digman et al. 2008; Hanley and Clayton 2005; Jameson et al. 1984; Redford and Clegg 2005).

$$x = \sum_s^S \frac{\alpha_s}{\sqrt{1 + (\omega_E \tau_s)^2}} \cos(\varphi_{F,s}) = M_{TOT} \cos(\varphi_{TOT}) \tag{7.22}$$

$$y = \sum_s^S \frac{\alpha_s}{\sqrt{1 + (\omega_E \tau_s)^2}} \sin(\varphi_{F,s}) = M_{TOT} \sin(\varphi_{TOT}) \tag{7.23}$$

The x,y values of Equations 7.22 and 7.23 are the components of a two-dimensional vector that can be plotted on an x,y coordinate system. They represent the response of the fluorescence system to a pure sinusoidal excitation of frequency (ω_E) (of course, any of the Fourier frequency components (ω_n) could be used).

The polar plot is a model-independent way to present and analyze FLIM data. Even when the lifetime components are not known, the measured point on the polar plot is a unique classifier. Each fluorescent

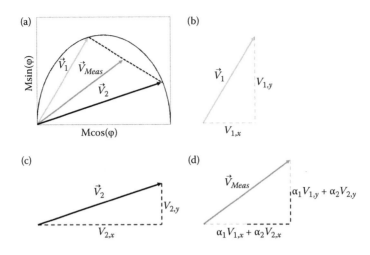

Figure 7.5 A two-component lifetime projected on the polar plot. (a) The measured polar coordinate $\left(\vec{V}_{Meas}\right)$ is shown together with the polar coordinates of its constituent single-lifetime components $\left[\left(\vec{V}_1\right) \text{and} \left(\vec{V}_2\right)\right]$. Any polar coordinates that lie along the dotted line represent the locations of samples containing linear combinations of the two single lifetime components. (b) The polar coordinate of the first lifetime component $\left(\vec{V}_1\right)$ has been decomposed to show its projections on the x-axis and y-axis separately. (c) In black, the x and y components comprising the polar coordinate of the second lifetime component $\left(\vec{V}_2\right)$ are drawn. (d) On the polar plot, the measured polar coordinate $\left(\vec{V}_{Meas}\right)$ is the sum of the two polar coordinates $\left(\vec{V}_1\right)$ and $\left(\vec{V}_2\right)$, each weighted by their corresponding contributions to the steady-state intensity (α_1) and (α_2). On the polar plot, this relationship can be expressed by simple vector addition along the x-axis and y-axis.

species contributes a certain fraction to the measured polar coordinate, and hence, the polar plot can be used easily to monitor variations in fluorescent populations.

On the polar plot, a universal semicircle defines the positions of all single-lifetime systems at a given modulation frequency (Figure 7.5). The semicircle is centered at $(x, y) = (0.5, 0)$, with a radius of 0.5. Because the polar coordinate of a multicomponent fluorescence emission is a weighted sum of the contributing single-lifetime polar vectors, it will lie inside the semicircle (Digman et al. 2008; Hanley and Clayton 2005; Jameson et al. 1984; Redford and Clegg 2005). In certain cases where one or more lifetime components are known, both their intensity fractions can be extracted directly from the polar plot. (See also Chapter 10 for details of the use of the polar plots). If the intensity fractions (α_s) and the lifetimes (τ_s) are known, the species fractions (a_s) (see Equation 7.2) can be determined.

If the fluorescent molecule becomes excited through an excited-state reaction, such as the excitation of a fluorescent acceptor from a donor in FRET, the point on the polar plot will lie outside the semicircle (Chen and Clegg 2011; Forde and Hanley 2006; Lakowicz and Balter 1982a, b).

When distributions of single lifetimes are present, the polar coordinate may lie inside the semicircle. Distributions of lifetimes may be caused, for example, by samples containing a mixture of donor and acceptor distances in the case of FRET. However, in general, even for relatively broad distributions, the polar plot points do not lie far inside the semicircle (Chen and Clegg 2011; Redford and Clegg 2005). Long lifetime components contribute more intensity than short components for identical species concentrations. Therefore, in a polar plot, the single component lifetimes of a symmetrical distribution (such as a Gaussian), which are located slightly inside the semicircle, will typically be skewed toward longer lifetimes (Chen and Clegg 2011). If two lifetime components are present in the signal, each with a distribution, the point on the polar point will lie on a straight line between the two polar plot points of the separate distributions; the position of the point on this straight line depends on the relative intensities of the two lifetime components, as in the case of two singular lifetimes discussed in the last paragraph.

7.4 INSTRUMENTATION

7.4.1 EXCITATION SOURCES

7.4.1.1 External modulators

Frequency-domain lifetime systems, both single-channel and FLIM, often use continuous light sources with external modulators, such as acousto-optic modulators (AOMs) (Piston et al. 1989; Sapriel 1979) or electro-optic modulators (EOMs) (Holshouser et al. 1961; Kaminow and Turner 1966).

7.4.1.1.1 Acousto-optic modulators

An AOM is an optically transparent material with a mechanically coupled piezoelectric transducer (Piston et al. 1989; Sapriel 1979). Upon receiving a radio-frequency electrical input, the piezoelectric transducer vibrates, creating oscillating periodic changes in the index of refraction in the transparent medium. Light passing through this device will be diffracted depending on the gradient of index of refraction, which creates a time-dependent grating. The output beam forms an interference pattern where the intensity of the central interference beam is modulated at twice the driving frequency. The principle drawback of using AOMs is that effective modulation of the incident light's intensity is only accessible in a narrow range near the resonance frequencies of the specific AOM. Also, the frequency modes of AOMs are very temperature dependent.

7.4.1.1.2 Electro-optic modulators

EOMs are another class of external modulators that rotate the polarization of incident light based on a voltage applied to them (Kaminow and Turner 1966). For frequency-domain lifetime measurements, the applied voltage is usually a DC-biased radio-frequency signal at the desired frequency of intensity modulation. By placing crossed polarizers on either side of an EOM, an incident polarization of light can be rotated to a new orientation, which is passed through an exit polarizer. Hence, by modulating the polarization of the light beam passing through the EOM, intensity modulation can be achieved. Some examples of EOMs are the Kerr cell (Holshouser et al. 1961; Kaminow and Turner 1966) and the Pockels cell (Kaminow and Turner 1966); the Kerr cell is no longer used, but Pockels cells are common on frequency-domain FLIM instruments. Although EOMs do not have the restrictions in modulation frequencies of AOMs, EOMs typically require larger input voltages for effective operation.

7.4.1.2 Pulsed lasers

Several instruments for lifetime imaging in the frequency domain use pulsed lasers as excitation sources (Hanson and Clegg 2005; Hanson et al. 2002; Murray et al. 1986; Piston et al. 1992; So et al. 1995). Particularly in the areas of tissue and live cell imaging, 2-photon excitation, using the high pulse repetition frequencies of Ti:sapphire lasers (usually 80–100 MHz), are commonly used for FLIM in the frequency domain. The near-infrared (IR) wavelength of a Ti:sapphire laser output also increases the depth of imaging compared to visible light. The exceptionally short pulses generated by these lasers (typically 100–200 fs) contain many higher harmonics of the repetition frequency, which are useful for multifrequency analysis, as described in Section 6.2.2.

7.4.1.3 Light-emitting diodes and laser diodes

More recently, light-emitting diodes (LEDs) and laser diodes (LDs) have been used for lifetime imaging systems (Herman et al. 2001; Lakowicz 2006; Sipior et al. 1997; Thompson et al. 1992). The intensity of light is modulated directly, eliminating the need for external modulators. Benefits of these units include low power consumption, large modulation depths, and availability of a wide range of modulation frequencies at relatively low cost. The LEDs and LDs available presently emit throughout the visible and ultraviolet spectrum. In addition, many of the currently available LDs have output power in the milliwatt range.

7.4.2 DETECTORS

7.4.2.1 Photomultiplier tubes

Photomultiplier tubes (PMTs) have been used for a long time as detectors for frequency-domain lifetime instruments, as well as for FLIM measurements with laser scanning confocal microscope systems. For direct heterodyne or homodyne detection, the gain of the PMT is modulated by coupling a radio-frequency alternating current (AC) voltage to one of the dynodes (Gratton and Limkeman 1983; Spencer and Weber 1969). The PMTs have the time resolution (response times) necessary so that the modulation of the gain can be performed up to frequencies on the order of several hundreds of megahertz (Hammamatsu Photonics 2007; Yamazaki and Tamai 1985).

7.4.2.2 Image intensifiers

Wide-field FLIM, whereby the whole image is acquired simultaneously, requires modulation of the image acquisition system at high frequencies for homodyne (or heterodyne) data acquisition. This is usually accomplished in the frequency domain with voltage-gated image intensifiers (Gadella et al. 1993; Lakowicz and Berndt 1991; Schneider and Clegg 1997). The intensifier is placed between the emission port of the microscope and a charge-coupled device (CCD) camera. The image of the fluorescent specimen is focused on the cathode of the intensifier. The ejected photoelectrons are then accelerated by an applied voltage through the channels of a microchannel plate (MCP; the amplification within each channel of the MCP is similar to that of the dynode stages in a PMT). The photoelectrons ejected from the output of the MCP are accelerated through a large potential drop between the back end of the MCP and a phosphor screen. When the electrons hit the phosphor screen, they are converted to photons (inverse of the photoelectron effect). The image on the phosphor screen is finally focused onto, and detected by, a CCD. The intensifier acts as the modulated detector, as described in Section 7.3.1. The gain of the image intensifier is modulated by varying the voltage across the intensifier cathode (Clegg 1996) or by alternatively varying the voltage across the MCP with an applied radio frequency (AC voltage) (Clegg 1996). Intensifiers add a significant amount of noise to images and are susceptible to damage from exposure to high light levels.

7.4.2.3 Direct modulation of a camera

It would be preferable to modulate the CCD directly because CCDs have much lower noise than image intensifiers. However, directly modulating the gain or, equivalently, the sensitivity of CCD cameras at frequencies necessary to resolve nanosecond lifetimes still remains difficult (Mitchell et al. 2002; Morgan et al. 1997). Attempts have been made by periodically dumping the accumulated charge in a row of pixels instead of moving it to the transfer registers during image acquisition with interline charge transfer CCDs (Morgan et al. 1997). Another reported technique selectively varied the number of active voltage gates associated with a row of pixels to periodically change the sensitivity of a frame-transfer CCD (Mitchell et al. 2002). At this time, modulation frequencies of CCDs on the order of tens of kilohertz are possible with these techniques.

In a recent article, lifetime images have been collected at modulation frequencies up to 20 MHz by adjusting the signal applied to activate pixels of CMOS cameras (Esposito et al. 2006). Each pixel on these cameras consists of multiple gates, which are controlled by the cameras' electronics. Therefore, the modulation of the detector's gain can be accomplished by changing the phase of the voltage (signal) applied to the gates, which activate a specific pixel. A reference (clock) signal describing this gain modulation can be outputted so that the precise synchronization of the signal modulating the intensity of light emitted from the excitation source and the signal modulating the voltages on the gates of the camera's pixels can be maintained during an experiment.

7.5 ADVANCED FEATURES OF FLIM

7.5.1 MORPHOLOGY, WAVELETS, AND DENOISING

7.5.1.1 Background and overview of analysis with wavelets

For any lifetime imaging instrument, the relevant features of an acquired image are often accompanied by noise from the detectors, unwanted background signals, and extraneous fluorescence. Noise specific to both the instrument and photon counting as well as extraneous fluorescence reduce the reliability of the data and obscure features in an image.

In images generated by FLIM, the lifetimes are frequently correlated with specific objects or morphological features of interest. Because objects of different sizes and shapes are associated with different spatial frequencies (i.e., changes in some spatial signal—e.g., intensity or color—over a certain distance), it is useful to determine the spatial frequencies in an image that correspond to selected morphologies. A spatial Fourier transformation is commonly used to identify and select spatial aspects of an image. However, a spatial Fourier transform is a global analysis and cannot identify the spatial frequencies at specific localizations within an image. Confining the Fourier transform to specific locations leads to well-known artifacts (Walker 1997). Wavelets are useful in this regard. Unlike the basis functions of Fourier, the basis functions for wavelets have selected windows of spatial frequencies that can be associated with specific locations of interest in an image. In addition, extended wavelet analysis can also be used to reduce noise of a particular type (e.g., Poisson or Gaussian) (Buranachai et al. 2008; Spring and Clegg 2009).

Wavelet functions are localized waveforms (not the harmonic waves of Fourier) that can be spatially stretched (scaled) and translated throughout an image to provide information about spatial frequencies present at localized areas (Farge 1992; Hubbard 1998; Rao and Bopardikar 1998). The scaled and translated wavelets derived from a "mother wavelet" comprise a set of basis functions with which an image or signal can be decomposed. The wavelet transform process is fully reversible (back transform), and different spatial windows (spatial frequency bands) can be combined. Some examples of applications of wavelets are image compression (Antonini et al. 1992), noise removal (Spring and Clegg 2009; Willet and Nowak 2004), and general analysis of spatial frequencies. To decompose an image into constituent spatial frequency parts, the continuous wavelet transform (CWT) can be applied by shifting wavelet basis functions at various scales throughout an image.

The discrete wavelet transform (DWT), by contrast, provides an equivalently accurate but less computational expensive means to decompose a signal by sampling the scales and positions on which the wavelet basis function operates by powers of two (Farge 1992; Hubbard 1998; Rao and Bopardikar 1998). Essentially, the DWT converts an image using successive iterations through a set of low-pass and band-pass filters into a series of approximations and detail coefficient matrices. In the DWT, the scaling of the wavelet in discrete steps creates a series of band-pass filters through which the image is converted into a set of approximation matrices. Primarily, the presence of low spatial scales (high spatial frequencies) in the image is determined by the scaling function in the DWT. The scaling function contains a low-pass spectrum (with respect to scale) detecting HF details in the image.

7.5.1.2 Wavelets applied to FLIM for background removal

The removal of background from the intensity images collected on a homodyne FLIM instrument can be achieved by incorporating the procedures of DWT into the multiresolution analysis described by Mallat (1989). Each approximation of the image contains the band pass of frequencies corresponding to the wavelet that generates the approximation at the particular scale (spatial resolution). If two approximations are taken at distinct scales (resolutions) and subtracted, the resulting difference image contains only the spatial frequencies in the region of frequencies between the two approximations. Although distinct from the process of purely reconstructing a signal from the entire set of approximation and detail coefficients derived from the DWT decomposition, multiresolution analysis can specifically distinguish HF noise generated from background factors and data acquisition noise from meaningful low-frequency (high-scale) morphological features of the sample. As has been shown, multiresolution analysis can substantially eliminate background when applied to FLIM (Buranachai et al. 2008). The optimization of removing background from the intensity images with the DWT and multiresolution analysis requires both the selection of the appropriate wavelet and the scales to be studied.

7.5.2 PHASE SUPPRESSION APPLIED TO FLIM

7.5.2.1 Introduction to phase suppression

When lifetime images are acquired by homodyning, both the contrast between distinct fluorescent species in an image and acquisition speed can be enhanced with phase suppression (Clegg 1996; Lakowicz and

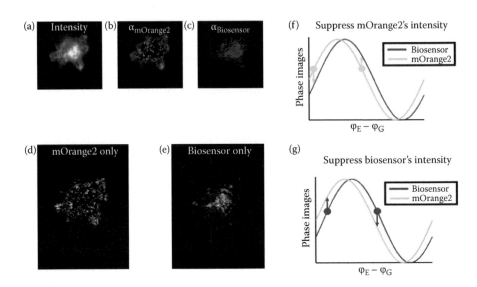

Figure 7.6 Phase suppression applied to study live cells. (a) The intensity image of an HT1080 cell expressing the MT1-MMP biosensor containing the mOrange2 and mCherry fluorescent proteins is shown. (b–c) The same HT1080 cell in (a) is shown as an intensity-masked image. The color in each pixel was determined by whichever conformation (green = cleaved biosensor [mOrange2], red = intact biosensor) of the biosensor had the largest fractional intensity contribution to the average intensity (intensity fraction) in that pixel. (d) Two detector phase angles were chosen to suppress the intensity from the intact biosensor. When the images corresponding to these detector phase angles were subtracted, that which remains should only be intensity from the mOrange2 (cleaved biosensor). (e) Two other detector phase angles (different from those chosen in (d)) were selected to suppress the intensity from the mOrange2 (cleaved biosensor). Following the subtraction of the two corresponding phase images, the intensity from the intact biosensor was clearly enhanced over the intensity from the mOrange2 (cleaved biosensor). (f) Two simulated curves collected by homodyning are shown (green = cleaved biosensor, red = intact biosensor). The points indicated on the green curve show a set of points that could be used to suppress the intensity from the cleaved biosensor. The remaining intensity from the intact biosensor would be the result of the subtraction of the intensities on the red curve indicated by the arrows. (g) The same curves from (f) are plotted but with the intensity from the intact biosensor suppressed. (The curves shown do not represent the actual difference in phase delay for the cleaved and intact biosensor.)

Berndt 1991; Lakowicz et al. 1992). This method is an extension of the use of phase suppression to isolate fluorescence spectra from solutions of mixed fluorophores with different lifetimes (Lakowicz and Cherek 1981, 1982). For phase suppression, two images acquired at two different phases are subtracted (Figure 7.6f and g). The two phases (differing by π) are selected in order to suppress the contribution of a selected fluorescence lifetime component in the difference image corresponding to a particular phase delay. As a result, with only two exposures of the sample, a spatial mapping of fluorescent species based on their phase delay (corresponding to their lifetimes) can be constructed. Although not frequently reported, the rapid image acquisition and selectivity offered by phase suppression has definite advantages for imaging samples with rapidly changing fluorescence signals, such as in live cells.

7.5.2.2 Example of phase suppression applied to live cell imaging with FLIM

Recently, we have applied phase suppression to study the dynamics of a biosensor that monitors the collagenase membrane type 1 matrix metalloproteinase (MT1-MMP) in live fibrosarcoma (HT1080) cells. The MT1-MMP biosensor contains the substrate for the catalytic domain of MT1-MMP suspended between the mOrange2 and mCherry fluorescent proteins (Ouyang et al. 2008, 2010; Shaner et al. 2004, 2008). When expressed in a cell, the mOrange2 and mCherry proteins are in close proximity. In this conformation, FRET can take place, resulting in emission from mCherry when the mOrange2 is excited. When an active MT1-MMP enzyme cleaves the substrate, the mOrange2 and mCherry proteins are permanently separated, eliminating all energy transfer in the system.

The MT1-MMP biosensor was expressed in HT1080 cells and imaged both with phase suppression and by determining the phase delay and modulation ratio by homodyning (Figure 7.6). HT1080 cells endogenously produce the MT1-MMP enzyme, and hence, there is a mixed population of intact and cleaved biosensors in the cell. Images were collected with phase suppression to highlight the intensity from the intact and cleaved biosensor separately (Figure 7.6d and e). The intensity fractions of each conformation of the biosensor (Figure 7.6b and c) closely paralleled the corresponding intensities resulting from the phase suppression (Figure 7.6d and e). For example, the green region in Figure 7.6b indicating the areas where the intensity fraction from the cleaved biosensor is greater than that of the intact biosensor corresponds to the intensities highlighted when the intact biosensor was suppressed (Figure 7.6d). Therefore, phase suppression distinguishes well the two conformations of the MT1-MMP biosensor and reduces our acquisition time to less than one-tenth of the time required for the traditional lifetime imaging.

7.5.3 SPECTRAL FLIM

Spectral FLIM is an extension of FLIM where lifetime-resolved images are collected at a series of different wavelengths. The spectral resolution adds an additional parameterization for distinguishing fluorescent species by correlating spectra with lifetimes. It also increases the reliability of differentiating, identifying, or discovering lifetimes in a FLIM image with several fluorescent species. Recently, spectral imaging has been applied to both time- and frequency-domain FLIM systems (Bird et al. 2004; Chen and Clegg 2009, 2011; De Beule et al. 2007; Hanley and Clayton 2005; Hanley et al. 2002; Rueck et al. 2007; Strat et al. 2011).

The simplest way to achieve some degree of spectral resolution is to acquire FLIM images through a set of emission band-pass filters and subsequently determine the separate (known) spectral component contributions by linear unmixing. This has been done in steady-state intensity fluorescence imaging for some time and has also been applied for FLIM (Chen and Clegg 2009, 2011; Chen et al. 2010; Forde and Hanley 2006; Garini et al. 2006; Hanley et al. 2002; Haraguchi et al. 2002; Rueck et al. 2007; Strat et al. 2011; Thaler et al. 2005; Tsurui et al. 2000; Zimmermann 2005). Samples containing multiple fluorophores emitting with similar but not identical spectra are sometimes difficult to analyze reliably with only a few filters. To dissect more reliably the fluorescence emission into separate constituent spectra of the fluorescent components, diffraction gratings, prisms, or a series of filters at many wavelengths can be used.

The contribution of each component spectrum to the measured signal can be determined by spectral linear unmixing (Zimmermann 2005). The unmixing algorithm estimates the best statistical fit of the measured intensity $M(\lambda)$ to the sum of the known reference spectra of each fluorophore $f_s(\lambda)$ in the sample multiplied by the corresponding intensity contribution of the fluorophore (c_s), as indicated in Equation 7.24,

$$M(\lambda) = \sum_s^S c_s f_s(\lambda) \tag{7.24}$$

Iterative fitting procedures and singular value decomposition (SVD) algorithms are used to perform the unmixing at every pixel (Chen and Clegg 2009, 2011; Chen et al. 2010; Tsurui et al. 2000; Zimmermann 2005).

A common application of spectral FLIM is to estimate the efficiency of FRET, which is particularly advantageous for homodyne full-field FLIM. The FLIM measurements can be carried out at many different wavelengths throughout the spectral region of both the donor and acceptor fluorescence. For homodyne FLIM, the fluorescence spectra from the donor and acceptor can be unmixed in each phase image individually; then the series of phase images can be analyzed to determine the lifetime parameters (modulation ratio and phase delay) of only the donor and the acceptor separately, without problems of overlapping spectra. Alternatively, the modulation ratio and phase delay of the combined donor and acceptor fluorescence can be determined at every different wavelength (no spectral unmixing). In this latter analysis, as the spectrum progresses from lower wavelengths (more donor, less acceptor) to longer wavelengths (more acceptor and less donor), the intensity fractions of fluorescence of donor and acceptor change, as do the corresponding points on the polar plot. Therefore, the polar plot coordinates of the points measured at the different wavelengths will lie on a straight line between the two spectrally distinct distributions of donors and acceptors. By extrapolation of this straight line to the semicircle of the polar plot, estimates can be made of the acceptor-alone fluorescence lifetime (this is usually a check on the validity of the analysis) and the lifetime of the donor in the presence of the acceptor (from which the FRET efficiency can be determined). The location of the measured polar plot points for the spectrally separated acceptor is also on the straight line but will lie outside the semicircle (because the acceptor is excited by a decaying ensemble of donors). The extent to which the acceptor points lie outside the semicircle depends on the relative lifetimes of the acceptor without donor and the donor in the absence of the acceptor, as well as on the amount of free acceptor (i.e., with no donor).

From the latter type of analysis, one can estimate not only the efficiency of transfer, but also, one can establish the presence and amount of interacting donors and acceptors in the sample, and the amount of donor and acceptor that are not undergoing FRET (i.e., free donor and acceptor). Such analysis with spectral FLIM has been shown to be particularly useful for determining FRET efficiencies in cases where the labeling of donors and acceptors is not 1:1 (Chen and Clegg 2009, 2011). Furthermore, the resolution of specific fluorescing species emitting at low intensity (such as dim acceptor fluorescence) but with overlapping spectra can be significantly improved with spectral FLIM (Chen and Clegg 2009, 2011).

7.6 HISTORICAL DEVELOPMENT OF FREQUENCY-DOMAIN FLIM SYSTEMS

The motivation for the development of lifetime imaging was mainly originated from the need to probe the complex environments of living systems such as cells and tissues. The earliest FLIM system reported in 1959 determined effective lifetimes in the frequency domain by comparing the phase delay of the intensity-modulated transmitted light to that of the sample's emitted fluorescence (Venetta 1959). Without frequency mixing, specimens, including suspensions of stained tumor cells, were imaged at specific locations on a microscope-based imaging system. Interest in measuring the lifetimes of fluorophores in live cells continued throughout the 1970s, with several instruments reported that detected lifetimes in images in the time domain (Arndt-Jovin et al. 1979; Loeser and Clark 1972; Loeser et al. 1972). In 1986, effective lifetimes were also recorded in the frequency domain at single points within images by taking the Fourier transform of data collected by photon counting (Murray et al. 1986). This system by Murray et al. (1986)was one of the earliest descriptions of short pulses of light used as an excitation source for FLIM in the frequency domain (the use of the frequency overtones of short pulses of light is an established technique used earlier for single-channel cuvette-type frequency-domain lifetime measurements). Later improvements in detectors led to two-dimensional lifetime imaging in 1989 with the incorporation of an image-dissecting tube with heterodyning in the detection system (Wang et al. 1989).

In the early 1990s, FLIM systems in the frequency domain sampled more of the illuminated field offered by conventional microscopes. In 1991, a system was designed where the focused output of a pulsed laser could be moved across the field of view with a beam-steering device (Verkman et al. 1991). Although not confocal, this system applied cross-correlation detection to determine the phase delay and modulation ratio of the sample at the locations where the focused laser was exciting the sample. Full-field FLIM systems were later reported in 1991 by Lakowicz and Berndt and again in 1993 by Gadella et al. Both systems detected the intensity-modulated emission of a microscope-based sample from every pixel simultaneously by homodyning using a gain-modulated intensifier and CCD camera.

Improvements in the available spatial resolution of frequency-domain FLIM systems have also been made throughout the 1990s and into the 2000s. Single-photon and two-photon scanning confocal techniques (Buurman et al. 1992; Morgan et al. 1992; Piston et al. 1992; So et al. 1995) have made accurate optical sectioning and three-dimensional imaging possible with frequency-domain FLIM. Full-field FLIM systems operating in the frequency domain have been recently equipped with spinning disk attachments in order to perform high-speed confocal imaging (Buranachai et al. 2008; van Munster et al. 2007).

A variety of frequency-domain FLIM systems are commercially available and span many of the modalities of the previously developed systems. Full-field systems that use gain-modulated image intensifiers and CCD cameras as detectors are commercially available. Spinning disk attachments can also be incorporated into these systems for applications requiring optical sectioning. Scanning confocal FLIM systems with either single- or two-photon excitation are available. Alternatively, any of the FLIM systems that measure lifetimes in the time domain can be converted into the frequency domain by simply taking the Fourier transform of the decay.

7.7 CONCLUSIONS

Mapping lifetimes in an image quantitatively classifies fluorescent species without the need for extensive corrections that are typically used to translate steady-state measurements into definite physical parameters, such as FRET efficiency. Furthermore, the analysis of a sample's time-dependent fluorescence (necessary to compute lifetimes) on the polar plot eliminates the need for computationally intensive processes such as iterative fitting when classifying different fluorescent species in images. When imaging tissues or other complex biological systems, the lifetimes mapped in an image can therefore accurately distinguish between the fluorescence being emitted from molecules of interest such as fluorescently labeled proteins and unwanted light from either a tissue's autofluorescence or the media in which the sample is mounted. In order to take advantage of the quantitative insights offered by lifetime imaging when studying tissues, two-photon lasers can now be readily incorporated into lifetime imaging systems. These lasers can probe deep into tissues because of the reduced scattering inherent to the longer excitation wavelengths and also because the excitation of the sample is limited to only the focus of the objective.

The temporal resolution of lifetime imaging is, however, fundamentally degraded because multiple exposures of the sample are necessary to generate a single-lifetime image. Extended imaging sessions to observe the kinetics of an enzymatic process, for example, can be difficult, and very often, the samples can incur excessive photodamage. At the same time, using carefully designed fluorescent molecules or sensors to monitor the metabolic or enzymatic activity of cells cultured in tissue is one of the necessary steps to study the biology of a cell's decision making. As discussed in this chapter, the techniques that process data collected in the frequency domain (e.g., phase suppression) can lower the exposure time required when imaging so that dynamic events can be monitored for extended periods of time. In our lab, for example, we can monitor the activity of a protease contained within a cancer cell cultured in a three-dimensional matrix of collagen for more than an hour with phase suppression. In these experiments, nearly 20 images mapping our sensor's time-dependent fluorescence can be generated. With continual improvements being made in the excitation sources, detectors, and data acquisition systems, lifetime imaging will no doubt continue to be a very useful tool to understand dynamics and complexities of many biological systems.

APPENDIX 7.1: FOURIER ANALYSIS

The equations describing the excitation and relaxation responses of a fluorescence system are linear. In the frequency-domain experiment, the excitation of the fluorescence system is applied repetitively, and the ensuing fluorescence response has the same repetition frequency as the excitation. The most convenient means for solving such problems is to expand the repetitive waveforms in a sum of sinusoidal waveforms. In other words, the time-dependent excitation and emission signals are expanded in Fourier series (Bracewell 1978; Brigham 1988). The frequencies of the sine and cosine terms in the Fourier series are integer multiples of $\left(\frac{2\pi}{T}\right)$, that is, $\left(n\frac{2\pi}{T}\right)$, where (T) is the period of repetition of the signal and n is an integer ($n = 0$ refers to the steady-state [DC] component). For every sinusoidal component at frequency $\left(n\frac{2\pi}{T}\right)$ of the excitation waveform, there is a corresponding sinusoidal component for the fluorescence response; of course, the modulation depth and phase of the excitation and fluorescence are different due to the delayed response of fluorescence. All sine and cosine functions of frequencies $\left(n\frac{2\pi}{T}\right)$, where n is an integer, are orthogonal to each other when integrated over the repetition period and contribute independently of each other. For both the excitation and fluorescence waveform, the coefficient for every nth frequency of both the sine and cosine component represents the amplitude and phase contribution of that frequency component to the repetitive signal. Each sinusoidal component contributes autonomously a certain amount to the repetitive signal, independent of the other components. Because of linearity, the fluorescence response at frequency $\left(n\frac{2\pi}{T} \right)$ is coupled only to the identical frequency component of excitation. The contribution of each sinusoidal component of the Fourier expansion of the signals (i.e., the modulation and phase for both cosine and sine terms of frequency $\left(n\frac{2\pi}{T} \right)$) are determined separately and independently of all other frequencies by performing a digital discreet Fourier transform over the repetition period from 0 to T (Chen et al. 2010; Clegg and Schneider 1996). This is demonstrated for the simple case of only two frequency terms in Figure 7.3.

APPENDIX 7.2: FITTING DATA COLLECTED BY HOMODYNING

When acquiring lifetimes by homodyning, a series of images are collected at various phase shifts applied to the detector (φ_G) (referred to as detector phases) sampled evenly from 0 to 2π. The measured phase delay and modulation ratio are commonly determined on a pixel-wise basis by taking the discrete Fourier transform (DFT) of the set of images (Bracewell 1978; Brigham 1988; Schneider and Clegg 1997) or by iterative least square fitting the data to a sine wave. Iterative fitting and digital Fourier analysis (DFA) result in the same phase delay and modulation ratio (Hamming 1973); because DFA is faster, there is no need for iterative fitting techniques when determining the modulation ratio and the phase delay at any of the component frequencies.

To apply the DFA, the data corresponding to the function $[G(t) \cdot F(t)]_{LF,Homo}$ is sampled at evenly spaced phases by varying the phase of the modulation of the intensifier amplification (φ_G) from 0 to 2π.

$$[G(t)\cdot F(t)]_{LF,Homo} = G_o E_o \sum_s a_s \tau_s$$
$$+ 2G_\omega E_\omega \sum_s \frac{a_s \tau_s}{\sqrt{1+(\omega_E \tau_s)^2}} \cos\left(\varphi_G - \varphi_E + \varphi_{F,s}\right) \tag{7.B.1}$$

The DFA analysis of the fluorescence signal will produce a single measured phase delay (φ_{TOT}) and modulation ratio (M_{TOT}) (Equations 7.10 and 7.11). The function $[G(t) \cdot F(t)]_{LF,Homo}$ is normalized to its corresponding steady-state intensity (SS), as shown in Equation 7.B.2

$$\left(\frac{[G(t)\cdot F(t)]_{LF,Homo}}{SS} \right) = 1 + \frac{G_\omega E_\omega}{G_o E_o} M_{TOT} \cos(\varphi_G - \varphi_E + \varphi_{TOT}) \tag{7.B.2}$$

Experimentally, $\left(\dfrac{[G(t)\cdot F(t)]_{LF,Homo}}{SS} \right)$ is discretely sampled over a specific set of evenly spaced phase delays. The detector phases can be expressed as a function oscillating at a frequency that depends on the number of detector phases N sampled. If eight detector phases are used ($N = 8$) then, for every value of n, from $n = 0$ to $n = N - 1$,

$$\left(\frac{[G(t)\cdot F(t)]_{LF,Homo}}{SS} \right)_n = 1 + \frac{G_\omega E_\omega}{G_o E_o} M_{TOT} \cos\left(\varphi_G^0 + \left(\frac{\pi}{4}\right)n - \varphi_E + \varphi_{TOT} \right), \tag{7.B.3}$$

where φ_G^0 is the phase of the intensifier modulation when $n = 0$. Of course, N could be larger or smaller but must be at least 3. Fitting is carried out at every pixel (i) of an image based on Equation 7.B.3. The average intensity over the repetition period is also measured at every pixel.

The amplitude of oscillation $\left(\dfrac{G_\omega E_\omega}{G_o E_o} M_{TOT} \right)$ and the phase delay value $\left(\varphi_G^0 - \varphi_E + \varphi_{TOT} \right)$ are calculated by taking the digital sine and cosine transforms (Equations 7.B.4 and 7.B.5) using the measured values $\left(\dfrac{[G(t) \cdot F(t)]_{LF,Homo}}{SS} \right)_n$ over the n values of the measurement.

$$F_{\sin,i} = \sum_{n=0}^{N-1} S_i \sin\left(\left(\frac{\pi}{4} \right) n \right) \qquad (7.B.4)$$

$$F_{\cos,i} = \sum_{n=0}^{N-1} S_i \cos\left(\left(\frac{\pi}{4} \right) n \right) \qquad (7.B.5)$$

$\left(\varphi_G^0 \right)$ and (φ_E) are known, so (φ_{TOT}) can be determined.

REFERENCES

Antonini, M., M. Barlaud, P. Mathieu, and I. Daubechies. (1992). Image coding using wavelet transform. *IEEE Trans Image Process* 1(2):205–220.

Arndt-Jovin, D.J., S.A. Latt, G. Striker, and T.M. Jovin. (1979). Fluorescence decay analysis in solution and in a microscope of DNA and chromosomes stained with quinacrine. *J Histochem Cytochem* 27(1):87–95.

Bird, D.K., K.W. Eliceiri, C.H. Fan, and J.G. White. (2004). Simultaneous two-photon spectral and lifetime fluorescence microscopy. *Appl Opt* 43(27):5173–5182.

Birks, J. (1970). *Photophysics of Aromatic Molecules*. Wiley-Interscience, New York.

Boens, N., W. Qin, N. Basaric, J. Hofkens, M. Ameloot, J. Pouget, J.P. Lefevre, B. Valeur, E. Gratton, M. vandeVen, N.D. Silva, Jr., Y. Engelborghs, K. Willaert, A. Sillen, G. Rumbles, D. Phillips, A.J. Visser, A. van Hoek, J.R. Lakowicz, H. Malak, I. Gryczynski, A.G. Szabo, D.T. Krajcarski, N. Tamai, and A. Miura. (2007). Fluorescence lifetime standards for time and frequency domain fluorescence spectroscopy. *Anal Chem* 79(5):2137–2149.

Bracewell, R.A. (1978). *The Fourier Transformation and Its Applications*. McGraw-Hill Kogakusha, Hamburg.

Brigham, E.O. (1988). *The Fast Fourier Transform and Its Application*. Prentice Hall, Englewood Cliffs, NJ.

Buranachai, C., D. Kamiyama, A. Chiba, B.D. Williams, and R.M. Clegg. (2008). Rapid frequency-domain FLIM spinning disk confocal microscope: Lifetime resolution, image improvement and wavelet analysis. *J Fluoresc* 18(5):929–942.

Buurman, E.P., R. Sanders, A. Draaijer, H.C. Gerritsen, J.J.F. Van Veen, P.M. Houpt, and Y.K. Levine. (1992). Fluorescence lifetime imaging using a confocal laser scanning microscope. *Scanning* 14:155–159.

Chen, Y.C., and R.M. Clegg. (2009). Fluorescence lifetime-resolved imaging. *Photosynth Res* 102(2–3):143–155.

Chen, Y.C., and R.M. Clegg. (2011). Spectral resolution in conjunction with polar plots improves the accuracy and reliability of FLIM measurements and estimates of FRET efficiencies. *J Microsc* 244(1):21–37.

Chen, Y.-C., B.Q. Spring, C. Buranachai, B. Tong, G. Malachowski, and R.M. Clegg. (2010). *General Concerns of FLIM Data Representation and Analysis: Frequency-Domain Model-Free Analysis*. Periasamy A., Clegg R.M., editors. Chapman & Hall/CRC, New York.

Clegg, R.M. (1996). *Fluorescence Lifetime-Resolved Imaging Microscopy (FLIM)*. Verga Scheggi A.M., editor. Kluwer Academic Publishers, Dordrecht, Netherlands.

Clegg, R.M., and P.C. Schneider. (1996). Fluorescence lifetime-resolved imaging microscopy: A General description of lifetime-resolved imaging measurements. In *Fluorescence Microscopy and Fluorescent Probes*. Slavik J., editor. Plenum Press, New York. 15–33.

Colyer, R.A., C. Lee, and E. Gratton. (2008). A novel fluorescence lifetime imaging system that optimizes photon efficiency. *Microsc Res Tech* 71(3):201–213.

De Beule, P., D.M. Owen, H.B. Manning, C.B. Talbot, J. Requejo-Isidro, C. Dunsby, J. McGinty, R.K. Benninger, D.S. Elson, I. Munro, M. John Lever, P. Anand, M.A. Neil, and P.M. French. (2007). Rapid hyperspectral fluorescence lifetime imaging. *Microsc Res Tech* 70(5):481–484.

Digman, M.A., V.R. Caiolfa, M. Zamai, and E. Gratton. (2008). The phasor approach to fluorescence lifetime imaging analysis. *Biophys J* 94(2):L14–L16.

Esposito, A., H.C. Gerritsen, T. Oggier, F. Lustenberger, and F.S. Wouters. (2006). Innovating lifetime microscopy: A compact and simple too for life sciences, screening and diagnostics. *J Biomed Opt* 11(3):034016.

Farge, M. (1992). Wavelet transforms and their applications to turbulence. *Annu Rev Fluid Mech* 24:395–457.

Forde, T.S., and Q.S. Hanley. (2006). Spectrally resolved frequency domain analysis of multi-fluorophore systems undergoing energy transfer. *Appl Spectrosc* 60(12):1442–1452.

Gadella, T.W., Jr., T.M. Jovin, and R.M. Clegg. (1993). Fluorescence lifetime imaging microscopy (FLIM): Spatial resolution of microstructures on the nanosecond time scale. *Biophys Chem* 48:221–239.

Garini, Y., I.T. Young, and G. McNamara. (2006). Spectral imaging: Principles and applications. *Cytometry A* 69(8):735–747.

Gratton, E., D.M. Jameson, and R.D. Hall. (1984a). Multifrequency phase and modulation fluorometry. *Annu Rev Biophys Bioeng* 13:105–124.

Gratton, E., and M. Limkeman. (1983). A continuously variable frequency cross-correlation phase fluorometer with picosecond resolution. *Biophys J* 44:315–324.

Gratton, E., M. Limkeman, J.R. Lakowicz, B.P. Maliwal, H. Cherek, and G. Laczko. (1984b). Resolution of mixtures of fluorophores using variable-frequency phase and modulation data. *Biophys J* 46(4):479–486.

Hammamatsu Photonics (2007). *Photomultiplier Tubes: Basics and Applications*. Hamamatsu Photonics K.K. Electron Tube Divison.

Hamming, R.W. (1973). *Numerical Methods for Scientists and Engineers*. McGraw-Hill, New York.

Hanley, Q.S., D.J. Arndt-Jovin, and T.M. Jovin. (2002). Spectrally resolved fluorescence lifetime imaging microscopy. *Appl Spectrosc* 56(2):155–166.

Hanley, Q.S., and A.H. Clayton. (2005). AB-plot assisted determination of fluorophore mixtures in a fluorescence lifetime microscope using spectra or quenchers. *J Microsc* 218(Pt 1):62–67.

Hanson, K.M., M.J. Behne, N.P. Barry, T.M. Mauro, E. Gratton, and R.M. Clegg. (2002). Two-photon fluorescence lifetime imaging of the skin stratum corneum pH gradient. *Biophys J* 83(3):1682–1690.

Hanson, K.M., and R.M. Clegg. (2005). Two-photon fluorescence imaging and reactive oxygen species detection within the epidermis. *Methods Mol Biol* 289:413–422.

Haraguchi, T., T. Shimi, T. Koujin, N. Hashiguchi, and Y. Hiraoka. (2002). Spectral imaging fluorescence microscopy. *Genes Cells* 7(9):881–887.

Herman, P., B.P. Maliwal, H.J. Lin, and J.R. Lakowicz. (2001). Frequency-domain fluorescence microscopy with the LED as a light source. *J Microsc* 203(Pt 2):176–181.

Holshouser, D.F., H.V. Foerster, and G.L. Clark. (1961). Microwave modulation of light using the Kerr EFFECT. *J Opt Soc Am* 51(12):1360–1365.

Hubbard, B.B. (1998). *The World According to Wavelets: The Story of a Mathematical Technique in the Making*. A.K. Peters, Ltd., Natick, MA.

Jameson, D.M., E. Gratton, and R.D. Hall. (1984). The measurement and analysis of hetergeneous emissions by multifrrequency phase and modulation fluorometry. *Appl Spectrosc Rev* 20(1):55–106.

Kaminow, I.P., and E.H. Turner. (1966). Electrooptic light modulators. *Proc IEEE* 54(10):1374–1390.

Lakowicz, J.R. (2006). *Principles of Fluorescence Spectroscopy*. Springer, New York, NY.

Lakowicz, J.R., and A. Balter. (1982a). Analysis of excited-state processes by phase-modulation fluorescence spectroscopy. *Biophys Chem* 16(2):117–132.

Lakowicz, J.R., and A. Balter. (1982b). Theory of phase-modulation fluorescence spectroscopy for excited-state processes. *Biophys Chem* 16(2):99–115.

Lakowicz, J.R., and K.W. Berndt. (1991). Lifetime-selective fluorescence imaging using an rf phase-sensitive camera. *Rev Sci Instrum* 62(7):1727–1734.

Lakowicz, J.R., and H. Cherek. (1981). Phase-sensitive fluorescence spectroscopy: A new method to resolve fluorescence lifetimes or emission spectra of components in a mixture of fluorophores. *J Biochem Biophys Methods* 5(1):19–35.

Lakowicz, J.R., and H. Cherek. (1982). Resolution of heterogeneous fluorescence by phase-sensitive fluorescence spectroscopy. *Biophys J* 37(1):148–150.

Lakowicz, J.R., H. Szmacinski, K. Nowaczyk, and M.L. Johnson. (1992). Fluorescence lifetime imaging of free and protein-bound NADH. *Proc Natl Acad Sci USA* 89(4):1271–1275.

Loeser, C.N., and E. Clark. (1972). Intracellular fluorescence decay time of anilinonaphthalene sulfonate. *Exp Cell Res* 72:485–488.

Loeser, C.N., E. Clark, M. Maher, and H. Tarkmeel. (1972). Measurement of fluorescence decay time in living cells. *Exp Cell Res* 72(1972):480–484.

Mallat, S.G. (1989). A theory for multiresolution signal decomposition: The wavelet representation. *IEEE Trans Pattern Anal Mach Intell* 11(7):674–692.

Mitchell, A.C., J.E. Wall, J.G. Murray, and C.G. Morgan. (2002). Direct modulation of the effective sensitivity of a CCD detector: A new approach to time-resolved fluorescence imaging. *J Microsc* 206(3):225–232.

Morgan, C.G., A.C. Mitchell, and J.G. Murray. (1992). Prospects for confocal imaging based on nanosecond fluorescence decay time. *J Microsc* 165(1):49–60.

Morgan, C.G., A.C. Mitchell, J.G. Murray, and J.E. Wall. (1997). New approaches to lifetime-resolved luminescence imaging. *J Fluoresc* 7(1):65–73.

Murray, J.G., R.B. Cundall, C.G. Morgan, G.B. Evans, and C. Lewis. (1986). A single-photon-counting Fourier transform microfluorometer. *J Phys E: Sci Instrum* 19:349–355.

Noomnarm, U., and R.M. Clegg. (2009). Fluorescence lifetimes: Fundamentals and interpretations. *Photosynth Res* 101(2–3):181–194.

Ouyang, M., H. Huang, N.C. Shaner, A.G. Remacle, S.A. Shiryaev, A.Y. Strongin, R.Y. Tsien, and Y. Wang. (2010). Simultaneous visualization of protumorigenic Src and MT1-MMP activities with fluorescence resonance energy transfer. *Cancer Res* 70(6):2204–2212.

Ouyang, M., S. Lu, X.Y. Li, J. Xu, J. Seong, B.N. Giepmans, J.Y. Shyy, S.J. Weiss, and Y. Wang. (2008). Visualization of polarized membrane type 1 matrix metalloproteinase activity in live cells by fluorescence resonance energy transfer imaging. *J Biol Chem* 283(25):17740–17748.

Piston, D.W., G. Marriott, T. Radivoyevich, R.M. Clegg, T.M. Jovin, and E. Gratton. (1989). Wide-band acousto-optic light modulator for frequency domain fluorometry and phosphorimetry. *Rev Sci Instrum* 60(8):2596–2600.

Piston, D.W., D.R. Sandison, and W.W. Webb. (1992). Time-resolved fluorescence imaging and background rejection by two-photon excitation in laser scanning microscopy. *Proc SPIE* 1640:379–389.

Rao, R.M., and A.S. Bopardikar. (1998). *Wavelet Transforms: Introduction to Theory and Applications.* Addison-Wesley, Reading, MA.

Redford, G.I., and R.M. Clegg. (2005). Polar plot representation for frequency-domain analysis of fluorescence lifetimes. *J Fluoresc* 15(5):805–815.

Rueck, A., C. Huelshoff, I. Kinzler, W. Becker, and R. Steiner. (2007). SLIM: A new method for molecular imaging. *Microsc Res Tech* 70:485–492.

Sapriel, J. (1979). *Acousto-Optics.* Francis S., Kelly B., translators. John Wiley and Sons, New York.

Schlachter, S., A.D. Elder, A. Esposito, G.S. Kaminski, J.H. Frank, L.K. van Geest, and C.F. Kaminski. (2009). mhFLIM: Resolution of heterogeneous fluorescence decays in widefield lifetime microscopy. *Opt Express* 17(3):1557–1570.

Schneider, P., and R.M. Clegg. (1997). Rapid acquisition, analysis and display of fluorescence lifetime-resolved images for real-time applications. *Rev Sci Instrum* 68(11):4107–4119.

Shaner, N.C., R.E. Campbell, P.A. Steinbach, B.N. Giepmans, A.E. Palmer, and R.Y. Tsien. (2004). Improved monomeric red, orange and yellow fluorescent proteins derived from Discosoma sp. red fluorescent protein. *Nat Biotechnol* 22(12):1567–1572.

Shaner, N.C., M.Z. Lin, M.R. McKeown, P.A. Steinbach, K.L. Hazelwood, M.W. Davidson, and R.Y. Tsien. (2008). Improving the photostability of bright monomeric orange and red fluorescent proteins. *Nat Methods* 5(6):545–551.

Sipior, J., G.M. Carter, J.R. Lakowicz, and G. Rao. (1997). Blue light-emitting diode demonstrated as an ultraviolet excitation source for nanosecond phase-modulation fluorescence lifetime measurements. *Rev Sci Instrum* 68(7):2666–2670.

So, P.T.C., T.E. French, W. Yu, K.M. Berland, C.-Y. Dong, and E. Gratton. (1995). Time-resolved fluorescence microscopy using two-photon excitation. *Bioimaging* 3(2):49–63.

Spencer, R.D., and G. Weber. (1969). Measurements of subnanosecond fluorescence lifetimes with a cross-correlation phase fluorometer. *Ann NY Acad Sci* 158:361–376.

Spring, B.Q., and R.M. Clegg. (2009). Image analysis for denoising full-field frequency-domain fluorescence lifetime images. *J Microsc* 235(2):221–237.

Squire, A., P.J. Verveer, and P.I. Bastiaens. (2000). Multiple frequency fluorescence lifetime imaging microscopy. *J Microsc* 197(Pt 2):136–149.

Strat, D., F. Dolp, B. von Einem, C. Steinmetz, C.A. von Arnim, and A. Rueck. (2011). Spectrally resolved fluorescence lifetime imaging microscopy: Forster resonant energy transfer global analysis with a one- and two-exponential donor model. *J Biomed Opt* 16(2):026002.

Thaler, C., S.V. Koushik, P.S. Blank, and S.S. Vogel. (2005). Quantitative multiphoton spectral imaging and its use for measuring resonance energy transfer. *Biophys J* 89(4):2736–2749.

Thompson, R.B., J.K. Frisoli, and J.R. Lakowicz. (1992). Phase fluorometry using a continuously modulated laser diode. *Anal Chem* 64:2075–2078.

Tsurui, H., H. Nishimura, S. Hattori, S. Hirose, K. Okumura, and T. Shirai. (2000). Seven-color fluorescence imaging of tissue samples based on Fourier spectroscopy and singular value decomposition. *J Histochem Cytochem* 48(5):653–662.

Valeur, B. (2002). *Molecular Fluorescence: Principles and Applications.* Wiley-VCH, Weinheim, Germany.

van Munster, E.B., J. Goedhart, G.J. Kremers, E.M. Manders, and T.W. Gadella, Jr. (2007). Combination of a spinning disc confocal unit with frequency-domain fluorescence lifetime imaging microscopy. *Cytometry A* 71(4):207–214.

Venetta, B. (1959). Microscope phase fluorometer for determining the fluorescence lifetime of fluorochromes. *Rev Sci Instrum* 30(6):450–457.

Verkman, A.S., M. Armijo, and K. Fushimi. (1991). Construction and evaluation of a frequency-domain epifluorescence microscope for lifetime and anisotropy decay measurements in subcellular domains. *Biophys Chem* 40(1):117–125.

Walker, J.S. (1997). Fourier analysis and wavelet analysis. *Notices of the AMS* 44(6):658–670.

Wang, X.F., T. Uchida, and S. Minami. (1989). A fluorescence lifetime distribution measurement system based on phase-resolved detection using an image dissector tube. *Appl Spectrosc* 43(5):840–845.

Weber, G. (1981). Resolution of fluorescence lifetimes in a heterogenous system by phase and modulation measurements. *J Phys Chem* 85:949–953.

Willet, R.M., and R.D. Nowak. (2004). Fast multiresolution photon-limited image reconstruction. *IEEE Int Symp Biomed Imag* 2:1192–1195.

Yamazaki, I., and N. Tamai. (1985). Microchannel-plate photmultiplier applicability to the time-correlated photon-counting method. *Rev Sci Instrum* 56(6):1187–1194.

Zimmermann, T. (2005). Spectral imaging and linear unmixing in light microscopy. *Adv Biochem Eng Biotechnol* 95:245–265.

Fluorescence lifetime imaging techniques: Time-gated fluorescence lifetime imaging

8

James McGinty, Christopher Dunsby, and Paul M. W. French

Contents

8.1 INTRODUCTION

Wide-field time-gated fluorescence lifetime imaging (FLIM) essentially entails illuminating a sample with an ultrashort pulse of excitation radiation and sampling the resulting time varying fluorescence "image" following excitation by acquiring a series of gated fluorescence intensity images recorded at different relative delays with respect to the excitation pulse. This is represented schematically in Figure 8.1. In the simplest case, a map of the mean fluorescence decay times across the field of view is obtained. If the sampling of the fluorescence decay profiles is appropriately detailed, then the entire fluorescence decay profile for each image pixel can be acquired, and the resulting data set can be fitted to complex temporal decay models. For example, a double exponential decay model is frequently used to analyze data from Förster resonant energy transfer (FRET) experiments. The acquisition of time-gated fluorescence intensity images requires a 2-D detector, normally a charge-coupled device (CCD) camera, and some kind of fast "shutter" able to sample fluorescence decay profiles on subnanosecond timescales. Such a "shutter" function cannot be provided by mechanical means or yet by electronic circuitry and is typically provided by optical image intensifiers whose gain can be modulated by varying the applied voltage.

The first demonstrations of wide-field FLIM utilizing microchannel plate (MCP) optical image intensifiers with modulated gain were reported around 1990 and were frequency-domain instruments. These were initially read out using linear photodiode arrays (Gratton et al. 1990; Morgan et al. 1990), but these were soon replaced by CCD camera technology (Gadella et al. 1993; Lakowicz and Berndt 1991). Time-domain FLIM was demonstrated using a streak camera-based approach (Minami and Hirayama 1990) and by implementing time-gated imaging using gated MCP image intensifiers coupled to CCD cameras, for example, in the work of Cubeddu et al. (1993) and Oida et al. (1993). Initially, the shortest MCP gate widths that could be applied were over 5 ns, but the technology quickly developed to provide subnanosecond gating times (Wang et al. 1991) and then the use of a wire mesh proximity-coupled to the MCP photocathode led to devices with sub-100 ps resolution (Hares 1987). The gated optical imaging intensifier (GOI) has remained the key component for wide-field time-gated FLIM with essentially the same functionality.

Figure 8.2 shows a schematic of the GOI in the context of FLIM. While there have been different designs of GOI, for example, with one or two MCPs, different photocathodes and phosphors, and different

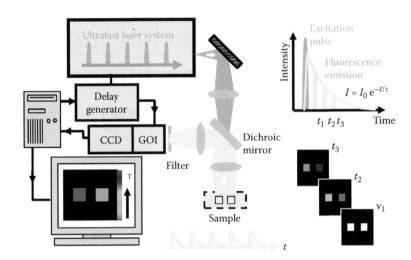

Figure 8.1 Schematic of wide-field time-gated FLIM. (Adapted from Elson, D.S. et al., *Photochem. Photobiol. Sci.*, 3, 2004.)

repetition rates, the main advances in time-gated FLIM have been the increase in GOI repetition rate from kilohertz to hundreds of megahertz, and the concurrent advances in excitation laser technology. Initially, time-gated FLIM instruments operated at relatively low repetition rates (up to approximately kilohertz), for example, in the work of Cubeddu et al. (1993), Elson et al. (2007), Oida et al. (1993), and Urayama et al. 2003), with excitation pulses from dye lasers pumped by excimer lasers or nitrogen lasers—or from these ultraviolet lasers directly. While nitrogen lasers in particular provided a convenient and low-cost pulsed excitation, the pulse durations of approximately nanoseconds are comparable to many fluorescence decay times and their relatively low repetition rate and pulse-to-pulse amplitude noise and temporal jitter compromise the instrumentation in terms of image acquisition time and signal-to-noise ratio. The development of mode-locked lasers provided excitation pulses of significantly shorter duration and increased amplitude stability. Initially, the limitation of the GOI to kilohertz repetition rates required the excitation pulse trains from mode-locked lasers to be reduced in repetition rate for wide-field time-gated FLIM, and this was realized by cavity dumping (Scully et al. 1996) and regenerative amplification (Dowling et al. 1997, 1998). The advent of mode-locked Ti:sapphire lasers in the 1990s, which could be frequency doubled to provide an almost ideal excitation source for FLIM over the range of ~350–520 nm (and provide direct two-photon excitation without second harmonic generation), stimulated

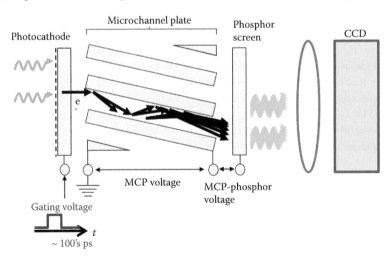

Figure 8.2 Schematic of a GOI employed in wide-field time-gated FLIM.

the development of higher-repetition-rate GOIs. The development of diode-pumped mode-locked solid-state lasers provided increasingly compact and practically deployable excitation sources, and these were complemented by the advances in ultrafast gain-switched diode laser technology. High-repetition-rate wide-field time-gated FLIM was realized using a picosecond diode laser excitation source (Elson et al. 2002), which was applied to microscopy and to a wide-field FLIM multiwell plate reader, and using a mode-locked Ti:sapphire laser, which was applied to FLIM microscopy and endoscopy, for example, in the work of Siegel et al. (2003).

Unfortunately, frequency doubling of Ti:sapphire lasers will not excite many important fluorophores above ~500 nm. While ultrashort pulses in this spectral region can be provided by a combination of optical parametric generation and second harmonic generation with mode-locked solid-state lasers, an increasingly popular alternative approach to provide tunable (and ultrafast) excitation over the visible and near-infrared (NIR) spectrum is to use ultrafast laser-pumped supercontinuum generation (Ranka et al. 2000) in microstructured optical fibers (Birk and Storz 2001; Deguil et al. 2004; Jureller et al. 2003; Kudlinski et al. 2006; McConnell 2004). This was demonstrated for FLIM using a femtosecond Ti:sapphire pump laser (Dunsby et al. 2004), but is now more widely implemented in microscopes using compact, robust, and relatively inexpensive ultrafast fiber lasers (Champert et al. 2002; Schreiber et al. 2003). FLIM can also be conveniently implemented at a range of discrete spectral wavelengths using relatively low-cost semiconductor diode laser and LED technology. Gain-switched picosecond diode lasers have been applied to time-domain FLIM in scanning (Böhmer et al. 2001) and wide-field (Elson et al. 2002) FLIM systems, although their relatively low average output power (typically <1 mW) and sparse spectral coverage limit their range of applications. It is possible to reach higher average powers using sinusoidal modulation for frequency-domain FLIM (Booth and Wilson 2004; Dong et al. 2001), and for wide-field FLIM where their low spatial coherence is an advantage, LEDs are increasingly interesting (van Geest and Stoop 2003), offering higher average powers and extended spectral coverage that potentially extends to the deep ultraviolet (e.g., 280 nm; McGuinness et al. 2004). An LED excitation source has also been used to realize frequency-domain FLIM in a scanning microscope (Herman et al. 2001).

8.2 HIGH-SPEED FLIM

The maximum imaging rate for FLIM is ultimately limited by the minimum number of detected photons/pixels required for an accurate determination of the fluorescence lifetime, which is typically a few hundred photons to reach ~10% accuracy fitting to a single exponential fluorescence decay model (Gerritsen et al. 2002; Köllner and Wolfrum 1992), and therefore will be a function of the sample brightness. This will depend on the excitation power, which in turn will be constrained by considerations of photobleaching and phototoxicity. Because the parallel pixel acquisition of wide-field frequency- and time-domain FLIM techniques entails much-reduced excitation intensity compared to scanning microscopy techniques, imaging rates can be significantly faster, with demonstrations of FLIM at rates of tens of hertz for tissue autofluorescence (Requejo-Isidro et al. 2004) and green fluorescent protein (GFP) labeled cells (Grant et al. 2007) and up to hundreds of hertz with dye fluorophores (Agronskaia et al. 2003) having been reported. This improvement over scanning techniques, for example, utilizing time-correlated single-photon counting (TCSPC), is in spite of the inherently reduced photon economy resulting from the time-varying gain applied to the MCP image intensifier, which can be particularly significant in the time domain if very short (~100 ps) time gates are applied to sample the fluorescence decay profile. In many situations, however, such short time gates are not necessary since the fast-rising and -falling edges of the time gate provide the required time resolution, so the gate can remain "open" for times comparable to the fluorescence decay time (Munro et al. 2005). Thus, the photon economy of wide-field time-gated imaging can approach that of wide-field frequency-domain FLIM.

For time-domain measurements, it is necessary to acquire at least three time-gated images in order to obtain a mean fluorescence lifetime in the presence of an offset (background signal). If the background can be determined in a separate measurement and then subtracted from subsequent acquisitions, it is possible to implement high-speed FLIM with only two time-gated images. In comparison, frequency-domain FLIM requires a minimum of three phase-resolved images. The required fluorescence lifetimes can be calculated analytically such that fluorescence lifetime images can be displayed in real time for both frequency- and time-domain FLIM. A frequency-domain FLIM microscope achieved a FLIM rate of 0.7 Hz for a field of view of 300 × 220 pixels and also provided "lifetime-resolved" images based on the difference between only two phase-resolved images at up to 55 Hz for 164 × 123 pixels (Holub et al. 2000). Frequency-domain FLIM was extended to FLIM endoscopy, acquiring phase-resolved images at 25 Hz to achieve real-time FLIM for a field of view of 32 × 32 pixels (Mizeret et al. 1999). In the time domain, rapid lifetime determination (RLD) is often implemented using just two time gates (with the background level assumed to be zero or subtracted), with an analytical approach to calculate the single exponential decay lifetime (Devries and Khan 1989). At Imperial, we demonstrated video-rate wide-field RLD FLIM using two sequentially acquired time-gated fluorescence images (subtracting a previously acquired background image) using a GOI-based system, which we also applied to FLIM endoscopy (Requejo-Isidro et al. 2004). RLD has been extensively studied, and if both gates are of equal width and contiguous, the minimum error in lifetime determination is found to be when the gate separation, Δt, is 2.5 times the lifetime being investigated (Ballew and Demas 1989). This optimized measurement is only accurate over a narrow range of sample fluorescence lifetimes, and to optimize RLD FLIM over a wider range of sample lifetimes, one can increase the width of the second time gate (Chan et al. 2001) or use three or more time gates, albeit at the expense of imaging speed. There are also analytic expressions for RLD of a single exponential decay with an unknown background (Chan et al. 2001) or a biexponential decay using four time-gated images (Sharman et al. 1999). Further optimization of RLD FLIM can exploit the potential to vary the CCD integration time in order to collect more photons at the later stages of the decay profile (Munro et al. 2005).

High-speed FLIM is important for application to dynamic or moving samples. Time-gated (or phase-resolved frequency-domain) wide-field imaging, however, is subject to severe artifacts if the sample moves between successive time-gated (or phase-resolved) acquisitions (Elson et al. 2004a). This problem is less severe for TCSPC FLIM, where sample motion during an image acquisition will degrade the spatial resolution but not drastically change the apparent fluorescence lifetimes. It is possible to avoid this issue of motion artifacts in RLD FLIM by acquiring the different time-gated (or phase-resolved) images simultaneously. This was first implemented in a "single-shot" FLIM system that utilized an optical image splitter to produce two images on the same GOI, but with the photons from one image being delayed with respect to the other by an optical relay (Agronskaia et al. 2003). This elegant approach was demonstrated at up to 100 Hz imaging rat neonatal myocytes stained with the calcium indicator Oregon green BAPTA-1. Its main drawbacks are that the field of view is reduced with respect to the conventional sequential time-gated imaging acquisition approach and that the complexity of the optical imaging delay line makes it difficult to adjust the delay between the parallel images in order to accommodate a range of sample fluorescence lifetimes. These issues can be addressed using multiple independently gated GOI detectors to maintain the field of view (Young et al. 1988) at the cost of significantly increased system size and complexity. Alternatively, the optical imaging delay line can be avoided by using a single GOI with a segmented photocathode providing multiple image channels that can be independently gated through the introduction of resistive elements. This approach was implemented as wide-field four-channel RLD FLIM at up to 20 Hz and was applied to label-free FLIM of *ex vivo* tissue (Elson et al. 2004a). We note that this parallel acquisition of multiple time-gated (phase-gated) images has recently been implemented for frequency-domain FLIM using a modulated complementary metal-oxide-semiconductor (CMOS) detector that can accumulate photoelectrons in two parallel image stores according to a modulated voltage (Esposito et al. 2005). This permits fast FLIM (with two phase-resolved images) using solid-state camera technology that could potentially be cheaper and faster than current GOI technology, although to date, it has only been realized with a 20 MHz modulation frequency and 124 × 160 pixels in each phase-resolved image.

While high-speed FLIM is the compelling application of wide-field time/phase-gated imaging, it is not always applied using RLD. For applications such as FRET, and also for imaging tissue autofluorescence, it is necessary to image samples exhibiting complex fluorescence decay profiles and to extract information concerning different components of the decay. In the time domain, this entails sampling the decay profiles with more time gates and usually fitting the data to a complex decay model using for example, a nonlinear least-squares Levenberg–Marquardt algorithm, although it is possible to use an analytical modified RLD approach to analyze double exponential decay profiles (Sharman et al. 1999). Sampling and fitting complex decay profiles, however, inevitably increases the FLIM acquisition time since significantly more detected photons are required for accurate lifetime determination (typically >10^4 detected photons to fit to a double exponential decay model; Köllner and Wolfrum 1992). For imaging biological tissue, it is often not possible to acquire so many photons/pixels given the issues of photobleaching and phototoxicity and the requirement for the sample to be stationary during an acquisition. One pragmatic approach to this issue is to approximate the complex fluorescence decay profiles to a single exponential decay model for which the resulting average fluorescence lifetime can still provide useful contrast since it will usually reflect changes in the decay times or relative contributions of different components. Of course, the interpretation of a change in the average lifetime of a complex fluorescence decay profile can be subject to ambiguity, and if quantitative multicomponent FLIM is necessary, for example, to quantify contributions from free and bound NADH, it is desirable to reduce the number of photons required to be detected using *a priori* knowledge or assumptions, for example, about the magnitude of one or more lifetime components or about their relative contributions. It can sometimes be useful to assume that some parameters are global, that is, they take the same value in each pixel of the image. Such global analysis (Verveer et al. 2000) is often used for the application of FLIM to FRET, as discussed further in Chapter 12, and increasingly finds application for autofluorescence. When such approaches are utilized, it is possible to determine multiple lifetime components with a similar number of detected photons per pixel (i.e., a few hundred) as is required for determining the mean lifetime by fitting to a monoexponential fluorescence decay model.

Wide-field time-gated FLIM has been applied to tissue imaging using photosensitizers in mice (Cubeddu et al. 1995) and humans (Cubeddu et al. 1999); utilizing autofluorescence in cells (Elson et al. 2004b; Schneckenburger et al. 1998) and tissue (Dowling et al. 1998; Elson et al. 2004b), including for studies of breast cancer (Tadrous et al. 2003), cartilage degeneration (Elson et al. 2004b; Talbot 2007), atherosclerosis (Elson et al. 2004b; Hegyi et al. 2006; Marcu 2010; Talbot 2007; Talbot et al. 2010), gastrointestinal (GI) and pancreatic cancer (McGinty et al. 2010), skin cancer (Galletly et al. 2008), and oral carcinoma (Sun et al. 2009); and guiding brain surgery (Sun et al. 2010). It has been implemented in microscopes and endoscopes, for which the potential for rapid imaging of large fields of view is important. Wide-field time-gated FLIM endoscopy has been implemented in a rigid ("Hopkins") arthroscope (Requejo-Isidro et al. 2004) and conventional forward-viewing flexible optical fiber bundle endoscopes (Munro et al. 2005). More recently, a special side-viewing wide-field endoscope was designed and demonstrated with time-gated FLIM (Elson et al. 2007) and applied *in vivo* in a hamster cheek pouch model and in humans (Sun et al. 2010). Figure 8.3 shows some label-free tissue autofluorescence images acquired by wide-field time-gated FLIM.

A key advantage of FLIM implemented in laser scanning microscopes is the ability to produce optically sectioned FLIM images, although the time to acquire a z-stack of fluorescence lifetime images on a multiphoton or laser scanning confocal microscope can be prohibitively long (or require prohibitively high excitation powers) and is rarely practical for live cells. It is possible, however, to obtain optically sectioned fluorescence lifetime images while benefitting from the parallel pixel acquisition of wide-field detection. This has been realized by scanning multiple excitation beams in parallel, and time-gated imaging has been combined with multibeam multiphoton microscopy (Benninger et al. 2005; Leveque-Fort et al. 2004; Straub and Hell 1998) and single-photon Nipkow disc confocal microscopy implemented with time-gated (Grant et al. 2005, 2007) and phase-gated imaging (van Munster et al. 2007). An alternative approach is to employ structured illumination microscopy (Neil et al. 1997) with wide-field time-gated detection (Cole et al. 2000), which entails acquiring three or more wide-field time-gated images and using these to calculate the optically sectioned (fluorescence lifetime) image in postprocessing. Unfortunately, this approach to optically sectioned FLIM suffers from reduced signal-to-noise ratio, since the out-of-focus light is recorded at the detector, and it is highly sensitive to motion artifacts. Other approaches to implementing optically sectioned FLIM with parallel pixel detection include multiple TCSPC channels combined with multiple-beam excitation (Kumar et al. 2007) and line-scanning FLIM, which can be implemented with a streak camera (Krishnan et al. 2003) or a GOI. In the latter case, a line of fluorescence emission can be relayed to the input slit of a spectrograph to facilitate "push-broom" hyperspectral FLIM, acquiring the emission spectral profile and fluorescence lifetime decay for each pixel on the line (De Beule et al. 2007; Owen et al. 2007).

Principles of fluorescence lifetime instrumentation

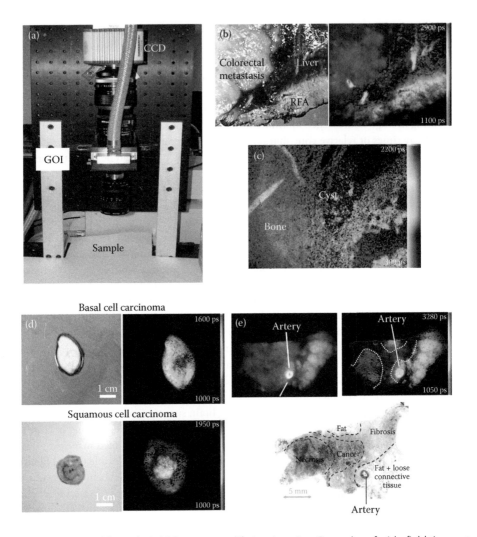

Figure 8.3 (a) Detector used for wide-field fluorescence lifetime imaging. Examples of wide-field time-gated FLIM of freshly resected tissue autofluorescence excited at 355 nm: (b) White light and FLIM images of liver tissue showing colorectal metastasis and RF ablation damage. (Adapted from McGinty, J. et al., *Biomed. Opt. Express*, 1, 2010.) (c) FLIM image of unstained fixed section of (human) femoral head showing degeneration and loss of articular cartilage with collapse of subchondral bone and subchondral cyst formation. (Adapted from Elson, D.S. et al., *Photochem. Photobiol. Sci.*, 3, 2004.) (d) White light and FLIM images of basal and squamous cell carcinomas and (e) fluorescence intensity and lifetime images of pancreatic tissue showing necrosis, cancer, fat, and loose connective tissue with an artery. (Adapted from Galletly, N.P. et al., *Br. J. Dermatol.*, 159, 2008, and McGinty, J. et al., *Biomed. Opt. Express*, 1, 2010.)

8.3 SIGNAL-TO-NOISE CONSIDERATIONS FOR WIDE-FIELD TIME-GATED FLIM

While wide-field time-gated FLIM can provide parallel image acquisition and higher speed imaging than single-beam laser scanning techniques, the requirement to sample the fluorescence decay results in a reduced photon economy. Furthermore, gated optical image intensifiers can introduce excess noise compared to TCSPC measurements, which are essentially shot-noise limited and represent the highest photon economy of any FLIM technique albeit being limited in speed by considerations of photon pileup and the constraints of the electronic circuitry. It is therefore useful to analyze the signal-to-noise performance of time-gated FLIM in order to optimize the data acquisition to maximize the accuracy of the lifetime determination for a given acquisition time.

Wide-field time-gated FLIM produces a fluorescence lifetime image from a series of gated intensity measurements, and the accuracy of the calculated lifetime is therefore determined by the signal-to-noise characteristics of the individual intensity measurements. At Imperial, we undertook a signal-to-noise characterization of two GOIs, one containing a single MCP and one a double MCP (McGinty et al. 2009a). The detectors were illuminated with a constant flux, determined from photon-counting measurements, with the average detected intensity and standard deviation calculated from 100 acquired images at a range of integration times and gain voltages. For the single MCP GOI, we found that the noise was shot noise–like, in that $\sigma_I^2 = E.I$, where I is the measured intensity, σ_I is the standard deviation, and E is the excess noise factor ($E = 1$ is the shot-noise–limited case). For the double MCP GOI, the noise statistics were also shot noise–like for gain voltages below 1100 V, but above this voltage, a higher-order dependence was observed.

Having determined the noise statistics for individual gated intensity measurements, it is possible to investigate how the acquisition parameters affected the accuracy of fluorescence lifetime measurements and therefore how to optimize the data acquisition. Assuming a single exponential fluorescence decay profile and using a linear least-squares approach, an analytic expression for the fluorescence lifetime can be derived for n equally spaced gated intensity measurements (Equation 8.1):

$$\tau = \frac{\left(\sum_{i=1}^{n} \frac{(i-1)s}{\sigma_i'^2}\right)^2 - \sum_{i=1}^{n} \frac{1}{\sigma_i'^2} \sum_{i=1}^{n}\left(\frac{(i-1)s}{\sigma_i'}\right)^2}{\sum_{i=1}^{n}\frac{1}{\sigma_i'^2}\sum_{i=1}^{n}\frac{(i-1)s\ln|I_i|}{\sigma_i'^2} - \sum_{i=1}^{n}\frac{(i-1)s}{\sigma_i'^2}\sum_{i=1}^{n}\frac{\ln|I_i|}{\sigma_i'^2}} \tag{8.1}$$

where I_i is the ith gated intensity measurement, s is the gate separation, and σ' the transformed noise characteristics (required by linearization). Unlike the situation for TCSPC, in wide-field time-gated FLIM, the user has control over the sampling of the fluorescence decay profile and can adjust the number of gates, their separation, width, and the integration time of the CCD. By inserting the appropriate expression for the noise in Equation 8.1, an error propagation analysis can be performed to determine the expected error in the fluorescence lifetime as the acquisition parameters are varied. For the case $n = 2$, the expression simplifies to the two-gate RLD (McGinty et al. 2009a; Munro et al. 2005), Equation 8.2, and its expected error, Equation 8.3.

$$\tau = \frac{s}{\ln\left|\frac{I_1}{I_2}\right|} \tag{8.2}$$

$$|\sigma_\tau| \propto \frac{\tau^2}{s}\left[\frac{1+\exp\left(-\frac{s}{\tau}\right)}{I_1}\right]^{\frac{1}{2}} \tag{8.3}$$

In these expressions, the gate separation is independent of the gate width and, by differentiating $|\sigma_\tau|$ with respect to s, the optimum gate separation of 2.22τ is found. This approach can be extended to more than two time gates (Sharman et al. 1999) and to sampling with different integration times at each time delay. For a study where the integration time was allowed to differ for two gated intensity measurements while keeping the total acquisition time constant, we plotted the expected error in the fluorescence lifetime as a 2-D surface, as shown in Figure 8.4. The minimum for this surface occurs for an integration ratio of 3.6 (i.e., where the integration time is 3.6 times longer for the second time gate at the later delay) and a gate separation of 2.56τ. In this case, the error in the fluorescence lifetime determination is reduced by ~12% compared to the case for equal integration times and the same total acquisition time.

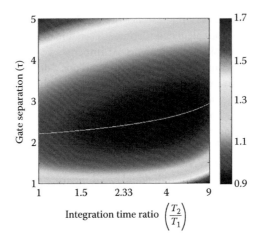

Figure 8.4 Surface showing the relative accuracy (given by color) of the two-gate RLD method with respect to the optimum separation with equal integration times. The white curve shows the optimum gate separation for different values of integration time ratio. $T_{1,2}$ are the integration times of the two gates such that $T_1 + T_2 =$ const. (Adapted from McGinty, J. et al., *J. Phys. D: Appl. Phys.*, 42, 2009.)

This analysis was undertaken with the assumption of monoexponential fluorescence decay profiles. A similar analysis can be made for the frequency-domain approach, where the acquisition can be optimized for excitation modulation frequency, number and position of phase measurements, and integration time (Esposito et al. 2007; Philip and Carlsson 2003). In principle, it should be possible to apply a similar approach to optimize the sampling and data acquisition of time-gated FLIM of samples exhibiting complex decay profiles, but this has not yet been done to our knowledge.

8.4 TOMOGRAPHIC FLIM

Wide-field time-gated imaging is also of increasing interest for time-resolved optical tomography, for which the parallel pixel measurement provides an important advantage with respect to imaging speed. Time-resolved detection has long been used to improve the reconstruction of tomographic images in highly scattering media (Hebden et al. 1997) utilizing transmitted light and fluorescence. Serendipitously, the same time-gated (phase-resolved) imaging technology used for diffuse optical tomography (DOT) and diffuse fluorescence tomography (DFT) to aid tomographic reconstruction of fluorescence intensity images by inverse scattering techniques (as discussed in chapter 8 of this volume), can also be applied to FLIM (Paithankar et al. 1997). Tomographic FLIM has been applied to scattering phantoms using frequency-domain (Godavarty et al. 2005; Shives et al. 2002) and time-domain (Kumar et al. 2005, 2008) approaches, and time-gated fluorescence molecular tomography (FMT) of live mice has recently been demonstrated *in vivo* (Niedre et al. 2008; Nothdurft et al. 2009). At Imperial College London, we have shown that we can reconstruct the fluorescence lifetime and other optical properties of strongly scattering samples with inclusions containing genetically expressed fluorophores exhibiting FRET (McGinty et al. 2009b) using the experimental setup shown in Figure 8.5a and employing a reconstruction algorithm based on the minimization of a cost functional between the measured data and a model of light diffusion in the Fourier domain (Soloviev et al. 2009). The sample is mounted on a rotation stage, and a fiber laser–pumped supercontinuum-based light is used to excite the fluorescence, which is imaged onto the photocathode of a GOI with the resulting wide-field time-gated and intensified fluorescence signals being recorded by a CCD camera. There is considerable interest in applying FLIM tomography to live mice labeled with genetically expressed fluorescent proteins (Kumar et al. 2009), particularly for FRET readouts. Accordingly, we applied this tomoFLIM apparatus to tomographic FLIM FRET of enhanced-GFP/mCherry (McGinty et al. 2011a) expressed in the leg muscles of live mice Figure 8.5b. This proof-of-principle demonstration highlights the prospect to noninvasively map FRET signals using tomographic FLIM in live disease models, which would provide a powerful means for *in vivo* localization of, for example, cell signaling events

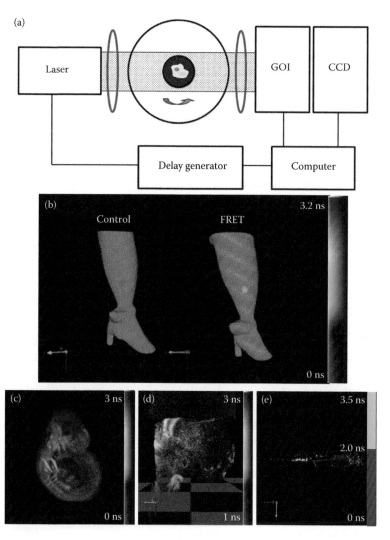

Figure 8.5 (a) Schematic of an experimental set-up for tomographic imaging based on rotational scanning. (b) 3-D fluorescence lifetime reconstructions of a mixture (control) and linked construct (FRET) of EGFP and mCherry expressed in a mouse leg *in vivo*. (Adapted from McGinty, J. et al., *Biomed. Opt. Express*, 2, 2011.) (c) 3-D fluorescence lifetime reconstructions of a fixed, optically cleared mouse embryo with an Alexa-488 labeled neurofilament. (Adapted from McGinty, J. et al., *J. Biophotonics*, 1, 2008.) (d) Autofluorescence FLIM reconstruction of a fixed, optically cleared lung sample. (e) 3-D fluorescence lifetime of a live lysC:GFP transgenic zebra fish embryo 3 days postfertilization. (Adapted from McGinty, J. et al., *Biomed. Opt. Express*, 2, 2011.)

for biomedical research and drug discovery, permitting longitudinal studies with a reduced number of animals. Ultimately, such readouts could be combined and correlated with label-free readouts of changes in metabolic pathways and tissue matrix properties. It remains challenging, however, to obtain tomographic fluorescence lifetime images of deep internal tissues owing to the high absorption, and scattering of visible radiation and the development of near infrared labels, particularly fluorescent proteins (Shu et al. 2009) is of critical importance for FLIM tomography of highly scattering disease models such as mice.

One way to avoid the difficulties associated with the high absorption and scattering of visible radiation by live animals such as mice is to avoid them by working with smaller, more transparent disease models such as *Drosophila, Caenorhabditis elegans,* and *Danio rerio* (zebra fish), which are optically accessible and can be genetically manipulated. This approach is of increasing importance for drug discovery and fundamental research since *in vivo* experiments provide a more relevant physiological context for compounds and molecular biology under study. Thus, high-content assays of biomolecular interactions could be translated from *in vitro* microscopy and analysis of cultured cells to imaging of live organisms and then to mice and higher disease models. The robust nature of fluorescence lifetime measurements makes FLIM an interesting readout for this vision, including of FRET, and wide-field time-gated imaging provides a means to rapidly acquire 3-D images of whole organisms. To this end, we have combined time-gated FLIM with optical projection tomography (OPT) (Sharpe et al. 2002), producing 3-D fluorescence lifetime images of transparent samples in the range ~0.1–1 cm (McGinty et al. 2008).

OPT is the optical equivalent of x-ray computed tomography (x-ray CT), for which a stack of X-Z slices are reconstructed from a series of transverse (X-Y) projections acquired at a number of projection angles. Due to its similarity to x-ray CT, many of the image reconstruction techniques, for example, in the work of Kak and Slaney (1988), can be directly applied, providing that the sample is treated (chemically cleared) to eliminate the optical scattering. The classic filtered back-projection technique assumes parallel ray paths, so optical beam propagation considerations limit its applicability samples that are smaller than the depth of focus of the imaging system (Sharpe 2004). For FLIM OPT, the experimental setup is essentially similar to that depicted in Figure 8.5a with the sample being mounted beneath a rotation stage and suspended in a reservoir of index-matching fluid. For FLIM OPT, the 3-D time-gated fluorescence intensity distribution is determined for each time delay, and then the fluorescence lifetime distribution is calculated for each voxel from the volumetric intensity decay data.

Figure 8.5c shows 3-D fluorescence lifetime reconstruction of an optically cleared fixed mouse embryo imaged with FLIM OPT (McGinty et al. 2008) that was fixed in formaldehyde and the neurofilament labeled with an Alexa-488-conjugated antibody. The Alexa-488-labeled neurofilament presented a lifetime of 1360 ± 180 ps, and autofluorescence, thought to be from elastin associated with the circulatory system, was also observed with a fluorescence lifetime of 1030 ± 135 ps. As well as imaging optically cleared or transparent disease models such as mouse embryos or zebra fish, OPT is also useful to map 3-D structures and molecular distributions in relatively large tissue samples that would be challenging for microscopy. Figure 8.5d shows an optically cleared excised lung sample for which FLIM OPT is providing label-free contrast based on autofluorescence lifetime. In particular, contrast between vessel walls and the surrounding stromal tissue is apparent. Such volumetric imaging of intact biopsy specimens is much faster and less likely to compromise the sample structure than the conventional approach of mechanically slicing the sample and reconstructing the volumetric image from many serial sections. We believe that FLIM OPT will be useful to image 3-D fluorescence lifetime distributions in histopathological samples and to provide contrast with both extrinsic labels and intrinsic autofluorescence signals. We have recently extended FLIM OPT to live zebra fish, and Figure 8.5e shows the 3-D rendered FLIM image of a transgenic (lysC:GFP) zebra fish embryo at 3 days postfertilization in which the myeloid cells (e.g., neutrophils and macrophages) express GFP. The GFP fluorescence is contrasted with autofluorescence by the binary lifetime color scale (McGinty et al. 2011b).

8.5 MULTIDIMENSIONAL FLUORESCENCE IMAGING

Increasingly, fluorescence instrumentation aims to provide more information than just the localization or distribution of specific fluorescent molecules, and this trend to higher content fluorescence readouts increasingly exploits multidimensional fluorescence imaging and measurement capabilities with instrumentation that resolves fluorescence lifetime together with other spectroscopic parameters such as excitation and emission wavelength and polarization, providing image information in two or three spatial dimensions as well as with respect to elapsed time. Analyzing fluorescence signals with respect to two or more spectral dimensions can increase the ability to contrast different fluorophores or their local environments. For example, utilizing the excitation-emission matrix (EEM) or applying spectrally resolved lifetime measurements can improve the capability to unmix signals from different fluorophores, particularly for autofluorescence-based experiments, and to improve quantitation of changes to the local molecular environment.

Such multidimensional fluorescence imaging and metrology approaches should be optimized to obtain as much useful information as possible from a sample in a given acquisition time since photobleaching/ damage or experimental considerations inevitably impose a limited photon "budget." This is particularly

(a)

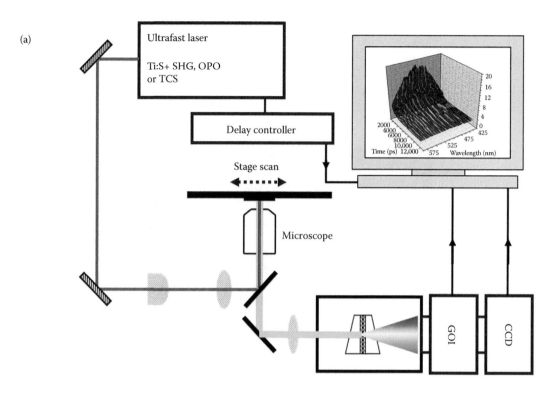

(b)

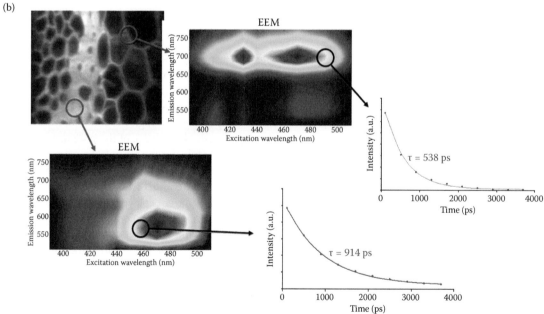

Figure 8.6 (a) Experimental setup for line-scanning hyperspectral FLIM. (b) Integrated intensity image of a sample with EEMs and decay curves for the two indicated regions. (c) Time-integrated central wavelength image of a frozen human artery exhibiting atherosclerosis and regions corresponding to medium and fibrous and lipid-rich plaques. (d) Spectrally integrated lifetime map of sample autofluorescence. (e) Autofluorescence lifetime-emission matrix. (Adapted from De Beule, P. et al., *Microsc. Res. Tech.*, 70, 2007.)

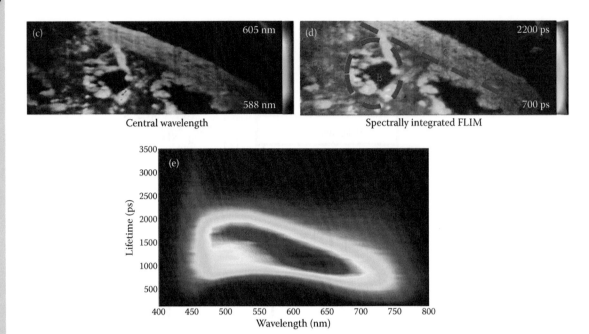

Figure 8.6 (Continued) (a) Experimental setup for line-scanning hyperspectral FLIM. (b) Integrated intensity image of a sample with EEMs and decay curves for the two indicated regions. (c) Time-integrated central wavelength image of a frozen human artery exhibiting atherosclerosis and regions corresponding to medium and fibrous and lipid-rich plaques. (d) Spectrally integrated lifetime map of sample autofluorescence. (e) Autofluorescence lifetime-emission matrix. (Adapted from De Beule, P. et al., *Microsc. Res. Tech.*, 70, 2007.)

important for real-time clinical diagnostic applications, as well as for higher-throughput assays and the investigation of dynamic biological systems. Signal-to-noise ratios will be compromised if the photon budget is allocated over more spectral channels than necessary, and optimal performance requires consideration of prior knowledge and the target information. In practice, for any fluorescence imaging experiment including FLIM, spectral discrimination is inherently applied through the choice of filters, dichroic beam splitters, and excitation wavelengths and comprehensive knowledge of the system, and sample spectral properties can facilitate better optimization, for example, for faster clinical measurements.

Wide-field time-gated FLIM provides three dimensions of detection that can be utilized for multi-dimensional fluorescence imaging (MDFI). Using a dichroic image splitter, FLIM can be combined with multispectral imaging using a tunable filter or a multichannel image splitter deployed in front of the GOI to acquire time-resolved images in a few discrete spectral channels. This has been combined with structured illumination to provide multispectral optically sectioned FLIM, resolving fluorescence with respect to five dimensions (Siegel et al. 2001). Alternatively, wide-field FLIM can be combined with hyperspectral imaging, for which the full time-resolved excitation or emission spectral profile is acquired for each image pixel. This can be realized by the so-called push-broom approach to hyperspectral imaging (Schultz et al. 2001) implemented in a line-scanning microscope. For hyperspectral FLIM, the fluorescence resulting from a line excitation is imaged to the entrance slit of a spectrograph to produce an $(x\text{-}\lambda)$ "subimage" that can be recorded on a wide-field FLIM detector. Stage scanning along the y axis then sequentially provides the full $(x\text{-}y\text{-}\lambda\text{-}\tau)$ data set. This approach is photon efficient, and the line-scanning microscope configuration provides "semiconfocal" optical sectioning.

Figure 8.6 shows an implementation of this push-broom line-scanning hyperspectral FLIM utilizing wide-field time-gated imaging (De Beule et al. 2007) together with data from its application to autofluorescence contrast in an unstained fixed section of human artery exhibiting atherosclerosis. This MDFI approach can be further extended by employing a tunable excitation laser to also resolve the autofluorescence with respect to excitation wavelength to realize six-dimensional imaging (Owen et al. 2007). Using spectral selection of a fiber laser-pumped supercontinuum source to provide tunable excitation from 390 to 510 nm, the line-scanning hyperspectral FLIM microscope can acquire the fluorescence excitation-emission-lifetime (EEL) matrix for each image pixel of a sample. Such a multidimensional data set can be used to subsequently reconstruct conventional EEMs for any image pixel and to obtain the fluorescence decay profile for any point in this EEM space. While this can provide exquisite sensitivity to perturbations in fluorescence emission and can enhance the ability to unmix signals from different fluorophores, the size and complexity of such high-content hyperspectral FLIM data sets is almost beyond the scope of *ad hoc* manual analysis. Sophisticated bioinformatics software tools are required to automatically analyze and present such data to identify trends and fluorescence "signatures." The combination of MDFI with image segmentation for high-content analysis should provide powerful tools, for example, for histopathology and for screening applications.

FLIM can also be combined with polarization-resolved imaging to provide information about rotational mobility, molecular clustering, and protein–protein interactions via homoFRET. While steady-state imaging of polarization-resolved fluorescence (i.e., fluorescence anisotropy) is a well-established technique (Lakowicz 2006) to obtain information concerning fluorophore orientation and to probe resonant energy transfer and rotational decorrelation, time-resolved imaging of fluorescence polarization anisotropy is less widely undertaken, although it has also been used to identify different components in the NADH signal from cardiomyocytes (Vishwasrao et al. 2005). Time-resolved, polarization-resolved imaging has been realized using wide-field time-gated imaging combined with a polarization-resolved image splitter to acquire the polarized images in parallel to avoid motion artifacts (Siegel et al. 2002).

8.6 SUMMARY

We have reviewed the development and application of wide-field time-gated imaging utilizing gated optical image intensifiers for FLIM. This permits rapid image acquisition for applications including the readout of signaling processes via FRET in live cells, high-content analysis with medium throughput, and tomographic FLIM of live organisms. Time-gated FLIM can also be combined with optical sectioning and with other spectroscopic imaging modalities, for example, resolving excitation and/or emission spectra and/or polarization, to realize multidimensional fluorescence imaging. By adjusting the gate delays and durations and the camera integration times, it is possible to optimize the precision in lifetime determination for a given photon budget or total acquisition time, although to date, this optimization has only been investigated for monoexponential decay profiles.

REFERENCES

Agronskaia, A. V., Tertoolen, L. and Gerritsen, H. C. 2003. High frame rate fluorescence lifetime imaging. *Journal of Physics D-Applied Physics* **36**:1655–1662.

Ballew, R. M. and Demas, J. N. 1989. An error analysis of the rapid lifetime determination method of single exponential decays. *Analytical Chemistry* **61**:30–33.

Benninger, R. K. P., Hofmann, O., McGinty, J. et al. 2005. Time-resolved fluorescence imaging of solvent interactions in microfluidic devices. *Optics Express* **13**:6275–6285.

Birk, H. and Storz, R. 2001. Illuminating device and microscope. US Patent 6,611,643. Leica Microsystems Heidelberg GmbH.

Böhmer, M., Pampaloni, F., Wahl, M., Rahn, H. J., Erdmann, R. and Enderlein, J. 2001. Time-resolved confocal scanning device for ultrasensitive fluorescence detection. *Review of Scientific Instruments* **72**:4145–4152.

Booth, M. J. and Wilson, T. 2004. Low-cost, frequency-domain, fluorescence lifetime confocal microscopy. *Journal of Microscopy-Oxford* **214**:36–42.

Champert, P. A., Popov, S. V., Solodyankin, M. A. and Taylor, J. R. 2002. Multiwatt average power continua generation in holey fibers pumped by kilowatt peak power seeded ytterbium fiber amplifier. *Applied Physics Letters* **81**:2157–2159.

Chan, S. P., Fuller, Z. J., Demas, J. N. and DeGraff, B. A. 2001. Optimized gating scheme for rapid lifetime determinations of single-exponential luminescence lifetimes. *Analytical Chemistry* **73**:4486–4490.

Cole, M. J., Siegel, S., Webb, S. E. D. et al. 2000. Whole-field optically sectioned fluorescence lifetime imaging. *Optics Letters* **25**:1361–1363.

Cubeddu, R., Taroni, P. and Valentini, G. 1993. Time gated imaging system for tumour diagnosis. *Optical Engineering* **32**:320–325.

Cubeddu, R., Pifferi, A., Taroni, P., Valentini, G. and Canti, G. 1995. Tumor detection in mice by measurements of fluorescence decay time matrices. *Optics Letters* **20**:2553–2555.

Cubeddu, R., Pifferi, A., Taroni, P., Torricelli, A., Valentini, G. and Sorbellini, E. 1999. Fluorescence lifetime imaging: An application to the detection of skin tumors. *IEEE Selected Topics on Quantum Electronics* **5**:923–929.

De Beule, P., Owen, D. M., Manning, H. B. et al. 2007. Rapid hyperspectral fluorescence lifetime imaging. *Microscopy Research and Technique* **70**:481–484.

Deguil, N., Mottay, E., Salin, F., Legros, P. and Choquet, D. 2004. Novel diode-pumped infrared tunable laser system for multi-photon microscopy. *Microscopy Research and Technique* **63**:23–26.

Devries, P. D. and Khan, A. A. 1989. An efficient technique for analyzing deep level transient spectroscopy data. *Journal of Electronic Materials* **18**:543–547.

Dong, C. Y., Buehler, C., So, P. T. C., French, T. and Gratton, E. 2001. Implementation of intensity-modulated laser diodes in time-resolved, pump-probe fluorescence microscopy. *Applied Optics* **40**:1109–1115.

Dowling, K., Hyde, S. C. W., Dainty, J. C., French, P. M. W. and Hares, J. D. 1997. 2-D fluorescence lifetime imaging using a time-gated image intensifier. *Optics Communications* **135**:27–31.

Dowling, K., Dayel, M. J., Lever, M. J., French, P. M. W., Hares, J. D. and Dymoke-Bradshaw, A. K. L. 1998. Fluorescence lifetime imaging with picosecond resolution for biomedical applications. *Optics Letters* **23**:810–812.

Dunsby, C., Lanigan, P. M. P., McGinty, J. et al. 2004. An electronically tunable ultrafast laser source applied to fluorescence imaging and fluorescence lifetime imaging microscopy. *Journal of Physics D-Applied Physics* **37**:3296–3303.

Elson, D. S., Siegel, J., Webb, S. E. D. et al. 2002. Fluorescence lifetime system for microscopy and multiwell plate imaging with a blue picosecond diode laser. *Optics Letters* **27**:1409–1411.

Elson, D. S., Munro, I., Requejo-Isidro, J. et al. 2004a. Real-time time-domain fluorescence lifetime imaging including single-shot acquisition with a segmented optical image intensifier. *New Journal of Physics* **6**:180.

Elson, D. S., Requejo-Isidro, J., Munro, I. et al. 2004b. Time domain fluorescence lifetime imaging applied to biological tissue. *Photochemical & Photobiological Sciences* **3**:795–801.

Elson, D. S., Jo, J. A. and Marcu, L. 2007. Miniaturized side-viewing imaging probe for fluorescence lifetime imaging (FLIM): Validation with fluorescence dyes, tissue structural proteins and tissue specimens. *New Journal of Physics* **9**:127.

Esposito, A., Oggier, T., Gerritsen, H. C., Lustenberger, F. and Wouters, F. S. 2005. All-solid-state lock-in imaging for wide-field fluorescence lifetime sensing. *Optics Express* **13**:9812–9821.

Esposito, A., Gerritsen, H. C. and Wouters, F. S. 2007. Optimizing frequency-domain fluorescence lifetime sensing for high-throughput applications: Photon economy and acquisition speed. *Journal of the Optical Society of America A* **24**:3261–3273.

Gadella, T. W. J., Jovin, T. M. and Clegg, R. M. 1993. Fluorescence lifetime imaging microscopy (FLIM)—Spatial resolution of structures on the nanosecond timescale. *Biophysical Chemistry* **48**:221–239.

Galletly, N. P., McGinty, J., Dunsby, C. et al. 2008. Fluorescence lifetime imaging distinguishes basal cell carcinoma from surrounding uninvolved skin. *British Journal of Dermatology* **159**:152–161.

Gerritsen, H. C., Asselbergs, M. A. H., Agronskaia, A. V. and Van Sark, W. G. J. H. M. 2002. Fluorescence lifetime imaging in scanning microscopes: Acquisition speed, photon economy and lifetime resolution. *Journal of Microscopy* **206**:218–224.

Godavarty, A., Sevick-Muraca, E. M. and Eppstein, M. J. 2005. Three-dimensional fluorescence lifetime tomography. *Medical Physics* **32**:992–1000.

Grant, D. M., Elson, D. S., Schimpf, D. et al. 2005. Optically sectioned fluorescence lifetime imaging using a Nipkow disc microscope and a tunable ultrafast continuum excitation source. *Optics Letters* **30**:3353–3355.

Grant, D. M., McGinty, J., McGhee, E. J. et al. 2007. High speed optically sectioned fluorescence lifetime imaging permits study of live cell signaling events. *Optics Express* **15**:15656–15673.

Gratton, E., Feddersen, B. and van de Ven, M. 1990. Parallel acquisition of fluorescence decay using array detectors. In *Time-Resolved Laser Spectroscopy in Biochemistry II*, ed. J. R. Lakowicz. Bellingham, WA: SPIE, 1204, pp. 21–25.

Hares, J. D. 1987. Advances in sub-nanosecond shutter tube technology and applications in plasma physics. In *X-Rays from Laser Plasmas*, ed. M. C. Richardson. Bellingham, WA: SPIE, 831, pp. 165–170.

Hebden, J. C., Arridge, S. R. and Delpy, D. T. 1997. Optical imaging in medicine: I. Experimental techniques. *Physics in Medicine and Biology* **42**:825–840.

Hegyi, L., Talbot, C., Monaco, C. et al. 2006. Fluorescence lifetime imaging of unstained human atherosclerotic plaques. *Atherosclerosis Supplements* **7**:587.

Herman, P., Maliwal, B. P., Lin, H.-J. and Lakowicz, J. R. 2001. Frequency-domain fluorescence microscopy with the LED as a light source. *Journal of Microscopy* **203**:176–181.

Holub, O., Seufferheld, M. J., Gohlke, C., Govindjee and Clegg, R. M. 2000. Fluorescence lifetime imaging (FLI) in real-time—A new technique in photosynthesis research. *Photosynthetica* **38**:581–599.

Jureller, J. E., Scherer, N. F., Birks, T. A., Wadsworth, W. J. and Russell, P. S. J. 2003. Widely tunable femtosecond pulses from a tapered fiber for ultrafast microscopy and multiphoton applications. In *Ultrafast Phenomena XIII*, eds. R. J. D. Miller, M. M. Murnane, N. F. Scherer, and A. M. Weiner. Berlin: Springer-Verlag, 684–686.

Kak, A. C. and Slaney, M. 1988. *Principles of Computerized Tomographic Imaging*. New York: IEEE Press.

Köllner, M. and Wolfrum, J. 1992. How many photons are necessary for fluorescence-lifetime measurements? *Chemical Physics Letters* **200**:199–204.

Krishnan, R. V., Saitoh, H., Terada, H., Centonze, V. E. and Herman, B. 2003. Development of a multiphoton fluorescence lifetime imaging microscopy system using a streak camera. *Review of Scientific Instruments* **74**:2714–2721.

Kudlinski, A., George, A. K., Knight, J. C. et al. 2006. Zero-dispersion wavelength decreasing photonic crystal fibers for ultraviolet-extended supercontinuum generation. *Optics Express* **14**:5715–5722.

Kumar, A. T. N., Skoch, J., Bacskai, B. J., Boas, D. A. and Dunn, A. K. 2005. Fluorescence-lifetime-based tomography for turbid media. *Optics Letters* **30**:3347–3349.

Kumar, A. T. N., Raymond, S. B., Dunn, A. K., Bacskai, B. J. and Boas, D. A. 2008. A time domain fluorescence tomography system for small animal imaging. *IEEE Transactions on Medical Imaging* **27**:1152–1163.

Kumar, A. T. N., Chung, E., Raymond, S. B. et al. 2009. Feasibility of in vivo imaging of fluorescent proteins using lifetime contrast. *Optics Letters* **34**:2066–2068.

Kumar, S., Dunsby, C., De Beule, P. A. A. et al. 2007. Multifocal multiphoton excitation and time correlated single photon counting detection for 3-D fluorescence lifetime imaging. *Optics Express* **15**:12548–12561.

Lakowicz, J. R. 2006. *Principles of Fluorescence Spectroscopy*, 3rd edition. New York: Springer.

Lakowicz, J. R. and Berndt, K. W. 1991. Lifetime-selective fluorescence imaging using an rf phase-sensitive camera. *Review of Scientific Instruments* **62**:1727–1734.

Leveque-Fort, S., Fontaine-Aupart, M. P., Roger, G. and Georges, P. 2004. Fluorescence-lifetime imaging with a multifocal two-photon microscope. *Optics Letters* **29**:2884–2886.

Marcu, L. 2010. Fluorescence lifetime in cardiovascular diagnostics. *Journal of Biomedical Optics* **15**:011106.

McConnell, G. 2004. Confocal laser scanning fluorescence microscopy with a visible continuum source. *Optics Express* **12**:2844–2850.

McGinty, J., Tahir, K. B., Laine, R. et al. 2008. Fluorescence lifetime optical projection tomography. *Journal of Biophotonics* **1**:390–394.

McGinty, J., Requejo-Isidro, J., Munro, I. et al. 2009a. Signal-to-noise characterization of time-gated intensifiers used for wide-field time-domain FLIM. *Journal of Physics D: Applied Physics* **42**:1–9.

McGinty, J., Soloviev, V. Y., Tahir, K. B. et al. 2009b. Three-dimensional imaging of Förster resonance energy transfer in heterogeneous turbid media by tomographic fluorescence lifetime. *Optics Letters* **34**:2772–2774.

McGinty, J., Galletly, N. P., Dunsby, C. et al. 2010. Wide-field fluorescence lifetime imaging of cancer. *Biomedical Optics Express* **1**:627–640.

McGinty, J., Stuckey, D. W., Soloviev, V. Y. et al. 2011a. In vivo fluorescence lifetime tomography of a FRET probe expressed in mouse. *Biomedical Optics Express* **2**:1907–1917.

McGinty, J., Taylor, H. B., Chen, L. et al. 2011b. In vivo fluorescence lifetime optical projection tomography. *Biomedical Optics Express* **2**:1340–1350.

McGuinness, C. D., Sagoo, K., McLoskey, D. and Birch, D. J. S. 2004. A new sub-nanosecond led at 280 nm: Application to protein fluorescence. *Measurement Science and Technology* **15**:L19–L22.

Minami, T. and Hirayama, S. 1990. High-quality fluorescence decay curves and lifetime imaging using an elliptic scan streak camera. *Journal of Photochemistry and Photobiology A:Chemistry* **53**:11–21.

Mizeret, J., Stepinac, T., Hansroul, M., Studzinski, A., van den Bergh, H. and Wagnieres, G. 1999. Instrumentation for real-time fluorescence lifetime imaging in endoscopy. *Review of Scientific Instruments* **70**:4689–4701.

Morgan, C. G., Mitchell, A. C. and Murray, J. G. 1990. Nanosecond time-resolved fluorescence microscopy: Principles and practice. *Transactions of the Royal Microscopy Society* **1**:463–466.

Munro, I., McGinty, J., Galletly, N. et al. 2005. Toward the clinical application of time-domain fluorescence lifetime imaging. *Journal of Biomedical Optics* **10**:051403.

Neil, M. A. A., Juškaitis, R. and Wilson, T. 1997. Method of obtaining optical sectioning by using structured light in a conventional microscope. *Optics Letters* **22**:1905–1907.

Niedre, M. J., de Kleine, R. H., Aikawa, E., Kirsch, D. G., Weissleder, R. and Ntziachristos, V. 2008. Early photon tomography allows fluorescence detection of lung carcinomas and disease progression in mice in vivo. *Proceedings of the National Academy of Sciences of the United States of America* **105**:19126–19131.

Nothdurft, R. E., Patwardhan, S. V., Akers, W., Ye, Y. P., Achilefu, S. and Culver, J. P. 2009. In vivo fluorescence lifetime tomography. *Journal of Biomedical Optics* **14**:024004.

Oida, T., Sako, Y. and Kusumi, A. 1993. Fluorescence lifetime imaging microscopy (flimscopy). Methodology development and application to studies of endosome fusion in single cells. *Biophysical Journal* **64**:676–685.

Owen, D. M., Auksorius, E., Manning, H. B. et al. 2007. Excitation-resolved hyperspectral fluorescence lifetime imaging using super-continuum generation. *Optics Letters* **32**:3408–3410.

Paithankar, D. Y., Chen, A. U., Pogue, B. W., Patterson, M. S. and Sevick-Muraca, E. M. 1997. Imaging of fluorescent yield and lifetime from multiply scattered light reemitted from random media. *Applied Optics* **36**:2260–2272.

Philip, J. and Carlsson, K. 2003. Theoretical investigation of the signal-to-noise ratio in fluorescence lifetime imaging. *Journal of the Optical Society of America A* **20**:368–379.

Ranka, J. K., Windeler, R. S. and Stentz, A. J. 2000. Visible continuum generation in air-silica microstructure optical fibers with anomalous dispersion at 800 nm. *Optics Letters* **25**:25–27.

Requejo-Isidro, J., McGinty, J., Munro, I. et al. 2004. High-speed wide-field time-gated endoscopic fluorescence lifetime imaging. *Optics Letters* **29**:2249–2251.

Schneckenburger, H., Gschwend, M. H., Sailer, R., Mock, H.-P. and Strauss, W. S. L. 1998. Time-gated fluorescence microscopy in molecular and cellular biology. *Cellular and Molecular Biology* **44**:795–805.

Schreiber, T., Limpert, J., Zellmer, H., Tunnermann, A. and Hansen, K. P. 2003. High average power supercontinuum generation in photonic crystal fibers. *Optics Communications* **228**:71–78.

Schultz, R. A., Nielsen, T., Zavaleta, J. R., Ruch, R., Wyatt, R. and Garner, H. R. 2001. Hyperspectral imaging: A novel approach for microscopic analysis. *Cytometry* **43**:239–247.

Scully, A. D., Mac Robert, A. J., Botchway, S. et al. 1996. Development of a laser-based fluorescence microscope with subnanosecond time resolution. *Journal of Fluorescence* **6**:119–125.

Sharman, K. K., Periasamy, A., Ashworth, H., Demas, J. N. and Snow, N. H. 1999. Error analysis of the rapid lifetime determination method for double-exponential decays and new windowing schemes. *Analytical Chemistry* **71**:947–952.

Sharpe, J. 2004. Optical projection tomography. *Annual Review of Biomedical Engineering* **6**:209–228.

Sharpe, J., Ahlgren, U., Perry, P. et al. 2002. Optical projection tomography as a tool for 3D microscopy and gene expression studies. *Science* **296**:541–545.

Shives, E., Xu, Y. and Jiang, H. 2002. Fluorescence lifetime tomography of turbid media based on an oxygen-sensitive dye. *Optics Express* **10**:1557–1562.

Shu, X., Royant, A., Lin, M. Z. et al. 2009. Mammalian expression of infrared fluorescent proteins engineered from a bacterial phytochrome. *Science* **324**:804–807.

Siegel, J., Elson, D. S., Webb, S. E. D. et al. 2001. Whole-field five-dimensional fluorescence microscopy combining lifetime and spectral resolution with optical sectioning. *Optics Letters* **26**:1338–1340.

Siegel, J., Suhling, K., Leveque-Fort, S. et al. 2002. Wide-field time-resolved fluorescence anisotropy imaging (TR-FAIM)—Imaging the mobility of a fluorophore. *Review of Scientific Instruments* **73**:182–192.

Siegel, J., Elson, D. S., Webb, S. E. D. et al. 2003. Studying biological tissue with fluorescence lifetime imaging: Microscopy, endoscopy and complex decay profiles. *Applied Optics* **42**:2995–3004.

Soloviev, V. Y., D'Andrea, C., Valentini, G., Cubeddu, R. and Arridge, S. R. 2009. Combined reconstruction of fluorescent and optical parameters using time-resolved data. *Applied Optics* **48**:28–36.

Straub, M. and Hell, S. W. 1998. Fluorescence lifetime three-dimensional microscopy with picosecond precision using a multifocal multiphoton microscope. *Applied Physics Letters* **73**:1769–1771.

Sun, Y., Phipps, J., Elson, D. S. et al. 2009. Fluorescence lifetime imaging microscopy: In vivo application to diagnosis of oral carcinoma. *Optics Letters* **34**:2081–2083.

Sun, Y. H., Hatami, N., Yee, M. et al. 2010. Fluorescence lifetime imaging microscopy for brain tumor image-guided surgery. *Journal of Biomedical Optics* **15**:56022.

Tadrous, P. J., Siegel, J., French, P. M. W., Shousha, S., Lalani, E. N. and Stamp, G. W. H. 2003. Fluorescence lifetime imaging of unstained tissues: Early results in human breast cancer. *Journal of Pathology* **199**:309–317.

Talbot, C. B. 2007. Temporal and spectral resolution of tissue autofluorescence. PhD Thesis, Imperial College London.

Talbot, C. B., McGinty, J., McGhee, E. et al. 2010. Fluorescence lifetime imaging and metrology for biomedicine. In *Handbook of Photonics for Biomedical Science*, ed. V. Tuchin. Boca Raton, FL: CRC Press, 159–196.

Urayama, P., Zhong, W., Beamish, J. A. et al. 2003. A UV-visible-NIR fluorescence lifetime imaging microscope for laser-based biological sensing with picosecond resolution. *Applied Physics B* **76**:483–496.

van Geest, L. K. and Stoop, K. W. J. 2003. FLIM on a wide field fluorescence microscope. *Letters in Peptide Science* **10**:501–510.

van Munster, E. B., Goedhart, J., Kremers, G. J., Manders, E. M. and Gadella, T. W. 2007. Combination of a spinning disc confocal unit with frequency-domain fluorescence lifetime imaging microscopy. *Cytometry A* **71**:207–214.

Verveer, P. J., Squire, A. and Bastiaens, P. I. H. 2000. Global analysis of fluorescence lifetime imaging microscopy data. *Biophysical Journal* **78**:2127–2137.

Vishwasrao, H. D., Heikal, A. A., Kasischke, K. A. and Webb, W. W. 2005. Conformational dependence of intracellular NADH on metabolic state revealed by associated fluorescence anisotropy. *Journal of Biological Chemistry* **280**:25119–25126.

Wang, X. F., Uchida, T., Coleman, D. M. and Minami, S. 1991. A two-dimensional fluorescence lifetime imaging system using a gated image intensifier. *Applied Spectroscopy* **45**:360–366.

Young, P. E., Hares, J. D., Kilkenny, J. D., Phillion, D. W. and Campbell, E. M. 1988. Four-frame gated optical imager with 120-ps resolution. *Review of Scientific Instruments* **59**:1457–1460.

Fluorescence lifetime imaging techniques: Time-correlated single-photon counting

Wolfgang Becker

Contents

9.1 INTRODUCTION

Fluorescence lifetime imaging (FLIM) techniques can be classified into time-domain and frequency-domain techniques, photon counting and analog techniques, and point-scanning and wide-field imaging techniques. Virtually all combinations are in use. This leads to a wide variety of instrumental principles, a number of which are described in other chapters of this book. Different principles differ in their photon efficiency, that is, in the number of photons required for a given lifetime accuracy (Ballew and Demas 1989; Gerritsen et al. 2002; Köllner and Wolfrum 1992; Philip and Carlsson 2003), the acquisition time required to record these photons, the photon flux they can be used at, their time resolution, their ability to resolve the parameters of multiexponential decay functions, multiwavelength capability, optical sectioning capability, and compatibility with different imaging techniques. An overview has been given by Becker and Bergmann (2008) and Becker (2012b). This chapter focuses on time-domain FLIM by time-correlated single-photon counting, or (TCSPC).

TCSPC FLIM uses a multidimensional TCSPC process (Becker 2005, 2012a), which is an extension of the classic TCSPC technique (O'Connor and Phillips 1984). The sample is scanned by the focused beam of a high-frequency pulsed laser. Data recording is based on detecting single photons of the fluorescence light emitted and determining the arrival times of the photons with respect to the laser pulses and the positions of the laser beam in the moment of photon detection. From these parameters, a photon distribution over the spatial coordinates, x, y, and the times of the photons, t, after the laser pulses is derived. The result is a three-dimensional data array that represents the pixels of the two-dimensional scan, with each pixel containing photons in a large number of time channels for consecutive times after the excitation pulses.

More parameters can be added to the photon distribution, such as the wavelength of the photons, or the time from a periodic stimulation of the sample. Because the technique builds up multidimensional photon distributions, it is called multidimensional TCSPC. Among the FLIM techniques described in this book, multidimensional TCSPC delivers the highest time resolution and the best lifetime accuracy, or photon efficiency, for a given number of photons detected from the sample (Becker 2005). TCSPC FLIM has a number of other features important to lifetime imaging of biological systems: It is able to resolve complex decay profiles, it is tolerant to dynamic changes in the fluorescence decay parameters during the acquisition (Becker 2005, 2012a), and it is perfectly compatible with confocal (Pawley 2006) and multiphoton (Denk et al. 1990; Diaspro 2001; Göppert-Mayer 1931) laser scanning systems. Because these systems provide optical sectioning, TCSPC FLIM delivers fluorescence decay data from an accurately defined plane of the sample, without contamination by out-of-focus fluorescence.

9.2 TECHNICAL PRINCIPLES OF TCSPC FLIM

9.2.1 SINGLE-PHOTON DETECTION

TCSPC makes use of the special properties of high-repetition-rate optical signals detected by a high-gain detector. Understanding these signals is the key to the understanding of TCSPC. The situation is illustrated in Figure 9.1.

Fluorescence of a sample is excited by a laser of 80 MHz pulse repetition rate (Figure 9.1a). The expected fluorescence waveform is shown in Figure 9.1b. However, the photon detection rate in biological applications of FLIM is on the order of 10^4 to 10^7 s^{-1}, which is far lower than the laser pulse repetition rate. Therefore, the detector signal (Figure 9.1c) has no similarity with the expected fluorescence waveform. Instead, it consists of individual pulses randomly spread over the time axis. The pulses represent the detection of single photons of the fluorescence signal. The fluorescence waveform (Figure 9.1b), therefore,

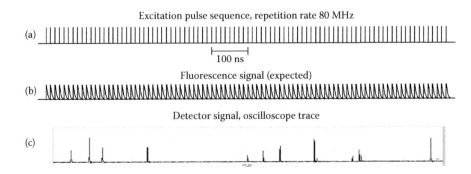

Figure 9.1 Detector signal for fluorescence detection at a pulse repetition rate of 80 MHz. Detection rate, 10^7 photons per second. (a) Excitation pulse sequence, (b) expected fluorescence waveform, and (c) detector signal.

has to be considered a probability distribution of the photons, not anything like a directly observable signal waveform.

9.2.2 PRINCIPLE OF TIME-CORRELATED SINGLE-PHOTON COUNTING

Figure 9.1 shows that, at the photon rates to be considered, the detection of a photon in a particular signal period is a relatively unlikely event. The detection of several photons in one signal period is so unlikely that it can be neglected.

If only one photon per excitation period needs to be considered, the buildup of a photon distribution over the time in the signal period is a relatively straightforward task; see Figure 9.2. Fluorescence is excited by a pulsed light source of high repetition rate. When a photon is detected, the time of the photon in the pulse period is measured. The time is used to address a histogram memory in which the photon detection events are accumulated. Thus, over a large number of excitation periods, the distribution of the photons over the detection time builds up. The procedure shown in Figure 9.2 is the classic TCSPC principle. It delivers a near-ideal signal-to-noise ratio (SNR; all detected photons appear in the result) and a very good time resolution (the resolution is limited only by the transit-time jitter in the detector, not by the width of the single-electron response). However, the principle also has a drawback: It is intrinsically one-dimensional, that is, it only records the waveform of a single optical signal. To apply the classic TCSPC procedure to FLIM, a slow-scan procedure had to be used: The sample had to be scanned at a rate slow enough to accumulate a full waveform, read the data from the histogram memory, and store them in a computer sequentially for each individual pixel. Such procedures have indeed been used (Bugiel et al. 1989) but are not compatible with the fast scan rates used in modern laser scanning microscopes and not appropriate for FLIM of biological samples.

The problem of fast scanning has been solved by a multidimensional TCSPC introduced by Becker & Hickl in 1993. Multidimensional TCSPC, for every photon, determines not only the time in the signal period but also additional parameters, such as the location, x and y, in an image area; the wavelength of the photon; the time from a stimulation of the sample; or the time within the period of an additional modulation of a pulsed laser. The recording process builds up a multidimensional photon distribution over these parameters. The principle is illustrated in Figure 9.3.

To record FLIM data with a fast scanner, the time of a photon in the excitation pulse period and the coordinates of the laser spot in the sample in the moment of its detection would be determined. These values would be used to build up the photon distribution over the images' coordinates and the time in the pulse period. Other parameters can be determined and added to the photon distribution: Adding the wavelength yields multiwavelength FLIM, adding the time from an additional modulation leads to phosphorescence lifetime imaging (PLIM), and replacing one spatial coordinate with a time from a periodical stimulation yields fluorescence lifetime-transient scanning (FLITS).

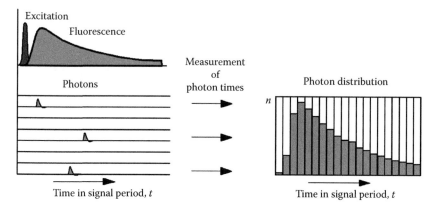

Figure 9.2 Principle of classic time-correlated single-photon counting.

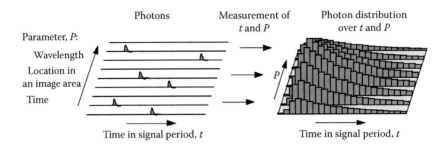

Figure 9.3 Multidimensional TCSPC. Several parameters are determined for every photon, and a multidimensional photon distribution is built up.

9.2.3 SINGLE-CHANNEL FLIM

The general architecture of a TCSPC FLIM system is shown in Figure 9.4. A laser scanning microscope scans the sample with a focused beam of a high-repetition-rate pulsed laser. Depending on the laser used, the fluorescence in the sample can be excited either by one photon or by multiphoton excitation. The FLIM detector is attached to either a confocal or non-descanned port of the scanning system. For every detected photon, the detector sends an electrical pulse into the TCSPC module. Moreover, the TCSPC module receives scan clock signals (pixel, line, and frame clock) from the scanning unit of the microscope.

For each photon pulse from the detector, the TCSPC module determines the time, t, within the laser pulse period (i.e., in the fluorescence decay) and the location within the scanning area, x and y. The photon times, t, and the spatial coordinates, x and y, are used to address a memory in which the detection events are accumulated. Thus, in the memory, the distribution of the photon density over x, y, and t builds up. The result is a data array representing the pixel array of the scan, with every pixel containing a large number of time channels with photon numbers for consecutive times after the excitation pulse. In other words, the result is an image that contains a fluorescence decay curve in each pixel (Becker et al. 2004a; Becker 2005).

The results are normally displayed as pseudo-color images. The brightness represents the number of photons per pixel. The color can be assigned to any parameter of the decay profile: the lifetime of a single-exponential approximation of the decay, the average lifetime of a multiexponential decay, the lifetime or amplitude of a decay component, or the ratio of such parameters (Becker & Hick GmbH 2012a, b).

An example is shown in Figure 9.5. The color shows the amplitude-weighted mean lifetime of a double-exponential decay. The data format is 512 × 512 pixels. Every pixel contains 256 time channels. Decay curves in two selected pixels are shown on the right. Similar curves are contained in any pixel of the image.

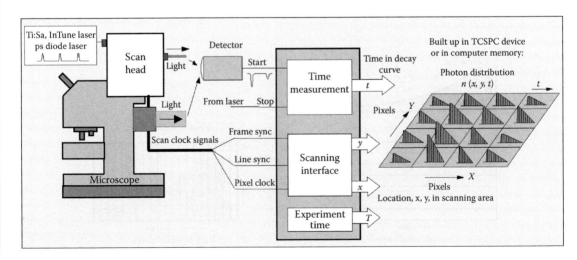

Figure 9.4 Multidimensional TCSPC architecture for FLIM.

Principles of fluorescence lifetime instrumentation

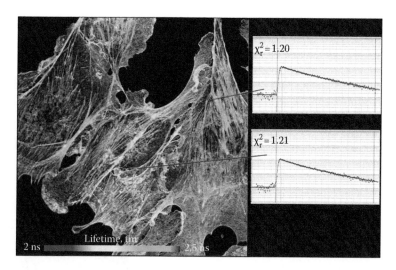

Figure 9.5 Lifetime image of a bovine pulmonary artery endothelia (BPAE) cell, stained with Alexa 488. FLIM data, 512 × 512 pixels, 256 time channels per pixel. Fluorescence decay shown for two selected pixels. (From Becker & Hickl GmbH, Zeiss LSM 710 Intune system with Becker & Hickl Simple-Tau 150 FLIM system, 2012.)

The FLIM data can be built up directly in the device memory of the TCSPC module, or the data of the individual photons and the scan clocks can be transferred into the computer and the photon distribution built up there.

The first technique has the advantage that the recording runs independently of the computer. It thus works up to extremely high count rates and scan speeds (Becker 2012a). The disadvantage is that the size of the FLIM data is limited by the onboard memory capacity of the TCSPC device.

The second technique has the advantage of having the large memory of the computer available for the buildup of the FLIM data. It thus delivers FLIM data with large numbers of pixels and time channels. Moreover, in addition to building up FLIM data, it keeps the full information about the individual photons available. Such "parameter tagged photon data" can be used in various ways, such as for multiparameter single-molecule spectroscopy (Prummer et al. 2004; Widengren et al. 2006), fluorescence correlation spectroscopy (FCS) (Becker et al. 2006; Felekyan et al. 2005), and PLIM (Becker et al. 2011a). The disadvantage is that a huge amount of data has to be transferred and processed. State-of-the-art FLIM modules therefore have both principles implemented; please see Becker (2012a) for details.

It should be explicitly noted that multidimensional TCSPC does not require that the scanner stays in one pixel until enough photons for a full fluorescence decay curve have been acquired. It is only necessary that the *total pixel time*, over a large number of subsequent frames, is large enough to record a reasonable number of photons per pixel. Thus, TCSPC FLIM works even at the highest scan rates available in laser scanning microscopes. At pixel rates used in practice, the recording process is more or less random: A photon is just stored in a memory location according to its time in the fluorescence decay, its detector channel number, and the location of the laser spot in the sample in the moment of detection.

9.2.4 PARALLEL-CHANNEL FLIM

There are a number of FLIM experiments that use several detection channels. Typical applications of dual-channel systems are anisotropy measurements where two channels for different polarization are used and autofluorescence measurements, which usually record simultaneously in the wavelength intervals of nicotinamide adenine dinucleotide (NADH) and adenine dinucleotide (FAD).

Simultaneous recording in several channels can be obtained by a "routing" technique that uses a single TCSPC channel for several detectors (Becker 2005, 2012a). Although an elegant solution for multiwavelength FLIM (see Figure 9.7), routing does not increase the counting capability and the throughput of a FLIM system. Systems with no more than four detector channels are therefore increasingly using parallel TCSPC channels; see Figure 9.6. Parallel systems deliver high throughput rates. Another

Principles of fluorescence lifetime instrumentation

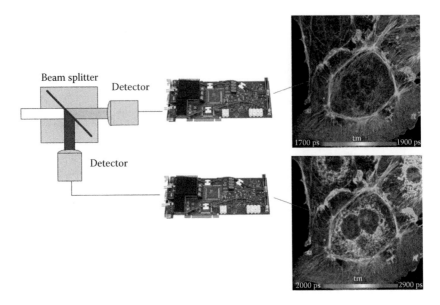

Figure 9.6 Parallel-channel TCSPC system. The light is split in two wavelength channels, the signals of which are recorded by parallel TCSPC FLIM modules.

advantage is that the channels are independent. If one channel overloads, the other ones still deliver correct data. Dual-channel systems have become standard in laser scanning FLIM microscopy (Becker 2012a; Becker & Hick GmbH 2012a, b). Four-channel systems are easily feasible (Becker et al. 2004b, Becker 2012a). An eight-channel parallel FLIM system has been demonstrated (Becker et al. 2009).

9.2.5 MULTIWAVELENGTH FLIM

The principle described in Figure 9.4 can be extended to simultaneously detecting in a large number of wavelength channels (Becker et al. 2002, 2007). As shown in Figure 9.1, the count rate of the detector is far lower than the laser pulse rate. Thus, the probability of detecting several photons per period is negligible. Now consider the case that the light signal delivered to the detector is split spectrally, and the spectrum spread over a one-dimensional array of detectors. The total intensity (or count rate) for the whole array is the same as for a single detector receiving the undispersed signal. Thus, it is also unlikely that the whole array will detect several photons per signal period. In particular, it is unlikely that several detectors of the array will detect a photon in the same signal period. This is the basic idea behind multiwavelength TCSPC: Although several detectors are *active simultaneously, they are unlikely to detect a photon in the same signal period*. The times of the photons detected in all detectors of the array can therefore be determined in a single TCSPC channel.

To obtain multiwavelength FLIM data, it is sufficient to spread a spectrum of the fluorescence light over an array of detector channels and determine the detection times, the channel numbers in the detector array, and the position, x, and y, of the laser spot for the individual photons. These pieces of information are used to build up a photon distribution over the time of the photons in the fluorescence decay, the wavelength, and the coordinates of the image. The technique is also known as "routing" because the "channel" signal routes the photons into different decay data blocks. The architecture of multiwavelength FLIM is shown in Figure 9.7.

As for single-wavelength FLIM, the result of the recording process is an array of pixels. However, the pixels now contain several decay curves for different wavelengths. Each decay curve contains a large number of time channels; the time channels contain photon numbers for consecutive times after the excitation pulse. A result of multiwavelength FLIM is shown in Figure 9.8. For other applications, please see Bird et al. (2004), Biskup et al. (2007), Chorvat and Chorvatova (2006, 2009), and Rück et al. (2005, 2007).

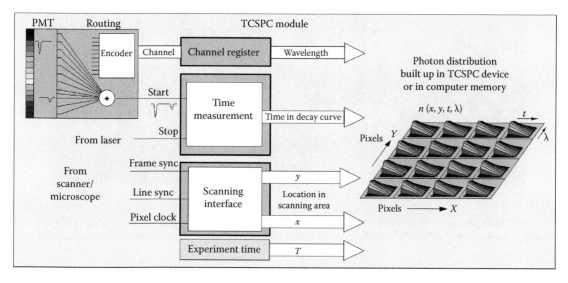

Figure 9.7 Principle of multiwavelength TCSPC FLIM.

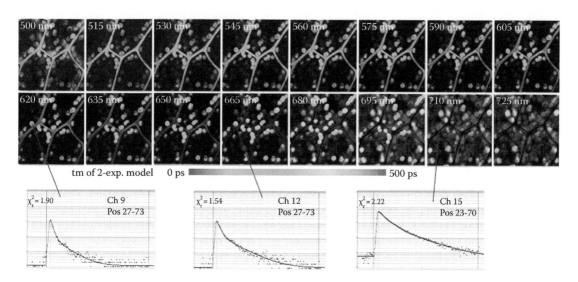

Figure 9.8 Multiwavelength FLIM of plant tissue, 16 wavelength intervals, 128 × 128 pixels, 256 time channels. Two-photon excitation at 850 nm, detection from 500 to 725 nm. (From Becker & Hickl GmbH, Zeiss LSM 710, bh Simple-Tau 150 FLIM system. Amplitude-weighted mean lifetime of double-exponential fit, normalized intensity, 2012.)

9.2.6 LASER WAVELENGTH MULTIPLEXING

Another extension of the TCSPC FLIM principle is obtained by using the laser wavelength as a dimension of the photon distribution (Becker 2005, 2012a). Several laser wavelengths are multiplexed, and the wavelength is used as a dimension of the recording process. Wavelength multiplexing is obtained either by multiplexing several lasers of different wavelength or by switching the wavelength of the acousto-optical filter of a supercontinuum laser. Multiplexing can be performed pulse by pulse, pixel by pixel, line by line, or frame by frame. For FLIM, pixel, line, or frame multiplexing is to be preferred because it avoids cross talk due to incomplete decay, reflections in optical fibers, or pileup effects.

An example of laser wavelength multiplexing is shown in Figure 9.9. Two diode lasers of 405 and 473 nm were multiplexed, and fluorescence was detected in two wavelength intervals from 432 to 510 nm and 510 to 560 nm, respectively.

Principles of fluorescence lifetime instrumentation

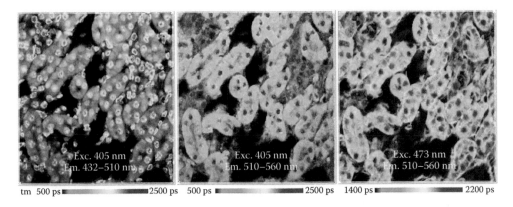

tm 500 ps ■━━━━━━2500 ps   500 ps ■━━━━━━2500 ps   1400 ps ■━━━━━━2200 ps

Figure 9.9 Multiplexing of diode lasers. Pixel-by-pixel multiplexing, picosecond diode lasers, 405 and 473 nm. Detection wavelength intervals indicated. (From Becker & Hickl GmbH, DCS-120 confocal FLIM system, 2012.)

9.2.7 PHOSPHORESCENCE LIFETIME IMAGING

TCSPC is able to simultaneously record fluorescence (FLIM) and phosphorescence lifetime images (PLIM). The technique is based on on-off modulating a high-frequency pulsed laser, and assigning two times to the individual photons. One is the time from the previous excitation pulse, the other a time from the modulation pulse (Becker et al. 2011; Becker 2012a).

The principle is shown in Figure 9.10. A high-frequency-pulsed laser is used for excitation. The laser is turned on for a short period of time, T_{on}, at the beginning of each pixel. For the rest of the pixel time, the laser is turned off. Within the on-time, the laser excites fluorescence and builds up phosphorescence. Within the rest of the pixel dwell time, T_{off}, pure phosphorescence is obtained. Photon times are determined both with respect to the laser pulses and with respect to the modulation pulse. The times from the laser pulse, t, are used to build up the fluorescence decay. The phosphorescence decay is built up from times from the modulation pulse, T-$T0$.

An example is shown in Figure 9.11. It shows yeast cells stained with tris(2,2′-bipyridyl) dichlororuthenium(II)hexahydrate. On the picosecond time scale, the yeast cells emit autofluorescence from NADH and FAD. On the microsecond time scale, phosphorescence of the ruthenium dye is emitted. The fluorescence lifetime image is shown in Figure 9.11a, the phosphorescence lifetime image in Figure 9.11b.

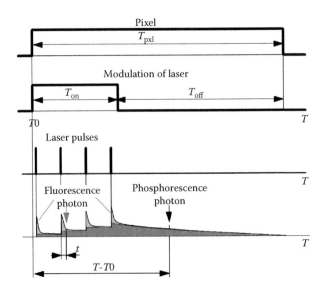

Figure 9.10 Principle of simultaneous phosphorescence and fluorescence lifetime imaging.

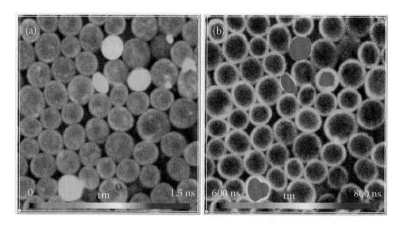

Figure 9.11 Fluorescence lifetime image (a) and phosphorescence lifetime image (b) of yeast cells stained with tris(2,2'-bipyridyl)dichlororuthenium(II)hexahydrate. Amplitude-weighted mean lifetime of double-exponential fit to decay data. Data analysis by Becker & Hickl SPCImage.

Figure 9.12 shows decay curves for a selected spot in the images. The dots are the photon numbers in the subsequent time channels, the solid curve is a fit with a double-exponential decay model, and the grey curve is the effective instrument response function (IRF). Please note that the IRF of the fluorescence decay is the laser pulse convoluted with the detector response, whereas the IRF of the phosphorescence decay is the waveform of the laser modulation.

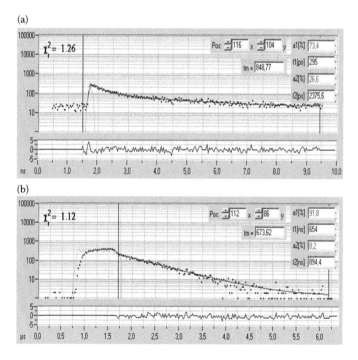

Figure 9.12 Decay curves in selected spot of Figure 9.11. (a) Fluorescence. (b) Phosphorescence. The dots are photon numbers in the time channels, the solid curve is a fit with double-exponential decay model, and the grey curve is the effective instrument response function. Inserts: Amplitude-weighted lifetime and double-exponential decay parameters. Decay parameters in picoseconds for fluorescence and in nanoseconds for phosphorescence.

Principles of fluorescence lifetime instrumentation

9.2.8 FLUORESCENCE LIFETIME-TRANSIENT SCANNING

Transient effects in the fluorescence lifetime of a sample can be recorded by time-series FLIM. Subsequent FLIM recordings would be performed, and the data saved into consecutive data files (Becker 2012a). Time series of FLIM images can be recorded at surprisingly a high rate, especially if readout times are avoided by dual-memory recording (Katsoulidou et al. 2007). Nevertheless, each step of a time series requires at least one complete x–y scan of the sample. With the typical frame rates of fast galvanometer scanners, lifetime changes can be recorded at a maximum resolution on the order of 1 s.

Faster effects can be resolved by line scanning. The technique has been named "FLITS," fluorescence *lifetime-transient* scanning. FLITS is based on building up a photon distribution over the distance along the line of the scan, the experiment time after a stimulation of the sample, and the arrival times of the photons after the excitation pulses (Becker et al. 2012). Technically, FLITS is obtained by replacing the frame clock of a FLIM system (see Figure 9.4) with a trigger pulse coincident with a stimulation of the sample. The resulting photon distribution is shown in Figure 9.13.

As long as the stimulation occurs only once, the recording process is simple: The sequencer of the TCSPC module starts to run with the stimulation and puts the photons in consecutive experiment-time channels along the T axis. The result is a time-series of line scans.

However, there is an important difference to a simple time series: The data are still in the memory (either in the onboard memory of the TCSPC module or in the computer) when the sequence is completed. Thus, the recording process can be made repetitive: The sample would be stimulated periodically, and the start of T triggered by the stimulation. The recording then runs along the T axis periodically, and the photons are accumulated into one and the same photon distribution.

With "triggered accumulation," it is no longer necessary that each T step acquires enough photons to obtain a complete decay curve in each pixel and T channel. No matter when and from where a photon arrives, it is assigned to the right location in x, the right experiment time T, and to the right arrival time, t, after the laser pulse. A desired SNR is obtained by simply running the acquisition process for a sufficiently long time. Obviously, the resolution in T is limited by the period of the line scan only, which is about 1 ms for the commonly used scanners.

Recordings of the "chlorophyll transients" (Govindjee 1995), that is, the change in the fluorescence lifetime in a plant leaf on illumination, are shown Figure 9.14. The horizontal axis is the distance along the line scanned; the vertical axis (bottom to top) is the experiment time, T, in this case, the time from the start of the illumination.

The nonphotochemical transient is shown in Figure 9.14a. The data were recorded in a single sweep over the T axis; the T scale is from 0 to 14 s. Decay curves for a selected pixel within the line are shown for $T = 0.5$, 7.5, and 13.4 s. The changes in the decay profiles are clearly visible. The amplitude-weighted lifetimes, t_m, obtained from a double-exponential fit, are 560, 385, and 311 ps, respectively.

The photochemical transient is shown in Figure 9.14b. It was obtained by turning on and off the laser periodically and recording the photons by triggered accumulation. The T scale is from 0 to 200 ms. Decay

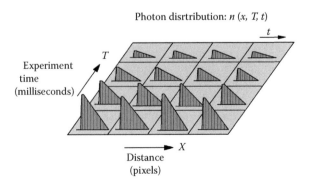

Figure 9.13 Photon distribution built up by FLITS.

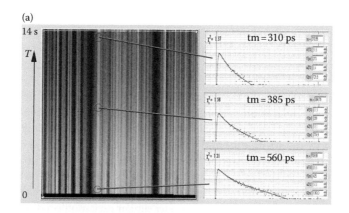

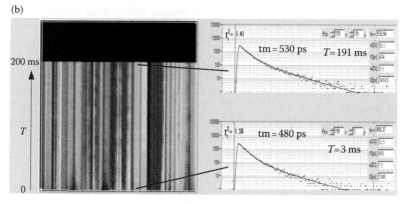

Figure 9.14 FLITS recording of chlorophyll transients in a plant leaf. Horizontal: Distance along the line scanned. Vertical, bottom to top: Time, T, after the start of illumination. Color: Amplitude-weighted lifetime of double-exponential fit. (a) Nonphotochemical transient, vertical time scale 0 to 14 s, decay curves shown for a selected pixel at T = 0.5, 7.5, and 13.4 s after turn-on of the laser. (b) Photochemical transient. Vertical time scale 0 to 200 ms, decay curves shown for a selected pixel at T = 3 ms and T = 191 ms into the laser-on phases. The amplitude-weighted lifetime increases from 480 to 530 ps.

curves are shown for a selected pixel within the line for $T = 3$ ms and $T = 191$ ms into the laser-on phases. The amplitude-weighted lifetime increases from 480 to 530 ps.

9.3 ACCURACY CONSIDERATIONS

9.3.1 SIGNAL-TO-NOISE RATIO OF FLIM

From a single-exponential fluorescence decay recorded under ideal conditions, the fluorescence lifetime can be obtained with SNR, relative standard deviation, or "coefficient of variation," CV_τ, of

$$CV_\tau = \frac{\sigma_\tau}{\tau} = \frac{1}{\sqrt{N}} \qquad (9.1)$$

with N = number of recorded photons (Ballew and Demas 1989; Gerritsen et al. 2002; Köllner and Wolfrum 1992). In other words, the fluorescence lifetime can be obtained at the same accuracy as the intensity. Measurement under ideal condition means the decay function is recorded
- With an IRF that is short compared to the decay time
- Into a large number of time channels
- Within a time interval several times longer than the decay time
- With negligible background of environment light and detector dark counts or detector after pulsing

Under these conditions, the equation given above can easily be verified: The average arrival time of the photons is equal to the decay time, and the variance of the arrival time of the individual photons is equal to the decay time. The variance of the average arrival time for N photons decreases with $N^{1/2}$. Consequently, the relative variance of the lifetime is $N^{-1/2}$. Correctly recorded TCSPC FLIM data come very close to the theoretical limit (Becker & Hickl GmbH 2012a, b).

Things become complicated when the fluorescence decay is no longer recorded under ideal conditions. An investigation of the SNR under nonideal conditions has been published by Köllner and Wolfrum (1992). It turns out that the factor impairing the lifetime accuracy most dramatically is background. Also, this can easily be explained: The arrival time of the background photons varies over the full recording time interval. This is usually 10 times longer than the fluorescence lifetime. That means background photons, on average, introduce 10 times more timing variation in the decay data than the fluorescence photons.

The fact that FLIM achieves the same SNR as intensity imaging is in apparent contradiction with common experience: From the same sample imaged under similar excitation conditions, intensity images are obtained in a far shorter acquisition time than FLIM images. Where is the mistake?

The mistake is to implicitly assume that "intensity imaging" (in the sense of spatial imaging) and FLIM are recording the same physical properties of the sample. This is wrong: Intensity imaging aims at resolving the spatial structure of the sample. To do so, it uses the fluorescence intensity of a fluorophore attached to the internal sample structures. In other words, intensity imaging records the concentration variation of a fluorophore in the sample. Relative concentrations in stained and unstained structures of the sample can vary by 1:100 and more.

FLIM aims at resolving changes in the fluorescence lifetime, that is, in the quantum efficiency of the fluorophore. Relative changes in these parameters are much smaller than the concentration changes. Lifetime changes are usually in the 10% range. Consequently, a higher SNR and a higher number of photons are needed to resolve the changes.

The differences are illustrated in Figure 9.15. The figure shows an intensity image (Figure 9.15a) and a lifetime image (Figure 9.15b) derived from the same TCSPC FLIM data. The number of photons, N, in the bright pixels is about 80. The SNR is about 9. The concentration changes are much larger than this. The spatial structure of the sample is therefore resolved very well. The SNR of the lifetime in the FLIM image is on the same order as the intensity noise in the intensity image. However, the lifetime variation in the sample is on the order of 10%, and thus at the limit of detectability.

The signal-to-noise problem is further enhanced by the fact that samples for FLIM normally have fluorophores linked to highly specific targets in the cells or do not contain exogenous fluorophores at all. That means the fluorophore concentration is low. At a given emission rate, the fluorophore molecules have to perform more excitation-emission cycles, and photobleaching becomes a problem. Photobleaching has little influence on the spatial structure of a sample. In FLIM, however, different fluorophores or fluorophore fractions may bleach at different rates. Severe photobleaching may therefore change the fluorescence decay parameters.

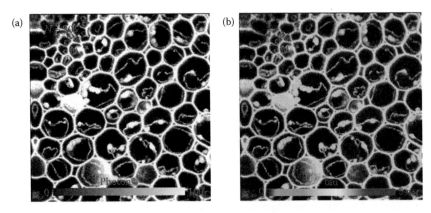

Figure 9.15 Intensity (a) and lifetime (b) images derived from the same TCSPC data.

It should be noted that "intensity imaging" may also be used to investigate molecular interactions of the fluorophores with their environment. In that case, the aim is to record variation in the fluorescence quantum efficiency of the fluorophores. The quantum efficiency and the fluorescence lifetime are directly connected. Consequently, the relative intensity changes to be recorded are the same as for the lifetime in a FLIM experiment. However, the intensity also depends on the concentration. The variation in concentration changes must be removed from the intensity data by reference measurements. These introduce additional noise. A practical evaluation of FLIM versus intensity measurement for FRET applications has been published by Pelet et al. (2006). The authors found that FLIM indeed delivered a better SNR than intensity-based FRET measurement.

9.3.2 ACQUISITION TIME

The equation given above can be used to estimate the acquisition time needed to acquire FLIM data for a given accuracy, a given number of pixels, and a given photon count rate. The required acquisition time is shown graphically in Figure 9.16.

For example, the acquisition time for an image of 256 × 256 pixels, a count rate of 10^6 s^{-1} and an accuracy of 10% would be acquired within 6 s. However, it would take almost 1 h to acquire a 512 × 512 pixel image at an accuracy of 1%.

Surprisingly, the acquisition time of FLIM for a given accuracy is often *shorter* than the times given in Figure 9.16. The explanation is that the lifetime analysis software uses, or can use, overlapping binning of the fluorescence decay data (Becker 2012a; Becker & Hick GmbH 2012a, b). That means the decay information for every pixel is taken not only from a single pixel but also from the pixels around it. This works because the images are usually oversampled. Of course, the number of photons and the number of pixels in the diagrams in Figure 9.16 refer to the binned lifetime data. With an oversampling factor of 5 and a binning of 5 × 5 pixels, a 512 × 512 pixel image is effectively reduced to 102 × 102 pixels. Consequently, a lifetime accuracy of 1% is reached in about 100 s of acquisition time. An accuracy of 10% is reached within a less than 2 s.

The acquisition time is a source of constant arguments between proponents of different FLIM techniques. On the one hand, it is commonly accepted that TCSPC has the highest photon efficiency, that is, reaches the best lifetime accuracy for a given number of detected photons. On the other hand, it is believed that TCSPC is slow in terms of acquisition time. This is a contradiction in itself and requires explanation.

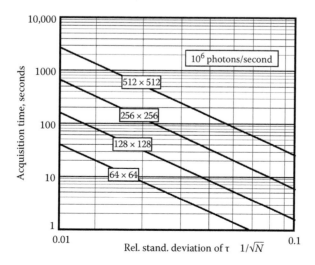

Figure 9.16 Acquisition time as a function of the desired lifetime accuracy for different image sizes. Count rate, 10^6 s^{-1}.

Consider TCSPC FLIM in a scanning microscope: It is correct that multidimensional TCSPC reaches the highest photon efficiency of all FLIM techniques. It does so up to a detector count rate of a few megahertz. Under these conditions, it reaches the shortest acquisition time of all techniques. However, if a sample delivers a detector count rate higher than about 10 MHz, TCSPC starts losing photons, the efficiency degrades by counting loss, and at even higher count rates, the detected lifetimes may be impaired by pileup effects; see Section 9.3.3. The excitation power then has to be reduced. That means that a less efficient technique, which is not based on photon counting, may deliver a result in a shorter acquisition time. However, in a laser scanning microscope, such count rates are obtained only from samples with high fluorophore concentration. However, in most FLIM experiments, the fluorophore concentration is low, and so is the count rate. This is exactly the situation where TCSPC delivers the shortest acquisition times of all techniques based on sample scanning.

Another argument against TCSPC is often that gated or modulated CCD cameras are faster because they record the data of all pixels in parallel. This is not entirely correct: Camera-based FLIM uses parallel recording in space but sequential recording in time or in phase (Becker and Bergmann 2008). Because the number of time or phase data points is usually smaller than the number of pixels, the camera can indeed record faster than TCSPC with spatial scanning. But this comes at a price: There is no depth resolution, scanning in time or phase is not artifact-free for samples with dynamic lifetime changes or in the presence of photobleaching, the photon efficiency is low, and the exposure of the sample is higher than for TCSPC (Becker and Bergmann 2008; Becker 2012a, b).

9.3.3 COUNTING LOSS AND PILE-UP EFFECTS

TCSPC is based on the assumption that the light intensity at the detector is low enough so that the detection of several photons per signal period is unlikely (Becker 2005; O'Connor and Phillips 1984). If the light intensity becomes too high, two instrumental effects have to be considered: The loss of a second photon detected *within the signal processing time* ("dead time") of a previous one, and the loss of a second photon detected *in the same signal period* with a previous one. The first effect causes "counting loss," that is, a nonlinearity in the recorded intensities. The second one is known as "pileup" and causes an error in the detected fluorescence lifetime.

Counting loss is illustrated in Figure 9.17. Figure 9.17a shows oscilloscope traces of laser pulses at 80 MHz repetition rate and single-photon pulses of a photomultiplier tube (PMT) at a count rate of 10 MHz. For a dead time of 100 ns, as it is typical for TCSPC, there is a 50% probability that a photon falls in the dead time caused by a previous one. The probability of losing a photon increases with the count rate. The result is a nonlinearity in the intensity scale, as shown in Figure 9.17b.

(a)
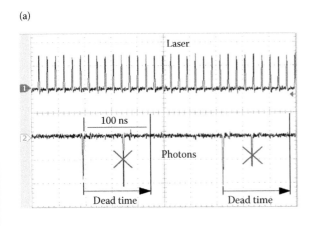

(b)
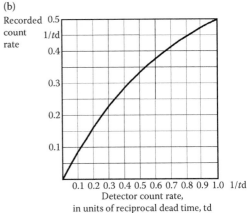

Figure 9.17 (a) Detection (and loss) of a photon within the dead time caused by a previous one. 40 ns per division, laser repetition rate 80 MHz, detector count rate 10 MHz. (b) Nonlinearity of recorded versus detected count rate.

Nonlinearity in the intensity scale is rarely a problem in FLIM experiments: FLIM is used to obtain molecular information independently of the concentration of the fluorophores. That means the intensity is not used as an analysis parameter. The effect of counting loss is merely that images taken at extremely high count rates look "flat" because bright pixels are recorded with a lower intensity than they actually have. An example is shown in Figure 9.18.

"Pileup" means the detection (and loss) of a second photon within the same laser period of a previous one. Pileup causes a distortion in the recorded decay profiles. The change in the lifetime of a single-exponential decay model is approximately

$$\tau_{meani} \approx \tau\,(1 - P/4)$$

where P = average number of photons per laser period, τ = fluorescence lifetime, and τ_{meani} = intensity-weighted lifetime of the measured decay profile.

Thus, even with 0.2 photons per laser period, the lifetime error is only 5%. A detailed derivation of the pileup distortion has been given by Becker (2005). A curve of the ratio of the detected lifetime and true lifetime as a function of the number of photons per signal period is shown in Figure 9.19.

A lifetime image taken at 4 MHz average count rate and 50 MHz laser repetition rate is shown in Figure 9.20. It was obtained from the same data as the intensity image shown in Figure 9.18, right. A single Becker & Hickl SPC-150 TCSPC FLIM module was used. The peak detector count rate of the brightest pixel was about 10 MHz, that is, 20% of the laser repetition rate. An uncorrected lifetime image is shown in Figure 9.20a, a lifetime image corrected for pileup in Figure 9.20b. The difference between the images is barely visible. This confirms that pileup in TCSPC FLIM is far less important than commonly believed.

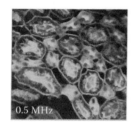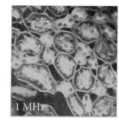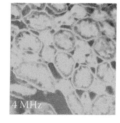

Figure 9.18 Intensity images taken at average count rates of 0.5, 1, and 4 MHz. Intensity normalized on brightest pixel.

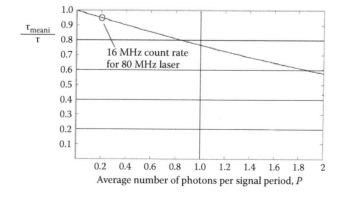

Figure 9.19 Ratio of detected lifetime and true lifetime as a function of the average number of photons per laser period. The error in the lifetime at 16 MHz detection rate and 80 MHz laser repetition rate is only 5%.

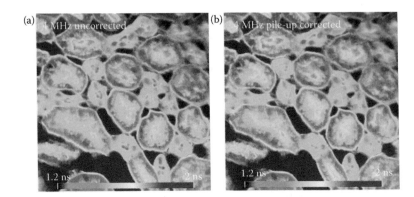

Figure 9.20 A lifetime image taken at 4 MHz average count rate. Same data as shown in Figure 9.18, right. Single SPC-150 module, laser 50 MHz, peak detector count rate of brightest pixels 12 MHz. (a) Uncorrected lifetime image. (b) Lifetime image with pileup corrected. The difference is barely visible.

9.4 APPLICATIONS

FLIM makes use of the fact that the fluorescence lifetime of a fluorophore depends on its molecular environment but not on the concentration. Molecular effects can thus be investigated independently of the unknown and usually variable fluorophore concentration (Berezin and Achilefu 2010; Lakowicz 2006; Roberts et al. 2011). Effects used in FLIM applications are

- Fluorescence quenching by various ions (Lakowicz 2006), in particular, Ca^{++} and Cl^-, both of which are important to the function of the neuronal system (Funk et al. 2008; Gilbert et al. 2007; Kaneko et al. 2004; Kuchibhotla et al. 2009).
- Fluorescence quenching by oxygen. Oxygen is an efficient fluorescence quencher for a large number of fluorophores (Lakowicz 2006), especially those of longer fluorescence lifetime. Strong oxygen quenching is observed for the phosphorescence of organic complexes of ruthenium, europium, platinum, and palladium.
- A strong oxygen effect exists on the fluorescence of the endogenous fluorophores NADH and FAD. Chance et al. (Chance 1976; Chance et al. 1979) defined a "redox ratio" that is a direct indicator of the amount of oxygen used in the mitochondria of the cells. Effects of oxygen possibly exist also for other endogenous fluorophores (Schweitzer et al. 2004).
- Binding to proteins, protein configuration. Fluorescence lifetimes change on binding of a fluorophore to a biological target. The most likely mechanism is a change in the configuration of the fluorophore, which, in turn, changes the rate of internal nonradiative decay. For almost all dyes used in cell biology, the fluorescence lifetime depends more or less on the binding to proteins, DNA, or lipids (Knemeyer et al. 2002; Lakowicz 2006; Van Zandvoort et al. 2002). When a fluorophore is bound to a protein, also, the protein configuration can have an influence on the fluorescence lifetime (Benesch et al. 2007; Treanor et al. 2005).
- NADH/FAD dynamics: For the endogenous fluorophores NADH and FAD, it is known that the lifetimes change on binding to proteins (Lakowicz et al. 1992; Paul and Schneckenburger 1996). The lifetime changes are substantial, and bound and unbound fractions can be easily distinguished by FLIM. Both the fluorescence lifetimes of the decay components and the relative amplitudes have been found sensitive to the metabolic state of cells and tissue (Bird et al. 2005; Chorvat and Chorvatova 2009; Ghukasyan and Kao 2009; Skala et al. 2007a, b).
- Local viscosity. Viscosity can be sensed by "molecular rotors." These are fluorophores that have a high degree of internal flexibility. Rotation inside the fluorophore provides for a radiationless decay path. The nonradiative decay rate changes with the viscosity of the solvent. Viscosity measurement by FLIM with a molecular rotor has been demonstrated by Kuimova et al. (2008) and Levitt et al. (2009).
- Local refractive index: The radiative decay rate of a fluorophore depends on the local refractive index (Strickler and Berg 1962). The refractive index varies within biological cells and tissue. The lifetime changes induced by these variations can be used to gain structural information about biological specimens (Tregido et al. 2008).

- pH: Many fluorescent molecules have a protonated and a deprotonated form. The equilibrium between both depends on the pH. If the protonated and deprotonated form have different lifetimes, the apparent lifetime is an indicator of the pH (Hanson et al. 2002; Lakowicz 2006; Sanders et al. 1995). An example is shown under "Measurement of pH and ion concentrations."
- Proximity to metal surfaces: Extremely strong effects on the decay rates must be expected if dye molecules are bound to metal surfaces, especially to metallic nanoparticles (Geddes et al. 2003; Malicka et al. 2003; Ritman-Meer 2007). Similar, yet less dramatic, effects have been reported also for nonmetallic nanoparticles (Muddana et al. 2009).
- Nanoparticles: FLIM helps identify and localize nanoparticles in biological tissue via their luminescence decay times or second-harmonic generation (SHG) emission. This is important when such particles are used as carriers of drug delivery (Prow et al. 2011) or when the diffusion of nanoparticles through skin has to be examined (Lin et al. 2011).
- Aggregates: The radiative and nonradiative decay rates depend on possible aggregation of the dye molecules. Aggregation is influenced by the local environment; the associated lifetime changes can be used as a probe function. Aggregation has been used to observe the internalization of dyes into cells (Kelbauskas and Dietel 2002).
- Förster resonance energy transfer: Förster resonance energy transfer (Förster 1948), or FRET, is an interaction of two molecules in which the emission band of one molecule overlaps the absorption band of the other. In this case, the energy from the first molecule, the donor, transfers immediately into the second one, the acceptor. FRET is used to investigate protein interactions and protein configuration. The measurement of FRET is the most frequent TCSPC FLIM application.

Typical applications of TCSPC FLIM are described below.

9.4.1 MEASUREMENT OF pH AND ION CONCENTRATIONS

An example of FLIM-based pH measurement is shown in Figure 9.21a. A skin sample was stained with 2′,7′-bis-(2-carboxyethyl)-5-(and-6)-carboxyfluorescein (BCECF), and a FLIM image was taken by a Zeiss LSM 510 NLO multiphoton microscope. BCECF has a protonated and a deprotonated form. In aqueous solution, the lifetimes for the deprotonated and the protonated form are 2.75 and 3.90 ns, respectively (Hanson et al. 2002). In the pH range from 4.5 to 8.5, both forms exist, and the fluorescence decay function is a mixture of both decay components. Thus, the lifetime of a single-exponential fit can be used as an indicator of the pH.

An example of Cl$^-$ concentration measurement is shown in Figure 9.21b. The image shows a spinal ganglion of a mouse. MQAE (6-methoxy-quinolyl acetoethyl ester) was used as a fluorescent probe. MQAE is quenched by Cl$^-$; thus, short lifetime means high Cl$^-$ concentration. Details have been described by Kaneko et al. (2004). By using two-photon excitation, the authors were able to obtain Z stacks of the Cl$^-$ concentration in dendrites over depth intervals up to 150 µm. Changes in the Cl$^-$ concentration during

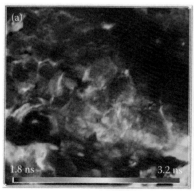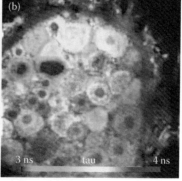

Figure 9.21 (a) Lifetime image of skin tissue stained with BCECF. The lifetime is an indicator of the pH. (Theodora Mauro, University of San Francisco, Zeiss LSM 510 NLO with bh SPC-830.) (b) Spinal ganglion of a mouse stained with MQAE. Short lifetime indicates high Cl$^-$ concentration. (Data courtesy of Thomas Gensch, Jülich Research Centre, Germany.)

chloride homeostasis in neurons were investigated by Gilbert et al. (2007), changes on inflammation by Funk et al. (2008).

Due to the large lifetime shifts, FLIM measurements of pH and ion concentrations are relatively easy as long as only relative concentration changes are considered. Absolute measurements are more difficult. The fluorescence lifetime of the dye in the biological environment is not necessarily the same as in aqueous solution. Therefore, calibrations in the expected biological environment are required.

9.4.2 FRET

Because of its dependence on the distance, FRET (Förster 1948) has become an important tool of cell biology (Periasamy 2001). Different proteins are labeled with the donor and the acceptor; FRET is then used to verify whether the proteins are physically linked and to determine distances on the nanometer scale.

The problem of steady-state FRET techniques is that the concentration of the donor and the acceptor changes throughout the sample. The techniques therefore depend on ratios of the donor and acceptor intensities. However, the FRET-excited acceptor intensity is not directly available. There is "donor bleed-through" due to the overlap of the donor fluorescence into the acceptor emission band. Moreover, some of the acceptor molecules are excited directly. Steady-state FRET techniques therefore require careful calibration, including measurements of samples containing only the donor and only the acceptor.

The correction problems can partially be solved by the acceptor photobleaching technique. An image of the donor is taken, then the acceptor is destroyed by photobleaching, and another donor image is taken. The increase in the donor intensity is an indicator of FRET. The drawback is that this technique is destructive and that it is difficult to use in live specimens. In fixed specimens, however, the protein structure is changed so that the results are not necessarily correct.

All steady-state techniques have the problem that there is usually a mixture of interacting and noninteracting proteins. Both the fraction of interacting proteins and the distance between the proteins influence the "FRET efficiency." It therefore cannot be told whether a variation in the FRET efficiency is due to a variation in the distance between donor and acceptor or in the fraction of interacting proteins.

9.4.2.1 FLIM FRET

The use of FLIM for FRET has the obvious benefit that the FRET intensity is obtained from a single lifetime image of the donor. Donor bleed-through and directly excited acceptor fluorescence therefore have no influence on FLIM FRET measurements. The only reference value needed is the donor lifetime in absence of the acceptor (Chen and Periasamy 2004; Duncan et al. 2004; Periasamy 2001; Periasamy and Clegg 2009; Peter and Ameer-Beg 2004). It will be shown later that FLIM FRET can work even without an external reference lifetime.

Figure 9.22 shows a single-exponential lifetime image of a cultured human embryonic kidney (HEK) cell expressing two interacting proteins labeled with CFP and YFP.

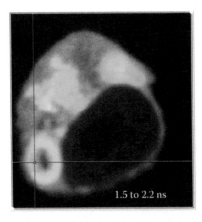

1.5 to 2.2 ns

Figure 9.22 FRET in an HEK cell. Interacting proteins were labeled with CFP (donor) and YFP (acceptor). Single-exponential lifetime image of the donor fluorescence, blue to red corresponds to a lifetime range of 1.5 to 2.2 ns. (Courtesy of Christoph Biskup, University Jena, Germany, LSM 510 NLO.)

Single-exponential lifetime images as the one shown in Figure 9.22 are very useful to locate the areas in a cell where the labeled proteins interact. It has been shown that for a given efficiency of the optical system and the detector and a given excitation power, FLIM-based FRET measurements give better accuracy than steady-state techniques (Pelet at al. 2006).

9.4.2.2 Double-exponential FRET analysis

Single-exponential decay measurements do not solve the general problem that not all of the donor molecules interact with an acceptor. An obvious reason for imperfect interaction is that the relative orientation of donor and acceptor is random. Donor molecules perpendicular to the acceptor do not interact. The problem is addressed by introducing the κ^2 factor in the calculation of the FRET efficiency (Lakowicz 2006).

There are also other reasons why a donor molecule may not interact. In the simplest case, a fraction of the donor molecules may not be linked to their targets, or not all of the acceptor targets may be labeled with an acceptor. This can happen especially in specimens with conventional antibody labeling. Even if the labeling is complete, by far not all of the labeled proteins in a cell may interact, and the fraction of interacting protein pairs varies throughout the cell.

TCSPC FLIM solves the problem of interacting and noninteracting donor by double-exponential decay analysis. The resulting donor decay functions can be approximated by a double-exponential model, with a slow lifetime component from the noninteracting (unquenched) and a fast component from the interacting (quenched) donor molecules. If the labeling is complete, as can be expected if the cell is expressing fusion proteins of the GFP variants, the decay components (corrected by the κ^2 factor) represent the fractions of interacting and noninteracting proteins. The composition of the donor decay function is illustrated in Figure 9.23.

Double-exponential decay analysis delivers the lifetimes, τ_0 and τ_{fret}, and the intensity factors (amplitudes), a and b, of the two decay components. From these parameters can be derived the true FRET efficiency for the interacting proteins, the ratio of the distance and the Förster radius, and the ratio of the number of interacting and noninteracting donor molecules. Please note that a reference lifetime from a donor-only cell is no longer required. The reference lifetime is the lifetime, τ_0, of the slow donor decay component. The advantage is that τ_0 comes from the same local environment as τ_{fret}. Double-exponential FRET is thus, within reasonable limits, independent of lifetime variations induced by variations in the local environment.

Figure 9.24 shows the result of a double-exponential analysis of the data shown in Figure 9.22. Figure 9.24a shows the ratio of the lifetimes of the noninteracting and interacting donor fractions, τ_0/τ_{fret}. The distribution of τ_0/τ_{fret} in different regions is shown far left. The locations of the maxima differ by only 10%, corresponding to a distance variation of only 2%. However, the variation in the intensity coefficients, a/b, is about 10:1, see Figure 9.24b.

The results show clearly that the variation in the single-exponential lifetime (Figure 9.22) is almost entirely caused by a variation in the fraction of interacting proteins, *not* by a change in distance. In other words, interpreting variations in the single-exponential lifetime (or classic FRET efficiencies from steady-state experiments!) as distance variations leads to wrong results.

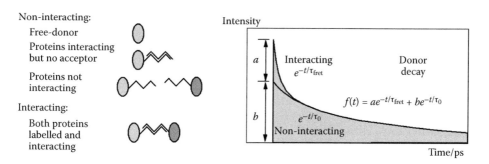

Figure 9.23 Fluorescence decay components in FRET systems.

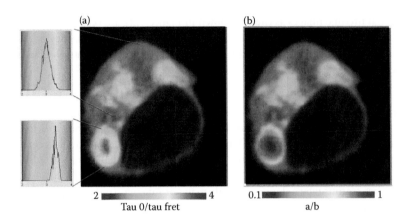

Figure 9.24 FRET results obtained by double-exponential lifetime analysis. (a) $\tau_0/\tau_{\text{fret}}$. (b) N_{fret}/N_0. (a) Ratio of lifetimes of noninteracting and interaction donor, and (b) ratio of intensity coefficients of interacting and noninteracting donor.

Similar double-exponential decay behavior is commonly found in FRET experiments based on multidimensional TCSPC (Bacskai et al. 2003; Becker et al. 2004a; Biskup et al. 2004a; Calleja et al. 2003; Duncan et al. 2004; Ellis et al. 2008; Peter and Ameer-Beg 2004). Double-exponential decay profiles have also been confirmed by streak-camera measurements (Biskup et al. 2004b). There is no doubt that the double-exponential behavior is due to the presence of an interacting and a noninteracting donor fraction.

The double-exponential decay profiles found in FRET experiments have an impact on the calculation of conventional (steady-state) FRET efficiencies. A double-exponential decay cannot be described by a single "fluorescence lifetime." The problem has already been mentioned by J. R. Lakowicz in his 1999 edition of *Principles of Fluorescence Spectroscopy* (Lakowicz 1999). To obtain the correct energy transfer efficiency, an *amplitude-weighted* average of the two lifetime components has to be used. The amplitude-weighted average is different from the lifetime obtained from a single-exponential fit. Thus, correct (conventional) FRET efficiencies are only obtained by double-exponential decay analysis.

It should be noted that there are a few more problems of FRET experiments. These are not especially related to FLIM FRET but can be better identified when FLIM is used.

There is the problem that a protein can be expressing (or be labeled with) several donor or acceptor molecules. Energy can then migrate between the donors, and an the energy transfer rate may increase due to interaction with several acceptor molecules. In FLIM experiments, interaction with several acceptors shows up by "impossibly" short interacting donor lifetimes. For physical background and possible pitfalls, please see Koushik and Vogel (2008) and Vogel et al. (2006).

It should also be mentioned here that double-exponential FRET measurements can face the problem that the fluorescence decay of the unquenched donor itself is multiexponential. This is reported especially for the CFPs and its latest variant, Cerulean. For CFP, multiexponential behavior is caused by heterogeneity in the CFP. Usually, there are two lifetime components of about 1.3 and 2.9 ns, with an amplitude of 20%–40% and 60%–80%, respectively (Becker et al. 2004a). The resulting decay function is still close to a single-exponential decay so that double-exponential FLIM FRET is possible. For Cerulean in live cells, we did not find noticeable double-exponential decay behavior. However, strong multiexponential behavior is induced by fixation procedures (Becker 2012a). We therefore recommend to run FRET measurements only on live specimens.

FRET experiments by any FRET technique must avoid that the results are impaired by photobleaching (Hoffmann et al. 2008). The noninteracting donor is likely to photobleach faster so that the apparent FRET efficiency increases. Photobleaching artifacts are likely to be reduced by new fluorescent proteins with better photostability (Day et al. 2008; Day and Schaufele 2008) and by FLIM systems with more efficient detectors (Becker et al. 2011b).

A huge number of FLIM FRET applications have been published in the last few years. Most of the FRET applications are more or less related to protein interactions or protein organization (Biskup et al. 2004a; Jiang et al. 2007; Klucken et al. 2006; Mandal et al. 2004, Nyborg et al. 2006; Peltan et al.

2006; Rickman et al. 2007; Sever et al. 2005, 2006; Stagi et al. 2006). It has also been shown that conformational changes of proteins in cells can be monitored by FLIM FRET (Calleja et al. 2003; Lleo et al. 2004). There are a number of FRET applications that are directly related to clinical research (Medine et al. 2007). Because FRET yields information about protein interaction, it can be used to investigate the formation—and possible dissolution—of amyloid plaques of Alzheimer's disease (Bacskai et al. 2003; Berezovska et al. 2003a, b, 2005; Lleo et al. 2004). Another potential application is the investigation of infection mechanisms. The technique was demonstrated by Ghukasyan et al. (2007) for the infection of HeLa cells with enterovirus 71. For an overview about the existing FLIM FRET literature, please see Becker (2012a).

9.4.3 AUTOFLUORESCENCE FLIM OF TISSUE

Biological tissue contains a variety of endogenous fluorophores (König and Riemann 2003; Richards-Kortum et al. 2003). The fluorescence decay profiles of most of them depend on the binding to proteins, the metabolic state of the tissue, the oxygen concentration, and other biologically relevant parameters. The decay functions of autofluorescence therefore contain information about the metabolic state and the constitution of the tissue.

Due to the presence of several fluorophores or fluorophore fractions , the fluorescence decay profiles of tissue autofluorescence decay profiles are multiexponential, with decay components from about 100 ps to several nanoseconds. The deviations from single-exponential decay are substantial; see Figure 9.25. Extracting meaningful decay parameters from the data therefore requires recording at high time resolution, sufficient number of time channels, and double- or triple-exponential decay analysis. To obtain clean decay data for multiexponential analysis, contamination of the decay data by out-of-focus signals must be avoided. Multiphoton or confocal laser scanning with TCSPC FLIM recording is therefore the method of choice.

9.4.3.1 NADH and FAD imaging

The use of fluorescence lifetime variations is especially promising for NADH and FAD. These coenzymes are involved in the electron transfer mechanism of the cell metabolism. Both NADH and FAD form redox pairs. NADH is fluorescent in its reduced form but loses fluorescence when oxidized. FAD is fluorescent when oxidized and loses fluorescence when reduced. The fluorescence intensity ratio of NADH and FAD therefore changes with the redox state of the tissue (Chance 1976; Chance et al. 1979). Measurements of the redox state are related to the "Warburg effect": Normal cells have an oxidative metabolism, and cancer cells have a reductive metabolism (Warburg 1956). For practical applications, please see Skala et al. (2010), Skala and Ramanujam (2010), and Walsh et al. (2012).

Moreover, the fluorescence lifetimes of NADH and FAD depend on the binding to proteins (Lakowicz et al. 1992; Lakowicz 2006; Paul and Schneckenburger 1996). The ratio of bound and unbound NADH depends on the metabolic state (Bird et al. 2005; Chia et al. 2008; Chorvat and Chorvatova 2006, 2009; Ghukasyan and Kao 2009; Skala et al. 2007a, b). FLIM data of NADH and FAD are therefore used to detect precancerous and cancerous alterations (Kantelhardt et al. 2007; Leppert et al. 2006; Skala et al. 2007a, b). It has also been shown that the fluorescence decay parameters of the NADH fluorescence change with maturation of cells, and during apoptosis and necrosis (Ghukasyan and Kao 2009; König and

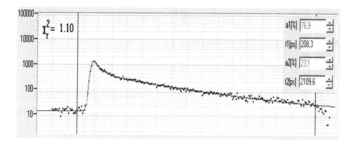

Figure 9.25 Typical decay curve of tissue autofluorescence. Stratum corneum of human skin, two-photon excitation at 800 nm.

Uchugonova 2009; König et al. 2011; Sanchez et al. 2010). Chorvat et al. (2010) used NADH FLIM data to detect early stages in the metabolism of isolated live cardiomyocytes in cases of rejection of transplanted hearts. The effect of ouabain, a pharmaceutical drug, on the NADH decay functions in live cadiomyocytes was studied by Chorvatova et al. (2012).

9.4.3.2 Ophthalmic FLIM

Significant changes in the decay parameters have been found for the autofluorescence of the human ocular fundus (Schweitzer et al. 2004). There are fluorescence components from a wide range of fluorophores, such as NADH, FAD, lipofuscin, collagen, melanin, and advanced glycation end products (AGE). Different decay times have been associated with different fluorophores and morphological structures of the retina (Schweitzer et al. 2007). Changes in the decay functions can result both from changes in the tissue constitution and from changes in the local environment of the fluorophores. Distinct signatures of the decay parameters were found for early and late age-related macula degeneration (AMD), for undersupplied regions after arterial branch occlusions, for metabolic alterations due to diabetes mellitus, and for extramacular drusen (Schweitzer et al. 2009, 2012; Schweitzer 2009, 2010).

Clinical FLIM of the human ocular fundus is obtained by combining an ophthalmic scanner with one or several picosecond diode lasers and a TCSPC FLIM system. The problems of ophthalmic FLIM are that the excitation power is limited by eye safety considerations and that motion of the eye during the image acquisition must be taken onto account. For technical details, please see Becker (2012a) and Schweitzer et al. (2004, 2007). Ophthalmic FLIM is currently in the state of clinical trials.

A few typical images are shown in Figure 9.26. The figure shows the fundus of a patient with wet age-related macula degeneration, AMD. The images were scanned by a modified HRA laser ophthalmoscope of Heidelberg Engineering. A Becker & Hickl BDL-445 SMC picosecond diode laser was used for excitation. The detection part had two spectral channels from 490 to 560 nm and 560 to 700 nm. The signals were detected by Hamamatsu R3809U-50 MCP PMTs and recorded by a Becker & Hickl SPC-150 TCSPC FLIM module. The images shown are from the 490 to 560 nm channel.

The decay data in the pixels of the image were analyzed by fitting a triple-exponential decay model. Figure 9.26a through d, shows a lifetime image of the amplitude-weighted average lifetime, t_m, and images of the relative contributions, $q1$, $q2$, and $q3$, of the fast, medium, and slow fluorescence components to the total intensity.

9.4.3.3 Multiphoton fluorescence tomography of skin

Human skin emits fluorescence from a wide variety of endogenous fluorophores, such as NADH, FAD, and melanin. It has been has shown that the decay times are different for normal skin, nevi, and malignant melanoma (Dimitrow et al. 2009). There is hope that the fluorescence spectrum and lifetime of melanin can be used as parameters for classification of skin lesions. Clear identification requires recording of spectral information, lifetime information, and morphological information on the cell level in different depth of the skin. To reach deep skin layers, two-photon excitation with non-descanned detection is used.

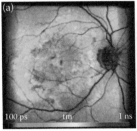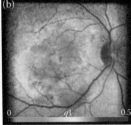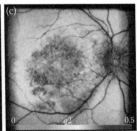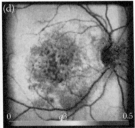

Figure 9.26 TCSPC lifetime images of the ocular fundus of a patient with wet AMD. (a–d) Amplitude-weighted lifetime, t_m, and relative intensity contributions of fast, medium, and slow fluorescence components, q1, q2, q3. Excitation wavelength 445 nm, detection wavelength interval from 490 to 560 nm. (Data courtesy of Dietrich Schweitzer, Friedrich Schiller University Jena, Germany.)

The technique is in the state of clinical evaluation (König and Riemann 2003; König 2008; Roberts et al. 2011); the first instruments are on the market (König 2008; König and Uchugonova 2009).

FLIM images obtained at human skin are shown in Figure 9.27. Figure 9.27a–d shows the stratum corneum (5 µm deep), Figure 9.27e–h, the stratum spinosum (50 µm deep). Double-exponential decay analysis was applied to the data. The color represents the fast lifetime component, τ_1, the slow lifetime component, τ_2, the ratio of lifetime components, τ_1/τ_2, and the ratio of the amplitudes of the components, a_1/a_2. The brightness of the pixels represents the intensity. The decay parameters show considerable variations throughout the image.

9.4.3.4 Combined fluorescence lifetime and SHG detection

Additional information about the constitution of tissue can be obtained from SHG signals (Masters and So 2008). The ratio of SHG and autofluorescence intensity in mammalian skin is an indicator of skin ageing (Koehler et al. 2006).

An example of detecting SHG signals in combination with fluorescence lifetime is shown in Figure 9.28. The images show a pig skin sample exited by two-photon excitation at 800 nm. Figure 9.28a shows the wavelength channel below 480 nm. This channel contains both fluorescence and SHG signals. The SHG fraction of the signal has been extracted from the FLIM data and displayed by color. Figure 9.28b is from the channel >480 nm. It contains only fluorescence; the color corresponds to the amplitude-weighted mean lifetime of a double-exponential decay model.

9.4.3.5 Z-stack recording

Measurements as the ones described above can favorably be performed by Z stack FLIM. Figure 9.29 shows the FLIM image of a pig skin sample excited at 800 nm in the wavelength channel above 480 nm. The color of the images represents the average (intensity-weighted) lifetime of a double-exponential fit to the decay data. The intensity of the images was normalized on the intensity of the brightest pixel.

Figure 9.30 shows the SHG intensity extracted from the FLIM data in the channel below 480 nm. Also here, the intensity of the images was normalized on the intensity of the brightest pixel.

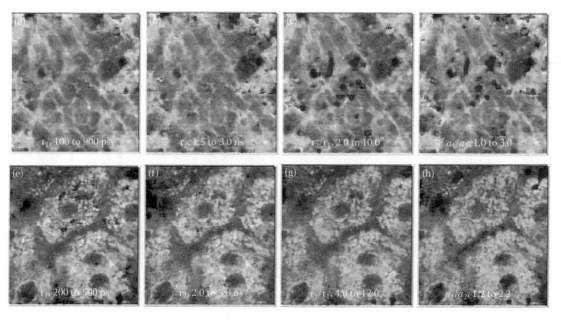

Figure 9.27 Time-resolved *in vivo* autofluorescence images of human stratum corneum (a–d, 5 µm deep) and stratum spinosum (e–h, 50 µm deep). Double-exponential analysis, (a, e) fast lifetime component, τ_1; (b, f) slow lifetime component, τ_2; (c, g) ratio of the lifetime components, τ_1/τ_2; (d, h) ratio of amplitudes, a_1/a_2. The indicated parameter range corresponds to a color range from blue to red. (Data courtesy of Karsten König and Iris Riemann, Fraunhofer Institute of Biomedical Techniques, St. Ingbert.)

Principles of fluorescence lifetime instrumentation

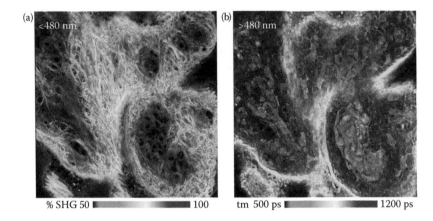

% SHG 50 ▬▬▬ 100   tm 500 ps ▬▬▬ 1200 ps

Figure 9.28 Two-photon FLIM of pig skin. Two-photon excitation, non-descanned detection. (a) Wavelength channel <480 nm, color shows percentage of SHG in the recorded signal. (b) Wavelength channel >480 nm, color shows amplitude-weighted mean lifetime. Microscope Zeiss LSM 710 NLO, with bh Simple-Tau 152 FLIM system.

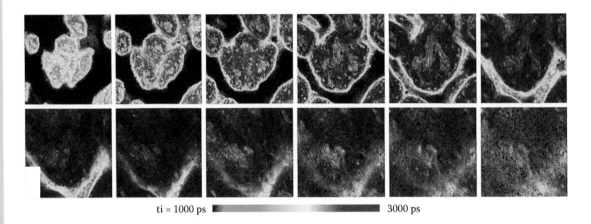

ti = 1000 ps ▬▬▬ 3000 ps

Figure 9.29 FLIM Z stack recorded at a pig skin sample, excitation at 800 nm, emission above 480 nm. Z step width 5.09 μm, scan area 212 × 212 μm. Images from 5 μm to about 60 μm below the surface. Color represents average (intensity-weighted) lifetime of a double-exponential fit. FLIM data format 256 × 256 pixels, 256 time channels. Normalized intensity. Microscope Zeiss LSM 710 NLO, with bh Simple-Tau 152 FLIM system.

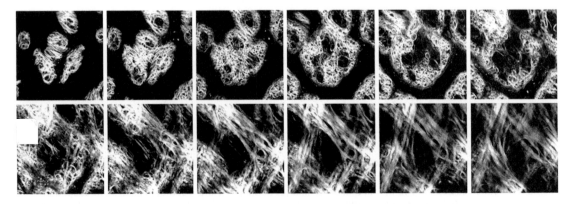

Figure 9.30 Z stack recorded at a pig skin sample, excitation at 800 nm, emission below 480 nm, SHG signal, extracted by selecting photons in an early time window. Z step width is 5.09 μm, scan area 212 × 212 μm. Images from 5 μm to about 60 μm depth. Normalized intensity.

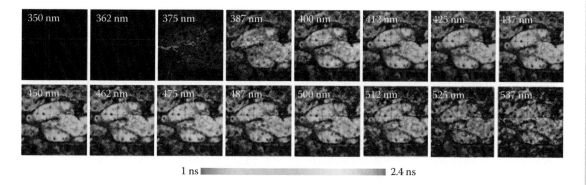

Figure 9.31 Multiwavelength fluorescence lifetime images of a mouse kidney section. Two-photon excitation at 750 nm. Images in subsequent wavelength intervals from 350 to 550 nm. (Courtesy of Christoph Biskup, University Jena, Germany. Zeiss LSM 510 NLO with Becker & Hickl MW FLIM detector and SPC-830 FLIM module.)

9.4.3.6 Multiwavelength FLIM of tissue

Due to the wide variety of fluorophores involved in autofluorescence, the FLIM data recorded are usually a mixture of the emission of several fluorophores. A way to better separate the signals is multiwavelength FLIM. Figure 9.31 shows autofluorescence images of a mouse kidney section recorded by a bh MW-FLIM system. The microscope was a Zeiss LSM 510 NLO. Two-photon excitation at a wavelength of 750 nm was used. Each of the images covers a wavelength range of 12.5 nm. The color of the images represents the lifetime obtained from a single-exponential fit.

An overview on autofluorescence multiwavelength FLIM is given by Chorvat and Chorvatova (2009). Skin measurements have been described by Dimitrow et al. (2009), applications related to PDT by Rück et al. (2005, 2007). Full exploitation of multiwavelength autofluorescence FLIM data requires a global fit with a suitable model that includes lifetimes and effective spectra of the expected fluorophores. The task is extremely difficult because most of the endogenous fluorophores are, in fact, mixtures of slightly different compounds, with different lifetimes and absorption and emission spectra. Moreover, the apparent fluorescence spectra may be changed by absorbers of unknown absorption spectra, inhomogeneous distribution, and unknown concentration in the tissue. Nevertheless, there are a number of approaches that are expected to solve at least a part of the problem (Chorvat and Chorvatova 2006, 2009; Chorvat et al. 2007; Provenzano et al. 2008).

9.5 SUMMARY

TCSPC FLIM is based on scanning a sample with a focused laser beam, detecting single photons of the fluorescence signal, and determining parameters for the individual photons. These parameters are used to build up photon distributions. Depending on which parameters are used, fluorescence lifetime images at a single wavelength or at multiple excitation or emission wavelengths and phosphorescence lifetime images can be built up. It should be noted that TCSPC FLIM is based on a sorting process, not on time gating or wavelength scanning. It therefore reaches an almost ideal efficiency of photon recording. Another advantage of TCSPC FLIM is that the data contain clearly resolved decay curves in the individual pixels. Biological information hidden in the composition of multiexponential decay profiles can therefore be extracted. Moreover, data in different wavelength channels and fluorescence and phosphorescence data are obtained simultaneously. The data therefore remain comparable, independently of possible dynamic effects in the sample during the acquisition time. TCSPC can even be used to record dynamic changes in the fluorescence decay functions within a sample. By building up a data array of decay functions over the distance in a line scan and the time after stimulation of a sample, dynamic changes can be resolved at a resolution down to milliseconds.

Typical applications of TCSPC FLIM are ion concentrations measurements, protein interaction measurements by FRET, and measurements of the autofluorescence of cells and tissue.

Ion concentration measurements by FLIM are based on the variation of the fluorescence lifetime with the local environment of the fluorophore molecules. The advantage over intensity-based measurements is that no special ratiometric fluorophores are needed. Therefore, a much wider selection of fluorescence markers can be used, and a wider range of cell parameters is accessible. A potential extension of ion concentration measurement is the measurement of dynamic changes on stimulation of a sample by FLITS.

FRET experiments are used to investigate protein interactions and protein configuration. FLIM FRET measurements are based on the change of the donor decay functions on interaction with an acceptor. FLIM-based FRET measurement does not have to cope with problems like donor bleed-through or directly excited acceptor fluorescence. This relaxes the requirements to the absorption and emission spectra of the donors and acceptors used. FRET measurements by TCSPC FLIM are able to distinguish interacting and noninteracting fractions of the donor. This yields independent information about distances and interacting and noninteracting protein fractions. It also yields correct classic FRET efficiencies in cases where not all donor molecules interact.

Autofluorescence FLIM measurements make use of the fact that the fluorescence decay times of most endogenous fluorophores vary with the binding state and with the redox state. Moreover, endogenous fluorophores are usually mixtures of slightly different compounds with different fluorescence lifetimes. By multiexponential analysis of the decay profiles, TCSPC FLIM delivers much better biological information than steady-state techniques, or techniques that record only the lifetime of a single-exponential approximation of the decay. A future extension of autofluorescence FLIM may be the combination with oxygen concentration measurement via the phosphorescence lifetime.

REFERENCES

Bacskai, B.J., Skoch, J., Hickey, G.A., Allen, R. & Hyman, B.T. (2003) Fluorescence resonance energy transfer determinations using multiphoton fluorescence lifetime imaging microscopy to characterize amyloid-beta plaques. *J. Biomed. Opt.* **8**, 368–375.

Ballew, R.M. & Demas, J.N. (1989) An error analysis of the rapid lifetime determination method for the evaluation of single exponential decays. *Anal. Chem.* **61**, 30–33.

Becker, W., Bergmann, A., Biskup, C., Zimmer, T., Klöcker, N. & Benndorf, K. (2002) Multi-wavelength TCSPC lifetime imaging. *Proc. SPIE* **4620**, 79–84.

Becker, W., Bergmann, A., Hink, M.A., König, K., Benndorf, K. & Biskup, C. (2004a) Fluorescence lifetime imaging by time-correlated single photon counting. *Microsc. Res. Tech.* **63**, 58–66.

Becker, W., Bergmann, A., Biscotti, G., König, K., Riemann, I., Kelbauskas, L. & Biskup, C. (2004b) High-speed FLIM data acquisition by time-correlated single photon counting. *Proc. SPIE* **5323**, 27–35.

Becker, W. (2005) *Advanced Time-Correlated Single-Photon Counting Techniques*. Springer, Berlin, Heidelberg, New York.

Becker, W., Bergmann, A., Haustein, E., Petrasek, Z., Schwille, P., Biskup, C., Kelbauskas, L., Benndorf, K., Klöcker, N., Anhut, T., Riemann, I. & König, K. (2006) Fluorescence lifetime images and correlation spectra obtained by multi-dimensional TCSPC. *Microsc. Res. Tech.* **69**, 186–195.

Becker, W., Bergmann, A. & Biskup, C. (2007) Multi-spectral fluorescence lifetime imaging by TCSPC. *Microsc. Res. Tech.* **70**, 403–409.

Becker, W. & Bergmann, A. (2008) Lifetime-resolved imaging in nonlinear microscopy. In: B.R. Masters, P.T.C. So, eds., *Handbook of Biomedical Nonlinear Optical Microscopy*. Oxford University Press, Oxford, New York.

Becker, W., Su, B. & Bergmann, A. (2009) Fast-acquisition multispectral FLIM by parallel TCSPC. *Proc. SPIE* **7183**, 718305.

Becker, W., Su, B., Bergmann, A., Weisshart, K. & Holub, O. (2011a) Simultaneous fluorescence and phosphorescence lifetime imaging. *Proc. SPIE* **7903**, 790320.

Becker, W., Su, B., Weisshart, K. & Holub, O. (2011b) FLIM and FCS detection in laser-scanning microscopes: Increased efficiency by GaAsP hybrid detectors. *Microsc. Res. Tech.* **74**, 804–811.

Becker, W., Su, B. & Bergmann, A. (2012) Spatially resolved recording of transient fluorescence lifetime effects by line-scanning TCSPC. *Proc. SPIE* **8226**, 82260C-1–82260C-6.

Becker, W. (2012a) *The bh TCSPC Handbook*, 5th edn. Becker & Hickl GmbH. Available from www.becker-hickl.com.

Becker, W. (2012b) Fluorescence lifetime imaging—Techniques and applications. *J. Microsc.* **247**, 119–136.

Becker & Hickl GmbH (2012a) DCS-120 Confocal scanning FLIM systems, user handbook. Becker & Hickl GmbH, Berlin. Available from www.becker-hickl.com.

Becker & Hickl GmbH (2012b) Modular FLIM systems for Zeiss LSM 510 and LSM 710 laser scanning microscopes, user handbook. Becker & Hickl GmbH, Berlin. Available from www.becker-hickl.com.

Benesch, J., Hungerford, G., Suhling, K., Tregidgo, C., Mano, J.F. & Reis, R.L. (2007) Fluorescence probe techniques to monitor protein adsorption-induced conformation changes on biodegradable polymers. *J. Colloid Interface Sci.* **312**, 193–200.

Berezin, M.Y. & Achilefu, S. (2010) Fluorescence lifetime measurement and biological imaging. *Chem. Rev.* **110**, 2641–2684.

Berezovska, O., Ramdya, P., Skoch, J., Wolfe, M.S., Bacskai, B.J. & Hyman, B.T. (2003a) Amyloid precursor protein associates with a nicastrin-dependent docking site on the presenilin 1-γ-secretase complex in cells demonstrated by fluorescence lifetime imaging. *J. Neurosci.* **23**, 4560–4566.

Berezovska, O., Bacskai, B.J. & Hyman, B.T. (2003b) Monitoring proteins in intact cells. *Sci. Aging Knowl. Environ., SAGE KE* 14.

Berezovska, O., Lleo, A., Herl, L.D., Frosch, M.P., Stern, E.A., Bacskai, B.J. & Hyman, B.T. (2005) Familial Alzheimer's disease presenilin 1 mutations cause alterations in the conformation of presenilin and interactions with amyloid precursor protein. *J. Neurosci.* **25**, 3009–3017.

Bird, D.K., Eliceiri, K.W., Fan, C.-H. & White, J.G. (2004) Simultaneous two-photon spectral and lifetime fluorescence microscopy. *Appl. Opt.* **43**, 5173–5182.

Bird, D.K., Yan, L., Vrotsos, K.M., Eliceiri, K.E. & Vaughan, E.M. (2005) Metabolic mapping of MCF10A human breast cells via multiphoton fluorescence lifetime imaging of coenzyme NADH. *Cancer Res.* **65**, 8766–8773.

Biskup, C., Kelbauskas, L., Zimmer, T., Benndorf, K., Bergmann, A., Becker, W., Ruppersberg, J.P., Stockklausner, C. & Klöcker, N. (2004a) Interaction of PSD-95 with potassium channels visualized by fluorescence lifetime-based resonance energy transfer imaging. *J. Biomed. Opt.* **9**, 735–759.

Biskup, C., Zimmer, T. & Benndorf, K. (2004b) FRET between cardiac Na⁺ channel subunits measured with a confocal microscope and a streak camera. *Nat. Biotechnol.* **22**, 220–224.

Biskup, C., Zimmer, T., Kelbauskas, L., Hoffmann, B., Klöcker, N., Becker, W., Bergmann, A. & Benndorf, K. (2007) Multi-dimensional fluorescence lifetime and FRET measurements. *Microsc. Res. Tech.* **70**, 403–409.

Bugiel, I., König, K. & Wabnitz, H. (1989) Investigations of cells by fluorescence laser scanning microscopy with subnanosecond time resolution. *Laser Life Sci* **3**(1), 47–53.

Calleja, V., Ameer-Beg, S., Vojnovic, B., Woscholski, R., Downwards, J. & Larijani, B. (2003) Monitoring conformational changes of proteins in cells by fluorescence lifetime imaging microscopy. *Biochem. J.* **372**, 33–40.

Chance, B. (1976) Pyridine nucleotide as an indicator of the oxygen requirements for energy-linked functions of mitochondria. *Circ. Res.* **38**, 131–138.

Chance, B., Schoener, B., Oshino, R., Itshak, F. & Nakase, Y. (1979) Oxidation–reduction ratio studies of mitochondria in freeze-trapped samples. NADH and flavoprotein fluorescence signals. *J. Biol. Chem.* **254**, 4764–4771.

Chen, Y. & Periasamy, A. (2004) Characterization of two-photon excitation fluorescence lifetime imaging microscopy for protein localization. *Microsc. Res. Tech.* **63**, 72–80.

Chia, T.H., Williamson, A., Spencer, D.D. & Levene, M.J. (2008) Multiphoton fluorescence lifetime imaging of intrinsic fluorescence in human and rat brain tissue reveals spatially distinct NADH binding. *Opt. Express* **16**, 4237–4249.

Chorvat, D. & Chorvatova, A. (2006) Spectrally resolved time-correlated single photon counting: A novel approach for characterization of endogenous fluorescence in isolated cardiac myocytes. *Eur. Biophys. J.* **36**, 73–83.

Chorvat, D., Mateasik, A., Kirchnerova, J. & Chorvatova, A. (2007) Application of spectral unmixing in multi-wavelength time-resolved spectroscopy. *Proc. SPIE* **6771**, 677105-1–677105-12.

Chorvat, D. & Chorvatova, A. (2009) Multi-wavelength fluorescence lifetime spectroscopy: A new approach to the study of endogenous fluorescence in living cells and tissues. *Laser Phys. Lett.* **6**, 175–193.

Chorvat, D. Jr., Mateasik, A., Cheng, Y., Poirier, N., Miro, J., Dahdah, N.S. & Chorvatova, A. (2010) Rejection of transplanted hearts in patients evaluated by the component analysis of multi-wavelength NAD(P)H fluorescence lifetime spectroscopy. *J. Biophotonics* **3**, 646–652.

Chorvatova, A., Elzwiei, F., Mateasik, A. & Chorvat, D. (2012) Effect of ouabain on metabolic oxidative state in living cardiomyocytes evaluated by time-resolved spectroscopy of endogenous NAAD(P)H fluorescence. *J. Biomed. Opt.* **17**, 101505-1–101505-7.

Day, R.N., Booker, C.F. & Periasamy, A. (2008) Characterization of an improved donor fluorescent protein for Förster resonance energy transfer microscopy. *J. Biomed. Opt.* **13**, 031203-1–031203-9.

Day, R.N. & Schaufele, F. (2008) Fluorescent protein tools for studying protein dynamics in living cells: A review. *J. Biomed. Opt.* **13**, 031202-1–031202-6.

Denk, W., Strickler, J.H. & Webb, W.W.W. (1990) Two-photon laser scanning fluorescence microscopy. *Science* **248**, 73–76.

Diaspro, A. (ed.) (2001) *Confocal and Two-Photon Microscopy: Foundations, Applications and Advances.* Wiley-Liss, Inc., New York.

Dimitrow, E., Riemann, I., Ehlers, A., Koehler, M., Norgauer, J., Elsner, P., König, K. & Kaatz, M. (2009) Spectral fluorescence lifetime detection and selective melanin imaging by multiphoton laser tomography for melanoma diagnosis. *Exp. Dermatol.* **18**, 509–515.

Duncan, R.R., Bergmann, A., Cousin, M.A., Apps, D.K. & Shipston, M.J. (2004) Multi-dimensional time-correlated single-photon counting (TCSPC) fluorescence lifetime imaging microscopy (FLIM) to detect FRET in cells. *J. Microsc.* **215**, 1–12.

Ellis, J.D., Llères, D., Denegri, M., Lamond, A.I. & Cáceres, J.F. (2008) Spatial mapping of splicing factor complexes involved in exon and intron definition. *J. Cell Biol.* **181**, 921–934.

Felekyan, S., Kühnemuth, R., Kudryavtsev, V., Sandhagen, C., Becker, W. & Seidel, C.A.M. (2005) Full correlation from picoseconds to seconds by time-resolved and time-correlated single photon detection. *Rev. Sci. Instrum.* **76**, 083104.

Förster, Th. (1948) Zwischenmolekulare Energiewanderung und Fluoreszenz. *Ann. Phys.* (Serie 6) **2**, 55–75.

Funk, K., Woitecki, A., Franjic-Würtz, C., Gensch, Th., Möhrlein, F. & Frings, S. (2008) Modulation of chloride homeostasis by inflammatory mediators in dorsal ganglion neurons. *Mol. Pain* **4**, 32.

Geddes, C.D., Cao, H., Gryczynski, I., Fang, J. & Lakowicz, J.R. (2003) Metal-enhanced fluorescence (MEF) due to silver colloids on a planar surface: Potential applications of indocyanine green to in vivo imaging. *J. Phys. Chem. A* **107**, 3443–3449.

Gerritsen, H.C., Asselbergs, M.A.H., Agronskaia, A.V. & van Sark, W.G.J.H.M. (2002) Fluorescence lifetime imaging in scanning microscopes: Acquisition speed, photon economy and lifetime resolution. *J. Microsc.* **206**, 218-224.

Ghukasyan, V., Hsu, Y.-Y., Kung, S.-H. & Kao, F.-J. (2007) Application of fluorescence resonance energy transfer resolved by fluorescence lifetime imaging microscopy for the detection of enterovirus 71 infection in cells. *J. Biomed. Opt.* **12**, 024016-1–024016-8.

Ghukasyan, V. & Kao, F.-J. (2009) Monitoring cellular metabolism with fluorescence lifetime of reduced nicotinamide adenine dinucleotide. *J. Phys. Chem. C* **113**, 11532–11540.

Gilbert, D., Franjic-Würtz, C., Funk, K., Gensch, Th., Frings, S. & Möhrlen, F. (2007) Differential maturation of chloride homeostasis in primary afferent neurons of the somatosensory system. *Int. J. Dev. Neurosci.* **25**, 479–489.

Göppert-Mayer, M. (1931) Über Elementarakte mit zwei Quantensprüngen. *Ann. Phys.* **9**, 273–294.

Govindjee (1995) Sixty-three years since Kautsky: Chlorophyll α fluorescence. *Aust. J. Plant Physiol.* **22**, 131–160.

Hanson, K.M., Behne, M.J., Barry, N.P., Mauro, T.M. & Gratton, E. (2002) Two-photon fluorescence imaging of the skin stratum corneum pH gradient. *Biophys. J.* **83**, 1682–1690.

Hoffmann, B., Zimmer, T., Klöcker, N., Kelbauskas, L., König, K., Benndorf, K. & Biskup, C. (2008) Prolonged irradiation of enhanced cyan fluorescent protein or Cerulean can invalidate Förster resonance energy transfer measurements. *J. Biomed. Opt.* **13**, 031250-1–031250-9.

Jiang, Y., Borrelli, L.A., Kanaoka, Y., Bacskai, B.J. & Boyce, J.A. (2007) Cys LT2 receptors interact with CysLT1 receptors and down-modulate cysteinyl leukotriene–dependent mitogenic responses of mast cells. *Blood* **110**, 3263–3270.

Kaneko, H., Putzier, I., Frings, S., Kaupp, U.B. & Gensch, Th. (2004) Chloride accumulation in mammalian olfactory sensory neurons. *J. Neurosci.* **24**, 7931–7938.

Kantelhardt, S.R., Leppert, J., Krajewski, J., Petkus, N., Reusche, E., Tronnier, V.M., Hüttmann, G. & Giese, A. (2007) Imaging of brain and brain tumor specimens by time-resolved multiphoton excitation microscopy ex vivo. *Neuro Oncol* **9**, 103–112.

Katsoulidou, V., Bergmann, A. & Becker, W. (2007) How fast can TCSPC FLIM be made? *Proc. SPIE* **6771**, 67710B-1–67710B-7.

Kelbauskas, L. & Dietel, W. (2002) Internalization of aggregated photosensitizers by tumor cells: Subcellular time-resolved fluorescence spectroscopy on derivates of pyropheophorbide-a ethers and chlorin e6 under femtosecond one- and two-photon excitation. *Photochem. Photobiol.* **76**, 686–694.

Klucken, J., Outeiro, T.F., Nguyen, P., McLean, P.J. & Hyman, B.T. (2006) Detection of novel intracellular-synuclein oligomeric species by fluorescence lifetime imaging. *FASEB J.* **20**, 2050–2057.

Knemeyer, J.-P., Marmé, N. & Sauer, M. (2002) Probes for detection of specific DNA sequences at the single-molecule level. *Anal. Chem.* **72**, 3717–3724.

Koehler, M.J., König, K., Elsner, P., Bückle, R. & Kaatz, M. (2006) In vivo assessment of human skin aging by multiphoton laser scanning tomography. *Opt. Lett.* **31**, 2879–2881.

Köllner, M. & Wolfrum, J. (1992) How many photons are necessary for fluorescence-lifetime measurements? *Phys. Chem. Lett.* **200**, 199–204.

König, K. & Riemann, I. (2003) High-resolution multiphoton tomography of human skin with subcellular spatial resolution and picosecond time resolution. *J. Biom. Opt.* **8**, 432–439.

König, K. (2008) Clinical multiphoton tomography. *J. Biophoton.* **1**, 13–23.

König, K. & Uchugonova, A. (2009) Multiphoton fluorescence lifetime imaging at the dawn of clinical application. In: A. Periasamy, R.M. Clegg, eds., *FLIM Microscopy in Biology and Medicine.* CRC Press, Boca Raton, FL.

König, K., Uchugonova, A. & Gorjup, E. (2011) Multiphoton fluorescence lifetime imaging of 3D-stem cell spheroids during differentiation. *Microsc. Res. Tech.* **74**, 9–17.

Koushik, S.V. & Vogel, S.S. (2008) Energy migration alters the fluorescence lifetime of Cerulean: Implications for fluorescence lifetime imaging Forster resonance energy transfer measurements. *J. Biomed. Opt.* **13**, 031204-1–031204-9.

Kuchibhotla, K.V., Lattarulo, C.R., Hyman, B. & Bacskai, B.J. (2009) Synchronous hyperactivity and intercellular calcium waves in astrocytes in Alzheimer mice. *Science* **323**, 1211–1215.

Kuimova, M.K., Yahioglu, G., Levitt, J.A. & Suhling, K. (2008) Molecular rotor measures viscosity of live cells via fluorescence lifetime imaging. *J. Am. Chem. Soc.* **130**, 6672–6673.

Lakowicz, J.R., Szmacinski, H., Nowaczyk, K. & Johnson, M.L. (1992) Fluorescence lifetime imaging of free and protein-bound NADH. *PNAS* **89**, 1271–1275.

Lakowicz, J.R. (1999) *Principles of Fluorescence Spectroscopy*, 2nd edn. Kluwer Academic/Plenum Publishers, Boston.

Lakowicz, J.R. (2006) *Principles of Fluorescence Spectroscopy*, 3rd edn. Springer, Berlin, Heidelberg, New York.

Leppert, J., Krajewski, J., Kantelhardt, S.R., Schlaffer, S., Petkus, N., Reusche, E., Hüttmann, G. & Giese, A. (2006) Multiphoton excitation of autofluorescence for microscopy of glioma tissue. *Neurosurgery* **58**, 759–767.

Levitt, J.A., Kuimova, M.K., Yahioglu, G., Chung, P.-H., Suhling, K. & Phillips, D. (2009) Membrane-bound molecular rotors measure viscosity in live cells via fluorescence lifetime imaging. *J. Phys. Chem. C* **113**, 11634–11642.

Lin, L.L., Grice, J.E., Butler, M.K., Zvyagin, A.V., Becker, W., Robertson, T.A., Soyer, H.P., Roberts, M.S. & Prow, T.W. (2011) Time-correlated single photon counting for simultaneous monitoring of zinc oxide nanoparticles and NAD(P)H in intact and barrier-disrupted volunteer skin. *Pharm. Res.* **28**, 2920–2930.

Lleo, A., Berezovska, O., Herl, L., Raju, S., Deng, A., Bacskai, B.J., Frosch, M.P., Irizarry, M. & Hyman, B.T. (2004) Nonsteroidal anti-inflammatory drugs lower Aβ42 and change presenilin 1 conformation. *Nat. Med.* **10**, 1065–1066.

Malicka, J., Gryczynski, I., Geddes, C.D. & Lakowicz, J.R. (2003) Metal-enhanced emission from indocyanine green: A new approach to in vivo imaging. *J. Biomed. Opt.* **8**, 472–478.

Mandal, A.K., Skoch, J., Bacskai, B.J., Hyman, B.T., Christmas, P., Miller, D., Yamin, T.D., Xu, S., Wisniewski, D., Evans, J.F. & Soberman, R.J. (2004) The membrane organization of leukotriene synthesis. *PNAS* **101**, 6587–6592.

Masters, B.R. & So, P.T.C. (2008) *Handbook of Biomedical Nonlinear Optical Microscopy*. Oxford University Press, Oxford, New York.

Medine, C.N., McDonald, A., Bergmann, A. & Duncan, R. (2007) Time-correlated single photon counting FLIM: Some considerations for physiologists. *Microsc. Res. Tech.* **70**, 421–425.

Muddana, H.S., Morgan, T.T., Adair, J.H. & Butler, P.J. (2009) Photophysics of Cy3-encapsulated calcium phosphate nanoparticles. *Nano Lett.* **9**, 1559–1556.

Nyborg, A.C., Herl, L., Berezovska, O., Thomas, A.V., Ladd, T.B., Jansen, K., Hyman, B.T. & Golde, T.E. (2006) Signal peptide peptidase (SPP) dimer formation as assessed by fluorescence lifetime imaging microscopy (FLIM) in intact cells. *Mol. Neurodegener.* **1**, 16.

O'Connor, D.V. & Phillips, D. (1984) *Time-Correlated Single Photon Counting*. Academic Press, London.

Paul, R.J. & Schneckenburger, H. (1996) Oxygen concentration and the oxidation-reduction state of yeast: Determination of free/bound NADH and flavins by time-resolved spectroscopy. *Naturwissenschaften* **83**, 32–35.

Pawley, J. (ed.) (2006) *Handbook of Biological Confocal Microscopy*, 3rd edn. Springer Science+Business Media, New York.

Pelet, S., Previte, M.J.R. & So, P.T.C. (2006) Comparing the quantification of Förster resonance energy transfer measurement accuracies based on intensity, spectral, and lifetime imaging. *J. Biomed. Opt.* **11**, 034017-1–034017-11.

Peltan, I.D., Thomas, A.V., Mikhailenko, I., Strickland, D.K., Hyman, B.T. & von Arnim, C.A.F. (2006) Fluorescence lifetime imaging microscopy (FLIM) detects stimulus-dependent phosphorylation of the low density lipoprotein receptor-related protein (LRP) in primary neurons. *Biochem. Biophys. Res. Commun.* **349**, 24–30.

Periasamy, A. (2001) *Methods in Cellular Imaging*. Oxford University Press, Oxford, New York.

Periasamy, A. & Clegg, R.M. (2009) *FLIM Microscopy in Biology and Medicine*. CRC Press, Boca Raton, FL.

Peter, M. & Ameer-Beg, S.M. (2004) Imaging molecular interactions by multiphoton FLIM. *Biol Cell* **96**, 231–236.

Philip, J.P. & Carlsson, K. (2003) Theoretical investigation of the signal-to-noise ratio in fluorescence lifetime imaging. *J. Opt. Soc. Am.* **A20**, 368–379.

Provenzano, P.P., Rueden, C.T., Trier, S.M., Yan, L., Ponik, S.M., Inman, D.R., Keely, P.J. & Eliceiri, K.W. (2008) Nonlinear optical imaging and spectral-lifetime computational analysis of endogenous and exogenous fluorophores in breast cancer. *J. Biomed. Opt.* **13**, 031220-1–031220-11.

Prow, T.W., Grice, J.E., Lin, L.L., Faye, R., Butler, M., Becker, W., Wurm, E.M.T., Yoong, C., Robertson, T.A., Peter Soyer, H. & Roberts, M.S. (2011) Nanoparticles and microparticles for skin drug delivery. *Adv. Drug Deliv. Rev.* **63**, 470–491.

Prummer, M., Sick, B., Renn, A. & Wild, U.P. (2004) Multiparameter microscopy and spectroscopy for single-molecule analysis. *Anal. Chem.* **76**, 1633–1640.

Richards-Kortum, R., Drezek, R., Sokolov, K., Pavlova, I. & Follen, M. (2003) Survey of endogenous biological fluorophores. In: M.-A. Mycek, B.W. Pogue, eds., *Handbook of Biomedical Fluorescence*. Marcel Dekker Inc., New York, 237–264.

Rickman, C., Medine, C.N., Bergmann, A. & Duncan, R.R. (2007) Functionally and spatially distinct modes of MUNC18-syntaxin 1 interaction. *JBC* **282**, 12097–12103.

Ritman-Meer, T., Cade, N.I. & Richards, D. (2007) Spatial imaging of modifications to fluorescence lifetime and intensity by individual Ag nanoparticles. *Appl. Phys. Lett.* **91**, 123122.

Roberts, M.S., Dancik, Y., Prow, T.W., Thorling, C.A., Li, L., Grice, J.E., Robertson, T.A., König, K. & Becker, W. (2011) Non-invasive imaging of skin physiology and percutaneous penetration using fluorescence spectral and lifetime imaging with multiphoton and confocal microscopy. *Eur. J. Pharm. Biopharm.* 77, 469–488.

Rück, A., Dolp, F., Hülshoff, C., Hauser, C. & Scalfi-Happ, C. (2005) FLIM and SLIM for molecular imaging in PDT. *Proc. SPIE* **7500**, 182–189.

Rück, A., Hülshoff, Ch., Kinzler, I., Becker, W. & Steiner, R. (2007) SLIM: A new method for molecular imaging. *Microsc. Res. Tech.* **70**, 403–409.

Sanchez, W.Y., Prow, T.W., Sanchez, W.H., Grice, J.E. & Roberts, M.S. (2010) Analysis of the metabolic deterioration of ex vivo skin, from ischemic necrosis, through the imaging of intracellular NAD(P)H by multiphoton tomography and fluorescence lifetime imaging microscopy (MPT-FLIM). *J. Biomed. Opt.* **15**, 0046008-1–046008-11.

Sanders, R., Draaijer, A., Gerritsen, H.C., Houpt, P.M. & Levine, Y.K. (1995) Quantitative pH Imaging in cells using confocal fluorescence lifetime imaging microscopy. *Anal. Biochem.* **227**, 302–308.

Schweitzer, D., Hammer, M., Schweitzer, F., Anders, R., Doebbecke, T., Schenke, S. & Gaillard, E.R. (2004) In vivo measurement of time-resolved autofluorescence at the human fundus. *J. Biomed. Opt.* **9**, 1214–1222.

Schweitzer, D., Schenke, S., Hammer, M., Schweitzer, F., Jentsch, S., Birckner, E. & Becker, W. (2007) Towards metabolic mapping of the human retina. *Microsc. Res. Tech.* **70**, 403–409.

Schweitzer, D. (2009) Quantifying fundus autofluorescence. In: N. Lois, J.V. Forrester, eds., *Fundus Autofluorescence*. Wolters Kluwer, Lippincott Willams & Wilkins, Philadelphia, PA.

Schweitzer, D., Quick, S., Schenke, S., Klemm, M., Gehlert, S., Hammer, M., Jentsch, S. & Fischer, J. (2009) Vergleich von Parametern der zeitaufgelösten Autofluoreszenz bei Gesunden und Patienten mit früher AMD. *Der Ophthalmol.* **8**, 1–8.

Schweitzer, D. (2010) Metabolic mapping. In: F.G. Holz, R.F. Spaide, eds., *Medical Retina, Essentials in Ophthalmology*. Springer, Heidelberg, Germany.

Schweitzer, D., Gaillard, E.R., Dillon, J., Mullins, R.F., Russell, S., Hoffmann, B., Peters, S., Hammer, M. & Biskup, C. (2012) Time-resolved autofluorescence imaging of human donor retina tissue from donors with significant extramacular drusen. *IOVS* **53**, 3376–3386.

Sever, S., Skoch, J., Bacskai, B.J. & Newmyer, S. (2005) Assays and functional properties of auxilin–dynamin interactions. *Methods Enzymol.* **404**, 570–585.

Sever, S., Skoch, J., Newmyer, S., Ramachandran, R., Ko, D., McKee, M., Bouley, R., Ausiello, D., Hyman, B.T. & Bacskai, B.J. (2006) Physical and functional connection between auxilin and dynamin during endocytosis. *EMBO J.* **25**, 4163–4174.

Skala, M.C., Riching, K.M., Bird, D.K., Dendron-Fitzpatrick, A., Eickhoff, J., Eliceiri, K.W., Keely, P.J. & Ramanujam, N. (2007a) In vivo multiphoton fluorescence lifetime imaging of protein-bound and free nicotinamide adenine dinucleotide in normal and precancerous epithelia. *J. Biomed. Opt.* **12**, 02401-1–02401-10.

Skala, M.C., Riching, K.M., Gendron-Fitzpatrick, A., Eickhoff, J., Eliceiri, K.W., White, J.G. & Ramanujam, N. (2007b) In vivo multiphoton microscopy of NADH and FAD redox states, fluorescence lifetimes, and cellular morphology in precancerous epithelia. *PNAS* **104**, 19494–19499.

Skala, M.C., Fontanella, A., Lan, L., Izatt, J.A. & Dewhirst, M.W. (2010) Longitudinal optical imaging of tumor metabolism and hemodynamics. *J. Biomed. Opt.* **15**, 011112-1–011112-8.

Skala, M.C. & Ramanujam, N. (2010) Multiphoton redox ratio imaging for metabolic monitoring in vivo. *Methods Mol. Biol.* **594**, 155–162.

Stagi, M., Gorlovoy, P., Larionov, S., Takahashi, K. & Neumann, H. (2006) Unloading kinesin transported cargoes from the tubulin track via the inflammatory c-Jun N-terminal kinase pathway. *FASEB J.* **20**, E1–E12.

Strickler, S.J. & Berg, R.A. (1962) Relationship between absorption intensity and fluorescence lifetime of molecules. *J. Chem. Phys.* **37**, 814–822.

Treanor, B., Lanigan, P.M.P., Suhling, K., Schreiber, T., Munro, I., Neil, M.A.A., Phillips, D., Davis, D.M. & French, P.M.W. (2005) Imaging fluorescence lifetime heterogeneity applied to GFP-tagged MHC protein at an immunological synapse. *J. Microsc.* **217**, 36–43.

Tregido, C., Levitt, J.A. & Suhling, K. (2008) Effect of refractive index on the fluorescence lifetime of green fluorescent protein. *J. Biomed. Opt.* **13**, 031218-1–031218-8.

Van Zandvoort, M.A.M.J., de Grauw, C.J., Gerritsen, H.C., Broers, J.L.V., Egbrink, M.G.A., Ramaekers, F.C.S. & Slaaf, D.W. (2002) Discrimination of DNA and RNA in cells by a vital fluorescent probe: Lifetime imaging of SYTO13 in healthy and apoptotic cells. *Cytometry* **47**, 226–232.

Vogel, S.S., Thaler, C. & Koushik, S.V. (2006) Fanciful FRET. *Sci. STKE* **2006**, re2 1–8.

Walsh, A., Cook, R.C., Rexer, B., Arteaga, C.L. & Skala, M.C. (2012) Optical imaging of metabolism in HER2 overexpressing breast cancer cells. *Biomed. Opt. Express* **3**, 75–85.

Warburg, O. (1956) On respiratory impairment in cancer cells. *Science* **124**, 269–270.

Widengren, J., Kudryavtsev, V., Antonik, M., Berger, S., Gerken, M. & Seidel, C.A.M. (2006) Single-molecule detection and identification of multiple species by multiparameter fluorescence detection. *Anal. Chem.* **78**, 2039–2050.

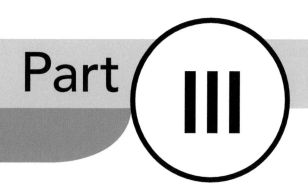

Part III

Analysis of fluorescence lifetime data

The phasor approach to fluorescence lifetime imaging: Exploiting phasor linear properties

10

Michelle A. Digman and Enrico Gratton

Contents

10.1 INTRODUCTION

The phasor approach to fluorescence lifetime imaging (FLIM) is emerging as a practical method for data visualization and analysis. The main feature of the phasor approach is that it is a "fit-free" analysis tool that provides quantitative results about mixtures of fluorophores, Förster resonant energy transfer (FRET), and autofluorescence [1–8]. Since phasor FLIM analysis is relatively simple, fast and accurate, its use is becoming more frequent [1,9]. A less emphasized feature of the phasor approach is that it provides a substantial new concept for FLIM analysis, which arises from the linear transformation of the fluorescence decay at each pixel. In this work, we explore consequences of using this linear transformation for improving the signal-to-noise ratio of FLIM data by applying filtering methods to the images of phasor components. It is well established that linear addition of phasors is crucial for facilitating data analysis and interpretation. It is less known that the linear phasor transformation opens up the possibility of decreasing the variance of FLIM images using simple data filtering. We also show that linear combination of phasors allows color coding of FLIM images on the basis of the fraction of molecular species without using fitting procedures. Finally, we show that image segmentation based on algorithms exploiting the intensity information present in the FLIM image is simplified since the result of the segmentation can be seen real time in the FLIM image.

The phasor approach to fluorescence lifetime analysis was first introduced by Jameson et al. [10] in the context of single-point lifetime measurements. Recently, the phasor idea was proposed for the analysis of multiexponential decaying systems from cuvette measurements [11,12]. In 2008, Digman et al. [5] used the phasor approach to analyze FLIM data. In the Digman et al. paper, three fundamental new ideas were introduced: (1) phasor fingerprinting for the identification of molecular species, (2) linear combination of molecular species (as opposed to linear combination of exponential components), and (3) FRET analysis in FLIM using trajectories in the phasor plot. More importantly, in the work of Digman et al., it was emphasized that each molecular species has a specific location in the phasor plot and that the identification of a species in a sample does not require the resolution of the decay into exponential components.

10.2 THE PHASOR "LINEAR TRANSFORMATION"

In the FLIM analysis method, a fluorescence decay curve is collected at every pixel of an image. The specific method of data collection can be both in the time and in the frequency domain [13]. Here we will assume that data are collected in the time domain under the form of the time-correlated single-photon counting (TCSPC) decay histogram at each pixel [14]. However, the derivations and discussions of this chapter apply to data collected in the frequency domain as well. Therefore, we start from the TCSPC collection of fluorescence decay curves at each pixel of an image. In the phasor approach, the decay $I(t)$ at each pixel is transformed into two coordinates in a Cartesian plot according to the following equations:

$$s_i(\omega) = \int_0^\infty I(t)\sin(n\omega t)\,dt \bigg/ \int_0^\infty I(t)\,dt, \quad g_i(\omega) = \int_0^\infty I(t)\cos(n\omega t)\,dt \bigg/ \int_0^\infty I(t)\,dt \qquad (10.1)$$

where, $g_i(\omega)$ and $s_i(\omega)$ are the x and y coordinates of the phasor in the phasor plot, respectively; ω is the angular repetition frequency of the excitation source; and n is the harmonic frequency (typically 1 or 2).

In the frequency domain, the phases (ϕ) and the modulations (m) of the signal are measured at several harmonic frequencies. In this case, the phasor coordinates are given by

$$s_i(\omega) = m\sin(\phi), \, g_i(\omega) = m\cos(\phi) \qquad (10.2)$$

In the phasor plot, the values of $g_i(\omega)$ and $s_i(\omega)$ are thought of as coordinates of vectors with origin in the (0,0) point. Phasors are normalized so that the coordinates have no units. In the phasor plot, the horizontal axis is used for the g (or cosine) transform. The values of g are between 0 and 1. The vertical axis is for s (the sine transform), which has a value between 0 and 0.5. These phasors follow the normal vector algebra. Their coordinates can be added or subtracted.

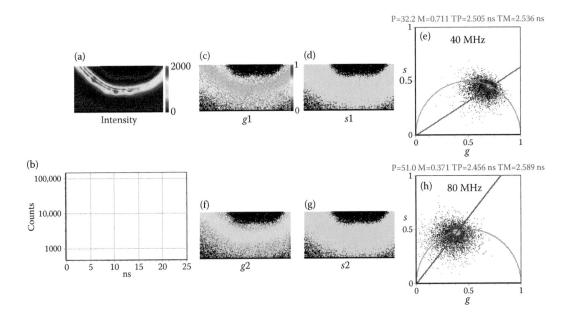

Figure 10.1 (a) Intensity image. The fluorescence is from a protein construct containing GFP. (b) Average decay (all pixels). The laser repetition frequency is 40 MHz. (c) g1 image. The color code (from 0 to 1) is shown in the side bar of this panel. (d) s1 image, same color scale as in (c). (e) Phasor histogram of the first harmonics at 40 MHz corresponding to images g1 and s1. (f) g2 image at 80 MHz, same color scale as for (c). (g) s2 image at 80 MHz, same color scale as for (c). (h) Phasor histogram of the second harmonics at 80 MHz. Data shown in this figure were collected with a Picoquant HARP300 system in the laboratory of Professor Claus Siedel, Dusseldorf, Germany. Excitation was from a picosecond laser at 488 nm, and emission was measured using a hybrid photomultiplier detector and a band-pass filter in the region 510–530 nm.

After the phasor transformation, for each pixel, we store five numbers, which are the total intensity, $g1$, $g2$, $s1$, and $s2$. The g and s are the coordinates of the phasor transform at two harmonics of the laser repetition frequency, respectively, for each pixel of the image. The total intensity at each pixel must be stored separately so that the sum implicit in the terms of Equation 10.1 can be used to combine the decay from adjacent pixels such as in binning. These five sets of numbers provide five matrices (images), which are combined to provide the phasor histogram at each harmonic, as shown in Figure 10.1. As the phasor approach is gaining popularity, images at more than two harmonics are stored, which enables better separation of molecular species when their locations in the phasor plot are very close to each other (Figure 10.1e and h).

The decay at each pixel (Figure 10.1b) is transformed using Equation 10.1 in matrices of pixels (images) corresponding to the coordinates of the phasors (Figure 10.1c and d). The phasors are then plotted in the phasor histogram, as shown in Figure 10.1e. A phasor transformation is also performed for the second harmonics of the laser repetition frequency (Figure 10.1f and g), and a second phasor plot is generated at the second selected harmonic (Figure 10.1h). In principle, we could have as many harmonics as the data

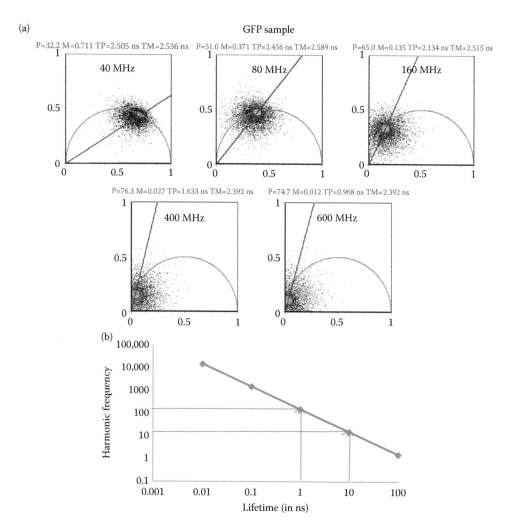

Figure 10.2 (a) For the GFP sample shown in Figure 19.1, the phasor histogram is shown for five selected harmonics from 40 MHz (fundamental) to 600 (MHz), the 15th harmonics. (b) Plot of the frequency that provides the best resolution as a function of the lifetime of the probe. For lifetime values in the range of 1–10 ns, the best frequency is in the range of 16–160 MHz.

support, that is, equal to half the number of bins used to collect the data in the TCSPC technique. In the following example, we show calculations of five harmonics up to 600 MHz. The phasor histogram does not broaden very much as the frequency is increased for the example above. As the frequency increases (from 40 to 600 MHz), the phasor histogram shifts along the universal circle toward the (0,0) coordinate. The best frequency at which we will have the maximal sensitivity to changes in the phasor location is when the phasor is located in the central region of the phasor plot (Figure 10.2, between 40 and 80 MHz) [15]. For example, for a probe with a lifetime of 2.5 ns (GFP) the best modulation frequency is about 60 MHz (Figure 10.2b).

Generally, a laser repetition rate in the range of 20–40 MHz is ideal for FLIM analysis of biological samples. Note that the optimal condition for FLIM data collection for typical biological sample is realized if both the laser pulse and the detector response are in the subnanosecond range. On the contrary, if the repetition rate is too high, the phasors of various molecular species will be crowed in a small area of the phasor plot, and their separation becomes problematic.

10.3 THE MEDIAN FILTER APPLIED TO PHASOR COMPONENT IMAGES

The phasor transformation provides "images" of the decay for the g and s components. These images can be refined using various image processing tools, providing an entirely new approach to FLIM data analysis. Note that this image refinement approach is not available in the common time-domain data analysis based on fit methods of data reduction.

In FLIM data collection, the calculation of the lifetime at one pixel has generally a relatively large error due to the limited number of photons that are collected at any one pixel [13]. For example, if at one pixel, 100 photons are collected, error propagation shows that the uncertainty in the measured lifetime, assuming that the decay is exponential, is about 20% of the value of the lifetime [13]. Since the variance (square of the error) decreases linearly with the number of photons collected, it is a common procedure in FLIM to average neighboring pixels, for example, in a 3 × 3 pixel region. This is done by summing the decay histograms in nine adjacent pixels and then assigning this average histogram to the center of this 3 × 3 pixel region. The binning procedure decreases the spatial resolution of the image, which is a problem.

Figure 10.3 shows an experiment of FLIM data collection for a solution of rhodamine 110. The average pixel count is 114, and the image has 256 × 256 pixels (Figure 10.3b). The average lifetime of rhodamine 110 is 4.0 ns, and the standard deviation of the average pixel lifetime determination is 0.648 ns (16% error). By binning pixels three by three, the standard deviation decreases to 0.216 ns (5.4% error). However, this operation cannot be repeated since the spatial resolution is also strongly degraded. The binning procedure gives pixilated images, and it becomes difficult to distinguish specific cellular features to be associated with lifetime values.

We describe here a procedure used in the phasor approach that maintains the resolution of the original image while dramatically improving the determination of the phasor in one pixel. In the phasor approach, we can follow a totally different procedure to reduce the noise of the phasor location. Since the coordinates of the phasor are additive, we treat the coordinates g_i and s_i as two images. Filtering procedures are then applied to these two images, resulting in a dramatic reduction of the variance of the position of the phasors (Figure 10.3d through f). Generally, we use a 3 × 3 convolution filter based on the median operation. The median filter does not reduce the resolution, but it tends to sharpen the borders, as will be shown later. The filter can be applied successively to decrease the uncertainty of the phasor location. The inverse of the variance of the phasor determination changes linearly with the number of filter applications (Figure 10.3c).

In Figure 10.4, we describe the operation of the median filter since this is a relatively unknown method to this field. We show in Figure 10.4 the principle of the general operation of a convolution filter. For a discussion of filters applied to images, see the tutorial at http://micro.magnet.fsu.edu/primer/java/digitalimaging/processing/medianfilter/index.html.

For the purpose of illustration, we describe a convolution with a 3 × 3 matrix, but other sizes of matrices can be used as well, such as 5 × 5 or larger matrices. The median filter assigns the median value of the nine pixels to the central pixel of the 3 × 3 region (Figure 10.4). The nine values are sorted in order of increasing value, and the central value is selected.

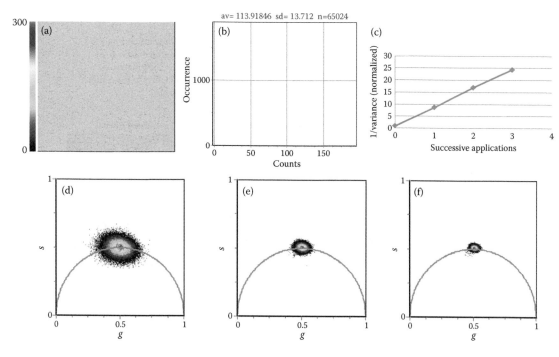

Figure 10.3 Successive applications of the median filter to the g and s images reduce the variance of the phasor location. (a) Image of a slide of a solution of rhodamine 110. (b) Histogram of pixel counts. The average is about 114 counts. (c) The inverse of the variance changes linearly with the number of filter applications. (d) Original phasor distribution. (e) One application of the median filter. (f) Three applications of the median filter. In the phasor plot, some points appear outside the main distribution after filtering. These phasors are at the image border that is not filtered in the convolution procedure.

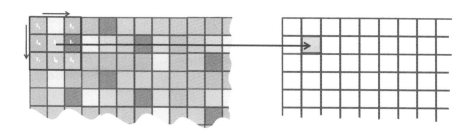

Figure 10.4 Operation of the convolution filter. An image is analyzed using a moving window of 3 × 3 pixels. In the resulting image on the right, a pixel corresponding in location to the central pixel of the 3 × 3 moving window is assigned with a value according to a logical or arithmetic operation. In the median filter, the pixel value is the median of the values of the nine pixels in the convolution window (red box). In the conventional smoothing filter, the pixel value is the arithmetic weighted mean of weights assigned to each value of the exploring matrix. By moving the window one pixel at a time both in the x and y direction, we obtain a filtered image. The border of the image is not filtered.

10.4 MAINTAINING THE SPATIAL RESOLUTION

Figure 10.5a schematically shows the operation of the median filter and comparison with the common smoothing filter. The specific weights used for the smoothing filter in this example are shown in Figure 10.5a. The image in Figure 10.5b (fluorescence from a tissue sample) shows bright fluorescent granules as well as broad structures that correspond to cells. After three applications of the median filter, the image

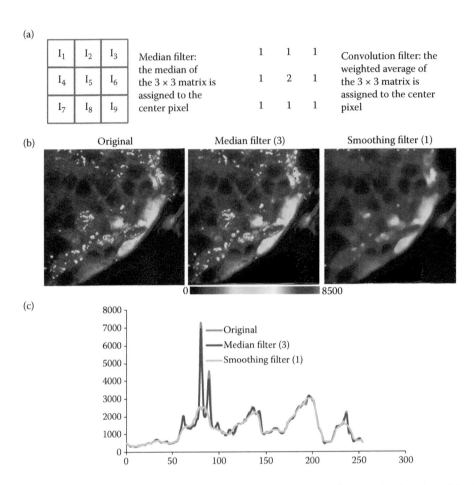

Figure 10.5 (a) Schematic representation of the 3 × 3 median and smoothing filters. A 3 × 3 region of the image is systematically explored. The value of the center pixels of the 3 × 3 pixel region is substituted with the median value (median filter) or with the weighted average of the intensity values of the 3 × 3 region each multiplied by the corresponding weighted matrix value (smoothing filter). (b) Comparison of the original image, the image after three applications of the median filter, and after one application of the smoothing filter. (c) Profile of the horizontal line 128 of the image in (b).

resolution is not affected, and the granules are still visible. Figure 10.5c shows the profile of the horizontal line 128 of the image. The original image and the image processed with the median filter have the same resolution, and the bright spots are maintained. Instead, after application of the weighted moving average filter, the image resolution is greatly degraded. The bright spots have disappeared, and only the broad features are maintained (Figure 10.5c).

10.5 FILTERING IN THE PHASOR PLOT REVEALS HIDDEN COMPONENTS

Successive application of the median filter to the g and s images sharpens the phasor plot. Features that were invisible in the raw phasor plot could become distinct after this filtering procedure. Figure 10.6 shows a FLIM measurement of a cell in which the membrane is labeled with a protein construct that contains the GFP protein (same sample as in Figure 10.1). The count histogram of the original image shows that there is a bright region of the image (the cell wall) with about 1200–1400 counts per pixel (Figure 10.6b). The average lifetime image histogram shows a complex distribution, with a relatively narrow lifetime distribution component arising from the bright regions of the image and a broad pedestal arising from the

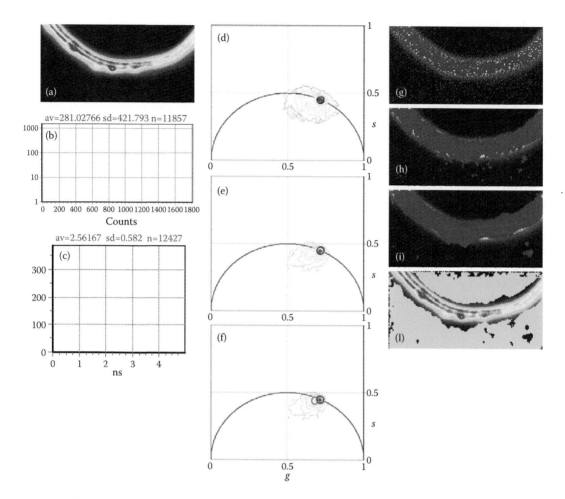

Figure 10.6 Successive applications of the median filter reveal the origin of hidden components. (a) Image of GFP construct in a cell wall. (b) The histogram of counts shows that at the cell wall, there are about 1200–1400 counts. (c) Average lifetime distribution. (d) Raw phasor distribution. (e) Phasor distribution after one application of the median filter. (f) Phasor distribution after three applications of the median filter. (g) Selecting pixels within the red cursor in the phasor plot (d); mainly selects pixel of the cell wall but also other pixels. (h) After one application of the median filter (e), the phasor distributions becomes sharper, and most of the pixels of the cell wall are selected by the small red cursor. (i) After three applications of the median filter (f), the distribution becomes even sharper. A second minor contribution is now appearing as shown in Figure 10.3f by the green circle. (j) The phasor in this region corresponds to pixels in the region of the cell away from the membrane yet populated by the GFP construct. The lifetime in this region is slightly modified by the linear combination with a background, which, in this case, has a phasor at (0,0).

dim parts of the image (Figure 10.6c). The lifetime appears to be uniform everywhere, and the different counts statistics determine the broadening of the average lifetime distribution.

The phasor histogram shows a phasor distribution centered around 2.56 ns; the phasor histogram is not symmetric but elongated toward the origin of the phasor plot (Figure 10.6d). Successive applications of the median filter show that the phasor distribution is made of two components, which can be identified with the bright and dim parts of the image, respectively (Figure 10.6f, i, and j). In the dim parts, the phasor distribution is shifted toward the (0,0) point. This indicates that the "component" that is added to the GFP phasor is unmodulated background. This component can be easily separated, its location can be identified, and its relative contribution can be quantified at each pixel.

Also, the successive application of the median filter "packs" the phasor distribution without affecting the resolution of the image, as previously shown in Figure 10.5 for the tissue sample.

10.6 LINEAR COMBINATIONS OF MOLECULAR SPECIES IN THE SAME PIXEL: QUANTITATION OF FRET

Perhaps the most emphasized property of the phasor analysis is the linear combination of molecular species. Here, we show the graphical selection of pixels that are a linear combination of two phasors. The linear combination of the phasors corresponding to the two species can be used to map their distribution in an image using the phasor plot. In this example, we use the GTPase Rac1 biosensor construct from the Hahn lab [16,17]. The Rac1 is a dual-chain biosensor that consists of the CyPet-Rac1 protein and its binding domain, YPet-PBD. Upon activation with epidermal growth factor (EGF), the two proteins come together, and FRET between the CyPet and Y-Pet fluorescent proteins can occur. COS7 cells were transfected with both donor (CyPet-Rac1) and acceptor (YPet-PBD) measured before and after EGF stimulation (donor channel). In the inactive state, the biosensor displays no FRET (the low FRET [L-FRET] state), while in the activated state, the biosensor shows high FRET (H-FRET) (Figure 10.7). In this experiment, a FLIM image (image 1 in Figure 10.7a) is measured before stimulation of the cell with epithelial growth factor (EGF). Then the cell is stimulated at frame 2, and subsequent frames were acquired every 10 s (frames 3–12). The phasor histogram in Figure 10.7b is the compound histogram from FLIM images 1–12. The fraction of FRET depends on the fractional intensity of the molecules that are in the active state. Experimentally, phasors distribute along a line joining the L-FRET and the H-FRET states (Figure 10.7b).

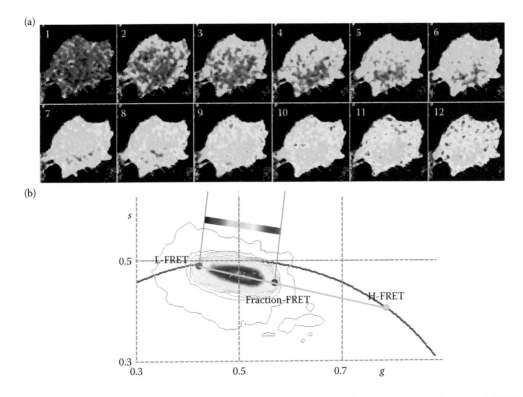

Figure 10.7 (a) Sequence of FLIM images colored according to the scale of fraction of molecules in the L-FRET and H-FRET state obtained in (b). Image 1 is collected before activation with EGF. From 2 to 12, images were recorded every 10 s. (b) Compound phasor histogram of all images 1–12 of (a). The phasor histogram is elongated from L-FRET to H-FRET. Extrapolation of the points intercepting the universal circle of the best line through the histogram gives the L-FRET (blue point) and H-FRET (green point) state of the Rac1 biosensor. The maximal fraction H-FRET in a pixel (red point) is about 45%. The FLIM image is color coded according to the scale from L-FRET to the maximal fraction of FRET.

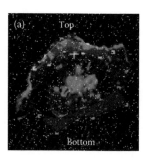 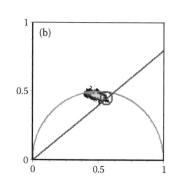 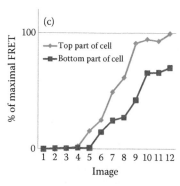

Figure 10.8 (a) Cell measured in Figure 19.7, analyzed for pixels reaching maximal FRET in the top and bottom part of the cell. (b) Pixels in the red circle are considered as pixels with maximal FRET. (c) Fraction of pixels reaching maximal FRET in the FLIM image sequence of Figure 10.7.

Note that the experimental phasor distribution is along a line rather than along the universal circle. This implies that each pixel in the image contains a combination of L-FRET and H-FRET molecules. For this biosensor, there is no gradual change in FRET efficiency that would have given phasors on the universal circle. Instead, the intercept of the green line in Figure 10.7b with the universal circle gives the value of the H-FRET state (extrapolated lifetime of 1.10 ns), which, in this case, corresponds to a FRET efficiency of 54% (L-FRET state lifetime is 2.40 ns). The maximum fraction measured is about 45% (red point at fraction of FRET in Figure 10.7b).

Since the phasor distribution is along the line of linear combination of the L-FRET and H-FRET states, we color coded pixels in the image according to the color scale shown in Figure 10.7b for the FRET fraction range measured in this experiment. Note that there are no assumptions about the two FRET states or their locations. All points (red, blue, and green in Figure 10.7b) are directly derived from the phasor plot using graphical constructions. We can unequivocally derive the value of the maximal FRET efficiency of the biosensor and the fractional intensity contribution of H-FRET molecules in each pixel. To obtain the fraction of molecules (concentration), we must account for the relative quantum yield of the two states, which we assume are proportional to their lifetimes (the extrapolated points on the universal circle for this example). The formula for converting fractional intensities in concentration ratios is given in the work of Celli et al. [18].

In the example of the Rac biosensor, the FLIM images also provide the distribution of the active biosensor as a function of time. If we spatially section the image according to the regions outlined in Figure 10.8a in red, we can count the number of pixels that are in the active state by plotting their corresponding phasor distribution and selecting the maximal H-FRET state (red circle in Figure 10.8b) for the image sequence (1–12) in Figure 10.7a.

Pixels with maximal activation of the Rac biosensor start at frame 4–5 after stimulation with EGF and reach saturation at about frame 9–10. Note that in the bottom part of the cell, there are fewer pixels with maximal activation.

10.7 EXPLOITING THE LINEAR PROPERTY TO SEGMENT IMAGES

The linear property of the phasor representations is crucial for the separation and assignment of phasors to molecular species. In the following example, we show how to recognize features in the FLIM image characterized by different phasor clusters. Figure 10.9 shows a portion of tissue from a seminiferous mouse tube where stem cells express the Oct4-eGFP genetic fluorescent marker. In the intensity image (Figure 10.9a), we distinguish some broad features that could be identified with cells, a background where it is difficult to recognize any specific feature, and bright puncti covering various parts of the image. In

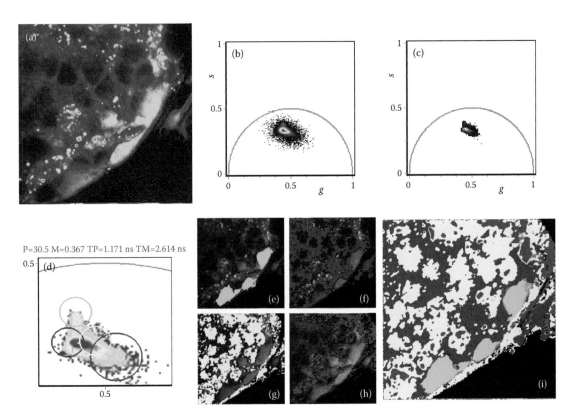

Figure 10.9 (a) Image of an excised seminiferous tube tissue of a mouse with stem cells expressing the GFP-Oct4 gene using 900 nm excitation in the two-photon microscope. (b) Phasor histogram, showing a broad distribution. (c) After application of the median filter, the phasor histogram reveals clusters that can be identified with specific features in the tissue. (d) Zoomed-in region of the phasor plot in which four regions are selected. (e–h) The color-coded region selected in (d). (i) Compound image of the color-coded regions according to the selection in part (d).

Figure 10.9b we show the phasor plot of this image at 80 MHz obtained with a two-photon excitation at 900 nm. As previously noted, the cluster of phasors is rather broad, and it is impossible to assign specific phasors to each feature in the image. At this level of representation of the cluster of phasors, we need to unscramble the cluster to determine the contributions of the various molecular species to the overall compound phasor distribution. In Figure 10.9c, after the application of the median filter, we can distinguish several features in the phasor plot that we would like to identify with features in the image. In Figure 10.9d, we zoom in the phasor plot, and we select four different regions indicated by the four colored circles. The regions of the image selected on the basis of the clusters in the phasor plot are shown in Figure 10.9e through h, using the same color code in Figure 10.9d. Now we can clearly distinguish the stem cells that express the GFP in the Oct4 complex (Figure 10.9e), at least two cellular or extracellular compartments colored in yellow and pink, and another component at shorter lifetime in blue that could arise from blood vessels. The color-coded image in Figure 10.9i is the superposition of Figure 10.9e through h where the precedence of color (in case the selected regions superimpose) is from green to pink to yellow to blue, with the blue having the lower precedence.

If we try identifying the cellular features colored in pink and in yellow, the intensity image shows that the pink-colored pixels are located at the cell borders in the tissues, while the yellow pixels are located in the cell interior. For this experiment, we collected FLIM images using two excitation wavelengths in the two-photon microscope, at 790 and 900 nm, respectively. Figure 10.10 shows four different regions of the seminiferous tube of the same mouse. In Figure 10.10a through d, the excitation wavelength is 900 nm,

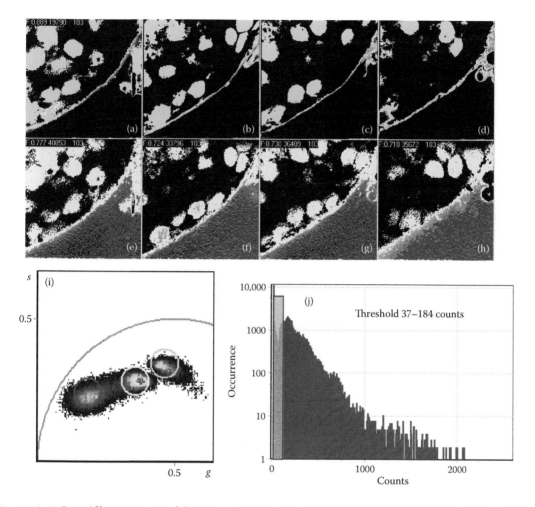

Figure 10.10 Four different regions of the seminiferous tube of the same mouse tissue shown in Figure 10.9. (a–d) Excitation at 900 nm, (e–h) Excitation at 790 nm. (i) Compound phasor plot of the eight images (at both wavelengths) with the applied threshold. (j) Only pixels within a given value of intensity (low intensity) are selected, shown in the green rectangle.

and in Figure 10.10e through f, the excitation is at 790 nm. The compound phasor plot of the images is shown in Figure 10.10i. To better select the dim parts of the tissues corresponding to cells, we apply a threshold based on the intensity, as shown in the green bar in Figure 10.10j. The median filter was applied to these images. Clearly, both excitation wavelengths excite the same cells, but the emission must be from different molecular species since the phasor position is quite different when we select the same cells. Note that the stem cells and the bright spots have been excluded for the analysis of this image since they are bright.

In order to select the bright regions, we must plot the phasors corresponding to a higher threshold. In Figure 10.11 we select only the part of the image with high intensity. Images in Figure 10.11a through d are obtained exciting at 900 nm, and images in Figure 10.10e through h are excited at 790 nm. Figure 10.11i shows the compound phasor plot. This time, the cluster selected by the yellow circle in Figure 10.10i shows that the bright spots have the same phasor location at the two excitation wavelengths. By applying a threshold in regions of low and high intensity, we can identify and map different molecular species in great detail. The linear property of the phasors allows us to separate and navigate the phasor plot at ease.

Analysis of fluorescence lifetime data

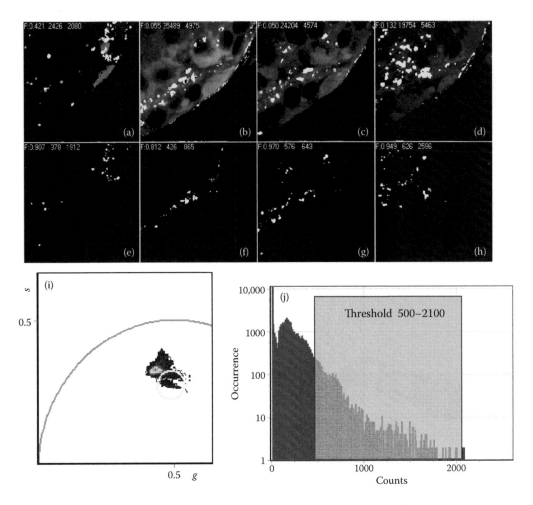

Figure 10.11 Four different regions of the seminiferous tube of the same mouse. (a–d) Excitation at 900 nm. (e–h) Excitation at 790 nm. (i) Compound phasor plot of the eight images. (j) Only pixels within a given value of intensity (high intensity) are selected.

10.8 CELL PHASOR ANALYSIS

The cell phasor analysis is another segmentation approach that maps phasor contribution of individual cells and shows how neighboring cells in a tissue have diverse metabolic states [19]. In Figure 10.12, we show the cell phasor segmentation analysis of the seminiferous tube. In this example, we segmented the image manually by tracing the contour of a cell by thresholding the intensity as to select only the fluorescent cells. We then calculate the average phasor of the selected region yielding one point on the phasor plot (noted by a star in Figure 10.11c). We can see that each cell has a slightly different position in the phasor plot. We note that the error of each cell phasor is very small since each region contains thousands of phasors. The position between cells becomes significant. Three of the cells (cells 2, 4, and 6) form a cluster in the direction of the GFP phasor. These cells will be identified as stem cells expressing the GFP-Oct4 genes, the true stem cells. The remaining cells (cells 1, 3, and 5) are bright, emitting in the green, but have a lifetime that is incompatible with the expression of GFP. On the basis of this segmentation, we were able to distinguish undifferentiated stem cells from differentiated stem cells [19].

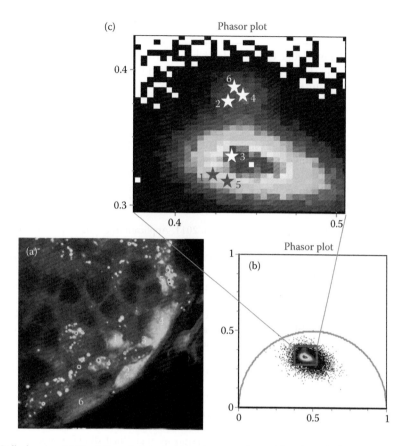

Figure 10.12 Cell phasors. (a) Pixels in cells labeled 1 to 6 were selected for the calculation of the average value of the phasor in the cell region. (b) The phasor plot shows a superposition of many pixels. (c) Each cell has a different average phasor. Larger cells (2, 4, and 6) are the stem cells expressing the GFP-Oct4 gene. Cells 1, 3, and 5 are smaller cells.

10.9 CONCLUSIONS

The linear property of the phasor transformation is a powerful feature that allows a series of simple tools to be used to explore the phasor space. First, the most important point in this chapter on phasor analysis is the description of the filtering tool (image smoothing) that greatly decreases the noise in the phasor spread. This decrease in spread, based on the vector addition law, allows better identification of pixels with similar decay and better identification of linear combination of species. As a consequence of this law, if only two species are present in a pixel, all phasors must be contained in a narrow region of the phasor plot along the line joining the phasors of the pure species. If more (than two) species are present, we can use phasor plots at different harmonic frequencies to better separate their contributions. Since phasors add in the compound phasor histogram of many images, we can select portions of the phasor histogram based on the intensity image histogram. This property allows separation of apparently superimposing constitutions to the phasor plot.

All data analyses shown in this work have been performed using the SimFCS program that is available at the Laboratory for Fluorescence Dynamics website (http://www.lfd.uci.edu/). A number of tutorials that explain the use of this software also are available at this website.

Analysis of fluorescence lifetime data

ACKNOWLEDGMENTS

The authors would like to acknowledge Claus Seidel, Chiara Stringari, and Elizabeth Hinde for providing some of the FLIM data used in this work. This work is supported in part by NIH-P41 P41-RRO3155, NIH P50-GM076516, and U54 GM064346.

REFERENCES

1. Colyer, R., O. Siegmund, A. Tremsin, J. Vallerga, S. Weiss, and X. Michalet. 2009. Phasor-based single-molecule fluorescence lifetime imaging using a wide-field photon-counting detector. In *Single Molecule Spectroscopy and Imaging II,* edited by Jörg Enderlein, Zygmunt K. Gryczynski, Rainer Erdmann. *Proc. of SPIE* Vol. 7185, 71850T.

2. Grecco, H. E., P. Roda-Navarro, A. Girod, J. Hou, T. Frahm, D. C. Truxius, R. Pepperkok, A. Squire, and P. I. Bastiaens. 2010. In situ analysis of tyrosine phosphorylation networks by FLIM on cell arrays. *Nat Methods* 7:467–472.

3. James, N. G., J. A. Ross, M. Stefl, and D. M. Jameson. 2011. Applications of phasor plots to in vitro protein studies. *Anal Biochem* 410:70–76.

4. Sanchez, S., L. Bakas, E. Gratton, and V. Herlax. 2011. Alpha hemolysin induces an increase of erythrocytes calcium: A FLIM 2-photon phasor analysis approach. *PloS One* 6:e21127.

5. Digman, M. A., V. R. Caiolfa, M. Zamai, and E. Gratton. 2008. The phasor approach to fluorescence lifetime imaging analysis. *Biophys J* 94:L14–L16.

6. Chen, Y. C., and R. M. Clegg. 2009. Fluorescence lifetime-resolved imaging. *Photosynth Res* 102:143–155.

7. Matsubara, S., Y. C. Chen, R. Caliandro, Govindjee, and R. M. Clegg. 2011. Photosystem II fluorescence lifetime imaging in avocado leaves: Contributions of the lutein-epoxide and violaxanthin cycles to fluorescence quenching. *J Photochem Photobiol* 104:271–284.

8. Spring, B. Q., and R. M. Clegg. 2009. Image analysis for denoising full-field frequency-domain fluorescence lifetime images. *J Microsc* 235:221–237.

9. Behne, M. J., S. Sanchez, N. P. Barry, N. Kirschner, W. Meyer, T. M. Mauro, I. Moll, and E. Gratton. 2011. Major translocation of calcium upon epidermal barrier insult: Imaging and quantification via FLIM/Fourier vector analysis. *Arch Dermatol Res* 303:103–115.

10. Jameson, D. M., E. Gratton, and R. Hall. 1984. The measurement and analysis of heterogeneous emissions by multifrequency phase and modulation fluorometry. *App Spec Rev* 20:55–106.

11. Redford, G. I., and R. M. Clegg. 2005. Polar plot representation for frequency-domain analysis of fluorescence lifetimes. *J Fluoresc* 15:805–815.

12. Clayton, A. H., Q. S. Hanley, and P. J. Verveer. 2004. Graphical representation and multicomponent analysis of single-frequency fluorescence lifetime imaging microscopy data. *J Microsc* 213:1–5.

13. Gratton, E., S. Breusegem, J. Sutin, Q. Ruan, and N. Barry. 2003. Fluorescence lifetime imaging for the two-photon microscope: Time-domain and frequency-domain methods. *J Biomed Opt* 8:381–390.

14. Becker, W., A. Bergmann, M. A. Hink, K. Konig, K. Benndorf, and C. Biskup. 2004. Fluorescence lifetime imaging by time-correlated single-photon counting. *Microsc Res Tech* 63:58–66.

15. Lakowicz, J. R., G. Laczko, H. Cherek, E. Gratton, and M. Limkeman. 1984. Analysis of fluorescence decay kinetics from variable-frequency phase shift and modulation data. *Biophys J* 46:463–477.

16. Hodgson, L., O. Pertz, and K. M. Hahn. 2008. Design and optimization of genetically encoded fluorescent biosensors: GTPase biosensors. *Methods Cell Biol* 85:63–81.

17. Hodgson, L., F. Shen, and K. Hahn. 2010. Biosensors for characterizing the dynamics of rho family GTPases in living cells. *Curr Protoc Cell Biol* Chapter 14 (Unit 14. 11):1–26.

18. Celli, A., S. Sanchez, M. Behne, T. Hazlett, E. Gratton, and T. Mauro. 2010. The epidermal Ca(2+) gradient: Measurement using the phasor representation of fluorescent lifetime imaging. *Biophys J* 98:911–921.

19. Stringari, C., A. Cinquin, O. Cinquin, M. A. Digman, P. J. Donovan, and E. Gratton. 2011. Phasor approach to fluorescence lifetime microscopy distinguishes different metabolic states of germ cells in a live tissue. *Proc Natl Acad Sci USA* 108:13582–13587.

Analysis of time-domain fluorescence measurements using least-squares deconvolution

11

Jing Liu, Daniel S. Elson, and Laura Marcu

Contents

11.1 INTRODUCTION

A main challenge in the analysis of fluorescence lifetime measurements is the ability to deconvolve the experimental data to obtain results that are free of distortions introduced by the instrument used in the experiment. The purpose of deconvolution is to free the fluorescence emission decay transients from the main sources of distortion. For time-domain fluorescence measurements, this includes distortions due to the finite rise time, width, and decay of the laser excitation system; distortions introduced by the limited frequency response of detector electronics; and distortions introduced by the light dispersion in optical fiber when fiber-optic probes are used for exciting and collecting the fluorescence emission. This chapter describes the use of the least-squares method as means of deconvolving the intrinsic fluorescence impulse response function (fIRF) from the measured fluorescence pulse transients.

Mathematically, the measured fluorescence emission as a function of time, $y(t)$ (that is, fluorescence decay transient) is a *convolution* of tissue intrinsic fIRF, $h(t)$, and instrumental response, $I(t)$ (Oppenheim et al. 1999). The time-domain measurements consist of sampling the fluorescence decay profile at discrete times. In the discrete-time representation, for N equally spaced sampling time points, $t_i = i\delta t$, $i = 0,\ldots,N-1$, and sampling time interval δt, the measured fluorescence decay is given by the following equation:

$$y(k) = \sum_{i=0}^{k} I(k-i) \cdot h(k) + \varepsilon_k \qquad (11.1)$$

for $k = 0,\ldots,N-1$, where $I(k)$ and $h(k)$ are the discrete-time representation of instrumental response and fIRFs, respectively. ε_k denotes the additive measurement noise. The *deconvolution* of fluorescence signals consists of the recovery of intrinsic fIRF, $h(t)$, from the noise-corrupted measurements and the measured instrument response.

Central to any deconvolution technique is the choice of models for which the underlying fIRF is parameterized. Two categories of parameterizations were generaly used in defining the fIRF models: (1) *Parametric* models, which rely on the *a priori* knowledge of the characteristics of the fluorescent system under study, for example, the number of fluorescent constituents and the decay characteristics of each fluorophore. In this case, model parameters that are often related to physical quantities provide the parameterization of fIRFs. Perhaps the most widely used parametric fIRF model is the summation of multiple exponential decay components. Specifically, the *a priori* assumptions are that the fluorescent system consists of a discrete number of fluorophores, each with a specific number of decay components, and the decay dynamics are governed by the first-order differential equation. When a continuous distribution of fluorophores is assumed, stretched-exponential (SE) models for fIRFs are often adopted (Lee et al. 2001); and (2) *Nonparametric* models do not make any *a priori* assumption of the physical behavior of fluorescent systems in response to optical excitation. Rather, these models are often parameterized solely for the purpose of mathematical convenience. For example, the fIRF can be expanded onto a set of orthogonal basis functions and parameterized by the resulting set of expansion coefficients. Several commonly used basis sets are Fourier series (Andre et al. 1979), Laplace series (Gafni et al. 1975), exponential series (Ware et al. 1973), and Laguerre series (Maarek et al. 2000).

The parametric models are most appropriate if they describe the physics of underlying fluorescent systems accurately. In fact, the prevalent use of multiple exponential fIRF models is due to the fact that many physical phenomena follow first-order differential equations (Istratov and Vyvenko 1999), especially when considering identification of known fluorescent species in more controllable situations such as in cell imaging. However, for complex systems, such as biological tissues, the number and type of fluorophores are rarely known *a priori*. It is often hard to justify parametric models from a physics perspective. Except for a few limited situations, parameters associated with parametric models cannot be interpreted in terms of underlying fluorophore content. On the contrary, *nonparametric* models are capable of representing fIRFs without *a priori* assumption of the the number of fluorescent constituents and their underlying decay function characteristics. By carefully choosing the expansion basis set, nonparametric models are often mathematically more easy and numerically more efficient to evaluate. However, without constraints from underlying physical principles, nonparametric models are often sensitive to measurement noise and may result in decay functions (i.e., fIRFs) that are obviously *unphysical* in the presence of high noise levels. For example, it is well known that Fourier series expansion may induce oscillation behavior (bumps) in fIRFs that do not match well with the smooth intensity decay function as expected (Andre et al. 1979). We emphasize that the distinction between parametric and nonparametric models is purely based on whether the models are parameterized with a physical model in mind. Usually, the decay dynamics from any given physical system are not entirely unknown. In particular, the fIRFs are considered as functions that smoothly decay to zero at long-enough time delays. Recently, a *semiparametric* approach has been developed (Liu et al. 2012) to provide an efficient way of combining *a priori* knowledge into nonparametric models. To the extent to which they are controlled by additional constraints, such semiparametric fIRF models can be considered a trade-off between parametric and nonparametric models.

A number of deconvolution methods have been developed in accordance with different fIRF models. In principle, Fourier transform and Laplace transform approaches are natural choices for nonparametric fIRF models. These integral transform deconvolution techniques are also suitable for parametric models, providing that the analytical form for the corresponding integral transforms can be easily computed. The method of moments (Isenberg et al. 1973), expectation–maximization (Ng et al. 2009), Prony method (Zhang et al. 1996), and method of modulating functions (Valeur 1978) have been adapted specific to parametric fIRF models in the form of multiple exponential summations. Over the past few decades, least-squares deconvolution (LSD) methods (or more generally maximum likelihood methods) have become popular for analyzing time-domain fluorescence signals. LSD is not limited to the functional form of fIRF models and has been commonly adopted when models (either parametric or nonparametric) other than multiple component exponentials are of interest (Lee et al. 2001; Murata et al. 1995; Ware et al. 1973). In addition, LSD has been shown to be reliable and robust to measurement noise in a variety of simulation and experimental settings (McKinnon et al. 1977; O'Connor et al. 1979). Generally, LSD searches for the fIRF that best fits the observed data after the fIRF has been reconvolved with the instrument response.

In this chapter, we will focus on time-domain deconvolution techniques based on LSD. First, we will present the mathematical concept underlying the LSD method. Second, we will demostrate LSD for two cases of parametric fIRF models, that is, multiexponential and SE models and a nonparametric fIRF model, that is, Laguerre series expansion. A variant to the LSD with nonparametric fIRF models (constrained LSD) that incorporates *a priori* information on the shape of fIRFs (or a semiparametric model) is also described. Finally, we show the performance of these three approaches in deconvolving time-domain fluorescence signals from both time-resolved fluorescence lifetime spectroscopy (TRFS) and fluorescence lifetime imaging (FLIM) with *ex vivo* tissue data.

11.2 LEAST-SQUARES DECONVOLUTION METHOD

In general, intrinsic fIRFs parameterized by an L-dimensional real vector $\mathbf{c} \in \mathbb{R}^L$, in a discrete-time representation, can be formally expressed as a column vector $\mathbf{h}(\mathbf{c}) = [h(0;\mathbf{c}),\ldots,h(N-1;\mathbf{c})]^T$. By defining a measurement data vector, $\mathbf{y} = [y(0),\ldots,y(N-1)]^T$, and the Toeplitz convolution matrix,

$$\mathbf{T} = \begin{pmatrix} I(0) & 0 & 0 & \cdots & 0 \\ I(1) & I(0) & 0 & \cdots & 0 \\ I(2) & I(1) & I(0) & \cdots & 0 \\ \vdots & \vdots & \vdots & \ddots & \vdots \\ I(N-1) & I(N-2) & I(N-3) & \cdots & I(0) \end{pmatrix} \tag{11.2}$$

the discrete-time convolution (Equation 11.1) can be recast into a matrix-vector form,

$$\mathbf{y} = \mathbf{Th}(\mathbf{c}) + \boldsymbol{\varepsilon} \tag{11.3}$$

where $\boldsymbol{\varepsilon} = [\varepsilon_0,\ldots,\varepsilon_{N-1}]^T$ is multivariate random noise, with zero mean and covariance matrix $\sum_{\varepsilon} = \boldsymbol{\varepsilon}\boldsymbol{\varepsilon}^T$. Most generally, LSD searches for an estimate of $\mathbf{c}$ that minimizes the *generalized* sum of squares error (SSE),

$$\Omega(\mathbf{c}) = \frac{1}{2}(\mathbf{y} - \mathbf{Th}(\mathbf{c}))^T \sum_{\varepsilon}^{-1} (\mathbf{y} - \mathbf{Th}(\mathbf{c})) \tag{11.4}$$

It should be noted that representation of the discrete-time data vector by a vector-valued deterministic function plus an additive error (Equation 11.3) is general enough to describe fluorescence measurements from almost all practical applications. In fact, the measurement noise may (1) depend on data and model parameters, that is, $\sum_{\varepsilon} = \sum_{\varepsilon}(\mathbf{y},\mathbf{c})$, or (2) have unequal variance and autocovariance structures, that is, $\sum_{\varepsilon}$ has nonconstant diagonal elements and nonzero off-diagonal elements. In practice, the random nature of noise is intrinsic to most measurement configurations. Additional assumptions on the structure of the noise covariance matrix will significantly reduce the complexity of subsequent numerical analysis. When measurement noise is mainly due to instrumental electronic noise (e.g., in time-domain signals from pulse sampling), it is often assumed that the noise is independent and that there are identically distributed random variables with equal variance, such that $\sum_{\varepsilon}$ is diagonal with constant diagonal entries. On the other hand, in time-correlated single photon counting (TCSPC) measurements, noise is often assumed to be independent and to follow a Poisson probability distribution due to the counting nature of the TCSPC method. In the case of a large number of photon counts at each sampling time, t_k [i.e., $y(k)$ is large], the variance of Poisson noise (at t_k) can be directly estimated by the measured $y(k)$, such that $\sum_\varepsilon = \text{diag}\{\mathbf{y}\}$.

11.2.1 LINEAR LSD

Minimization of (Equation 11.4) is achieved by setting its first derivative with respect to the parameters to zero, that is,

$$J(c) = \frac{\partial \Omega}{\partial \mathbf{c}} = \left(\frac{\partial \Omega}{\partial c_1}, \frac{\partial \Omega}{\partial c_2}, \ldots, \frac{\partial \Omega}{\partial c_L} \right)^T = 0 \tag{11.5}$$

Generally, the first derivative of a multivariate function, that is, $\mathbf{J}$, is also known as the *Jacobian*.

If the model for the fIRF, $\mathbf{h(c)}$, is linear in its parameters, that is,

$$\mathbf{h(c)} = \mathbf{Bc},$$

where $\mathbf{B}$ is the system matrix specific to the chosen model (to be discussed later), the Jacobian is also linear in $\mathbf{c}$, such that Equation 11.5 becomes

$$\mathbf{B}^T \mathbf{T}^T \sum_\varepsilon^{-1} (\mathbf{y} - \mathbf{TBc}) = 0,$$

which has a closed-form solution to $\mathbf{c}$,

$$\mathbf{c} = \left(\mathbf{B}^T \mathbf{T}^T \sum_\varepsilon^{-1} \mathbf{TB} \right)^{-1} \mathbf{B}^T \mathbf{T}^T \sum_\varepsilon^{-1} \mathbf{y}$$

11.2.2 NONLINEAR LSD

In the case that models for fIRFs are nonlinear in their parameters, the Jacobian matrix is no longer linear in model parameters. A common strategy for solving Equation 11.5 is to use Newton-type iterative methods (Nocedal and Wright 2006). The basic idea behind Newton's method is to break the nonlinear estimation Equation 11.5 into a sequence of linear equations and solve them iteratively. More specifically, let the estimated parameter at the mth iteration to be $\mathbf{c}^{(m)}$. A Taylor expansion of the Jacobian matrix around $\mathbf{c}^{(m)}$ gives

$$J(c) \approx J(c^{(m)}) + \frac{\partial J(c^{(m)})}{\partial \mathbf{c}} (\mathbf{c} - \mathbf{c}^{(m)}) \tag{11.6}$$

The derivative of the Jacobian, also known as a *Hessian* matrix, is the second derivative of the generalized SSE (Equation 11.4), that is,

$$\frac{\partial J}{\partial \mathbf{c}} = \begin{pmatrix} \dfrac{\partial^2 \Omega}{\partial c_1^2} & \dfrac{\partial^2 \Omega}{\partial c_1 \partial c_2} & \cdots & \dfrac{\partial^2 \Omega}{\partial c_1 \partial c_L} \\ \dfrac{\partial^2 \Omega}{\partial c_2 \partial c_1} & \dfrac{\partial^2 \Omega}{\partial c_2^2} & \cdots & \dfrac{\partial^2 \Omega}{\partial c_2 \partial c_L} \\ \vdots & \vdots & \ddots & \vdots \\ \dfrac{\partial^2 \Omega}{\partial c_L \partial c_1} & \dfrac{\partial^2 \Omega}{\partial c_L \partial c_2} & \cdots & \dfrac{\partial^2 \Omega}{\partial c_L^2} \end{pmatrix} \tag{11.7}$$

To simplify notation, let $\mathbf{J}^{(m)} = \mathbf{J}(\mathbf{c}^{(m)})$ and $\mathbf{H}^{(m)} = \dfrac{\partial \mathbf{J}(\mathbf{c}^{(m)})}{\partial \mathbf{c}}$. By keeping only the linear term in Equation 11.6, the Jacobian function near $\mathbf{c}^{(m)}$ is approximated by its tangential hyperplane. The updated estimate of $\mathbf{c}$ is found by solving a linear equation,

$$\mathbf{J}^{(m)} + \mathbf{H}^{(m)}(\mathbf{c} - \mathbf{c}^{(m)}) = 0, \tag{11.8}$$

which gives the estimated parameter for the $(m+1)$th iteration,

$$\mathbf{c}^{(m+1)} = \mathbf{c}^{(m)} - (\mathbf{H}^{(m)})^{-1} \mathbf{J}^{(m)} \tag{11.9}$$

Formally, the Jacobian and Hessian matrices can be found as

$$\mathbf{J}^{(m)} = -\left(\frac{\partial \mathbf{h}(\mathbf{c}^{(m)})}{\partial \mathbf{c}}\right)^{T} \mathbf{T}^{T} \mathbf{S}_{e}^{-1}(\mathbf{y} - \mathbf{Th}(\mathbf{c}^{(m)})) \tag{11.10}$$

$$\mathbf{H}^{(m)} = -\left((\mathbf{y} - \mathbf{Th}(\mathbf{c}^{(m)}))^{T} \sum_{\varepsilon}^{-1} \mathbf{T} \otimes \mathbf{I}\right)\left(\frac{\partial^{2} \mathbf{h}(\mathbf{c}^{(m)})}{\partial \mathbf{c} \partial \mathbf{c}'}\right)^{T}$$
$$-\left(\frac{\partial \mathbf{h}(\mathbf{c}^{(m)})}{\partial \mathbf{c}}\right)^{T} \mathbf{T}^{T} \sum_{\varepsilon}^{-1} \mathbf{T}\left(\frac{\partial \mathbf{h}(\mathbf{c}^{(m)})}{\partial \mathbf{c}}\right), \tag{11.11}$$

where $\mathbf{I}$ denotes an $L \times L$ identity matrix and $\otimes$ denotes the matrix Kronecker product. Here, the first and second derivative of the fIRFs with respect to model parameters should be understood as

$$\frac{\partial \mathbf{h}(\mathbf{c})}{\partial \mathbf{c}} = \begin{pmatrix} \dfrac{\partial \mathbf{h}(t_1; \mathbf{c})}{\partial c_1} & \dfrac{\partial \mathbf{h}(t_1; \mathbf{c})}{\partial c_2} & \cdots & \dfrac{\partial \mathbf{h}(t_1; \mathbf{c})}{\partial c_L} \\[2ex] \dfrac{\partial \mathbf{h}(t_2; \mathbf{c})}{\partial c_1} & \dfrac{\partial \mathbf{h}(t_2; \mathbf{c})}{\partial c_2} & \cdots & \dfrac{\partial \mathbf{h}(t_2; \mathbf{c})}{\partial c_L} \\[2ex] \cdots & \cdots & \ddots & \vdots \\[2ex] \dfrac{\partial \mathbf{h}(t_N; \mathbf{c})}{\partial c_1} & \dfrac{\partial \mathbf{h}(t_N; \mathbf{c})}{\partial c_2} & \cdots & \dfrac{\partial \mathbf{h}(t_N; \mathbf{c})}{\partial c_L} \end{pmatrix} \tag{11.12}$$

and

$$\frac{\partial^{2} \mathbf{h}(\mathbf{c})}{\partial \mathbf{c} \partial \mathbf{c}'} = \left(\frac{\partial^{2} \mathbf{h}(\mathbf{c})}{\partial \mathbf{c} \partial c_1}, \ldots, \frac{\partial^{2} \mathbf{h}(\mathbf{c})}{\partial \mathbf{c} \partial c_L}\right) \tag{11.13}$$

with

$$\frac{\partial^{2} \mathbf{h}(\mathbf{c})}{\partial \mathbf{c} \partial c_i} = \begin{pmatrix} \dfrac{\partial^{2} \mathbf{h}(t_1; \mathbf{c})}{\partial c_1 \partial c_i} & \dfrac{\partial^{2} \mathbf{h}(t_1; \mathbf{c})}{\partial c_2 \partial c_i} & \cdots & \dfrac{\partial^{2} \mathbf{h}(t_1; \mathbf{c})}{\partial c_L \partial c_i} \\[2ex] \dfrac{\partial^{2} \mathbf{h}(t_2; \mathbf{c})}{\partial c_1 \partial c_i} & \dfrac{\partial^{2} \mathbf{h}(t_2; \mathbf{c})}{\partial c_2 \partial c_i} & \cdots & \dfrac{\partial^{2} \mathbf{h}(t_2; \mathbf{c})}{\partial c_L \partial c_i} \\[2ex] \cdots & \cdots & \ddots & \vdots \\[2ex] \dfrac{\partial^{2} \mathbf{h}(t_N; \mathbf{c})}{\partial c_1 \partial c_i} & \dfrac{\partial^{2} \mathbf{h}(t_N; \mathbf{c})}{\partial c_2 \partial c_i} & \cdots & \dfrac{\partial^{2} \mathbf{h}(t_N; \mathbf{c})}{\partial c_L \partial c_i} \end{pmatrix} \tag{11.14}$$

for $i = 1, \ldots, L$.

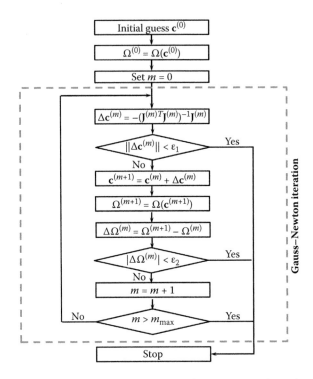

Figure 11.1 Gauss–Newton method in nonlinear LSD. ε_1, ε_2, and m_{max} are preselected constants defining the stopping criteria for the iteration.

It should be noted that iterative searches for model parameters using Newton's method with Equation 11.9 involves computing second-order derivatives of the objective function (i.e., the Hessian matrix) and its inverse. Unless the analytical forms are easily available, direct calculation of the Hessian and its inverse are often avoided in practice. Alternatively, various approximations to the Hessian matrix (and its inverse) have been proposed. Specific to the least-squares minimization as in LSD, the *Gauss–Newton* method and its variants are commonly used, where the Hessian matrix is approximated as

$$\mathbf{H} \approx \mathbf{J}^T\mathbf{J} \tag{11.15}$$

such that the updated Equation 11.9 is

$$\mathbf{c}^{(m+1)} = \mathbf{c}^{(m)} - \left(\mathbf{J}^{(m)T}\mathbf{J}^{(m)}\right)^{-1}\mathbf{J}^{(m)} \tag{11.16}$$

The Gauss–Newton method is illustrated in Figure 11.1.

11.2.3 CONSTRAINED LSD

In many practical applications, it is necessary to restrict the model parameters (i.e., $\mathbf{c}$) within a subset of $\mathbb{R}^L$. For example, for multiexponential fIRF models, all parameters are necessarily positive—it is not realistic to have negative lifetime and fractional contribution values. For nonparametric fIRF models, model parameters should guarantee the decay functions $\mathbf{h}(\mathbf{c})$ are smoothly and strictly decaying. However, such properties cannot be specified in the framework of unconstrained LSD, although this may not be a problem for deconvolution of signals with high signal-to-noise ratio. In general, one may explicitly search the solution to the LSD subject to various forms of constraints without loss of generality, for example,

$$\begin{aligned} &\text{minimize} \quad \Omega(\mathbf{c}) \\ &\text{subject to} \quad g_i(\mathbf{c}) \geq 0 \quad \text{for } i = 1,\ldots,l. \end{aligned} \tag{11.17}$$

where $\Omega(\mathbf{c})$ is defined in Equation 11.4 and $g_i(\mathbf{c})$ specifies the ith functional form of the constraint on parameters. It should be noted that constraints are often linear in parameters, such that $g_i(\mathbf{c}) = \mathbf{a}_i^T \mathbf{c}$. The solution to the constrained LSD is out of the scope of this chapter, and we refer the interested reader to a few reference books (Boyd and Vandenberghe 2004; Nocedal and Wright 2006).

11.3 MODELS OF FLUORESCENCE DECAY

The intrinsic fIRF of a fluorescent system, $h(t)$, is the rate of photon emission as a function of time after the system is excited by an impulse optical stimulus. This is closely related to the time evolution of electron relaxation from excited states to ground states. In this section, we describe parametric and nonparametric models that are commonly used for representing fIRFs.

11.3.1 MULTIPLE EXPONENTIAL EXPANSION MODELS

The relaxation of excited electrons is often assumed to follow a first-order differential equation. For a simple fluorescent system with one fluorophore, the decay rate of electrons in excited state is proportional to the number of electrons in excited states, such that

$$\frac{dN(t)}{dt} = -\lambda N(t), \tag{11.18}$$

where λ is a time constant characterizing the decay rate of the relaxation process. The solution to Equation 11.18 is given by an exponential function,

$$N(t) = N_0 \exp(-\lambda t) \tag{11.19}$$

Since photon emission intensity is proportional to the electronic decay rate, fIRF is given by

$$h(t) = A \exp\left(-\frac{t}{\tau}\right), \tag{11.20}$$

where $\tau = 1/\lambda$, which is often referred to as fluorescence *lifetime*. In this equation, A is the initial emission intensity at time zero, or the *amplitude*. This is the conventional exponential model for fIRFs from single-species systems. If multiple fluorophores are simultaneously contributing to the fluorescence signal, by assuming a noninteracting environment, fIRFs are often expressed as a summation of exponential decay terms,

$$h(t) = \sum_{i=1}^{M} A_i \exp\left(-\frac{t}{\tau_i}\right), \tag{11.21}$$

where M is the total number of fluorophores. A_i and τ_i are amplitude and lifetime for the ith fluorophore. The parameter for a multiple exponential expansion fIRF model, $h(t;\mathbf{c})$ with M exponential terms, is $\mathbf{c} = [\{A_i, \tau_i, i = 1,\ldots,M\}]^T$. Note that in case of multiple exponential fIRF models, the Jacobian matrix and Hessian matrix are easily computed.

Since model parameters completely specify the functional form of the fIRF, in practice, only amplitude and lifetime values from individual components are reported instead of the fIRF itself. The *averaged lifetime*, which summarizes the information within fIRF, is given by

$$\tau_{avg} = \frac{\displaystyle\int_{t=0}^{\infty} t h(t)\, dt}{\displaystyle\int_{t=0}^{\infty} h(t)\, dt} = \frac{\displaystyle\sum_{i=1}^{M} A_i \tau_i^2}{\displaystyle\sum_{i=1}^{M} A_i \tau_i} \tag{11.22}$$

In addition, by rearranging Equation 11.22, we have

$$
\tau_{\text{avg}} = \sum_{i=1}^{M} \left(\frac{A_i \tau_i}{\sum\limits_{i=1}^{M} A_i \tau_i} \right) \cdot \tau_i = \sum_{i=1}^{M} f_i \tau_i, \tag{11.23}
$$

where $f_i = A_i \tau_i \Big/ \sum\limits_{i=1}^{M} A_i \tau_i$ is also known as *fractional contribution* from the ith fluorophore.

In fact, Equation 11.21 defines a class of fIRF models with different numbers of exponential terms. Except for limited cases where the number of noninteracting fluorophores is known *a priori*, it is required to select models with appropriate M. While models with fewer terms may not fit to data sufficiently, models with more terms may increase the risk of fitting-to-noise and computation burden. The Bayesian information criterion (BIC) has been used for selecting M "objectively" (Ng et al. 2009). Other *information-theoretic* metrics, such as minimum description length (MDL) and Akaike information criterion (AIC), can also be used (Akaike 1974; Ljung 1999; Rissanen 1978). However, model selection using information-theoretic metrics introduces another level of computation complexity, and using different criteria may result in different models. Moreover, models selected in this way cannot be justified from a physics perspective, that is, parameters associated with each exponential component cannot be interpreted in terms of underlying fluorophore content (James and Ware 1985). In practice, it has been observed that fIRFs from many fluorescent systems (e.g., biological tissues) are sufficiently represented by a two-term exponential model, for which the parameters can be attributed to a "fast" component and a "slow" component.

11.3.2 STRETCHED-EXPONENTIAL MODELS

The multiple exponential fIRF models fundamentally assume a discrete number of fluorophores in the system. However, there are many situations in which one cannot expect a finite number of fluorescent species with discrete lifetime values; for example, a fluorophore in a mixture of solvents may be subject to various relaxation processes, which may result in fluorescence exponential decays with a range of lifetime values (Lakowicz 2006). In such cases, it is natural to model the component-wise fluorescence amplitudes (i.e., A in Equation 11.21) by a distribution function in terms of lifetime values. Two common choices for such distribution functions are the *Gaussian* function

$$
A(\tau; c) = \frac{A_0}{\sqrt{2\pi}\sigma} \exp\left(\frac{(\tau - \bar{\tau})^2}{2\sigma^2} \right) \tag{11.24}
$$

and the *Cauchy* function

$$
A(\tau; c) = \frac{A_0}{\pi} \frac{\sigma}{(\tau - \bar{\tau})^2 + \sigma^2} \tag{11.25}
$$

By analog to the discrete multiple exponential model (Equation 11.21), models for fIRFs with continuous distribution of lifetime values can be expressed by

$$
h(t; c) = \int_{\tau=0}^{\infty} A(\tau; c) \exp\left(-\frac{t}{\tau} \right) d\tau \tag{11.26}
$$

It should be noted that, in cases where the amplitude distributions are modeled by either Gaussian or Cauchy functions, the fIRFs are parameterized by $c = (A_0, \bar{\tau}, \sigma)^T$. Here, A_0 measures the peak value of the fluorescence intensity. $\bar{\tau}$ and σ quantify the central values and the width of the distributions, respectively.

However, the practical uses of these two distribution functions are limited due to the following reasons. Firstly, the choice of distribution functions is often arbitrary in practice. Secondly, neither function provides a closed form of fIRF function, the LSD of which requires numerical evaluation of integrals. More importantly, both distributions (as functions of τ) are symmetric around their corresponding central values with no protection against negative lifetime values.

A seemingly complicated distribution that is only defined over positive values (therefore realistic for fluorescence lifetimes) does give closed functional forms of fIRF models. This distribution is given by

$$A(\tau; c) = \frac{A_0 \tau_0}{\pi \tau} \sum_{k=0}^{\infty} \frac{(-1)^k}{k!} \sin(\pi \beta k) \Gamma(\beta k + 1) \left(\frac{\tau}{\tau_0} \right)^{\beta k},$$

(11.27)

where $\Gamma(\cdot)$ is the Gamma function and $0 \leq \beta \leq 1$. By inserting Equation 11.27 into 11.26, it can be shown that

$$h(t; c) = \int_{\tau=0}^{\infty} -\frac{A_0 \tau_0}{\pi \tau} \sum_{k=0}^{\infty} \frac{(-1)^k}{k!} \sin(\pi \beta k) \Gamma(\beta k + 1) \left(\frac{\tau}{\tau_0} \right)^{\beta k} \exp\left(-\frac{t}{\tau} \right) d\tau$$

$$= A_0 \exp\left(-\left(\frac{t}{\tau_0} \right)^{\beta} \right)$$

(11.28)

The proof is nontrivial, and we refer the interested reader to derivations in the work of Berberan-Santos et al. (2005) and Lindsey and Patterson (1980). Following the definition for averaged lifetime values in Equation 11.22, we have

$$\tau_{\text{avg}} = \frac{\int_{t=0}^{\infty} t \exp\left(-(t/\tau)^{\beta} \right) dt}{\int_{t=0}^{\infty} \exp\left(-(t/\tau)^{\beta} \right) dt} = \tau_0 \frac{\Gamma(2/\beta)}{\Gamma(1/\beta)}$$

(11.29)

The fIRF model in the form of Equation 11.28 is also known as the SE (or *Kohlrausch–Williams–Watts*) model for fIRF. The SE model is completely specified by parameter $c = (A_0, \tau_0, \beta)^T$. Note that the width of the lifetime distribution is indicated by the parameter β, and for $\beta = 1$, we recover the single exponential decay function. It should be clear that the underlying *a priori* assumption for an SE fIRF model is that there exist a range of exponential decays, whose amplitude distribution function (ADF) is governed by the distribution function (Equation 11.27).

11.3.3 LAGUERRE EXPANSION MODELS

As noted above, the nonparametric fIRF models do not require an *a priori* assumption of underlying physical processes. The fIRFs are typicaly expanded onto a set of basis functions, preferably *ordered* orthonormal functions, and are parameterized by the corresponding expansion coefficients. In this section, we present one of such orthonormal basis sets consisting of Laguerre basis functions (LBFs), for which the $(l + 1)$th element can be defined as in following equation (Abramowitz and Stegun 1973):

$$b_l(t) = p^{1/2} \exp\left(-\frac{pt}{2} \right) L_l(pt)$$

(11.30)

where $l = 0, \ldots, L - 1$ is the order of the LBF, $p > 0$ is a scaling factor for continuous-time LBFs, and

$$L_l(t) = \frac{\exp(t)}{l!} \frac{d^l}{dt^l}\left(\exp(-t)t^l\right) \tag{11.31}$$

is the lth order Laguerre polynomial. The orthonormality of LBFs follows directly from the orthonormality of Laguerre polynomials. LBFs of the form in Equation 11.30 are also known as *continuous-time* LBFs. In terms of LBFs, fIRFs are formally given by

$$h(t;\mathbf{c}) = \sum_{l=0}^{L-1} c_l b_l(t) \tag{11.32}$$

with model parameters $\mathbf{c} = (c_0,\ldots,c_{L-1})^T$. Note that fIRFs are linear in parameters for the given Laguerre basis set.

11.3.3.1 Discrete-time representation

By sampling at a discrete-time grid (i.e., $\{t_i = i\delta t, i = 0,\ldots,N-1\}$), continuous-time LBFs (Equation 11.30) can be directly used as an expansion basis for representating discrete-time time-domain signals. It should be noted that the mutual orthonormality between "sampled" continuous-time LBFs is only approximately satisfied due to numerical errors. In practice, the *discrete-time* version of LBFs can be generated directly using a recursive relation (Marmarelis 1993) in order to preserve their mutual orthonormality,

$$b_0(k) = \sqrt{\alpha}\,b_0(k-1) + \sqrt{(1-\alpha)}\,\delta_{k0}$$
$$b_l(k) = \sqrt{\alpha}\,b_l(k-1) + \sqrt{a}\,b_{l-1}(k) - b_{l-1}(k-1) \tag{11.33}$$

with $b_l(-1) = 0$. Formally, the discrete-time LBFs have the form

$$b_l(k;\alpha) = \alpha^{\frac{k-l}{2}}(1-\alpha)^{\frac{1}{2}}\sum_{i=0}^{l}(-1)^i \binom{k}{i}\binom{l}{i}\alpha^{l-i}(1-\alpha)^i \tag{11.34}$$

It can be shown that continuous-time LBFs are, in fact, the limit of Equation 11.34 by letting $\alpha = \exp(-p\delta t)$ and $\delta t \to 0$.

Note that LBFs are argumented by a *scale* parameter (α in discreate case, or p in continuous case) that controls the overall decay rate of the set of LFBs. In principle, if the fIRF is expanded on an infinite number of LBFs ($L \to \infty$) over infinite long time intervals, the choice of scale parameter is arbitrary. In practice, only a finite number of expansion terms (i.e., L is finite) can be used, and the measured time series are always truncated for numerical analysis. Thus, appropriate selection of a Laguerre basis set is critical for practical use of the Laguerre expansion (LE) of fIRFs. The optimal selection of both parameters is out of the scope of this chapter. Interested readers are referred to the method described in Liu et al. (2012).

11.3.3.2 Constraint on $h(t;\mathbf{c})$

In general, the fIRFs of physical fluorescent systems, $h(t;\mathbf{c})$, are considered "decay" functions that smoothly decay to zero at long-enough time delay following the excitation pulse. However, such a property is not guaranteed by the form of the fIRF model (Equation 11.32). In fact, at least two properties of LBFs may give rise to unphysical behavior of the fluorescence decay functions, especially when higher-order LBFs are used in the expansion, firstly, the built-in oscillatory terms in LBFs, and secondly, the fact that all of LBFs contribute to fIRF values near time zero. Therefore, additional constraints on the functional form of $h(t;\mathbf{c})$

are necessary such that it is positive, monotonically decreases, is strictly convex, and asymptotically goes to zero. Mathematically (Liu et al. 2012), this is equivalent to the conditions that

1. $\lim_{t \to \infty} h(t; \mathbf{c}) = 0$ and
2. $h(t; \mathbf{c}) > 0$, $h'(t; \mathbf{c}) > 0$ and $h''(t; \mathbf{c}) > 0$.

It should also be noted that the second derivative of $h(t; \mathbf{c})$ is required to be strictly positive so that there are no "flat" (i.e., zero curvature) line segments on a decay profile. Although these conditions are not able to capture all features of a smooth "decay" function, they allow us to limit our search for the fIRF to a smaller subspace of functions that are more physically realistic. For practical use, simplied conditions have been derived such that

1. $\lim_{t \to \infty} h'(t; \mathbf{c}) = 0$, $\lim_{t \to \infty} h''(t; \mathbf{c}) = 0$, $\lim_{t \to \infty} h'''(t; \mathbf{c}) = 0$ and
2. $h'''(t; \mathbf{c}) \le 0$.

Note that for parametric models of fIRF such as multiexponential and SE, the conditions 1 and 2 are automatically satisfied. For $h(t; \mathbf{c})$ expanded on the set of LBFs (Equation 11.32), its asymptotic behavior is determined by the expansion basis functions. Following the definition of LBFs in Equation 11.30, it is easy to verify that all orders of derivatives of LBFs go to zero for $t \ge 0$. Thus, in principle, condition 1 is also satisfied for LE of $h(t; \mathbf{c})$. In practice, time-resolved measurements for evaluating the fluorescence decays are always truncated to a finite range. It is important to choose an appropriate Laguerre basis set such that all its basis functions and corresponding derivatives decay "sufficiently close" to zero at the end of the measured time series. In other words, condition 1 is imposed implicitly in choosing the expansion basis set. In addition, condition 2 can be formulated in a constrained optimization problem (Equation 11.17), in which $g(\mathbf{c}) = -h'''(t; \mathbf{c})$. A detailed discussion of choosing LBFs and the implementation of constrained LSD can be found in Liu et al. (2012).

11.4 GOODNESS OF FIT

Quantitative assessment of the deconvolution quality in practice is often based on the analysis of residuals (i.e., unfitted components) between the measured time-domain signals and the fitted values. Specifically, the *fitted* values of time-resolved signals in the time domain are the reconvolution of the estimated fIRF, $h(t; \mathbf{c})$, and the instrumental response,

$$\hat{y}(k) = \sum_{i=0}^{k} I(k-i)\hat{h}(i) \tag{11.35}$$

and the *residuals* are given by

$$r(k) = y(k) - \hat{y}(k) \tag{11.36}$$

In practice, we consider a deconvolution "good enough" if the fitting residuals follow closely the underlying noise model and exhibit no autocorrelation structures. Perhaps the most widely used diagnostic statistic is based on χ^2 goodness of fit. Ideally, the residuals should not contain any unfitted component other than random noise. A direct measure of the unfitted component against noise is based on the *reduced* χ^2 goodness-of-fit statistic (Bevington and Robinson 2003),

$$\chi_\nu^2 = \frac{1}{(N-L)} \sum_{k=0}^{N-1} \frac{r(k)^2}{\sigma_k^2}, \tag{11.37}$$

where N is the total number of time sampling points, L is the number of fitting parameters, and σ_k^2 is the variance of noise. Note that χ_ν^2 estimated from residuals is a random variable. For an ideal fit, χ_ν^2 is close to 1, reflecting sample variance of residuals is comparable to noise. More rigorously, if a good estimate of noise variance is available, under the assumption of the normal distribution noise model, the random variable χ_ν^2 follows $\chi^2(\nu)$ distribution normalized to the value of $N - L$ (also known as degrees of freedom).

Ljung–Box test statistic. The independence of residuals is tested using the Ljung–Box test statistic, which is given by (Brockwell and Davis 2002)

$$T_{LB} = N(N+2) \sum_{k=1}^{K} \frac{\rho(k)^2}{N-k}, \tag{11.38}$$

where $\rho(k)$ is the sample autocorrelation at time lag t_k, defined as

$$\rho(k) = \frac{\sum_{i=0}^{N-1-k} [r(i+k) - \bar{r}][r(i) - \bar{r}]}{\sum_{i=0}^{N-1} [r(i) - \bar{r}]^2}$$

with $\bar{r} = \sum_{i=0}^{N-1} r(i)/N$. The Ljung–Box test measures the distance from zero of autocorrelation functions averaged over a number of time lags. Under the null hypothesis that residuals are independently distributed, T_{LB} follows a χ^2 distribution with degree of freedom (d.o.f.) K. Detailed analysis of the Ljung–Box summary statistics is out of the scope of this chapter. Such analysis can be found in the work of Tsay (2005). It should be noted that although formal summary statistics provide an efficient way of detecting anomalies within deconvolution residuals (i.e., unfitted components other than measurement noise), no real measurement noise strictly follows an independent normal distribution. Non-normality and correlation can either be caused by correlated random noise, which may be due to limited bandwidth of detector frequency responses, or be intrinsic to signal sampling schemes, such as those based on TCSPC techniques. Therefore, deconvolution residuals should be appropriately weighted (e.g., in case of Poisson distribution of measurement noise) or down-sampled (e.g., in case sampling frequency is higher than maximum detector frequency response) before being tested against the hypotheses of normality and independence.

11.5 COMPARISON OF fIRF MODELS

In this section, we demonstrate the use of LSD with fIRF models based on (1) biexponential (BE), (2) SE and (3) LE using computer-simulated data and experimental data from *ex vivo* tissue samples. The fIRF models are compared in terms of their goodness-of-fit statistics and the accuracy in estimating average lifetimes for a variety of simulation and experiment settings.

11.5.1 COMPARISON BASED ON SIMULATED DATA

11.5.1.1 fIRF with discrete lifetime components

In simulation, the underlying fIRF models are known, and the data are generated from the convolution of fIRF models with the instrumental response (i.e., Equation 11.1 or 11.3). The additive random noise level was set to be 40 dB (Gaussian white noise, but Poissonian white noise could also be implemented). The simulated data set consisted of randomly generated signals using multiexponential fIRF models with M exponential components. For the mth component, a random fractional contribution (uniformly distributed between 0% and 100%) and a random lifetime value (uniformly distributed between $\frac{10(m-1)}{M} + 1$ and $\frac{10m}{M} + 1$ ns) were assigned, such that the lifetime values from all M individual components had sufficient coverage over the range 1–11 ns. This simulation was repeated for $M = 2, \ldots, 8$, and for each M, 1000 simulated decays were generated. It is expected that the simulated signals should mimic, as closely as possible, the data generated from biological tissues, where the number of components, component-wise fractional contribution, and lifetime values are unknown.

The performance of the deconvolution based on the three fIRF models noted above is compared here in terms of their χ^2 and *Ljung–Box* test statistics. Specifically, for each M, all 1000 signals were deconvolved with the three fIRF models, and both test statistics were computed from the resulting residuals. Figure 11.2a shows the proportions (out of 1000 residuals) that failed the test of autocorrelation (i.e., *Ljung–Box* test) at a significance level of .05, for data sets simulated with different numbers of exponential components. These are also the proportions of signals that were not sufficiently fitted according to the *Ljung–Box* test.

As expected for nonparametric models, LSD-LE performed equally well regardless of the "ground truth" models; the proportion that were rejected was close to 5% for all M values. LSD-BE (fIRF with two exponential components) was able to fit signals that were generated from fIRFs with more than two components, although the rejected proportion increased with the increasing number of exponential components as a result of model mis-specification. In the worst case with $M = 8$, about 12% of the signals were not adequately fitted using a BE model. However, the fact that over 80% of the signals with multiexponential components can be deconvolved solely based on the BE model is strong evidence that it is not always possible to differentiate the number of components in a fluorescence system purely based on the number of exponential terms used. The SE model was shown to be extremely restrictive. For M values greater than 3, over 70% of the deconvolutions were rejected, and for $M = 2$, about 55% were rejected. This suggests that the SE fIRF model was not appropriate for signals generated from exponential fIRFs with a discrete number of components. Admittedly, there were cases where the SE model could be used, especially when a long lifetime component dominated other decay components. However, in the presence

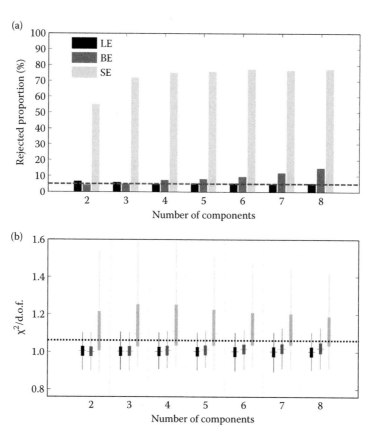

Figure 11.2 Assessment of deconvolution qualities using LE, BE, and SE models for data sets simulated from multiexponential fIRFs. (a) Comparison of rejected proportions from Ljung–Box test. The deconvolution is not adequate (i.e., *rejected*) if the probability of getting a deconvolution residual that is more extreme than the resulting deconvolution residual is less than 0.05. (b) Distributions of χ^2 goodness-of-fit values. Dashed line indicates the critical value at a significance level of 0.05.

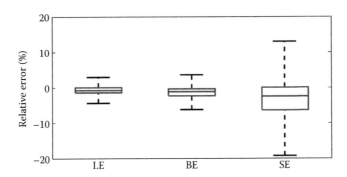

Figure 11.3 Distribution of relative errors in average lifetime values from LE, BE, and SE models. Data were simulated from multiexponential fIRF models with one to eight components.

of fluorescence systems with unknown fluorescence decay properties, such as biological tissues, the SE model could result in high risk for deconvolution errors.

These observations were confirmed in Figure 11.2b, where the distributions of χ_v^2 values computed from deconvolution residuals using three fIRF models were shown. For an ideal fit, the χ_v^2 values should follow a χ_v^2 distribution (normalized to the d.o.f.) with an expected value of 1. For both LSD-BE and LSD-LE, most χ_v^2 values fell below the critical value (0.05 level), suggesting a good fit. On the contrary, the majority of χ_v^2 values from LSD-SE deconvolution were above the critical value, which also suggests that the SE model was not adequate where multiple discreet lifetime values are expected.

In the context of TRFS and FLIM, of particular interest is the average lifetime. Thus, it is also interesting to compare the relative error in estimating average lifetime values from these three fIRF models. Figure 11.3 summarizes the distribution of relative errors for all the above simulated data using LSD-LE, LSD-BE, and LSD-SE models. As expected from the poor deconvolution quality, average lifetime values estimated from LSD-SE resulted in errors greater than 18% (absolute values). On the other hand, the estimation error of average lifetime from LSD-LE and LSD-BE were within the range of –5% to +5% of the ground truth.

11.5.1.2 fIRF with continuous lifetime components

It has been observed that LSD using an SE fIRF model is appropriate when one expects a distributions of relaxation times. In this second simulation study, we considered signals generated from an underlying fIRF consisting of components with a range of lifetime values and distributed amplitudes. In particular, the ADFs were chosen to be beta functions (Abramowitz and Stegun 1973). We will not go into details of the functional form of beta functions but only note that for certain parameter values, a beta function is unimodal and only has nonzero values over a closed interval. One thousand random ADFs were generated over the lifetime range of 1–10 ns. Two examples of such ADFs are shown in Figure 11.4a, and the corresponding signals generated with these two ADFs are given in Figure 11.4b. The additive noise level was set to be 40 dB.

Simulated data were deconvolved with LSD based on LE, BE, and SE fIRF models. In addition, a fIRF model with a monoexponential (ME) component was also used. Figure 11.5a shows the proportions of deconvolution residuals that failed the *Ljung–Box* test on a significance level of 0.05. The distributions of corresponding χ_v^2 values are given in Figure 115b. The relative errors in estimating average lifetime values from different fIRF models are compared in Figure 11.6. The ME fIRF model was found not appropriate for deconvolution of systems with distributed lifetime values. In fact, over 80% of simulated signals were not sufficiently deconvolved with the ME fIRF model, according to both test statistics. As expected, the performance of LSD-SE was improved in the case of distributed lifetime systems; less than 10% deconvolution residuals were signaled to be abnormal according to the *Ljung–Box* test, and majority of χ_v^2 values fell below the critical value. It should be noted that for fluorescence systems with distributed lifetime values, both LSD-LE and LSD-BE were able to provide sufficient deconvolution. This also confirms the fact that it is often impossible to identify the distribution of fluorophores and the underlying physical processes purely based on the fIRF models used.

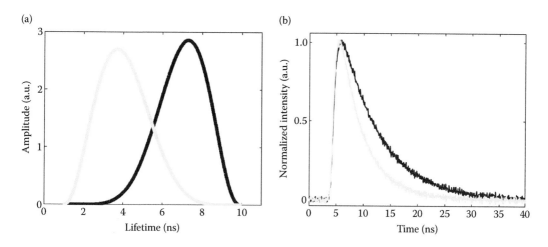

Figure 11.4 (a) Examples of amplitude distribution functions. (b) Simulated signals using fIRFs with distributed amplitude and lifetime components specified in (a).

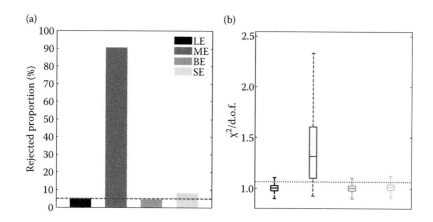

Figure 11.5 Assessment of deconvolution qualities using LE, ME, BE, and SE models for data sets simulated from fIRF with continuous lifetime components. (a) Comparison of rejected proportions from Ljung–Box test. Dashed line indicates 5% level. (b) Distributions of χ^2_v goodness-of-fit values. Dashed line indicates the critical value at a significance level of 0.05.

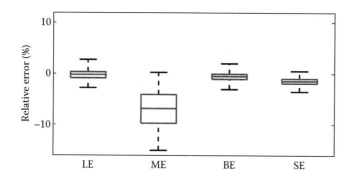

Figure 11.6 Distribution of relative errors in average lifetime values from LE, ME, BE, and SE models. Data were simulated from fIRF models with continuous lifetime components.

11.5.2 COMPARISON BASED ON EXPERIMENTAL DATA

We consider the deconvolution of TRFS and FLIM measurement of *ex vivo* human atherosclerotic plaques using data acquired with a system similar to that described in Chapters 4 and 19.

11.5.2.1 Deconvolution of TRFS data

The TRFS apparatus measures the sample fluorescence pulse transients (decay) over a broad range of emission wavelengths. Figure 11.7 shows the average lifetime values over the 360–550 nm wavelength range estimated from LSD with four fIRF models, that is, LE, multiexponential with two and

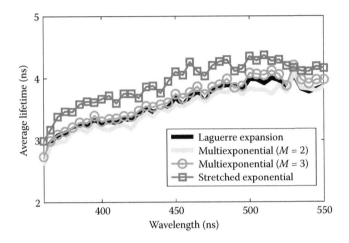

Figure 11.7 Average lifetime spectrum (i.e., average lifetime over wavelength range 360–550 nm) from LSD with fIRF models of Laguerre expansion, multiexponential with two components, multiexponential with three components, and stretched exponential.

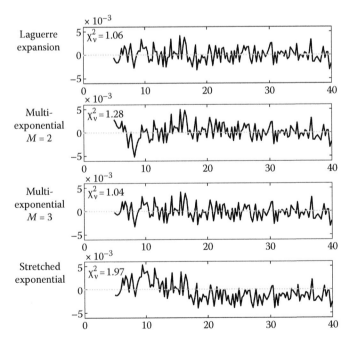

Figure 11.8 Deconvolution residuals of time-domain measurement at wavelength 455 nm, using LSD with fIRF models of Laguerre expansion, multiexponential with two components, multiexponential with three components, and stretched exponential (from top to bottom).

three components, and SE. The estimated average lifetime within the wavelength range was found to be almost identical for LE and three-component multiexponential fIRF models. When compared with Laguerre and triexponential models, the estimates from the two-component exponential model presented slightly lower values, while the average lifetime values from the SE model were found to be higher.

Therefore, it is critical to choose the appropriate average lifetime values over wavelengths for which the fIRF models provided adequate deconvolution for the measured data. The deconvolution residuals for time-domain signals measured at 455 nm with the above four fIRF models and their corresponding χ_ν^2 statistics are given in Figure 11.8. It is clear that the measured signal was accurately deconvolved using both the LE model (nonparametric) and the three-component multiexponential fIRF model (parametric). In contrast, there were apparent unfitted components when the two-component multiexponential and SE models were used. The χ_ν^2 and Ljung–Box test statistics were computed for deconvolution residuals over all wavelengths (Figure 11.9). For LE and three-component exponential models, both test statistics fell below their corresponding critical values, indicating a good deconvolution over the entire wavelength range. However, both test statistics indictate an inaccurate deconvolution for the two-component multiexponential and SE models, especially for wavelengths below 500 nm. It should be noted that while both test statistics were consistent in detecting an inaccurate deconvolution, the χ_ν^2 relies on a good estimate of the noise variance (i.e., σ_k^2 in Equation 11.37). Here, the noise variance was estimated using deconvolution residuals from the nonparametric model, LE. In the case that noise variance is hard to estimate, χ_ν^2 may provide misleading results, and thus, the Ljung–Box test statistic is recommended.

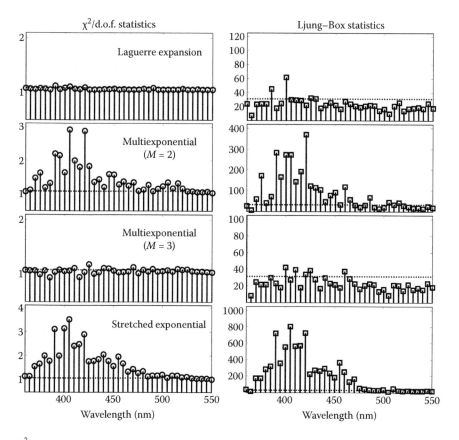

Figure 11.9 χ_ν^2 statistics (left panel) and Ljung–Box statistics (right panel) over wavelengths, calculated from deconvolution residuals using LSD with fIRF models of Laguerre expansion, multiexponential with two components, multiexponential with three components, and stretched exponential (from top to bottom). Dashed lines are the critical value at a significance level of 0.05.

Analysis of fluorescence lifetime data

11.5.2.2 Deconvolution of FLIM data

The FLIM apparatus measures the fluorescence emission (often at a limited number of wavelengths) to generate a spectroscopic image of a region of interests (ROI). Figure 11.10a shows the region of a FLIM scan measurement on the surface of the *ex vivo* atherosclerotic sample. Fluorescence transients measured at all spatial locations were then deconvolved using LSD with fIRF models of LE, multiexponential with two and three components, and stretched exponential. Respective average lifetime distributions over the measured region are shown in Figure 11.10c through f. LSD with LE and three-component multiexponential fIRF models gave similar estimated average lifetime values. The estimates from the two-component exponential model were slightly lower, and the average lifetime values from the SE model were higher than those estimated from Laguerre and three-component multiexponential models. It is noted that these results are quite similar to the deconvolution of TRFS (Section 5.2.1).

The performance of LSD with all fIRF models was compared based on Ljung–Box test on the fitting (i.e., deconvolution) residuals. For the deconvolution residual of the fluorescence signal at each spatial location, we computed the test statistic and corresponding P value. The P values were mapped on to the same ROI for each model (Figure 11.11) to show regions where the deconvolution is insufficient (that is, $P < .05$). A completely randomly spatial distribution of P values from the LE model (Figure 11.11a) and three-component exponential model (Figure 11.11c) indicates that the measured data over the ROI were adequately deconvolved. Note that 5% of P values should be allowed to be less than .05 purely by chance. On the other hand, LSD with a two-component exponential (Figure 11.11b) and an SE model (Figure 11.11d) showed connected regions where P values were less than .05, indicating inadequate deconvolution of fluorescence transients within those regions. It is noted that these (rejected) regions corresponded to those with high average lifetime values (in Figure 11.10).

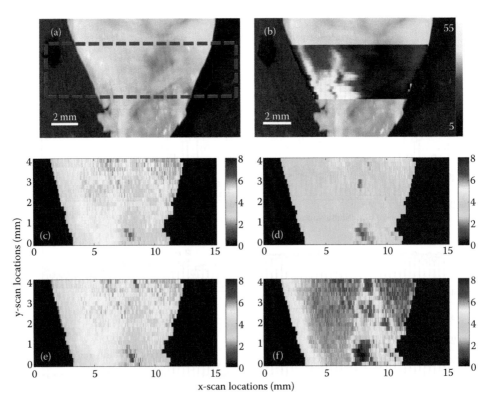

Figure 11.10 FLIM measurement on an *ex vivo* human cardiovascular plaque at 450 nm emission wavelength. (a) FLIM line-scan region of interests (red dashed box). (b) Integrated fluorescence intensity arbitrary unit (a.u.) overlayed on the tissue sample. (c–f) Average lifetime images (nanoseconds) from LSD with fIRF models of Laguerre expansion, multiexponential with two components, multiexponential with three components, and stretched exponential, respectively.

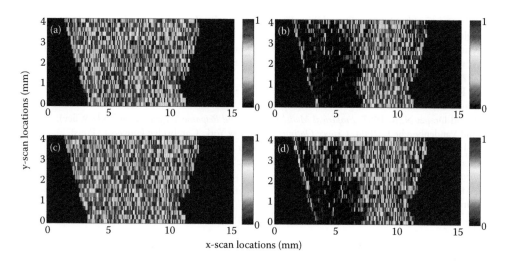

Figure 11.11 (a–d) P-value maps of Ljung–Box test on residuals from LSD with fIRF models of Laguerre expansion, multiexponential with two components, multiexponential with three components, and stretched exponential, respectively.

11.6 CONCLUSION

In this chapter, we demonstrated the use of LSD for recovering intrinsic fluorescence IRFs from time-resolved fluorescence measurements. In essence, LSD searches for the fIRF, within the model space, that has the minimal least-squares distance to the observed data after the fIRF has been reconvolved with the instrumental response. Both parametric and nonparametric fIRF models were implemented and compared.

Parametric models rely on *a priori* assumptions of the underlying constituents and decay dynamics of fluorescent systems. While multiple exponential fIRF models assume that the fluorescence decay consists of a discrete number of decay dynamics, the stretched exponential model assumes a distribution of such decays. It was found that parametric models are most appropriate if they describe the physics of underlying fluorescent systems accurately. In particular, the SE model could only be used when a distribution of fluorescence lifetimes is expected. Interestingly, fIRFs from some fluorescent systems (with discrete or continuous in lifetime values) are well represented by a two-term exponential model, for which the parameters can be attributed to a "fast" component and a "slow" component. However, for more complex fluorescent systems (e.g., biological tissue), additional exponential terms are often needed for the adequate representation of fIRFs.

Nonparametric models are capable of representing fIRFs regardless of the fluorescent system physics. Due to their built-in decay characteristics and orthonrmality, LBFs provide an efficient (nonparametric) representation for fIRFs from compelx fluorescence systems. However, without constraints from underlying physical principles, nonparametric models are often sensitive to measurement noise and may result in decay functions that are obviously *unphysical* in the presence of high noise levels. Additional constraints on fIRFs could be easily incorporated in the constrained LSD framework.

We conclude this chapter by quoting Dr. George Box (Box and Draper 1987): "Essentially, all models are wrong, but some are useful." Given the fact that measurements are sufficiently fitted (or deconvolved), there is no more information that can be extracted from measured data for differentiation between models. The choice of model is then a trade-off between numerical efficiency and physical interpretation and is a matter of convenience.

REFERENCES

Abramowitz M. and Stegun I. A. 1973. *Handbook of Mathematical Functions: With Formulas, Graphs and Mathematical Tables* (New York: Dover Publications).

Akaike H. 1974. New look at statistical-model identification. *IEEE Transactions on Automatic Control* **AC19** 716–23.

Andre J. C., Vincent L. M., Oconnor D. and Ware W. R. 1979. Applications of fast fourier-transform to deconvolution in single photon-counting. *Journal of Physical Chemistry* **83** 2285–94.

Berberan-Santos M. N., Bodunov E. N. and Valeur B. 2005. Mathematical functions for the analysis of luminescence decays with underlying distributions 1. Kohlrausch decay function (stretched exponential). *Chemical Physics* **315** 171–82.

Bevington P. R. and Robinson D. K. 2003. *Data Reduction and Error Analysis for the Physical Sciences* (Boston, MA: McGraw-Hill).

Box G. E. P. and Draper N. R. 1987. *Empirical Model-Building and Response Surfaces* (New York: Wiley).

Boyd S. P. and Vandenberghe L. 2004. *Convex Optimization* (New York: Cambridge University Press).

Brockwell P. J. and Davis R. A. 2002. *Introduction to Time Series and Forecasting* (New York: Springer).

Gafni A., Modlin R. L. and Brand L. 1975. Analysis of fluorescence decay curves by means of Laplace transformation. *Biophysical Journal* **15** 263–80.

Isenberg I., Mullooly J. P., Dyson R. D. and Hanson R. 1973. Studies on analysis of fluorescence decay data by method of moments. *Biophysical Journal* **13** 1090–115.

Istratov A. A. and Vyvenko O. F. 1999. Exponential analysis in physical phenomena. *Review of Scientific Instruments* **70** 1233–57.

James D. R. and Ware W. R. 1985. A fallacy in the interpretation of fluorescence decay parameters. *Chemical Physics Letters* **120** 455–9.

Lakowicz J. R. 2006. *Principles of Fluorescence Spectroscopy* (New York: Springer).

Lee K. C. B., Siegel J., Webb S. E. D., Leveque-Fort S., Cole M. J., Jones R., Dowling K., Lever M. J. and French P. M. W. 2001. Application of the stretched exponential function to fluorescence lifetime imaging. *Biophysical Journal* **81** 1265–74.

Lindsey C. P. and Patterson G. D. 1980. Detailed comparison of the Williams-Watts and Cole-Davidson functions. *Journal of Chemical Physics* **73** 3348–57.

Liu J., Sun Y., Qi J. and Marcu L. 2012. A novel method for fast and robust estimation of fluorescence decay dynamics using constrained least-squares deconvolution with Laguerre expansion. *Physics in Medicine and Biology* **57** 843–65.

Ljung L. 1999. *System Identification Theory for the User*. Thomas Kailath, ed. (Upper Saddle River, NJ: Prentice Hall).

Maarek J. M. I., Marcu L., Snyder W. J. and Grundfest W. S. 2000. Time-resolved fluorescence spectra of arterial fluorescent compounds: Reconstruction with the Laguerre expansion technique. *Photochemistry and Photobiology* **71** 178–87.

Marmarelis V. Z. 1993. Identification of nonlinear biological-systems using Laguerre expansions of kernels. *Annals of Biomedical Engineering* **21** 573–89.

McKinnon A. E., Szabo A. G. and Miller D. R. 1977. Deconvolution of photoluminescence data. *Journal of Physical Chemistry* **81** 1564–70.

Murata S., Matsuzaki S. Y. and Tachiya M. 1995. Transient effect in fluorescence quenching by electron-transfer. 2. Determination of the rate parameters involved in the Marcus equation. *Journal of Physical Chemistry* **99** 5354–8.

Ng B. K., Fu C. Y. and Razul S. G. 2009. Fluorescence lifetime discrimination using expectation–maximization algorithm with joint deconvolution. *Journal of Biomedical Optics* **14064009**.

Nocedal J. and Wright S. J. 2006. *Numerical Optimization* (New York: Springer).

O'Connor D. V., Ware W. R. and Andre J. C. 1979. Deconvolution of fluorescence decay curves—Critical comparison of techniques. *Journal of Physical Chemistry* **83** 1333–43.

Oppenheim A. V., Schafer R. W. and Buck J. R. 1999. *Discrete-Time Signal Processing* (Upper Saddle River, NJ: Prentice Hall).

Rissanen J. 1978. Modeling by shortest data description. *Automatica* **14** 465–71.

Tsay R. S. 2005. *Analysis of Financial Time Series*, vol. 543 (Hoboken, NJ: Wiley-Interscience).

Valeur B. 1978. Analysis of time-dependent fluorescence experiments by method of modulating functions with special attention to pulse fluorometry. *Chemical Physics* **30** 85–93.

Ware W. R., Doemeny L. J. and Nemzek T. L. 1973. Deconvolution of fluorescence and phosphorescence decay curves—Least-squares method. *Journal of Physical Chemistry* **77** 2038–48.

Zhang Z. Y., Grattan K. T. V., Hu Y. L., Palmer A. W. and Meggitt B. T. 1996. Prony's method for exponential lifetime estimations in fluorescence-based thermometers. *Review of Scientific Instruments* **67** 2590–4.

Global analysis of FLIM-FRET data

Hernán E. Grecco and Peter J. Verveer

Contents

12.1 INTRODUCTION

To understand how biological function arises from dynamic biochemical networks, it is essential to quantify the molecular state of proteins in living cells. Optical microscopy plays an important role in this strategy as it can visualize the spatiotemporal dynamics of cellular processes in single intact cells. However, the typical length scale of proteins lies well below the resolution limit of standard optical microscopy, and even current super-resolution methods fall short of resolving conformational changes and interactions (Grecco and Verveer 2011). In other words, two molecules can be within the same resolvable element of the optical system but still far apart in molecular terms.

Fluorescence resonance energy transfer (FRET), the nonradiative transfer of energy between two fluorescent species in close proximity (~5 nm), provides a way to reduce the coincidence detection volume to a molecular scale. One of the most robust ways to detect FRET is to monitor the reduction in fluorescence lifetime, that is, the average time that a fluorophore spends in the excited state after absorption of a photon. Fluorescence lifetime imaging microscopy (FLIM) provides an effective way to quantify FRET in intact, living cells by estimating the fluorescence lifetimes of the donor and of the donor/acceptor species and their relative fluorescence intensities (Bastiaens and Squire 1999; Gadella et al. 1993). Provided the FLIM system is properly calibrated for the instrument response, no additional measurements with reference samples are needed. This is an advantage over intensity-based FRET methods that require separate calibration samples (Gordon et al. 1995). Moreover, unlike intensity-based FRET methods, the recovered lifetimes yield an estimate of the quantum yield of both the free donor and of the donor/acceptor complexes, which makes it possible to renormalize the estimated fractional fluorescence to true relative concentrations (Verveer et al. 2000, 2001).

In FLIM-FRET experiments, the specific aim is to resolve the fluorescence lifetimes of the donor and donor–acceptor complex, and the fraction of donor in complex with the acceptor. For a given set of values for these parameters, an error measure for the difference between model and observation can be calculated. Varying these values yields an error landscape, and fitting the data to the model corresponds to finding its global minimum. However, determining the optimal values for the parameters of nonlinear multidimensional models is often hampered by codependence of the variables and by experimental noise, which creates multiple local minima and renders the error landscape shallow.

Global analysis uses *a priori* information to identify common (global) parameters across experimental data sets and theoretical models, which are then simultaneously fitted. By solving an overdetermined system, the error landscape is effectively sharpened. The linkage between parameters reduces covariance "valleys," increasing the robustness and the stability of the fitting procedure (Knutson et al. 1983). Early on, global analysis was used to analyze fluorescence measurements in cuvettes (Beechem 1992; Beechem and Brand 1986; Beechem et al. 1983, 1985; Gratton et al. 1984; Knutson et al. 1983; Lakowicz et al. 1984). Global analysis has also been applied to the recovery of kinetic parameters of molecular interactions, for example, by analyzing data from different surface densities of surface plasmon resonance biosensors (Myszka et al. 1996). It has also been used to distinguish between similar physical models for the association of biomolecules using time-resolved fluorescence anisotropy data (Bialik et al. 1998). Global analysis has also been used to implement spectrally resolved fluorescence correlation spectroscopy (FCS) (Previte et al. 2008).

In this chapter, we will focus on the application of global analysis to FLIM data, in particular, in the context of FRET microscopy. There are two main experimental techniques, usually named time-domain FLIM and frequency-domain FLIM, which are described in depth elsewhere in this book. Briefly, in time-domain FLIM, a train of sharp pulses is used to excite the sample, and the fluorescence decay is resolved using box-car detection with slow detectors or using continuous detection with fast detectors (Figure 12.1a). On the other hand, in frequency-domain FLIM, the demodulation and the phase shift of the fluorescence with respect to a modulated excitation source are measured using phase-sensitive detection such as phase-stepping or by employing a lock-in amplifier (Figure 12.1b). Both approaches exploit the response to repetitive excitation (Figure 12.1), and there is a clear theoretical correspondence between them, which can be understood through the duality of the Fourier transform. For example, the time stepping in box-car detection is equivalent to the phase-stepping method in the frequency domain. The information content of both methods is, in principle, equivalent, and in both cases, the data can be fit to obtain the fluorescence lifetimes of the donor in absence and presence of FRET, and the relative concentration of complexes.

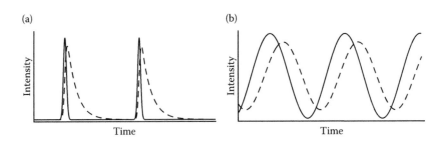

Figure 12.1 Basic principles of FLIM. (a) In time-domain FLIM measurements, a train of narrow pulses is used to excite the sample, and the fluorescence kinetics response is directly fitted to obtain the fluorescence lifetimes. (b) In frequency-domain methods, the sample is excited by a sinusoidal wave, and the fluorescence lifetimes are derived from the phase shift and demodulation of the response. In practice, the pulses used in the time domain may be broad, while in the frequency domain, the signals are not necessarily purely sinusoidal. The two methods have a repetitive excitation in common and can therefore be analyzed in the same way using Fourier methods.

In the case of FLIM, data are acquired in a spatially resolved manner, either in parallel in wide-field microscopy or sequentially by scanning and using a point detector. Traditionally, the data in each pixel have been fit independently, which is a challenging problem since the signal-to-noise ratios of the individual pixels may be low. However, if the fluorescence lifetime is resilient to environmental changes, the lifetimes of the free and complexed donor can be regarded as global parameters that are not expected to depend on spatial location, whereas the relative fraction of complexes may be different in each pixel. This makes FLIM-FRET data an ideal candidate for global analysis methods. In this chapter, we will describe how time- and frequency-domain data sets can be analyzed using global analysis, show some applications to biological problems, and discuss problems and pitfalls of the method.

12.2 FOURIER DESCRIPTION OF FLIM DATA

The decay kinetics of a mixture of fluorophores are given by a sum of Q exponentials with fluorescence lifetime τ_q:

$$F(t) = \sum_{q=1}^{Q} \frac{\alpha_q}{\tau_q} \exp(-t / \tau_q),$$

(12.1)

where the sum of the fractional contributions to the fluorescence, α_q, is normalized to 1, that is, $\sum_{q=1}^{Q} \alpha_q = 1$.

12.2.1 TIME-RESOLVED FLUORESCENCE RESPONSE

In fluorescence lifetime measurements, the excitation is not a single event but a periodic waveform (frequency-domain approach) or a train of narrow pulses (time-domain approach) with period T, which can be written as a Fourier series:

$$E(t) = \sum_{n=-\infty}^{\infty} E_n \exp(-jn\omega t),$$

(12.2)

where $\omega = 2\pi/T$. The time-resolved response to the excitation defined by Equation 12.2 is given by

$$S(t) = S_T \int_{-\infty}^{t} E(u)F(t-u)\,du,$$

(12.3)

where S_T is the total fluorescence signal. This can be written as

$$S(t) = S_T \sum_{n=-\infty}^{\infty} \sum_{q=1}^{Q} \frac{E_n D_n \alpha_q \exp(-jn\omega t)}{1 - jn\omega\tau_q},$$

(12.4)

where the extra term D_n has been incorporated to take into account the limited bandwidth of the detector. The product of $E_n D_n$ represents the Fourier transform of the response to an ideal pulse. This response is known as the instrument response function (IRF) and can be obtained with a calibration measurement.

It is convenient to define the contribution of the monoexponential species q to the harmonic n as

$$R_{n,q} = \frac{1}{1 - jn\omega\tau_q},$$

(12.5)

and rewrite Equation 12.4 in the form of a Fourier series:

$$S(t) = S_T \sum_{n=-\infty}^{\infty} E_n D_n R_n \exp(-jn\omega t), \qquad (12.6)$$

with

$$R_n = \sum_{q=1}^{Q} \alpha_q R_{n,q}. \qquad (12.7)$$

Figure 12.2a shows a geometric representation of Equation 12.7 in the complex plane. All possible values of $R_{n,q}$ fall on a semicircle of radius 0.5 centered at (0.5, 0). These represent the values of R_n for pure species with their monoexponential fluorescence lifetimes increasing counterclockwise. All possible values of R_n for arbitrary mixtures of lifetimes lie within the semicircle. This representation is convenient for the analysis of FLIM data, since it can be used to visually distinguish clusters of pixels that correspond to certain states, such as the occurrence of FRET (Digman et al. 2007).

12.2.2 HOMODYNE AND HETERODYNE DETECTION

In FLIM methods with continuous detection such as time-correlated single-photon counting FLIM (TCSPC-FLIM), $F(t)$ is measured directly, and R_n can be obtained by a Fourier analysis of the time-resolved data, after correcting for the instrument response $E_n D_n$. In frequency-domain methods, the detector is periodically modulated at a frequency equal to ω and integrated into a constant signal. In this case, the signal is sampled by varying the relative phase of the detector modulation with respect to the excitation. The modulation of the detector can be written as a Fourier series:

$$M(t) = \sum_{n=-\infty}^{\infty} M_n \exp(-jn\omega t) \exp(-jn\Delta\phi), \qquad (12.8)$$

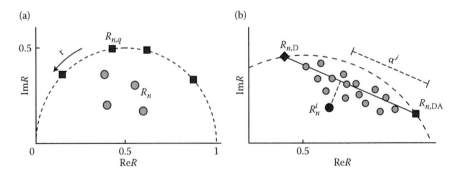

Figure 12.2 Principle of Fourier domain analysis of FLIM data. (a) The Fourier coefficients of pure lifetime species (monoexponential decays) lie on a semicircle if their imaginary part is plotted against their real part. Mixtures with multiple lifetime components are found within this semicircle. (b) The Fourier components of mixtures (R_n^i) are linear combinations of the Fourier components of the pure species. In the case of two components, all mixtures are found on a straight line, which intersects the semicircle at the positions of the pure components ($R_{n,D}$, $R_{n,DA}$ in the case of FRET). The relative fluorescence of each species α^i for a given mixture is found by the normalized distance along the connecting straight line. Global analysis consists of finding the intersections of the straight line with the semicircle, which requires sufficient variation in the distribution of R_n^i (or equivalently, in α^i).

where $\Delta\phi$ is a tunable delay with respect to the excitation. Instrumentally, this is the phase shift in frequency-domain FLIM or the variable delay in box-car detection time-domain FLIM. The measured signal is given by the multiplication of $S(t)$ and $M(t)$ and retaining only constant terms:

$$S(\Delta\phi) = S_T \sum_{n=-\infty}^{\infty} \sum_{q=1}^{Q} \frac{E_n D_n M_n \alpha_q \exp(-jn\Delta\phi)}{1 - jn\omega\tau_q}, \tag{12.9a}$$

$$= S_T \sum_{n=-\infty}^{\infty} E_n D_n M_n R_n \exp(-jn\Delta\phi), \tag{12.9b}$$

where the R_n are given by Equation 12.7. These can be estimated by Fourier analysis of the phase-dependent data, after correction for $E_n D_n M_n$. If the excitation or the detector modulation has only a single harmonic frequency, Equation 12.9 reduces to the more familiar cosine form that is employed in single-frequency FLIM. This approach is called *homodyne* detection, where the modulation signal has the same frequency as the excitation. A less common variation of this technique is *heterodyne* detection where the modulation of the detector differs slightly from the excitation, leading to a slowly modulated response of the same form as Equation 12.6.

12.2.3 MEASUREMENT OF INSTRUMENT PARAMETERS

In time-domain measurements, the $E_n D_n$ parameters are the Fourier coefficients of IRF, which can be measured directly as the response to a sample that has a lifetime equal to zero (for instance, a reflective sample) or derived from a sample with a known lifetime. Similarly, in a frequency-domain system, $E_n D_n G_n$ follows from the phase shift and demodulation of the system response to a zero-lifetime sample (Gadella et al. 1993; Squire and Bastiaens 1999). It is important to note that for detection systems with wavelength-dependent response properties, such as avalanche photodiodes, a reflective sample will provide a biased estimation of the IRF. Instead, a fluorescent sample with the same spectral properties as the specimen of interest must be used.

12.2.4 EQUIVALENCE OF TIME- AND FREQUENCY-DOMAIN FLIM

The analysis above shows that both in time-domain and frequency-domain methods, the data can be described as a Fourier series with the Fourier components given by Equation 12.7, after correction for $E_n D_n$ (Equation 12.6) or $E_n D_n M_n$ (Equation 12.9). All information on the lifetimes τ_q and the relative fluorescence fractions α is contained in the corrected Fourier components R_n. Therefore, Fourier data analysis methods can be applied to both time-domain and frequency-domain data.

Fourier analysis methods are commonly applied to frequency-domain data, but time-domain data are usually fitted directly in the time domain. However, in the next section, we will show that the problem of global analysis of FLIM data is greatly simplified after applying Fourier analysis, and that this is of particular advantage for time-domain data.

12.3 GLOBAL ANALYSIS OF FLIM-FRET DATA

12.3.1 SINGLE-FREQUENCY GLOBAL ANALYSIS

Global analysis of FLIM-FRET data was introduced by Verveer et al. (2000) who noted that, in many cases, the fluorescence lifetime values in each pixel are given by the lifetime of the donor τ_D and of the donor in complex with the acceptor τ_{DA}. Moreover, these values can be assumed to be the same in each pixel, if, apart from FRET, there are no other photophysical processes that influence the fluorescence lifetime. The relative fraction of the donor/acceptor complex α is the only parameter that is different in each pixel. In such a case, the two lifetimes can be considered global parameters, whereas the α values are local parameters. It was shown that it was possible to fit τ_D and τ_{DA} together with all values of α to single-frequency data, which is not possible if all pixels are analyzed individually.

It was shown by Clayton et al. (2003) and later confirmed by others (Esposito et al. 2005; Redford and Clegg 2005) that the problem is linear, which considerably simplifies the calculation. As linear least squares regression can be computed element-by-element, there is no need to store the complete data set in memory, enabling the simultaneous fit of an unlimited number of images, which can be of great advantage, for instance, in high-throughput screening applications (Grecco et al. 2010).

The linear calculation is based on Equation 12.7, which can be simplified to the sum of only two species in the case of FLIM-FRET measurements:

$$R_n^i = (1-\alpha^i)R_{n,D} + \alpha^i R_{n,DA} \tag{12.10}$$

where the superscript i has been added to number the pixels, and $R_{n,D}$ and $R_{n,DA}$ are given by Equation 12.5 with $\tau_1 = \tau_D$ and $\tau_2 = \tau_{DA}$. Figure 12.2b shows a geometric representation of this equation. The values of R_n^i fall on a straight line, which intersects the semicircle at $R_{n,D}$ and $R_{n,DA}$, and α_i is the normalized distance from R_n^i to $R_{n,DA}$. Global analysis of the data consists of finding the intersections of the straight line with the semicircle by fitting the values of R_n^i to a linear equation in the complex plane:

$$\Im R_n^i = u_n + v_n \Re R_n^i, \tag{12.11}$$

where $\Im R_n^i$ and $\Re R_n^i$ are the imaginary and real components of the complex number R_n^i, respectively. The fluorescence lifetimes of the mixture then follow from the fitted values of u_n and v_n by applying Equation 12.5:

$$\tau_D, \tau_{DA} = \frac{1 \pm \sqrt{1 - 4u_n(u_n + v_n)}}{2n\omega u_n} \tag{12.12}$$

Note that the data used for fitting the straight line do not need to be limited to a single FLIM data set. Multiple sets with multiple conditions can be pooled, as long as it can be guaranteed that the different conditions only affect the fraction of molecules that participate in FRET, but not the fluorescence lifetimes τ_D and τ_{DA}. In this way, the accuracy of the fit can be improved by increasing the number of data points, and it can be ensured that there is sufficient variation in α^i to sample the straight line properly.

The corresponding fractional fluorescence emitted by the complexed donor molecules in each pixel (α^i) is obtained by the normalized scalar projection of R_n^i on the line connecting $R_{n,D}$ and $R_{n,DA}$. This turns out to be equivalent to a least squares estimation of α^i (Verveer and Bastiaens 2003):

$$\alpha^i = \frac{n\omega(\tau_{DA} + \tau_D)\Re R_n^i + (n^2\omega^2\tau_{DA}\tau_D - 1)\Im R_n^i - n\omega\tau_{DA}}{n\omega(\tau_D - \tau_{DA})}. \tag{12.13}$$

It is important to realize that α^i is not equal to the relative amount of interacting donor, but that it represents the fraction of the total fluorescence it emits. To obtain the true relative concentration of the interacting donor, the fractional fluorescence of free and complexed donor must be divided by their quantum yields and then renormalized to unity. In a FRET system, the quantum yields of the free and complexed donor species are proportional to their lifetimes, τ_D and τ_{DA}, respectively. For the relative concentration of donor in complex c^i, we then find

$$c^i = \frac{\alpha^i \tau_D}{\alpha^i(\tau_D - \tau_{DA}) + \tau_{DA}}. \tag{12.14}$$

Single-frequency global analysis is an attractive approach, in particular, for single-frequency FLIM data, where it allows the estimation of α^i without prior knowledge of the fluorescence lifetimes. In a single pixel τ_D, τ_{DA} and α^i cannot be resolved, since only two measurements are available (phase and demodulation).

However, if N pixels are analyzed together with global analysis, $2N$ measurements are available, but only $N + 2$ parameters need to be estimated. The only requirement is that there is sufficient variation in the values of α^i to guarantee that the linear relation given by Equation 12.11 can be sampled and fit accurately (see also Figure 12.2).

As described in Section 12.2, time-domain data can also be analyzed by Fourier methods. In this case, additional information is found in the higher harmonics of the data, and in principle, a single pixel analysis can resolve τ_D, τ_{DA}, and α^i. However, single-frequency global analysis can also be reliably applied to time-domain data by analyzing the first harmonic of the Fourier transform of the time-domain data (Grecco and Verveer 2009). In this case, the analysis still benefits from the reduction in the number of parameters that need to be estimated, yielding more precise results.

12.3.2 PREFILTERING OF DATA FOR GLOBAL ANALYSIS

FLIM data can be very noisy, particularly in the case of time-domain data acquired by confocal systems that employ time-correlated photon counting approaches. To account for this, pixels that are too noisy should be excluded from the global analysis. A simple prefiltering step excludes pixels that have an intensity that is lower than some preset threshold. For the purpose of global analysis, it can also be of advantage to exclude pixels that correspond to physically impossible lifetime values, such as negative values. Grecco and Verveer (2009) devised a filtering method that excludes all pixels that lead to values of R_n^i that fall significantly outside the semicircle, or that would fall on a hypothetical line that intersects with the semicircle at points that correspond to negative lifetimes. This was shown to considerably improve the estimation of the fluorescence lifetimes in FRET applications using data acquired by a time-domain confocal FLIM system (Grecco and Verveer 2009).

12.3.3 ESTIMATION OF INSTRUMENT PARAMETERS

Fourier analysis of either frequency-domain or time-domain data, which is a prerequisite for single-frequency global analysis, requires the measurement of instrument correction factors (see Section 12.2). Only the correction factor for the first harmonic is necessary. In frequency-domain acquisition, this is equivalent to the phase and the modulation of the system, which is usually measured using a reflective or low-lifetime sample (Gadella et al. 1993; Squire and Bastiaens 1999). In time-domain acquisition, the correction is the first Fourier component of the IRF, which can be measured directly, and corresponds roughly to the delay and width of the IRF. Alternatively, the IRF can be approximated by a Gaussian shape, which works well for estimating the correction for the first harmonic, and can be extracted reliably from the higher harmonics of the decay curve (Grecco and Verveer 2009).

12.3.4 ALTERNATIVE APPROACHES

Extensions of the method above to multiple-frequency FLIM data (Schlachter et al. 2009; Squire et al. 2000) are possible but have not been employed in practical applications yet. For time-domain approaches, several alternative methods have been described (Barber et al. 2008; Laptenok et al. 2010; Pelet et al. 2004). The most common approach to global analysis in time-domain FLIM data sets involves fitting a sum of the exponentials plus the background in each pixel, linking the fluorescence lifetimes across them. The main drawback of this method stems from the nonlinear nature of the fitting function, which leads to high memory and computational requirements, and sensitivity of the result to the initial guess of the fitting parameters. Several approaches have been described to minimize such problems. For example, starting from rationally chosen parameters can speed up the convergence of the fit. The rapid lifetime determination method (Sharman et al. 1999) can be used to efficiently compute initial estimations of the global lifetimes. Other methods pool the pixels into multiple sets and use the fitted parameters for each set to constrain the initial guess. Automatic image segmentation has been used to pool together pixels from the same type of structures to derive local initial estimations (Pelet et al. 2004). Additionally, other methods such as maximum likelihood (Bajzer et al. 1991), maximum entropy (Livesey and Brochon 1987), and maximum fidelity (Walther et al. 2011) have been used to provide a better estimation of the fitting parameters.

These methods are designed to utilize all information in the data, in contrast to the single-frequency global analysis method, which only uses the first harmonic of the data. It is straightforward to extend frequency-domain global analysis to include multiple harmonics (Schlachter et al. 2009). Intuitively, using all harmonics suggests better results, since all information is utilized. However, Grecco and Verveer (2009) argue that this is not necessarily the case, since the first harmonic contains most of the information and is generally estimated most reliably. At the current state of the art, the single-frequency global analysis method appears to provide results of similar quality to the more involved nonlinear methods at a significantly lower computational cost.

An important consideration in all time-domain FLIM analysis methods is the need to account for the shape of the IRF. When the width of the IRF is comparable to the lifetimes to be resolved, it is necessary to deconvolve its effects from the raw data. As the IRF is a low-pass filter, the inverse filtering enhances the high-frequency noise contribution. Most fitting methods use iterative reconvolution algorithms in which the IRF shape is either measured separately or obtained from the data and then convolved with the fitted exponential decay. This leads to higher computational costs and to an increased potential for introducing noise and systematic errors in the higher harmonics of the data. Single-frequency global analysis only needs an estimation of the first harmonic of the IRF, which is more easily obtained in a reliable fashion. This may be one of the reasons for the good performance of single-frequency global analysis methods on FLIM data (Grecco and Verveer 2009).

12.4 APPLICATIONS

Global analysis of FLIM data has been used successfully in biological applications to quantitatively determine the fraction of interacting molecules with FRET assays (Grecco et al. 2010; Ng et al. 2001; Offterdinger and Bastiaens 2008; Reynolds et al. 2003; Rocks et al. 2005; Verveer et al. 2000; Wimmer-Kleikamp et al. 2004; Xouri et al. 2007; Yudushkin et al. 2007). Here we show examples of two applications: phosphorylation of the epidermal growth factor receptor (EGFR) in fixed cells and the activation of the small GTPase Ras in living cells.

The prototype application of global analysis to FLIM-FRET data is the quantification of tyrosine phosphorylation of signaling molecules, such as the EGFR, in single cells (Wouters and Bastiaens 1999). Global analysis was applied to frequency-domain FLIM-FRET data (Verveer et al. 2000), which made it possible to visualize and quantify the relative fraction of active receptors in single intact cells. Figure 12.3a

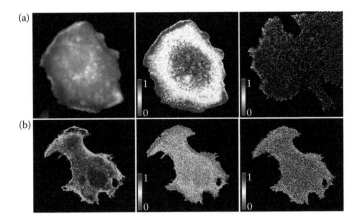

Figure 12.3 Application of global analysis to measure EGFR phosphorylation, detected by FRET. (a) Global analysis of FLIM-FRET data acquired with a wide-field frequency domain system. Left: GFP intensity image. Middle: Relative concentration of an active receptor after 5 min of stimulation with EGF, as measured by FRET between the GFP tag fused to EGFR and a Cy3-labeled antibody against phosphotyrosine. Right: Relative concentration of active receptor without stimulation. (Adapted from Verveer, P. J. et al., *Science 290*, 1567–1570, 2000.) (b) Global analysis of FLIM-FRET data acquired with a confocal time-domain system. Left: Citrine intensity image. Middle: Relative concentration of an active receptor after 5 min of stimulation with EGF, as measured by FRET between the Citrine tag fused to EGFR and Cy3.5 labeled antibody. Right: Same data fitted on a pixel-by-pixel basis to a two-component exponential model. (Adapted from Grecco, H. E. and P. J. Verveer, *Opt. Express 17*, 6493–6508, 2009.)

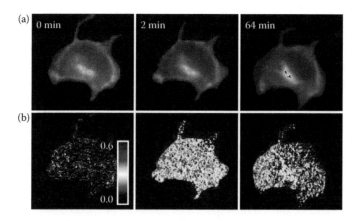

Figure 12.4 Application of global analysis to measure Ras activation in living cells. Ras activation measured using a wide-field frequency-domain FLIM system, by detecting FRET between YFP-H-Ras and RafRBD-dhcRed, as a function of time. (a) YFP intensity images. (b) Relative concentration of active H-Ras. (Adapted from Rocks, O. et al., *Science 307*, 1746–1752, 2005.)

shows the application of global analysis to FLIM data, acquired from samples expressing EGFR-GFP and incubated with Cy3 labeled PY72, a generic antibody against phosphotyrosine. This assay can also be employed in a time-domain setup, yielding similar results at a higher spatial resolution (Grecco and Verveer 2009). Figure 12.3b shows results obtained with a confocal time-domain FLIM setup, from samples expressing EGFR-Citrine, fixed and incubated with Cy3.5 labeled PY72. In this case, a standard pixel-by-pixel analysis of the data does not show the correct spatial distribution due to the low photon counts, but the global analysis shows that only the peripheral receptors are activated.

The quantification of protein phosphorylation using antibodies against phosphotyrosine can only be done in fixed cells. Although this is a valuable tool, it is highly desirable to follow the dynamics of activation in living cells. This can be achieved by measuring FRET between two fluorescent proteins. The usual pair of CFP/YFP is not optimal for use with global analysis (see Section 12.5), but in recent years, a wide selection of red fluorescence proteins has become available that can be used as an acceptor together with green or yellow donor fluorophores. This was first employed in combination with global analysis by Rocks et al. (2005) to measure the activity of Ras in living cells. Figure 12.4 gives an example that clearly shows the initial activation of Ras at the plasma membrane, while at later time points, Ras is shown to be mostly active at the Golgi apparatus. This reflects the dynamic distribution of Ras that is controlled by an acylation cycle, which determines the spatial distribution of Ras.

12.5 PROBLEMS AND PITFALLS

The basic requirement for global analysis is that linked parameters need to be invariant across data sets. For FLIM, this means that the fluorescence lifetimes are only affected by FRET and are insensitive to other environmental conditions. Moreover, the rate of energy transfer needs to be the same everywhere, a condition that is not always fulfilled, but which can be reasonably assumed in tight protein complexes.

In current single-frequency FLIM global analysis methods, each species should have monoexponential fluorescence decay kinetics. This condition is not always fulfilled; for instance, the commonly used donor CFP has a distinct biexponential decay and therefore its use should be avoided. On the other hand, GFP shows only slight biexponential behavior and therefore can be used with only a minor effect: a small positive bias in the estimation of the relative concentrations of complexes. Also, the second component must be monoexponential, which in the case of a FRET system means that the donor should still have monoexponential kinetics if bound to an acceptor. This condition can be fulfilled for a tight fixed geometry distribution of acceptor (or acceptors) or if it leads to a narrow distribution of FRET efficiencies, and thereby of the lifetimes. While failure to fulfill this condition leads to a wrong estimation of the parameters, for FRET efficiencies higher than 50%, the value of α is not significantly affected

(Grecco et al. 2010). Although multiexponential decays can be fitted using multiple harmonics information (time-domain FLIM or multifrequency-domain FLIM) to resolve any of these potential issues, the difficulties to obtain accurate results with a limited amount of photons has hindered its application so far.

In practice, the constraints on the FLIM data are not limiting, and many successful applications have been reported. In particular, the important application of measuring the phosphorylation of proteins using FRET between a fluorescence protein and a labeled antibody has been validated experimentally using an *in vitro* system (Grecco et al. 2010). The precision and accuracy of this assay were demonstrated using purified EPHA3 in which the fraction of phosphorylated protein can be accurately controlled (Figure 12.5). Different ratios of phosphorylated and dephosphorylated protein were mixed and then measured. The increasing value of the fraction of phosphorylated protein is clearly seen in the linear procession of the values within the semicircle (Figure 12.5a), and the measured values of the relative concentration of phosphorylated EPHA3 correspond well to the expectations (Figure 12.5b, c).

The conditions of invariance and a monoexponential decay can be checked to some extent using samples with only one of the species, that is, donor-only samples in the case of a FRET system. For example, in frequency-domain FLIM, a fluorophore can be regarded as monoexponential if the lifetimes derived from phase and modulation are equal within the experimental uncertainty. The invariance of the donor fluorescence lifetime can be tested by pooling pixels from different cellular compartments and comparing the resulting fluorescence lifetimes.

Within the family of fluorescent proteins, we have found that the various versions of YFP, in particular, mCitrine, are very good donors as they have monoexponential decay kinetics and are robust against fusion. Recent work has delivered bright and monoexponential CFP replacements (mTFP and mTurquoise), leaving only the red part of the spectra without suitable donors for FRET applications.

Global analysis methods will have difficulties if there is little variation in the sample. For instance, if there is no FRET at all, global analysis is meaningless since no second lifetime is present. The variation in the set of samples in a FRET assay can, however, often be extended experimentally, for instance, by using

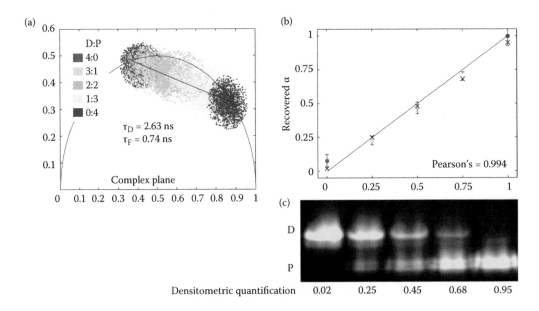

Figure 12.5 Experimental validation of global analysis of FLIM-FRET data. (a) Complex plane representation of the *in vitro* phosphorylation of purified mCit-EPHA3c as measured by FLIM. Each gray level represents a different mixing ratio between dephosphorylated (D) and phosphorylated (P) mCit-EPHA3c. (b) Average recovered phosphorylated fraction (α) as a function of the prepared fraction of phosphorylated mCit-EPHA3c. Error bars, s.d. of five fields analyzed. (c) Native gel shift assay of mCit-EPH3Ac phosphorylation, with densitometric quantification of phosphorylated protein shown below the gel. (Adapted from Grecco, H. E. et al., *Nat. Methods 7*, 467–472, 2010.)

different stimuli to induce more or less binding, by adding measurements of a sample without acceptor, or by photobleaching the acceptor.

12.6 CONCLUSIONS

In this chapter, we have described the application of global analysis techniques to FLIM data. For single-frequency FLIM, this has the crucial advantage that a mixture of two species can be resolved without further prior knowledge. This makes it possible to calculate the relative fraction of fluorescence contributed by each species. Importantly, in a FRET system, this fraction can be renormalized to the relative concentration of complexes, which in most biological applications is the ultimate parameter of interest. This level of quantification is not possible with intensity-based FRET systems and is a unique advantage of measuring FRET by FLIM. Single-frequency FLIM has been available for some time now, and global analysis has been, and continues to be, used in biological applications successfully.

The global analysis method that we described in detail is equally applicable to time-domain data. In principle, such data contain sufficient information to resolve the mixture on a pixel-by-pixel basis. This approach is popular in confocal laser scanning systems, where a point detector with single-photon counting can be employed to precisely sample the fluorescence response. In practice, the signal-to-noise ratios of these data are not good enough for a pixel-by-pixel analysis, and global analysis allows reliable analysis of these data. In addition to the Fourier analysis method presented here, several other approaches for global analysis methods have been described, which, at least in theory, should perform at least equally well. So far, global analysis methods for time-domain data have not been widely applied in biological systems, but we expect that this will change in the near future, since confocal FLIM systems have become widely available in recent years.

There are several prerequisites for the successful application of global analysis, which therefore should be used with care. It is important that the data are well understood, for instance, by applying standard FLIM analysis methods before proceeding with global analysis. However, so far, this has not inhibited the successful use of global analysis of FLIM data to several relevant biological questions. These methods have been mainly applied to microscopy data of single cells, but they should be equally applicable to fitting a double-exponential decay model to tissue autofluorescence decay data. We expect that future developments, both in FLIM instrumentation and in data analysis methods, will widen the range of biological applications where global analysis can be applied successfully to quantify protein interactions in living cells.

REFERENCES

Bajzer, V., T. M. Therneau, J. C. Sharp, and F. G. Prendergast (1991). Maximum likelihood method for the analysis of time-resolved fluorescence decay curves. *Eur. Biophys. J. 20*, 247–262.

Barber, P. R., S. M. Ameer-Beg, J. Gilbey, L. M. Carlin, M. Keppler, T. Ng, and B. Vojnovic (2008). Multiphoton time-domain fluorescence lifetime imaging microscopy: Practical application to protein–protein interactions using global analysis. *J. R. Soc. Interface 6*, S93–S105.

Bastiaens, P. I. H. and A. Squire (1999). Fluorescence lifetime imaging microscopy: Spatial resolution of biochemical processes in the cell. *Trends Cell Biol. 9*, 48–52.

Beechem, J. M. (1992). Global analysis of biochemical and biophysical data. *Methods Enzymol. 210*, 2127–2137.

Beechem, J. M., M. Ameloot, and L. Brand (1985). Global and target analysis of complex decay phenomena. *Anal. Instrum. 14*, 379–402.

Beechem, J. M. and L. Brand (1986). Global analysis of fluorescence decay: Applications to some unusual experimental and theoretical studies. *Photochem. Photobiol. 44*, 323–329.

Beechem, J. M., J. Knutson, B. A. Ross, B. W. Turner, and L. Brand (1983). Global resolution of heterogeneous decay by phase/modulation fluorometry: Mixtures and proteins. *Biochemistry 22*, 6054–6058.

Bialik, C. N., B. Wolf, E. L. Rachofsky, J. B. Ross, and W. R. Laws (1998). Dynamics of biomolecules: Assignment of local motions by fluorescence anisotropy decay. *Biophys. J. 75*, 2564–2573.

Clayton, A. H. A., Q. S. Hanley, and P. J. Verveer (2003). Graphical representation and multicomponent analysis of single-frequency fluorescence lifetime imaging microscopy data. *J. Microsc. 213*, 1–5.

Digman, M., V. R. Caiolfa, M. Zamai, and E. Gratton (2007). The phasor approach to fluorescence lifetime imaging analysis. *Biophys. J. 94*, L14–L16.

Esposito, A., H. C. Gerritsen, and F. S. Wouters (2005). Fluorescence lifetime heterogeneity resolution in the frequency domain by lifetime moments analysis. *Biophys. J. 89*, 4286–4299.

Gadella, T. W. J., T. M. Jovin, and R. M. Clegg (1993). Fluorescence lifetime imaging microscopy (FLIM)—Spatial resolution of microstructures on the nanosecond time-scale. *Biophys. Chem. 48*, 221–239.

Gordon, G. W., B. Chazotte, X. F. Wang, and B. Herman (1995). Analysis of simulated and experimental fluorescence recovery after photobleaching. Data for two diffusing components. *Biophys. J. 68*, 766–778.

Gratton, E., M. Limkeman, J. R. Lakowicz, B. P. Maliwal, H. Cherek, and G. Laczko (1984). Resolution of mixtures of fluorophores using variable-frequency phase and modulation data. *Biophys. J. 46*, 479–486.

Grecco, H. E., P. Roda-Navarro, A. Girod, J. Hou, T. Frahm, D. C. Truxius, R. Pepperkok, A. Squire, and P. I. H. Bastiaens (2010). *In situ* analysis of tyrosine phosphorylation networks by FLIM on cell arrays. *Nat. Methods 7*, 467–472.

Grecco, H. E. and P. J. Verveer (2009). Global analysis of time correlated single photon counting FRET-FLIM data. *Opt. Express 17*, 6493–6508.

Grecco, H. E. and P. J. Verveer (2011). FRET in cell biology: Still shining in the age of super-resolution? *Chem. Phys. Chem. 12*, 484–490.

Knutson, J., J. Beechem, and L. Brand (1983). Simultaneous analysis of multiple fluorescence decay curves: A global approach. *Chem. Phys. Lett. 102*, 501–507.

Lakowicz, J. R., G. Laczko, H. Cherek, E. Gratton, and M. Limkeman (1984). Analysis of fluorescence decay kinetics from variable-frequency phase shift and modulation data. *Biophys. J. 46*, 463–477.

Laptenok, S. P., J. W. Borst, K. M. Mullen, I. H. M. van Stokkum, A. J. W. G. Visser, and H. van Amerongen (2010). Global analysis of Förster resonance energy transfer in live cells measured by fluorescence lifetime imaging microscopy exploiting the rise time of acceptor fluorescence. *Phys. Chem. Chem. Phys. 12*, 7593–7602.

Livesey, A. and J. Brochon (1987). Analyzing the distribution of decay constants in pulse-fluorimetry using the maximum entropy method. *Biophys. J. 52*, 693–706.

Myszka, D. G., P. R. Arulanantham, T. Sana, Z. Wu, T. A. Morton, and T. L. Ciardelli (1996). Kinetic analysis of ligand binding to interleukin-2 receptor complexes created on an optical biosensor surface. *Protein Sci. 5*, 2468–2478.

Ng, T., M. Parsons, W. E. Hughes, J. Monypenny, D. Zicha, A. Gautreau, M. Arpin, S. Gschmeissner, P. J. Verveer, P. I. Bastiaens, and P. J. Parker (2001). Ezrin is a downstream effector of trafficking PKC-integrin complexes involved in the control of cell motility. *EMBO J. 20*, 2723–2741.

Offterdinger, M. and P. I. H. Bastiaens (2008). Prolonged EGFR signaling by ERBB2-mediated sequestration at the plasma membrane. *Traffic 9*, 147–155.

Pelet, S., M. J. R. Previte, L. H. Laiho, and P. T. C. So (2004). A fast global fitting algorithm for fluorescence lifetime imaging microscopy based on image segmentation. *Biophys. J. 87*, 2807–2817.

Previte, M. J. R., S. Pelet, K. H. Kim, C. Buehler, and P. T. C. So (2008). Spectrally resolved fluorescence correlation spectroscopy based on global analysis. *Anal. Chem. 80*, 3277–3284.

Redford, G. I. and R. M. Clegg (2005). Polar plot representation for frequency-domain analysis of fluorescence lifetimes. *J. Fluoresc. 15*, 805–815.

Reynolds, A. R., C. Tischer, P. J. Verveer, O. Rocks, and P. I. H. Bastiaens (2003). EGFR activation coupled to inhibition of tyrosine phosphatases causes lateral signal propagation. *Nat. Cell Biol. 5*, 447–453.

Rocks, O., A. Peyker, M. Kahms, P. J. Verveer, C. Koerner, M. Lumbierres, J. Kuhlmann, H. Waldmann, A. Wittinghofer, and P. I. H. Bastiaens (2005). An acylation cycle regulates localization and activity of palmitoylated ras isoforms. *Science 307*, 1746–1752.

Schlachter, S., A. D. Elder, A. Esposito, G. S. Kaminski, J. H. Frank, L. K. van Geest, and C. F. Kaminski (2009). mhFLIM: Resolution of heterogeneous fluorescence decays in widefield lifetime microscopy. *Opt. Express 17*, 1557–1570.

Sharman, K. K., A. Periasamy, H. Ashworth, and J. N. Demas (1999). Error analysis of the rapid lifetime determination method for double-exponential decays and new windowing schemes. *Anal. Chem. 71*, 947–952.

Squire, A. and P. I. H. Bastiaens (1999). Three dimensional image restoration in fluorescence lifetime imaging microscopy. *J. Microsc. 193*, 36–49.

Squire, A., P. J. Verveer, and P. I. H. Bastiaens (2000). Multiple frequency fluorescence lifetime imaging microscopy. *J. Microsc. 197*, 136–149.

Verveer, P. J. and P. I. H. Bastiaens (2003). Evaluation of global analysis algorithms for single frequency fluorescence lifetime imaging microscopy data. *J. Microsc. 209*, 1–7.

Verveer, P. J., A. Squire, and P. I. Bastiaens (2000). Global analysis of fluorescence lifetime imaging microscopy data. *Biophys. J. 78*, 2127–2137.

Verveer, P. J., A. Squire, and P. I. H. Bastiaens (2001). Frequency-domain fluorescence lifetime imaging microscopy: A window on the biochemical landscape of the cell. In A. Periasamy (Ed.), *Methods in Cellular Imaging*, pp. 273–292. Oxford: Oxford University Press.

Verveer, P. J., F. S. Wouters, A. R. Reynolds, and P. I. H. Bastiaens (2000). Quantitative imaging of lateral ErbB1 receptor signal propagation in the plasma membrane. *Science 290*, 1567–1570.

Walther, K. A., B. Papke, M. B. Sinn, K. Michel, and A. Kinkhbwala (2011). Precise measurement of protein interacting fractions with fluorescence lifetime imaging microscopy. *Mol. BioSyst. 7*, 322–336.

Wimmer-Kleikamp, S. H., P. W. Janes, A. Squire, P. I. H. Bastiaens, and M. Lackmann (2004). Recruitment of Eph receptors into signaling clusters does not require ephrin contact. *J. Cell Biol. 164*, 661–666.

Wouters, F. S. and P. I. H. Bastiaens (1999). Fluorescence lifetime imaging of receptor tyrosine kinase activity in cells. *Curr. Biol. 9*, 1127–1130.

Xouri, G., A. Squire, M. Dimaki, B. Geverts, P. J. Verveer, S. Taraviras, H. Nishitani, A. B. Houtsmuller, P. I. H. Bastiaens, and Z. Lygerou (2007). Cdt1 associates dynamically with chromatin throughout G1 and recruits geminin onto chromatin. *EMBO J. 26*, 1303–1314.

Yudushkin, I. A., A. Schleifenbaum, A. Kinkhabwala, B. G. Neel, C. Schultz, and P. I. H. Bastiaens (2007). Live-cell imaging of enzyme-substrate interaction reveals spatial regulation of PTP1B. *Science 315*, 115–119.

Analysis of fluorescence lifetime data

Fluorescence lifetime imaging in turbid media

Vadim Y. Soloviev, Teresa M. Correia, and Simon R. Arridge

Contents

13.1 INTRODUCTION

Fluorescence optical tomography (fDOT) is a functional and molecular imaging modality, which consists of illuminating the object of study using a light source in the visible or near-infrared (NIR) wavelength range to excite a fluorescent probe and measuring emitted fluorescence light, and possibly excitation light, in order to reconstruct spatial maps of fluorescence parameters, such as lifetime and fluorescence yield, which is linearly related to the quantum yield and absorption coefficient of a fluorophore (Gao et al. 2010; Ntziachristos et al. 2005). Light in the NIR region is preferred because it is known to penetrate deeper into biological tissue than visible light due to the low absorption by the main tissue chromophores, such as water, oxyhemoglobin, and deoxyhemoglobin. Usually, fDOT systems with the ability to measure fluorescent lifetimes are known as fluorescence lifetime imaging microscopy (FLIM) systems.

This chapter introduces source–detector geometries and strategies for data acquisition in fDOT (FLIM). Furthermore, the techniques used to generate 3D images of fluorescence parameters from measured data are described. This image reconstruction problem is known as the inverse problem. It is essential to first establish a forward model that accurately describes the propagation of light through biological tissue, which is a highly scattering medium. The radiative transfer equation (RTE) is a general model for describing light propagation in the presence of multiple scattering. The RTE can be solved numerically using Monte Carlo (MC) techniques, which are computationally very expensive, analytically for simple cases and geometries or approximations to the RTE. The aim of this chapter is to describe state-of-the-art methods used in FLIM to obtain data, generate the forward model, and reconstruct images.

13.2 INSTRUMENTATION

The three main imaging technologies used in fDOT are continuous wave (CW) (Soubret et al. 2005), time domain (TD) (D'Andrea et al. 2003; McGinty et al. 2011; Venugopal et al. 2010b), and frequency domain (FD) (Godavarty et al. 2003; O'Leary et al. 1996). The CW systems use steady-state light sources and can only be used to recover fluorescence yield. The TD and FD systems provide the additional possibility of measuring the lifetime parameter. Fluorescence lifetimes are of the order of picoseconds to nanoseconds, and therefore, fast detection schemes must be used.

In fDOT, there are different types of illumination and detection strategies: point illumination and point detection (Milstein et al. 2003; Wu et al. 1997), where multiple fiber optics-based sources and detectors are coupled to the object of study, point illumination, and wide-field detection (Kumar et al. 2008; Soubret et al. 2005), which uses a noncontact detection scheme and multiple fiber-based sources, and wide-field illumination and detection (D'Andrea et al. 2010; Thompson et al. 2003), which uses a noncontact illumination and detection setup. In point detection technologies, light is usually collected by a fast detector such as a photomultiplier tube (PMT) or avalanche photodiode (APD), whereas in wide-field detection, a charge-coupled device (CCD) camera records projection images. Noncontact imaging systems that acquire projection images over 360°, by either rotating a laser source and detector around the source or by rotating the object, have been recently developed (McGinty et al. 2011; Schulz et al. 2010).

13.2.1 TIME DOMAIN

TD systems employ a short pulse (order of picoseconds) of light and measure the time of flight of photons through tissue. A fluorophore embedded in the tissue is excited by the light pulse and subsequently emits a fluorescence pulse with a characteristic lifetime decay.

Fiber-based TD systems typically use a detection method known as time-correlated single-photon counting (TCSPC; Gao et al. 2010; Montcel and Poulet 2006). In the TCSPC technique, the time delay between the excitation pulse and the detection of a single emission photon is recorded. Multiple pulses are used to construct a histogram of the number of detected photon counts as a function of time, providing a temporal distribution known as fluorescence temporal point spread function (TPSF; Delpy et al. 1988). The absorbing and highly scattering nature of biological tissue leads to broadening and amplitude attenuation of the TPSF. The multiple scattering events increase the optical path length and, hence, the mean time of flight. Therefore, early-arriving photons will deviate less from their initial trajectory and have shorter times of flight than late-arriving photons. TCSPC systems have high time resolution at the expense of long data acquisition times. In the simplest case, the fluorescence decay curve is a monoexponential process:

$$I(t) = I_0 \exp^{-\frac{t}{\tau}} \tag{13.1}$$

where I_0 is the fluorescence intensity immediately after excitation and τ is the fluorescence lifetime. Nevertheless, most fluorescence decays are multiexponential. Furthermore, the observed decay is the convolution of the true TPSF with the instrument response function (IRF; Han et al. 2008; Venugopal et al. 2010a). An ideal system should have an IRF with the width of the incident pulse. However, nonideal real detection systems and electronics result in the broadening of the IRF.

Another detection method used in fluorescence lifetime imaging is *time gating*. After the excitation pulse, only the photons that arrive within specified time windows are detected. Time-gating detection can be achieved by a picosecond gated intensified CCD (ICCD) camera (D'Andrea et al. 2003; Venugopal et al. 2010b). In this case, multiple projection images are recorded sequentially at different time delays after the excitation pulse. Hence, each pixel of the image contains information about fluorescence lifetime.

Time-gated detection of early-arriving photons can provide images of fluorescence objects with high spatial resolution but low signal-to-noise ratio (SNR), since very few photons reach the detector without scattering (Wu et al. 1997).

Alternatively, strobe pulses can be used to turn on the detector for very narrow time windows, which is known as the *stroboscopic* technique (James et al. 1992). The fluorescence TPSF can be obtained by using multiple excitation pulses and moving the time window after each pulse within the desired time range.

13.2.2 FREQUENCY DOMAIN

In FD systems, the light source is amplitude-modulated at frequencies of a few hundreds of megahertz (MHz). The emitted fluorescence signal is compared to the excitation reference signal, and the phase shift and amplitude are recorded. Fluorescence light has the same modulation frequency as excitation light; however, the lifetime of a fluorophore introduces a phase shift in the signal. Two fluorophores with different lifetimes cause different phase shifts in the signals. Fluorescence lifetime is related to the phase shift ϕ by the following expression:

$$\tau_p = \frac{1}{\omega} \tan(\phi) \tag{13.2}$$

where ω is the modulation frequency, and is related to the demodulation measurements by

$$\tau_m = \frac{1}{\omega} \sqrt{\frac{1}{m^2} - 1}, \tag{13.3}$$

where

$$m = \frac{I_{fmax}/I_{fmin}}{I_{emax}/I_{emin}} \tag{13.4}$$

is the demodulation calculated from the maximum and minimum intensity of the fluorescence signal, I_{fmax} and I_{fmin}, respectively, and the maximum and minimum intensity of the emission signal, I_{emax} and I_{emin}, respectively. In the case of monoexponential decay, $\tau_p = \tau_m = \tau$, whereas for multiexponential decay, $\tau_p < \tau_m$.

FD measurements are related to TD measurements via Fourier transform (Nothdurft et al. 2009). However, TD systems acquire measurements over a broad range of frequencies simultaneously. FD typically uses a single or a few discrete frequencies ranging from about 100 MHz up to 1 GHz. An advantage of FD systems is that ambient light can be filtered by rejection of unmodulated light.

FD systems operating at a high modulation frequency are required to detect fluorophores with short fluorescence lifetimes. However, detection of high-frequency signals can be challenging due to the loss of modulation and decreased performance of detectors. In FD systems, homodyne or heterodyne detection methods are used to convert high-frequency signals to low-frequency signals or constant (DC) signal (Chance et al. 1998). In *homodyne* detection, the detected light has the same modulation frequency as the excitation light, but it is out of phase. After frequency mixing the two signals, a DC signal is generated from which information about phase shift and demodulation can be extracted. In *heterodyne* detection, the detected signal is modulated at a slightly different frequency than the excitation signal, which after frequency mixing results in a low-frequency signal of a few kHz.

Measurement of multiple lifetimes requires several measurement using different excitation frequencies. Single frequency measurements can be recorded sequentially for a number of frequencies, which is a slow process. Fast data acquisition using *multiple frequencies* can be achieved by using the harmonics of a laser pulse train (Chance et al. 1998). For example, if a pulse train has a frequency ω, then the harmonics have frequencies of 2ω, 3ω, 4ω, and so forth, up to a high frequency of the order of Ghz.

Noncontact CCD-based imaging systems record amplitude and phase offsets at all pixels simultaneously.

13.2.3 ILLUMINATION AND DETECTION

Fiber-based imaging systems rely on point illumination and point detection techniques. In point illumination, only a small volume of the object of study is sampled by the excitation light, and therefore, the fluorescence target may not be excited, leading to sparse data sets. Image reconstruction from sparse data sets is challenging, and images generally exhibit artifacts and low resolution. These systems have high temporal resolution. However, they have the disadvantage of requiring either the optical fibers to be in contact with the surface of the object or coupling liquids surrounding the object with matching optical properties. Moreover, fiber-based systems have rigid geometries and use a limited number of fibers.

On the other hand, wide-field illumination techniques sample a large surface area simultaneously, which is clearly faster than using sequential point-source illumination strategies to sample the same area. Furthermore, a large volume of the object is sampled, and noncontact detection schemes can be used to measure light emerging from a large surface area of the object, providing measurements with improved information content. These systems produce large data sets that can provide reconstructed images with improved spatial resolution. However, it can be challenging and computationally consuming to obtain image reconstructions from large data sets.

Recently, wide-field illumination methods using structured light or illuminations patterns have been proposed in order to reduce both acquisition and computational times. The main idea behind this approach derives from the fact that tissue behaves like a low-pass filter for spatial Fourier components. Therefore, a small number of low-frequency illumination patterns can be used, without the risk of losing significant information content (D'Andrea et al. 2010; Ducros et al. 2011). Illumination patterns can be produced using a digital micromirror device (Ducros et al. 2011).

13.3 LIGHT TRANSPORT MODELS

Recovery of spatial and time-dependent information relies on clear understanding of the physics behind light transport in turbid media. In this section, we describe a general model based on the RTE. Only a limited set of solutions for the RTE are available analytically. Here, we describe methods for solving it numerically. Due to the computational overhead of solving the RTE, simplifications are used, which are valid under certain assumptions. We discuss the most popular approximations, viz., the diffusion and the telegraph equations. We also describe MC methods. Comparisons of different approaches are given.

13.3.1 RADIATIVE TRANSFER EQUATION

The main quantity characterizing the radiation is the intensity I, which can be introduced as follows. Let us consider an infinitesimal area dS perpendicular to the direction of radiation along the unit vector $\mathbf{s}$ at a particular point in the scattering medium $\mathbf{r}$. For the interval of frequencies of radiation $(\nu, \nu + d\nu)$ one would obtain the energy of radiation dE passing through dS within an infinitesimal solid angle $d^2\mathbf{s}$ within a time interval dt as

$$dE = I dS d^2 \mathbf{s} d\nu dt. \tag{13.5}$$

Therefore, the intensity of radiation is the coefficient of proportionality in this equation. Note that the intensity in Equation 13.5 is proportional to the photon distribution function, which is not the same as the intensity introduced as the time averaged Poyting vector (Born and Wolf 1968). Physical dimensions of the intensity can be easily determined from Equation 13.5 depending on the choice of the metric system. Some useful quantities characterizing the radiation can be introduced as well. The most common are (i) the average intensity

$$u(\mathbf{r}) = \frac{1}{4\pi} \int_{(4\pi)} I(\mathbf{r}, \mathbf{s}) d^2 \mathbf{s} \tag{13.6}$$

and (ii) the flux

$$q(r) = \frac{1}{4\pi} \int_{(4\pi)} s I(r,s) d^2 s. \tag{13.7}$$

Another quantity characterizing the radiation is the energy density ϱ, which relates to the average intensity as $\varrho = 4\pi u/c$, where c is the speed of light in the medium.

As a beam of radiation of intensity I travels through the medium in a particular direction s, it loses energy due to absorption of radiation and scattering to other directions s' and gains energy by emission and scattering from every direction s' to the given direction s. The differential form of the RTE is written in the form (Chandrasekhar 1960; Sobolev 1963)

$$\frac{1}{c}\frac{dI}{dt} = \frac{1}{c}\frac{\partial I}{\partial t} + s \cdot \nabla I = -\mu I + \mu_s B. \tag{13.8}$$

In Equation 13.8, (i) μI is the energy loss due to scattering and absorption, where μ is the transport coefficient, which is a sum of the scattering, μ_s, and absorption, μ_a, coefficients, and (ii) $\mu_s B$ is the energy gain due to scattering and emission. The function B in the source term reads

$$B = \int_{(4\pi)} p(s,s') I(s') d^2 s' + p(s,s_0) I_0, \tag{13.9}$$

where I_0 denotes the incident radiation, and the phase function $p(s,s')d^2 s'$ describes a probability for a photon incident in the direction s' to be scattered into the direction s. The RTE is the linear variant of the Boltzmann equation. Methods of solution of the RTE mainly depend on a particular form of the phase function.

In general, the phase function is anisotropic. Because it is a probability distribution function, it integrates over the unit sphere to unity

$$\int_{(4\pi)} p(s,s') d^2 s' = 1, \tag{13.10}$$

which also follows from the energy conservation low. In many physically interesting cases, the phase function possesses the azimuthal symmetry depending only on a scalar product $s \cdot s' = \cos \vartheta$. In such case, the phase function can be expanded over Legendre polynomials:

$$p(\cos \vartheta) = \sum_{l=0}^{\infty} p_l P_l(\cos \vartheta). \tag{13.11}$$

This expansion of the phase function leads to well-known P_N approximation to the RTE by truncating series 13.11 at $l = N$. The application of the addition theorem (Lebedev 1972) to Legendre polynomials results in the expansion of the phase function over spherical harmonics $Y_l^m(\theta,\phi)$, where $0 \leq l \leq N$, and $-l \leq m \leq l$. The intensity I belongs to a five-dimensional space, which can be decomposed into the coordinate space r and the velocity space $c = cs$. The velocity space is two-dimensional due to locally constant speed of light, c. By introducing the spherical system of coordinates (c, θ, ϕ) in the velocity space, one can expand the intensity over spherical harmonics (Jackson 1998; Lebedev 1972)

$$I(r,s) = \sum_{l=0}^{N} \sum_{m=-l}^{l} I_l^m(r) Y_l^m(\theta,\phi). \tag{13.12}$$

Substitution of Equations 13.11 and 13.12 into the RTE and use of the orthogonality property of spherical harmonics on the unit sphere results in the system of equation for unknown coefficients $I_l^m(\mathbf{r})$.

The most popular phase function used in modeling of the scattering phenomenon is the Henyey–Greenstein phase function

$$p(\cos\vartheta) = \frac{1}{4\pi}\frac{1-g^2}{(1+g^2-2g\cos\vartheta)^{3/2}}, \tag{13.13}$$

where g is the parameter controlling the shape of the phase function and satisfies $|g| < 1$. The case $g = 0$ gives isotropic scattering, while the case $g \to 1$ describes sharply peaked forward scattering.

Another approximation to the phase function was suggested by Reynolds and McCormick (1980). This approximation is useful for modeling super sharply peaked scattering and is based on the two-parameter formula

$$p(\cos\vartheta) = \frac{1}{4\pi}\frac{K}{(1+g^2-2g\cos\vartheta)^{\alpha+1}}, \tag{13.14}$$

where $\alpha > -1/2$ and

$$K = \frac{4\alpha g(1-g^2)^{2\alpha}}{(1+g)^{2\alpha}-(1-g)^{2\alpha}}. \tag{13.15}$$

Clearly, $\alpha = 1/2$ gives the Henyey–Greenstein phase function. In the case of the Reynolds-McCormick phase function, however, it is more convenient to use the Gegenbauer polynomials as the basis functions, and, therefore, the P_N approximation should be modified accordingly.

There are several exact and approximate analytical methods developed for solving the RTE. Most of them are covered in classical monographs by Chandrasekhar (1960) and Sobolev (1963). Analytical solutions of the RTE are found for simple geometries such as a layer, half-space, a cylinder, and a sphere. Constant scattering, μ_s, and absorption, μ_a, coefficients are usually assumed. The most commonly considered phase functions are (i) isotropic or (ii) a simple anisotropic function of the form

$$p(\cos\vartheta) = \frac{1}{4\pi}(1+\epsilon_1\cos\vartheta), \tag{13.16}$$

where $\epsilon_1 \in [-1,1]$. All these cases play a role of "a hydrogen atom" and are important for better understanding of the light scattering phenomenon. In many cases, the Rayleigh phase function plays an important role. The choice of the Rayleigh phase function will be described below (see Equation 13.50).

13.3.2 MC METHODS

Due to multiple scattering, the coherence property of the light is destroyed, and photons can be described as noninteracting particles scattered on heavy centers. That is, photons in the medium form a Lorentz gas. Moreover, considering a single scattering event, the probability for a photon to be scattered in a particular direction depends only on its current direction and, in contrast to the theory of the Boltzmann equation, does not depend on the distribution of other photons in the medium. Therefore, the scattering process of every single photon defines a Markov chain and can be efficiently modeled by methods of MC.

MC is one of the earliest numerical methods used for modeling light transport in biological tissues. The main idea behind MC is fairly simple. Photons are launched one by one into the medium from the source position. The source can be the point source, collimated source, or any arbitrary source distribution. The distance between consequent scattering events is sampled according to the photon mean free path length $l = 1/\mu$ (Wang et al. 1995). Each photon is regarded as a ray, or a packet, with the initial energy = 1. When scattering event occurs, a photon changes its direction of flight according to the phase function. The energy of the photon is attenuated as $\exp(-\mu_a L)$ along its path, where L is the total distance along the photon trajectory. The attenuated energy is deposited in each grid element visited by the photon. Sometimes there is no need to store photon energy inside the entire domain. In this case, photons are accumulated only on the boundary where a camera or detectors are located.

MC has been used in many applications involving light transport. There exists a vast body of literature devoted to development of MC codes. We could not cover MC methods in every detail. From our point of view, the most interesting approaches involve the modeling of polarized light transport (Wang et al. 2003) and, especially, recent advances in development of parallel MC algorithms accelerated by graphics processing units (GPUs; Alerstam et al. 2010; Fang and Boas 2009).

13.3.3 DIFFUSION AND TELEGRAPH EQUATION APPROXIMATIONS

An application of the diffusion approximation (DA) to the RTE in early 1990s revolutionized the whole field of biomedical optics (Arridge and Hebden 1997; Arridge and Schweiger 1995; Arridge et al. 1992, 1993, 1995; Schweiger et al. 1993). The DA is a simple and adequate description of the light scattering in large objects when the scattering dominates the absorption. The DA assumes the simplest anisotropic phase function, Equation 13.16, and results from the first two moments of the RTE, Equations 13.6 and 13.7. Sometimes it is termed as the P_1 approximation and can be derived by following the standard procedure by expanding the phase function and the intensity over the spherical harmonics. Here we present an analogous but slightly simpler derivation of the DA together with its companion, the telegraph equation approximation (TE).

The intensity is approximated by the expression

$$I \simeq u + 3\mathbf{s} \cdot \mathbf{q}, \tag{13.17}$$

where the average intensity, u, and the flux, $\mathbf{q}$, are defined according to Equations 13.6 and 13.7. Next, we introduce a spherical system of coordinates in the velocity space (c, θ, ϕ), where θ and ϕ are the polar and azimuthal angles, respectively. Therefore, $\mathbf{s} = (\sin\theta\cos\phi, \sin\theta\sin\phi, \cos\theta)^T$, where components of $\mathbf{s}$ can be considered as linear combinations of spherical harmonics $Y_1^m(\theta, \phi)$, where $-1 \leq m \leq 1$. It is seen that Equations 13.6 and 13.7 are recovered by multiplying Equation 13.17 by 1 and by $\mathbf{s}$ and integrating over the whole solid angle. Therefore, inserting Equation 13.17 into the RTE Equations 13.8 and 13.9 and integrating the RTE over the solid angle with weights 1 and $\mathbf{s}$, one would arrive at the system of two coupled first-order partial differential equations:

$$\left(\mu_a + \frac{1}{c}\frac{\partial}{\partial t} \right) u + \nabla \cdot \mathbf{q} = \mu_s u_0, \tag{13.18}$$

$$\left(1 + \frac{3\varkappa}{c}\frac{\partial}{\partial t} \right) \mathbf{q} + \varkappa \nabla u = \frac{\epsilon_1}{4\pi} \varkappa \mu_s \mathbf{s}_0 I_0, \tag{13.19}$$

where

$$u_0 = \frac{1}{4\pi} I_0, \tag{13.20}$$

and the diffusion coefficient is

$$\varkappa = (3\mu - \epsilon_1\mu_s)^{-1} = [3(\mu_a + \mu_s')]^{-1}, \tag{13.21}$$

where $\mu_s' = \mu_s(1 - \epsilon_1/3)$ is the reduced scattering coefficient. The source term in Equation 13.19 can be neglected, which results from the following consideration. Bearing in mind a weak formulation of the direct problem, we integrate Equation 13.18 over an infinitesimal volume and apply the Gauss theorem to the second term on the left-hand side. This gives $\int \mathbf{q} \cdot \mathbf{n} \, dS$, where the integration is performed over the surface enclosing the infinitesimal volume, and $\mathbf{n}$ is the outward surface normal. Then, making use of Equation 13.19 for computing $\int \mathbf{q} \cdot \mathbf{n} \, dS$, we see that the surface integral $\int \mu_s \varkappa I_0 \mathbf{s}_0 \cdot \mathbf{n} \, dS$ can be approximated by $\mu_s \varkappa I_0 \int \mathbf{s}_0 \cdot \mathbf{n} \, dS$ in regions where the optical parameters are continuous and smooth. Then, it is seen that the integral $\int \mathbf{s}_0 \cdot \mathbf{n} \, dS$ vanishes.

Applying the operator $1 + (3\varkappa/c)\partial/\partial t$ to Equation 13.18, one would obtain

$$\frac{1 + 3\varkappa\mu_a}{c}\frac{\partial u}{\partial t} + \frac{\varkappa}{c^2}\frac{\partial^2 u}{\partial t^2} - \nabla \cdot \varkappa\nabla u + \mu_a u = \mu_s\left(1 + \frac{3\varkappa}{c}\frac{\partial}{\partial t}\right)u_0. \tag{13.22}$$

Furthermore, in the case $\mu \gg \mu_a$, the term $3\varkappa\mu_a\partial u/\partial t$ can be neglected because $\varkappa\mu_a \ll 1$. The DA is obtained from Equation 13.22 by making one more step further by neglecting terms with $\partial^2 u/\partial t^2$ and $\partial u_0/\partial t$. This can be done when $c\bar{t} \gg \varkappa$, where $\bar{t}$ denotes the characteristic timescale of the problem. This results in the diffusion equation

$$\frac{1}{c}\frac{\partial u}{\partial t} - \nabla \cdot \varkappa\nabla u + \mu_a u = \mu_s u_0. \tag{13.23}$$

In the literature, Equation 13.22 is called the telegraph equation (TE) (Arridge 1999; Durian and Rundick 1997). The TE has a broader range of applicability than the DA (Equation 13.23) but more complex in the TD. Asymptotic solution of the TE is found for the cases of a half-space and a layer (Soloviev et al. 2003, 2004) with constant optical parameters. It is instructive to analyze the solution of the TE. Let us assume for the sake of simplicity an infinite space and isotropic scattering. Then, the asymptotic expression at long times and small absorption of the TE Green function reads

$$G_{TE} = C_0 \frac{c(t - t_0) + L}{2L^{5/2}}\exp\left\{-\mu c(t - t_0) + \frac{1}{2}\mu_s[c(t - t_0) + L]\right\}, \tag{13.24}$$

where C_0 is some dimensional constant and the interval L is given by

$$L = \left[c^2(t - t_0)^2 - 3|\mathbf{r} - \mathbf{r}_0|^2\right]^{1/2}. \tag{13.25}$$

The interval, L, is the real valued function. That is, when $|\mathbf{r} - \mathbf{r}_0| \geq (c/\sqrt{3})(t - t_0)$, the Green function is 0. This means that the average intensity propagates with the finite speed $c/\sqrt{3}$. The finite propagation speed is physically meaningful in comparison to the infinite propagation speed resulting from the DA. This is because the finite propagation speed is provided by the term $(\varkappa/c^2)\partial^2 u/\partial t^2$, which is neglected in the DA.

Therefore, in contrast to the DA, the TE allows studying very fast processes such as transition processes from ballistic regime (not scattered photon) to diffusive ones.

In the case

$$c(t - t_0) \gg \sqrt{3} |\mathbf{r} - \mathbf{r}_0|, \tag{13.26}$$

G_{TE} transforms to the solution of the corresponding diffusion equation (Equation 13.23). Thus, expanding the interval L into the Taylor series and retaining the first two terms in the exponent, we arrive at

$$G_{DA} \simeq \frac{C_0}{[c(t - t_0)]^{3/2}} \exp\left\{ -\mu_a c(t - t_0) - \frac{3\mu_s |\mathbf{r} - \mathbf{r}_0|^2}{4c(t - t_0)} \right\}. \tag{13.27}$$

A direct substitution of Equation 13.27 into the diffusion Equation 13.23 shows that, indeed, G_{DA} solves Equation 13.23 approximately when $\mu_s \simeq \mu$ (or $\varkappa \simeq 1/3\mu_s$). For fluorescence lifetime imaging in turbid media, the characteristic timescale, $\bar{\tau}$, is about a few nanoseconds, and physical dimensions vary from millimeters to centimeters. That is, the inequality equation 13.26 always holds implying applicability of the DA. Nevertheless, the TE is the method of choice in the case of short times and strong absorption (Durian and Rundick 1997).

The lifetime reconstruction requires time-dependent information describing evolution of a physical system. Acquired time-dependent data can be Fourier-transformed with respect to time to give the equivalent Fourier domain data at multiple harmonic samples. Lifetime reconstruction in the Fourier domain has significant advantages over the TD reconstruction due to its simplicity (Soloviev et al. 2007b, 2008, 2009). Therefore, it is sufficient to solve the TE or the DA in the Fourier domain, which is much simpler than solving them in the TD. The TE in the Fourier domain turns to the Helmholtz equation:

$$\Lambda u = \mu_s u_0, \tag{13.28}$$

where the differential operator is given by

$$\Lambda = -\nabla \cdot \kappa \nabla + \mu_a (1 + i\omega/\mu_a c), \tag{13.29}$$

the complex diffusion coefficient is defined as

$$\kappa = (3\tilde{\mu} - \mu_s \epsilon_1)^{-1}, \tag{13.30}$$

and the complex extinction (attenuation) coefficient is

$$\tilde{\mu} = \mu(1 + i\omega/\mu c). \tag{13.31}$$

Equation 13.28 must be supplied with boundary conditions, which follow from Equation 13.17 and the Fourier analog of Equation 13.19. That is, $I(\mathbf{s} \cdot \mathbf{n} < 0) = \gamma I(\mathbf{s} \cdot \mathbf{n} > 0)$ at the open boundary of the scattering domain, where $\mathbf{n}$ is the surface normal, and γ is a constant depending on the refractive index mismatch (Schweiger et al. 1995). The Fourier analog of the DA follows from Equation 13.28 by neglecting dimensionless parameter $\omega/\mu c$ in the complex diffusion coefficient (Equations 13.30 and 13.31), when $\omega/\mu c \ll 1$. In the Fourier domain, both approximations give almost the same results at low frequencies ω and for a weak absorption.

It is also interesting to notice that the Helmholtz equation remains formally the same if we set $\epsilon_1 = 0$, which affects only the value of the diffusion coefficient. This admits the following interpretation. Recalling that the transport coefficient, μ, is reciprocal to photon's mean path length l, the efficient mean path length can be introduced as $l_{eff} = 1/\mu_{eff}$, where $\mu_{eff} = \mu(1 - \lambda \epsilon_1/3)$, and $\lambda \in [0,1]$ is the albedo of a single scattering event. Therefore, the light diffusion can be considered as a quasi-isotropic scattering process with longer photon's mean path length.

Green's function is a solution of the corresponding equation with the point-like source, which is usually modeled by the δ-function. Solution of the DA or TE for arbitrary distributed sources is found by convolving a Green's function with the source distribution, $f(\mathbf{r}',t')$, over space and time

$$U(\mathbf{r},t) = \int_{t_0}^{t} dt' \int_{V} G(\mathbf{r} - \mathbf{r}', t - t') f(\mathbf{r}',t') d^3\mathbf{r}'. \tag{13.32}$$

To illustrate the application of the above formula, we compute the energy density in the infinite space with the source term in the form of a spherical wave:

$$f(\mathbf{r},t) = \mu_s I_0 \frac{\exp(-\mu|\mathbf{r} - \mathbf{r}_0|)}{4\pi|\mathbf{r} - \mathbf{r}_0|^2} \delta\left(t - t_0 - |\mathbf{r} - \mathbf{r}_0|/c\right). \tag{13.33}$$

The convolution of the this source term with the Green's function of the telegraph equation was computed in the Fourier domain and contains two terms:

$$U = C\tilde{\mu} \exp(-i\omega t_0)\left[\frac{1}{2\eta} \ln\left(\frac{\tilde{\mu} + \eta}{\tilde{\mu} - \eta}\right)G + D\right]. \tag{13.34}$$

Here, G is the Fourier image of the Green function:

$$G = \frac{\exp(-\eta|\mathbf{r} - \mathbf{r}_0|)}{4\pi|\mathbf{r} - \mathbf{r}_0|}, \tag{13.35}$$

and the term D is given by

$$D = \frac{1}{4\pi|\mathbf{r} - \mathbf{r}_0|} \int_{\tilde{\mu}}^{\infty} \frac{\exp(-\nu|\mathbf{r} - \mathbf{r}_0|)}{\eta^2 - \nu^2} d\nu, \tag{13.36}$$

where

$$\eta^2 = 3\tilde{\mu}(\tilde{\mu} - \mu_s). \tag{13.37}$$

The obtained solution (Equation 13.34) can be analyzed for several values of parameters $\tilde{\mu}$ and η depending on the value of the albedo λ, the transport coefficient μ, and ω. In the case $|\tilde{\mu}| \gg |\eta|$, the term D behaves as $\exp(-\tilde{\mu}|\mathbf{r} - \mathbf{r}_0|)/|\mathbf{r} - \mathbf{r}_0|$ and can be neglected. Therefore, the energy density can be described by the Green's function G alone:

$$U\big|_{|\tilde{\mu}| \gg |\eta|} \simeq C \exp(-i\omega t_0)G. \tag{13.38}$$

It is clear that Equation 13.38 describes the diffusion regime, which corresponds to long times and distances. The case $|\tilde{\mu}| \ll |\eta|$ is the opposite one. The second term D dominates and the contribution to the energy density from the Green's function can be ignored, which results in

$$U \big|_{|\tilde{\mu}| \ll |\eta|} \simeq \frac{C\tilde{\mu}}{2\eta^2} \frac{\exp\left(-\tilde{\mu}|\mathbf{r} - \mathbf{r}_0| - i\omega t_0\right)}{4\pi|\mathbf{r} - \mathbf{r}_0|^2}. \tag{13.39}$$

Equation 13.39 is applicable to very short times and distances and propagates with speed of light c in the TD. Probably the most interesting case is the transition regime from the ballistic to the diffusion one, which corresponds to $|\tilde{\mu}| \sim |\eta|$ and results from removing logarithmic singularity in Equation 13.34:

$$U \big|_{|\tilde{\mu}| \sim |\eta|} \simeq \frac{C\tilde{\mu}}{2\eta} \exp(-i\omega t_0) \ln\left[(\tilde{\mu} + \eta)|\mathbf{r} - \mathbf{r}_0|\right] G. \tag{13.40}$$

Noticeably, this solution contains the factor $\ln\left[(\tilde{\mu} + \eta)|\mathbf{r} - \mathbf{r}_0|\right]/\eta$, which is responsible for quite different spatial, frequency, and temporal behavior than the behavior of the Green's function alone.

13.3.4 FOKKER–PLANCK EQUATION

It is possible to replace the RTE with the Fokker–Planck equation in the case of a sharply peaked phase function (Kim 2004; Kim and Keller 2003; Kim and Moscoso 2004; Larsen 1999; Lehtikangas et al. 2010). Here, we briefly describe this approach. The collision term in Equation 13.9 is simplified by expansion of the intensity in a Taylor series about $\mathbf{s}$. Thus, $\mathbf{s}' = \mathbf{s} + \Delta\mathbf{s}$, where $\Delta\mathbf{s} = \mathbf{s}' - \mathbf{s}$. Retaining the first three terms in the intensity, we have

$$I(\mathbf{s}') \simeq I(\mathbf{s}) + \Delta s_i \frac{\partial I}{\partial s_i} + \frac{1}{2}\Delta s_i \Delta s_j \frac{\partial^2 I}{\partial s_i \partial s_j}, \tag{13.41}$$

where a summation over repeated indices is assumed, and s_i and Δs_i, $i = 1,2,3$, are components of $\mathbf{s}$ and $\Delta\mathbf{s}$, respectively. Direct substitution of Equation 13.41 into Equations 13.8 and 13.9 results in

$$\frac{1}{c}\frac{\partial I}{\partial t} + \mathbf{s} \cdot \nabla I + \mu_a I = \mu_s B, \tag{13.42}$$

where the function B becomes

$$B \simeq p(\mathbf{s} \cdot \mathbf{s}_0)I_0 + (v_i - s_i)\frac{\partial I}{\partial s_i}$$
$$+ \frac{1}{2}(s_i s_j - s_i v_j - s_j v_i + w_{ij})\frac{\partial^2 I}{\partial s_i \partial s_j}, \tag{13.43}$$

and the tensors v_i and w_{ij} are given by

$$v_i = \int_{(4\pi)} s_i' p(\mathbf{s} \cdot \mathbf{s}') d^2\mathbf{s}', \tag{13.44}$$

$$w_{ij} = \int_{(4\pi)} s_i' s_j' p(\mathbf{s} \cdot \mathbf{s}') d^2\mathbf{s}'. \tag{13.45}$$

In Equations 13.43–13.45, integration should be performed over small solid angle about **s**. However, we made use of sharpness of the phase function and extended integration over the whole solid angle. The function B (Equation 13.43) can be significantly simplified. As mentioned above, photons are distributed on a sphere (c, θ, ϕ) in the velocity space $c\mathbf{s}$. From this, it follows that vectors **s** and $\partial I/\partial \mathbf{s}$ are orthogonal, that is, $s_i \partial I/\partial s_i = 0$. Furthermore, as it is seen from Equation 13.44, v is the mean value of $\mathbf{s}'$. Therefore, v points in the same direction as **s**, and the term $v_i \partial I/\partial s_i$ vanishes as well. Then, B simplifies to

$$B \simeq \frac{1}{2} w_{ij} \frac{\partial^2 I}{\partial s_i \partial s_j} + p(\mathbf{s} \cdot \mathbf{s}_0) I_0. \tag{13.46}$$

Equation 13.42 with the source term in the form of Equation 13.46 is known as the Fokker–Planck equation. The tensor w_{ij} can be computed if the phase function is known. In the literature, mostly the Henyey–Greenstein phase function is considered. However, we believe that the two-parameter Reynolds–McCormick phase function (Equation 13.14) is a more general and, therefore, preferable model of the sharply-peaked phase function. Here, we present a brief and rather sketchy outline of further simplification of Equation 13.46.

First we recognize that the term $(1 + g^2 - 2g\cos\vartheta)^{-\alpha-1}$ in the Reynolds–McCormick phase function is the generating function for the Gegenbauer polynomials, $C_l^{\alpha+1}(\cos\vartheta)$ (Lebedev 1972). It can be shown by making use of the power series of these polynomials, for instance, that only diagonal elements of w_{ij} survive, while nondiagonal elements are 0 ($i \neq j$). Then, the terms with $\partial^2 I/\partial s_i^2$ only remain, which can be expressed in the spherical coordinates (θ, ϕ) (Larsen 1999).

The Fokker–Planck equation was considered in the context of its applicability to mesoscopic imaging of live objects. That is, light transport in a live object usually involves light scattering, whose influence depends strongly on the object size in terms of the photon mean path length. For very small objects scattering may be negligible, whereas for large objects, scattering is so dominant that light propagation can be considered as a diffusion process. In between these extremes, for objects varying in size from several hundred micrometers to several millimeters, multiple scattering is present but not fully diffuse; this regime is known as the mesoscopic scale.

Although the Fokker–Planck equation was already used in imaging of small biological samples (Vinegoni et al. 2008), its applicability to mesoscopic imaging remains questionable. Thus, an application of the Fokker–Planck equation assumes the presence of snake photons, which travel close to the shortest path in the medium. However, in the case of collimated intensity measurements, ballistic and diffusive photons are mostly collected by the CCD array, while snake photons are clipped out of recorded intensity. Secondly, the choice of a phase function adequately describing scattering properties of biological tissues at the mesoscopic scale remains uncertain. It was found earlier (Soloviev et al. 2011a) from a simple angularly resolved imaging experiment of a weakly scattering medium that the Rayleigh phase function with dominant isotropic part could be a reasonable choice of the law of scattering for materials such as Intralipid.

13.3.5 MESOSCOPIC-SCALE APPROXIMATIONS

The DA or the TE does not consider the propagation of light according to geometric optics since it assumes that all photons are scattered from their original ballistic trajectories. The DA or the TE is not a good approximation for scattering objects from which photons have a finite probability to leave without any scattering events. Realistic light transport models for small or weakly scattering objects should take account of both photon diffusion and geometrical optics. These properties are described by the RTE (Sobolev 1963), but the numerical solution of the RTE is highly challenging with prohibitive computational overheads for most applications. In this section, we consider novel, computationally inexpensive approximations to the RTE, which combine light diffusion and ray optics.

Including light propagation along rays into the light transport model is important for interpretation of images taken by the CCD camera. Images are formed from the energy of absorbed photons by pixels of the CCD array within the camera's exposure time. Photons reach the CCD array by propagating along rays.

Thus, the intensity of light rays per pixel area and per exposure time gives the absorbed energy by a pixel of the CCD array (Equation 13.5). Absorbed energy is further transformed into an image. Therefore, the time-gating technique relies on intensity-based collimated measurements. The DA, for instance, describes the average intensity (or energy density), which is the first moment of the intensity (Equation 13.6). The method of Schwarzschild–Schuster (Sobolev 1963) can be utilized for an image's transformation from intensity measurements to energy density measurements (Soloviev et al. 2010). However, this method has its limitations. Thus, a model describing the intensity of light rather than its moments is preferable.

Let us start with the simplest case of the isotropic phase function and consider the problem in the Fourier domain (ω). In this case, the function B (Equation 13.9) simplifies to

$$B = u + u_0, \tag{13.47}$$

where u and u_0 are given by Equations 13.6 and 13.20, respectively. When the average intensity is found by solving the TE or the DA, then the solution of the RTE can be computed by integrating along ray paths:

$$I(\mathbf{r},\mathbf{s}) = I_0 + \int_0^{l_{max}} \mu_s(\mathbf{r} - \mathbf{s}l)$$

$$\times B(\mathbf{r} - \mathbf{s}l) \exp\left(-\int_0^l \tilde{\mu}(\mathbf{r} - \mathbf{s}l')dl'\right) dl, \tag{13.48}$$

where the integration is performed from the observation point $\mathbf{r}$ in the reverse direction of light propagation $-\mathbf{s}$, and l_{max} denotes the maximum distance of a ray's path contributing to the intensity. If the distribution of sources is known, the intensity of the direct radiation I_0 is given by a convolution of the Green function with the source distribution. The Green function, I_0, satisfies the equation

$$\mathbf{s} \cdot \nabla I_0 + \tilde{\mu} I_0 = Q_0 \delta(\mathbf{r} - \mathbf{r}_0)\delta(\mathbf{s} - \mathbf{s}_0), \tag{13.49}$$

where Q_0 is the source amplitude at $\mathbf{r}_0$. Equation 13.48 connects ray optics and light diffusion as follows. Let us assume that μ is large enough, providing a very fast decay of the exponent. Then, in highly scattering media, $\mu_s B$ is a slowly varying function in comparison to the exponent and can be taken out from the integral at the observation point $\mathbf{r}$. Applying the Laplace method (Olver 1974) and neglecting ballistic and singly scattered photons, one obtains approximately $I \simeq \lambda u$, where λ is the albedo.

The methodology of combining ray optics and light diffusion can be extended to the case of the simplest form of anisotropic phase function (Equation 13.16; Soloviev and Arridge 2011a, b). However, it is more interesting to consider the phase function containing the first three terms in its series over the Legendre polynomials (Equation 13.11; Soloviev et al. 2011a). Therefore, we write the phase function as

$$p(\cos\vartheta) \simeq \frac{1}{4\pi}\left(\frac{3}{3+\epsilon_2} + \epsilon_1 \cos\vartheta + \frac{3\epsilon_2}{3+\epsilon_2}\cos^2\vartheta\right), \tag{13.50}$$

where the coefficient $3/(3+\epsilon_2)$ results from the normalization condition (Equation 13.10), and $\epsilon_2 \geq 0$. Thus, the case $\epsilon_1 \neq 0$ and $\epsilon_2 = 0$ provides the simplest form of the anisotropic phase function, while the case $\epsilon_1 = 0$ and $\epsilon_2 = 1$ gives the well-known Rayleigh phase function (Chandrasekhar 1960; Sobolev 1963; van de Hulst 1981).

Now, the intensity is approximated by three terms:

$$I \simeq u + 3\mathbf{s} \cdot \mathbf{q} + \frac{15}{2}\overset{\circ}{\mathbf{s}\mathbf{s}} : \hat{\mathbf{g}}, \tag{13.51}$$

Analysis of fluorescence lifetime data

where $\overset{\circ}{s}s$ denotes the nondivergent dyadic tensor built from $\mathbf{s}$

$$
\overset{\circ}{s}s = \begin{pmatrix} \sin^2\theta\cos^2\phi - 1/3 & \dfrac{1}{2}\sin^2\theta\sin 2\phi & \dfrac{1}{2}\sin 2\theta\cos\phi \\[2mm] \dfrac{1}{2}\sin^2\theta\sin 2\phi & \sin^2\theta\sin^2\phi - 1/3 & \dfrac{1}{2}\sin 2\theta\sin\phi \\[2mm] \dfrac{1}{2}\sin 2\theta\cos\phi & \dfrac{1}{2}\sin 2\theta\sin\phi & \cos^2\theta - 1/3 \end{pmatrix}.
\tag{13.52}
$$

The symbol " : " denotes the double product of two tensors $\mathbf{a}$ and $\mathbf{b}$ such as $\mathbf{a}:\mathbf{b} = \sum_{ij} a_{ij}b_{ji}$. The average intensity u and the flux $\mathbf{q}$ in Equation 13.51 are defined by Equations 13.6 and 13.7. The new quantity $\hat{\mathbf{g}}$ in Equation 13.51 is an analog of the photon pressure tensor:

$$
\hat{\mathbf{g}} = \frac{1}{4\pi}\int_{(4\pi)} \overset{\circ}{s}s I(\mathbf{s})d^2\mathbf{s}.
\tag{13.53}
$$

Note that the nondivergent dyadic tensor $\overset{\circ}{s}s$ contains only five linearly independent entries. Therefore, the expansion of I over the orthogonal basis $(1,\mathbf{s},\overset{\circ}{s}s)$ (Equation 13.51) is completely analogous to the expansion of I over the spherical harmonics $Y_l^m(\theta,\phi)$, where $0 \le l \le 2$, and $-l \le m \le l$, due to the linear dependence of functions forming the basis $(1,\mathbf{s},\overset{\circ}{s}s)$ and $Y_l^m(\theta,\phi)$.

The substitution of the approximate intensity into the function B (Equation 13.9) gives

$$
\begin{aligned}
B = u &+ \epsilon_1 s_i q_i + p(\mathbf{s}\cdot\mathbf{s}_0)I_0 \\
&+ \frac{3\epsilon_2}{3+\epsilon_2}\left(s_i s_j - \frac{1}{3}\delta_{ij}\right)\left(g_{ij} - \frac{g}{3}\delta_{ij}\right).
\end{aligned}
\tag{13.54}
$$

Here, and in the rest of this section, a summation over repeated indices is assumed, and $g/3 = g_{ii}/3$.

Next, we find equations satisfied by the average intensity u, the flux $\mathbf{q}$, and the tensor $\hat{\mathbf{g}}$, by computing the moments of the RTE (Equation 13.8). Multiplying the RTE consequently by 1, $\mathbf{s}$, and $\overset{\circ}{s}s$ and integrating over the whole solid angle, we arrive at the system of first-order partial differential equations:

$$
\frac{\partial q_i}{\partial x_i} + (\mu_a + i\omega/c)u = \mu_s u_0,
\tag{13.55}
$$

$$
q_i = -\kappa\frac{\partial u}{\partial x_i} - 3\kappa\frac{\partial}{\partial x_j}\left(g_{ij} - \frac{g}{3}\delta_{ij}\right),
\tag{13.56}
$$

$$
g_{ij} - \frac{g}{3}\delta_{ij} = -\frac{\sigma}{2}\left(\frac{\partial q_i}{\partial x_j} + \frac{\partial q_j}{\partial x_i}\right) + \frac{\sigma}{3}\delta_{ij}\frac{\partial q_k}{\partial x_k},
\tag{13.57}
$$

where the complex diffusion coefficient κ is the same as Equation 13.30, and an analog of the viscosity coefficient is defined by

$$\sigma = \left(\frac{5}{2}\tilde{\mu} - \frac{\mu_s \epsilon_2}{3 + \epsilon_2} \right)^{-1}. \tag{13.58}$$

The source term in Equations 13.56 and 13.57 are neglected for the reasons explained right after Equations 13.18 and 13.19 in Section 13.3.3.

It is worth solving the system (Equations 13.55 through 13.57) numerically. However, solving the system requires sufficient amount of memory for allocation of a nine-dimensional vector at every point of the domain. Moreover, in the context of the inverse problem, solving the system could be exceedingly expensive. Therefore, we seek an approximate solution of this system. Departing from the TE, we define the zero-order approximation for the flux as

$$q_j^{(0)} \simeq -\kappa \frac{\partial u}{\partial x_j}, \tag{13.59}$$

which results in the zero-order approximation for the tensor $\hat{\mathbf{g}}$

$$g_{ij}^{(0)} - \frac{g^{(0)}}{3}\delta_{ij} \simeq \frac{\sigma}{2}\left(\frac{\partial}{\partial x_j}\kappa\frac{\partial u}{\partial x_i} + \frac{\partial}{\partial x_i}\kappa\frac{\partial u}{\partial x_j} \right)$$
$$- \frac{\sigma}{3}\delta_{ij}\frac{\partial}{\partial x_k}\kappa\frac{\partial u}{\partial x_k}. \tag{13.60}$$

Equations 13.59 and 13.60 are substituted into the function B (Equation 13.54). Performing summation, we take into account that the terms $\partial/\partial x_j \kappa \partial u/\partial x_i$ are counted twice. Then, making use of the identity $s_i \partial u/\partial x_i = \partial u/\partial l$, we arrive at

$$B \simeq u - \epsilon_1 \kappa \frac{\partial u}{\partial l} + p(\mathbf{s}\cdot\mathbf{s}_0)I_0$$
$$+ \frac{3\epsilon_2\sigma}{3+\epsilon_2}\frac{\partial}{\partial l}\kappa\frac{\partial u}{\partial l} - \frac{\epsilon_2\sigma}{3+\epsilon_2}\nabla\cdot\kappa\nabla u. \tag{13.61}$$

Furthermore, the last term $-\nabla\cdot\kappa\nabla u$ in Equation 13.61 is replaced with $\lambda\mu u_0 - (\mu_a + i\omega/c)u$ in accordance with Equation 13.55.

Summarizing, in order to compute the intensity, I, we first compute I_0 and use it for computing the source term in the TE (Equation 13.28). Solving the TE, we find the function B, and finally compute the intensity according to Equation 13.48. The most expensive step here is the numerical solution of the TE, while the ray integration is inexpensive if performed by using the Siddon algorithm (Siddon 1985).

It is instructive to compare light scattering according to the Rayleigh phase function ($\epsilon_1 = 0$ and $\epsilon_2 = 1$) against scattering according to the simplest anisotropic phase function Equation 13.16 ($\epsilon_1 = 1$ and $\epsilon_2 = 0$). For this purpose, we construct a numerical phantom consisting of weakly and highly scattering regions, whose transport coefficients differ by an order of magnitude. We further assume that recorded photons coming from weakly scattering regions are scattered only once, that is, $1/\mu$ is a length scale on the order of physical dimensions of the scattering domain. For such case, the method of successive approximations (Sobolev 1963) provides the function B in the form of the scattered once direct radiation $p(\mathbf{s}\cdot\mathbf{s}_0)I_0$.

Therefore, in weakly scattering regions, terms containing u in Equation 13.61 are neglected. In our simulations, a cylinder with the background transport coefficient $\mu = 0.1$ mm^{-1} and albedo $\lambda = 0.9$ serves as a weakly scattering medium. A highly scattering object is embedded within a weakly scattering cylinder, which is a knot made of two scattering tangled tori, whose values of the transport coefficient and albedo are set to $\mu = 0.75$ mm^{-1} and $\lambda = 0.9$, respectively. In the middle of the knot, two bent absorbing rods with reduced values of albedo, $\lambda = 0.5$, are inserted. The direct light I_0 is taken in the form of parallel rays entering the domain along the unit vector $\mathbf{s}_0 = 2^{-1/2}(1,0,-1)^T$ (Equation 13.49), with $\mathbf{r}_0$ lying on a source plane $\mathbf{r}_0 \cdot \mathbf{s}_0 = $ const. The amplitude of the direct radiation in the Fourier domain was set to 10 in energy units per square millimeter. A CCD camera rotates around the weakly scattering cylinder starting from its initial position with $\mathbf{n} = (1,0,0)^T$, where $\mathbf{n}$ is the outward camera's normal.

The CCD camera images are shown in Figure 13.1. The difference in observed scattered light depending on the phase function is clearly illustrated in these images. The upper row in Figure 13.1 shows real parts of the camera images of scattered light in accordance with the Rayleigh phase function, $\epsilon_1 = 0$ and $\epsilon_2 = 1$. The second row (Figure 13.1e through h) shows the camera images of scattered light according to the phase function 13.16 ($\epsilon_1 = 1$ and $\epsilon_2 = 0$). Images are shown for the camera rotation angles {0°, 90°, 180°, 270°}. Computations were performed on 100 × 100 × 200 Cartesian grid corresponding to the physical

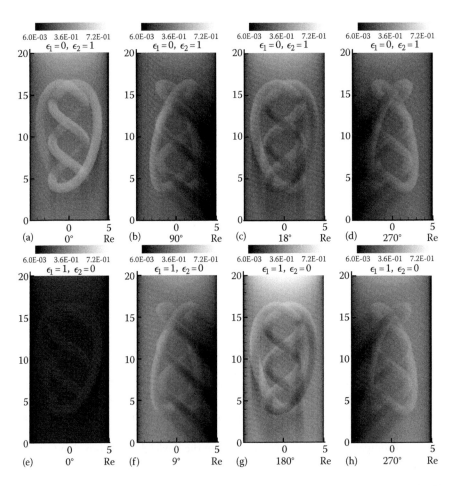

Figure 13.1 Camera images displaying real parts of Fourier transformed intensity scattered on a highly scattering knot embedded within a weakly scattering cylinder. The upper row (a–d) shows real parts of camera images of scattered light according to the Rayleigh phase function, $\epsilon_1 = 0$ and $\epsilon_2 = 1$. The second row (e–h) shows camera images of scattered light according to the simplest anisotropic phase function, $\epsilon_1 = 1$ and $\epsilon_2 = 0$. Images are shown for camera rotation angles {0°, 90°, 180°, 270°}.

dimensions of the computational domain $10 \times 10 \times 20$ mm. Free space propagation is assumed outside the weakly scattering medium with values of the transport coefficient and the albedo set to 0 and 1, respectively. Images illustrate the importance of knowledge of the phase function in optical tomographic imaging.

13.3.6 NUMERICAL METHODS

Numerical methods of solving the light transport equations are a growth area in biomedical optics. In this section, we cover a few numerical techniques for solving the TE and the DA including a finite volume (FV) numerical scheme, a finite element method (FEM), and a discontinuous Galerkin method (DG). We do not cover other methods such as the boundary element method (Elisee et al. 2011), meshless methods, and many other numerical methods.

The simplest and very robust numerical scheme, which is widely used for solving partial differential equations, is the FV. The FV assumes constant values of unknowns at each computational cell of an appropriately discretized computational domain. Its derivation is fairly simple for a Cartesian grid. Thus, integrating Equation 13.55 over a cell of the volume ΔV, we obtain

$$\sum_k \mathbf{q} \cdot \mathbf{n}_k (\Delta S_k / \Delta V) + (\mu_a + i\omega/c)u = \mu_s u_0, \tag{13.62}$$

where ΔS_k is an area of the kth cell's interface, $\mathbf{n}_k$ is the cell's interface normal, and $\mathbf{q}_k$ is understood here as an interface flux. Denoting jumps and averages of some quantity a across the kth cell's interface as $[a]_k = a - a'$ and $\{a\} = 1/2(a + a')$, respectively, where a' denotes the value of quantity a in the neighboring cell, we integrate Equation 13.59 across the kth cell's interface and arrive at

$$\mathbf{q}_k \cdot \mathbf{n}_k \simeq \{\kappa\}_k [u]_k (\Delta S_k / \Delta V). \tag{13.63}$$

Insertion of Equation 13.63 into Equation 13.62 results in the FV numerical scheme. Because Equation 13.63 contains $[u]_k = u - u'$, the interface flux $\mathbf{q}_k$ is split into outgoing and incoming fluxes, which enter to the system matrix.

The FV can be easily generalized for adaptive meshes. Consider a computational domain, which is initially represented by a single cell, which could be a cube, hexahedra, or tetrahedra. Adaptive meshes are generated recursively by refining cells into child cells. For the hexahedral and tetrahedral meshes, the most convenient data structure supporting this recursive refinement is an octal tree (Carey 1997). This technique is computationally effective because meshes can evolve during computation (Soloviev 2006; Soloviev and Krasnosselskaia 2006). Adaptive meshes introduce two cases to consider while computing interface fluxes: (i) neighboring cells are finer cells, and (ii) the neighboring cell is a coarser cell. In the first case, the incoming flux is the sum of fluxes. In the second case, neighboring coarser cells can be refined up to the level the current cell. In order to avoid information loss, coarsening of finer cells is not performed but rather the volumetric average is computed.

Another very popular numerical method is the FEM. The FEM is widely used for solving parabolic and elliptic differential equations in complex geometries and covered in many textbooks (see, for instance, Hughes 2003; Zienkiewicz et al. 2005). Here we provide a brief overview of the FEM for the Fourier domain (ω). The entire computational domain is divided into elements (or cells), and all functions entering the Helmholtz equation (Equations 13.28 and 13.29) are expanded over the shape functions $\phi_i(\mathbf{r})$ satisfying the completeness condition $\sum_i \phi_i = 1$ inside a cell. It is convenient to introduce the notation $\tilde{\mu}_a = \mu_a + i\omega/c$. Then

$$u = u_i \phi_i, \quad u_0 = u_{0,i} \phi_i. \tag{13.64}$$

$$\kappa = \kappa_k \phi_k, \quad \tilde{\mu}_{a,k} = \tilde{\mu}_{a,k} \phi_k, \tag{13.65}$$

where a summation over repeated indices is assumed. Substitution of these expansions into the Helmholtz equation (Equation 13.28) and consequent integration of the equation multiplied by $\phi_j(\mathbf{r})$ over the entire computational domain result in

$$\{a_{ijk}\kappa_k + b_{ijk}\tilde{\mu}_{a,k} + c_{ij}\}u_i = b_{ijk}\mu_{s,k}u_{0,i}, \tag{13.66}$$

where

$$a_{ijk} = \int_V (\nabla\phi_i \cdot \nabla\phi_j)\phi_k d^3\mathbf{r}, \tag{13.67}$$

$$b_{ijk} = \int_V \phi_i\phi_j\phi_k d^3\mathbf{r}, \tag{13.68}$$

$$c_{ij} = \frac{1}{3}\frac{1-\gamma}{1+\gamma}\oint_{\partial V} \phi_i\phi_j d^2\mathbf{r}. \tag{13.69}$$

In Equation 13.66, the term c_{ij} (Equation 13.69) results from the boundary conditions

$$\kappa u_i\mathbf{n}\cdot\nabla\phi_i = -\frac{1}{3}\frac{1-\gamma}{1+\gamma}\phi_i u_i, \tag{13.70}$$

where $\mathbf{n}$ is the outward surface normal, and γ is a constant depending on the refractive index mismatch.

Equation 13.66 can be written in matrix notation as $\mathbf{Au} = \mathbf{u}_0$, whose solution involves sparse matrix inversion. FEM can be further extended to dynamically adaptive meshes as was done (Joshi et al. 2006a, b). However, implementation of the mesh adaptation technique is much simpler when FV or DG numerical schemes are employed.

The DG is a relatively new numerical technique (Rivière 2008). Therefore, we provide a rather detailed overview of this technique in the Fourier domain (ω) following our earlier publication (Soloviev et al. 2010). The DG is the combination of the FV and the FEM. The most valuable feature of DG is the ease of mesh adaptation inherited from FV, which is important for three-dimensional problems. Secondly, the DG method effectively deals with discontinuities in the solution, which may be present in media having, for instance, internal refractive index mismatches (Mohan et al. 2011). On the other hand, an application of the DG numerical scheme to inverse problems has several disadvantages. One of them is the rather high computational cost due to the higher degree of freedom it offers.

We start with the Helmholtz equation written as a system of two first-order equations:

$$\nabla\cdot\mathbf{q} + \tilde{\mu}_a u = \mu_s u_0, \tag{13.71}$$

$$\mathbf{q} = -\kappa\nabla u. \tag{13.72}$$

As above, the computational domain is divided into cells, where the solution of Equations 13.71 and 13.72 is expanded over shape functions $\phi_i(\mathbf{r})$. For the sake of computational performance, we represent optical parameters as piecewise constant functions, following the FV framework, while all other functions are expanded over a piecewise linear basis. Thus, we assume that the optical parameters are slowly varying functions inside computational cells but may suffer jumps across cells' interfaces. Therefore, only the energy density and the source term are represented by Equation 13.64. Multiplying Equation 13.55 by ϕ_j and integrating over the cell's volume, we arrive at a weak formulation of the direct problem in the local form:

$$w_j + \kappa a_{ij}u_i + \tilde{\mu}_a b_{ij}u_i = \mu_s b_{ij}u_{0,i} \tag{13.73}$$

where matrix elements are given by

$$w_j = \oint_{\partial V} (\mathbf{q} \cdot \mathbf{n})\phi_j d^2\mathbf{r}, \tag{13.74}$$

$$a_{ij} = \int_V (\nabla\phi_i \cdot \nabla\phi_j) d^3\mathbf{r}, \tag{13.75}$$

$$b_{ij} = \int_V \phi_i\phi_j d^3\mathbf{r}. \tag{13.76}$$

Here V denotes the cell's volume having an outward normal $\mathbf{n}$, and ∂V denotes the cell's interface. In contrast to FEM, at this stage, the local form (Equation 13.73) consists of a system of uncoupled equations. Coupling between cells is provided by replacing $\mathbf{q}$ with the interface flux, which is derived below.

A sum of fluxes at cells' interfaces (Equations 13.45) together with the observation that $\mathbf{n} = -\mathbf{n}'$ results in

$$w_j + w_{j'} = \oint_{\partial V} [\mathbf{q}\phi_j] \cdot \mathbf{n} d^2\mathbf{r}. \tag{13.77}$$

This expression is further transformed into the following:

$$w_j + w_{j'} = \oint_{\partial V} (\{\mathbf{q}\}[\phi_j] + [\mathbf{q}]\{\phi_j\}) \cdot \mathbf{n} d^2\mathbf{r}, \tag{13.78}$$

where [·] and {·} denote a jump and an average across a cell interface defined above, respectively. Noting that the condition $[\mathbf{q}] \cdot \mathbf{n} = 0$ is satisfied for the exact solution and taking into account the flux equation $\mathbf{q} = -\kappa u_i \nabla\varphi_i$ simplifies Equation 13.74 to

$$w_j = u_i f_{ij} + u_{i'} f_{i'j}, \tag{13.79}$$

where the interface flux contribution to the jth row and the ith and i'th columns of the system matrix is

$$f_{ij} = -\frac{1}{2}\kappa \oint_{\partial V} \phi_j \mathbf{n} \cdot \nabla\phi_i d^2\mathbf{r},$$

$$f_{i'j} = -\frac{1}{2}\kappa' \oint_{\partial V} \phi_j \mathbf{n} \cdot \nabla\phi_i d^2\mathbf{r}. \tag{13.80}$$

Expression 13.79 has a simple meaning as a sum of outgoing and incoming fluxes through the shared cell's interface.

The numerical flux $\hat{\mathbf{q}} = -\{\kappa u_i \nabla\phi_i\}$ in Equation 13.78, which replaces $\mathbf{q}$ in Equation 13.45, is the so-called Bassi–Rebay interface flux (Bassi and Rebay 1997). The scheme with the Bassi–Rebay flux is stable for the polynomial interpolation of shape functions ϕ_i of degree higher than 1. For a linear interpolation, the scheme is unstable and must be regularized. For this purpose, we impose a set of constraints that the solution is continuous across the cell's interfaces, that is, $[u] = 0$, and add the following "zero term" to the right-hand side of Equation 13.78:

$$v_j + v_{j'} = \beta \oint_{\partial V} \{\kappa\nabla\phi_j\} \cdot \mathbf{n}[u] d^2\mathbf{r} + \delta \oint_{\partial V} [\phi_j][u] d^2\mathbf{r}, \tag{13.81}$$

where parameters $\beta = \{-1,0,1\}$ and $\delta \in \mathbb{R}^+$ are penalty parameters. In the same way as before, we identify v_j in this expression as

$$v_j = u_i e_{ij} + u_i e_{i'j},$$ (13.82)

where

$$e_{ij} = \frac{\beta}{2} \kappa \oint_{\partial V} \phi_i \mathbf{n} \cdot \nabla \phi_j d^2 \mathbf{r} + \delta \oint_{\partial V} \phi_i \phi_j d^2 \mathbf{r},$$

$$e_{i'j} = -\frac{\beta}{2} \kappa \oint_{\partial V} \phi_i \mathbf{n} \cdot \nabla \phi_j d^2 \mathbf{r} - \delta \oint_{\partial V} \phi_i \phi_j d^2 \mathbf{r}.$$ (13.83)

These penalty terms improve the condition number of the system matrix.

Figure 13.2 illustrates the application of the DG discretization scheme. Thus, Figure 13.2a shows experimentally recorded intensity of 633 nm wavelength leaving the surface of a solid phantom excited from the opposite side by a laser beam. The phantom is a solid cylinder 40 mm in diameter and 100 mm in height. The phantom was made of toner and TiO_2 powder dispersed in the hardener. It has tissue-like optical parameters $\mu_a = 0.01$ mm^{-1} and $\mu_s' = 0.83$ mm^{-1}. Figure 13.2b shows the image of the average intensity on the surface of the phantom obtained from the intensity map. Figure 13.2c displays computed average intensity by using the DG implementation of the TE. As it is seen by comparing Figure 13.2b and c, the TE and the DA are very accurate models of the light transport in highly scattering media with negligible absorption. On the other hand, comparison of Figure 13.2a and b or c indicates a significant difference between distributions of intensity and average intensity on the phantom surface. This difference deserves some explanations. As was emphasized above, the CCD camera records the absorbed energy of the radiation, where the intensity enters as a coefficient of proportionality. The intensity is easily found when

Analysis of fluorescence lifetime data

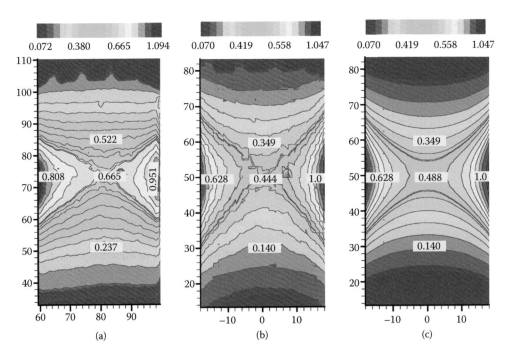

(a) (b) (c)

Figure 13.2 (a) Image recorded by the CCD camera; (b) corrected image; (c) computed average intensity on the surface of a homogeneous cylinder. Intensities of all images are scaled by 10^5. Images are shown at $\omega = 0$. The source is located on the opposite side of the visible part of the surface.

the pixel area and the camera exposure time are known. Then, the surface distribution of experimentally measured intensity must be transformed into the average intensity according to the angular dependence of radiation leaving the surface. The method of Schwarzschild–Shuster (Sobolev 1963; Soloviev et al. 2010) provides the following relationship between the intensity and the average intensity on the phantom surface

$$u\mid_{\mathbf{r}\in\partial V} \sim \frac{I(\mathbf{r},\mathbf{s}_0)}{1/2 + \mathbf{n}\cdot\mathbf{s}_0}\mid_{\mathbf{r}\in\partial V}, \tag{13.84}$$

where $I(\mathbf{r},\mathbf{s}_0)$ is the intensity of radiation leaving the surface toward the CCD camera in the direction $\mathbf{s}_0$, and $\mathbf{n}$ is the phantom surface normal. This expression (Equation 13.84) was used for transforming the image in Figure 13.2a into the image in Figure 13.2b.

13.4 INVERSE PROBLEM

In this section, we provide a detailed treatment of the inverse problem in diffusion imaging. The basic tool is the development of a relationship between changes in optical parameters and changes in measured data. This linear relationship can be inverted directly or used as the basis for a nonlinear reconstruction technique. Reported methods differ in regard to the number of parameters reconstructed and the assumptions made with regard to *a priori* knowledge. We will explain the meaning of ill-posedness and methods for using priors and/or regularization to overcome this difficulty.

13.4.1 LINEAR INVERSE PROBLEMS IN FLUORESCENCE OPTICAL IMAGE

In the absence of scattering, inverse problems in biomedical optics can be formulated as the inversion of the Radon transform or its analog such as the attenuated radon transform. For instance, optical projection tomography (OPT) was developed as an optical analog of x-ray computerized tomography (CT) and utilizes the inverse radon transform. The attenuated Radon transform found its applications in positron emission tomography (PET) and single-photon emission computed tomography (SPECT). In the presence of scattering, inverse problems are more complex and mainly considered as inverse scattering problems. The inverse scattering problem is the problem of determining the characteristics of an object from measurement data of radiation scattered from the object. Except in special cases, reconstruction algorithms typically call for the solution of an optimization problem of an appropriately constructed cost functional.

In homogeneous media, the problem of reconstruction of the quantum yield η and the lifetime τ is a linear problem. These two parameters enter to the source term, and therefore this problem is known as a source reconstruction problem. Assume that the size of the scattering medium, L, satisfies the condition $\mu L \gg 1$ together with $\mu_s \gg \mu_a$. Therefore, the light transport model is the diffusion process governed by the TE or the DA. Assume for the sake of simplicity the same optical parameters for excitation and fluorescent light. Fluorescence is considered as a re-emission of the excitation radiation, which was absorbed by fluorescent particles. Then, the transport of the excitation and fluorescent light is governed by two coupled Helmholtz equations:

$$\Lambda u = \mu_s u_0, \tag{13.85}$$

$$\Lambda u' = \frac{\mu_a \eta}{1 + i\omega\tau} u, \tag{13.86}$$

where the differential operator Λ is given by Equation 13.86, and u and u' are the excitation and fluorescence energy densities, respectively. The source term in Equation 13.86 has a simple physical meaning: $\mu_a u$ is the absorbed energy density of the excitation radiation, and the quantum yield η ($0 \le \eta < 1$) describes the efficiency of the re-emission process. The factor $(1 + i\omega\tau)^{-1}$ is the Fourier transform of the exponential distribution, where the lifetime τ is the mean time of the spontaneous emission.

Next, it is convenient to introduce the Green function g as

$$\Lambda g = C\delta(\mathbf{r} - \mathbf{r}'), \tag{13.87}$$

where C is some dimensional constant. Then, the solution of Equation 13.86 is given in the form of spatial convolution:

$$u'(\mathbf{r}) = C(\omega)\int_V u(\mathbf{r}' - \mathbf{r}_0)g(\mathbf{r} - \mathbf{r}')q(\mathbf{r}')d^3\mathbf{r}', \tag{13.88}$$

where the complex source function $q(\mathbf{r}')$ is

$$q(\mathbf{r}') = \frac{\mu_a\eta(\mathbf{r}')}{1 + i\omega\tau(\mathbf{r}')}. \tag{13.89}$$

In Equation 13.88, we let $C(\omega)$ depend on the Fourier parameter ω because now it may contain the CCD camera gating function and the amplitude modulation function of the excitation radiation. In Equation 13.88, only the complex source function $q(\mathbf{r}')$ is unknown. That is, Equation 13.88 is a linear Fredholm integral equation of the first kind.

It is instructive to compare Equation 13.88 with its TD analog. As can be seen, Equation 13.88 in the TD results in a double time convolution when the CCD camera gating function and the amplitude modulation function are neglected. Practically, however, the width of a gate of the CCD camera is about a nanosecond, which has a significant effect. This leads to a triple time convolution. That is to say, the inverse problem is three-dimensional in time (Soloviev et al. 2007b).

Next, we let $\mathbf{r}$ and $\mathbf{r}_0$ vary on the surface of the scattering object and replace $u'(\mathbf{r}, \mathbf{r}_0)$ on the left-hand side of Equation 13.88 with experimental data sets, denoted by $u'_e(\mathbf{r})$. However, this replacement must be accompanied by a transformation of the experimentally collected data set into the average intensity $u'_e(\mathbf{r})$. As was illustrated above (Equation 13.84), the data set collected by the CCD camera should be multiplied by an image correction factor. It is not always possible to find such a simple relationship as Equation 13.84. In any case, this problem can be avoided by introducing the so-called Born ratio. Thus, denoting the experimentally measured excitation average intensity by $u_e(\mathbf{r} - \mathbf{r}_0)$, we rewrite Equation 13.88 in the form

$$\frac{u'_e(\mathbf{r})}{u_e(\mathbf{r} - \mathbf{r}_0)} = \frac{C(\omega)}{u(\mathbf{r} - \mathbf{r}_0)}\int_V u(\mathbf{r}' - \mathbf{r}_0)g(\mathbf{r} - \mathbf{r}')q(\mathbf{r}')d^3\mathbf{r}'. \tag{13.90}$$

Because excitation and fluorescent data sets are collected in the same way, any factors in u_e and u'_e cancel each other on the left-hand side of Equation 13.90.

In practice, the inverse problem posed as the Fredholm integral equation of the first kind (Equation 13.42 or 13.90) is solved numerically. To pursue a numerical solution, Equation 13.42 or 13.90 is discretized. Depending on the discretization scheme, computational cells in the case of the FV or nodes in the cases of FEM or DG are enumerated. Enumeration results in the unknown vector $x_i = q(\mathbf{r}'_i)$ and the source vector $y_k = u'(\mathbf{r}_h, \mathbf{r}_{0,m})$, where $0 \le i < N$; $0 \le m < M$; $0 \le h < H$; and $k = hm$. Here N is the dimension of the vector $\mathbf{x}$, M is the number of source-camera positions, and H is the number of discrete points on the surface of the scattering object where measurements are taken. Discretization of the kernel of the integral equation results in the matrix $\mathbf{A}$, which has MH rows and N columns. Therefore, Equation 13.88 or 13.90 in discrete form reads

$$\mathbf{y} = \mathbf{A}\mathbf{x}. \tag{13.91}$$

Because dimensions of $\mathbf{x}$ and $\mathbf{y}$ do not match, the solution of the linear system (Equation 13.91) is sought by applying the least squares method, which results in the normal equation

$$\mathbf{A}^T\mathbf{A}\mathbf{x} = \mathbf{A}^T\mathbf{y}. \tag{13.92}$$

Unfortunately, for all problems involving light diffusion, the matrix $\mathbf{A}^T\mathbf{A}$ is rank deficient. Because of the rank deficiency, a solution is not unique, that is, any vector from the null space of the matrix $\mathbf{A}^T\mathbf{A}$ can be added to the true solution $\mathbf{x}$, producing an infinite set of solutions all satisfying Equation 13.92. Therefore, regularization must be used to correct the rank deficiency of linear operator $\mathbf{A}^T\mathbf{A}$. For this purpose, the quasi-solution $\mathbf{x}^\alpha$ is introduced (Tikhonov and Arsenin 1977), which solves the optimization problem:

$$\min_{\mathbf{x}^\alpha}\left\{\|\mathbf{y} - \mathbf{A}\mathbf{x}^\alpha\|^2 + \alpha\Upsilon(\mathbf{x}^\alpha)\right\}. \tag{13.93}$$

In Equation 13.93, $\Upsilon(\mathbf{x}^\alpha)$ is a positively defined regularization functional and α is a regularization parameter. The parameter α depends on the norm of the experimental error δ in such a way that $\alpha \to 0$ when $\delta \to 0$. The quasi-solution $\mathbf{x}^\alpha$ has the following properties:

$$\mathbf{x}^\alpha \to 0; \quad \alpha \to \infty, \tag{13.94}$$

$$\|\mathbf{y} - \mathbf{A}\mathbf{x}^\alpha\| \le \delta; \quad \alpha \to 0. \tag{13.95}$$

The choice of the regularization functional Υ in the form of $\|\mathbf{x}^\alpha\|^2$ leads to the classic Tikhonov regularization (Kirsch 1996; Tikhonov and Arsenin 1977). In this case, variation of Equation 13.93 results in

$$(\mathbf{A}^T\mathbf{A} + \alpha\mathbf{I})\mathbf{x}^\alpha = \mathbf{A}^T\mathbf{y}. \tag{13.96}$$

The optimal value of the regularization parameter α depends not only on the experimental error δ but also on the class of *a priori* conditions imposed on the solution $\mathbf{x}$. Thus, in the case $\|\mathbf{A}^{-1}\mathbf{x}\| \le C$, where C is a constant, the optimal α is given by

$$\alpha_{opt} = \delta/C. \tag{13.97}$$

Equation 13.96 is usually solved by use of iterative methods such as the Landweber method, the steepest descent, the conjugate gradient (CG) method, or generalized minimal residual (GMRES) method (Faddeev and Faddeeva 1963; Golub and Loan 1996). Iterative techniques have a remarkable property. It was noticed that earlier termination of the iterative process has a regularization effect. This effect is known as a discrepancy principal, in accordance to which iterations are terminated when

$$\|\mathbf{y} - \mathbf{A}\mathbf{x}^\alpha\| \simeq a\delta, \tag{13.98}$$

where $a \ge 2$. In many practical cases, the constant a is found empirically.

Figure 13.3 serves as an illustration of the linear inverse problem in the form of the integral equation, Equation 13.88 or 13.90. The lifetime and the quantum yield were reconstructed from experimental time-gated data (Soloviev et al. 2007b). For this experiment, a solid phantom of slab geometry $120 \times 75 \times 40$ mm shown in Figure 13.3a was built using a solution of epoxy and hardener mixed with the scattering particles TiO_2. The following values of optical parameters were measured $\mu_s \simeq 2.002$mm^{-1}, $\mu_a \simeq 0.001$mm^{-1}, and $\epsilon_1 \simeq 0.75$. The slab contains four wells 8 mm in diameter, each with the same optical parameters but mixed with Rhodamine 6G. The FV discretization scheme in Equation 13.96 was used on

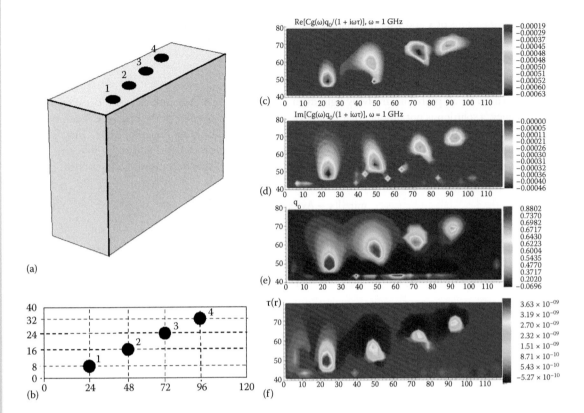

Figure 13.3 (a) Phantom. (b) Cross section showing the positions of tubes filled with Rhodamine 6G. (c, d) Real and imaginary parts of the source function q. (e) Reconstructed quantum yield η. (f) Reconstructed lifetime τ. Reconstruction of the lifetime is performed at ω = 1 GHz. The quantum yield is reconstructed at ω = 0.

the hexahedral adaptive mesh. Quantum yield was reconstructed from the case ω = 0. Slice at $z = 25$ mm height is shown in Figure 13.3e. Lifetime was computed from the reconstructed complex-valued source function, whose real and imaginary parts are shown in Figure 13.3c and d at ω = 1 GHz. Reconstructed lifetime is displayed in Figure 13.3f. Note that the yield can be found from the reconstructed complex valued source function (Figure 13.3c, d) in a similar way as the lifetime. However, the case ω = 0 gives less noisy reconstruction results. This approach was further extended to the case of a strong background fluorescence (Soloviev et al. 2007a).

Some nonlinear inverse problems can be posed in terms of a linear integral equation as a result of linearization. For instance, reconstruction of perturbation of the absorption coefficient $\delta\mu_a \ll \mu_a$ results in the Fredholm integral equation of the first kind (Konecky et al. 2008). As before, we assume a highly scattering medium, $\mu_s \gg \mu_a$, such that the constant diffusion coefficient κ can be assumed. Therefore, representing $\mu_a = \mu_a^{(0)} + \delta\mu_a$, we introduce two Green functions satisfying the following:

$$\begin{cases} -\kappa\Delta g_0(\mathbf{r} - \mathbf{r}_1) + \mu_a^{(0)} g_0(\mathbf{r} - \mathbf{r}_1) = \delta(\mathbf{r} - \mathbf{r}_1), \\ -\kappa\Delta g(\mathbf{r} - \mathbf{r}_2) + \mu_a g(\mathbf{r} - \mathbf{r}_2) = \delta(\mathbf{r} - \mathbf{r}_2). \end{cases} \tag{13.99}$$

Multiplying the first equation by $g(\mathbf{r} - \mathbf{r}_2)$, the second by $g_0(\mathbf{r} - \mathbf{r}_1)$, subtracting them, and integrating the result over the whole domain, we arrive at

$$g_0(\mathbf{r}_1 - \mathbf{r}_2) - g(\mathbf{r}_1 - \mathbf{r}_2) = \int_V \delta\mu_a g_0(\mathbf{r} - \mathbf{r}_1) g(\mathbf{r} - \mathbf{r}_2) d^3\mathbf{r}. \tag{13.100}$$

Due to the same boundary conditions satisfied by g and g_0, the term $g\Delta g_0 - g_0\Delta g$ in Equation 13.100 vanishes. As can be seen, Equation 13.100 is the Fredholm integral equation of the first kind for unknown perturbation $\delta\mu_a$, which can be solved by using the method above (Markel et al. 2003; Schotland and Markel 2001).

13.4.2 ANISOTROPIC DIFFUSION REGULARIZATION

The incorporation of *a priori* information can improve the reconstruction quality. Anisotropic diffusion regularization can be used to incorporate in the reconstruction process prior knowledge about the smoothness/sharpness of the solution and anatomical information. The concept of anisotropic diffusion was introduced in image processing by Perona and Malik (1990). This technique has the ability to preserve and enhance edges present in the images while simultaneously removing image noise.

The anisotropic diffusion functional is given by (Correia et al. 2011):

$$\Upsilon(\mathbf{x}^{\alpha}, \mathbf{x}^{ref}) = \int_V W(|\nabla\mathbf{x}^{ref}|)\psi(|\nabla\mathbf{x}^{\alpha}|), \tag{13.101}$$

where W is a weighting factor calculated from an anatomical image $\mathbf{x}^{ref}$ obtained from another medical imaging modality, and ψ is an image to image mapping. The application of standard variational techniques leads to

$$\Upsilon'(\mathbf{x}^{\alpha}, \mathbf{x}^{ref}) = \nabla \cdot [W(|\nabla x^{ref}|)\kappa(|\nabla\mathbf{x}^{\alpha}|)\nabla\mathbf{x}^{\alpha}] \tag{13.102}$$

where

$$\kappa(|\nabla\mathbf{x}^{\alpha}|) = \frac{\psi'(|\nabla\mathbf{x}^{\alpha}|)}{|\nabla\mathbf{x}^{\alpha}|}, \tag{13.103}$$

is a diffusivity (edge-preserving) function. Perona and Malik (1990) proposed the following function:

$$\psi(|\nabla\mathbf{x}^{\alpha}|) = \frac{T^2}{2}\{1 - \exp[-(|\nabla\mathbf{x}^{\alpha}|/T)^2]\}, \tag{13.104}$$

and, therefore,

$$\kappa(|\nabla\mathbf{x}^{\alpha}|) = \exp[-(|\nabla\mathbf{x}^{\alpha}|/T)^2], \tag{13.105}$$

where T is the threshold parameter, which determines which image gradients are assumed to be edges and, hence, preserved. In order to obtain the weighting factor W, one can apply the edge preserving function κ to the anatomical image, which returns an image where homogeneous regions have larger weight than close to edges, where the weight approaches zero. In regions where the anisotropic diffusion function has a small value, the diffusion is blocked across boundaries, whereas where it has a large value, diffusion is isotropic. Therefore, the function defines which regions are to be enhanced or smoothed.

Images were reconstructed using the anisotropic diffusion regularization from *in vivo* mouse data. The mouse had a brain tumour marked by a fluorescent dye. Data were collected using a hybrid fDOT-CT system, which had an x-ray tube and a detector mounted onto a rotating gantry and, on the perpendicular axis, a laser and CCD camera (Schulz et al. 2010). The gantry rotates around the object placed in the center. fDOT data consisted of 16 projections obtained at evenly spaced positions over the full 360° range. Data compression is used to improve computational performance. The CT images were used to generate a finite element mesh with 75,845 nodes and 350,411 elements and to obtain the anatomical information.

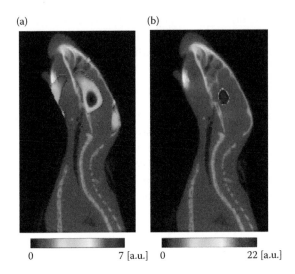

(a) (b)

0 7 [a.u.] 0 22 [a.u.]

Figure 13.4 (a) Reconstruction of the fluorescence yield of a tumor targeting fluorescent dye using Tikhonov regularization and (b) anisotropic diffusion regularization. (Data courtesy of Prof. Vasilis Ntziachristos.)

Figure 13.4 shows the reconstructed images of the fluorochrome overlayed on the CT image. Figure 13.4a shows the reconstructed image of the fluorescence yield obtained using Tikhonov regularization, which is highly contaminated with noise. Figure 13.4b shows the reconstructed image using anisotropic diffusion regularization, where it is visible that this method accentuates the tumor edges and smooths the remaining regions.

13.4.3 VARIATIONAL METHODS

In general, the inverse problem of reconstruction of optical and fluorescent parameters is nonlinear. Nonlinearity manifests itself through the cross-dependence of optical and fluorescent parameters. Moreover, the diffusive nature of light transport in turbid media such as biological tissues presents significant difficulties in image reconstruction problems and requires development of more sophisticated algorithms than those used in CT or in OPT. Thus, the well-known back-projection algorithm cannot be applied for the diffuse imaging. Nevertheless, an analog of the back-projection algorithm for diffusive imaging can be derived by constructing an appropriate cost functional.

We consider a simple experimental setup wherein positions of the laser source and the CCD camera are fixed, but the object under study is rotated. Computationally, it is much simpler to rotate the source and camera with fixed spatial orientation of the object. Let us set the unit vector **s** to point in the direction along the excitation laser beam. We set further **n** = −**s**, where **n** is the camera normal. Then, the variational problem is formulated as a minimization problem of the cost functional $\mathcal{F}$:

$$\mathcal{F} = \int \xi(\mathbf{s})(\mathcal{E} + \mathcal{L})d^2\mathbf{s} + \Upsilon, \tag{13.106}$$

where the error norm, $\mathcal{E}$, is given as

$$\mathcal{E} = \int_{-\infty}^{\infty} \varsigma(\omega)d\omega \int_{V} \chi(\mathbf{r})\left(\left|u_e - u\right|^2 + \left|u_e' - u'\right|^2\right)d^3\mathbf{r}. \tag{13.107}$$

The function u, u_e, u', and u'_e depend on the direction of the applied excitation radiation **s**. The function $\xi(\mathbf{s})$ is introduced for convenience and for emphasizing similarity with the back-projection operator. It samples directions of the laser beam and the position of the camera:

$$\xi(\mathbf{s}) = \sum_{0 \le n < N} \delta(\mathbf{s} - \mathbf{s}_n), \tag{13.108}$$

where N is the number of source-camera positions. Similarly, the functions χ and ς represent sampling of measurements in space and frequency:

$$\chi(\mathbf{r}) = \sum_{0 \le m < M} a_m \delta(\mathbf{r} - \mathbf{r}_m), \quad \varsigma(\omega) = \sum_{0 \le l < L} \delta(\omega - \omega_l), \tag{13.109}$$

where M is the number of discrete points on the imaged phantom's surface; L denotes the number of samples in the Fourier domain (ω); the vector $\mathbf{r}_m$ denotes the surface points visible by the CCD camera corresponding to the direction of the excitation beam, **s**. Factors a_m are surface areas around $\mathbf{r}_m$ such that $\int \chi(\mathbf{r}) d^3 \mathbf{r}$ gives the total visible area. The form of $\mathcal{F}$ is chosen in order to simplify a variational procedure. Thus, the function χ allows replacing a sum over surface points visible by the CCD camera with a volume integral. Analogously, the function ς replaces a sum over samples in the Fourier domain with an integral.

The Lagrangian terms in Equation 13.25 are denoted by $\mathcal{L}$ and explicitly given by

$$\mathcal{L} = \mathrm{Re} \int_{-\infty}^{\infty} \varsigma(\omega) \langle \psi, \Lambda u - \mu_s u_0 \rangle d\omega$$

$$+ \mathrm{Re} \int_{-\infty}^{\infty} \varsigma(\omega) \langle \phi, \Lambda u' - qu \rangle d\omega. \tag{13.110}$$

In expression 13.110, $\langle \cdot, \cdot \rangle$ denotes the inner product; the functions ψ and ϕ are Lagrange multipliers satisfying the same boundary conditions as the average intensities.

Next, define a vector containing unknown optical and fluorescent parameters by $\mathbf{x} = \left(\mu'_s, \mu_a, \eta \mu_a, \tau \right)^T$ at every point of the domain and choose a dynamic form of the regularization term, $\Upsilon(\mathbf{x})$, depending on the kth iteration as

$$\Upsilon(\mathbf{x}) = \sum_{1 \le i \le 4} \alpha_i \int_{-\infty}^{\infty} \varsigma(\omega) \| x_{k+1}^i - x_k^i \|^2 \, d\omega, \tag{13.111}$$

where x_k^i is the ith component of $\mathbf{x}_k$, and α_i are the Tikhonov regularization parameters.

Usually, the cost functional $\mathcal{F}$ is minimized by applying the Gauss–Newton algorithm. Alternatively, $\mathcal{F}$ can be minimized by setting first variations with respect to unknown functions to zero, which results in the system of partial differential equations for average intensities, adjoint average intensities (Lagrange multipliers), and optical and fluorescence parameters (Soloviev et al. 2008, 2009; Tadi 1997). This system consists of Equations 13.85 and 13.86 and the following equations:

$$\Lambda\phi^* = 2\chi(\mathbf{r})\left(u_e'^* - u'^*\right), \tag{13.112}$$

$$\Lambda\psi^* = 2\chi(\mathbf{r})\left(u_e^* - u^*\right) + q\phi^*, \tag{13.113}$$

$$x_{k+1}^i = x_k^i + (1/\alpha_i)\int \xi(\mathbf{s}) f_i(\mathbf{s}, \mathbf{x}_k) d^2\mathbf{s}, \tag{13.114}$$

where the asterisk denotes complex conjugation, and functions $f_i(\mathbf{s}, \mathbf{x}_k)$ are given by

$$f_1 = 3\mathrm{Re}[\kappa^2(\nabla\phi^* \cdot \nabla u' + \nabla\psi^* \cdot \nabla u) + u_0/(3 - \epsilon_1)], \tag{13.115}$$

$$f_2 = -\mathrm{Re}(\phi^* u' + \psi^* u) + \eta f_3, \tag{13.116}$$

$$f_3 = \mathrm{Re}\Phi, f_4 = \omega\mathrm{Im}(q\Phi), \tag{13.117}$$

$$\Phi = (1 + i\omega\tau)^{-1} \phi^* u. \tag{13.118}$$

Insertion of the function $\xi(\mathbf{s})$ (Equation 13.108) into Equation 13.114 allows us to rewrite this equation in the form

$$x_{k+1}^i = \left[x_{s,k}^i + (1/\alpha_i) f_i(\mathbf{s}_s, \mathbf{x}_k) \right] + (1/\alpha_i)\sum_{n=s+1}^{N-1} f_i(\mathbf{s}_n, \mathbf{x}_k), \tag{13.119}$$

where $x_{0,k}^i = x_k^i$, and define a subsequence of the kth iteration by letting $\mathbf{x}_k$ depend on the laser beam direction $\mathbf{s}_s$ as

$$x_{s+1,k}^i = x_{s,k}^i + (1/\alpha_i) f_i(\mathbf{s}_s, \mathbf{x}_{s,k}). \tag{13.120}$$

Next, we let the index s run over directions $\mathbf{s}_s$ in an arbitrary order and associate the index k with samples in the Fourier domain. In this form, Equation 13.120 presents a variant of the Landweber–Kaczmarz method (Kaczmarz 1993).

Equations 13.85, 13.86, and 13.112 through 13.114 are the Karush–Kuhn–Tucker conditions for optimization of $\mathcal{F}$ with conditions $\mathcal{L}$ (Nocedal and Wright 1999). Equations satisfied by average intensities, ϕ^* and ψ^*, are Helmholtz equations (Fourier images of corresponding telegraph equations), with source terms placed on the surface of a scattering object visible by the CCD camera. Sources' amplitudes are differences between recorded and computed average intensities. Equations 13.115 through 13.118 are used for reconstruction of optical and fluorescent parameters. The physical meaning of these equations can be given by introducing the most probable (averaged) photons' propagation paths. These photons' paths provide the largest contribution to recorded or computed average intensities on the surface of a scattering object from a given source. For instance, for a single source–detector pair, averaged photons' propagation paths can be visualized as banana-like distributions connecting the source and the detector (Arridge and Schweiger 1995; Colak et al. 1997). Thus, parameters are updated iteratively by summing up back-projected differences between recorded and computed average intensities on the visible part of the surface while stepping through laser beam directions, $\mathbf{s}_n$, and frequencies, ω_j. Parameters $1/\alpha_i$ in Equation 13.120 are computed at each iteration step as described in Soloviev et al. (2009). Iterations are terminated when $\mathcal{E} + \Upsilon$ attains its global minimum.

This reconstruction algorithm was applied to the experimental data set acquired from a highly scattering phantom containing fluorescent inclusions and utilized the DG discretization scheme. The phantom diagram is shown in Figure 13.5. It is a solid homogeneous cylinder 40 mm in diameter and 100 mm in height. The phantom was made of toner and TiO_2 powder dispersed in the hardener with $\mu_a = 0.01$ mm^{-1} and $\mu'_s = 0.83$ mm^{-1}. The phantom has three tubes; two of them were filled with fluorophore (Nile Blue dissolved in methanol) and one was filled with an absorber only. One fluorescent tube is truncated, which makes the inverse problem three-dimensional. All tubes are placed 10 mm off the phantom's axis. Tubes A and C contain fluorophore. The concentration of fluorophore in tube C was four times higher than in tube A (10^{-5} M in tube C and 2.5×10^{-6} M in tube A). However, the quantum yield is the same in both tubes. Tube B was filled with absorber ($\mu_a = 0.04$ mm^{-1}), and its reduced scattering coefficient, μ'_s, has been set to background. Tube A has twice lower value of μ'_s than the background. Tubes A and B have height 100 mm and diameter 4 mm. Tube C has smaller diameter 3 mm and shorter height 50 mm. Its volume is approximately 3.6 times smaller than the volume of tube A, and therefore, in spite of higher fluorophore concentration, its brightness is comparable with brightness of tube A upon excitation. The phantom was probed at three different heights $y = \{42.5; 50.0; 67.5\}$ mm. At each height, the phantom was rotated by $\pi/6$ and imaged. For each camera's position, 41 time windows were acquired.

Reconstruction results are shown in Figure 13.6. Each row displays (i) the reduced scattering coefficient μ'_s (in mm^{-1}); (ii) the absorption coefficient μ_a (in mm^{-1}); (iii) the product $\eta\mu_a$ (in mm^{-1}), and (iv) the lifetime τ (in ns). Each column displays slices at three different heights $y = 40, 50$, and 60 mm. Two frequencies were used in reconstruction: 500 and 750 MHz. The value of the quantum yield η can be estimated by dividing reconstructed fluorescence efficiency by μ_a. It is seen that the reduced scattering coefficient was reconstructed relatively well. Its minimal value in tube A is about 0.45 mm^{-1} at height $y = 40$ mm, which is slightly higher than the true value 0.415 mm^{-1}. The value of μ'_s slightly increases with height achieving 0.52 mm^{-1} at $y = 60$ mm. As usual, reconstruction artifacts are also present. Reconstruction of the absorption coefficient μ_a is far less accurate. Its value in tube B is roughly 1.5 lower than it should. Thus, its maximum reconstructed value is 0.027 mm^{-1} while the true value is 0.04 mm^{-1}. Tubes A and C, which were filled with fluorophore, also appear in reconstruction. This is an expected result due to absorbing properties of fluorophores. Localization of the fluorescent efficiency appears to be relatively good. It is clearly seen that two tubes appear at height $y = 40$ mm and only one at heights 50 and 60 mm. The value of the quantum yield is lower than expected. Thus, dividing $\eta\mu_a$ by reconstructed μ_a, we obtain approximately [0.18; 0.2] at the center of tube A, while the true value is about [0.26; 0.27]. The lifetime distribution τ, the last column in Figure 13.6, has quite high contrast and is well localized. Slices showing the lifetime have background value almost 0. Reconstructed lifetime values are close to the true

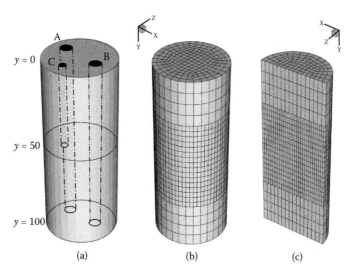

Figure 13.5 (a) Phantom. (b) Surface mesh. (c) Mesh slice showing internal mesh structure.

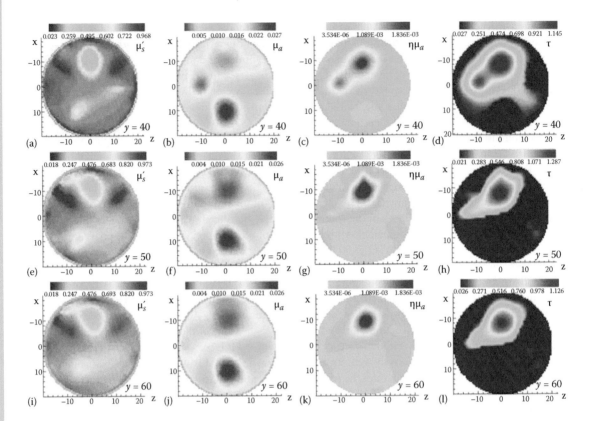

Figure 13.6 Reconstruction results. First (a–d), second (e–h), and third (i–l) rows show slices at y = 40, 50, and 60 mm, respectively. First column (a, e, i) shows reconstructed reduced scattering coefficient μ′$_s$, second column (b, f, j) shows the absorption coefficient μ$_a$, third column (c, g, k) shows the fluorescence efficiency ημ$_a$, and fourth column (d, h, l) shows the lifetime τ.

value of Nile Blue fluorophore, which is 1.2 ns. Thus, (i) the maximum lifetime is 1.17 ns at $y = 40$ mm; (ii) 1.29 ns at $y = 50$ mm; and (iii) 1.14 ns at $y = 60$ mm.

Previous examples tell us that the image resolution achievable in the diffusion imaging is not high. It is well known that diffusion processes are accompanied by an increase in the entropy, which can be interpreted as an increase in the missing information. Therefore, it should be expected that in the case of moderate or weak scattering, the resolution of reconstructed images increases. It is instructive to compare the inverse problem in the diffusion imaging with the inverse problem of imaging objects embedded in a weakly scattering medium. Such a situation is quite common in nature. For instance, the atmosphere can be considered mostly nonscattering or weakly scattering with clouds presenting highly scattering inclusions. A live embryo or fetus is mostly transparent, with internal organs being highly scattering. Let us consider again the case of a highly scattering knot embedded in a weakly scattering cylinder (see Figure 13.1). We insert two bent fluorescent rods in the middle of the knot and use this object to simulate camera's images, which serve as an experimental data set. Before proceeding with the inverse problem, we provide the fluorescent light transport model first.

It is more convenient to use the transport coefficient μ and the albedo λ in place of the scattering and absorption parameters μ_s and μ_a for image reconstruction in a weakly scattering medium. The set of optical parameters $\{\lambda, \mu\}$ is completely analogous to the conventional set $\{\mu_s, \mu_a\}$ and relates to the latter as $\mu_s = \lambda\mu$ and $\mu_a = (1 - \lambda)\mu$. Moreover, exploiting the physical meaning of the transport coefficient as the reciprocal quantity to the mean free path length, we see that μ describes the scattering properties of the medium, while the albedo controls the absorption. In this case, the scattering and absorption properties are described separately from each other by two independent parameters. When the scattering medium is a mixture of two types of particles with different scattering properties μ_j and λ_j ($j = 1, 2$), then the resulting transport coefficient and the albedo are found as $\mu = \mu_1 + \mu_2$ and $\lambda = \lambda_1 \lambda_2 / \bar{\lambda}$, respectively, where $\bar{\lambda} = (\lambda_1 + \lambda_2)/2$ is the average albedo. Parameters of a mixture of more than two types of scattering particles are computed recursively.

Fluorescent light transport is analogous to the excitation one. Usually fluorescence is considered as a re-emission of the excitation radiation, which was absorbed by a fluorophore. However, in accordance with our approximation, it can be viewed as the resonantly scattered excitation radiation. The resonant scattering by fluorescent particles is isotropic and accompanied by Stokes shift. The transport of such scattered excitation radiation is governed by the equation

$$\mathbf{s} \cdot \nabla I_0' + \tilde{\mu} I_0' = \frac{1}{4\pi} B_0',$$

(13.121)

where

$$B_0'(\mathbf{r}) = \frac{1 - \lambda}{1 + i\omega\tau} \mu\eta(u + I_0).$$

(13.122)

Next, we assume that fluorescent intensity I_0' propagating toward each pixel of the CCD camera results from coherently influenced waves resonantly scattered by the fluorescent particles in a voxel of the domain, which is much larger than several central Fresnel zones (Born and Wolf 1968; van de Hulst 1981). This assumption implies rectilinear propagation of the fluorescent light in accordance with Huygens' principle, which is completely analogous to propagation of a singly scattered excitation radiation described above (Equation 13.9). Therefore, we have

$$I_0'(\mathbf{r}, \mathbf{s}) = \frac{1}{4\pi} \int_0^{l_{max}} B_0'(\mathbf{r} - \mathbf{s}l) \exp\left(-\int_0^l \tilde{\mu}(\mathbf{r} - \mathbf{s}l') \, dl' \right) dl,$$

(13.123)

Multiply scattered fluorescent intensity, I', coming from highly scattering regions to the CCD array satisfies the transport equation

$$\mathbf{s} \cdot \nabla I' + \tilde{\mu} I' = \lambda\mu B',$$

(13.124)

where the function B' reads

$$B' \simeq u - \epsilon_1 \kappa \frac{\partial u'}{\partial l} + p(\mathbf{s} \cdot \mathbf{s}_0) I_0'$$

$$+ \frac{3\epsilon_2 \sigma}{3 + \epsilon_2} \frac{\partial}{\partial l} \kappa \frac{\partial u'}{\partial l} - \frac{\epsilon_2 \sigma}{3 + \epsilon_2} \nabla \cdot \kappa \nabla u'.$$

(13.125)

In Equation 13.125, u' is the fluorescence average intensity, which is defined analogously to Equation 13.64 and satisfies the Helmholtz equation

$$\Lambda u' = B_0'.$$

(13.126)

Finally, denoting the total fluorescent intensity by I'', we arrive at

$$I''(\mathbf{r},\mathbf{s}) = I_0'(\mathbf{r},\mathbf{s}) + \int_0^{l_{max}} \lambda(\mathbf{r}-\mathbf{s}l)\mu(\mathbf{r}-\mathbf{s}l)$$

$$\times B'(\mathbf{r}-\mathbf{s}l,\mathbf{s})\exp\left(-\int_0^l \tilde{\mu}(\mathbf{r}-\mathbf{s}l')\,dl'\right)dl,$$

(13.127)

where B' is set to 0 in weakly scattering regions.

The CCD camera images displaying fluorescent intensity resonantly scattered by two bent fluorescent rods embedded inside a highly scattering knot are shown in Figure 13.7. The excitation light enters the domain in the same way as shown in Figure 13.1. The phase function is chosen in the form of Equation 13.50, where $\epsilon_1 = 0.9$ and $\epsilon_2 = 0$.

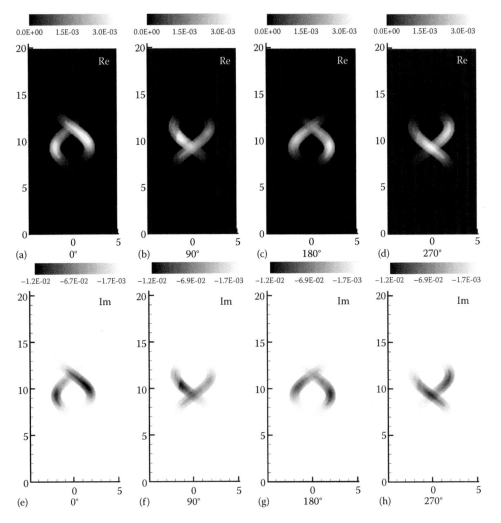

Figure 13.7 Camera images displaying fluorescent intensity resonantly scattered by two bent fluorescent rods embedded inside a highly scattering knot. Values of the quantum yield and the lifetime are set to 0.1 and 1.0 ns, respectively. The same values of optical parameters governing fluorescent and excitation light transport are assumed. The upper row (a–d) shows real parts of camera images for rotation angles {0°, 90°, 180°, 270°}. The lower row (e–h) displays imaginary parts of images. Images are taken at ω = 500 MHz.

Analogously to the diffusion case, we formulate the inverse problem as a variational problem. However, in contrast to the diffusion imaging, we do not consider the inverse problem as inverse scattering but rather pose it as a tomographic reconstruction problem. Technically, we introduce only adjoint intensities satisfying adjoint transport equations without an introduction of the adjoint average intensities. Thus, we need to solve the diffusion equation for the excitation and fluorescent light omitting solution of corresponding adjoint equations. This simplification makes the inverse problem less expensive computationally. Nevertheless, it can be extended by adding corresponding terms to the cost functional if necessarily (Soloviev et al. 2011b).

The variational problem is formulated as a minimization problem of the cost functional given by Equation 13.106. Because of the intensity-based measurements, the error norm now reads

$$\mathcal{E} = \int \xi(\mathbf{s}) d^2 \mathbf{s} \int_V \chi(\mathbf{r}) \left(\left| I_E - I \right|^2 + \left| I_F - I'' \right|^2 \right) d^3 \mathbf{r}, \tag{13.128}$$

where I_E and I are experimentally recorded and computed excitation intensities in the direction $\mathbf{s}$, respectively, and I_F and I' are recorded and computed fluorescent intensities, correspondingly. Lagrangian terms in the functional $\mathcal{F}$ (Equation 13.106) are replaced with

$$\mathcal{L} = \mathrm{Re} \int \xi(\mathbf{s}) \left\langle J, \mathbf{s} \cdot \nabla I + \tilde{\mu} I - \lambda \mu B \right\rangle d^2 \mathbf{s}$$
$$+ \mathrm{Re} \int \xi(\mathbf{s}) \left\langle H, \mathbf{s} \cdot \nabla I'' + \tilde{\mu} I'' - \lambda \mu B' - \frac{1}{4\pi} B_0' \right\rangle d^2 \mathbf{s}, \tag{13.129}$$

where $\langle \cdot, \cdot \rangle$ denotes the inner product, and J and H are excitation and fluorescent adjoint intensities, respectively. Then, we define a vector of four unknown parameters at every point of the scattering domain as

$$\mathbf{x} = (\mu, \lambda, \eta, \tau)^T, \tag{13.130}$$

and introduce the regularization term by Equation 13.111.

The reconstruction algorithm is based on the condition that the first variation $\delta \mathcal{F}(I, I'', J, H, \mathbf{x})$ vanishes. To avoid unnecessary complexity, some simplifying assumptions are made. Firstly, having assumed the same optical parameters governing the excitation and fluorescent light transport, it is sufficient to use only excitation data sets for reconstruction of the transport coefficient and albedo. Secondly, in many practical cases, $\omega/\mu c \ll 1$ and terms containing this parameter are neglected. Therefore, the resulting system is given by

$$\mathbf{s}_n \cdot \nabla I + \tilde{\mu} I = \lambda \mu B, \tag{13.131}$$

$$\mathbf{s}_n \cdot \nabla I'' + \tilde{\mu} I'' = \lambda \mu B' + \frac{1}{4\pi} B_0', \tag{13.132}$$

$$-\mathbf{s}_n \cdot \nabla J + \tilde{\mu} J = 2\chi(\mathbf{r})(I_E - I), \tag{13.133}$$

$$-\mathbf{s}_n \cdot \nabla H + \tilde{\mu} H = 2\chi(\mathbf{r})(I_F - I''). \tag{13.134}$$

$$x_{k+1}^i = x_k^i + (1/\alpha_i) \int \xi(\mathbf{s}) f_i(\mathbf{s}, \mathbf{x}_k) d^2 \mathbf{s}, \tag{13.135}$$

where the back-projection operators are given by

$$f_1 \simeq \lambda\Psi - \text{Re}(J^*I),$$ (13.136)

$$f_2 \simeq \mu\Psi - 3\epsilon\text{Re}\{(\mu\kappa)^2 J^* \mathbf{s}\cdot\nabla u\},$$ (13.137)

$$f_3 \simeq (1-\lambda)\mu\text{Re}\left\{\frac{\Theta}{1+i\omega\tau}\right\},$$ (13.138)

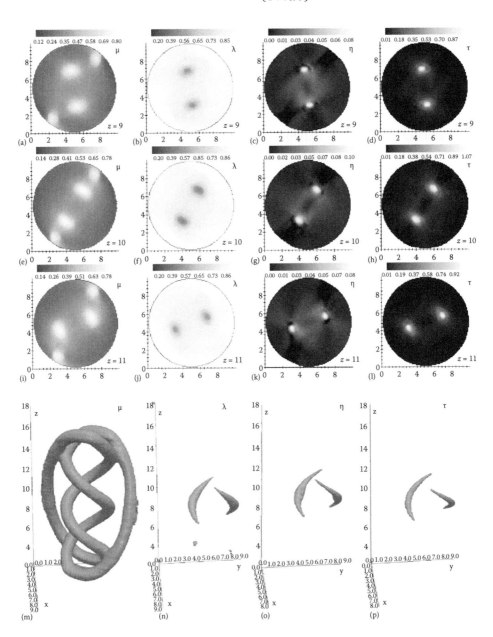

Figure 13.8 Slices showing reconstruction results of the knot embedded in the weakly scattering cylinder at three different heights. The first (a–d), second (e–h), and third (i–l) rows display the transport coefficient μ, albedo λ, quantum yield η, and lifetime τ at $z = \{9, 10, 11\}$ mm, respectively. (m) Isosurface of the transport coefficient μ. (n) Isosurface of the albedo λ. (o) Isosurface of the quantum yield η. (p) Isosurface of the lifetime τ in nanoseconds.

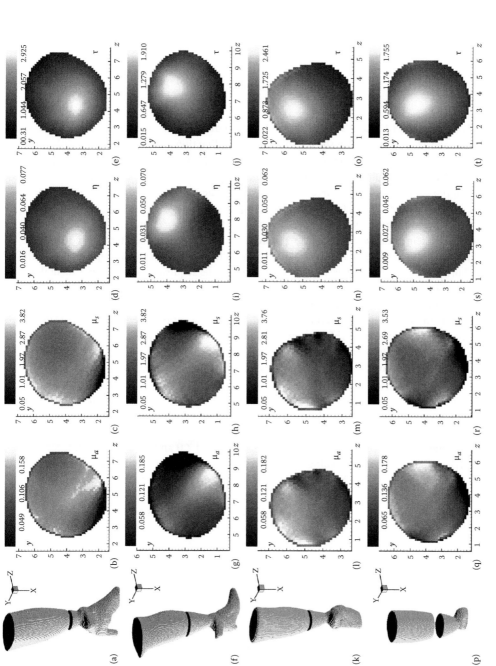

Figure 13.9 Reconstruction results based on the hybrid model. The first column (a, f, k, p) displays meshes of four legs with cuts at heights where slices are taken. The first (b–e) and third (l–o) rows show reconstruction of parameters of legs transfected with unlinked eGFP and mCherry. The second (g–j) and fourth (q–t) rows display parameters of legs transfected with FRET construct. The second (b, g, l, q) and third (c, h, m, r) columns show the absorption and scattering coefficients in mm⁻¹. The fourth (d, i, n, s) columns display the quantum yield and lifetime reconstructions. The lifetime is given in nanoseconds.

Analysis of fluorescence lifetime data

$$f_4 \simeq \omega(1-\lambda)\mu\eta \, \mathrm{Im}\left\{\frac{\Theta}{(1+i\omega\tau)^2}\right\}, \tag{13.139}$$

where the asterisk denotes complex conjugation, and the functions Ψ and Θ are defined as

$$\Psi = \mathrm{Re}[u + p(\mathbf{s} \cdot \mathbf{s}_0)I_0]J^*, \tag{13.140}$$

$$\Theta = \frac{1}{4\pi}(u + I_0)H^*. \tag{13.141}$$

Notice that in Equations 13.133 and 13.134, the propagation direction is reversed, $\mathbf{s} = -\mathbf{s}_n$. Therefore, adjoint intensities J and H propagate from the CCD array in the direction of its normal $\mathbf{n}$.

Reconstruction results are shown in Figure 13.8. Slices displaying parameters at $z = \{9, 10, 11\}$ mm are shown in Figure 13.8a through l. Figure 13.8m through p displays isosurfaces of parameters.

Recently, this approach was applied to three-dimensional lifetime imaging of Förster resonance energy transfer in mice *in vivo* (Soloviev et al. 2011b). Data were acquired by imaging two mice with right hind legs cotransfected by electroporation with plasmids for enhanced green fluorescence protein (eGFP) and mCherry fluorescent proteins that were expressed separately and two mice with legs transfected with plasmids encoding an eGFP–mCHerry FRET construct in which eGFP (donor) was coupled to mCherry (acceptor) by a short peptide flexible linker. More information on the acquisition of the experimental data can be found in McGinty et al. (2011).

Reconstruction results of four legs are presented in Figure 13.9. First and third rows (Figure 13.9a through e and k through o) show reconstruction of optical and fluorescent parameters of two control mouse legs transfected with unlinked eGFP and mCherry. Second and fourth rows (Figure 13.9f through j and p through t) present mouse legs transfected with FRET construct. The first column shows meshes of four mouse legs. Meshes are cut at heights where slices are taken. At these heights, the legs were expressing fluorophore, and therefore, slices contain highest contrast in the parameter reconstructions. The second column presents reconstruction of the absorption coefficient, μ_a, while the third column displays the scattering coefficient, μ_s, in mm^{-1}. Reconstructed quantum yield, η, and lifetime, τ, are displayed in the fourth and fifth columns. The lifetime is given in nanoseconds (ns). The maximum values of reconstructed lifetimes are 2.97 and 2.54 for control mice legs, and 2.0 and 1.86 for legs transfected with FRET construct. Reference values obtained from cytosol preparations of lifetimes are approximately 2.61 ns for unlinked eGFP and 2.13 ns for eGFP linked with mCherry (McGinty et al. 2011). Therefore, reconstructions demonstrate a consistent lifetime contrast for FRETing and non-FRETing cases. Reconstructed quantum yield, however, does not present such contrast. Reconstructed values of the yield vary within the (0.6, 0.8) interval. Moreover, accuracy in lifetime reconstruction indicates that the error in reconstructed values can reach 15%–20% or more. Inaccuracy of reconstruction of fluorescent parameters is mostly attributed to the ill-posed nature of the inverse problem and autofluorescence present in the visible part of the light spectrum (480 nm).

13.5 LIMITATIONS

The major limitation of lifetime imaging in turbid media is the severe ill-posedness mentioned above. Direct consequences of complexity of lifetime imaging in turbid media are the lack of spatial resolution and the lack of accuracy of lifetime estimation. Localization depends on the knowledge of optical parameters such as absorption and scattering coefficient. Therefore, in the general case, these parameters must be reconstructed together with the quantum yield and lifetime. Unknown optical parameters results in non-uniqueness of the inverse problem, when the number of unknowns is greater than the number of equations. This problem can be addressed by using time gating imaging technique and the prior information coming from other imaging modalities such as CT or MRI imaging. However, additional spatial information coming from priors cannot resolve all difficulties in lifetime estimation. For instance, multiple lifetimes of the fluorescence probe cannot be recovered due to the non-uniqueness of this inverse problem.

13.6 SUMMARY

In this chapter, experimental, mathematical, and computational aspects of FLIM are considered. The first and the second sections of this chapter cover illumination and detection technique of biological samples. Imaging techniques such as CW, TD, frequency domain, and Fourier domain are presented and discussed.

In Section 13.3, the light transport models applicable to light scattering biological tissues are introduced. The most general description of the light transport is given by the RTE. The methods of solution of the RTE mostly depend on the law of scattering, which is described by the phase functions. Models of the phase functions are also discussed in this section. In general, solving the RTE is hard and, therefore, approximation to the RTE is used. We cover the Markov chain MC methods, the TE, the DA, and the Fokker–Planck equation. The relatively novel mesoscopic-scale approximation to the RTE is also introduced and discussed. Numerical methods of solving RTE approximations are covered next. They include the FV, FEM, and DG methods.

The last part of this chapter is devoted to the inverse problem, which appears in FLIM. We start with a general introduction to the inverse problem theory and derive the Fredholm integral equation of the first kind, which is used in the lifetime image reconstruction. The use of anatomical priors is beneficial and can be incorporated into the image reconstruction problem to improve the image quality. This methodology is presented in Section 13.4.2. Variational methods of minimization of an appropriately constructed cost functional are analogous to the well-known back-projection algorithms used in CT. They are introduced in the last section of this chapter. As an illustration of reconstruction algorithms, images obtained from experimentally acquired data are presented and discussed.

REFERENCES

Alerstam, E., W. C. Y. Lo, T. D. Han, J. Rose, S. Andersson-Engels, and L. Lilge. "Next-generation acceleration and code optimization for light transport in turbid media using GPUs." *Biomed. Opt. Express* 1: (2010) 658–675.

Arridge, S. R. "Photon measurement density functions. Part I: Analytical forms." *Appl. Opt.* 34: (1995) 7395–7409.

Arridge, S. R. "Optical tomography in medical imaging." *Inverse Probl* 15: (1999) R41–R93.

Arridge, S. R., M. Cope, and D. T. Delpy. "Theoretical basis for the determination of optical pathlengths in tissue: Temporal and frequency analysis." *Phys. Med. Biol.* 37: (1992) 1532–1560.

Arridge, S. R., and J. C. Hebden. "Optical imaging in medicine II: Modelling and reconstruction." *Phys. Med. Biol.* 42: (1997) 841–853.

Arridge, S. R., and M. Schweiger. "Photon measurement density functions. Part II: Finite element results." *Appl. Opt.* 34: (1995) 8026–8037.

Arridge, S. R., M. Schweiger, M. Hiraoka, and D. T. Delpy. "A finite element approach to modelling photon transport in tissue." *Med. Phys.* 20: (1993) 299–309.

Bassi, F., and S. Rebay. "A high-order accurate discontinuous finite elements method for the numerical solution of the compressible Navier-Stokes equations." *J. Comp. Phys.* 131: (1997) 267–279.

Born, M., and E. Wolf. *Principles of Optics.* Oxford: Pergamon Press, 1968.

Carey, G. F. *Computational Grids: Generation and Adaptation and Solution Strategies.* Washington, DC: Taylor & Francis, 1997.

Chance, B., M. Cope, E. Gratton, N. Ramanujam, and B. Tromberg. "Phase measurement of light absorption and scatter in human tissue." *Rev. Sci. Instrum.* 69, 10: (1998) 3457–3481.

Chandrasekhar, S. *Radiative Transfer.* New York: Dover Publications, 1960.

Colak, S. B., D. G. Papaioannou, G. W. t Hooft, H. Schomberg, M. B. van der Mark, J. C. J. Paasschens, J. B. M. Melissen, and N. A. A. J. van Asten. "Tomographic image reconstruction from optical projection in light diffusing media." *Appl. Opt.* 36: (1997) 180–213.

Correia, T., J. Aguirre, A. Sisniega, J. Chamorro-Servent, J. J. Vaquero, J. Abascal, M. Desco, V. Kolehmainen, and S. Arridge. "Split operator method for fluorescence diffuse optical tomography using anisotropic diffusion regularisation with prior anatomical information." *Biomed. Opt. Express* 2, 9: (2011) 2632–2648.

D'Andrea, C., D. Comelli, A. Pifferi, A. Torricelli, G. Valentini, and R. Cubeddu. "Time-resolved optical imaging through turbid media using a fast data acquisition system based on a gated CCD camera." *J. Phys. D: Appl. Phys.* 36, 14: (2003) 1675–1681.

D'Andrea, C., N. Ducros, A. Bassi, S. Arridge, and G. Valentini. "Fast 3D optical reconstruction in turbid media using spatially modulated light." *Biomed. Opt. Express* 1, 2: (2010) 471–481.

Delpy, D. T., M. Cope, P. van der Zee, S. R. Arridge, S. Wray, and J. S. Wyatt. "Estimation of optical pathlength through tissue from direct time of flight measurement." *Phys. Med. Biol.* 33: (1988) 1433–1442.

Ducros, N., A. Bassi, G. Valentini, M. Schweiger, S. Arridge, and C. D'Andrea. "Multiple-view fluorescence optical tomography reconstruction using compression of experimental data." *Opt. Lett.* 36: (2011) 1377–1379.

Durian, D. J., and J. Rundick. "Photon migration at short times and distances and in cases of strong absorption." *J. Opt. Soc. Am. A* 14: (1997) 235–245.

Elisee, J., A. Gibson, and S. Arridge. "Diffuse optical cortical mapping using the boundary element method." *Biomed. Opt. Express* 2: (2011) 568–578.

Faddeev, D. K., and V. N. Faddeeva. *Computational Methods of Linear Algebra*. New York: W. H. Freeman and Company, 1963.

Fang, Q., and D. A. Boas. "Monte Carlo simulation of photon migration in 3D turbid media accelerated by graphics processing units." *Opt. Express* 17: (2009) 20178–20190.

Gao, F., J. Li, L. Zhang, P. Poulet, H. Zhao, and Y. Yamada. "Simultaneous fluorescence yield and lifetime tomography from time-resolved transmittances of small-animal-sized phantom." *Appl. Opt.* 49: (2010) 3163–3172.

Godavarty, A., M. J. Eppstein, C. Zhang, S. Theru, M. Gurfinkel, A. B. Thompson, and E. M. Sevick-Muraca. "Fluorescence-enhanced optical imaging in large tissue volumes using a gain-modulated ICCD camera." *Phys. Med. Biol.* 48: (2003) 1701–1720.

Golub, G. H., and C. F. Van Loan. *Matrix Computations*. Baltimore: The Johns Hopkins University Press, 1996.

Han, S., S. Farshchi-Heydari, and D. J. Hall. "Analysis of the fluorescence temporal point-spread function in a turbid medium and its application to optical imaging." *J. Biomed. Opt.* 13, 6: (2008) 064038.

Hughes, T. J. R. *The Finite Element Method*. New York: Dover Publications, 2003.

Jackson, J. D. *Classical Electrodynamics*. New York: John Wiley & Sons, 1998.

James, D. R., A. Siemiarczuk, and W. R. Ware. "Stroboscopic optical boxcar technique for the determination of fluorescence lifetimes." *Rev. Sci. Instrum.* 63, 2: (1992) 1710–1716.

Joshi, A., W. Bangerth, K. Hwan, J. C. Rasmussen, and E. M. Sevick-Muraca. "Fully adaptive FEM based fluorescence tomography from time-dependant measurements with area illumination and detection." *Med. Phys.* 33: (2006a) 1299–1310.

Joshi, A., W. Bangerth, and E. M. Sevick-Muraca. "Non-contact fluorescence optical tomography with scanning patterned illumination." *Opt. Express* 14: (2006b) 6516–6534.

Kaczmarz, S. "Approximate solution of system of linear equations." *Int. J. Control* 57: (1993) 1269–1271.

Kim, A. D. "Transport theory for light propagation in biological tissue." *J. Opt. Soc. Am. A* 21: (2004) 820–827.

Kim, A. D., and J. B. Keller. "Light propagation in biological tissue." *J. Opt. Soc. Am. A* 20: (2003) 92–98.

Kim, A. D., and M. Moscoso. "Beam propagation in sharply peaked forward scattering media." *J. Opt. Soc. Am. A* 21: (2004) 797–803.

Kirsch, A. *An Introduction to the Mathematical Theory of Inverse Problems*. New York: Springer, 1996.

Konecky, S. D., G. Y. Panasyuk, K. Lee, V. A. Markel, J. C. Schotland, and A. G. Yodh. "Imaging complex structures with diffuse light." *Opt. Express* 16: (2008) 5048–5060.

Kumar, A. T. N., S. B. Raymond, A. K. Dunn, B. J. Bacskai, and D. A. Boas. "A time domain fluorescence tomography system for small animal imaging." *IEEE Trans. Med. Imaging* 27, 8: (2008) 1152–1163.

Larsen, E. W. "The linear Boltzmann equation in optically thick systems with forward-peaked scattering." *Prog. Nucl. Energy* 34: (1999) 413–423.

Lebedev, N. N. *Special Functions & Their Applications*. New York: Dover Publications, 1972.

Lehtikangas, O., T. Tarvainen, V. Kolehmainen, A. Pulkkinen, S. R. Arridge, and J. P. Kaipio. "Finite element approximation of the Fokker-Planck equation for diffuse optical tomography." *J. Quant. Spectrosc. Radiat. Transfer* 111: (2010) 1406–1417.

Markel, V. A., V. Mital, and J. C. Schotland. "The inverse problem in optical diffusion tomography III: Inversion formulas and singular value decomposition." *J. Opt. Soc. Am. A* 20: (2003) 890–902.

McGinty, J., D. W. Stuckey, V. Y. Soloviev, R. Laine, D. J. Wells, M. Wylezinska-Arridge, S. R. Arridge, P. M. W. French, J. V. Hajnal, and A. Sardini. "*In vivo* fluorescence lifetime tomography of a FRET probe expressed in mouse." *Biomed. Opt. Express* 2, 7: (2011) 1907–1917.

Milstein, A. B., S. Oh, K. J. Webb, C. A. Bouman, Q. Zhang, D. A. Boas, and R. P. Millane. "Fluorescence optical diffusion tomography." *Appl. Opt.* 42, 16: (2003) 3081–3094.

Mohan, P. S., V. Y. Soloviev, and S. R. Arridge. "Discontinuous Galerkin method for the forward modelling in optical diffusion tomography." *Int. J. Numer. Meth. Eng.* 85: (2011) 562–574.

Montcel, B., and P. Poulet. "An instrument for small-animal imaging using time-resolved diffuse and fluorescence optical methods." *Nucl. Instrum. Methods Phys. Res. A* 569: (2006) 551–556.

Nocedal, J., and S. J. Wright. *Numerical Optimization*. New York: Springer-Verlag, 1999.

Nothdurft, R. E., S. V. Pawardhan, W. Akers, Y. Ye, S. Achilefu, and J. P. Culver. "*In vivo* fluorescence lifetime tomography." *J. Biomed. Opt.* 14, 2: (2009) 024004.

Ntziachristos, V., J. Ripoll, L. V. Wang, and R. Weissleder. "Looking and listening to light: The evolution of whole-body photonic imaging." *Nat. Biotechnol.* 23: (2005) 313–320.

O'Leary, M. A., D. A. Boas, B. Chance, and A. G. Yodh. "Fluorescence lifetime imaging in turbid media." *Opt. Lett.* 21, 2: (1996) 158–160.

Olver, F. W. J. *Asymptotics and Special Functions.* London: Academic Press, 1974.

Perona, P., and J. Malik. "Scale-space and edge detection using anisotropic diffusion." *IEEE Trans. Pattern Anal. Mach. Intell.* 12: (1990) 629–639.

Reynolds, L. O., and N. J. McCormick. "Approximate two-parameter phase function for light scattering." *J. Opt. Soc. Am.* 70: (1980) 1206–1212.

Rivière, B. *Discontinuous Galerkin Methods for Solving Elliptic and Parabolic Equations.* Philadelphia, PA: SIAM, 2008.

Schotland, J. C., and V. A. Markel. "Inverse scattering with diffusing waves." *J. Opt. Soc. Am. A* 18: (2001) 2767–2777.

Schulz, R. B., A. Ale, A. Sarantopoulos, M. Freyer, E. Sowhngen, M. Zientkowska, and V. Ntziachristos. "Hybrid system for simultaneous fluorescence and X-ray computed tomography." *IEEE Trans. Med. Imaging* 29, 2: (2010) 465–473.

Schweiger, M., S. R. Arridge, and D. T. Delpy. "Application of the finite element method for the forward and inverse problem in optical tomography." *J. Math. Imaging Vis.* 3: (1993) 263–283.

Schweiger, M., S. R. Arridge, M. Hiraoka, and D. T. Delpy. "The finite element method for the propagation of light in scattering media: Boundary and source conditions." *Med. Phys.* 22: (1995) 1779–1792.

Siddon, R. L. "Fast calculation of the exact radiological path for a three-dimensional CTH array." *Med. Phys.* 12: (1985) 252–255.

Sobolev, V. V. *A Treatise on Radiative Transfer.* Princeton, NJ: Van Nostrand, 1963.

Soloviev, V. Y. "Mesh adaptation technique for Fourier-domain fluorescence lifetime imaging." *Med. Phys.* 33: (2006) 4176–4183.

Soloviev, V. Y., and S. R. Arridge. "Fluorescence lifetime optical tomography in weakly scattering media in the presence of highly scattering inclusions." *J. Opt. Soc. Am. A* 28: (2011a) 1513–1523.

Soloviev, V. Y., and S. R. Arridge. "Optical tomography in weakly scattering media in the presence of highly scattering inclusions." *Biomed. Opt. Express* 2: (2011b) 440–451.

Soloviev, V. Y., A. Bassi, L. Fieramonti, G. Valentini, C. D'Andrea, and S. R. Arridge. "Angularly selective mesoscopic tomography." *Phys. Rev. E* 84: (2011a) 051915.

Soloviev, V. Y., C. D'Andrea, M. Brambilla, G. Valentini, R. Cubeddu, R. B. Schulz, and S. R. Arridge. "Adjoint time domain method for fluorescent imaging in turbid media." *Appl. Opt.* 47: (2008) 2303–2311.

Soloviev, V. Y., C. D'Andrea, P. S. Mohan, G. Valentini, R. Cubeddu, and S. R. Arridge. "Fluorescence lifetime optical tomography with discontinuous Galerkin discretisation scheme." *Biomed. Opt. Express* 1: (2010) 998–1013.

Soloviev, V. Y., C. D'Andrea, G. Valentini, R. Cubeddu, and S. R. Arridge. "Combined reconstruction of fluorescent and optical parameters using time resolved data." *Appl. Opt.* 48: (2009) 28–36.

Soloviev, V. Y., and L. V. Krasnosselskaia. "Dynamically adaptive mesh refinement technique for image reconstruction in optical tomography." *Appl. Opt.* 45: (2006) 2828–2837.

Soloviev, V. Y., J. McGinty, D. W. Stuckey, R. Laine, D. J. Wells, M. Wylezinska-Arridge, A. Sardini, J. V. Hajnal, P. M. W. French, and S. R. Arridge. "Förster resonance energy transfer imaging in vivo with approximated radiative transfer equation." *Appl. Opt.* 50: (2011b) 6583–6590.

Soloviev, V. Y., J. McGinty, K. B. Tahir, M. A. A. Neil, J. V. Hajnal, A. Sardini, S. R. Arridge, and P. M. W. French. "Fluorescence lifetime tomography of live cells expressing enhanced green fluorescent protein embedded in a scattering medium exhibiting background autofluorescence." *Opt. Lett.* 32: (2007a) 2034–2036.

Soloviev, V. Y., K. B. Tahir, J. McGinty, D. S. Elson, P. M. W. French, M. A. A. Neil, and S. R. Arridge. "Fluorescence lifetime imaging by using time gated data acquisition." *Appl. Opt.* 46: (2007b) 7384–7391.

Soloviev, V., D. Wilson, and S. Vinogradov. "Phosphorescence lifetime imaging in turbid media: The forward problem." *Appl. Opt.* 42: (2003) 113–123.

Soloviev, V. Y., D. F. Wilson, and S. A. Vinogradov. "Phosphorescence lifetime imaging in turbid media: The inverse problem and experimental image reconstruction." *Appl. Opt.* 43: (2004) 564–574.

Soubret, A., J. Ripoll, and V. Ntziachristos. "Accuracy of fluorescent tomography in the presence of heterogeneities: Study of the normalized born ratio." *IEEE Trans. Med. Imaging* 24, 10: (2005) 1377–1386.

Tadi, M. "Inverse heat conduction based on boundary measurement." *Inverse Probl.* 13: (1997) 1585–1605.

Thompson, A. B., D. J. Hawrysz, and E. M. Sevick-Muraca. "Near-infrared fluorescence contrast-enhanced imaging with area illumination and area detection: The forward imaging problem." *Appl. Opt.* 42, 19: (2003) 4125–4136.

Tikhonov, A. N., and V. Y. Arsenin. *Solution of Ill-Posed Problems.* Washington, D.C.: Winston, 1977.

van de Hulst, H. C. *Light Scattering by Small Particles.* New York: Dover Publications, 1981.

Venugopal, V., J. Chen, and X. Intes. "Development of an optical imaging platform for functional imaging of small animals using widefield excitation." *Biomed. Opt. Express* 1, 1: (2010a) 143–156.

Venugopal, V., J. Chen, F. Lesage, and X. Intes. "Full-field time-resolved fluorescence tomography of small animals." *Opt. Lett.* 35, 19: (2010b) 3189–3191.

Vinegoni, C., C. Pitsouli, D. Razansky, N. Perrimon, and V. Ntziachristos. "*In vivo* imaging of Drosophila melanogaster pupae with mesoscopic fluorescence tomography." *Nat. Methods* 5: (2008) 45–47.

Wang, L. H., S. L. Jacques, and L. Q. Zheng. "MCML-Monte Carlo modeling of light transport in multi-layered tissues." *Comput. Methods Programs Biomed.* 47: (1995) 131–145.

Wang, X., L. V. Wang, C.-W. Sun, and C.-C. Yang. "Polarized light propagation through scattering media: Time-resolved Monte Carlo simulations and experiments." *J. Biomed. Opt.* 8: (2003) 608–617.

Wu, J., L. Perelman, R. R. Dasari, and M. S. Feld. "Fluorescence tomographic imaging in turbid media using early-arriving photons and Laplace transforms." *Proc. Natl. Acad. Sci. U. S. A.* 94: (1997) 8783–8788.

Zienkiewicz, O. C., R. L. Taylor, and J. Z. Zhu. *The Finite Element Method: Its Basis and Fundamentals.* Barcelona: Elsevier Butterworth-Heinemann, 2005.

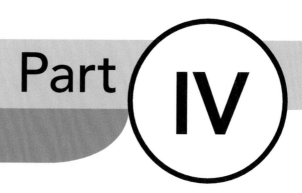

Part IV

Tissue autofluorescence lifetime spectroscopy and imaging
Applications

Oncology applications: Optical diagnostics of cancer

Alzbeta Chorvatova and Dusan Chorvat

Contents

14.1 INTRODUCTION

Cancers develop from premalignant to malignant in multiple steps in the process of carcinogenesis. Most human cancers arise from the epithelium, which undergoes transformations in the precancerous stage, including increase in the metabolic activity, cell density and nuclear variability, as well as low cellular maturation and thickening of the epithelial layer (Evan and Vousden 2001). The presence of epithelial precancer is also accompanied by architectural changes in the underlying stroma and submucosa, including neovascularization and slow destruction of the collagen cross-link by protease. In the course of cancer progression, chronic alterations of fuel metabolism and oxidative stress status are factors that could impair the capacity of the mitochondria to fulfill their crucial role in energy production and thus contribute to the activation of pathways governing cell death by apoptosis and/or necrosis. Modifications in metabolic functions almost always accompany pathological processes, as they are closely related to the energy need of the altered cellular processes machinery during disease. In cells and tissues, changes during physiological and pathological processes result in modifications of the amount and distribution of endogenous fluorophores and chemical–physical properties of their environment. Sensing of such biochemical and morphological changes in cell homeostasis using endogenous fluorescence or autofluorescence (AF) therefore provides valuable information for the optical diagnostics of cancers at their early stages.

Evaluation of the oxidative phosphorylation cycle is an attractive way how to differentiate cancerous from noncancerous tissue, as well as to investigate invasion and metastasis (Sud et al. 2006), This includes monitoring of key biomolecules, such as NAD(P)H, flavoproteins (Figures 14.1 and 14.2a, b), oxygen,

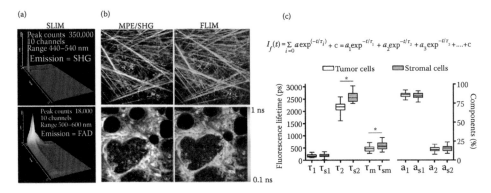

$$I_f(t) = \sum_{i=0} a \exp^{(-t/\tau_i)} + c = a_1 \exp^{-t/\tau_1} + a_2 \exp^{-t/\tau_2} + a_3 \exp^{-t/\tau_3} + \ldots + c$$

Figure 14.1 FLIM and spectral lifetime imaging microscopy (SLIM) analysis of mammary tumors. (a) Multiphoton SLIM analysis of the emission spectrum from endogenous fluorescence resulting from excitation at 890 nm. Emission from collagen (at half of the input wavelength) showed a very strong and sharp signal with a no appreciable decay (lifetime) confirming the SHG nature of the collagen signal (top). Emission spectra of endogenous fluorescence from tumor and stromal cells showed that the only substantial emission signal is at 530 nm, indicating that the source of the autofluorescence signal is FAD. (b) Multiphoton intensity and FLIM images of the stroma near a tumor (top) and the tumor and stromal components (bottom) from wild-type tumors. Note the increased intensity and fluorescent lifetimes of stromal cells (quantified in (c)) and the low lifetime of collagen. The color map in (b) represents the weighted average of the two-term model components ($\tau_m = (a_1\tau_1 + a_2\tau_2)/(a_1 + a_2)$) using the equation shown in (c). (c) Quantitative analysis of fluorescent lifetime components from tumor and stromal (subscript s) cells using the equation shown. * indicates a statistically significant ($p < 0.05$) difference following analysis with one-way analysis of variance (ANOVA) with a post-hoc Tukey–Kramer test. (Reproduced in the original form from the BioMed Central from Figure 6 of Provenzano PP et al., *BMC Med* 6, 11, 2008.)

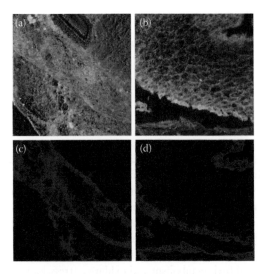

Figure 14.2 Fluorescence confocal microscopy measurements in the biopsy from colon of patients. Images of endogenous fluorescence of flavins (a, b) and PPIX (c, d), recorded *ex vivo* using laser scanning microscope in the biopsy of the healthy part of the colon (a, c) and of the cancerous colon (b, d) of patients. Excitation with 488 nm laser, emission at >500 nm for flavins or >633 nm for PPIX. One pixel corresponds to 0.22 μm. Patients received 5 mg/kg of ALA prior the measurement. Note an increase in the accumulation of the fluorescence to the colon wall with cancer.

and ATP consumption/generation. Other markers of cancer invasiveness include changes in the tumor microenvironment, namely, distribution of collagens of extracellular matrix (ECM) (Figure 14.1) and cellular protoporphyrins (Figure 14.2c, d). Monitoring of cell death by noninvasive AF measurements is another approach that can be used in the early identification and study of cancer (Wang et al. 2008). Indeed, preclinical and clinical studies have shown that the detectable levels of apoptosis or programmed cell death in cancerous tissue correlate with tumor progression.

This chapter is the first part of oncological applications, and its focus is on the methods employed in optical diagnostics of cancer, including laser-induced detection, time-resolved spectroscopy studies, and non-linear methods. Endogenous fluorophores present in cells and tissues, which are employed in cancer detection and treatment are also discussed,as summarized in the Table 14.1.

14.2 OVERVIEW OF METHODS FOR OPTICAL DIAGNOSTICS OF CANCER

14.2.1 FLUORESCENCE SPECTROSCOPY FOR EARLY CANCER DIAGNOSTICS

The potential to use fluorescence for tumor detection is known since the early 20th century (Policard 1924). Quantitative fluorescence spectroscopy approaches were first applied to discriminate normal and malignant tissue in 1965 (Lycette and Leslie 1965). Since 1980, advances in fiber optics and detectors enabled development of clinical instrumentation for *in vivo* measurements of fluorescence spectra from cancerous and normal tissues (Ramanujam 2000; Wagnieres et al. 1998). A large number of *in vitro* and *in vivo* investigations have demonstrated the diagnostic potential of fluorescence spectroscopy for detecting early lesions in variety of organ sites *in vivo* and *ex vivo* (Ramanujam 2000).

Increased nuclear size and nuclear-to-cytoplasmic ratio, increased microvascularization, degradation of stromal collagen, and changes in the concentration of mitochondrial fluorophores such as NAD(P)H and flavins lead to changes in optical scattering, absorption, and AF characteristics (Fryen et al. 1997; Gillenwater et al. 19998, 2006; Muller et al. 2008). Tissue AF based on the endogenous fluorescence of collagens and NAD(P)H was proposed as a real-time diagnostic tool for analyzing surgical tissue specimens of meningioma, based on variabilities in both spectral and time-resolved emission characteristics (Butte et al. 2005). Differences in fluorescence lifetimes between normal and cancerous tissue have been described in the oral cavity (Chang et al. 2004; Chen et al. 2005), esophagus (DaCosta et al. 2002; Pfefer et al. 2003), brain (Butte et al. 2005; Katz et al. 2002; Leppert et al. 2006; Marcu et al. 2004), and breast (Bird et al. 2005; Tadrous et al. 2003; Yang et al. 1997; Zipfel et al. 2003) (see Chapters 15 through 18 for more details). Fluorescence emission spectroscopy was also investigated for the recognition of cervical cancers and precancers *in vitro* (Lohmann et al. 1989). At 365 nm excitation, fluorescence of normal and abnormal cervix exhibited a single emission peak at 475 nm. The fluorescence intensity increased with the degree of progress from normal to precancer, proportionally to the degree of dysplasia. In contrast, the fluorescence of tumors is very small, with a rise in the intensity at the border between malignant and healthy tissue. Others (see Richards-Kortum and Sevick-Muraca 1996 for review) have reported consistent differences in the fluorescence of nonmalignant and malignant tissues, attributed to either an increase in NADH or a decrease in collagen or elastin fluorescence, indicating that measuring fluorescence improves the contrast in diagnostics of cervical cancers.

Probing of cell metabolism by detecting mitochondrial NAD(P)H and flavins and monitoring collagen fluorescence in stroma or melanin in melanocytes are just some examples of the approaches used in optical diagnostics of cancer. Others include comparative methods related to changes in multiple parameters simultaneously; for example, when compared to normal tissues, high-grade malignant colonic tissues exhibited low NAD(P)H and FAD fluorescence, but high endogenous fluorescence of amino acids porphyrins and protoporphyrin IX and changes in collagen distribution (reviewed in Richards-Kortum and Sevick-Muraca 1996; see also Chapter 17 for more details). Consequently, the use of comparative ratios of endogenous fluorescence measurements was introduced, namely, the ratio of SHG signal to AF signal on one hand (Cicchi et al. 2009) or Red-Ox ratio measurements, defined as NADH/FAD ratio and/or free/bound NAD(P)H component ratio (Bird et al. 2005; Provenzano et al. 2009; Wu and Qu 2006; Zhang et al. 2004) on the other hand.

Table 14.1 Fluorescence parameters of endogenous fluorescence in cancerous tissues and cell lines

CELL TYPE/TISSUE	FLUOROPHORE	EXCITATION (nm)	EMISSION (nm)	FLUORESCENCE LIFETIME (ns)	CHANGES IN CANCER	REFERENCES
Cancerous Tissues						
Normal living human esophageal HET-1	NAD(P)H	364	435–485	2.51	↓ τ in Barrett's adenocarcinoma SEG-1 cells: 2.21 ns	Sud et al. 2006
Glioblastoma multiform by endoscopy from brain tumor patients	NAD(P)H	337	460	1.59	Weaker fluorescence and ↑τ in glioblastoma multiform than in normal cortex (1.28 ns)	Sun et al. 2010
Meningioma from patients undergoing brain tumor surgery	NAD(P)H	337	460	5.1	Similar τ as in dura but larger than cortex, significantly lower fluorescence intensity when compared to dura and cortex	Butte et al. 2005
Hamster cheek pouch *in vivo*	NAD(P)H	780 (2p)		0.29 2.03	↓ τ2 in precancerous tissues: 1.58 ns (low-grade precancer) 1.83 ns (high-grade precancer)	Skala et al. 2007a
Breast (mouse mammary tissue, stromal cells)	FAD	890 (2p)		Not quantified	↑ <τ> in invading cells vs. primary tumor cells due to ↑ in τ1 and ↑ in τ2	Provenzano et al. 2008a
Melanoma human patients *in vivo*	NAD(P)H and melanin	760 and 800 (2p)		0.461 2.312	In keratinocytes ↓ τ1 and τ2 in melanocytes to 0.146 and 1.0970	Dimitrow et al. 2009a
Meningioma from patients undergoing brain tumor surgery	Collagen types I and III	337	385–390	2.04	Strong intensity emission when compared with cortex	Butte et al. 2005

Tissue	Fluorophore	Excitation	Emission	Lifetime	Findings	Reference
Breast (mouse mammary tissue, stromal cells)	Collagen	890 (2p)		Not quantified	↓ <τ> due to ↓ τ2 in tumor cells	Provenzano et al. 2006
Normal urothelium (human bladder)	Porphyrin PPIX	425		15	No significant change in carcinoma *in situ*	Glanzmann et al. 1996
Normal oral mucosa (human patients)	Porphyrin PPIX	410	633	1.76 8.14	↓ τ1 and ↑ τ2 in oral premalignant lesions: 1.76 ns and 12.59 ns (verrucous hyperplasia) 1.65 and 12.97 ns (epithelial hyperplasia) 1.43 and 10.66 ns (epithelial dysplasia)	Chen et al. 2005
Basal cell carcinoma	Unspecified tissue AF	355	375 (also 455)	1.551 1.617 1.567 (<τ>)	↓ <τ> in basal cell carcinomas 1.398 ns 1.617 ns 1.417 ns (<τ>)	Galletly et al. 2008
Nonmelanoma skin cancer	Unspecified tissue AF					Yaroslavsky et al. 2007
Lung cancer bronchoscopy in patients	Unspecified tissue AF	405	Long pass 450	0.17 2.02 6.84	No change for healthy mucosa and preneoplastic or neoplastic lesions	Uehlinger et al. 2009
Colonic cancer						
Cervical cancer						

(continued)

Table 14.1 (Continued) Fluorescence parameters of endogenous fluorescence in cancerous tissues and cell lines

CELL TYPE/TISSUE	FLUOROPHORE	EXCITATION (nm)	EMISSION (nm)	FLUORESCENCE LIFETIME (ns)	CHANGES IN CANCER	REFERENCES
Cell Lines and Cells in Culture						
3T3-L1 adipocytes and fibroblasts	NAD(P)H	370	420–480	0.58 2.46 9.00 7.23 ($<\tau>$)	↓ $<\tau>$ to 6.73 in high glucose	Evans et al. 2005
Endothelial cells from calf aorta BKEz-7	NAD(P)H	375	440–445 475–480	0.15–0.2 0.4–0.5 2.0–2.5		Schneckenburger et al. 2004
Hepatocytes (primary culture from C57BL/6 mice)	NAD(P)H	750 and 790 (2p)	Short pass 640	0.45 2.3		Ramanujan et al. 2005
Cultured HEK293T cells	NAD(P)H	750 and 790 (2p)	Short pass 640	0.5 2.4		Ramanujan et al. 2005
HeLa	NAD(P)H	750 and 790 (2p)	Short pass 640	0.48 2.5		Ramanujan et al. 2005
HeLa	NAD(P)H	750 (2p)	Band pass 450 ± 40	1.3 ($<\tau>$)	↑ $<\tau>$ after apoptotic treatment to 3.54 ns	Wang et al. 2008
Epithelial cells (cheek pouches)	NAD(P)H	800 (2p)	490 short pass	2.35	↓ τ in precancerous tissues: 2.35 and 2.25 (low-grade precancer) 2.25–2.35 and 2.15 ns (high-grade precancer)	Skala et al. 2007b

Epithelial cells (cheek pouches)	FAD	890 (2p)	490 short pass	1.65–1.75 2.05	↑ τ1 and ↓ τ2 in precancerous tissues: 1.65–1.75 and 1.95 ns (low-grade precancer) 1.70–1.75 and 1.95 ns (high-grade precancer)	Skala et al. 2007b
Rat epithelial cells in culture	Porphyrin PPIX	398	LP590	3.6 7.4		Kress et al. 2003
Glioma U87 cell line (also G112 cell line)	Unspecified tissue AF	750–770 (2p)	Short pass	1.181 (<τ>) cytosol	↓ <τ> in nucleus to 0.775 ns	Leppert et al. 2006
Epithelial tissue	Unspecified tissue AF	405		0.4–0.6 3–4	τ1 dominant in nonkeratinized tissue, τ2 in keratinized epithelium and stroma	Wu et al. 2007
Bronchial mucosa epithelium	Unspecified tissue AF		430–680	0.2 2.0 6.9		Glanzmann et al. 2001

Note: Changes in the endogenous fluorescence in cancer tissues and/or cell lines and cells in culture derived from cancer. <τ> mean lifetime, 2p: two-photon excitation, PPIX: protoporphyrin, ↑: increase, ↓: decrease.

Nowadays, fluorescence and reflectance spectroscopy provide complementary information useful for precancer diagnosis (Sokolov et al. 2002). Fiber-optic confocal reflectance microscope was used to differentiate between normal, preneoplastic, and neoplastic oral tissue *in situ* (Maitland et al. 2008). In addition, near-infrared (NIR) absorption and scattered tomography imaging was used in the diagnostics of different types of cancers, namely, the presence of glioma tumor with endogenous fluorescence of PPIX was demonstrated using a noncontact single-photon counting fan-beam acquisition system interfaced with microCT imaging (Kepshire et al. 2009). Perspectives of AF detection for cancer diagnostics and treatment include the use of laser-induced AF imaging (detection) of early carcinoma in tissues.

14.2.2 LASER-INDUCED AF FOR CANCER DETECTION

Optical methods for cancer diagnostics and treatment include the use of laser-induced AF as a sensitive technique for detection of early carcinoma in tissues, such as the case of colonic cancer. Colonic cancer (see Chapter 17 for details) is an important public health concern in industrialized countries and around the world. Present detection methods do not provide accurate diagnosis in the disease's early stages. Diagnosis and localization of early carcinoma, therefore, play an important role in the prevention and curative treatment of this type of cancer (Li et al. 2005). In colon neoplasms, primary differences between the fluorescence of neoplastic and nonneoplastic tissues were attributed to collagen fluorescence and hemoglobin reabsorption (Schomacker et al. 1992a, b). In addition, Mycek et al. (1998) performed a time-resolved AF spectroscopy study of colonic polyp types *in vivo*. Pancreatic adenocarcinoma being one of the leading causes of cancer death, with a 5-year survival rate of only 4%, fluorescence spectroscopy was also investigated as a potential diagnostic tool to differentiate between diseased and normal pancreatic tissue states (Chandra et al. 2007), and also between pancreatic cancer (adenocarcinoma) and nonneoplastic inflammation (pancreatitis).

Optical methods are now also widely used for the recognition of cancer vs. noncancer in the case of surgical resection and/or ablation of cancer, as an appropriate safe surgical resection is a prerequisite to its optimal therapy. Taking into consideration that distinct excitation profiles and lifetimes of endogenous fluorescence were identified for specific brain regions (Kantelhardt et al. 2007; Leppert et al. 2006), the imaging of brain and brain tumor specimens was based on monitoring of endogenous molecular fluorophores that can be selectively isolated accordingly to their excitation spectra and fluorescence lifetimes.

14.2.3 TIME-RESOLVED AF IN CANCER DIAGNOSTICS

Fluorescence lifetime imaging microscopy (FLIM) of endogenous fluorescence (Figure 14.1) has been widely explored for cancer diagnostics purposes. Currently, there are no other known diagnostic techniques able to detect such early tissue transformations. Time-resolved fluorescence spectroscopy of endogenous fluorophores is an ideal technique for these purposes, because of its ability to examine tissue surfaces together with functional changes and its adaptability to endoscopic devices. For these reasons, this technique was widely tested in different *in vivo* and *ex vivo* conditions, as well as in cell cultures (for review of the use of steady-state endogenous fluorescence in cancer diagnostics, see Ramanujam 2000).

Time-resolved methods allow detection of epithelial precancerous stages, particularly early neoplasia, which may not be apparent using conventional diagnostic imaging techniques, resulting in an increase in the cancer stage classification and also better localization and resection of cancer (Ramanujam 2000). Such diagnostics is based on the inherent difference between fluorescence spectral and lifetime characteristics in normal vs. precancerous lesions. When combined with multiphoton endoscopy or confocal endoscopy, FLIM or fiber-based fluorescence lifetime spectroscopy provides a powerful tool in clinical optical diagnostics of cancer (Cicchi et al. 2009, 2010).

Steady-state and time-resolved fluorescence spectroscopy was employed to measure epithelial cell AF (Konig et al. 1993). AF properties of human epithelial cells were extensively studied because carcinoma, the most common form of human cancer, is derived from epithelial cells. Lifetime imaging of endogenous AF also allowed structural imaging of tumors and the central nervous system architecture at the subcellular level (Leppert et al. 2006), as excitable endogenous fluorophores in cells have been shown to be derived from tumors of different histotypes with individual fluorescence lifetime profiles for distinct cell types. In addition, fluorescence measurements are used to increase generally poor contrast of the legions, and thus tumor localization and precise excision (Yaroslavsky et al. 2007); this technique was used, for example, to identify nonmelanoma skin cancers, which include basal cell and squamous cell carcinomas and are more common than all other human cancers, as every fourth Caucasian develops at least one lesion during his/her lifetime and most of which are curable by surgery if detected and treated early (see Chapter 16 for more details). As the extent of tumor resection directly correlates with the prevention of recurrence, evaluation of excised tumor specimens is particularly important. AF lifetime allowed to distinguish meningioma, slow growing lesions accounting for 15%–20% of primary brain tumors and primarily treated by surgical resection (Butte et al. 2005), from normal dura and cortex and thus contributed to the accuracy of tissue classification. Consequently, time-resolved spectroscopy was proposed as a tool for the identification of meningiomas, which can serve for the development of real-time diagnostic tools analyzing brain tumor tissue surgical specimens.

14.2.4 NONLINEAR OPTICAL METHODS IN CANCER DIAGNOSTICS

Nonlinear optical methods have recently become an important add-on to the wide spectrum of methods used for cancer diagnostics. As an example, multiphoton excitation applied for AF imaging, or second harmonic generation (SHG) technique used for imaging of collagen fiber networks, offers a broad range of new biomedical applications, established as noninvasive and painless methods for *in vivo* and *ex vivo* examination of tissues, thus holding great promise for cancer diagnostics (Konig et al. 2008).

Methods employing multiphoton excitation allow sensitive detection of endogenous fluorophores in distinct tissue regions, namely, it enables functional imaging of deep-tissue cells and their cellular compartments. A well-studied example is skin cancer (see Chapter 16 for details), which accounts for 40% of annually detected cancer and which early diagnosis is essential for successful cure. Using 2P FLIM endoscopic tomography, differentiated keratinocytes, basal stem cells, melanocytes, and macrophages can be detected and distinguished simultaneously by their AF properties (Konig 2008). In addition, this technique can also be used for probing and characterization of the network of elastin fibers and collagen bundles in healthy tissue and their changes in cancer. In general, fluorescence tomography provides a tool for preclinical molecular contrast agent assessment in oncology, taking into consideration the fact that this system has capacity of noncontact imaging, automated boundary recovery, and inclusion of sophisticated internal tissue shapes in the recovered images (Choi et al. 2005; Kepshire et al. 2009; Ntziachristos 2006; Ntziachristos et al. 2002, 2004).

SHG method is another well-suited microscopy technique to analyze connective tissues due to the significant modification/changes of second-order nonlinear susceptibility of collagen, observed in the peritumoral stroma (Cicchi et al. 2008; Lin et al. 2005). SHG was therefore used for investigating collagen–fiber orientation and their structural changes in the tumor microenvironment (Brown et al. 2003; Cicchi et al. 2009; Lin et al. 2006; Provenzano et al. 2009).

14.3 CANCER DIAGNOSTICS USING ENDOGENOUS FLUOROPHORES

14.3.1 TRACKING CHANGES IN METABOLIC ACTIVITY OF CANCER CELLS WITH METABOLITES, ENZYMES, AND COENZYMES

14.3.1.1 NAD(P)H and FAD fluorescence

Change in NAD(P)H and FAD fluorescence is considered a metabolic signature of cancers (Figures 14.1. and 14.2a, b). Increased FAD fluorescence intensity and lifetime were associated with invading metastatic cells (Provenzano et al. 2008a; Table 14.1). Indeed, cancer metastasis involves complex cell behavior and interaction with the ECM by metabolically active cells. AF consistent with NADH and flavins was found decreased, potentially as a result of changes in the redox state of the fluorophores in human epithelial cell lines, both an immortalized and a distinct tobacco-carcinogen-transformed human bronchial epithelial cell lines, used as useful models for studies seeking to distinguish between normal and transformed human bronchial epithelial cells (Pitts et al. 2001). In this study, a loss in AF was observed in carcinogen-transformed bronchial epithelial cells and proposed to result from changes in fluorophore oxidation state (Pitts et al. 2001). AF from bronchial tissue has been investigated as a means of lung cancer detection.

NAD(P)H fluorescence, measured in human epithelial tissues *in vivo*, was proposed as a quantitative fluorescence biomarker for the *in vivo* detection of dysplasia in the esophagus (see Chapter 17 for details), when esophageal cancer invasiveness was tested by FLIM (Sud et al. 2006). AF was also used for diagnostics of bladder cancers: in malignant tumors, significant decrease in collagen, NADH, and FAD fluorescence was observed and attributed to change in tissue morphology, namely, increased thickness of the urothelium in malignant lesions (Zheng et al. 2003). A clinical study investigating the course of esophageal cancer progression has applied the time-resolved optical molecular imaging to study the underlying biological basis of endogenous fluorescence changes (Sud et al. 2006). Higher intracellular oxygen and NADH levels, attributed to altered metabolic pathways in malignant cells in Barret's adenocarcinoma cells, compared to normal living human esophageal cells, were also revealed by FLIM (Table 14.1; Sud et al. 2006). Several other cultured cells were tested for metabolic redox imaging using fluorescence lifetimes, including hepatocytes (Ramanujan et al. 2005), osteosarcomas (Wang et al. 2008), and HeLa cells (Ramanujan et al. 2005; Wang et al. 2008).

In the cancerous tissue, a shift of NADH from the bound to the free form (Colasanti et al. 2000) was documented together with a release of FAD (Skala et al. 2007b). In these studies, time-resolved spectroscopy showed its potential to help a complete resection of a tumor, thus reducing its recurrence rate and improve a patient's survival. The ratio of relative contributions of "free" and "bound" NADH lifetimes served for metabolic mapping of human breast cells (Bird et al. 2005). NADH decay was found modified in cancer cell lines, as nonmetastatic cells consistently showed higher average lifetime than metastatic and normal ones (Pradhan et al. 1995). This observation was related to a more hydrophobic environment of NADH in nonmetastatic cell lines. In addition, Tadrous et al. (2003) applied FLIM to study AF changes in fixed human breast tissues and found significant modification of AF lifetimes among stroma, malignancy-associated stroma, blood vessels, and malignant epithelium, suggesting the possibility of establishing a histological map of breast tissues with lifetime measurements. NADH/NADPH fluorescence was also used as an important tool for image-guided surgery of glioblastoma multiforme (Sun et al. 2010), as well as meningioma (Butte et al. 2010) brain tumors (see also Chapter 15 for details). Indeed, in the former study, NAD(P)H fluorescence at 460 nm exhibited weaker fluorescence intensity and longer lifetime in glioblastoma when compared to the normal cortex, while in the latter study, the short-lived (<1.5 ns) fluorescence emission (peak at 460 nm) was proposed to correspond to the NAD(P)H in meningioma, facilitating their resection. NADH decay was also found modified in cancer cell lines, as nonmetastatic cells consistently showed higher average lifetime than metastatic and normal ones (Pradhan et al. 1995). Finally, NADH lifetime measurements were also used as a noninvasive tool to detect apoptosis (cell death) and to distinguish apoptosis from necrosis in cell cultures of human osteosarcome and in HeLa cells (Wang et al. 2008).

14.3.1.2 NADH/FAD ratio

Metabolic imaging (NADH, NADPH, flavins) revealed loss of fluorescence in carcinogen-transformed cells—the result of an alteration in the redox state due to changes in the mitochondrial morphology—and became important for the preneoplasia detection in the lungs (Pitts et al. 2001). In precancerous epithelium, an increase in NADH fluorescence and a decrease in the redox ratio defined as FAD fluorescence over the sum of FAD and NADH fluorescence were found (Georgakoudi et al. 2002; Pradhan et al. 1995; Richards-Kortum and Sevick-Muraca 1996; Wu and Qu 2006). Time-resolved optical imaging was used to show altered metabolic functions in cancer cells and cell lines (Ramanujan et al. 2005; Sud et al. 2006; Wang et al. 2008). The fluorescence lifetimes of NADH and FAD and particularly the redox ratio of these two metabolites were altered in transformed cells and/or epithelial tissues (Bird et al. 2005; Provenzano et al. 2009; Wu and Qu 2006; Zhang et al. 2004). Variability in the metabolic redox ratio, based on reduced NADH and oxidized FAD imaging of fluorescence lifetime, was also evaluated in precancerous epithelial tissues (Skala et al. 2007a, b), where precancerous cells were showed to have changed NADH fluorescence when compared to normal epithelial cells. Consequently, NADH and FAD lifetimes were proposed to serve as fingerprints of cancer invasiveness (reviewed in Provenzano et al. 2009).

14.3.2 VITAMINS AND PTERINS IN CANCER DIAGNOSTICS

A part of changes described above, there is little information on changes of the overall content in individual vitamins or pterins in the course of cancer progression. Vitamin B3 or cholecalciferol (synthesized in the skin under the action of the UV light) has endogenous fluorescence after excitation with UV light in the range 380–460 nm (Zipfel et al. 2003), but despite the fact that vitamin B3 was proved to be used in the colon cancer prevention, so far, no information is available on its potential use in cancer diagnostics. Vitamin B6 was also noted disturbed in cancers (see Chapter 15 for details).

Folate is another potentially autofluorescent molecule that can help in cancer diagnostics. Folate receptor is overexpressed in a number of tumors (Garin-Chesa et al. 1993; Toffoli et al. 1997; Weitman et al. 1992). Folate receptor-positive cancers were showed to be optically imaged by administration of folate-conjugated fluorescent dye: folate-linked fluorophore can be used for optical imaging of tumors that express folate receptor (Kennedy et al. 2003), and its use was proposed to help the tumor localization during tumor resection. However, although pterins were used as sensitizers in photochemical reactions inducing DNA damage (Lorente et al. 2004), neither their fluorescence nor that of folate is employed for cancer diagnostics.

14.3.3 MONITORING ARCHITECTURAL CHANGES IN TUMOR MICROENVIRONMENT USING STRUCTURAL PROTEINS

14.3.3.1 Collagens

ECM proteins, namely collagen, are an important part of the tumor microenvironment (reviewed in Provenzano et al. 2009; Figure 14.1), together with a wide range of tumor-associated cells, including fibroblasts, macrophages, neurophils, and the vasculature. Understanding modification in this multicellular environment is therefore a prerequisite to comprehend a precise role that these components play in supporting tumor growth and, ultimately, the initial metastatic stages. Light-induced AF spectroscopy is able to detect premalignant and malignant lesions because normal, premalignant, and malignant tissues contain different amounts of collagens. Furthermore, in various carcinogenesis stages, cells also possess different local environments, thus exhibiting different fluorescence lifetimes. Nonlinear optical measurements by SHG and time-resolved measurements (Figure 14.1) are a valuable tool helping to identify cancerous invasions by studying their collagen network (Provenzano et al. 2008a). In this study, augmented collagen density has been determined by the sFLIM method in a mammary tumor initiation. Higher collagen density enhances tumorigenesis, local invasion, and metastasis, casually linking stromal collagen elevation to tumor formation and progression.

Collagen has become an important biomarker for optical cancer diagnostic methods. ECM provides a structural lattice for cells in tissues, and it also facilitates cellular communication, while being a key component during processes such as angiogenesis and neoplasia (Kirkpatrick et al. 2006). Collagen is an important structural protein of tissues providing their strength; there are numerous types of collagens, of which types I, II, and III constitute about 80%–90% of the body's collagen (Lodish et al. 2000). Type I collagen, one of the main components of the interstitial matrix, exhibits both optical scattering and endogenous fluorescence (Kirkpatrick et al. 2006). Both properties can be exploited to evaluate noninvasively changes in the ECM. Increased stromal collagen has been found to significantly accentuate with tumor formation and thus to result in a significantly more invasive phenotype; spectrally resolved lifetime detection was therefore proposed as a powerful tool to evaluate the invasiveness of tumor cells in breast cancer (Provenzano et al. 2008b).

More than 85% of cancers arise in epithelial tissues and are preceded by a precancerous stage in which neoplastic cells are confined to the epithelium. Fluorescence spectroscopy is used for detection of epithelial precancer and cancers (Nath et al. 2004). AF properties of human epithelial cells were extensively studied with the aim to battle carcinoma, the most common form of human cancer derived from epithelial cells. The detection of otherwise invisible early neoplastic growth in epithelial tissue sites is one of the most widely explored applications of fluorescence-based diagnostics of cancer (Ramanujam 2000). Such early neoplastic growth refers to premalignant changes such as dysplasia and carcinoma *in situ*, which precede malignancy, for example, invasive carcinoma. Development of epithelial precancer and cancer leads to well-documented molecular and structural changes in the epithelium, including loss of normal epithelial architecture (Arifler et al. 2007). Neoplasia is associated with alterations in both epithelial cells and the supporting stroma. In the stroma underlying epithelium, collagen fibers are the main scatterers. As a well-known source of AF and nonlinear emission, collagens were proposed to be employed for detecting such neoplastic changes.

Collagens and collagen cross-links in the stroma are also altered in head and neck tumors (Gillenwater et al. 1998, 2006; Muller et al. 2008; see also Shin et al. 2010 for review and Chapter 18 for details). Collagen is often decreased in stroma in oral cancer (Pavlova et al. 2008). Collagen fluorescence was also used, together with that of NADH, to record clinically observable differences between normal and dysplastic tissue spectra in cervical tissue (Drezek et al. 2001, 2003).

Fluorescence lifetime monitoring of collagen modifications was proposed to help to evaluate the presence of early stages of cancerous processes in the brain (see Chapter 15 for details). The fluorescence of connective tissue proteins, in particular, collagens, was showed to dominate the fluorescence emission of dura and meningioma (Butte et al. 2005). Interestingly, different decay characteristics were described in the two tissues, and these differences were attributed to distinct types of collagens and their cross-links: namely, time-resolved spectra of meningiomas at the blue-shifted wavelengths were proposed to be dominated by two types of collagen, collagen type I (with peak emission at 380 nm) and type III (peak emission at 390 nm). Larger relative collagen content was also detected in adenocarcinoma and pancreatitis (Chandra et al. 2007). The thickening of the epithelium and a slow destruction of the cross-link in collagen associated with the process of carcinogenesis resulted in the decrease in collagen fluorescence from the connective tissue under the epithelium (Wu and Qu 2006).

Basal cell carcinoma, the most common form of cancer worldwide, was also tested by FLIM in human biopsies to be distinguished from the surrounding uninvolved healthy skin (Galletly et al. 2008). Mean fluorescence lifetimes were significantly reduced in areas of carcinomas, as opposed to surrounding uninvolved skin, without significant change in AF intensity, suggesting possible use of the FLIM for early diagnosis of skin cancer based on collagen disruption. Collagen cross-linking also proved to be reduced in conditions such as photodamaged skin due to a posttranscriptionally reduced collagen synthesis (Lutz et al. 2012; see Chapter 16 for more details).

14.3.3.2 Elastin

AF from the elastin network in the submucosa provided the major fluorescence signal following 405 nm excitation and was proposed to be used to distinguish tumoral tissue in the human bronchial early cancer (Uehlinger et al. 2009); however, these authors did not identify changes in the fluorescence lifetimes in the diseased tissue.

14.3.4 AMINO ACIDS IN CANCER DETECTION

Tryptophan (Trp). An increase in the fluorescence of Trp residues (emission peak at 350 nm, excitation under 300 nm) was noted in cancerous tissues, namely, in bladder tumors (Zheng et al. 2003). Trp change was also reported in the malignant tissues of the colon (Ramanujam 2000). However, no change in Trp fluorescence was observed between normal and transformed lumen bronchial epithelial cells (Pitts et al. 2001). In addition, Trp residues were evaluated in metastatic and nonmetastatic cell lines, where an increase in Trp fluorescence was detected in nonmetastatic cells, but in these experiments, no change in average fluorescence lifetime for Trp decay was noted (Pradhan et al. 1995), suggesting a rise in the Trp concentration when the cells progress from a nonmetastatic or low metastatic to metastatic state. On the other hand, an increase in the fluorescence of Trp residues observed in bladder tumors was proposed to be linked to hyperactivity or to urothelial hyperproliferation in tumor. To discriminate tumor from normal bladder tissue, intensity ratios (at I350/I470 [excitation 280 nm] and at I390/I470 [excitation 330 nm]) were employed as diagnostic algorithms (Zheng et al. 2003). A combination of the fluorescence measurements, recording the Trp emission with the fluorescence lifetime evaluation of NADH, was proposed to be used as an internal standard to evaluate normal from cervical cancer cells (Li et al. 2009).

14.3.5 DETECTION OF MELANOMA USING PIGMENTS

14.3.5.1 Melanin

Time-resolved fluorescence of melanin was proposed as an additional mean to support diagnostic decision and improve the process of noninvasive early detection of melanoma (see Chapter 16 for details). In the case of melanoma, the tissue cell architecture and cell morphologies exhibited significant differences when compared to normal nevi (Konig et al. 2008). Time-resolved fluorescence is now used as additional information for melanoma detection in comparison with common melanocytic nevi (Dimitrow et al. 2009a), based on the observation that the fluorescence lifetime distribution is correlating with the intracellular amount of melanin. Malignant melanoma is a type of skin cancer which incidence has dramatically increased over the past decades, while the patient's outcome and curability depends on diagnosis and excision at early stages of tumor progression, prompting search for noninvasive techniques of melanoma detection (Dimitrow et al. 2009b). The AF intensity was found increased in melanocytic compound nevi when compared to the normal one, while the lifetime was decreased (Dimitrow et al. 2009a). Time-resolved fluorescence was used for noninvasive early detection of melanoma, malign cancerous lesions (Dimitrow et al. 2009a, b), related to differences found in the lifetime behavior of keratinocytes in contrast with melanocytes, as keratinocyte lifetime values were proposed to correspond to NAD(P)H and melanocyte lifetime to endogenous melanin. However, in these studies, it was not clear whether the modification of the fluorescence lifetime in the tissue is caused by changes in the melanin structure or NAD(P)H binding.

14.3.5.2 Mucin

Mucin is another pigment that can be a source of scarce tissue AF following excitation with UV/VIS light (Castillo et al. 1986). Mucin, a glycoprotein expressed on most epithelial cell surfaces, was confirmed as an epithelial tumor marker and thus a biomarker for the diagnosis of early cancers (Cheng et al. 2009).

14.3.6 CANCER DIAGNOSTICS AND THERAPY EMPLOYING PORPHYRINS

Protoporphyrin IX (PPIX). Red fluorescence specific for advanced ulcerated squamous cell carcinoma, ascribed to porphyrins, is known for nearly 100 years (Policard 1924). As described in Chapter 3, enzymatic differences between tumor and normal cells culminate in higher PPIX concentration in cancer (Figure 14.2c, d), allowing the use of PPIX fluorescence lifetimes to discriminate carcinoma tissues, including the human bladder (Glanzmann et al. 1996), or premalignant lesions in the oral mucosa (Chen et al. 2005). In the colon cancer, an increase in the accumulation of the endogenous flavin and PPIX fluorescence to the colon wall is noted in cancer as opposed to noncancerous tissues in biopsies from patients that were administered with ALA (Figure 14.2; see also Chapter 23 for details).

Long fluorescence lifetime of porphyrins was also proposed to have the potential to be used for early cancer diagnosis, as well as delineation of the boundaries of the cancerous tumor (Berezin and Achilefu 2010). The origin of the protoporphyrin in ulcerated squamous cell carcinomas was related to several sources (Richards-Kortum and Sevick-Muraca 1996). Different species of tumor-localizing porphyrins, particularly monomeric and aggregated porphyrin molecules, as well as ionic species located at different cellular sites, can be distinguished by their fluorescence lifetimes (Kress et al. 2003; Schneckenburger et al. 1995; Strauss et al. 1997). Fluorescence lifetime of PPIX has been used to differentiate malignant tissue from the healthy one in the brain (Kantelhardt et al. 2007, 2008), as well as in oral mucosa (Chen et al. 2005), but not in esophagus (Glanzmann et al. 1999).

Significant endogenous fluorescence from PPIX (which is enhanced when the subject imaged has been administered with ALA; Cubeddu et al. 1995; Kennedy and Pottier 1992; Kriegmair et al. 1994; Marcus et al. 1996) is provided by glioma tumors. Fluorescence lifetime imaging of glioma tissue was deployed to visualize solid tumors, the tumor–brain interface, as well as single invasive tumor cells. The analysis of fluorescence decays allows one to discriminate between tumor glioma cells and normal brain parenchyma, as shortening of the fluorescence lifetime was found in glioma when compared to normal brain tissue (Leppert et al. 2006). Time-resolved laser-induced fluorescence spectroscopy of endogenous fluorophores also demonstrated strong potential as a diagnostic tool in brain tumor operations (Sun et al. 2010). Measurement of fluorescence decays was, therefore, proposed as a technology that can provide noninvasive optical tissue assessment that could potentially be applied to intraoperative procedures, such as the detection of residual tumor tissue in glioma surgery (see Chapter 15 for more details). Using endogenous contrast from PPIX, fluorescence tomography system coupled to microCT was used to illustrate diagnostic detection of glioma tumors (Kepshire et al. 2009), which was enhanced when the subject imaged has been administered with ALA. To provide maximum sensitivity, time-correlated single-photon counting (TCSPC) was used in this fluorescence tomography system.

Laser-induced fluorescence of endogenous PPIX was investigated in malignant and normal tissues in mice (Nilsson et al. 1994), as well as in rats (Johansson et al. 1997). Comparable studies have been performed in patients in connection with photodynamic therapy (PDT) of basal cell carcinomas and adjacent normal skin following topical application of 5-Ala in order to study PPIX build-up (Klinteberg et al. 1999; Palsson et al. 2003), as well as in various kinds of malignant, premalignant, and benign lesions in the head and neck region (Wang et al. 1999; see also Andersson-Engels et al. 2009 and Celli et al. 2010 for review and Chapters 16 through 18 and 23 for more details).

14.4 CONCLUSIONS

Optical diagnostics of cancer, utilizing optical methods detecting changes in endogenous cell and tissue fluorophores, has a great potential to provide noninvasive screening and detection of surface tumors for clinical applications, particularly when combined with endoscopy. This chapter provides an overview of optical techniques that proved to be useful in the search for cancer biomarkers in cells, as well as in assessing the histological architecture of cancer tissue, in early cancer diagnostics, or in surgical resection and delineation of the tumor from the normal tissue. Employment of these methods leads to early detection, diagnostics, and monitoring of pathological conditions and thus more effective treatment of this critical disease. This chapter also gives examples of the use of endogenous fluorophores in cancer diagnostics and treatment. Identification of an optically discernible mechanism that is consistently altered in a malignant process could prove very useful not only for further developing of therapeutic targets but also for understanding disease pathogenesis and designing minimally invasive optical technologies for early detection.

ACKNOWLEDGMENTS

We would like to acknowledge support from the research grant agency of the Ministry of Education, Science, Research and Sport of the Slovak Republic VEGA No. 1/0296/11 and the Slovak Research and Development Agency APVV-0242-11. We would also like to acknowledge support from Integrated Initiative of European Laser Infrastructures LaserLab Europe III (EC's FP7 under grant agreement no. 284464). We would like to thank Dr. A. Mateasik and Prof. P. Mlkvy from ILC, Bratislava, for providing Figure 14.2.

REFERENCES

Andersson-Engels S., Johansson J., Svanberg K., & Svanberg S. (2009). Fluorescence imaging in medical diagnostics. In: Fujimoto, J. G. & Farkas, D. L. (eds) *Biomedical Optical Imaging*. Oxford: Oxford University Press, pp. 265–305.

Arifler D., Pavlova I., Gillenwater A., & Richards-Kortum R. (2007). Light scattering from collagen fiber networks: Micro-optical properties of normal and neoplastic stroma. *Biophys J* 92, 3260–3274.

Berezin M. Y., & Achilefu S. (2010). Fluorescence lifetime measurements and biological imaging. *Chem Rev* 110, 2641–2684.

Bird D. K., Yan L., Vrotsos K. M., Eliceiri K. W., Vaughan E. M., Keely P. J., White J. G., & Ramanujam N. (2005). Metabolic mapping of MCF10A human breast cells via multiphoton fluorescence lifetime imaging of the coenzyme NADH. *Cancer Res* 65, 8766–8773.

Brown E., McKee T., diTomaso E., Pluen A., Seed B., Boucher Y., & Jain R. K. (2003). Dynamic imaging of collagen and its modulation in tumors in vivo using second-harmonic generation. *Nat Med* 9, 796–800.

Butte P. V., Fang Q. Y., Jo J. A., Yong W. H., Pikul B. K., Black K. L., & Marcu L. (2010). Intraoperative delineation of primary brain tumors using time-resolved fluorescence spectroscopy. *J Biomed Opt* 15, 027008-1–027008-8.

Butte P. V., Pikul B. K., Hever A., Yong W. H., Black K. L., & Marcu L. (2005). Diagnosis of meningioma by time-resolved fluorescence spectroscopy. *J Biomed Opt* 10, 064026-1–064026-9.

Castillo E. J., Koenig J. L., Anderson J. M., & Jentoft N. (1986). Protein adsorption on soft contact-lenses. 3. Mucin. *Biomaterials* 7, 9–16.

Celli J. P., Spring B. Q., Rizvi I., Evans C. L., Samkoe K. S., Verma S., Pogue B. W., & Hasan T. (2010). Imaging and photodynamic therapy: Mechanisms, monitoring, and optimization. *Chem Rev* 110, 2795–2838.

Chandra M., Scheiman J., Heidt D., Simeone D., McKenna B., & Mycek M. A. (2007). Probing pancreatic disease using tissue optical spectroscopy. *J Biomed Opt* 12, 060501-1–060501-3.

Chang C. L., You C., Chen H. M., Chiang C. P., Chen C. T., & Wang C. Y. (2004). Autofluorescence lifetime measurement on oral carcinogenesis. *Conf Proc IEEE Eng Med Biol Soc* 4, 2349–2351.

Chen H. M., Chiang C. P., You C., Hsiao T. C., & Wang C. Y. (2005). Time-resolved autofluorescence spectroscopy for classifying normal and premalignant oral tissues. *Lasers Surg Med* 37, 37–45.

Cheng A. K. H., Su H. P., Wang A., & Yu H. Z. (2009). Aptamer-based detection of epithelial tumor marker mucin 1 with quantum dot-based fluorescence readout. *Anal Chem* 81, 6130–6139.

Choi H. K., Yessayan D., Choi H. J., Schellenberger E., Bogdanov A., Josephson L., Weissleder R., & Ntziachristos V. (2005). Quantitative analysis of chemotherapeutic effects in tumors using in vivo staining and correlative histology. *Cell Oncol* 27, 183–190.

Cicchi R., Crisci A., Nesi G., Cosci A., Giancane S., Carini M., & Pavone F. S. (2009). *Multispectral Multiphoton Lifetime Analysis of Human Bladder Tissue*. SPIE-Int Soc Optical Engineering, Bellingham, WA.

Cicchi R., Kapsokalyvas D., De Giorgi V., Maio V., Van Wiechen A., Massi D., Lotti T., & Pavone F. S. (2010). Scoring of collagen organization in healthy and diseased human dermis by multiphoton microscopy. *J Biophotonics* 3, 34–43.

Cicchi R., Sestini S., De Giorgi V., Massi D., Lotti T., & Pavone F. S. (2008). Nonlinear laser imaging of skin lesions. *J Biophotonics* 1, 62–73.

Colasanti A., Kisslinger A., Fabbrocini G., Liuzzi R., Quarto M., Riccio P., Roberti G., & Villani F. (2000). MS-2 fibrosarcoma characterization by laser induced autofluorescence. *Lasers Surg Med* 26, 441–448.

Cubeddu R., Canti G., Taroni P., & Valentini G. (1995). Delta-aminolevulinic-acid induced fluorescence in tumor-bearing mice. *J Photochem Photobiol B-Biol* 30, 23–27.

DaCosta R. S., Wilson B. C., & Marcon N. E. (2002). New optical technologies for earlier endoscopic diagnosis of premalignant gastrointestinal lesions. *J Gastroenterol Hepatol* 17 Suppl, S85–S104.

Dimitrow E., Riemann I., Ehlers A., Koehler M. J., Norgauer J., Elsner P., Konig K., & Kaatz M. (2009a). Spectral fluorescence lifetime detection and selective melanin imaging by multiphoton laser tomography for melanoma diagnosis. *Exp Dermatol* 18, 509–515.

Dimitrow E., Ziemer M., Koehler M. J., Norgauer J., Konig K., Elsner P., & Kaatz M. (2009b). Sensitivity and specificity of multiphoton laser tomography for *in vivo* and *ex vivo* diagnosis of malignant melanoma. *J Invest Dermatol* 129, 1752–1758.

Drezek R. A., Richards-Kortum R., Brewer M. A., Feld M. S., Pitris C., Ferenczy A., Faupel M. L., & Follen M. (2003). Optical imaging of the cervix. *Cancer* 98, 2015–2027.

Drezek R., Sokolov K., Utzinger U., Boiko I., Malpica A., Follen M., & Richards-Kortum R. (2001). Understanding the contributions of NADH and collagen to cervical tissue fluorescence spectra: Modeling, measurements, and implications. *J Biomed Opt* 6, 385–396.

Evan G. I., & Vousden K. H. (2001). Proliferation, cell cycle and apoptosis in cancer. *Nature* 411, 342–348.

Evans N. D., Gnudi L., Rolinski O. J., Birch D. J., & Pickup J. C. (2005). Glucose-dependent changes in NAD(P)H-related fluorescence lifetime of adipocytes and fibroblasts in vitro: Potential for non-invasive glucose sensing in diabetes mellitus. *J Photochem Photobiol B* 80, 122–129.

Fryen A., Glanz H., Lohmann W., Dreyer T., & Bohle R. M. (1997). Significance of autofluorescence for the optical demarcation of field cancerisation in the upper aerodigestive tract. *Acta Otolaryngol* 117, 316–319.

Galletly N. P., McGinty J., Dunsby C., Teixeira F., Requejo-Isidro J., Munro I., Elson D. S., Neil M. A., Chu A. C., French PM, & Stamp GW (2008). Fluorescence lifetime imaging distinguishes basal cell carcinoma from surrounding uninvolved skin. *Br J Dermatol* 159, 152–161.

Garin-Chesa P., Campbell I., Saigo P. E., Lewis J. L., Jr., Old L. J., & Rettig W. J. (1993). Trophoblast and ovarian cancer antigen LK26. Sensitivity and specificity in immunopathology and molecular identification as a folate-binding protein. *Am J Pathol* 142, 557–567.

Georgakoudi I., Jacobson B. C., Muller M. G., Sheets E. E., Badizadegan K., Carr-Locke D. L., Crum C. P., Boone C. W., Dasari R. R., Van Dam J., & Feld M. S. (2002). NAD(P)H and collagen as in vivo quantitative fluorescent biomarkers of epithelial precancerous changes. *Cancer Res* 62, 682–687.

Gillenwater A., Jacob R., Ganeshappa R., Kemp B., El-Naggar A. K., Palmer J. L., Clayman G., Mitchell M. F., & Richards-Kortum R. (1998). Noninvasive diagnosis of oral neoplasia based on fluorescence spectroscopy and native tissue autofluorescence. *Arch Otolaryngol Head Neck Surg* 124, 1251–1258.

Gillenwater A., Papadimitrakopoulou V., & Richards-Kortum R. (2006). Oral premalignancy: New methods of detection and treatment. *Curr Oncol Rep* 8, 146–154.

Glanzmann T. M., Ballini J. P., Jichlinski P., van den Bergh H., & Wagnieres G. A. (1996). Tissue characteristics by time-resolved fluorescence spectroscopy of endogenous and exogenous fluorophores. *Proc SPIE: Int Soc Opt Eng* 2926, 41–50.

Glanzmann T. M., Ballini J. P., van den Bergh H., & Wagnieres G. (1999). Time-resolved spectrofluorometer for clinical tissue characterization during endoscopy. *Rev Sci Instrum* 70, 4067–4077.

Glanzmann T. M., Uehlinger P., Ballini J. P., Radu A., Gabrecht T., Monnier P., van den Bergh H., & Wagnieres G. A. (2001). Tissue characteristics by time-resolved fluorescence spectroscopy of endogenous and exogenous fluorophores. *Proc SPIE: Int Soc Opt Eng* 4432, 199–209.

Johansson J., Berg R., Svanberg K., & Svanberg S. (1997). Laser-induced fluorescence studies of normal and malignant tumour tissue of rat following intravenous injection of delta-amino levulinic acid. *Lasers Surg Med* 20, 272–279.

Kantelhardt S. R., Diddens H., Leppert J., Rohde V., Huttmann G., & Giese A. (2008). Multiphoton excitation fluorescence microscopy of 5-aminolevulinic acid induced fluorescence in experimental gliomas. *Lasers Surg Med* 40, 273–281.

Kantelhardt S. R., Leppert J., Krajewski J., Petkus N., Reusche E., Tronnier V. M., Huttmann G., & Giese A. (2007). Imaging of brain and brain tumor specimens by time-resolved multiphoton excitation microscopy ex vivo. *Neuro Oncol* 9, 103–112.

Katz A., Savage H. E., Schantz S. P., McCormick S. A., & Alfano R. R. (2002). Noninvasive native fluorescence imaging of head and neck tumors. *Technol Cancer Res Treat* 1, 9–15.

Kennedy J. C., & Pottier R. H. (1992). Endogenous protoporphyrin-IX, a clinically useful photosensitizer for photodynamic therapy. *J Photochem Photobiol B-Biol* 14, 275–292.

Kennedy M. D., Jallad K. N., Thompson D. H., Ben Amotz D., & Low P. S. (2003). Optical imaging of metastatic tumors using a folate-targeted fluorescent probe. *J Biomed Opt* 8, 636–641.

Kepshire D. S., Gibbs-Strauss S. L., O'Hara J. A., Hutchins M., Mincu N., Leblond F., Khayat M., Dehghani H., Srinivasan S., & Pogue B. W. (2009). Imaging of glioma tumor with endogenous fluorescence tomography. *J Biomed Opt* 14, 030501.

Kirkpatrick N. D., Hoying J. B., Botting S. K., Weiss J. A., & Utzinger U. (2006). *In vitro* model for endogenous optical signatures of collagen. *J Biomed Opt* 11, 054021-1–054021-8.

Klinteberg C., Enejder A. M., Wang I., Andersson-Engels S., Svanberg S., & Svanberg K. (1999). Kinetic fluorescence studies of 5-aminolaevulinic acid-induced protoporphyrin IX accumulation in basal cell carcinomas. *J Photochem Photobiol B* 49, 120–128.

Konig K. (2008). Clinical multiphoton tomography. *J Biophotonics* 1, 13–23.

Konig K., Schneckenburger H., Ruck A., Steiner R., & Walt H. (1993). Laser-induced autofluorescence of cells and tissue. *Proc SPIE* 1887, 213–221.

Konig K., Weinigel M., Hoppert D., Buckle R., Schubert H., Kohler M. J., Kaatz M., & Elsner P. (2008). Multiphoton tissue imaging using high-NA microendoscopes and flexible scan heads for clinical studies and small animal research. *J Biophotonics* 1, 506–513.

Kress M., Meier T., Steiner R., Dolp F., Erdmann R., Ortmann U., & Ruck A. (2003). Time-resolved microspectrofluorometry and fluorescence lifetime imaging of photosensitizers using picosecond pulsed diode lasers in laser scanning microscopes. *J Biomed Opt* 8, 26–32.

Kriegmair M., Ehsan A., Baumgartner R., Lumper W., Knuechel R., Hofstadter F., Steinbach P., & Hofstetter A. (1994). Fluorescence photodetection of neoplastic urothelial lesions following intravesical instillation of 5-aminolevulinic acid. *Urology* 44, 836–841.

Leppert J., Krajewski J., Kantelhardt S. R., Schlaffer S., Petkus N., Reusche E., Huttmann G., & Giese A. (2006). Multiphoton excitation of autofluorescence for microscopy of glioma tissue. *Neurosurgery* 58, 759–767.

Li D., Zheng W., & Qu J. Y. (2009). Imaging of epithelial tissue in vivo based on excitation of multiple endogenous nonlinear optical signals. *Opt Lett* 34, 2853–2855.

Li Q. B., Xu Z., Zhang N. W., Zhang L., Wang F., Yang L. M., Wang J. S., Zhou S., Zhang Y. F., Zhou X. S., Shi J. S., & Wu J. G. (2005). *In vivo* and *in situ* detection of colorectal cancer using Fourier transform infrared spectroscopy. *World J Gastroenterol* 11, 327–330.

Lin S. J., Jee S. H., Kuo C. J., Wu R. J., Lin W. C., Chen J. S., Liao Y. H., Hsu C. J., Tsai T. F., Chen Y. F., & Dong C. Y. (2006). Discrimination of basal cell carcinoma from normal dermal stroma by quantitative multiphoton imaging. *Opt Lett* 31, 2756–2758.

Lin S. J., Wu R., Jr., Tan H. Y., Lo W., Lin W. C., Young T. H., Hsu C. J., Chen J. S., Jee S. H., & Dong C. Y. (2005). Evaluating cutaneous photoaging by use of multiphoton fluorescence and second-harmonic generation microscopy. *Opt Lett* 30, 2275–2277.

Lodish H. F., Beck A., Zipursky S. L., Matsudaira P., Baltimore D., & Darnell J. (2000). *Molecular Cell Biology*, 4th ed. New York: W.H. Freeman, pp. 979–985.

Lohmann W., Mussmann J., Lohmann C., & Kunzel W. (1989). Native fluorescence of the cervix uteri as a marker for dysplasia and invasive carcinoma. *Eur J Obstet Gynecol Reprod Biol* 31, 249–253.

Lorente C., Capparelli A. L., Thomas A. H., Braun A. M., & Oliveros E. (2004). Quenching of the fluorescence of pterin derivatives by anions. *Photochem Photobiol Sci* 3, 167–173.

Lutz V., Sattler M., Gallinat S., Wenck H., Poertner R., & Fischer F. (2012). Impact of collagen crosslinking on the second harmonic generation signal and the fluorescence lifetime of collagen autofluorescence. *Skin Res Technol* 18, 168–179.

Lycette R. M., & Leslie R. B. (1965). Fluorescence of malignant tissue. *Lancet* 2, 436.

Maitland K. C., Gillenwater A. M., Williams M. D., El-Naggar A. K., Descour M. R., & Richards-Kortum R. R. (2008). In vivo imaging of oral neoplasia using a miniaturized fiber optic confocal reflectance microscope. *Oral Oncol* 44, 1059–1066.

Marcu L., Jo J. A., Butte P. V., Yong W. H., Pikul B. K., Black K. L., & Thompson R. C. (2004). Fluorescence lifetime spectroscopy of glioblastoma multiforme. *Photochem Photobiol* 80, 98–103.

Marcus S. L., Sobel R. S., Golub A. L., Carroll R. L., Lundahl S., & Shulman D. G. (1996). Photodynamic therapy (PDT) and photodiagnosis (PD) using endogenous photosensitization induced by 5-aminolevulinic acid (ALA): current clinical and development status. *J Clin Laser Med Surg* 14, 59–66.

Muller S., Pan Y., Li R., & Chi A. C. (2008). Changing trends in oral squamous cell carcinoma with particular reference to young patients: 1971–2006. The Emory University experience. *Head Neck Pathol* 2, 60–66.

Mycek M. A., Schomacker K. T., & Nishioka N. S. (1998). Colonic polyp differentiation using time-resolved autofluorescence spectroscopy. *Gastrointest Endosc* 48, 390–394.

Nath A., Rivoire K., Chang S., West L., Cantor S. B., Basen-Engquist K., dler-Storthz K., Cox D. D., Atkinson E. N., Staerkel G., Macaulay C., Richards-Kortum R., & Follen M. (2004). A pilot study for a screening trial of cervical fluorescence spectroscopy. *Int J Gynecol Cancer* 14, 1097–1107.

Nilsson H., Johansson J., Svanberg K., Svanberg S., Jori G., Reddi E., Segalla A., Gust D., Moore A. L., & Moore T. A. (1994). Laser-induced fluorescence in malignant and normal tissue in mice injected with 2 different carotenoporphyrins. *Br J Cancer* 70, 873–879.

Ntziachristos V. (2006). Fluorescence molecular imaging. *Annu Rev Biomed Eng* 8, 1–33.

Ntziachristos V., Schellenberger E. A., Ripoll J., Yessayan D., Graves E., Bogdanov A., Jr., Josephson L., & Weissleder R. (2004). Visualization of antitumor treatment by means of fluorescence molecular tomography with an annexin V-Cy5.5 conjugate. *Proc Natl Acad Sci U S A* 101, 12294–12299.

Ntziachristos V., Tung C. H., Bremer C., & Weissleder R. (2002). Fluorescence molecular tomography resolves protease activity in vivo. *Nat Med* 8, 757–760.

Palsson S., Gustafsson L., Bendsoe N., Thompson M. S., Andersson-Engels S., & Svanberg K. (2003). Kinetics of the superficial perfusion and temperature in connection with photodynamic therapy of basal cell carcinomas using esterified and non-esterified 5-aminolaevulinic acid. *Br J Dermatol* 148, 1179–1188.

Pavlova I., Williams M., El-Naggar A., Richards-Kortum R., & Gillenwater A. (2008). Understanding the biological basis of autofluorescence imaging for oral cancer detection: High-resolution fluorescence microscopy in viable tissue. *Clin Cancer Res* 14, 2396–2404.

Pfefer T. J., Paithankar D. Y., Poneros J. M., Schomacker K. T., & Nishioka N. S. (2003). Temporally and spectrally resolved fluorescence spectroscopy for the detection of high grade dysplasia in Barrett's esophagus. *Lasers Surg Med* 32, 10–16.

Pitts J. D., Sloboda R. D., Dragnev K. H., Dmitrovsky E., & Mycek M. A. (2001). Autofluorescence characteristics of immortalized and carcinogen-transformed human bronchial epithelial cells. *J Biomed Opt* 6, 31–40.

Policard A. (1924). Etude sur les aspects offerts par des tumeurs expérimentales examinées à la lumière de Wood. *CR Soc Biol* 91, 1423–1424.

Pradhan A., Pal P., Durocher G., Villeneuve L., Balassy A., Babai F., Gaboury L., & Blanchard L. (1995). Steady state and time-resolved fluorescence properties of metastatic and non-metastatic malignant cells from different species. *J Photochem Photobiol B* 31, 101–112.

Provenzano P. P., Eliceiri K. W., Campbell J. M., Inman D. R., White J. G., & Keely P. J. (2006). Collagen reorganization at the tumor-stromal interface facilitates local invasion. *BMC Med* 4, 38.

Provenzano P. P., Eliceiri K. W., & Keely P. J. (2009). Multiphoton microscopy and fluorescence lifetime imaging microscopy (FLIM) to monitor metastasis and the tumor microenvironment. *Clin Exp Metastasis* 261, 357–370.

Provenzano P. P., Inman D. R., Eliceiri K. W., Knittel J. G., Yan L., Rueden C. T., White J. G., & Keely P. J. (2008a). Collagen density promotes mammary tumor initiation and progression. *BMC Med* 26, 357–370.

Provenzano P. P., Rueden C. T., Trier S. M., Yan L., Ponik S. M., Inman D. R., Keely P. J., & Eliceiri K. W. (2008b). Nonlinear optical imaging and spectral-lifetime computational analysis of endogenous and exogenous fluorophores in breast cancer. *J Biomed Opt* 13, 031220-1–031220-4.

Ramanujam N. (2000). Fluorescence spectroscopy of neoplastic and non-neoplastic tissues. *Neoplasia* 2, 89–117.

Ramanujan V. K., Zhang J. H., Biener E., & Herman B. (2005). Multiphoton fluorescence lifetime contrast in deep tissue imaging: Prospects in redox imaging and disease diagnosis. *J Biomed Opt* 10, 051407-1–051407-11.

Richards-Kortum R., & Sevick-Muraca E. (1996). Quantitative optical spectroscopy for tissue diagnosis. *Annu Rev Phys Chem* 47, 555–606.

Schneckenburger H., Gschwend M. H., Sailer R., Ruck A., & Strauss W. S. (1995). Time-resolved pH-dependent fluorescence of hydrophilic porphyrins in solution and in cultivated cells. *J Photochem Photobiol B* 27, 251–255.

Schneckenburger H., Wagner M., Weber P., Strauss W. S., & Sailer R. (2004). Autofluorescence lifetime imaging of cultivated cells using a UV picosecond laser diode. *J Fluoresc* 14, 649–654.

Schomacker K. T., Frisoli J. K., Compton C. C., Flotte T. J., Richter J. M., Deutsch T. F., & Nishioka N. S. (1992a). Ultraviolet laser-induced fluorescence of colonic polyps. *Gastroenterology* 102, 1155–1160.

Schomacker K. T., Frisoli J. K., Compton C. C., Flotte T. J., Richter J. M., Nishioka N. S., & Deutsch T. F. (1992b). Ultraviolet laser-induced fluorescence of colonic tissue: Basic biology and diagnostic potential. *Lasers Surg Med* 12, 63–78.

Shin D., Vigneswaran N., Gillenwater A., & Richards-Kortum R. (2010). Advances in fluorescence imaging techniques to detect oral cancer and its precursors. *Future Oncol* 6, 1143–1154.

Skala M. C., Riching K. M., Bird D. K., Gendron-Fitzpatrick A., Eickhoff J., Eliceiri K. W., Keely P. J., & Ramanujam N. (2007a). In vivo multiphoton fluorescence lifetime imaging of protein-bound and free nicotinamide adenine dinucleotide in normal and precancerous epithelia. *J Biomed Opt* 12, 024014-1–024014-10.

Skala M. C., Riching K. M., Gendron-Fitzpatrick A., Eickhoff J., Eliceiri K. W., White J. G., & Ramanujam N. (2007b). In vivo multiphoton microscopy of NADH and FAD redox states, fluorescence lifetimes, and cellular morphology in precancerous epithelia. *Proc Natl Acad Sci U S A* 104, 19494–19499.

Sokolov K., Follen M., & Richards-Kortum R. (2002). Optical spectroscopy for detection of neoplasia. *Curr Opin Chem Biol* 6, 651–658.

Strauss W. S., Sailer R., Schneckenburger H., Akgun N., Gottfried V., Chetwer L., & Kimel S. (1997). Photodynamic efficacy of naturally occurring porphyrins in endothelial cells in vitro and microvasculature *in vivo*. *J Photochem Photobiol B* 39, 176–184.

Sud D., Zhong W., Beer D. G., & Mycek M. A. (2006). Time-resolved optical imaging provides a molecular snapshot of altered metabolic function in living human cancer cell models. *Opt Express* 14, 4412–4426.

Sun Y., Hatami N., Yee M., Phipps J., Elson D. S., Gorin F., Schrot R. J., & Marcu L. (2010). Fluorescence lifetime imaging microscopy for brain tumor image-guided surgery. *J Biomed Opt* 15, 056022-1–056022-5.

Tadrous P. J., Siegel J., French P. M., Shousha S., Lalani E., & Stamp G. W. (2003). Fluorescence lifetime imaging of unstained tissues: Early results in human breast cancer. *J Pathol* 199, 309–317.

Toffoli G., Cernigoi C., Russo A., Gallo A., Bagnoli M., & Boiocchi M. (1997). Overexpression of folate binding protein in ovarian cancers. *Int J Cancer* 74, 193–198.

Uehlinger P., Gabrecht T., Glanzmann T., Ballini J. P., Radu A., Andrejevic S., Monnier P., & Wagnieres G. (2009). In vivo time-resolved spectroscopy of the human bronchial early cancer autofluorescence. *J Biomed Opt* 14, 024011-1–024011-9.

Wagnieres G. A., Star W. M., & Wilson B. C. (1998). In vivo fluorescence spectroscopy and imaging for oncological applications. *Photochem Photobiol* 68, 603–632.

Wang H. W., Gukassyan V., Chen C. T., Wei Y. H., Guo H. W., Yu X. J., & Kao F. J. (2008). Differentiation of apoptosis from necrosis by dynamic changes of reduced nicotinamide adenine dinucleotide fluorescence lifetime in live cells. *J Biomed Opt* 13, 054011.

Wang I., Clemente L. P., Pratas R. M. G., Cardoso E., Clemente M. P., Montan S., Svanberg S., & Svanberg K. (1999). Fluorescence diagnostics and kinetic studies in the head and neck region utilizing low-dose delta-aminolevulinic acid sensitization. *Cancer Lett* 135, 11–19.

Weitman S. D., Lark R. H., Coney L. R., Fort D. W., Frasca V., Zurawski V. R., Jr., & Kamen B. A. (1992). Distribution of the folate receptor GP38 in normal and malignant cell lines and tissues. *Cancer Res* 52, 3396–3401.

Wu Y., & Qu J. Y. (2006). Autofluorescence spectroscopy of epithelial tissues. *J Biomed Opt* 11, 054023-1–054023-11.

Wu Y., Zheng W., & Qu J. Y. (2007). Time-resolved confocal fluorescence spectroscopy reveals the structure and metabolic state of epithelial tissue. *Proc SPIE: Int Soc Opt Eng* 6430, 643013.

Yang Y., Katz A., Celmer E. J., Zurawska-Szczepaniak M., & Alfano R. R. (1997). Fundamental differences of excitation spectrum between malignant and benign breast tissues. *Photochem Photobiol* 66, 518–522.

Yaroslavsky A. N., Salomatina E. V., Neel V., Anderson R., & Flotte T. (2007). Fluorescence polarization of tetracycline derivatives as a technique for mapping nonmelanoma skin cancers. *J Biomed Opt* 12, 014005-1–014005-9.

Zhang Z., Li H., Liu Q., Zhou L., Zhang M., Luo Q., Glickson J., Chance B., & Zheng G. (2004). Metabolic imaging of tumors using intrinsic and extrinsic fluorescent markers. *Biosens Bioelectron* 20, 643–650.

Zheng W., Lau W., Cheng C., Soo K. C., & Olivo M. (2003). Optimal excitation-emission wavelengths for autofluorescence diagnosis of bladder tumors. *Int J Cancer* 104, 477–481.

Zipfel W. R., Williams R. M., Christie R., Nikitin A. Y., Hyman B. T., & Webb W. W. (2003). Live tissue intrinsic emission microscopy using multiphoton-excited native fluorescence and second harmonic generation. *Proc Natl Acad Sci U S A* 100, 7075–7080.

Oncology applications: Brain

Pramod V. Butte, Adam N. Mamelak, and Laura Marcu

Contents

15.1 INTRODUCTION

15.1.1 BRAIN TUMORS: RESEARCH AND CLINICAL CHALLENGES

Primary brain tumors are the tumors that originate in the cranial cavity from various tissues like glial cells, meninges, and neural tissue. Primary and benign brain tumors account for 19.34 cases per 100,000 person-years, and it is estimated that 64,530 new cases of primary brain tumors are diagnosed every year (Central Brain Tumor Registry of the United States [CBTRUS] 2012). Most common types of primary brain tumors are gliomas and meningiomas. Glioblastoma multiforme (GBM) is very difficult to treat with a median survival of 12–18 months. Conventional imaging techniques such as contrast-enhanced computed topography (CT) and magnetic resonance imaging (MRI) are helpful in diagnosing the tumors. MRI provides superior anatomic details for grading and preoperative decision making. Once diagnosed, the standard of care includes surgical resection. Although surgery is not curative, it is associated with significant survival benefit either independently or by decreasing tumor burden and thereby enhancing the additional therapeutic effects. Most studies have shown the advantage of aggressive tumor resection (Devaux et al. 1993; Hess 1999; Lacroix et al. 2001; Laws et al. 2003; Sanai and Berger 2008; Sanai et al. 2011; Stummer et al. 2011; Stupp et al. 2005).

Intraoperatively, gliomas are typically identified based on color, texture, and vascularity, but the margins between the tumor and normal white matter are less distinct. Currently, intraoperative MRI and stereotactic guidance are used to localize the tumor and determine the margins. But it is well recognized that the change in anatomic lesion (brain shift) due to removal of space occupying lesions such as glioma makes presurgical images less reliable. Thus, newer optical techniques such as fluorescence spectroscopy are being researched to aid the surgeon in near complete resection.

15.1.1.1 Glioma

Gliomas are malignant lesions accounting for 27% of primary brain tumors, which arise from glial cells and include astrocytoma, glioblastoma, mixed gliomas, malignant gliomas not otherwise specified (NOS), and neuroepithelial tumors. Gliomas are the second most frequently reported histology and represent 44% of all brain tumors. Among these, astrocytoma and glioblastomas are the most common type of malignant brain tumor in adults accounting to more than 75% of gliomas. Generally, the survival rate of malignant brain tumors by histology for 5 and 10 years is 28% and 24%, respectively. However, there is a large variation in survival rates between different tumor histologies. For example, the 5-year survival rate exceeds 85% for pilocytic astrocytoma but is less than 5% for glioblastomas (CBTRUS 2012), suggesting that tumor phenotype plays an important role in defining the patient outcome. Histologically, gliomas are categorized in low-grade (grade I), intermediate-grade (grades II and III), and high-grade gliomas or GBM (grade IV) (Louis et al. 2007). Grade I is considered as benign astrocytoma lesions, grade II is considered to be low malignancy but with good prognosis, grade III is anaplastic astrocytomas with cellular atypia, and grade IV or GBM shows necrosis, endothelial proliferation, and mitotic activity. As presence of necrosis separates GBM from anaplastic astrocytomas (Bruner 1994), a review of 251 cases found no statistically significant difference in the rate of survival between grades II and III (Kim et al. 1991). Necrosis was found to be a significant predictor of short survival time (Nelson et al. 1983). Low-grade gliomas constitute about 10%–20% of all adult primary brain tumors. The majority are astrocytomas; approximately 5% are oligodendrogliomas or mixed oligoastrocytomas. Pathologically, they are well differentiated and lack the cellular features such as high cellularity, pleomorphism, mitoses, vascular endothelial proliferation, and necrosis. In patients whose low-grade glioma has been completely resected, radiation does not appear to increase the survival rate, making complete surgical resection the most important factor. In case of high-grade astrocytoma, the prognosis of anaplastic astrocytomas is superior to that of GBM (Byar et al. 1983). Surgery being the treatment of choice, maximal surgical resection improves the results of subsequent radiation and chemotherapy. Clinically, it was shown that extent of tumor resection was a significant independent variable for survival in patients treated for malignant glioma (Berger 1994; Winger et al. 1989).

15.1.1.2 Meningioma

Meningiomas arise from the arachnoidal cells in the meninges. Overall, they are the most common type of benign tumor and account for 27% of all the primary tumors of brain (CBTRUS 2012). Over 90% of the meningiomas are benign in nature. They are most common in older adults (median age at diagnosis 65 years). The rate of meningioma increases dramatically after age of 65 years and continues to be high among the population of 85 and older (CBTRUS 2012). Histologically, meningiomas are characterized in many groups such as meningothelial meningioma, which show presence of meningothelial cells; fibroblastic meningioma, which show presence of fibrous tissue with spindle-shaped cells; psammomatous meningioma, which has psammoma bodies, which are islands of calcified tissue with aggregate of meningothelial cells into whorls and lobules around the island; and angiomatous or angioblastic meningioma, which resemble a vascular malformation.

It was reported that even after a perceived "total resection," the disease recurrence was 9%. A review from Massachusetts General Hospital showed that a "total resection" is followed by 7%, 20%, and 32% recurrence rate at 5, 10, and 15 years, respectively (Mirimanoff et al. 1985). Extent of tumor resection directly correlates with prevention of recurrence. Intraoperative biopsy specimens are routinely removed by the neurosurgeon from the main tumor mass and possibly one to several dural margins. Neuropathologic evaluation for rapid intraoperative preliminary diagnosis is often requested. The standard intraoperative tissue specimen neuropathologic evaluation has been the "frozen section." Certain intracranial characteristics of meningiomas, moreover, can increase the difficulty of achieving a complete surgical resection. The "en plaque" variety and petroclival location of meningiomas are examples of meningiomas that often present difficulty in excising dura involved with tumor. In addition, some meningiomas can invade the subjacent brain parenchyma, usually the cerebral cortex.

15.1.2 ROLE OF IMAGING TECHNIQUES IN RESEARCH AND DIAGNOSIS OF BRAIN TUMORS. TARGETS: COMPOSITIONAL, STRUCTURAL, AND FUNCTIONAL FEATURES. PREOPERATIVE AND INTRAOPERATIVE IMAGING

In spite of the advent of radiotherapy as well as chemotherapy, surgery remains the fastest way to reduce the tumor bulk. Current neuroimaging techniques for preoperative and intraoperative evaluation of brain tumor are done using MRI or ultrasound imaging (USI). MRI performed 36 h prior to the surgery provides valuable information during surgery regarding localization of the tumor. Multiple MRI images are also used for stereotactic surgery, thus aiding the surgeon during resection. A shift in the brain after opening the cranial cavity reduces the accuracy of preoperative MRI. Intraoperative MRI-guided surgery can assist in resection but currently is prohibitively expensive, requires special setup, and can be used only on limited locations. Additionally, the degree to which a complete resection can be achieved in the brain is limited primarily by the difficulty of visually detecting differences between normal brain and malignant tissue during surgery; thus, neuropathologic evaluation for rapid intraoperative preliminary diagnosis is often requested. The standard intraoperative tissue specimen neuropathologic evaluation has been the "frozen section." This process of freezing the tissue, slicing the frozen specimen with a microtome, and staining/analysis can, at minimum, take 15 to 25 min. Multiple intraoperative "frozen section" specimen requests can lead to lengthy increases in operative time, as well as a taxing increase in workload for the surgical pathologist. Thus, patients with malignant gliomas often have a subtotal resection. In addition, resection of the normal tissue at the tumor boundary may increase the risk of neurological morbidity. By providing intraoperative biochemical diagnosis of the brain tumors, fluorescence spectroscopy can improve the resection of brain tumor.

15.1.3 ADVANTAGES OF FLUORESCENCE LIFETIME TECHNIQUES

The presence of intrinsic fluorophores such as amino acids, structural proteins, and enzyme cofactors (nicotinamide adenine dinucleotide, flavins) in human tissue offers the potential to probe biochemical, morphological, and physiological changes occurring in diseased tissues. The various alterations of these tissue fluorophores can be correlated by analyzing the changes of the relative contribution of each fluorescent constituent to the overall fluorescence emission. Either steady-state or TRFS techniques can be used to measure or monitor these changes (Marcu et al. 2001, 2004). The time-resolved measurement resolves fluorescence intensity decay in terms of lifetimes and thus provides additional information about the underlying fluorescence dynamics and improves its specificity.

Several groups, including ours, reported studies demonstrating the potential of both steady-state laser-induced fluorescence spectroscopy (LIFS) and time-resolved LIFS (TR-LIFS) of endogenous fluorophores (autofluorescence) for diagnosis of brain cancers (Bottiroli et al. 1995, 1998; Chung et al. 1997; Croce et al. 2003; Poon et al. 1992; Wagnieres et al. 1998). These include studies of glioblastoma (Butte et al. 2010; Croce et al. 2003; Lin et al. 2001; Marcu et al. 2004), astrocytoma (Lin et al. 2001), oligodendroglioma (Lin et al. 2001), and metastatic carcinoma (Lin et al. 2001).

15.2 ORIGINS OF FLUORESCENCE CONTRAST IN BRAIN TISSUE

Some of the most important fluorophores in the biological system are vitamins and metabolites. They are sensitive to changes in the biological systems and thus can be excellent indicators for underlying pathological condition. Some of the fluorophores significant in the brain are pyridoxine (vitamin B6), which is required for monoamine neurotransmitters such as dopamine, serotonin (5-hydroxytryptamine), glycine, D-serine, glutamate, γ-aminobutyrate (GABA); histamine (Bowling 2011); glutamate decarboxylase (GAD), which is a coenzyme required for converting glutamate in γ-aminobutyric acid (GABA); NADH/NAD(P)H, which is a basic metabolite in formation of ATP from glucose; and derivatives of vitamin B_2 such as flavin-adenine dinucleotide (FAD) and flavin mononucleotide (FMN; Table 15.1).

15.2.1 PYRIDOXINE

Vitamin B6 exists in three forms: pyridoxal and pyridoxamine from the animals and pyridoxine from plants (Sharma and Dakshinamurti 1992). Active forms of the vitamin, which form the cofactor in the phosphorylase *b* enzyme, are pyridoxal-5'-phosphate (PLP) and pyridoxamine-5'-phosphate (PMP) (Honikel and Madsen 1972; Sharma and Dakshinamurti 1992). PLP is synthesized in the body from pyridoxal (PL) by enzyme pyridoxal kinase (PK) and from pyridoxamine, pyridoxine by pyridoxine-5'-phosphate oxidase (PNPO) (Ngo et al. 1998). Pyridoxamine fluoresces with peak emission at 390 nm of wavelength (Zipfel et al. 2003). PLP mechanism is known to be disturbed in the cancers. PLP is converted from pyridoxine phosphate (PN) and pyridoxamine phosphate (PM) by PNPO, which is absent in neoplatic cells in rodents (Ngo et al. 1998). It has been shown by Ngo et al. (1998) that PNPO activity is lower in the fetal brain as well as in the neuronally derived tumors in rats. They also found lower PNPO activity in some of the human cell lines.

15.2.2 GLUTAMATE DECARBOXYLASE

GAD is an important enzyme that is required for reversible conversion of glutamate to GABA. GABA is the most important inhibitory neurotransmitter found in abundance in the brain. It is known that GABA is present in abundance in the cerebellar cortex (Flace et al. 2004). GABAA-benzodiazepine receptors are present in one-third of all brain, and the α1 subunit of a GABAA receptor is expressed throughout the brain (Wolf et al. 1997). The absorption maxima of GAD are 330 and 420 nm of wavelengths. The emission maxima occur at 380 nm when 335 nm light is used for activation and at 490 nm when the activation wavelength is 420 nm (Shukuya and Schwert 1960). Most of the GAD is present in the neurons

Table 15.1 **List of potential endogenous fluorophores**

ENDOGENOUS FLUOROPHORES	EXCITATION MAXIMA (NM)	EMISSION MAXIMA (NM)	LIFETIME (NS)
PLP-GAD (Shukuya and Schwert 1960)	320	380–400	NA
Pyridoxamine/pyridoxine (Greenaway and Ledbetter 1987; Honikel and Madsen 1972; Kempe and Stark 1975)	33	390	0.8–1.6
NADH/NADPH (free) (Jameson et al. 1989; Scott et al. 1969, 1970)	336	440–464	0.8
NADH/NADPH (bound)	336	464	1.2
Flavins (Wahl et al. 1974)	337–449	Broad around 530	2.4–2.7
Lipopigments (Crespi et al. 2004; Croce et al. 2003)	440	580	NA
Endogenous porphyrins (Schneckenburger and Konig 1992; Seidlitz et al. 1990)	440	630–690	1.5–3.0 (dimer) 10–14 ns (monomers)

at the synapses. It has been shown that neuroblastoma cells have GAD, and the activity of this enzyme increases with the differentiation. Activities of GAD are much lower in the C6 glioblastoma and C-1300 neuroblastoma cells in lower than normal mouse cerebral cortex (Ossola et al. 1979). It has been suggested that C6 glioma cells may represent a neuroglial precursor at an early developmental stage suggested by low levels of GAD_{65} mRNA and expression of GFAP. This may suggest an absence of glutamate decarboxylase of its active form in neoplastic tissue.

15.2.3 NADH

NADH fluorescence has long been used as an indicator of cellular metabolic state (Pradhan et al. 1995; Reinert et al. 2004; Zipfel et al. 2003). NADH is significant in characterizing brain tumors as the utilization of glucose in most of the tumors is through the anaerobic respiration, which leads to the decrease in the NADH levels in mitochondrion and can be a significant factor in the detection of tumor tissue. Additionally, the difference in peak fluorescence emission and the lifetime of free and bound form of NADH can be an indicator of the metabolic state in the cells. In gliomas with mutation of IDH1, IDH2 (isocitrate dehydrogenase) enzyme, which converts NADH to NAD+, leads to disturbance of NADH in cytoplasm. IDH1 and IDH2 are important for shuttling electrons between the mitochondria and the cytosol. Although mutations in IDH1 are expected to hinder these processes, newly described work (Dang et al. 2009) proposes a new gain-of-function role for glioma-associated mutants of IDH1. R132 mutations of IDH1 generate a new enzyme with a-ketoglutarate reductase activity that produces 2-hydroxyglutarate, and increased 2-hydroxyglutarate strongly correlates with cancer formation.

15.2.4 FAD/FMN

The second important set of fluorophores in the brain is FAD/FMN (Croce et al. 2003). These metabolites fluoresce when oxidized rather than reduced unlike NADH. FAD oxidizes by reducing NADH. Thus, when there is a decrease in the NADH, there will be a decrease in FAD+. During anaerobic respiration, there is higher lactic acid produced, which may lead to a decrease in NADH fluorescence and an increase in FAD+ fluorescence.

15.3 FLUORESCENCE LIFETIME INSTRUMENTATION USED IN BRAIN TISSUE/TUMOR CHARACTERIZATION

15.3.1 TIME-RESOLVED FLUORESCENCE SPECTROSCOPY

A few studies reported the time-resolved fluorescence of brain tumors recorded intraoperatively (Butte et al. 2010, 2011; Marcu et al. 2004; Pradhan et al. 1995; Yong et al. 2006). These experiments involved the use of a TRFS apparatus based on pulse sampling technique (see Chapter 4). Briefly, this consisted of a pulsed nitrogen laser (λ = 337 nm, pulse width = 700 ps FWHM) used as the excitation source, a custom-made sterilizable bifurcated fiber-optic probe described below, an imaging spectrometer/monochromator (F/4.4, 600 gr/mm grating, blazed at 450 nm), a gated multichannel plate photomultiplier tube (MCP-PMT) with a rise time = 180 ps and a fast preamplifier (1.5 GHz), a digital phosphor oscilloscope (5 Gsamples/s), a computer workstation, and peripheral electronics. The instrument allowed portability as it was contained in a standard endoscopic cart modified to accommodate the individual devices. Light delivery and collection were implemented with a custom-made bifurcated sterilizable probe (nonsolarizing silica/silica step index fibers of 0.11 numerical aperture). It had a central excitation fiber of 600 μm core diameter, surrounded by a collection ring of twelve 200 μm core diameter fibers. The collection fibers were beveled at a 10° angle in order to improve excitation/collection overlap for small tissue-to-probe distances. The center-to-center separation between the excitation and collection fibers was 480 μm. The probe was flexible throughout its entire length (3 m) except for a 7 cm distal part that consisted of a rigid stainless-steel tube. This facilitated the mounting and micromanipulation of the probe (Figure 15.1).

15.3.2 FLUORESCENCE LIFETIME IMAGING MICROSCOPY

A wide-field fluorescence lifetime imaging microscopy (FLIM) apparatus was also used in clinical setting (Sun et al. 2010). Tissue autofluorescence was induced by a pulsed nitrogen laser (337 nm, 700 ps) using

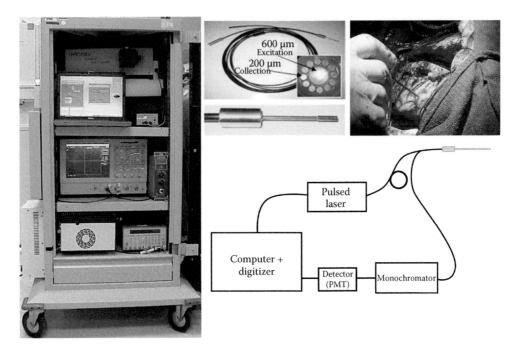

Figure 15.1 Schematic showing the setup for TRFS. Inset on the left shows the portable TRFS system and on the right shows the positioning of TRFS probe on the brain tissue.

a customized semiflexible endoscope probe that remotely delivered the excitation light. The fluorescence light was collected using a gradient index (GRIN) lens (NA of 0.5, 0.5 mm diameter, 4 mm field of view) cemented to a fiber image guide (0.6 mm diameter, 2 m long, 10,000 fibers). The fluorescence emitted from the proximal end of the fiber bundle was projected to a fast-gated ICCD with the gating time down to 200 ps. Two band-pass filters were used throughout this study (390/70 and 450/60 nm, central wavelength/bandwidth). Data acquisition time for each measurement was ~2 min including one steady-state image and a series of up to 29 time-gated images (0.5 ns gating time, 0.5 ns time interval; Figure 15.2).

15.3.3 TRFS AND FLIM DATA ANALYSIS

Analysis of TRFS data is performed in order to determine intrinsic fluorescence decay or fluorescence impulse response function (FIRF). FIRF is derived by deconvolving the laser signal used to excite the tissue from the fluorescence signal. For these studies, a method based on expansion of discrete time Laguerre basis as a means of retrieving the fluorescence impulse function (Jo et al. 2004) was employed. This approach is described in detail in Chapter 11.

15.4 *EX VIVO* APPLICATION OF TRFS TECHNIQUE

15.4.1 GLIOMAS

The first TRFS study of brain tumor was performed in specimens diagnosed as GBM Grade IV obtained during craniotomy (Marcu et al. 2004). This study included 23 point measurements (9 from GBM, 9 from normal cortex, and 5 from normal white matter) and was the first to demonstrate that fluorescence characteristics of brain tissue are both time- and wavelength-dependent. The average lifetime values of brain tumor tissue were found longer when compared with normal white matter and cortex across all emission wavelengths (360–520 nm). The lifetime values (~1.3 ns in tumor) at 460 nm emission suggest that the GBM fluorescence is likely dominated by the bound form of NAD(P)H fluorescence, whereas

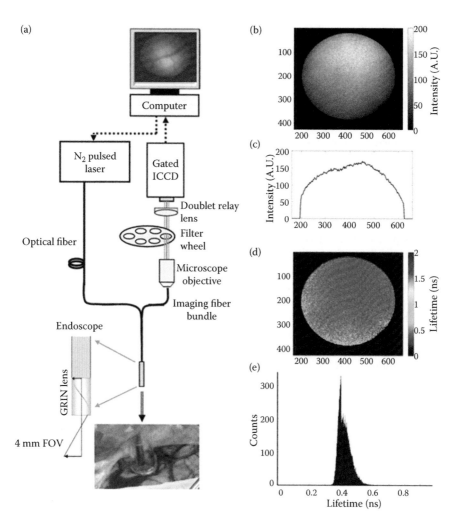

Figure 15.2 (a) Schematic of the FLIM apparatus, including a photo of the tip of the imaging bundle positioned on the interrogated area of the cortex. (b) Fluorescence intensity image of rhodamine B and (c) a cross section of the intensity distribution profile. Note that the fluorescence intensity is higher in the center of the image when compared to the edge of the image. (d) Fluorescence average lifetime image and (e) fluorescence average lifetime histogram of the rhodamine B solution measured and analyzed with the FLIM system. The pixel coordinates are shown on the left and bottom side of the images in (b) and (d). (Adapted from Sun, Y. et al., *J Biomed Opt* 15 (5):056022, 2010.)

the relatively shorter lifetime values (~1 ns) of the normal tissue are generated by a larger contribution of free-form NAD(P)H. In addition, this study showed that upon 337 nm excitation, both tumor and normal tissue exhibit a second peak fluorescence centered at about 390 nm. This peak was most prominent in GBM.

A more extensive evaluation of TRFS's ability to distinguish between distinct types of brain tumor tissues was reported in a subsequent study (Butte et al. 2010; Yong et al. 2006). This included differences between low-grade glioma (LGG), high-grade glioma (HGG), and high-grade glioma with necrotic changes (HGGN) from normal cerebral cortex (NC) and white matter (NWM). As determined in the earlier study, two wavelength ranges enabled discrimination of glioma tissue from normal brain tissue, specifically 370–400 nm (the region of the main peak emission for HGGN and normal white matter) and 440–480 nm (the region of the main peak emission for the normal cortex as well as LGG and HGG). The LGG (~1.1 ns average lifetime at 460 nm) exhibited a faster decay dynamics than HGG (~1.3 ns).

Table 15.2 **Calculated specificity and sensitivity in detecting *ex vivo* tissues under investigation**

	NORMAL CORTEX (%)	NORMAL WHITE MATTER (%)	LOW-GRADE GLIOMA (%)	HIGH-GRADE GLIOMA (%)	HGG + NECROSIS (%)
Sensitivity	83.33	87.50	81.82	76.92	100.00
Specificity	100.00	94.44	93.75	93.33	97.37

The fluorescence characteristics of LGG measured *ex vivo*, however, were found similar with those of NC but very different from those of NWM. The latter presented a much slower decay dynamics (~1.8 ns at 460 nm). These features were also associated with the NAD(P)H emission (Konig et al. 1997). In both the spectrum and time domain, the HGGN presented emission characteristics that were very distinctive from all other tissue types. Their emission was dominated by a strong peak centered at about 380–390 nm associated with an average lifetime of ~2 ns. These characteristics were associated with the emission of connective tissue proteins (e.g., collagen) that typically is formed in response to previous radiation treatment. The strong collagen fluorescence emission is likely to dominate the overall fluorescence emission of these tumors. A multivariate statistical analysis applied to the TRFS-derived parameters had shown that different grades of gliomas, normal white matter, and cortex tissues can be differentiated using a relatively limited number of predictor variables (five in total) from the two spectral ranges noted above. Parameters obtained from both spectral (intensity values) and time-resolved emission contributed to the accuracy of tissue classification. Although a small database was used in this study, the reported results indicate that parameters obtained from both spectral (intensity values) and time-resolved emission contributed allowed for a good discrimination of LGG from NWM (sensitivity 90%, specificity 100%) as well as a very good delineation of brain tissue exposed to radiation therapy (sensitivity 83%, specificity 100%; Table 15.2).

15.4.2 MENINGIOMAS

The separate TRFS study was designed to evaluate the use of TRFS as an adjunctive tool for the intraoperative rapid evaluation of tumor specimens and delineation of tumor from surrounding normal tissue. The study was conducted in tissue specimens from patients undergoing surgery for removal of tumor. Tissue samples include various types of meningioma, adjacent normal dura mater, and cerebral cortex (26 patients, 97 sites). Fluorescence parameters retrieved from both spectral and time domains were used for tissue characterization and classification. The results revealed that meningioma is characterized by unique fluorescence characteristics that enable discrimination of tumor from normal tissue with high sensitivity (>89%) and specificity (100%). The time-resolved fluorescence emission of meningioma samples presented a relatively narrow emission bandwidth characterized by a well-defined peak at a 385–390 nm wavelength with a long-lasting emission (>15 ns). The dynamics of the fluorescence decay was found wavelength-dependent with higher values in the region of peak fluorescence (τ_{390} = 2.04 ± 0.72 ns) when compared with the longer wavelengths (τ_{460} = 1.2 ± 0.46 ns). In contrast, the fluorescence emission of the dura mater was characterized by a broad wavelength band with two peaks emissions. The main peak is centered at about 385 to 390 nm and the second at about 440 nm. The fluorescence lifetime was found slightly longer in the region of main peak emission (τ_{390} = 1.7 ± 0.52 ns) when compared with the red-shifted wavelengths (τ_{460} = 1.4 ± 0.36 ns). Figure 15.3 summarizes these trends. Importantly, the use of both sets of spectroscopic parameters (spectral and time-resolved) resulted in improved overall accuracy of meningioma detection (91.8%) when compared with the case when only spectral variables were employed (63.9%). These findings showed the feasibility of TRFS for diagnosis of meningiomas.

The fluorescence emission of dura and meningioma is most likely due to autofluorescence of connective tissue proteins, in particular, collagens. High concentration of various types of collagen was reported in these tissues (Ng et al. 2010) and confirmed by the present study. The results showed that meningioma and dura were characterized by distinct fluorescence decay dynamic, albeit having common spectroscopic features such as an intense fluorescence emission in the 370–400 nm wavelength range. These differences suggest that fluorescence of meningioma and dura originates from distinct types of collagens and their cross-links. Studies of fluorophores from collagen-rich fractions of human dura mater using chromatography have showed that two fluorophores are associated with dura fluorescence emission (Sell

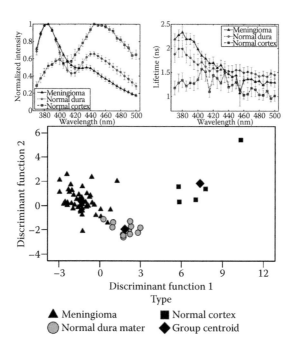

Figure 15.3 Results demonstrating the intensity and average lifetime derived from the time-resolved spectrum of *ex vivo* samples of normal cortex, normal dura mater, and meningioma. The below figure shows the clustering of these tissues using discriminant function analysis. Note the increased intensity at blue-shifted wavelength of 390 nm with long lifetime in both dura mater and meningioma due to presence of connective tissue such as collagen.

and Monnier 1989). These are a "P" fluorophore (a pyridinium compound similar to pyridinoline collagen cross-link, peak excitation/emission: 335/385 nm) and an "M" fluorophore (an age-related fluorophore due to acceleration of collagen browning; peak excitation/emission: 370/440 nm). The TRFS measurement of dura mater tissue corroborates these early studies by revealing two peak emissions, one at about 390 nm and the other at 440 nm. In addition, the change in the fluorescence decay dynamics with emission wavelength not only confirms the presence of at least two fluorophores in the dura mater but also indicates that these fluorophores are characterized by distinct lifetimes. In addition, several immunocytochemical studies have demonstrated the presence in meningioma of collagen I, collagen III, collagen IV, pro-collagens, laminin, and vimentine in meningiomas (Ng et al. 2010; Nitta et al. 1990). Evidence has also shown that meningioma cells are derived closely from the leptomeningeal arachnoid cells and are "fibrous response" of the leptomeninges to trauma, infection, or other pathologies such as tumor infiltration (Gill et al. 1990). The time-resolved spectra of meningiomas in the current study, characterized by a relatively narrow-band blue-shifted emission, closely resembled collagen type I (peak emission 380 nm) and collagen type III (peak 390 nm) fluorescence (Gill et al. 1990; Marcu et al. 2004; Richards-Kortum and Sevick-Muraca 1996). Such trends indicate that these two types of collagen most likely dominated the fluorescence emission of meningioma at the blue-shifted wavelengths (Table 15.3).

Table 15.3 **Specificity and sensitivity values from discriminant function analysis prediction**

	MENINGIOMA (%)	DURA MATER (%)	NORMAL CORTEX (%)
(a) Using Time-Resolved Parameters in Addition to Intensity Parameters for Discriminant Analysis			
Sensitivity	100	86.67	100
Specificity	90.00	100	100
(b) Using Only the Intensity Parameters for Discriminant Analysis			
Sensitivity	92.86	69.23	55.56
Specificity	68.18	86.27	100

Tissue autofluorescence lifetime spectroscopy and imaging: Applications

15.5 *IN VIVO* APPLICATION OF TRFS AND FLIM TECHNIQUES

15.5.1 TRFS OF GLIOMAS

The intraoperative application of TRFS was shown in two studies (Butte et al. 2010, 2011). These clinical studies were conducted in 17 and 42 brain cancer patients, respectively, scheduled for surgical removal of tumor. The emission characteristics of several types of tumor were investigated. This included LGG (oligodendroglioma, oligodendrocytoma, diffuse astrocytoma), intermediate grade (anaplastic astrocytoma), and HGG (anaplastic oligodendroglioma, anaplastic oligoastrocytoma, and glioblastoma multiforme). The intraoperative measurements were conducted using TRFS instrumentation (pulse-sampling) as described in Chapter 4. Areas for TRFS investigation were first identified based on preoperative MRI and surgeon experience. When feasible, tissue biopsy was conducted from areas where TRFS data were acquired. The biopsied tissue was subsequently histopathologically analyzed and used for TRFS measurement validation. With the instrumentation and data analysis method described above, such data set was acquired in less than 30 s and processed in less than 2 s. In the initial study, a Greenburg retractor was used to position the fiber-optic probe above the areas of interest in order to minimize the probe-positioning artifacts. In the second study, a spacer with two slits on the opposite sites was added in front of the distal end of the probe. This configuration enabled direct contact between probe and tissue while maintaining a fixed probe-to-tissue distance. Moreover, the slit permitted suction of fluids in the optical field of view.

As demonstrated by both studies, the LGG was characterized by distinctive emission characteristics enabling their discrimination from NC and NWM. Typically, the LGG (oligoastrocytoma and oligodendroglioma; Butte et al. 2010) presents a main peak emission at about 460 nm (Figure 15.4) that indicates NAD(P)H as the main biological fluorophore contributing to LGG fluorescence emission. This is contrasting with normal tissue that presents a secondary peak at about 390 nm. Moreover, LGG was characterized by a faster fluorescence decay dynamics (~1 ns at 390 nm and ~0.6 ns at 460 nm) relative

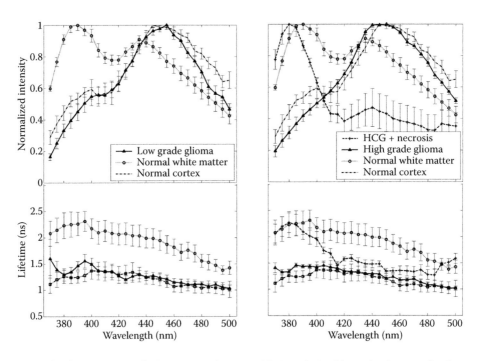

Figure 15.4 Results demonstrating the intensity and average lifetime derived from the time-resolved spectrum of *ex vivo* samples of normal cortex, normal white matter, LGG, and HGG. Both low- and high-grade tumors show reduced fluorescence at 390 nm wavelength while showing a slight increase in average lifetime when compared with normal tissues. The tumors with necrotic tissue demonstrated increased intensity at 390 nm wavelength with longer average lifetime possibly due to collagen deposition.

to NC and NWM (~2 ns at 390 nm and ~1 ns at 460 nm). A more detailed evaluation of LGG based on tumor genetic subclassification (Butte et al. 2011) including diffuse astrocytomas showed that this tumor type presents a longer-lasting decay dynamics (e.g., ~1.5 ns at 390 nm) compared to oligoastrocytoma and oligodendroglioma, results suggesting that the LGG fluorescence emission varies with tumor phenotype or genetic subclassification.

The TRFS studies of HGG (Butte et al. 2011) showed that this group displays a broad range of fluorescence emission characteristics. The spectral and temporal emission characteristics appeared to vary not only with tumor phenotype (e.g., glioblastoma multiforme and anaplastic oligodendroglioma) as observed for LGG but also with the location within the tumor from where the data were recorded. It was noted that HGG tumors are highly heterogeneous, and in many instances, the spectroscopic characteristics retrieved from the center of the tumor are very different from those obtained from tumor margins (Figure 15.5). In addition to the main peak emission at 460 nm, HGG consistently presented an additional peak emission centered at about 380–390 nm similar with that observed for NC and NWM. This peak was associated with two fluorophores known to emit fluorescence within this wavelength band specifically PMP and glutamate decarboxylase (GAD) (Butte et al. 2010; Yong et al. 2006). These biological fluorophores are known to be metabolically affected in gliomas (Honikel and Madsen 1972; Kempe and Stark 1975; Ngo et al. 1998; Shukuya and Schwert 1960). The HGG average lifetime was found longer (e.g., >1 ns

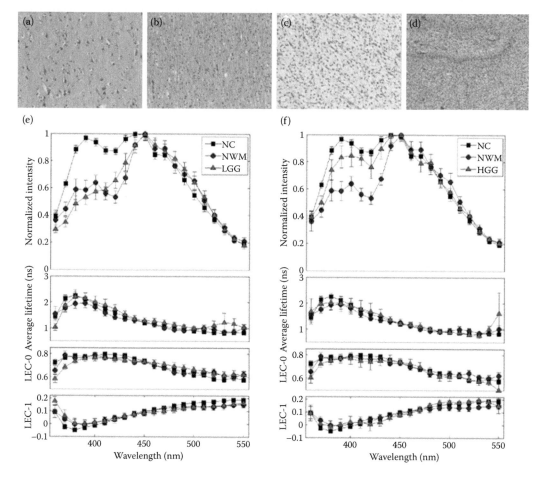

Figure 15.5 Comparison of average fluorescence values mean ± SE of the spectroscopic parameters, emission spectra intensity, average lifetime, average Laguerre coefficients (LEC-0, LEC-1) across the emission wavelengths for distinct brain tissue types (e) normal cortex, normal white matter and low grade glioma, (f) normal cortex, normal white matter and high grade glioma. Histology examples from the biopsy samples of (a) normal cortex, (b) normal white matter, (c) low grade glioma and (d) high grade glioma (glioblastoma).

Tissue autofluorescence lifetime spectroscopy and imaging: Applications

at 460 nm) than that of LGG. Separation of HGG and normal brain was primarily based on the use of Laguerre coefficients that provided a more accurate representation of the fluorescence decay dynamics than the conventional average fluorescence lifetime parameter. Measurements conducted in recurrent glioma with radiation necrosis also demonstrated very distinctive characteristics. Specifically, this tumor displayed a single emission peak at 390 nm suggesting the emission of fibrotic constituents such as collagen.

In addition, a combination of spectroscopic parameters derived from both the emission spectrum (ratios of distinct spectral intensities) and decay dynamics (average lifetimes, Laguerre coefficient values) utilized in a linear discriminant algorithm enabled the classification of LGG with 100% sensitivity and 98% specificity and that of HGG with 47% sensitivity and 94% specificity (Butte et al. 2011). Differentiating between LGG and NWM is of importance when trying to achieve a complete excision at the margins, as the tumor will be surrounded by NWM. Detection of LGG represents a major challenge to current tumor resection techniques. LGGs lack the vascularity of HGG and are often visually bland. As such, they are far more difficult to differentiate from surrounding normal brain than HGG. Further, these tumors typically infiltrate into surrounding tissues that maintain function, often resulting in surgeons performing only biopsies or limited resections. Several recent studies have demonstrated that the extent of resection of LGG correlates with long-term survival and the potential for cure (Sanai and Berger 2008; Smith et al. 2008). These findings are especially important in light of the fact that these tumors are more prevalent in younger patients, for whom increased life expectancy and cure are critical. Consequently, one of the great potential applications of TRFS is the ability to differentiate LGG from normal brain. As such, it may prove to be an extremely useful adjunct to increasing the extent of resection in LGG surgery.

The low sensitivity obtained for HGG was attributed to the high degree of heterogeneity of these tumors. Interestingly, the HGGs were classified with high specificity above 90% that indicates minimum false-positive results; however, current results show very low sensitivity below 50%. The latter was attributed to the high variability in the TRFS signals obtained from various subclasses of HGG. Such variability was due to the high degree of heterogeneity of these tumors. Such differences may also be attributed to the large variance in the protein expression (Umesh et al. 2009) within the same HGG tumor. Generally, HGG cells are more pleomorphic when compared with LGG. Thus, an important aspect to be investigated in future studies is to determine how well heterogeneities in HGG can be distinguished using TRFS-derived parameters. It is anticipated that a more comprehensive classification based on the biochemical information and immunohistochemistry, rather than structural features (H&E staining), will help in classifying the HGG with higher accuracy. The small sample size available for this study most likely also contributed to the low sensitivity values. In order to obtain an unbiased estimation of the classification accuracy, a larger set of data needs to be obtained, and the training set and the test set used have to be completely independent. Consequently, a higher number of HGGs need to be investigated along with the biochemical heterogeneities within the same tumor for a more comprehensive assessment of the ability of TRFS to distinguish HGG tissue from normal brain tissue and, in particular, to improve the sensitivity of the measurement.

Differences between *in vivo* and *ex vivo* TRFS measurements of brain tumors were also reported (Butte 2010). Although the exact cause of such differences is still to be elucidated, it has been suggested that such changes are due to differences in the environment and/or metabolic changes that may occur after tissue excision.

15.5.2 FLIM OF GLIOMAS

A FLIM technique for intraoperative delineation of brain tumor margins was also recently reported (Sun et al. 2010). The study employed a prototype apparatus as described above. Intraoperative measurements were conducted in three patients (13 sites) diagnosed with HGG (glioblastoma multiforme). This study was focused on the evaluation of whether fluorescence lifetime contrast can be achieved between normal brain and brain tumor areas previously identified by conventional diagnostic methods during neurosurgical procedures such as preoperative MRI images and neurosurgeon experience. Since the fluorescence lifetime images are minimally affected by factors that often confound point spectroscopic analysis, including nonuniform illumination, irregular tissue surfaces, and endogenous absorbers such as blood hemoglobin

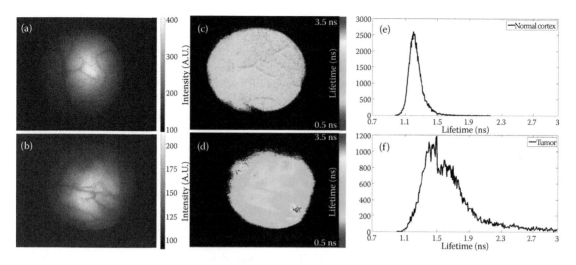

Figure 15.6 Representative fluorescence intensity and lifetime images. (a, b) Intensity images. (c, d) Fluorescence average lifetime images. (e, f) Fluorescence lifetime histograms. For each image, the average lifetime value was retrieved from the binning pixel of four original pixels (2 × 2 square). All average lifetime values for each binning pixel in the region of interest (ROI) were plotted together to show a histogram of average lifetime distribution. (Adapted from Sun, Y. et al., *J Biomed Opt* 15 (5):056022, 2010.)

in the operative field, FLIM rather than intensity-based fluorescence measurements was considered in particular to be appropriate for intraoperative use. The study employed a band-pass filter with a center wavelength of 460 nm and a bandwidth of 50 nm corresponding to NAD(P)H fluorescence. Data acquisition time for each measurement was ~2 min, including one steady-state image and a series of up to 29 time-gated images (0.5 ns gating time and 0.5 ns relative delay-time interval). During imaging, the probe was gently positioned perpendicular to the tissue surface and held with a Greenberg device to minimize the moving artifacts (Figure 15.1).

The results from FLIM measurements were found in agreement with those reported previously for TRFS measurements. Findings further emphasize that the fluorescence lifetime contrast generated by the changes in free vs. bound NAD(P)H plays an important role in the delineation of glioma tumor from normal surrounding tissue. The slightly longer fluorescence lifetime for GBM (~1.6 ns) when compared with normal cortex (~1.3 ns) suggests a higher contribution of bound-NAD(P)H to tissue autofluorescence (Figure 15.6).

15.6 FUTURE OF TRFS AND FLIM IN BRAIN TUMOR DIAGNOSIS

The validation and ultimately the use of fluorescence lifetime techniques such as TRFS and FLIM in clinical diagnostics would have to account for several challenges. These are related to numerous factors including the biological diversity of the brain tumors, ability to validate the spectroscopic technology against local pathophysiologic transformation in brain tissue, and label-free fluorescence-based technology to operate in clinical environment. Some of these challenges were analyzed previously by our group (Butte et al. 2011) and are summarized in the following.

First, the primary brain tumors are characterized by a broad biochemical, molecular, and metabolic diversity. Consequently, these tumors are likely to have a broad range of autofluorescent signatures. This is an important challenge as conventional histopathology does not describe the tumor heterogeneity in terms of metabolic states. As noted above, the intrinsic fluorophores potentially being responsible for the distinct fluorescence emission spectrum include NAD(P)H with a peak emission at ~460 nm and GAD or PMP with a reported peak emission at ~390 nm (Butte et al. 2010). The emission characteristic of such fluorophores most likely is distinctly modulated by various tumor pathophysiologies and metabolic states.

Thus, advancement and accurate validation of the fluorescence lifetime techniques depend on the ability to immunohistopathologically classify and subclassify the tumor as a function of their metabolic activity.

Second, another major challenge in the validation of these lifetime techniques is the correlation of TRFS point measurement or FLIM wide-field image with the pathology or biochemical characteristics of the biopsy tissue specimen acquired from the same site (TRFS case) or with the tissue surface from where the image was actually collected (FLIM case). For the TRFS validation, the size of the optically interrogated volume (~1.2 mm surface diameter × 250–400 μm penetration depth), as defined by the fiber probe excitation-collection geometry and the UV light penetration depth, is overall smaller than the actual size of the biopsied tissue (~3–4 mm³ depth). The orientation of the biopsy also cannot be determined, making it difficult to determine the exact tissue volume excited by the laser. Also, while numerous TRFS measurements can be taken from a single patient, only a limited number of biopsies can be obtained; thus, extensive validation of TRFS measurements against local tissue pathology is limited. Studies to determine *in vivo* the intrapatient signal variability are therefore difficult to perform. For the FLIM validation, the surface of optical interrogation is substantially larger. Only a few physical biopsies can be obtained from this area; thus, a complete validation of FLIM data with histopathologic data is virtually impractical.

Third, maximizing the signal-to-noise ratio (SNR) for improved detection sensitivity and minimizing the time of data acquisition are important features of TRFS and FLIM systems capable of providing diagnostic information based on endogenous contrast and of operating in an intraoperative setting. The described TRFS experimental apparatus with a relatively long data acquisition time of ~30 s posed a challenge related with the variable amount of blood in the field of measurement. This results in the distortion of the fluorescence intensity and reduction of the SNR. It was noticed that the use of a spacer between the optical probe and tissue and presence of a slit in the spacer, which enabled suction to remove the blood in the optical path during the experiments, have minimized the fluorescence signal reabsorption. Maintaining the constant excitation-collection geometry via the spacer, which stabilizes the probe on the brain tissue, has also minimized the distortion of the fluorescence emission spectral shape. Alternatively, TRFS or FLIM techniques that enable reduction of data acquisition time to a few seconds are needed. Recent developments in this area (Sun et al. 2008, 2011) can potentially address this challenge.

15.7 CONCLUSION

While at the current stage of research, TRFS and FLIM do not compare with the existing standard of diagnosis using histopathology, present results suggest that these techniques have the potential to delineate brain tumors from normal cortex to achieve improved tumor excision. The main goal of developing these techniques is not to completely replace the existing gold standard of histopathology but to provide a guide to the neurosurgeon during tumor resection. Acquisition of additional data, in order to differentiate the subclasses of tumors, represents an important next step in establishing the full potential of these fluorescence lifetime techniques. It is similar to all the current techniques deployed in the operating room (e.g., stereotactic navigation, intraoperative ultrasound, and frozen section), which provide guidance to the surgeon regarding the extent of the brain tumor, and it is the task of the surgeon to interpret the results based on full visual, tactile, imaging, and spatial data. The techniques mentioned above provide the surgeon with the structural information; TRFS or FLIM can represent additional techniques, which will enhance the neurosurgeon's ability to further evaluate the tissue by providing biochemical and metabolic information. The most important advantage of TRFS or FLIM is the speed at which they can provide the diagnosis, eliminating the requirement to repeat lengthy frozen section. The speed will also allow the surgeon to inspect multiple sites during the surgery, thus decreasing the chances of residual tumor and ensuring more complete resection.

REFERENCES

Berger, M. S. 1994. Malignant astrocytomas: Surgical aspects. *Semin Oncol* 21 (2):172–85.
Bottiroli, G., A. C. Croce, D. Locatelli, R. Marchesini, E. Pignoli, S. Tomatis, C. Cuzzoni, S. Di Palma, M. Dalfante, and P. Spinelli. 1995. Natural fluorescence of normal and neoplastic human colon: A comprehensive "ex vivo" study. *Lasers Surg Med* 16 (1):48–60.

Bottiroli, G., A. C. Croce, D. Locatelli, R. Nano, E. Giombelli, A. Messina, and E. Benericetti. 1998. Brain tissue autofluorescence: An aid for intraoperative delineation of tumor resection margins. *Cancer Detect Prev* 22 (4):330–9.

Bowling, F. G. 2011. Pyridoxine supply in human development. *Semin Cell Dev Biol* 22 (6):611–18.

Bruner, J. M. 1994. Neuropathology of malignant gliomas. *Semin Oncol* 21 (2):126–38.

Butte, P. V., Q. Fang, J. A. Jo, W. H. Yong, B. K. Pikul, K. L. Black, and L. Marcu. 2010. Intra-operative delineation of primary brain tumors using time-resolved fluorescence spectroscopy. *J Biomed Opt* 15 (2):027008.

Butte, P. V., A. N. Mamelak, M. Nuno, S. I. Bannykh, K. L. Black, and L. Marcu. 2011. Fluorescence lifetime spectroscopy for guided therapy of brain tumors. *Neuroimage* 54 Suppl 1:S125–35.

Byar, D. P., S. B. Green, and T. A. Strike. 1983. Prognostic factors for malignant glioma. In M. D. Walker (ed.), *Oncology of the Nervous System, Cancer Treatment and Research*, vol. 12. New York: Springer, pp. 379–95.

Central Brain Tumor Registry of the United States. 2012. CBTRUS Statistical Report: Primary Brain and Central Nervous System Tumors Diagnosed in the United States in 2004–2008. Central Brain Tumor Registry of the United States. Available on http://www.cbtrus.org/2012-NPCR-SEER/CBTRUS_Report_2004-2008_3-23-2012.pdf.

Chung, Y. G., J. A. Schwartz, C. M. Gardner, R. E. Sawaya, and S. L. Jacques. 1997. Diagnostic potential of laser-induced autofluorescence emission in brain tissue. *J Korean Med Sci* 12 (2):135–42.

Crespi, F., A. C. Croce, S. Fiorani, B. Masala, C. Heidbreder, and G. Bottiroli. 2004. Autofluorescence spectrofluorometry of central nervous system (CNS) neuromediators. *Lasers Surg Med* 34 (1):39–47.

Croce, A. C., S. Fiorani, D. Locatelli, R. Nano, M. Ceroni, F. Tancioni, E. Giombelli, E. Benericetti, and G. Bottiroli. 2003. Diagnostic potential of autofluorescence for an assisted intraoperative delineation of glioblastoma resection margins. *Photochem Photobiol* 77 (3):309–18.

Dang, L., D. W. White, S. Gross, B. D. Bennett, M. A. Bittinger, E. M. Driggers, V. R. Fantin, H. G. Jang, S. Jin, M. C. Keenan, K. M. Marks, R. M. Prins, P. S. Ward, K. E. Yen, L. M. Liau, J. D. Rabinowitz, L. C. Cantley, C. B. Thompson, M. G. V. Heiden, and S. M. Su. 2009. Cancer-associated IDH1 mutations produce 2-hydroxyglutarate. *Nature* 462 (7274):739–44.

Devaux, B. C., J. R. O'Fallon, and P. J. Kelly. 1993. Resection, biopsy, and survival in malignant glial neoplasms. A retrospective study of clinical parameters, therapy, and outcome. *J Neurosurg* 78 (5):767–75.

Flace, P., V. Benagiano, L. Lorusso, F. Girolamo, A. Rizzi, D. Virgintino, L. Roncali, and G. Ambrosi. 2004. Glutamic acid decarboxylase immunoreactive large neuron types in the granular layer of the human cerebellar cortex. *Anat Embryol (Berl)* 208 (1):55–64.

Gill, S. S., D. G. Thomas, N. Van Bruggen, D. G. Gadian, C. J. Peden, J. D. Bell, I. J. Cox, D. K. Menon, R. A. Iles, D. J. Bryant, and G. A. Coutts. 1990. Proton MR spectroscopy of intracranial tumours: In vivo and in vitro studies. *J Comput Assist Tomogr* 14 (4):497–504.

Greenaway, F. T., and J. W. Ledbetter. 1987. Fluorescence lifetime and polarization anisotropy studies of membrane surfaces with pyridoxal 5'-phosphate. *Biophys Chem* 28 (3):265–71.

Hess, K. R. 1999. Extent of resection as a prognostic variable in the treatment of gliomas. *J Neurooncol* 42 (3):227–31.

Honikel, K. O., and N. B. Madsen. 1972. Comparison of the absorbance spectra and fluorescence behavior of phosphorylase b with that of model pyridoxal phosphate derivatives in various solvents. *J Biol Chem* 247 (4):1057–64.

Jameson, D. M., V. Thomas, and D. M. Zhou. 1989. Time-resolved fluorescence studies on NADH bound to mitochondrial malate dehydrogenase. *Biochim Biophys Acta* 994 (2):187–90.

Jo, J. A., Q. Fang, T. Papaioannou, and L. Marcu. 2004. Fast model-free deconvolution of fluorescence decay for analysis of biological systems. *J Biomed Opt* 9 (4):743–52.

Kempe, T. D., and G. R. Stark. 1975. Pyridoxal 5'-phosphate, a fluorescent probe in the active site of aspartate transcarbamylase. *J Biol Chem* 250 (17):6861–9.

Kim, T. S., A. L. Halliday, E. T. Hedley-Whyte, and K. Convery. 1991. Correlates of survival and the Daumas-Duport grading system for astrocytomas. *J Neurosurg* 74 (1):27–37.

Konig, K., M. W. Berns, and B. J. Tromberg. 1997. Time-resolved and steady-state fluorescence measurements of beta-nicotinamide adenine dinucleotide-alcohol dehydrogenase complex during UVA exposure. *J Photochem Photobiol B* 37 (1–2):91–5.

Lacroix, M., D. Abi-Said, D. R. Fourney, Z. L. Gokaslan, W. Shi, F. DeMonte, F. F. Lang, I. E. McCutcheon, S. J. Hassenbusch, E. Holland, K. Hess, C. Michael, D. Miller, and R. Sawaya. 2001. A multivariate analysis of 416 patients with glioblastoma multiforme: Prognosis, extent of resection, and survival. *J Neurosurg* 95 (2):190–8.

Laws, E. R., I. F. Parney, W. Huang, F. Anderson, A. M. Morris, A. Asher, K. O. Lillehei, M. Bernstein, H. Brem, A. Sloan, M. S. Berger, and S. Chang. 2003. Survival following surgery and prognostic factors for recently diagnosed malignant glioma: Data from the Glioma Outcomes Project. *J Neurosurg* 99 (3):467–73.

Lin, W. C., S. A. Toms, M. Johnson, E. D. Jansen, and A. Mahadevan-Jansen. 2001. In vivo brain tumor demarcation using optical spectroscopy. *Photochem Photobiol* 73 (4):396–402.

Tissue autofluorescence lifetime spectroscopy and imaging: Applications

Louis, D. N., H. Ohgaki, O. D. Wiestler, W. K. Cavenee, P. C. Burger, A. Jouvet, B. W. Scheithauer, and P. Kleihues. 2007. The 2007 WHO classification of tumours of the central nervous system. *Acta Neuropathol* 114 (2): 97–109.

Marcu, L., M. C. Fishbein, J. M. Maarek, and W. S. Grundfest. 2001. Discrimination of human coronary artery atherosclerotic lipid-rich lesions by time-resolved laser-induced fluorescence spectroscopy. *Arterioscler Thromb Vasc Biol* 21 (7):1244–50.

Marcu, L., J. A. Jo, P. V. Butte, W. H. Yong, B. K. Pikul, K. L. Black, and R. C. Thompson. 2004. Fluorescence lifetime spectroscopy of glioblastoma multiforme. *Photochem Photobiol* 80:98–103.

Mirimanoff, R. O., D. E. Dosoretz, R. M. Linggood, R. G. Ojemann, and R. L. Martuza. 1985. Meningioma: Analysis of recurrence and progression following neurosurgical resection. *J Neurosurg* 62 (1):18–24.

Nelson, J. S., Y. Tsukada, D. Schoenfeld, K. Fulling, J. Lamarche, and N. Peress. 1983. Necrosis as a prognostic criterion in malignant supratentorial, astrocytic gliomas. *Cancer* 52 (3):550–4.

Ng, W. H., K. Mukhida, and J. T. Rutka. 2010. Image guidance and neuromonitoring in neurosurgery. *Childs Nerv Syst* 26 (4):491–502.

Ngo, E. O., G. R. LePage, J. W. Thanassi, N. Meisler, and L. M. Nutter. 1998. Absence of pyridoxine-5'-phosphate oxidase (PNPO) activity in neoplastic cells: Isolation, characterization, and expression of PNPO cDNA. *Biochemistry* 37 (21):7741–8.

Nitta, H., T. Yamashima, J. Yamashita, and T. Kubota. 1990. An ultrastructural and immunohistochemical study of extracellular matrix in meningiomas. *Histol Histopathol* 5 (3):267–74.

Ossola, L., M. Maitre, J. M. Blindermann, and P. Mandel. 1979. Enzymes of GABA metabolism in tissue culture. *Adv Exp Med Biol* 123:139–57.

Poon, W. S., K. T. Schomacker, T. F. Deutsch, and R. L. Martuza. 1992. Laser-induced fluorescence: Experimental intraoperative delineation of tumor resection margins. *J Neurosurg* 76 (4):679–86.

Pradhan, A., P. Pal, G. Durocher, L. Villeneuve, A. Balassy, F. Babai, L. Gaboury, and L. Blanchard. 1995. Steady state and time-resolved fluorescence properties of metastatic and non-metastatic malignant cells from different species. *J Photochem Photobiol B* 31 (3):101–12.

Reinert, K. C., R. L. Dunbar, W. Gao, G. Chen, and T. J. Ebner. 2004. Flavoprotein autofluorescence imaging of neuronal activation in the cerebellar cortex in vivo. *J Neurophysiol* 92 (1):199–211.

Richards-Kortum, R., and E. Sevick-Muraca. 1996. Quantitative optical spectroscopy for tissue diagnosis. *Annu Rev Phys Chem* 47:555–606.

Sanai, N., and M. S. Berger. 2008. Glioma extent of resection and its impact on patient outcome. *Neurosurgery* 62 (4):753–64; discussion 764–6.

Sanai, N., M. Y. Polley, M. W. McDermott, A. T. Parsa, and M. S. Berger. 2011. An extent of resection threshold for newly diagnosed glioblastomas. *J Neurosurg* 115 (1):3–8.

Schneckenburger, H., and K. Konig. 1992. Fluorescence decay kinetics and imaging of Nad(P)H and flavins as metabolic indicators. *Opt Eng* 31 (7):1447–51.

Scott, T. G., R. D. Spencer, N. J. Leonard, and G. Weber. 1969. Emission properties of NADH. Fluorescence lifetimes and quantum efficiencies of NADH and simplified synthetic models. *Abstr Papers Am Chem Soc* (Sep):Bi39.

Scott, T. G., R. D. Spencer, N. J. Leonard, and G. Weber. 1970. Emission properties of NADH. Studies of fluorescence lifetimes and quantum efficiencies of NADH, AcPyADH, and simplified synthetic models. *J Am Chem Soc* 92 (3):687–95.

Seidlitz, H. K., H. Schneckenburger, and K. Stettmaier. 1990. Time-resolved polarization measurements of porphyrin fluorescence in solution and in single cells. *J Photochem Photobiol B* 5 (3–4):391–400.

Sell, D. R., and V. M. Monnier. 1989. Isolation, purification and partial characterization of novel fluorophores from aging human insoluble collagen-rich tissue. *Connect Tissue Res* 19 (1):77–92.

Sharma, S. K., and K. Dakshinamurti. 1992. Determination of vitamin B6 vitamers and pyridoxic acid in biological samples. *J Chromatogr* 578 (1):45–51.

Shukuya, R., and G. W. Schwert. 1960. Glutamic acid decarboxylase. II. The spectrum of the enzyme. *J Biol Chem* 235:1653–7.

Smith, J. S., E. F. Chang, K. R. Lamborn, S. M. Chang, M. D. Prados, S. Cha, T. Tihan, S. Vandenberg, M. W. McDermott, and M. S. Berger. 2008. Role of extent of resection in the long-term outcome of low-grade hemispheric gliomas. *J Clin Oncol* 26 (8):1338–45.

Stummer, W., M. J. van den Bent, and M. Westphal. 2011. Cytoreductive surgery of glioblastoma as the key to successful adjuvant therapies: New arguments in an old discussion. *Acta Neurochir (Wien)* 153 (6):1211–8.

Stupp, R., W. P. Mason, M. J. van den Bent, M. Weller, B. Fisher, M. J. Taphoorn, K. Belanger, A. A. Brandes, C. Marosi, U. Bogdahn, J. Curschmann, R. C. Janzer, S. K. Ludwin, T. Gorlia, A. Allgeier, D. Lacombe, J. G. Cairncross, E. Eisenhauer, and R. O. Mirimanoff. 2005. Radiotherapy plus concomitant and adjuvant temozolomide for glioblastoma. *N Engl J Med* 352 (10):987–96.

Sun, Y., A. J. Chaudhari, M. Lam, H. Xie, D. R. Yankelevich, J. Phipps, J. Liu, M. C. Fishbein, J. M. Cannata, K. K. Shung, and L. Marcu. 2011. Multimodal characterization of compositional, structural and functional features of human atherosclerotic plaques. *Biomed Opt Express* 2 (8):2288–98.

Sun, Y., N. Hatami, M. Yee, J. Phipps, D. S. Elson, F. Gorin, R. J. Schrot, and L. Marcu. 2010. Fluorescence lifetime imaging microscopy for brain tumor image-guided surgery. *J Biomed Opt* 15 (5):056022.

Sun, Y., R. Liu, D. S. Elson, C. W. Hollars, J. A. Jo, J. Park, and L. Marcu. 2008. Simultaneous time- and wavelength-resolved fluorescence spectroscopy for near real-time tissue diagnosis. *Opt Lett* 33 (6):630–2.

Umesh, S., A. Tandon, V. Santosh, B. Anandh, S. Sampath, B. A. Chandramouli, and V. R. Sastry Kolluri. 2009. Clinical and immunohistochemical prognostic factors in adult glioblastoma patients. *Clin Neuropathol* 28 (5):362–72.

Wagnieres, G. A., W. M. Star, and B. C. Wilson. 1998. In vivo fluorescence spectroscopy and imaging for oncological applications. *Photochem Photobiol* 68 (5):603–32.

Wahl, P., J. C. Auchet, A. J. Visser, and F. Muller. 1974. Time resolved fluorescence of flavin adenine dinucleotide. *FEBS Lett* 44 (1):67–70.

Winger, M. J., D. R. Macdonald, and J. G. Cairncross. 1989. Supratentorial anaplastic gliomas in adults. The prognostic importance of extent of resection and prior low-grade glioma. *J Neurosurg* 71 (4):487–93.

Wolf, H. K., R. Buslei, I. Blümcke, O. D. Wiestler, and T. Pietsch. 1997. Neural antigens in oligodendrogliomas and dysembryoplastic neuroepithelial tumors. *Acta Neuropathol* 94 (5):436–43.

Yong, W. H., P. V. Butte, B. K. Pikul, J. A. Jo, Q. Fang, T. Papaioannou, K. Black, and L. Marcu. 2006. Distinction of brain tissue, low grade and high grade glioma with time-resolved fluorescence spectroscopy. *Front Biosci* 11:1255–63.

Zipfel, W. R., R. M. Williams, R. Christie, A. Y. Nikitin, B. T. Hyman, and W. W. Webb. 2003. Live tissue intrinsic emission microscopy using multiphoton-excited native fluorescence and second harmonic generation. *Proc Natl Acad Sci U S A* 100 (12):7075–80.

16 Oncology applications: Skin cancer

Rakesh Patalay, Paul M. W. French, and Christopher Dunsby

Contents

16.1 INTRODUCTION

The human skin is a layered tissue comprising an upper layer of keratin, a cellular epidermis (keratinocytes and melanocytes) containing NAD(P)H, keratin, flavins, and melanin, and the dermis consisting mainly of collagen and elastin. The presence of these endogenous fluorophores provides opportunities to use autofluorescence lifetime to contrast changes in skin caused by disease. Fluorescence lifetime measurements of human skin have been carried out using a wide range of instruments, from nonimaging fiber-optic probe-based point measurements using single-photon excitation through to multiphoton microscopy and tomography providing subcellular resolution. This chapter aims to review fluorescence and fluorescence lifetime measurements of skin according to the type of instrumentation used and the type of disease studied. The initial Sections 16.2 and 16.3 cover results employing single-photon excitation, while Section 16.4 onward consider multiphoton imaging.

While this book focuses on the application of fluorescence lifetime measurements, it is important to consider these in the context of the extensive work on steady-state and spectrally resolved measurements of autofluorescence of human skin. A number of studies have shown that the autofluorescence characteristics of neoplastic skin differ from those of normal tissue, as discussed in several reviews (e.g., Ramanujam 2000; Richards-Kortum et al. 1996; Wagnieres et al. 1998; Zeng and MacAulay 2003).

Using point-based measurements, Lohmann and Paul (1988) reported an increase in 366 nm excited autofluorescence from melanoma compared to nevi in a study of 82 patients. However, later work using the same instrument did not contrast dysplastic nevi from melanomas (Lohmann et al. 1991). Chwirot et al. (1998) used a digital imaging approach with the same excitation wavelength and collecting autofluorescence at 475 nm to distinguish melanomas from common and dysplastic nevi.

A number of groups have applied point spectroscopy of autofluorescence in basal cell carcinomas (BCCs) with broadband near-ultraviolet excitation (350–375 nm). Both Na et al. (2001) and Brancaleon et al. (2001) have reported that the autofluorescence intensity from BCCs at these excitation wavelengths is significantly lower than that from normal skin. It has been suggested that this may be due to the degradation of dermal collagen by BCC tumor matrix metalloproteinases, resulting in reduced collagen fluorescence from the upper dermis (Brancaleon et al. 2001). Conversely, Sterenborg et al. (1994) demonstrated no significant differences between the autofluorescence from BCCs and normal skin. Other groups have successfully used longer (410–440 nm) (Panjehpour et al. 2002; Zeng and MacAulay 2003) or shorter (295 nm) (Brancaleon et al. 2001) wavelengths of excitation light to differentiate BCCs from normal skin.

The group of Rajaram et al. (2010a) has recently applied an instrument combining diffuse reflectance spectroscopy and fluorescence spectroscopy to the study of nonmelanoma skin cancers. The diffuse reflectance spectroscopy provided information on blood volume fraction, oxygen saturation, blood vessel size, tissue microarchitecture, and melanin content, and the fluorescence spectroscopy used excitation at 337 and 445 nm to provide native fluorophore contributions of NADH, collagen, and FAD. This instrument was used to study nonmelanoma skin cancer in 48 lesions from 40 patients and demonstrated classification of BCCs from normal skin with a sensitivity and specificity of 94% and 89%, respectively (Rajaram et al. 2010b).

Most of these studies cited above have discriminated between normal and neoplastic tissue by detecting differences in the measured steady-state fluorescence intensities or spectra. However, fluorescence intensity measurements are sensitive to fluctuations in excitation intensity, tissue scattering, and absorption and are difficult to quantify and compare between samples. Spectrally resolved measurements provide additional information, but fluorescence emission spectra of many tissue fluorophores are broad and overlap significantly (Cubeddu et al. 1999). Therefore, it is interesting to explore the potential of autofluorescence lifetime measurements to provide a further opportunity to detect and contrast different tissue fluorophores and their relative distributions for different disease states.

16.2 FIBER-OPTIC POINT-PROBE-BASED FLUORESCENCE LIFETIME SPECTROSCOPY OF SKIN

Pitts and Mycek (2001) developed a fiber-optic point-probe device able to collect spectrally and temporally resolved fluorescence decays, and this instrument was applied to measurements of human skin autofluorescence in 2001. The forearm from a healthy volunteer was excited using light from a pulsed nitrogen laser at 337 nm with a repetition rate of 10 Hz, and fluorescence was collected above 460 nm. Fluorescence decays were fitted using a double-exponential decay model, which yielded decay components of 0.938 and 5.3 ns from the skin surface, which were thought to be compatible with the expected fluorescence decay of collagen.

Katika et al. (2006) used a pulsed LED excitation source at 375 nm and collected the fluorescence decays using time-correlated single-photon counting (TCSPC) in four emission bands between 442 and 496 nm. The skin from both palms was recorded from 35 volunteers, and various body sites were examined in one patient. Their aim was to compare differences in the fluorescence lifetime with age, sex, and skin type. A triexponential decay model was used to analyze the data, and decay components at 0.4, 2.7, and 9.7 ns were observed with the 442 nm emission filter. While the two shorter fluorescence decay components were thought to be due to NAD(P)H, the longest component was postulated to arise from advanced glycation end-products (AGEs). The authors found no correlation of the fluorescence lifetime with skin type and gender. The fluorescence lifetime parameters were not found to correlate with age ($R^2 <$ 0.03).

AGEs are the final product of chemical reactions between sugars, proteins, lipids, and nucleic acids. Although these accumulate naturally with age, the rate increases in diabetics. Some AGEs, such as pentosidine, fluoresce, and therefore, there is the potential that fluorescence spectroscopy can be used for noninvasive monitoring. Blackwell et al. (2008) used the same fiber-optic probe instrument described in the previous paragraph to collect time-resolved fluorescence decays *in vivo* from the skin with the aim to screen for type 2 diabetes. The authors found that the greatest signal was obtained from the palm and that the signal was independent of skin complexion, and so they chose this body site in their study. Data were collected from 38 diabetic and 37 nondiabetic patients, and a triple exponential decay model was employed for the fluorescence decay analysis. Repeated measurements over a 1 month period show a coefficient of variation of <5%, indicating that lifetimes were a sufficiently reproducible parameter. No statistical differences in lifetime with age, gender, or skin type were observed. No statistically significant differences were found in the lifetimes of the decay components between the two patient populations, but two of the fractional amplitudes were found to be different. The mean decay component lifetimes were 0.5, 2.6, and 9.2 ns for normal skin. The shorter two decay components were thought consistent with that of NADH. A comparison of the fluorescence decay from diabetic patients with foot ulcers (a marker for microvascular disease) versus age-matched healthy controls showed a change in the short fluorescence lifetime component τ_1 (477 to 510 ps, $p = 0.038$) and the fractional component a_1 (0.769 to 0.761, $p = 0.012$), with the most significant change seen for a_2 (0.196 to 0.217, $p = 0.002$). The authors speculated that the change was related to differences in the metabolic rate of the tissues rather than the presence of AGEs.

De Beule et al. (2007) presented another instrument for measuring time (4096 time bins) and spectrally resolved (16 spectral channels) fluorescence decays in skin. This instrument was used to measure the fluorescence from freshly excised lesions. Lesions were excited at both 355 and 435 nm using pulsed laser diodes, and fluorescence decays were collected using TCSPC and were fitted using a double-exponential decay model. Measurements of BCCs with excitation at 435 nm yielded a mean fluorescence lifetime of 2.44 ns, which was shorter than the lifetime of 2.96 ns measured for surrounding uninvolved skin. Six of the lesions contained a sufficiently wide margin of normal skin to permit paired lesional and perilesional measurements, which showed a decrease in mean fluorescence lifetime of 620 ps for BCCs ($p < 0.03$) using 435 nm excitation. The fluorescence lifetimes collected using 355 nm excitation showed little contrast between benign and malignant skin. Further spectrally resolved lifetime histograms for squamous cell carcinomas (SCCs), nevi, melanomas, and seborrheic keratosis were presented, but there were insufficient patient numbers to support further statistical analysis.

Thompson et al. (2012) applied similar equipment as De Beule et al. (except for the second laser wavelength being changed from 435 to 445 nm) to measure the fluorescence from a number of lesions on the skin *in vivo*, including 10 BCCs. They found that the spectrally averaged, intensity-weighted mean fluorescence lifetimes were consistently shorter for BCCs than paired perilesional normal skin (2240 ± 480 ps BCC vs. 3130 ± 413 ps normal) when excited at 445 nm. This shift of 886 ps was statistically significant ($p = 0.002$) and greater than the average variability found in the surrounding normal skin of 175 ps. UV excitation at 355 nm did not show a consistent change in lifetime between normal and BCC, and this is consistent with the data collected by De Beule et al. from *ex vivo* samples using the same equipment.

16.3 WIDE-FIELD FLIM OF SKIN

Galletly et al. (2008) published work evaluating 25 freshly excised BCCs and their margins by collecting FLIM images from the tissue surface. Tissue was excited at 355 nm using a frequency tripled mode-locked Nd:YVO$_4$ laser with a repetition rate of 80 MHz, and fluorescence was collected at wavelengths >375 and >445 nm using a wide-field time-gated FLIM system. Each pixel was defined manually as lesional or perilesional normal skin using the reflectance white light image, and a single-exponential decay was fitted to the data. Unlike many previous studies (e.g., Brancaleon et al. 2001), no consistent changes were seen in fluorescence intensity between normal and lesional skin. A significant difference was found in the fluorescence lifetime between BCCs and normal perilesional skin (means of 1.4 ns vs. 1.6 ns, respectively,

$p < 0.001$). As discussed above, neither De Beule et al. (2007) nor Thompson et al. (2012) found a significant change in autofluorescence lifetime at this excitation wavelength. This may be explained by the differences in excitation/collection geometry between fiber-optic point-probe and wide-field imaging systems and the differences in detection wavelength bands employed.

16.3.1 WIDE-FIELD FLIM OF SKIN WITH PHOTOSENSITIZER

Wide-field FLIM has also been applied to skin following treatment with photosensitizers. Cubeddu et al. (1999) and Andersson-Engels et al. (2000) published studies investigating the potential to increase the contrast of fluorescence from BCC/SCCs by applying the exogenous photosensitizer ALA, which causes PPIX to be preferentially accumulated in cancerous tissue compared to normal tissue. FLIM of the skin surface was implemented with excitation from a pulsed dye laser operating at 405 nm that provided sub-1 ns pulses. Subsequently, an *in vivo* study of 48 lesions in 34 patients (benign and malignant, mainly BCC) reported that, following application of ALA, tumors were found to have a longer lifetime of 18 ns compared to normal skin of 10 ns (Cubeddu et al. 2002).

16.4 MULTIPHOTON IMAGING OF NORMAL SKIN

The first publication of multiphoton imaging of human skin *in vivo* was published in 1997 by Masters et al. (1997). This work contrasted autofluorescence spectra and lifetimes of cells from two depths in the skin (0–50 and 100–150 µm), exciting fluorescence at 730 and 960 nm using a tunable femtosecond pulsed Ti:sapphire laser. An emission peak at 445–460 nm consistent with that of NAD(P)H was found to dominate images at both depths when excited at 730 nm. When exciting at 960 nm, a spectral peak centered at 520 nm consistent with the emission spectrum of flavins was found. Another peak at 450 nm was found but not identified. Mean fluorescence lifetimes were calculated for each pixel using frequency-domain FLIM, and values ranging from 0.5 to 3 ns were observed.

König et al. (König 2000; König and Riemann 2003; König et al. 2002) developed a multiphoton tomograph (DermaInspect®, JenLab GmbH) capable of measuring both fluorescence intensity and lifetimes in optically sectioned images of skin both *ex vivo* and *in vivo*. A tunable mode-locked Ti:sapphire laser was used to excite the tissue, and imaging with submicron lateral spatial resolution and TCSPC FLIM with 250 ps temporal resolution was implemented. In this initial work, *ex vivo* samples from a variety of dermatoses including psoriasis, nevi, a melanoma, and fungal infection were studied, and the forearms of two volunteers were also examined *in vivo*. The fluorescence decays acquired at each pixel were fitted to a double-exponential decay model, and the mean fluorescence lifetime of cellular autofluorescence was found to be in the range of 1.8–2.4 ns. The DermaInspect instrument was subsequently CE-marked for clinical use *in vivo*.

16.4.1 SPECTRALLY RESOLVED MULTIPHOTON IMAGING

Early spectrally resolved multiphoton imaging of the dermal layer of *ex vivo* human skin was carried out by Buehler et al. (2005) using a 16-channel multianode photomultiplier tube, which allowed the emission spectra from collagen-rich and elastin-rich regions of the image to be compared.

Laiho et al. (2005) imaged five samples of normal skin that had been stored at –3°C before use. Each sample was excited at 730, 780, and 830 nm using a tunable mode-locked Ti:sapphire laser, and the fluorescence was detected in 10 spectral channels between 375 and 600 nm using a filter wheel. The emission spectra from the whole sample and from four depths within the skin using all three excitation wavelengths were analyzed to study the presence of endogenous fluorophores, for example, tryptophan, NAD(P)H, melanin, collagen, and elastin. Fluorescence lifetime measurements were also undertaken using TCSPC for which the decay profile from each pixel was fitted to a double-exponential decay model. This allowed second harmonic generation (SHG) in collagen to be distinguished from elastin autofluorescence.

Pena et al. (2005) investigated autofluorescence from keratin using single- and two-photon microscopy and spectroscopy. They presented multiphoton imaging of autofluorescence and SHG from fixed sectioned normal human skin excited at 860 nm, and their spectral analysis of autofluorescence indicated that keratin and elastin spectra are similar (with spectral peaks at 477 and 471 nm, respectively, when excited at 760 nm).

Chen et al. (2006) performed multiphoton imaging with a mode-locked Ti:sapphire laser excitation source and a 32-channel hyperspectral detector. Emission spectra of both epidermis and dermis were recorded using two-photon excitation wavelengths in the range 790–830 nm. The maximum fluorescence signal from both epidermis and dermis was obtained using the shortest excitation wavelength, and the maximum SHG signal from dermis was obtained using 800 nm excitation. Zhuo et al. (2006) used the same instrument to study the dermis of fresh frozen human skin.

The group of Gerritsen presented spectrally resolved multiphoton imaging of mouse skin *in vivo* using a prism-based spectrometer readout by a high-speed EMCCD camera providing 100 spectral channels over the range 350–600 nm (Palero et al. 2006, 2008). Images were acquired using an excitation wavelength of 760 nm and were converted into an RGB scale for real-color visualization. The authors performed a detailed analysis of the spectral signatures of mouse skin (Palero et al. 2007) and determined the relative contributions and emission spectra of keratin, melanin, NAD(P)H, flavins, collagen, and elastin. Subsequent work included a study of the effect of ischemia on the emission spectrum of NAD(P)H (Palero et al. 2011) and of the two-photon excited emission spectra of human skin *in vivo* (Bader et al. 2011). The images of normal skin from two people with different skin types clearly demonstrated the high spatial resolution of the system and its ability to distinguish collagen, elastin, NAD(P)H/FAD, and melanin based on spectra alone.

Subsequently, a prism-based spectrometer incorporating a fiber-bundle spot-to-line converter was incorporated into a DermaInspect-based instrument and used to acquire spectrally resolved images of human skin *ex vivo* from a suspected nodular BCC with excitation at 760 and 840 nm (Talbot et al. 2011).

Breunig et al. (2010) used the DermaInspect to characterize the influence of the excitation wavelength on the fluorescence emission intensity from various depths within normal skin *in vivo*. Intracellular autofluorescence reduces while collagen SHG signal increases in intensity as the excitation wavelength is increased from 720 to 880 nm, and the authors discussed the optimum excitation wavelength for each of the fluorescent species investigated.

16.4.2 MULTIPHOTON IMAGING OF TISSUE STRUCTURE

Koehler et al. (2010) used the DermaInspect to study epidermal thickness. Image depth stacks were acquired from the dorsal forearm and the dorsum of the hand in 30 volunteers. Skin was excited at 800 nm, and two spectral channels with filters were used to collect the SHG using a 400 nm BPF and the second channel collecting autofluorescence between 410 and 490 nm. They described an objective method of distinguishing the levels in the skin using the number of photon counts per image of the autofluorescence and SHG signals. Using the peak in the SHG signal to signify the papillary dermis, they used ~9000 images to calculate the epidermal thickness from the acquired image depth stacks. The dorsum of the hand was found to have thicker total epidermis, cellular epidermis, and stratum corneum and higher depths of the papillary dermis than the forearm. No thickness differences were found with age. No significant variation was observed in the undulations of the dermo-epidermal junction with age. The same authors (Koehler et al. 2011) also present a study of 30 patients with normal skin, discussing tissue morphology and quantifying three parameters for two different body sites (dorsal forearm and dorsum of the hand) in three age groups (young, mid-aged, and old). The parameters investigated were the number of papillae per square millimeter and, in the stratum granulosum, the mean number of keratinocytes per unit area and the mean nuclear area. The total number of papillae was found to be significantly decreased in the old group compared with the young, and keratinocytes were found to be smaller for the dorsum of the hand compared with the dorsal forearm.

16.4.3 MULTIPHOTON FLIM OF NORMAL HUMAN SKIN

Breunig et al. (2010) acquired two FLIM images of the same field of view in the papillary dermis using 710 and 800 nm excitation. Fluorescence decays for each pixel were fitted using a double-exponential decay model, and the results highlighted that changes in the excitation wavelength lead to different measured fluorescence lifetimes due to the changing contributions from the different autofluorescent species present.

Sugata et al. (2010) applied the DermaInspect to evaluate the fluorescence lifetime of melanin within normal skin *in vivo* and compared it to melanin found in the hair bulb and from melanocytic 3D cell cultures. Ultrafast excitation pulses at 760 and 800 nm were employed, and fluorescence decays collected using TCSPC were fitted to a double-exponential decay model. Fluorescence lifetimes in the basal epidermal layer of τ_1 = 132 ps and τ_2 = 1762 ps were obtained using 760 nm excitation. The authors concluded that, for all samples, the short τ_1 component was dominated by melanin fluorescence while the longer τ_2 component mainly represented other fluorophores. The authors then fixed the lifetimes of the two decay components (to 120 and 1100 ps) in the fitting software and used the resulting ratio maps of the amplitude of the short and long fluorescence decay components (a_1/a_2) to produce maps of melanin concentration.

Benati et al. (2011) have published a comprehensive *in vivo* study of the fluorescence lifetimes from normal skin using multiphoton imaging. Forty-nine patients in total from two age groups had images taken from different body sites and depths. A DermaInspect was used with 760 nm excitation, and fluorescence decays were collected using TCSPC and fitted using a single-exponential decay model to estimate the mean fluorescence lifetime. The fluorescence lifetimes in the skin were found to vary with body site, increase with age, and shorten with tissue depth. The shorter lifetimes lower in the skin were partially attributable to the higher melanin content in the basal layers of the skin.

Many studies of diseased tissue have studied normal tissue as part of the investigation. In order to avoid repetition, these papers are discussed in Sections 16.5–16.7 covering diseased tissue.

16.4.4 MULTIPHOTON IMAGING OF PENETRATION OF PARTICLES AND CHEMICALS INTO HUMAN SKIN

Multiphoton imaging has been used to assess the penetration of a range of fluorescent chemicals and luminescent particles into skin. Yu et al. (2001, 2002) looked at the penetration of sulforhodamine B and rhodamine B hexyl esters in excised human cadaveric skin and used this approach to assess oleic acid–induced transdermal diffusion pathways (Yu et al. 2003). The distribution of sulforhodamine B was also studied in *ex vivo* skin by Ericson et al. (2008). Celli et al. (2010) incubated *ex vivo* human skin with Calcium Green and used multiphoton imaging to quantify changes in calcium concentration via the fluorescence decay dynamics of the calcium dye (changes on binding) using the phasor approach to analyze the fluorescence decay data.

The dermal penetration of nanoparticle-borne drugs in excised human skin has been assessed by multiphoton imaging where the nanoparticles were covalently labeled with fluorescein and where Texas Red was used as a drug model and dissolved in the particles to be released (Stracke et al. 2006). Zvyagin et al. (2008) used multiphoton imaging to investigate the penetration of zinc oxide nanoparticles into human skin in both *ex vivo* skin samples and healthy volunteers. Using a DermaInspect instrument, they excited fluorescence at 740 nm and monitored the luminescence intensity from ZnO particles in a spectral channel at 385 nm and tissue autofluorescence in a second broader channel. Subsequent work by the same group investigated the use of fluorescence lifetime information to study the position of the ZnO particles (Roberts et al. 2008). They also used multiphoton imaging and FLIM to study the penetration of ZnO in tape-stripped normal skin and in patients with atopic dermatitis and psoriasis *in vivo* (Lin et al. 2011).

16.4.5 MULTIPHOTON IMAGING OF SKIN TO STUDY THE EFFECTS OF AGING

The SHG signal to autofluorescence (AF) aging index of dermis (SAAID) has been defined as SAAID = (SHG − AF)/(SHG + AF), which provides a useful numerical index to assess the age of skin using multiphoton imaging (Lin et al. 2005). This index is related to the histological observation that collagen is progressively replaced by elastic fibers in older skin. Lin et al. (2005) investigated three samples of formalin fixed sections of facial skin. Multiphoton images were obtained exciting the tissue at 760 nm and collecting the fluorescence from areas in the superficial dermis at 380 nm for SHG and >435 nm for autofluorescence. The SAAID was found to decrease with age to −0.93 in a 70-year-old patient and correlated with an increase in elastin in the dermis in the form of solar elastosis and a decrease in the SHG signal from collagen causing a decrease in the collagen/elastin ratio.

Koehler et al. (2006) looked at the skin from the inside forearm from 18 volunteers *in vivo*. The skin was excited at 820 nm with the SHG collected at 410 nm and autofluorescence collected above 470 nm. Five fields of view were acquired from each patient from the upper dermis, and the SAAID was calculated for each pixel. The results showed that SAAID generally reduced with age (linear regression $R^2 = 0.57$), but there was large interpatient variability. This drop was greater for women ($R^2 = 0.89$) compared with men ($R^2 = 0.68$). In subsequent work, the same group (Koehler et al. 2008) assessed quantitatively the dermal matrix composition and showed characteristic changes with aging. In methodology identical to their 2006 study, 18 volunteers had the inside of their forearm imaged using the DermaInspect, and five images were acquired from the upper dermis. The dermis was evaluated using a multiphoton imaging-based dermis morphology score (MDMS). Images were evaluated for eight parameters in the following categories: fiber spread and fiber aspect, network pattern and image homogeneity (SHG images only), and clot formation (autofluorescence images only). A single MDMS score was calculated, which correlated strongly (linear regression $R^2 = -0.9$) with age and less strongly with SAAID ($R^2 = 0.66$), and did not show gender-specific differences.

Koehler et al. (2009) looked at the skin of 60 healthy volunteers including the young, old, and women using tanning salons. The skin elasticity was measured using both mechanical methods and SAAID. The skin was excited at 800 nm from the volar and dorsal aspect of the forearm, and SHG was collected at 400 nm and fluorescence collected from 410 to 490 nm. Sixteen regions at approximately 180 µm below the stratum granulosum were measured and averaged for each patient. The authors found that older patients had lower SAAID scores. Sun-exposed areas had a lower SAAID (significant in old men, old women, and young women using tanning but not in young men and young women). Mechanical measurements of skin elasticity (cytometry) indicated loss of elasticity with age, and a higher elasticity in the dorsum forearm than the ventral was found in all groups. The procedures used in this study reduced the intrapatient and interpatient variability compared to the 2006 and 2008 studies, and this study confirmed that SAAID decreased with age in a larger sample size. The authors found a lower SAAID in sun-exposed dorsal forearm compared to the sun-protected volar forearm in the old and in young women using tanning salons, but were not able to detect any damage caused by tanning beds.

Kaatz et al. (2010) examined the influence of depth and epidermal thickness on the SAAID index. The forearms of 30 healthy volunteers divided into three age groups were examined *in vivo*. Skin was excited at 800 nm using a DermaInspect, and SHG was collected at 400 nm and autofluorescence at 410–490 nm. The fluorescence intensities from the images were then analyzed based on changes in the autofluorescence and SHG intensity with depth, and the authors determined that the papillary dermis was reached at 91 ± 16 µm. The maximum penetration depth was found to be 130–180 µm based on the plateau in the autofluorescence and SHG intensities below this depth. The position of the basal layer and the presence of solar elastosis in older patients could be seen through small peaks in the plots of tissue autofluorescence intensities as a function of depth. A sharper decline in the SHG signal after the dermis is reached was found in the elderly compared to the young. The SHG peak in the upper dermis is due to the high content of fibrillar collagen (types I and III). The most significant differences between ages were found in the SAAID index when a constant depth below the surface was assessed (below the SHG peak and above the penetration depth), for example, 150 µm.

16.4.6 EFFECTS OF TISSUE EXCISION ON AUTOFLUORESCENCE

In 2011, Sanchez et al. (2010) outlined the effect of excision, temperature, and the use of culture media on the autofluorescence from NAD(P)H in freshly excised human skin with measurements performed on a daily basis. The results showed that intracellular fluorescence was maintained for days, even for samples stored at room temperature. The use of media hastened tissue degradation, but lowering the temperature to 4°C could maintain viability for 7 days despite this. Fitting the measured decay from each pixel to a double-exponential model, a peak around 800 ps appeared in histogram of τ_1 shortly after excision in addition to a persistent peak at 400 ps. The peak at 2500 ps in τ_2 broadened with time from excision.

Palero et al. (2011) studied the fluorescence emission spectrum of mouse skin during ischemia over 3.3 h. The authors focused on identifying the contributions of free and protein-bound NADH and used linear spectral unmixing to separate and quantify the spectrally shifted (~15 nm) emission spectra of these two moieties. The total fluorescence intensity increased compared to pre-ischemic baseline by an average of 71% when measured over 50–80 min post-ischemia. After 80 min, the fluorescence intensity then steadily dropped to below pre-ischemic levels. Following the onset of ischemia, the ratio of free/bound NADH decreased below initial levels at 50 min and then increased to above initial levels from 75 min, which was interpreted in terms of changes in the cellular redox state.

Patalay et al. (2012) studied the changes in the fluorescence lifetime of cellular autofluorescence in *ex vivo* human skin biopsies in four spectral detection windows for excitation at 760 nm over a 3 h period post-excision. The mean fluorescence lifetime averaged over four separate samples decreased from the initial value by 7.2%, 1.6%, 2.9%, and 8.6% in the blue (360–425 nm), green (425–515 nm), yellow (515–620 nm), and red (620–640/655 nm) spectral channels, respectively. These changes were all much smaller than the changes seen between normal and BCC diagnostic categories. The authors also compared the median cell mean fluorescence lifetime of normal skin imaged *ex vivo* to normal skin imaged *in vivo* and found no statistically significant changes (Wilcoxon rank sum test, $p = 0.98, 0.70, 0.63$, and 0.23 in the blue, green, yellow, and red spectral channels, respectively).

16.5 MULTIPHOTON IMAGING OF AUTOFLUORESCENCE IN NONMELANOMA SKIN CANCER

In one of the first studies of skin cancer using multiphoton imaging (Lin et al. 2006), autofluorescence intensity images were collected from nine fixed, sectioned slices from BCCs. Serially acquired images were montaged together to generate an autofluorescence image covering a large area crossing tumor margins, and this was presented with an exact correlative histopathological image. The SHG from dermal collagen was spectrally filtered from the emitted fluorescence, and the authors used an index of multiphoton-excited autofluorescence to SHG (MFSI, equivalent to the negative of the SAAID index; see Section 16.4.5) to distinguish cancer cells/stroma from normal dermis in BCCs. The authors found that the MFSI ratio was greatest within the tumor, lower in cancer stroma, and lowest in normal dermal stroma.

Paoli et al. (2008) published a study of the morphological features observed in multiphoton-excited autofluorescence intensity images. Fourteen freshly excised specimens of nonmelanoma skin cancer— including SCC *in situ*, superficial BCCs, and nodular BCCs)—were imaged using 780 nm excitation. They looked for the presence of morphological features in SCC *in situ* and in superficial BCCs. Their study confirmed that many of the established histopathological features could be seen using the *en face* multiphoton imaging images, although image features characteristic of disease were only found in one of three nodular BCCs due to the limited penetration depth.

The group of Pavone has applied *en face* multiphoton imaging and FLIM to assess freshly excised BCCs (Cicchi et al. 2007, 2008; De Giorgi et al. 2009). Using autofluorescence excitation at 740 nm and SHG imaging at 840 nm, they presented examples of the morphological features seen in multiphoton imaging images of BCCs and compared them directly to correlative histology. The SAAID index (see Section 16.4.5) was used to distinguish BCC tumors (negative index), stromal interface (marginally positive index), and the dermis (negative index). The autofluorescence emission spectrum of four BCCs was studied and found to be shifted toward the blue for BCC compared to normal skin, with this shift being the greatest for depths in the range 30–50 μm. They also studied fluorescence lifetime histograms obtained from FLIM images acquired at different depths and, in a study of four samples, found the fluorescence lifetime to be longer for BCC compared to normal skin tissue, with this difference also being the greatest (80 ps) for imaging depths in the range 30–50 μm.

Patalay et al. (2011) evaluated BCCs *ex vivo* using a DermaInspect that acquired FLIM data in two emission spectral channels (300–500 and 500–640 nm). Whole excised, fresh tissue samples were imaged using excitation at 760 nm. Each cell in every image was identified manually and all pixels within each region of interest (ROI) summed to provide a single fluorescence decay per cell. This approach increased the number of photons available in order to fit a double-exponential decay model to the data. In total, fluorescence decays from 615 cells from three nodular BCCs and 566 cells from four nevi were defined and fitted. The mean fluorescence lifetimes for nevi vs. nodular BCCs were calculated as 2516 vs. 2786 ps (<500 nm channel) and 1334 vs. 2085 ps (>500 nm channel). Shorter fluorescence lifetimes were observed in both spectral channels for nevi, and this was attributed to the increased melanin content in the nevi. Fitting fluorescence decays on a cell-by-cell basis allowed populations of nevi cells to be contrasted with nodular BCCs via their lifetime, and it was suggested that the variation in the melanin content in cells could be responsible for the lifetime histograms obtained. This study also indicated a large interpatient variability in fluorescence lifetimes of skin lesions.

Seidenari et al. (2013) investigated a range of morphological image descriptors—building on those identified by Paoli et al. (2008)—with the aim to optimize the differentiation of BCC from normal skin using multiphoton imaging. Initially, training image stacks acquired at 10 depths from one field of view per sample were acquired *ex vivo*, imaging 24 BCCs and 24 samples of normal skin. These multiphoton imaging stacks were used to identify nine morphological criteria/descriptors. In the main study, image stacks from a further 66 BCCs, 66 other lesions (23 nevi, 8 melanomas, 17 skin tumors, and 18 other skin lesions), and 66 normal regions of skin were acquired. The resulting images were then assessed by three independent observers who assessed each sample for the presence or absence of each descriptor. The mean number of descriptors per lesion was 2.64 for BCCs, 0.17 in other lesions, and 0 for normal skin. The presence of "Aligned elongated cells" was found in 73% of BCCs but in only 5% of other lesions. An overall sensitivity/specificity of 95%/89% was achieved when using the presence of one or more BCC descriptor(s) as the diagnostic criterion, whereas 83%/95% was obtained when using the presence of two or more descriptors. These results indicate the potential utility of multiphoton imaging as a diagnostic tool.

Seidenari et al. (2012) extended this approach to the combination of multiphoton intensity imaging and FLIM for the diagnosis of BCC, distinguishing them from a range of other skin lesions. A preliminary study of 35 BCCs and 35 healthy skin samples were imaged, and morphological descriptors were identified using the resulting FLIM images. In the main study, 63 BCCs, 66 other skin lesions (24 nevi, 8 melanomas, 15 inflammatory lesions, and 19 skin tumors), and 63 samples of healthy skin were imaged and the presence or absence of each descriptor assessed by three independent observers. Fifteen FLIM images were acquired at different depths for each sample using an excitation wavelength of 760 nm. Cells with a longer fluorescence lifetime were always observed for BCCs, and this descriptor provided the highest sensitivity/specificity (100%/70%) of all of the descriptors investigated. The mean fluorescence lifetime values, calculated on three representative cells in the lower layers of each of the 98 (preliminary and main study samples combined) healthy skin lesions and each of the 98 BCC lesions, were found to be 1012 and 1475 ps, respectively (a significant difference at $p < 0.001$), in agreement with earlier work on a smaller number of samples (Cicchi et al. 2008). The mean number of BCC descriptors per lesion was higher in BCC (3.86 ± 1.45) compared to miscellaneous lesions (0.54 ± 0.86). The presence of at least one BCC descriptor was observed in all BCCs but only in 36% of other skin samples. An overall sensitivity for the diagnosis of BCC from other lesions and healthy skin of 97% was achieved when the diagnostic criterion was chosen to be the presence of two or more descriptors, and this produced no false positives, either in other lesions or healthy skin. A specificity of 100% was obtained when considering the presence of five descriptors or more.

Patalay et al. (2012) studied BCCs using multiphoton imaging that combined morphological descriptors with multispectral FLIM data acquired using DermaInspect modified to provide TCSPC detection in four spectral channels (360–425, 425–515, 515–620, and 620–640/655 nm). A key step to realize robust multispectral FLIM data was the use of broadband ultrafast luminescence from gold from nanorods in order to obtain the system instrument response function in all spectral channels simultaneously (Patalay et al. 2011). Nineteen freshly excised BCCs and 27 samples of normal skin (imaged *in vivo* and *ex vivo*) were imaged with excitation at 760 nm and an exemplar dataset is shown in Figure 16.1. All FLIM images were assessed by a dermatologist for the presence of at least one of the multiphoton imaging diagnostic morphological descriptors for BCC described previously by Seidenari et al. (2012, 2013) or a new diagnostic descriptor "merging cells." This yielded a sensitivity/specificity of 79%/93% for the diagnosis of BCC from normal skin. The intracellular autofluorescence from each cell was identified using the ROI approach described previously and then fitted using a double-exponential decay model (Patalay et al. 2011). First, the ROIs were used to assess the morphology of each cell and its morphology with respect to its nearest neighbors (cellular morphology parameters). Second, the relative spectral contribution in each spectral channel and all of the fluorescence decay parameters for the double-exponential fit to each ROI were calculated (spectroscopic parameters). All of the parameters were then assessed for their ability to differentiate BCC from normal skin and ranked according to the area under the curve (AUC) of the receiver operating characteristic. The fluorescence decay parameters provided the greatest discrimination,

Tissue autofluorescence lifetime spectroscopy and imaging: Applications

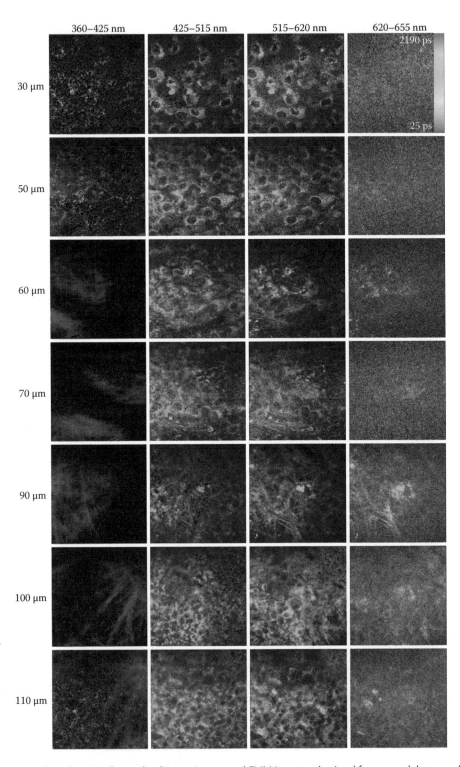

Figure 16.1 Depth and spectrally resolved intensity merged FLIM images obtained from a nodular superficial BCC. SHG from collagen appears as a short fluorescence lifetime in the 360–425 nm spectral channel. The 425–515 nm channel is mainly due to cellular NAD(P)H fluorescence with longer fluorescence lifetimes arising from elastin fibers in deeper layers. The 515–620 nm channel detects a mixture of flavin, NAD(P)H, and melanin fluorescence. Melanin fluorescence with a short lifetime dominates the 620–650 nm channel. (Reproduced from Patalay R. et al., 2012 *PLoS One* 7(9): e43460. doi:10.1371/journal.pone.0043460.)

with τ_1 from the red and blue channels providing AUCs of 0.82 and 0.80, respectively. All parameters were then input into a diagnostic algorithm using principal component analysis-based dimensionality reduction followed by linear discriminant analysis. The percentage of cells/ROIs classified as normal or BCC by the algorithm was then calculated for each patient, and the separation of these variables' distributions yielded an AUC of 0.83. As an example, if the threshold for classifying a patient as having BCC is set at having >30% of cells classified as BCC, then this yielded a sensitivity/specificity of 89%/73%. Fully automated analysis approaches—without the need for manual segmentation of cellular ROIs—were also investigated and found to provide a similar level of performance. The authors also demonstrated the ability of the multiphoton multispectral FLIM system to image fields of view larger than 1 mm through the montaging of serially acquired images.

16.6 MULTIPHOTON SPECTROSCOPY AND IMAGING OF PIGMENTED LESIONS

Teuchner et al. (1999) presented emission spectra and fluorescence lifetime measurements of synthetic melanin and melanin in *ex vivo* specimens of normal human skin reporting a complex fluorescence decay profile with a dominant short decay component at about 200 ps. Melanin excitation can also occur via a stepwise multiphoton excitation (Teuchner et al. 1999), which allows for preferential excitation of melanin, particularly when using longer (nanosecond) NIR excitation pulses (Eichhorn et al. 2009). In subsequent work, Teuchner et al. (2000) studied the emission spectra of normal skin, nevi, and melanoma and noted a shift in the peak of the fluorescence emission spectrum of the melanoma sample (~550 nm) compared to nevi (~520 nm) and normal skin (~500 nm). The same group observed the same trends in a larger study of 10 specimens in each of the three lesion types (Hoffmann et al. 2001).

Eichhorn et al. (2009) exploited selective excitation of melanin using 2.5 ns excitation pulses at a wavelength of 810 nm to study the fluorescence emission spectra of melanin in formalin-fixed paraffin-embedded samples. A total of 27 lesions were studied, including 9 malignant melanomas. For benign melanocytic nevi, spectral emission peaks of ~480 and ~490 nm were observed, respectively. For malignant melanoma, the autofluorescence was characterized by a peak at ~600 nm. A dysplastic compound nevus exhibited both peaks. Similar spectra were observed for a fresh unfixed specimen of malignant melanoma. These results were confirmed in a more detailed subsequent study by Leupold et al. (2011), where the authors measured the emission spectra of 167 cases of nevi and melanomas. Melanoma gave a characteristic emission peak at 640 nm with benign melanocytic nevi presenting a peak at 590 nm. This difference was thought to be due to changes in the ratio of eumelanin to pheomelanin. For each sample, the emission spectrum was acquired for an array of 12 × 9 positions on the surface of the specimen that were spaced laterally in steps of 50 μm. The fluorescence signal at each point was compared in two bands (483–520 and 585–620 nm) in order to classify the measurement as benign or malignant. Overall, this study reported a sensitivity, specificity, and diagnostic accuracy of 94%, 80%, and 83%, respectively, for distinguishing malignant melanoma from benign nevi.

Ehlers et al. (2007) used a DermaInspect to investigate the fluorescence lifetime of human hair, which is rich in melanin. Fluorescence lifetime analysis of hair of different colors revealed differences in the lifetime between blond (0.4, 2.2 ns) and black hair (0.2, 1.3 ns), and these lifetimes were compared to pure melanin and melanin measured *in vivo* from a mole. When a fast multichannel plate photomultiplier tube providing a 24 ps response time was used to detect the fluorescence decays, eumelanin in black hair (0.03, 0.8 ns) could be distinguished from pheomelanin in red/blonde hair (0.34, 2.3 ns) and white hair (0.3, 2.1 ns).

Cicchi et al. (2008) and De Giorgi et al. (2009) applied spectrally resolved multiphoton imaging and FLIM to study two melanoma specimens with excitation at 740 nm. Melanomas were found to have a shorter fluorescence lifetime than normal skin and BCCs but were not compared to other pigmented lesions. Spectral analysis showed a reduction in fluorescence intensity below 520 nm for melanoma compared to normal skin. However, the shift toward longer emission wavelengths for melanoma compared to normal skin reported by Teuchner et al. (2000) was not seen.

Dimitrow et al. applied the DermaInspect to study pigmented lesions. In their first study (Dimitrow et al. 2009b), they initially studied 15 *ex vivo* specimens (healthy skin, melanocytic lesions, and melanoma) using multiphoton imaging with excitation at 760 nm and identified six morphological features to distinguish melanomas from nevi. A further 83 lesions (42 both *in vivo* and *ex vivo*, 11 only *in vivo*, and 30 only *ex vivo*) were then imaged and the data assessed by four independent observers for the occurrence of each feature. The feature of "large intercellular distance" provided the greatest sensitivity of 80% and "dendritic cells" provided the greatest specificity of 96% (when considering all lesions). Binary logistic regression showed that the features of "architectural disarray," "poorly defined keratinocyte cell borders," "pleomorphic cells," and "dendritic cells" should also be included in diagnostic decisions, which yielded an overall diagnostic accuracy of 85%/97% for analysis of images acquired *in vivo/ex vivo*.

In a second study, Dimitrow et al. (2009a) examined 13 nevi and 10 melanomas, mostly from *ex vivo* samples. In this study, they found that multiphoton excitation at 800 nm enhanced many of the morphological features in the intensity images associated with melanoma, compared to excitation at 760 nm. Fluorescence lifetimes were calculated for a total of 84 manually selected pixels of the cytoplasm from either keratinocytes or melanocytes, and a biexponential decay model was fitted to the data. The authors found that both the short/long decay components were shorter for melanocytes than keratinocytes (140/1076 ps vs. 445/2269 ps, respectively) and that melanocytes had a more dominant short lifetime component than keratinocytes (93%/7% and 76%/24%, respectively). The differences in the fluorescence decay parameters between the two cell populations were attributed to a dominant NAD(P)H fluorescence in keratinocytes and a dominant melanin fluorescence in melanocytes. The observed fluorescence lifetimes were able to distinguish keratinocytes from melanocytes but not nevi from melanomas. The authors also investigated the fluorescence emission spectrum of four lesions (two of normal skin, one nevus, and one melanoma) and found a peak at 550 nm occurring in the melanoma sample, but not the nevus or normal skin. This peak was more pronounced when exciting at 800 nm compared to 760 nm.

16.7 MULTIPHOTON IMAGING OF OTHER SKIN DISEASES

16.7.1 ACTINIC KERATOSIS

Koehler et al. (2011) performed an *in vivo* study using the DermaInspect to compare normal skin from 30 patients with actinic keratosis (AK) with that from 27 patients. The skin was excited at 760 and 820 nm for autofluorescence intensity (470 nm long-pass filter) and SHG imaging, respectively. The authors observed wider intercellular spaces and the presence of a fluorescence perinuclear rim for AK. A number of cellular-size parameters were compared, but only the nuclear-to-keratinocyte-size ratio was deemed significantly different with AK. They also found an increase in the SAAID index with AK. In addition, they observed increased thicknesses of the total and viable epidermis consistent with epidermal acanthosis.

16.7.2 ATOPIC DERMATITIS

Huck et al. (2010) presented results comparing the changes in autofluorescence intensity images and lifetimes seen with the inflammatory dermatosis "atopic eczema" compared with normal skin *in vivo* using a DermaInspect with an excitation wavelength of 710 nm. In three patients with control, mild, and severe atopic dermatitis, they found that the peak mean lifetime lengthened from ~1150 to ~1300 to ~1500 ps, respectively. In addition, nonlesional skin in atopic dermatitis patients was found to have a longer mean lifetime than patients with normal skin (Huck et al. 2011).

16.7.3 SCLERODERMA

Lu et al. (2009) studied sclerodermatous skin *ex vivo* using frozen vertically sectioned tissue. Three samples from scleroderma patients and two from normal skin were imaged with multiphoton excitation at 810 nm, with SHG detected between 393 and 414 nm and autofluorescence between 430 and 650 nm. Intensity images from the dermis were analyzed, and general morphological differences were discussed. The net orientation of collagen bundles was assessed via the fast Fourier transform of the images and found to be higher in scleroderma compared to normal skin. In addition, the spacing of collagen fibrils and the epidermal thickness was found to be reduced in scleroderma. The SHG/autofluorescence ratio (similar to the SAAID index) was found to be increased in the lower dermis in scleroderma patients.

16.7.4 KELOIDS AND HYPERTROPHIC SCARS

Brewer et al. (2004) compared multiphoton excited image depth stacks of an excised keloid scar with normal skin. Cicchi et al. (2008) acquired five images for each of four different regions in one keloid specimen, showing that regions of fibroblastic proliferation and keloid demonstrated a strongly positive SAAID index compared to normal tissue.

Chen et al. (2009) applied multiphoton imaging with 32-spectral channel detection to acquire multispectral image depth stacks from five hypertrophic scars and one sample of normal skin in frozen sectioned tissue excited at 850 nm. They recorded the SHG between 414 and 436 nm and elastin autofluorescence between 457 and 714 nm, and presented the morphological features of the collagen and elastin fibers in normal and scar tissue.

16.7.5 MULTIPHOTON IMAGING OF SKIN WITH PHOTOSENSITIZER

Cicchi et al. (2007, 2008) applied multiphoton imaging and FLIM to study four BCC specimens treated with the photosensitizer ALA to induce the production of PPIX. A triple exponential model was required to fit the decay data, and ALA-treated BCC tissue showed an increased fluorescence signal and a much longer mean fluorescence lifetime (>8 ns) compared to tissue autofluorescence (typically ~2 ns).

16.8 SUMMARY

Autofluorescence provides a rich range of contrast parameters for studying diseased human skin, which can be complemented by exogenous contrast agents such as ALA. These fluorescence signals have been investigated using a range of instrumentation including single-point fiber probes, wide-field imaging with single-photon excitation, and multiphoton imaging. Autofluorescence intensity can be complemented by measurements using multiple excitation wavelengths, multiple emission detection bands, and measurements of fluorescence lifetime. A comprehensive picture of the characteristics of the main autofluorescence components of skin is emerging, and this is contributing to a deeper understanding of the mechanisms behind the contrast observed. Key challenges are now to design and construct instrumentation that is more compact and lower cost, in order to aid the translation of such devices into multicenter clinical trials.

REFERENCES

Andersson-Engels, S., G. Canti, R. Cubeddu et al. 2000. Preliminary evaluation of two fluorescence imaging methods for the detection and the delineation of basal cell carcinomas of the skin. *Lasers Surg Med* 26:76–82.

Bader, A. N., A.-M. Pena, C. Johan van Voskuilen et al. 2011. Fast nonlinear spectral microscopy of in vivo human skin. *Biomed Opt Express* 2:365–373.

Benati, E., V. Bellini, S. Borsari et al. 2011. Quantitative evaluation of healthy epidermis by means of multiphoton microscopy and fluorescence lifetime imaging microscopy. *Skin Res Technol* 17:295–303.

Blackwell, J., K. M. Katika, L. Pilon, K. M. Dipple, S. R. Levin, and A. Nouvong. 2008. In vivo time-resolved autofluorescence measurements to test for glycation of human skin. *J Biomed Opt* 13:014004.

Brancaleon, L., A. J. Durkin, J. H. Tu, G. Menaker, J. D. Fallon, and N. Kollias. 2001. In vivo fluorescence spectroscopy of nonmelanoma skin cancer. *Photochem Photobiol* 73:178–183.

Breunig, H. G., H. Studier, and K. König. 2010. Multiphoton excitation characteristics of cellular fluorophores of human skin in vivo. *Opt Express* 18:7857–7871.

Brewer, M. B., A. T. Yeh, B. Torkian, C.-H. Sun, B. J. Tromberg, and B. J. Wong. 2004. Multiphoton imaging of excised normal skin and keloid scar: Preliminary investigations. In *Proc. SPIE 5312, Lasers in Surgery: Advanced Characterization, Therapeutics, and Systems XIV*, 204–208.

Buehler, C., K. H. Kim, U. Greuter, N. Schlumpf, and P. T. C. So. 2005. Single-photon counting multicolor multiphoton fluorescence microscope. *J Fluoresc* 15:41–51.

Celli, A., S. Sanchez, M. Behne, T. Hazlett, E. Gratton, and T. Mauro. 2010. The epidermal Ca^{2+} gradient: Measurement using the phasor representation of fluorescent lifetime imaging. *Biophys J* 98:911–921.

Chen, G., J. Chen, S. Zhuo et al. 2009. Nonlinear spectral imaging of human hypertrophic scar based on two-photon excited fluorescence and second-harmonic generation. *Br J Dermatol* 161:48–55.

Tissue autofluorescence lifetime spectroscopy and imaging: Applications

Chen, J. X., S. M. Zhuo, T. S. Luo, X. S. Jiang, and J. J. Zhao. 2006. Spectral characteristics of autofluorescence and second harmonic generation from ex vivo human skin induced by femtosecond laser and visible lasers. *Scanning* 28:319–326.

Chwirot, B. W., S. Chwirot, J. Redzinski, and Z. Michniewicz. 1998. Detection of melanomas by digital imaging of spectrally resolved ultraviolet light-induced autofluorescence of human skin. *Eur J Cancer* 34:1730–1734.

Cicchi, R., D. Massi, S. Sestini et al. 2007. Multidimensional non-linear laser imaging of basal cell carcinoma. *Opt Express* 15:10135–10148.

Cicchi, R., S. Sestini, V. De Giorgi, D. Massi, T. Lotti, and F. S. Pavone. 2008. Nonlinear laser imaging of skin lesions. *J Biophotonics* 1:62–73.

Cubeddu, R., D. Comelli, C. D'Andrea, P. Taroni, and G. Valentini. 2002. Clinical system for skin tumour detection by fluorescence lifetime imaging. In Engineering in Medicine and Biology, 2002. 24th Annual Conference and the Annual Fall Meeting of the Biomedical Engineering Society EMBS/BMES Conference, 2002. Proceedings of the Second Joint, vol. 2293, 2295–2296.

Cubeddu, R., A. Pifferi, P. Taroni, A. Torricelli, G. Valentini, and E. Sorbellini. 1999. Fluorescence lifetime imaging: An application to the detection of skin tumors. *IEEE J Sel Topics Quantum Electron* 5:923–929.

De Beule, P. A., C. Dunsby, N. P. Galletly et al. 2007. A hyperspectral fluorescence lifetime probe for skin cancer diagnosis. *Rev Sci Instrum* 78:123101.

De Giorgi, V., D. Massi, S. Sestini, R. Cicchi, F. S. Pavone, and T. Lotti. 2009. Combined non-linear laser imaging (two-photon excitation fluorescence microscopy, fluorescence lifetime imaging microscopy, multispectral multiphoton microscopy) in cutaneous tumours: First experiences. *J Eur Acad Dermatol Venereol* 23:314–316.

Dimitrow, E., I. Riemann, A. Ehlers et al. 2009a. Spectral fluorescence lifetime detection and selective melanin imaging by multiphoton laser tomography for melanoma diagnosis. *Exp Dermatol* 18:509–515.

Dimitrow, E., M. Ziemer, M. J. Koehler et al. 2009b. Sensitivity and specificity of multiphoton laser tomography for in vivo and ex vivo diagnosis of malignant melanoma. *J Invest Dermatol* 129:1752–1758.

Ehlers, A., I. Riemann, K. König, and M. Stark. 2007. Multiphoton fluorescence lifetime imaging of human hair. *Microsc Res Tech* 70:154–161.

Eichhorn, R., G. Wessler, M. Scholz et al. 2009. Early diagnosis of melanotic melanoma based on laser-induced melanin fluorescence. *J Biomed Opt* 14:034033.

Ericson, M. B., C. Simonsson, S. Guldbrand, C. Ljungblad, J. Paoli, and M. Smedh. 2008. Two-photon laser-scanning fluorescence microscopy applied for studies of human skin. *J Biophotonics* 1:320–330.

Galletly, N. P., J. McGinty, C. Dunsby et al. 2008. Fluorescence lifetime imaging distinguishes basal cell carcinoma from surrounding uninvolved skin. *Br J Dermatol* 159:152–161.

Hoffmann, K., M. Stucker, P. Altmeyer, K. Teuchner, and D. Leupold. 2001. Selective femtosecond pulse-excitation of melanin fluorescence in tissue. *J Invest Dermatol* 116:629–630.

Huck, V., C. Gorzelanny, K. Thomas, V. Niemeyer, T. A. Luger et al. 2010. Intravital multiphoton tomography as a novel tool for non-invasive in vivo analysis of human skin affected with atopic dermatitis. In *Proc. SPIE* 7548, *Photonic Therapeutics and Diagnostics VI*, 75480B–75486R.

Huck, V., C. Gorzelanny, K. Thomas, V. Niemeyer, T. A. Luger et al. 2011. Intravital multiphoton tomography as an appropriate tool for non-invasive in vivo analysis of human skin affected with atopic dermatitis. In *Proc. SPIE* 7883, *Photonic Therapeutics and Diagnostics VII*, 78830R–78836R.

Kaatz, M., A. Sturm, P. Elsner, K. König, R. Bückle, and M. J. Koehler. 2010. Depth-resolved measurement of the dermal matrix composition by multiphoton laser tomography. *Skin Res Technol* 16:131–136.

Katika, K. M., L. Pilon, K. Dipple, S. Levin, J. Blackwell, and H. Berberoglu. 2006. In vivo time-resolved autofluorescence measurements on human skin. In *Photonic Therapeutics and Diagnostics II*, N. Kollias, H. Zeng, B. Choi, R. S. Malek, B. J. Wong, J. F. R. Ilgner, E. A. Trowers, W. T. de Riese, H. Hirschberg, S. J. Madsen, M. D. Lucroy, L. P. Tate, K. W. Gregory, and G. J. Tearney, editors. SPIE, San Jose, 83–93.

Koehler, M. J., S. Hahn, A. Preller et al. 2008. Morphological skin ageing criteria by multiphoton laser scanning tomography: Non-invasive in vivo scoring of the dermal fibre network. *Exp Dermatol* 17:519–523.

Koehler, M. J., K. König, P. Elsner, R. Buckle, and M. Kaatz. 2006. In vivo assessment of human skin aging by multiphoton laser scanning tomography. *Opt Lett* 31:2879–2881.

Koehler, M. J., A. Preller, N. Kindler et al. 2009. Intrinsic, solar and sunbed-induced skin aging measured in vivo by multiphoton laser tomography and biophysical methods. *Skin Res Technol* 15:357–363.

Koehler, M. J., T. Vogel, P. Elsner, K. König, R. Bückle, and M. Kaatz. 2010. In vivo measurement of the human epidermal thickness in different localizations by multiphoton laser tomography. *Skin Res Technol* 16:259–264.

Koehler, M. J., S. Zimmermann, S. Springer, P. Elsner, K. Konig, and M. Kaatz. 2011. Keratinocyte morphology of human skin evaluated by in vivo multiphoton laser tomography. *Skin Res Technol* 17:479–486.

König, K. 2000. Multiphoton microscopy in life science. *J Microsc* 200:83–104.

König, K., and I. Riemann. 2003. High-resolution multiphoton tomography of human skin with subcellular spatial resolution and picosecond time resolution. *J Biomed Opt* 8:432–439.

König, K., U. Wollina, I. Riemann et al. 2002. Optical tomography of human skin with subcellular spatial and picosecond time resolution using intense near infrared femtosecond laser pulses. In *Proc. SPIE* 4620, *Multiphoton Microscopy in the Biomedical Sciences II*, 191–201.

Laiho, L. H., S. Pelet, T. M. Hancewicz, P. D. Kaplan, and P. T. So. 2005. Two-photon 3-D mapping of ex vivo human skin endogenous fluorescence species based on fluorescence emission spectra. *J Biomed Opt* 10:024016.

Leupold, D., M. Scholz, G. Stankovic et al. 2011. The stepwise two photon excited melanin fluorescence is a unique diagnostic tool for the detection of malignant transformation in melanocytes. *Pigment Cell Melanoma Res* 24:438–445.

Lin, L., J. Grice, M. Butler et al. 2011. Time-correlated single photon counting for simultaneous monitoring of zinc oxide nanoparticles and NAD(P)H in intact and barrier-disrupted volunteer skin. *Pharm Res* 28:2920–2930.

Lin, S. J., S. H. Jee, C. J. Kuo et al. 2006. Discrimination of basal cell carcinoma from normal dermal stroma by quantitative multiphoton imaging. *Opt Lett* 31:2756–2758.

Lin, S. J., R. Wu, Jr., H. Y. Tan et al. 2005. Evaluating cutaneous photoaging by use of multiphoton fluorescence and second-harmonic generation microscopy. *Opt Lett* 30:2275–2277.

Lohmann, W., M. Nilles, and R. H. Bodeker. 1991. In situ differentiation between naevi and malignant melanomas by fluorescence measurements. *Naturwissenschaften* 78:456–457.

Lohmann, W., and E. Paul. 1988. In situ detection of melanomas by fluorescence measurements. *Naturwissenschaften* 75:201–202.

Lu, K., J. Chen, S. Zhuo et al. 2009. Multiphoton laser scanning microscopy of localized scleroderma. *Skin Res Technol* 15:489–495.

Masters, B. R., P. T. So, and E. Gratton. 1997. Multiphoton excitation fluorescence microscopy and spectroscopy of in vivo human skin. *Biophys J* 72:2405–2412.

Na, R., I. M. Stender, and H. C. Wulf. 2001. Can autofluorescence demarcate basal cell carcinoma from normal skin? A comparison with protoporphyrin IX fluorescence. *Acta Derm Venereol* 81:246–249.

Palero, J. A., A. N. Bader, H. S. de Bruijn, A. van der Ploeg-van den Heuvel, H. J. C. M. Sterenborg, and H. C. Gerritsen. 2011. In vivo monitoring of protein-bound and free NADH during ischemia by nonlinear spectral imaging microscopy. *Biomed Opt Express* 2:1030–1039.

Palero, J. A., H. S. de Bruijn, A. van der Ploeg-van den Heuvel, H. J. Sterenborg, and H. C. Gerritsen. 2006. In vivo nonlinear spectral imaging in mouse skin. *Opt Express* 14:4395–4402.

Palero, J. A., H. S. de Bruijn, A. van der Ploeg van den Heuvel, H. J. Sterenborg, and H. C. Gerritsen. 2007. Spectrally resolved multiphoton imaging of in vivo and excised mouse skin tissues. *Biophys J* 93:992–1007.

Palero, J. A., G. Latouche, H. S. de Bruijn, A. van der Ploeg van den Heuvel, H. J. Sterenborg, and H. C. Gerritsen. 2008. Design and implementation of a sensitive high-resolution nonlinear spectral imaging microscope. *J Biomed Opt* 13:044019.

Panjehpour, M., C. E. Julius, M. N. Phan, T. Vo-Dinh, and S. Overholt. 2002. Laser-induced fluorescence spectroscopy for in vivo diagnosis of non-melanoma skin cancers. *Lasers Surg Med* 31:367–373.

Paoli, J., M. Smedh, A. M. Wennberg, and M. B. Ericson. 2008. Multiphoton laser scanning microscopy on non-melanoma skin cancer: Morphologic features for future non-invasive diagnostics. *J Invest Dermatol* 128:1248–1255.

Patalay, R., C. Talbot, Y. Alexandrov et al. 2011. Quantification of cellular autofluorescence of human skin using multiphoton tomography and fluorescence lifetime imaging in two spectral detection channels. *Biomed Opt Express* 2:3295–3308.

Patalay, R., C. Talbot, Y. Alexandrov et al. 2012. Multiphoton multispectral fluorescence lifetime tomography for the evaluation of basal cell carcinomas. *PLoS One* 7:e43460.

Pena, A., M. Strupler, T. Boulesteix, and M. Schanne-Klein. 2005. Spectroscopic analysis of keratin endogenous signal for skin multiphoton microscopy. *Opt Express* 13:6268–6274.

Pitts, J. D., and M.-A. Mycek. 2001. Design and development of a rapid acquisition laser-based fluorometer with simultaneous spectral and temporal resolution. *Rev Sci Instrum* 72:3061–3072.

Rajaram, N., T. J. Aramil, K. Lee, J. S. Reichenberg, T. H. Nguyen, and J. W. Tunnell. 2010a. Design and validation of a clinical instrument for spectral diagnosis of cutaneous malignancy. *Appl Opt* 49:142–152.

Rajaram, N., J. S. Reichenberg, M. R. Migden, T. H. Nguyen, and J. W. Tunnell. 2010b. Pilot clinical study for quantitative spectral diagnosis of non-melanoma skin cancer. *Lasers Surg Med* 42:716–727.

Ramanujam, N. 2000. Fluorescence spectroscopy of neoplastic and non-neoplastic tissues. *Neoplasia* 2:89–117.

Richards-Kortum, R., and E. Sevick-Muraca. 1996. Quantitative optical spectroscopy for tissue diagnosis. *Annu Rev Phys Chem* 47:555–606.

Roberts, M. S., M. J. Roberts, T. A. Robertson et al. 2008. In vitro and in vivo imaging of xenobiotic transport in human skin and in the rat liver. *J Biophotonics* 1:478–493.

Sanchez, W. Y., T. W. Prow, W. H. Sanchez, J. E. Grice, and M. S. Roberts. 2010. Analysis of the metabolic deterioration of ex vivo skin from ischemic necrosis through the imaging of intracellular NAD(P)H by multiphoton tomography and fluorescence lifetime imaging microscopy. *J Biomed Opt* 15:046008.

Seidenari, S., F. Arginelli, S. Bassoli et al. 2013. Diagnosis of BCC by multiphoton laser tomography. *Skin Res Technol* 19:e297–304.

Seidenari, S., F. Arginelli, P. French et al. 2012. Multiphoton laser tomography and fluorescence lifetime imaging of basal cell carcinoma: Morphologic features for non-invasive diagnostics. *Exp Dermatol* 21:831–836.

Sterenborg, H. J. C. M., M. Motamedi, R. F. Wagner, Jr., M. Duvic, S. Thomsen, and S. L. Jacques. 1994. In vivo fluorescence spectroscopy and imaging of human skin tumours. *Lasers Med Sci* 9:191–201.

Stracke, F., B. Weiss, C.-M. Lehr, K. Konig, U. F. Schaefer, and M. Schneider. 2006. Multiphoton microscopy for the investigation of dermal penetration of nanoparticle-borne drugs. *J Invest Dermatol* 126:2224–2233.

Sugata, K., S. Sakai, N. Noriaki, O. Osanai, T. Kitahara, and Y. Takema. 2010. Imaging of melanin distribution using multiphoton autofluorescence decay curves. *Skin Res Technol* 16:55–59.

Talbot, C. B., R. Patalay, I. Munro et al. 2011. A multispectral FLIM tomograph for in-vivo imaging of skin cancer. In *Proc. SPIE 7903, Multiphoton Microscopy in the Biomedical Sciences XI*, CA, 79032B–79039.

Teuchner, K., J. Ehlert, W. Freyer et al. 2000. Fluorescence studies of melanin by stepwise two-photon femtosecond laser excitation. *J Fluoresc* 10:275–281.

Teuchner, K., W. Freyer, D. Leupold et al. 1999. Femtosecond two-photon excited fluorescence of melanin. *Photochem Photobiol* 70:146–151.

Thompson, A. J., S. Coda, M. B. Sørensen et al. 2012. In vivo measurements of diffuse reflectance and time-resolved autofluorescence emission spectra of basal cell carcinomas. *J Biophotonics* 5:240–254.

Wagnieres, G. A., W. M. Star, and B. C. Wilson. 1998. In vivo fluorescence spectroscopy and imaging for oncological applications. *Photochem Photobiol* 68:603–632.

Yu, B., C.-Y. Dong, P. T. C. So, D. Blankschtein, and R. Langer. 2001. In vitro visualization and quantification of oleic acid induced changes in transdermal transport using two-photon fluorescence microscopy. *J Invest Dermatol* 117:16–25.

Yu, B., K. H. Kim, P. T. C. So, D. Blankschtein, and R. Langer. 2002. Topographic heterogeneity in transdermal transport revealed by high-speed two-photon microscopy: Determination of representative skin sample sizes. *J Invest Dermatol* 118:1085–1088.

Yu, B., K. H. Kim, P. T. C. So, D. Blankschtein, and R. Langer. 2003. Visualization of oleic acid-induced transdermal diffusion pathways using two-photon fluorescence microscopy. *J Invest Dermatol* 120:448–455.

Zeng, H., and C. MacAulay. 2003. Fluorescence spectroscopy and imaging for skin cancer detection and evaluation. In *Handbook of Biomedical Fluorescence*, M.-A. Mycek, and B. Pogue, editors. Marcel Dekker, New York, 315–360.

Zhuo, S., J. Chen, T. Luo, D. Zou, and J. Zhao. 2006. Multimode nonlinear optical imaging of the dermis in ex vivo human skin based on the combination of multichannel mode and Lambda mode. *Opt Express* 14:7810–7820.

Zvyagin, A. V., X. Zhao, A. Gierden, W. Sanchez, J. A. Ross, and M. S. Roberts. 2008. Imaging of zinc oxide nanoparticle penetration in human skin in vitro and in vivo. *J Biomed Opt* 13:064031.

17

Oncology applications: Gastrointestinal cancer

Sergio Coda, Paul M. W. French, and Christopher Dunsby

Contents

17.1 INTRODUCTION

Cancers of esophagus, stomach, and colon are among the most common cancers worldwide, accounting for a total of 2.2 million new cases each year (Boyle and Levin 2008). Prevention of these conditions is currently based on early detection of early-stage cancers or premalignant conditions during conventional white-light endoscopy (WLE). Today, there is a range of more sophisticated biophotonics techniques under development that aim to enhance the contrast of areas of concern beyond what is possible with conventional WLE. Commercially available techniques include high-definition endoscopy (HDE; Adler et al. 2009; Buchner 2010; Rex and Helbig 2007), narrow band imaging (NBI; Gono et al. 2004), magnifying chromoendoscopy (MCE; Kudo et al. 1996), autofluorescence (AF) imaging (AFI; Nakaniwa et al. 2005), and confocal laser endomicroscopy (CLE; Kiesslich et al. 2004; Wang et al. 2007).

NBI has been shown to be useful in differentiating neoplastic from non-neoplastic lesions in experienced hands (Curvers et al. 2009; van den Broek et al. 2009b), but the general confidence of clinicians with this system is still to be established (Adler et al. 2008, 2009; Kaltenbach et al. 2008; Rex and Helbig 2007). The picture is even less clear for AFI as only a few studies have been performed showing mixed results (Kara et al. 2005; Matsuda et al. 2008; van den Broek et al. 2009a). CLE has shown impressive contrast between normal and diseased tissue in several studies and of a range of disease (e.g., Buchner and Wallace 2008; Kiesslich et al. 2004, 2006), but an international multicenter, randomized, controlled trial in 101 patients with Barrett's esophagus using the probe-based version of CLE showed that the specificity was lower than that of WLE alone (Sharma et al. 2011).

There is therefore increasing interest in exploiting spectroscopic readouts of tissue AF, and, as we will discuss below, a number of groups have shown that gastrointestinal (GI) cancers exhibit different autofluorescent signals from those of normal tissue. It remains to be seen whether AF and its associated spectroscopic properties, including fluorescence lifetime, will be useful in clinical practice, and further *in vivo* research studies using these techniques are urgently needed. The following sections aim to review by organ type some representative *ex vivo* and *in vivo* studies using AF spectroscopy and imaging in GI dysplasia and cancer. As there are relatively few reports of AF lifetime (AFL) studies to date, we include spectrally resolved studies, the results of which should help with the design of future AFL studies.

17.2 ESOPHAGUS

A number of important papers published in the 1990s presented AF emission spectra that were collected *in vivo* from Barrett's esophagus, malignant and normal tissue at 410 nm excitation (Panjehpour et al. 1995, 1996; Vo-Dinh et al. 1995, 1998). These results showed that fluorescence of normal esophageal mucosa and nondysplastic Barrett's epithelium could be distinguished from that of dysplasia and cancer in patients with Barrett's esophagus.

Mayinger et al. (2000) used an innovative optical system delivering either white or violet-blue light for excitation of tissue AF during routine endoscopy. Nine patients displayed a spinocellular carcinoma and four had an adenocarcinoma of the esophagus. The result of the study was that both types of cancer exhibited different emission spectra compared to normal esophagus.

The combined use of three different spectroscopic techniques (reflectance, fluorescence, and light-scattering spectroscopy) has been shown to provide useful complementary information in patients with dysplastic Barrett's esophagus (Georgakoudi et al. 2001). The fluorescence spectra of NADH and collagen in dysplastic tissue appeared considerably different from nondysplastic mucosa (Georgakoudi et al. 2002); dysplastic epithelium was characterized by low collagen and high NADH fluorescence intensity compared to normal esophagus.

Further work to investigate the differences in the metabolic pathway between normal tissue and cancer was made using FLIM to measure intracellular oxygen concentration with a ruthenium-based fluorescent probe (RTDP) in cultured normal and adenocarcinoma cells (from Barrett's epithelium) of human esophagus (Sud et al. 2006). Higher intracellular levels of oxygen and NADH fluorescence intensity were sensed in the neoplastic cell line studied.

The clinical diagnosis of adenocarcinoma in patients with short-segment Barrett's esophagus, which is defined as intestinal metaplasia occurring in columnar epithelium <3 cm in length, is often challenging and controversial. Niepsuj et al. (2003) used a monochromatic blue light (425–455 nm) in 34 patients with documented short-segment Barrett's esophagus. In their study, AFI increased the detection rate of high-grade dysplasia when compared with white light endoscopy.

Lin et al. (2009) studied *ex vivo* biopsy specimens from 30 patients with Barrett's esophagus using high-resolution AFI under 266, 355, and 408 nm excitation along with an image acquired using the same setup under white light illumination. Distinct cellular organization and morphology related to disease progression could be identified, and the pathological evaluation was confirmed by expert pathologists.

AFI has been applied in combination with HDE and NBI in the so-called "trimodal imaging" to identify inconspicuous Barrett's neoplasia (Curvers et al. 2008) and assist endoscopic mucosal resection (EMR) of early neoplasia in Barrett's esophagus (Thomas et al. 2009). In 84 patients with Barrett's esophagus, AFI plus HDE increased the detection rate of early neoplasia within the Barrett segment, and additional viewing with NBI increased the overall specificity (Curvers et al. 2008).

17.3 STOMACH

Chwirot et al. (1997) reported the first study on AFI of gastric cancers *ex vivo*. In particular, they imaged 21 surgically resected specimens using excitation at 325 and 442 nm and measured the fluorescence emission in six regions of the visible spectrum. In all the spectral channels, the AF intensity of neoplastic tissue was found to be lower than that of areas of normal stomach in the same specimen.

Immediately after surgery, Abe et al. (2000) applied light-induced fluorescence endoscopy (LIFE), which employs endoscopic AFI with excitation at 442 nm and emission detected in two bands around 520 and >630 nm, to 61 gastric cancers resected from 50 patients. Comparison of these data with the histopathological diagnosis showed that depth of infiltration and mucosal thickening were correlated with the diagnosis of a differentiated cancer but not in the case of undifferentiated cancer, which displayed a scattered pattern of tumor infiltration without altering the mucosal thickness.

Ohkawa et al. (2004) found that LIFE applied to 79 patients with gastric cancer provided high sensitivity but poor specificity with limited clinical utility. Also, only a small degree of correlation was observed between histopathology and AF findings.

Ito et al. (2001) stained and imaged biopsy specimens of freshly resected stomachs from three patients with gastric cancer using an anticarcinoembryonic antigen antibody coupled with a fluorescent label and an infrared fluorescence endoscope. They found that there was a good correlation between the infrared fluorescent images with the cancerous sites, suggesting that this technique can spot cancer cells and generate a relatively strong fluorescent signal to detect small cancers.

Xiao et al. (2002a, b) acquired AF images of gastric cancer using a laser scanning confocal microscope with two detection channels in green (505–530 nm) and red (>580 nm). The two spectrally resolved image channels were then merged and presented to the clinician. Sixteen gastric cancer specimens and corresponding normal gastric tissue were imaged using both 488 and 543 nm excitation wavelengths. Normal stomach appeared green, whereas a red-brown image was found to be characteristic of all neoplastic specimens.

Mayinger et al. (2004) appraised light-induced AF spectroscopy for *in vivo* diagnosis of gastric cancer studying normal and cancerous gastric mucosa in 15 patients with adenocarcinoma and 16 patients with signet-ring cell gastric carcinoma. A system capable of exciting tissue AF alternately with white or violet-blue light was used. Endogenous fluorescence spectra emitted by the tissue were collected with a fiber-optic probe and analyzed with a spectrograph. Both types of cancers emitted a less intense signal compared to normal tissue (2300 vs. 3700 a.u. approximately), with a lower total intensity between 480 and 570 nm and a higher red-to-green ratio than normal gastric mucosa. The sensitivity of fluorescence spectroscopy for the diagnosis of carcinoma decreased by 30% with the presence of signet-ring cells, which may be explained by the scattered and frequently submucosal infiltration of this subtype of adenocarcinoma (diffuse type according to the Lauren classification; Lauren 1965).

Silveira et al. (2008) used fluorescence spectroscopy with excitation at 488 nm to detect benign and malignant lesions of *ex vivo* human gastric mucosa. Endoscopic biopsies of antral gastritis and cancer were collected from 35 patients undergoing endoscopy. AF spectra were collected from two random areas for each sample. Analysis of the fluorescence spectra was able to discriminate neoplastic from normal tissue with 100% of sensibility and specificity.

Kim et al. (2008) performed preoperative AFI in 20 patients with early gastric cancers and then compared the endoscopic characteristics with the histology after endoscopic submucosal dissection, which is a new and more radical treatment for early GI cancers providing high rates of "en bloc" resection compared with EMR (Coda et al. 2007). Categorization of AFI images of gastric cancers into four patterns proved to be useful to delineate the dissection margins in the majority of cases.

17.4 LIVER AND PANCREAS

Izuishi et al. (1999b) published the first pilot study using AF endoscopy in cancer of the biliary tract. They performed percutaneous transhepatic AF cholangioscopy in nine patients with bile duct cancer using LIFE. Normal mucosa appeared as light blue, whereas cancerous legions displayed a dark red fluorescence and also a white fluorescence in the majority of cases suggesting an additional contrast parameter for the diagnosis of bile duct cancer by AF endoscopy.

Chandra et al. (2007) observed significant differences between the fluorescence and reflectance properties of normal, pancreatitis, and adenocarcinoma tissues. Measurements were associated with a relatively larger collagen content detected in adenocarcinoma and pancreatitis than in normal tissue. Reflectance data indicated that adenocarcinoma had higher reflectance in the 430 to 500 nm range compared to normal and only inflamed tissues. Subsequently (Chandra et al. 2010), they used a fiber-optic probe-based spectroscope to assess the diagnostic accuracy in 50 freshly resected specimens of pancreatic adenocarcinoma and chronic pancreatitis. They developed a classification algorithm—the spectral areas and ratios classifier—which they showed could discriminate between healthy and diseased tissue and identify cancer with reasonable sensitivity and specificity.

Liang et al. (2009) evaluated the ability of two-photon microscopy (TPM) to detect the number of HCV-infected cells in frozen sections of liver biopsy specimens from nine patients with chronic HCV-related hepatitis. TPM was able to determine the number of infected hepatocytes with considerably high sensitivity than that achieved by biochemical detection of HCV proteins or viral RNA.

Preliminary data in *ex vivo* tissue from our group have shown a statistically significant AF lifetime contrast between cancerous and healthy tissue in unfixed liver containing metastatic colorectal carcinoma and unfixed pancreas containing an area of pancreatic adenocarcinoma using wide-field FLIM. The lifetime of neoplastic tissue AF induced by 355 nm UV excitation was shorter than AF from surrounding non-neoplastic tissue (McGinty et al. 2010).

In a recent study conducted on freshly excised GI tissue from 10 normal and 5 transgenic mice, excitation spectral signatures of epithelium, lamina propria, collagen, and lymphatic tissue were determined using hyperspectral TPM with the excitation laser tuned from 710 to 920 nm in 5 nm increments and using three emission channels (350–505, 505–560, and 560–650 nm). Based on the optically sectioned two-photon images acquired, the morphology of these four main tissue types was clearly visualized and compared to cross-sectional histology with a high degree of concordance (Grosberg et al. 2011).

17.5 COLON

A large body of literature exists on the use of AF techniques to investigate colonic polyps and adenocarcinomas, which is discussed in more detail below.

The fluorescent architecture of normal colon, colonic adenoma, and adenocarcinoma was studied by fluorescence microscopy in frozen tissue sections with excitation in the range 351 to 364 nm (Romer et al. 1995; Wang et al. 1999; Zonios et al. 1996). Collagen, present in all the layers of the GI wall, was responsible for the blue fluorescence emitted from normal tissues and adenomas, whereas a yellow-amber fluorescence was mainly attributed to the lamina propria. Interestingly, adenomas displayed cytoplasmic blue-green fluorescence increasing in intensity with the degree of epithelial dysplasia. This cytoplasmic fluorescence was not emitted from normal mucosal cells. The lamina propria of adenomas exhibited less fluorescence from collagen and was richer in blue-green fluorescent eosinophilic granules than that of normal colon (Romer et al. 1995). Dysplastic adenomas displayed higher fluorescence intensity than normal tissue and lower fluorescence intensity than adenocarcinomas (Wang et al. 1999). In summary, these papers found that colonic neoplasms have decreased AF intensity compared to that of normal colon. This was attributed to the decrease with neoplasia of mucosal collagen, which is the dominant fluorophore for UV excitation wavelengths, as a consequence of the enlargement of crypts that progressively displace the lamina propria. In addition, the submucosal contribution to the fluorescence in adenomatous tissue is reduced compared to that in normal colon due to the increased thickness of the polyp and absorption by hemoglobin, as a result of increased intratumoral microvessel density.

Conversely, Fiarman et al. (1995), using confocal microscopy in frozen tissue sections with a longer excitation wavelength of 488 nm, observed that the fluorescence from normal colonic mucosa arises primarily from the lamina propria, whereas the epithelium is the main source of fluorescence in adenomas and hyperplastic polyps. In normal tissue, the epithelium/lamina propria fluorescence intensity ratio was significantly lower than that of adenomas and hyperplastic polyps, and no statistically significant difference was seen between the ratios of both types of polyps.

Izuishi et al. (1999a) did not observe any significant differences between the fluorescence intensities of frozen sections of normal, adenomatous, and cancerous tissue obtained using 400–440 nm excitation and 520 nm emission in a wide-field fluorescence microscope. Likewise, the concentration of flavins measured by high-performance liquid chromatography was not significantly different between normal and diseased tissue. But interestingly, the invasion of the submucosal layer was associated with a decreased fluorescence emission, in agreement with the hypothesis that collagen is the main source of colon AF. The screening effect of mucosal thickening and replacement of submucosa by cancer cells was adduced as a reason for the decreased fluorescence observed in cancerous tissue.

Bottiroli et al. (1995) identified different patterns of fluorescence intensity and spectra of AF emission between neoplastic and non-neoplastic tissue using a microspectrofluorometer with 366 nm excitation. The different patterns were related to distinct layers of the GI wall, and in the 480–580 nm band in particular, the spectra of neoplastic epithelia displayed higher intensity than those of normal colon epithelia, suggesting the role of different arrangements in the neoplastic stromal compartment as a possible explanation for the differences observed.

Kapadia et al. (1990) measured the emission spectra obtained with 325 nm excitation from 35 normal samples and 35 adenomatous polyps *ex vivo*. Retrospectively, they developed a score to differentiate adenomatous changes from normal and hyperplastic tissue. Then, spectra from additional normal ($n = 34$), adenomatous ($n = 16$), and hyperplastic ($n = 16$) specimens were prospectively classified with this algorithm with 100% sensitivity and 98% specificity. A similar study was conducted *in vivo* by Cothren et al. (1990) at 370 nm excitation from 31 adenomas, 4 hyperplastic polyps, and 32 normal sites of the colon. Following the sequence normal to hyperplasia to adenoma, they identified a characteristic decrease in the maximum fluorescence intensity emitted at 460 nm and a relative increase in the fluorescence intensity at 680 nm. Using these spectral features, each tissue state was differentiated with high sensitivity and specificity. Later on, they also used an algorithm that was able to detect dysplasia *in vivo* during endoscopy with 90% sensitivity and 95% specificity (Cothren et al. 1996).

Ex vivo fluorescence emission spectra from normal colon and adenomatous polyps were also obtained at a variety of excitation wavelengths (Richards-Kortum et al. 1991). Changes in the spectral line shapes between the two tissue types for diagnosing the presence of adenoma were most evident at 330, 370, and 430 nm excitation.

In two consecutive studies by Schomacker et al. (1992a, b), emission spectra from normal sites ($n = 86$), hyperplastic ($n = 35$), adenomatous polyps ($n = 49$), and adenocarcinomas ($n = 7$) were collected both *in vivo* and *ex vivo* at 337 nm excitation. The spectra of neoplastic and non-neoplastic tissue were differentiated with a diagnostic accuracy of 87%, which is about the same as that of routine clinical pathology. Interestingly, the spectra emitted from *ex vivo* and *in vivo* tissues were also different, and this change was attributed primarily to NADH, whose fluorescence was shown to decay with a half-life of approximately 100 min after resection.

Yang et al. (1995) excited fresh biopsies of normal ($n = 39$) and malignant colonic tissue ($n = 35$) using a range of excitation (290–340 nm) and emission (360–450 nm) wavelengths. The ratios of 325/360 to 380 and 290 to 340/450 were found to be most specific and sensitive for distinguishing normal from neoplastic tissue.

Mycek et al. (1998) performed time-resolved AF spectroscopy *in vivo* on 24 polyps (13 adenomas, 11 non-adenomas) from 17 patients using excitation at 337 nm and detection in the range 530–570 nm. This was the first study in the colon that demonstrated differences in fluorescence lifetime decay between normal and dysplastic tissue. Adenomas showed faster average decays than those of non-adenomas (9.3 ± 0.4 ns vs. 10.5 ± 0.7 ns).

Chwirot et al. (1999) observed that the intensity of AF emission at 325 nm excitation from 50 surgically resected specimens of neoplastic and premalignant conditions was lower than that of contiguous normal colon.

Distinctive fluorescence emission spectra and Raman spectra have also been shown *in vivo* from 32 adenomas and 114 hyperplastic polyps at 337 nm excitation (Eker et al. 1999), and from 10 adenomas and 9 hyperplastic polyps at 785 nm, respectively (Molckovsky et al. 2003).

Compared to normal colon, longer AF lifetimes were found in cancerous colonic tissues at 397 nm excitation and 635 nm emission (4.32 vs. 18.45 ns), and this was attributed to the contribution of abnormal accumulation in neoplasia of PpIX with known long lifetime (Li et al. 2006).

Recently, our group at Imperial College London has shown a statistically significant AF lifetime contrast between cancerous and healthy colon tissue in a series of 18 unstained *ex vivo* colonic tumors surgically resected, using wide-field FLIM. The lifetime of neoplastic tissue AF induced by 355 nm UV excitation was longer than AF from surrounding non-neoplastic tissue (McGinty et al. 2010).

17.6 SUMMARY

We have reviewed the use of endogenous fluorescence for the detection and diagnosis of disease in the GI tract. This has included a wide range of methods including point-based spectroscopic techniques, wide-field techniques, and optically sectioning imaging techniques. It is clear from the literature that there are changes in both the relative contributions of different fluorescence species with disease and also changes in the fluorescence properties of individual species. Further *in vivo* clinical studies are required to establish the optimum modalities and parameters for spectroscopic contrast (e.g., diffuse reflectance spectroscopy, AFI, etc., excitation and emission wavelength(s) of AF and fluorescence lifetime) for different types of tissue and disease, and this must be coupled with further developments in the instrumentation necessary for such studies.

REFERENCES

Abe, S., K. Izuishi, H. Tajiri, T. Kinoshita, and T. Matsuoka. 2000. Correlation of in vitro autofluorescence endoscopy images with histopathologic findings in stomach cancer. *Endoscopy* 32:281–286.

Adler, A., J. Aschenbeck, T. Yenerim et al. 2009. Narrow-band versus white-light high definition television endoscopic imaging for screening colonoscopy: A prospective randomized trial. *Gastroenterology* 136:410–416 e411; quiz 715.

Adler, A., H. Pohl, I. S. Papanikolaou et al. 2008. A prospective randomised study on narrow-band imaging versus conventional colonoscopy for adenoma detection: Does narrow-band imaging induce a learning effect? *Gut* 57:59–64.

Bottiroli, G., A. C. Croce, D. Locatelli et al. 1995. Natural fluorescence of normal and neoplastic human colon: A comprehensive "ex vivo" study. *Lasers Surg Med* 16:48–60.

Boyle, P., and B. Levin. 2008. *World Cancer Report 2008*. Geneva: World Health Organization.

Buchner, A. M. 2010. High-definition colonoscopy detects colorectal polyps at a higher rate than standard white-light colonoscopy. *Clin Gastroenterol Hepatol* 8:364–370.

Buchner, A. M., and M. B. Wallace. 2008. Future expectations in digestive endoscopy: Competition with other novel imaging techniques. *Best Pract Res Clin Gastroenterol* 22:971–987.

Chandra, M., J. Scheiman, D. Heidt, D. Simeone, B. McKenna, and M. A. Mycek. 2007. Probing pancreatic disease using tissue optical spectroscopy. *J Biomed Opt* 12:060501.

Chandra, M., J. Scheiman, D. Simeone, B. McKenna, J. Purdy, and M. A. Mycek. 2010. Spectral areas and ratios classifier algorithm for pancreatic tissue classification using optical spectroscopy. *J Biomed Opt* 15:010514.

Chwirot, B. W., S. Chwirot, W. Jedrzejczyk et al. 1997. Ultraviolet laser-induced fluorescence of human stomach tissues: Detection of cancer tissues by imaging techniques. *Lasers Surg Med* 21:149–158.

Chwirot, B. W., Z. Michniewicz, M. Kowalska, and J. Nussbeutel. 1999. Detection of colonic malignant lesions by digital imaging of UV laser-induced autofluorescence. *Photochem Photobiol* 69:336–340.

Coda, S., S. Y. Lee, and T. Gotoda. 2007. Endoscopic mucosal resection and endoscopic submucosal dissection as treatments for early gastrointestinal cancers in Western countries. *Gut Liver* 1:12–21.

Cothren, R., R. Richards-Kortum, M. Sivak et al. 1990. Gastrointestinal tissue diagnosis by laser-induced fluorescence spectroscopy at endoscopy. *Gastrointest Endosc* 36:105–111.

Cothren, R., M. Sivak, J. Van Dam et al. 1996. Detection of dysplasia at colonoscopy using laser-induced fluorescence: A blinded study. *Gastrointest Endosc* 44:168–176.

Curvers, W. L., R. Singh, L. M. Song et al. 2008. Endoscopic tri-modal imaging for detection of early neoplasia in Barrett's oesophagus: A multi-centre feasibility study using high-resolution endoscopy, autofluorescence imaging and narrow band imaging incorporated in one endoscopy system. *Gut* 57:167–172.

Curvers, W. L., F. J. van den Broek, J. B. Reitsma, E. Dekker, and J. J. Bergman. 2009. Systematic review of narrow-band imaging for the detection and differentiation of abnormalities in the esophagus and stomach (with video). *Gastrointest Endosc* 69:307–317.

Eker, C., S. Montan, E. Jaramillo et al. 1999. Clinical spectral characterisation of colonic mucosal lesions using autofluorescence and delta aminolevulinic acid sensitisation. *Gut* 44:511–518.

Fiarman, G. S., M. H. Nathanson, A. B. West, L. I. Deckelbaum, L. Kelly, and C. R. Kapadia. 1995. Differences in laser-induced autofluorescence between adenomatous and hyperplastic polyps and normal colonic mucosa by confocal microscopy. *Dig Dis Sci* 40:1261–1268.

Georgakoudi, I., B. C. Jacobson, M. G. Muller et al. 2002. NAD(P)H and collagen as in vivo quantitative fluorescent biomarkers of epithelial precancerous changes. *Cancer Res* 62:682–687.

Georgakoudi, I., B. C. Jacobson, J. Van Dam et al. 2001. Fluorescence, reflectance, and light-scattering spectroscopy for evaluating dysplasia in patients with Barrett's esophagus. *Gastroenterology* 120:1620–1629.

Gono, K., T. Obi, M. Yamaguchi et al. 2004. Appearance of enhanced tissue features in narrow-band endoscopic imaging. *J Biomed Opt* 9:568–577.

Grosberg, L. E., A. J. Radosevich, S. Asfaha, T. C. Wang, and E. M. C. Hillman. 2011. Spectral characterization and unmixing of intrinsic contrast in intact normal and diseased gastric tissues using hyperspectral two-photon microscopy. *PLoS One* 6:e19925.

Ito, S., N. Muguruma, Y. Kusaka et al. 2001. Detection of human gastric cancer in resected specimens using a novel infrared fluorescent anti-human carcinoembryonic antigen antibody with an infrared fluorescence endoscope in vitro. *Endoscopy* 33:849–853.

Izuishi, K., H. Tajiri, T. Fujii et al. 1999a. The histological basis of detection of adenoma and cancer in the colon by autofluorescence endoscopic imaging. *Endoscopy* 31:511–516.

Izuishi, K., H. Tajiri, M. Ryu et al. 1999b. Detection of bile duct cancer by autofluorescence cholangioscopy: A pilot study. *Hepatogastroenterology* 46:804–807.

Kaltenbach, T., S. Friedland, and R. Soetikno. 2008. A randomised tandem colonoscopy trial of narrow band imaging versus white light examination to compare neoplasia miss rates. *Gut* 57:1406–1412.

Kapadia, C. R., F. W. Cutruzzola, K. M. O'Brien, M. L. Stetz, R. Enriquez, and L. I. Deckelbaum. 1990. Laser-induced fluorescence spectroscopy of human colonic mucosa. Detection of adenomatous transformation. *Gastroenterology* 99:150–157.

Kara, M. A., M. E. Smits, W. D. Rosmolen et al. 2005. A randomized crossover study comparing light-induced fluorescence endoscopy with standard video endoscopy for the detection of early neoplasia in Barrett's esophagus. *Gastrointest Endosc* 61:671–678.

Kiesslich, R., J. Burg, M. Vieth et al. 2004. Confocal laser endoscopy for diagnosing intraepithelial neoplasias and colorectal cancer in vivo. *Gastroenterology* 127:706–713.

Kiesslich, R., L. Gossner, M. Goetz et al. 2006. In vivo histology of Barrett's esophagus and associated neoplasia by confocal laser endomicroscopy. *Clin Gastroenterol Hepatol* 4:979–987.

Kim, W. J., J. Y. Cho, S. W. Jeong et al. 2008. Comparison of autofluorescence imaging endoscopic findings with pathologic findings after endoscopic submucosal dissection of gastric neoplasms. *Gut Liver* 2:186–192.

Kudo, S., S. Tamura, T. Nakajima, H. Yamano, H. Kusaka, and H. Watanabe. 1996. Diagnosis of colorectal tumorous lesions by magnifying endoscopy. *Gastrointest Endosc* 44:8–14.

Lauren, P. 1965. The two histological main types of gastric carcinoma: Diffuse and so-called intestinal-type carcinoma. An attempt at a histo-clinical classification. *Acta Pathol Microbiol Scand* 64:31–49.

Li, B., Z. Zhang, and S. Xie. 2006. Steady state and time-resolved autofluorescence studies of human colonic tissues. *Chin Opt Lett* 4:348–350.

Liang, Y., T. Shilagard, S. Y. Xiao et al. 2009. Visualizing hepatitis C virus infections in human liver by two-photon microscopy. *Gastroenterology* 137:1448–1458.

Lin, B., S. Urayama, R. M. Saroufeem, D. L. Matthews, and S. G. Demos. 2009. Real-time microscopic imaging of esophageal epithelial disease with autofluorescence under ultraviolet excitation. *Opt Express* 17:12502–12509.

Matsuda, T., Y. Saito, K. I. Fu et al. 2008. Does autofluorescence imaging video endoscopy system improve the colonoscopic polyp detection rate? A pilot study. *Am J Gastroenterol* 103:1926–1932.

Mayinger, B., P. Horner, M. Jordan et al. 2000. Light-induced autofluorescence spectroscopy for tissue diagnosis of GI lesions. *Gastrointest Endosc* 52:395–400.

Mayinger, B., M. Jordan, T. Horbach et al. 2004. Evaluation of in vivo endoscopic autofluorescence spectroscopy in gastric cancer. *Gastrointest Endosc* 59:191–198.

McGinty, J., N. P. Galletly, C. Dunsby et al. 2010. Wide-field fluorescence lifetime imaging of cancer. *Biomed Opt Express* 1:627–640.

Molckovsky, A., L. M. Song, M. G. Shim, N. E. Marcon, and B. C. Wilson. 2003. Diagnostic potential of near-infrared Raman spectroscopy in the colon: Differentiating adenomatous from hyperplastic polyps. *Gastrointest Endosc* 57:396–402.

Mycek, M. A., K. T. Schomacker, and N. S. Nishioka. 1998. Colonic polyp differentiation using time-resolved autofluorescence spectroscopy. *Gastrointest Endosc* 48:390–394.

Nakaniwa, N., A. Namihisa, T. Ogihara et al. 2005. Newly developed autofluorescence imaging videoscope system for the detection of colonic neoplasms. *Dig Endosc* 17:235–240.

Niepsuj, K., G. Niepsuj, W. Cebula et al. 2003. Autofluorescence endoscopy for detection of high-grade dysplasia in short-segment Barrett's esophagus. *Gastrointest Endosc* 58:715–719.

Ohkawa, A., H. Miwa, A. Namihisa et al. 2004. Diagnostic performance of light-induced fluorescence endoscopy for gastric neoplasms. *Endoscopy* 36:515–521.

Panjehpour, M., B. F. Overholt, J. L. Schmidhammer, C. Farris, P. F. Buckley, and T. Vo-Dinh. 1995. Spectroscopic diagnosis of esophageal cancer: New classification model, improved measurement system. *Gastrointest Endosc* 41:577–581.

Panjehpour, M., B. F. Overholt, T. Vo-Dinh, R. C. Haggitt, D. H. Edwards, and F. P. Buckley. 1996. Endoscopic fluorescence detection of high-grade dysplasia in Barrett's esophagus. *Gastroenterology* 111:93–101.

Rex, D. K., and C. C. Helbig. 2007. High yields of small and flat adenomas with high-definition colonoscopes using either white light or narrow band imaging. *Gastroenterology* 133:42–47.

Richards-Kortum, R., R. P. Rava, R. E. Petras, M. Fitzmaurice, M. Sivak, and M. S. Feld. 1991. Spectroscopic diagnosis of colonic dysplasia. *Photochem Photobiol* 53:777–786.

Romer, T. J., M. Fitzmaurice, R. M. Cothren et al. 1995. Laser-induced fluorescence microscopy of normal colon and dysplasia in colonic adenomas: Implications for spectroscopic diagnosis. *Am J Gastroenterol* 90:81–87.

Schomacker, K., J. Frisoli, C. Compton et al. 1992a. Ultraviolet laser-induced fluorescence of colonic polyps. *Gastroenterology* 102:1155–1160.

Schomacker, K. T., J. K. Frisoli, C. C. Compton et al. 1992b. Ultraviolet laser-induced fluorescence of colonic tissue: Basic biology and diagnostic potential. *Lasers Surg Med* 12:63–78.

Sharma, P., A. R. Meining, E. Coron et al. 2011. Real-time increased detection of neoplastic tissue in Barrett's esophagus with probe-based confocal laser endomicroscopy: Final results of an international multicenter, prospective, randomized, controlled trial. *Gastrointest Endosc* 74:465–472.

Silveira, L., Jr., J. A. Betiol Filho, F. L. Silveira, R. A. Zangaro, and M. T. Pacheco. 2008. Laser-induced fluorescence at 488 nm excitation for detecting benign and malignant lesions in stomach mucosa. *J Fluoresc* 18:35–40.

Sud, D., W. Zhong, D. G. Beer, and M.-A. Mycek. 2006. Time-resolved optical imaging provides a molecular snapshot of altered metabolic function in living human cancer cell models. *Opt Express* 14:4412–4426.

Thomas, T., R. Singh, and K. Ragunath. 2009. Trimodal imaging-assisted endoscopic mucosal resection of early Barrett's neoplasia. *Surg Endosc* 23:1609–1613.

van den Broek, F. J. C., P. Fockens, S. van Eeden et al. 2009a. Clinical evaluation of endoscopic trimodal imaging for the detection and differentiation of colonic polyps. *Clin Gastroenterol Hepatol* 7:288–295.

van den Broek, F. J., J. B. Reitsma, W. L. Curvers, P. Fockens, and E. Dekker. 2009b. Systematic review of narrow-band imaging for the detection and differentiation of neoplastic and nonneoplastic lesions in the colon (with videos). *Gastrointest Endosc* 69:124–135.

Vo-Dinh, T., M. Panjehpour, and B. F. Overholt. 1998. Laser-induced fluorescence for esophageal cancer and dysplasia diagnosis. *Ann N Y Acad Sci* 838:116–122.

Vo-Dinh, T., M. Panjehpour, B. F. Overholt, C. Farris, F. P. Buckley, 3rd, and R. Sneed. 1995. In vivo cancer diagnosis of the esophagus using differential normalized fluorescence (DNF) indices. *Lasers Surg Med* 16:41–47.

Wang, H. W., J. Willis, M. I. F. Canto, M. V. Sivak, and J. A. Izatt. 1999. Quantitative laser scanning confocal autofluorescence microscopy of normal, premalignant, and malignant colonic tissues. *IEEE Trans Biomed Eng* 46:1246–1252.

Wang, T. D., S. Friedland, P. Sahbaie et al. 2007. Functional imaging of colonic mucosa with a fibered confocal microscope for real-time in vivo pathology. *Clin Gastroenterol Hepatol* 5:1300–1305.

Xiao, S. D., Z. Z. Ge, L. Zhong, and H. Y. Luo. 2002a. Diagnosis of gastric cancer by using autofluorescence spectroscopy. *Chin Dig Dis* 3:99–102.

Xiao, S. D., L. Zhong, H. Y. Luo, X. Y. Chen, and Y. Shi. 2002b. Autofluorescence imaging analysis of gastric cancer. *Chin Dig Dis* 3:95–98.

Yang, Y., G. C. Tang, M. Bessler, and R. R. Alfano. 1995. Fluorescence spectroscopy as a photonic pathology method for detecting colon cancer. *Lasers Life Sci* 6:259–276.

Zonios, G. I., R. M. Cothren, J. T. Arendt et al. 1996. Morphological model of human colon tissue fluorescence. *IEEE Trans Biomed Eng* 43:113–122.

18 Oncology applications: Intraoperative diagnosis of head and neck carcinoma

D. Gregory Farwell and Laura Marcu

Contents

18.1 INTRODUCTION

Head and neck squamous cell carcinoma (HNSCC) is the sixth most common cancer worldwide (Howlader et al. 2011; Upile et al. 2009). The estimated new cases of oral cancer exceeded 40,000 in 2012 in the United States alone (NCI 2012). The known risk factors include chronic topical carcinogen exposure from tobacco, alcohol, and betel quid. Since these carcinogens are typically widely exposed to the tissues of the upper aerodigestive tract through smoke inhalation or alcohol ingestion, patients are predisposed to field cancerization and the development of new malignancies adjacent or remote to the initial tumor. A more recent association has been made in the oropharynx with remote high-risk HPV infection and DNA integration of viral oncogenes.

While more recent data would suggest HPV associated oropharynx cancer is associated with an improved prognosis, the reported five-year survival rate of patients diagnosed with HNSCC is only 50%–60% (Howlader et al. 2011). A majority of these patients typically present with advanced disease in the clinic (Gourin et al. 2005; Rahman et al. 2008). This is despite the fact that oral and pharyngeal tumors are often easily identifiable within the mouth and throat by direct visualization or simple awake endoscopy (Gourin et al. 2005). This delay in diagnosis may be attributed to several factors such as relative paucity of symptoms until the tumors are advanced. Other reasons for a delay in diagnosis include access to health care (Arbes et al. 1999; Gourin et al. 2005), genetic susceptibility (Day et al. 1993), neglect on the part of the patient (Peacock et al. 2008), and difficulty differentiating from other malignancies (Bagan et al. 2010) as well as benign lesions (Cabot et al. 2010; de Souza Azevedo et al. 2008). Additional challenges in improving disease outcomes include tumor recurrence at the local and regional level, distant metastasis, and second primary tumors. Being able to accurately detect and diagnose HNSCC early is therefore an important clinical goal.

In patients affected by HNSCC, the standard clinical work-up involves a diagnostic assessment of the extent of the tumor through preoperative physical examination, endoscopy, and radiologic evaluation with x-ray, computed tomography (CT), magnetic resonance imaging (MRI), or positron emission tomography (PET)/CT scans. Subsequently, definitive treatment includes complete surgical excision oftentimes with postoperative radiation with or without chemotherapy (Shah and Gil 2009). A combination of chemotherapy and radiation therapy has also been recently utilized in a neoadjuvant setting for organ conservation (Hainsworth et al. 2009). Despite these advances, current treatment approaches often result in significant morbidity and impairment of speech, swallowing, taste, and facial appearance (Carvalho et al. 2005). Consequently, there is an urgent clinical need for techniques that allow for early diagnosis, optimal staging and restaging, and monitoring response following therapy in HNSCC.

18.2 ORIGIN OF FLUORESCENCE CONTRAST IN HEAD AND NECK TUMORS

The development of oral cancer results in changes in tissue morphology and metabolism that affect the inherent optical properties of the tissue including the fluorescence. The main group of endogenous fluorophores related to HNSCC includes the enzyme cofactors involved in cellular metabolism: reduced nicotinamide adenine dinucleotide (NADH) and flavin adenine dinucleotide (FAD). Both of them are found in the epithelial layer of the tissue. Another major class of fluorophores concerns the structural proteins, in particular, collagen and collagen cross-links present in the stroma. Cancerous changes alter the tissue structure and the concentration of these fluorophores. Typically, such changes result in loss of fluorescence. It was shown that stromal fluorescence signal decreases in dysplasia and cancer due to thickening of the epithelium and a decrease in the fluorescence of stromal collagen (Pavlova et al. 2008a). It was also suggested that the reduction in stromal fluorescence in the presence of inflammation is linked to the displacement of structural fibers by the infiltrating lymphocytes, which are less fluorescent. Inflammation also promotes the expression of matrix-degrading proteases leading to the breakdown of collagen cross-links (Mignogna et al. 2004). Porphyrins were also identified as the fluorescent substance in the HNSCC (de Veld et al. 2005). Moreover, the elution patterns on high-performance liquid chromatography (HPLC) revealed some porphyrin compounds (protoporphyrin IX, coproporphyrin, and related components) as specific to oral cancer (Onizawa et al. 2003). These studies suggest that autofluorescence in oral carcinoma correlates well with carcinogenesis, and that fluorescent components such as protoporphyrin are produced in association with the progression to cancer. Moreover, the autofluorescence properties of oral tissue vary based on the anatomic site within the oral cavity and the pathologic diagnosis (de Veld et al. 2003, 2005).

18.3 FLUORESCENCE TECHNIQUES FOR DIAGNOSIS OF HEAD AND NECK TUMORS

Autofluorescence contrast as a means of diagnosis of HNSCC including the cancer of the oral cavity was broadly evaluated using a variety of spectroscopy and imaging techniques (Chaturvedi et al. 2010; Delank et al. 2000; Heintzelman et al. 2000; Lane et al. 2006; Muller et al. 2003; Pavlova et al. 2008a, b, 2009; Roblyer et al. 2008, 2009; Schwarz et al. 2008, 2009; Vedeswari et al. 2009). The direct accessibility of these tumor sites enabled the development of both point spectroscopy and wide-field imaging devices that can access the area of interest via flexible (fiber optic), semiflexible (light guide), or rigid (direct illumination) light guides (Marcu 2012). The availability of the commercial device VELscope (LED Dental, Inc; VELscope Homepage 2012) has also enabled clinicians to directly visualize fluorescence in oral cavity. However, most devices tested in patients were based on steady-state spectroscopy or intensity-based imaging. In order to improve the specificity of the fluorescence measurements, the clinical devices typically have incorporated complementary diagnostic techniques. This includes diffuse reflectance spectroscopy (Muller et al. 2003; Schwarz et al. 2009) and light scattering spectroscopy (Muller et al. 2003), depth-sensitive probes (Schwarz et al. 2008, 2009), advanced computational models of signal calibration, and digital imaging processing (Lane et al. 2006). Only a few reports (Chen et al. 2005; Meier et al. 2010) describe the use of TRFS point spectroscopy or fluorescence lifetime imaging microscopy (FLIM) systems for characterization and diagnosis of oral carcinoma in preclinical or clinical settings in patients. These studies are presented below.

18.4 PRECLINICAL STUDIES OF ORAL CARCINOMA USING FLUORESCENCE LIFETIME TECHNIQUES

Preclinical *in vivo* studies of oral carcinoma have been carried out primarily in a hamster buccal pouch carcinogenesis model. This model was shown to provide a means to evaluate the ability of fluorescence-based techniques to distinguish healthy, premalignant, and malignant epithelial cells (Meier et al. 2010).

In this model, a known carcinogen, 7,12-dimethylbenz(a)anthracene (DMBA), is applied to the buccal pouch of immunocompetent animals a few times per week. Repeated exposure to the carcinogen leads to consistent temporal development of precancerous lesions followed by cancerous growths. Thus, this model allows study of the sequence of events from healthy epithelium to carcinoma. After visible tumors are generated, the surrounding epithelium may be healthy or dysplastic. This allows sampling using fluorescence spectroscopy of a wide variety of histopathologic specimens (Meier et al. 2010; Skala et al. 2007a).

Skala et al. (2007a) reported on the multiphoton scanning FLIM of protein-bound and free NADH in normal and precancerous epithelia. The FLIM system was built around a conventional microscope (Nikon Diaphot 200). A titanium–sapphire laser (Coherent, Mira; 76 MHz, 120 fs, at 780 nm) pumped by a solid-state laser (Coherent, Verdi) was used as excitation source. FLIM data were collected using a scanning unit (BioRad, MRC-600), a fast photon-counting PMT (Becker & Hickl, PMH-100), and time-correlated single-photon counting (TCSPC) electronics (Becker & Hickl, SPC-830). The fluorescence decays were approximated by a double-exponential model. Under these experimental conditions, the cytoplasm of normal hamster cheek pouch epithelial cells presented a short lifetime component of 0.29 ± 0.03 ns and long lifetime component of 2.03 ± 0.06 ns. These were attributed to free and protein-bound NADH, respectively. The mild to moderate dysplasia as well as severe dysplasia and carcinoma *in situ* were discriminated from normal tissues by their decreased protein-bound NADH lifetime. These trends were related to the inhibition of cellular glycolysis and oxidative phosphorylation in cell monolayers that produced an increase and decrease, respectively, in the protein-bound NADH lifetime. A subsequent report by the same group (Skala et al. 2007b) described the use of the same FLIM experimental setup for studying the NADH and FAD redox states, fluorescence lifetimes, and cellular morphology in precancerous epithelia. Consistent with earlier reports, a significant decrease was observed in the contribution and lifetime of protein-bound NADH (averaged over the entire epithelium) in all (low and high grade) epithelial precancers compared with normal epithelial tissues. In addition, a significant increase in the protein-bound FAD lifetime and a decrease in the contribution of protein-bound FAD were observed in high-grade precancers only. Moreover, an increase in the intracellular variability in the redox ratio, NADH, and FAD fluorescence lifetimes was reported in precancerous cells compared with normal cells. The spatial and depth resolution of the multiphoton FLIM system used in these studies provided a unique approach to study the metabolic changes involved in progression of oral carcinoma *in vivo* in animal models of oral carcinoma.

Farwell et al. (2010) reported the application of TRFS to distinct stages of oral carcinoma in the hamster buccal pouch model using a pulse-sampling TRFS technique and Laguerre deconvolution technique (Jo et al. 2004; Liu et al. 2012) for analysis of the fluorescence decay parameters. These techniques are described in Chapters 4 and 11, respectively. The tissue fluorescence was induced with a pulsed nitrogen laser (337 nm, 700 ps pulse width) and measured with customized TRFS instrumentation similar with that described in Chapter 4. Spectral intensities and average lifetime values at three spectral bands (SB1 = 380 ± 10 nm, SB2 = 460 ± 10 nm, and SB3 = 635 ± 10 nm) allowed for discrimination among four conditions: healthy epithelium, dysplasia, carcinoma *in situ*, and invasive carcinoma. Figure 18.1 depicts results from this study. The fluorescence emission of the three main tissue fluorophores known to emit upon 337 nm excitation was observed. The fluorescence of health tissue appeared dominated by collagen (very strong peak emission at ~390 nm with a longer-lasting emission >3 ns), while the fluorescence of cancerous lesions appeared dominated by the NADH (peak emission at ~460 nm with a short-lasting emission <1.5 ns). Porphyrin fluorescence (peak emission at ~630 nm) was observed in invasive carcinoma cases. The fluorescence intensity measured in healthy tissue was consistently stronger than that of lesions. When using only time-resolved parameters, the most important values to distinguish lesions were in the NADH emission band (SB2). A shorter lifetime was typically found in tumor relative to healthy tissue.

A linear discriminant algorithm combining spectral fluorescence parameters derived from both spectral and time-domain parameters (peak intensities, average fluorescence lifetimes, and the Laguerre coefficient [zero-order]) for healthy epithelium, dysplasia, carcinoma *in situ*, and invasive carcinoma provided the best diagnostic discrimination, with a sensitivity of 100%, 100%, 69.2%, and 76.5% and a specificity of 100%, 92.2%, 97.1%, and 96.2%, respectively.

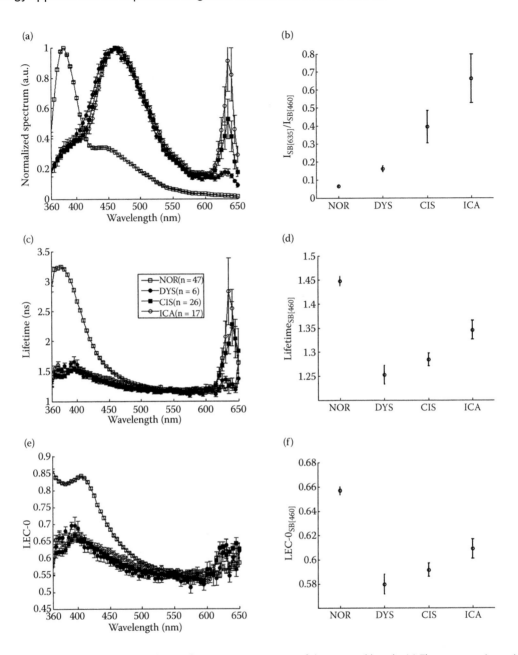

Figure 18.1 Spectral intensities and time-domain measurements of the spectral bands. (a) Fluorescence intensity spectra values of healthy cheek mucosa (NOR), dysplasia (DYS), carcinoma *in situ* (CIS), and carcinoma (CA). (b) Ratio of fluorescence intensities at 635 and 460 nm spectral bands ($I_{SB[635]}:I_{SB[460]}$) for each tissue type. (c) Lifetime values for each tissue type. (d) Average lifetime values at the 460 nm spectral band ($\tau_{SB[460]}$) for each tissue type. (e) Laguerre expansion coefficient, zero order (LEC-0) for each tissue type. (f) LEC-0 at the 460 nm spectral band ($\tau_{SB[460]}$) for each tissue type. Results are presented as mean ± SE of the data from each independent measurement. (Adapted from Farwell, D. G. et al., *Archives of Otolaryngology-Head & Neck Surgery* 136 (2):126–133, 2010.)

Sun et al. (2009) reported a FLIM study in the same hamster carcinoma model using a flexible imaging bundle (endoscopic-like). Similar to the TRFS technique described above, tissue autofluorescence was induced by pulsed nitrogen laser (337 nm, 700 ps pulse width) via a high numerical aperture fiber optic. The emitted light was collected using the gradient index (GRIN) lens (NA of 0.5, 0.5 mm diameter, 4 mm field of view) cemented to a fiber image guide (0.6 mm diameter, 2 m long, 10,000 fibers). A 20× microscope objective and

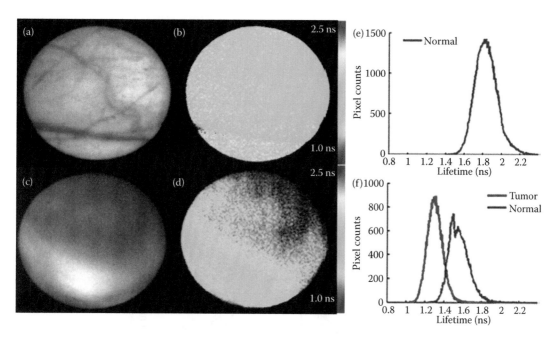

Figure 18.2 Fluorescence images of hamster cheek *in vivo*. Normal buccal pouch mucosa: (a) intensity image, (b) lifetime image, and (e) lifetime histogram. Carcinoma *in situ* with surrounding tissue: (c) intensity image, (d) lifetime map, and (f) its histogram. Two histogram regions were selected referring to the bright field and fluorescence intensity images. Each region covered about 60%–80% central area for both tumor and normal. (Adapted from Sun, Y. et al., *Optics Letters*, 34 (13):2081–2083, 2009.)

a 150 mm focal length achromatic lens were used to magnify fluorescence images to the intensified charge-coupled device (ICCD). The time-gated images were recorded for two emission bands (center wavelength/bandwidth): 390/70 and 450/65 nm. Figure 18.2 depicts representative results from this preclinical study.

The results of this study were in full agreement with those obtained using the point-spectroscopy TRFS technique described above (Farwell et al. 2010). The FLIM images demonstrate a significant contrast in fluorescence average lifetime between tumor (1.77 ± 0.26 ns) and normal (2.50 ± 0.36 ns) tissues at 450 nm and an over 80% intensity decrease at 390 nm emission in tumor versus normal areas. The results of this study have also underscored that the time-resolved images were minimally affected by tissue morphology, endogenous absorbers, and illumination.

18.5 APPLICATION OF TRFS TO INTRAOPERATIVE DIAGNOSIS OF HEAD AND NECK TUMORS

Clinical applications of fluorescence lifetime techniques are sparse and have evaluated, only two groups have evaluate the time-resolved fluorescence characteristics of the HNSCC in patients.

Chen et al. (2005) reported the time-resolved fluorescence emission of oral carcinoma at 633 nm corresponding to the porphyrin fluorescence. The experiments were conducted with a commercially available TCSPC system (HORIBA Jobin Yvon IBH). The experimental apparatus consisted of a short-pulsed laser diode (408 nm, 70 ps full width at half-maximum [FWHM] pulse width), a fast-rise photomultiplier, and a single-photon counting device. A customized handheld bifurcated fiber bundle probe was used for excitation and collection of the autofluorescence signal. The study was performed in 38 patients. The characteristics of fluorescence decay dynamics were fitted with a biexponential model and analyzed to determine the decay parameters allowing for discrimination of normal oral mucosa from oral premalignant lesions (verrucous hyperplasia, epithelial hyperplasia, and epithelial dysplasia). The fluorescence decay (lifetime) of normal mucosa at 633 nm emission was found significantly faster when

Tissue autofluorescence lifetime spectroscopy and imaging: Applications

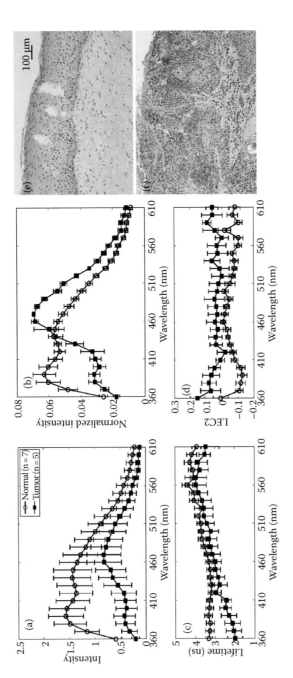

Figure 18.3 Mean values of TRFS features from vocal cord (normal tissue vs. tumor tissue): (a) absolute intensity spectra, (b) normalized spectrum, (c) average fluorescence lifetime, (d) second Laguerre coefficient. Error bars indicate standard errors. (e) H&E stained section of a normal vocal cord tissue sample, and (f) H&E stained section of a tumor located on the vocal cord.

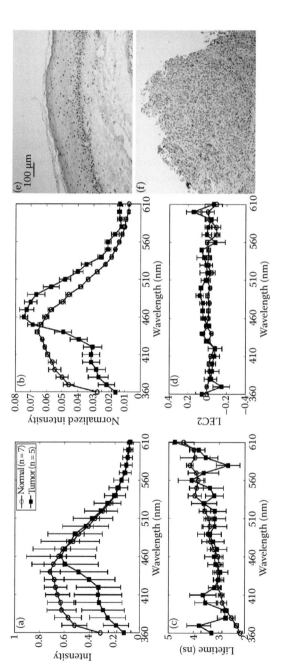

Figure 18.4 Mean values of TRFS features from tongue (normal tissue vs. tumor tissue): (a) absolute intensity spectra, (b) normalized spectrum, (c) average fluorescence lifetime, and (d) second Laguerre coefficient. Error bars indicate standard errors. (e) H&E stained section of a normal tongue tissue sample, and (f) H&E stained section of a tumor located on the tongue.

compared with that of the dysplastic tissue. Their results suggested that a good discrimination of distinct pathologies was possible based on the porphyrin fluorescence decay characteristics.

A more recent study (Marcu 2012; Meier et al. 2010) from our group employed a customized TRFS device to study the fluorescence decay characteristics of distinct types of tissues and sites of human tumors. These included not only the oral cavity but other locations within the upper aerodigestive tract such as the larynx (9 patients). The TRFS technique was based on pulse sampling technique (see Chapter 4) and fluorescence decay data analysis using Laguerre expansion method (see Chapter 11). This system allowed for the recording of time-resolved fluorescence emission spectra in the 360–560 nm range. Using a customized fiber optic probe (bifurcated configuration), various cancer sites were optically interrogated. This included buccal mucosa, tongue, vocal cord, and floor of the mouth. Representative results from TRFS point measurements are presented in Figure 18.3 for vocal cord and Figure 18.4 for tongue.

In agreement with previous reports (de Veld et al. 2005; Pavlova et al. 2008b, 2009), this study also showed that the fluorescence emission features are different for different sites not only in spectral domain but also in time domain. For example, the fluorescence characteristics of the normal vocal cord (Figure 18.3) with as main peak emission at about 390 nm and average lifetime longer than 3 ns in this spectral region are distinct from those of the normal tongue tissue (Figure 18.4) with a main peak emission at about 450 nm and average lifetime at wavelength below 400 nm shorter than 3 ns. In both cases, the spectral intensities, as well as the time-resolved data, provided means for discrimination between tumor and the surrounding normal tissue. Upon 337 nm excitation, the dynamics of the fluorescence decay in 460 ± 25 nm range allow for the discrimination of tumors (regardless their location) from normal tissue. Overall, the tumors exhibited a shorter average lifetime when compared with normal tissues. These differences are also attributed to different ratios of free- and bound-NAD(P)H in the cancerous vs. healthy tissue. When bound to proteins, the NAD(P)H fluorescence lifetime is known to increase multifold (Chorvat and Chorvatova 2009; Lakowicz 2006).

A small pilot study in patients diagnosed with cancer was carried out with the FLIM apparatus described above in the preclinical study (Sun et al. 2013). However, the fluorescence was excited and collected via a customized compact imaging probe suitable for clinical use. This included an excitation fiber with high numerical aperture (NA = 0.48), a collection fiber image bundle (10,000 fibers, ~0.6 mm total diameter), and a GRIN lens objective (FOV = 4 mm). The fluorescence emission was collected via a band-pass filter 460 ± 25 nm corresponding to NAD(P)H fluorescence emission. The fluorescence decays were also deconvolved using the Laguerre expansion technique (Jo et al. 2004). Similar to the TRFS study, the tumor area presented faster decay dynamics when compared with the healthy surrounding tissue. The fluorescence intensity also plays an important role in distinguishing tumor vs. normal within this spectral emission range with a lower intensity in tumor tissue vs. normal tissue. This trend is in agreement to steady-state fluorescence intensity studies reported by other groups (Chaturvedi et al. 2010; Heintzelman et al. 2000; Muller et al. 2003; Pavlova et al. 2008b, 2009; Roblyer et al. 2008; Schwarz et al. 2009; Vedeswari et al. 2009).

18.6 CONCLUSION

Head and neck cancer remains a common tumor worldwide. Treatment of these tumors is challenging with early diagnosis being the key to successful cure and optimizing prognosis. As these tumors become larger, the chance of cure goes down, and the morbidity associated with curative treatment becomes more debilitating. Cancers of the head and neck involve anatomic structures critical for many of the functions we value most as humans. Tumors involving the tongue, pharynx, and larynx may impede out ability to talk, breathe, and swallow. Additionally, treatment may disfigure patients as the face, head, and neck are not as easily concealed under clothing as other anatomical sites.

Speech and communication set us apart from other animals and are essential to our social nature. Swallowing is critical for maintaining nutrition and sustenance, but less appreciated is the impact that swallowing dysfunction has upon our social interactions. Much of modern society's social interactions take place around drinking and eating. Family meals at dinnertime, social interactions at a cocktail party, and having a cup of coffee with a peer are all challenges for patients who can no longer eat or drink normally from the side effects of treatments for head and neck cancers.

Technology that allows for the earlier diagnosis of tumors such as the TRFS and FLIM described above has the potential to improve head and neck cancer patient outcomes. Noninvasive or minimally invasive diagnosis via a direct probe or probe placed routinely through a flexible endoscope will allow a physician the ability to detect early fluorescent changes that have been shown to correlate with carcinogenesis. This will allow for targeted biopsies of subtle lesions, earlier detection of tumors, and subsequently improved prognosis. Additionally, earlier diagnosis allows for more targeted treatment. Smaller surgeries or less intensive nonsurgical treatment will preserve normal tissue and improve the ability of patients to continue with more normal swallowing, speech, and breathing and dramatically improve their quality of life.

The advantage of FLIM is the improved resolution of fluorescent characteristics in real-world situations such as surgery. Currently, assessment of the adequacy of surgical resection of tumor relies on the ability of the surgeon to visually remove all abnormal tissue and histopathologic sampling of the tissue at the periphery of the resection. These are imperfect techniques, and as a result, a significant number of patients suffer tumor recurrence. FLIM technology allows improved detection of tissue fluorescent characteristics despite irregular surfaces or intraoperative blood and allows for the potential to improve surgical margin control. Real-time evaluation of the surgical excision margins will potentially reduce the risk of tumor recurrence and ultimately improve survival.

Several studies have now demonstrated the potential of these technologies to accurately query tissue in the head and neck. The easy accessibility of these tissues in awake patients allows for the early implementation of this technology. As clinical acceptance proceeds, it is expected that these tools will play an increasing role in the early diagnosis and management of this challenging set of cancers.

ACKNOWLEDGMENT

The studies reviewed in this chapter were supported in part by the National Institutes of Health grants R01-HL67377, R21-RR0025818, and UL1 RR024146.

REFERENCES

Arbes, S. J., A. F. Olshan, D. J. Caplan, V. J. Schoenbach, G. D. Slade, and M. J. Symons. 1999. "Factors contributing to the poorer survival of black Americans diagnosed with oral cancer (United States)." *Cancer Causes and Control* 10 (6):513–523.

Bagan, J., G. Sarrion, and Y. Jimenez. 2010. "Oral cancer: Clinical features." *Oral Oncology* 46 (6):414–417.

Cabot, R. C., N. L. Harris, J. A. O. Shepard, E. S. Rosenberg, A. M. Cort, S. H. Ebeling, C. C. Peters, T. R. Flynn, G. J. Hunter, and M. M. Johnson. 2010. "Case 6–2010." *New England Journal of Medicine* 362 (8):740–748.

Carvalho, A. L., L. P. Kowalski, I. M. Agra, E. Pontes, O. D. Campos, and A. C. Pellizzon. 2005. "Treatment results on advanced neck metastasis (N3) from head and neck squamous carcinoma." *Otolaryngology-Head and Neck Surgery* 132 (6):862–868.

Chaturvedi, P., S. K. Majumder, H. Krishna, S. Muttagi, and P. K. Gupta. 2010. "Fluorescence spectroscopy for noninvasive early diagnosis of oral mucosal malignant and potentially malignant lesions." *Journal of Cancer Research and Therapeutics* 6 (4):497–502.

Chen, H. M., C. P. Chiang, C. You, T. C. Hsiao, and C. Y. Wang. 2005. "Time-resolved autofluorescence spectroscopy for classifying normal and premalignant oral tissues." *Lasers in Surgery and Medicine* 37 (1):37–45.

Chorvat, D., and A. Chorvatova. 2009. "Multi-wavelength fluorescence lifetime spectroscopy: A new approach to the study of endogenous fluorescence in living cells and tissues." *Laser Physics Letters* 6 (3):175–193.

Day, G. L., W. J. Blot, D. F. Austin, L. Bernstein, R. S. Greenberg, J. B. Schoenberg, D. M. Winn, J. K. McLaughlin, and J. F. Fraumeni. 1993. "Racial differences in risk of oral and pharyngeal cancer: Alcohol, tobacco, and other determinants." *Journal of the National Cancer Institute* 85 (6):465–473.

de Souza Azevedo, R., F. R. Pires, R. D. Coletta, O. P. de Almeida, L. P. Kowalski, and M. A. Lopes. 2008. "Oral myofibromas: Report of two cases and review of clinical and histopathologic differential diagnosis." *Oral Surgery, Oral Medicine, Oral Pathology, Oral Radiology, and Endodontology* 105 (6):e35–e40.

de Veld, D. C. G., M. Skurichina, M. J. H. Witjes, R. P. W. Duin, D. J. C. M. Sterenborg, W. M. Star, and J. L. N. Roodenburg. 2003. "Autofluorescence characteristics of healthy oral mucosa at different anatomical sites." *Lasers in Surgery and Medicine* 32 (5):367–376.

Tissue autofluorescence lifetime spectroscopy and imaging: Applications

de Veld, D. C. G., M. J. H. Witjes, H. J. C. M. Sterenborg, and J. L. N. Roodenburg. 2005. "The status of in vivo autofluorescence spectroscopy and imaging for oral oncology." *Oral Oncology* 41 (2):117–131.

Delank, W., B. Khanavkar, J. A. Nakhosteen, and W. Stoll. 2000. "A pilot study of autofluorescent endoscopy for the in vivo detection of laryngeal cancer." *Laryngoscope* 110 (3 Pt 1):368–373.

Farwell, D. G., J. D. Meier, J. Park, Y. Sun, H. Coffman, B. Poirier, J. Phipps, S. Tinling, D. J. Enepekides, and L. Marcu. 2010. "Time-resolved fluorescence spectroscopy as a diagnostic technique of oral carcinoma validation in the hamster buccal pouch model." *Archives of Otolaryngology-Head & Neck Surgery* 136 (2):126–133.

Gourin, C. G., W. J. McAfee, K. M. Neyman, J. W. Howington, R. H. Podolsky, and D. J. Terris. 2005. "Effect of comorbidity on quality of life and treatment selection in patients with squamous cell carcinoma of the head and neck." *Laryngoscope* 115 (8):1371–1375.

Hainsworth, J. D., D. R. Spigel, H. A. Burris III, T. M. Markus, D. Shipley, M. Kuzur, S. Lunin, and F. A. Greco. 2009. "Neoadjuvant chemotherapy/gefitinib followed by concurrent chemotherapy/radiation therapy/gefitinib for patients with locally advanced squamous carcinoma of the head and neck." *Cancer* 115 (10):2138–2146.

Heintzelman, D. L., U. Utzinger, H. Fuchs, A. Zuluaga, K. Gossage, A. M. Gillenwater, R. Jacob, B. Kemp, and R. R. Richards-Kortum. 2000. "Optimal excitation wavelengths for in vivo detection of oral neoplasia using fluorescence spectroscopy." *Photochemistry and Photobiology* 72 (1):103–113.

Howlader, N., A. Noone, M. Krapcho, N. Neyman, and W. W. Aminou. 2011. *SEER Cancer Statistics Review, 1975–2008*. Bethesda, MD: National Cancer Institute.

Jo, J. A., Q. Y. Fang, T. Papaioannou, and L. Marcu. 2004. "Fast model-free deconvolution of fluorescence decay for analysis of biological systems." *Journal of Biomedical Optics* 9 (4):743–752.

Lakowicz, J. R. 2006. *Principles of Fluorescence Spectroscopy*, 3rd ed. New York: Kluwer Academic/Plenum.

Lane, P. M., T. Gilhuly, P. Whitehead, H. S. Zeng, C. F. Poh, S. Ng, P. M. Williams, L. W. Zhang, M. P. Rosin, and C. E. MacAulay. 2006. "Simple device for the direct visualization of oral-cavity tissue fluorescence." *Journal of Biomedical Optics* 11 (2):024006.

Liu, J., Y. Sun, J. Y. Qi, and L. Marcu. 2012. "A novel method for fast and robust estimation of fluorescence decay dynamics using constrained least-squares deconvolution with Laguerre expansion." *Physics in Medicine and Biology* 57 (4):843–865.

Marcu, L. 2012. "Fluorescence lifetime techniques in medical applications." *Annals of Biomedical Engineering* 40 (2):304–331.

Meier, J. D., H. T. Xie, Y. Sun, Y. H. Sun, N. Hatami, B. Poirier, L. Marcu, and D. G. Farwell. 2010. "Time-resolved laser-induced fluorescence spectroscopy as a diagnostic instrument in head and neck carcinoma." *Otolaryngology-Head and Neck Surgery* 142 (6):838–844.

Mignogna, M. D., S. Fedele, L. Lo Russo, L. Lo Muzio, and E. Bucci. 2004. "Immune activation and chronic inflammation as the cause of malignancy in oral lichen planus: Is there any evidence." *Oral Oncology* 40 (2):120–130.

Muller, M. G., T. A. Valdez, I. Georgakoudi, V. Backman, C. Fuentes, S. Kabani, N. Laver, Z. Wang, C. W. Boone, R. R. Dasari, S. M. Shapshay, and M. S. Feld. 2003. "Spectroscopic detection and evaluation of morphologic and biochemical changes in early human oral carcinoma." *Cancer* 97 (7):1681–1692.

NCI. 2012. Oral Cancer Homepage: National Cancer Institute. National Cancer Institute, National Institutes of Health 2012 [cited November 2012]. Available from http://www.cancer.gov/cancertopics/types/oral.

Onizawa, K., N. Okamura, H. Saginoya, and H. Yoshida. 2003. "Characterization of autofluorescence in oral squamous cell carcinoma." *Oral Oncology* 39 (2):150–156.

Pavlova, I., C. R. Weber, R. A. Schwarz, M. Williams, A. El-Naggar, A. Gillenwater, and R. Richards-Kortum. 2008a. "Monte Carlo model to describe depth selective fluorescence spectra of epithelial tissue: Applications for diagnosis of oral precancer." *Journal of Biomedical Optics* 13 (6):064012.

Pavlova, I., M. Williams, A. El-Naggar, R. Richards-Kortum, and A. Gillenwater. 2008b. "Understanding the biological basis of autofluorescence imaging for oral cancer detection: High-resolution fluorescence microscopy in viable tissue." *Clinical Cancer Research* 14 (8):2396–2404.

Pavlova, I., C. R. Weber, R. A. Schwarz, M. D. Williams, A. M. Gillenwater, and R. Richards-Kortum. 2009. "Fluorescence spectroscopy of oral tissue: Monte Carlo modeling with site-specific tissue properties." *Journal of Biomedical Optics* 14 (1):014009.

Peacock, Z. S., M. A. Pogrel, and B. L. Schmidt. 2008. "Exploring the reasons for delay in treatment of oral cancer." *The Journal of the American Dental Association* 139 (10):1346–1352.

Rahman, M., P. Chaturvedi, A. M. Gillenwater, and R. Richards-Kortum. 2008. "Low-cost, multimodal, portable screening system for early detection of oral cancer." *Journal of Biomedical Optics* 13:030502.

Roblyer, D., C. Kurachi, V. Stepanek, M. D. Williams, A. K. El-Naggar, J. J. Lee, A. M. Gillenwater, and R. Richards-Kortum. 2009. "Objective detection and delineation of oral neoplasia using autofluorescence imaging." *Cancer Prevention Research* 2 (5):423–431.

Roblyer, D., R. Richards-Kortum, K. Sokolov, A. K. El-Naggar, M. D. Williams, C. Kurachi, and A. M. Gillenwater. 2008. "Multispectral optical imaging device for in vivo detection of oral neoplasia." *Journal of Biomedical Optics* 13 (2):024019.

Schwarz, R. A., W. Gao, D. Daye, M. D. Williams, R. Richards-Kortum, and A. M. Gillenwater. 2008. "Autofluorescence and diffuse reflectance spectroscopy of oral epithelial tissue using a depth-sensitive fiber-optic probe." *Applied Optics* 47 (6):825–834.

Schwarz, R. A., W. Gao, C. R. Weber, C. Kurachi, J. J. Lee, A. K. El-Naggar, R. Richards-Kortum, and A. M. Gillenwater. 2009. "Noninvasive evaluation of oral lesions using depth-sensitive optical spectroscopy." *Cancer* 115 (8):1669–1679.

Shah, J. P., and Z. Gil. 2009. "Current concepts in management of oral cancer-surgery." *Oral Oncology* 45 (4–5):394–401.

Skala, M. C., K. M. Riching, D. K. Bird, A. Gendron-Fitzpatrick, J. Eickhoff, K. W. Eliceiri, P. J. Keely, and N. Ramanujam. 2007a. "In vivo multiphoton fluorescence lifetime imaging of protein-bound and free nicotinamide adenine dinucleotide in normal and precancerous epithelia." *Journal of Biomedical Optics* 12 (2):024014.

Skala, M. C., K. M. Riching, A. Gendron-Fitzpatrick, J. Eickhoff, K. W. Eliceiri, J. G. White, and N. Ramanujam. 2007b. "In vivo multiphoton microscopy of NADH and FAD redox states, fluorescence lifetimes, and cellular morphology in precancerous epithelia." *Proceedings of the National Academy of Sciences of the United States of America* 104 (49):19494–194949.

Sun Y, J. E. Phipps, J. Meier, N. Hatami, B. Poirier, D. S. Elson, D. G. Farwell, and L. Marcu. 2013. Endoscopic fluorescence lifetime imaging for in vivo intraoperative diagnosis of oral carcinoma. *Microsc Microanal* 19 (4):791–798.

Sun, Y., J. E. Phipps, D. S. Elson, H. Stoy, S. Tinling, J. Meier, B. Poirier, F. S. Chuang, D. G. Farwell, and L. Marcu. 2009. "Fluorescence lifetime imaging microscopy: In vivo application to diagnosis of oral carcinoma." *Optics Letters* 34 (13):2081–2083.

Upile, T., W. Jerjes, H. J. Sterenborg, A. K. El-Naggar, A. Sandison, M. J. Witjes, M. A. Biel, I. Bigio, B. J. Wong, A. Gillenwater, A. J. MacRobert, D. J. Robinson, C. S. Betz, H. Stepp, L. Bolotine, G. McKenzie, C. A. Mosse, H. Barr, Z. Chen, K. Berg, A. K. D'Cruz, N. Stone, C. Kendall, S. Fisher, A. Leunig, M. Olivo, R. Richards-Kortum, K. C. Soo, V. Bagnato, L. P. Choo-Smith, K. Svanberg, I. B. Tan, B. C. Wilson, H. Wolfsen, A. G. Yodh, and C. Hopper. 2009. "Head & neck optical diagnostics: Vision of the future of surgery." *Head & Neck Oncology* 1:25.

Vedeswari, C. P., S. Jayachandran, and S. Ganesan. 2009. "In vivo autofluorescence characteristics of pre- and post-treated oral submucous fibrosis: A pilot study." *Indian Journal of Dental Research* 20 (3):261–267.

VELscope. 2012. LED Dental Inc. 2012 [accessed November 2012]. Available from http://www.leddental.com/.

Tissue autofluorescence lifetime spectroscopy and imaging: Applications

19 Fluorescence lifetime techniques in cardiovascular disease diagnostics

Jennifer E. Phipps, Yang Sun, and Laura Marcu

Contents

19.1 FLUORESCENCE IN ATHEROSCLEROTIC CARDIOVASCULAR DISEASE

Cardiovascular disease has long been the number one cause of death and morbidity for adults in developed countries (Roger et al. 2010). Coronary artery disease is the most common type of cardiovascular disease and is caused by the development of atherosclerotic plaques in the coronary arteries of the heart; plaque is composed of lipids, cholesterol, calcium, and other types of cells found circulating in the blood stream (Ross 1993). Heart attack and other acute coronary syndromes are caused by plaque rupture, and although there are oftentimes no precluding physical symptoms to this phenomenon, plaques with a higher risk of rupture are morphologically different than their more stable counterparts and are deemed high-risk or "vulnerable plaques" (Libby 2002; Virmani et al. 2000). There is currently no method clinically available for assessing risk of plaque rupture; thus, many research efforts are being expended to develop imaging techniques for this purpose. This chapter presents a review of the time-resolved fluorescence techniques applied to characterization and diagnosis of atherosclerosis.

19.1.1 ARTERIAL STRUCTURE AND PATHOGENESIS OF ATHEROSCLEROSIS

Arteries are composed of three layers: intima, media, and adventitia (Figure 19.1). The innermost layer, the intima, is composed of a single layer of endothelial cells that are in contact with the blood stream and cover a subendothelial matrix composed of elastin, collagen III, and a small amount of smooth muscle cells. The media, which is separated from the intima by internal elastic lamina, is composed of many layers of smooth muscle cells. The adventitia is the outermost layer of the artery and is separated from the media by external elastic lamina.

Atherosclerosis results from an inflammatory response in the artery after endothelial injury. The exact mechanism of endothelial injury is debatable (Packard and Libby 2008), but it is known that monocytes in the blood stream only adhere to an endothelium expressing adhesion molecules that are activated by an inflammatory response. Monocytes migrating into the intima begin the process of plaque formation. This occurs when ligands on the monocytes bind selectins expressed on the activated endothelium causing the monocytes to loosely roll along the endothelium. Then integrins, another type of adhesion molecule, foster stronger binding between monocytes and the endothelium, which causes chemokines within the inflamed intima to allow the intimal migration of the monocytes (Packard and Libby 2008). Once in the intima, monocytes begin to divide and develop into macrophages that accumulate cholesteryl esters, and other lipidic and cholesterol content through endocytosis becomes foam cells. The accumulation of foam cells and apoptotic debris begins the formation of a necrotic core within the plaque. A fibrotic cap will form on the luminal side of the necrotic core in an attempt to maintain the structural integrity of the vessel. Plaque

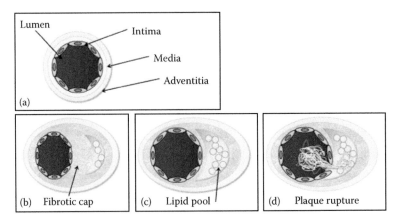

Figure 19.1 Artery diagrams. (a) Normal artery, (b) stable plaque, (c) rupture-prone place, and (d) plaque rupture. (Adapted from Libby, P., *Nature*, 420(6917), 868–874, 2002.)

will also develop without a necrotic core in which case it may become fibrotic, thickened by collagen I or calcified, thickened by calcium deposits, or both fibrotic and calcified. It is the rupture of the fibrous cap and the exposure of the thrombotic necrotic core to the luminal blood flow that cause the majority of acute coronary symptoms, including heart attack (Figure 19.1d).

Plaques are broadly categorized into one of several clinically relevant types that are described by Virmani et al. (2000): thickened intima, pathologic intimal thickening, fibrotic, calcified, fibrotic and calcified (fibrocalcified), and fibroatheroma with thick or thin fibrotic caps covering a necrotic/lipid core. The thin and thick cap fibroatheromas, pathologic intimal thickening, and calcified plaques with luminal calcified nodules are considered vulnerable pathologies.

Typically, three types of features are used to describe plaque vulnerability: structural (fibrous cap <65 μm, presence of lipid pool or necrotic core), biochemical (collagen, elastin, and lipid content), and functional (cellular activity). The current gold standard for imaging coronary plaques in patients is coronary angiography, despite the fact that this technique is unable to assess any of these three features. It works by acquiring x-ray images of contrast as it is injected into the coronary arteries, which allows for the visualization of plaques that occlude blood flow. Yet, the majority of acute coronary syndromes are shown to be caused by plaques that do not obstruct blood flow (Sharif and Murphy 2010). Intravascular ultrasound (IVUS) has been used in the past few years for assessing some structural features of the plaque including the presence of a lipid core, and an even newer technique, intravascular optical coherence tomography (OCT), accomplishes this with much higher resolution. Both IVUS and OCT are catheter-based techniques that are used in conjunction with coronary angiography. Molecular imaging techniques assess functional features by using contrast agents to target specific atherosclerotic processes such as the presence of matrix metalloproteinases (MMPs), the infiltration of inflammatory cells, or the activation of the endothelium (Libby et al. 2010). While these techniques have an advantage of high specificity, they have not been implemented clinically because the necessary contrast agents have yet to be approved for use in humans. Fluorescence has been used to study the biochemical composition of arteries and plaque since the 1960s. Thus far, the drawbacks of fluorescence techniques (challenges of intravascular light delivery, limited penetration depth, and signal acquisition through blood) outweighed the benefits (nondestructive, high resolution, biochemical analysis). However, more recently, researchers have developed methods to overcome these obstacles, and thus, fluorescence techniques are now considered to have high potential for future clinical applications.

19.1.2 ATHEROSCLEROTIC PLAQUE FLUOROPHORES

Upon ultraviolet (UV) light excitation, the normal and diseased arteries exhibit fluorescence properties (Baraga et al. 1989). This is due to distinct types of fluorescent biological molecules naturally occurring in arteries. These fluorescent constituents are also associated with distinct morphological features of the arteries. A healthy artery or very early lesion has a thin intima composed mainly of a single layer of endothelial cells (Figure 19.1a). Beneath this, the media is composed mostly of elastin, which has a broad emission band between 360 and 600 nm with a peak around 410 nm and average lifetime of 1.8 ns (e.g., upon 337 nm excitation; Marcu et al. 2003). A fibrotic plaque or thick cap fibroatheroma (Figure 19.1b) has an abundance of collagen I, which is characterized by an emission band between 360 and 510 nm with a peak at 390 nm and average lifetime of 3–4 ns (337 nm excitation; Marcu et al. 2003). The fluorescence emission of inflamed plaques with necrotic/lipid cores (Figure 19.1c) is likely generated by collagen I and lipid constituents. For example, ceroid is one type of lipopigment, which emits fluorescence between 375 and 590 nm with a peak at 450 nm (Verbunt et al. 1992). Very low-density lipoprotein (VLDL) and LDL exhibit peak fluorescence emission at 390 nm with an average lifetime shorter than collagen and elastin (337 nm excitation; Marcu et al. 2003). Cholesterol linoleate and oleate fluoresce at a wider bandwidth (390–430 nm) than LDL and VLDL with similar short average lifetimes (Marcu et al. 2003). The lipid content in macrophages has peak fluorescence emission at longer wavelengths (around 450 nm) and a shorter average lifetime (1.5 ns) than collagen and elastin (337 nm excitation) (Marcu et al. 2005). The differences in the emission bands and peaks of these different fluorophores enable the differentiation between normal artery/early lesions, fibrotic/thick cap fibroatheromas, and inflamed/thin cap fibroatheromas. Elastin has a shorter fluorescence lifetime than collagen, and cholesterol/lipid lifetime is generally shortest of all the arterial fluorophores. The fluorescence lifetime contrast generated by these molecules can be analyzed based on the time-resolved fluorescence measurements from an investigated volume of tissue (Marcu et al. 2003). Figure 19.2 summarizes the average lifetime values from the literature for various plaque components at three wavelength bands (Marcu et al. 2003).

The source of plaque autofluorescence from tissue can be validated with histology, a technique that analyzes the microstructure with various stains that allow pathologists to recognize certain features. For example, collagen, elastin, and smooth muscle can be visualized with a trichrome stain. Hematoxylin and eosin are used to stain cell nuclei, cytoplasm, and other structural components, and CD68 is an immunohistochemical stain used to identify macrophages. Figure 19.3 demonstrates four plaque types depicted by a trichrome stain.

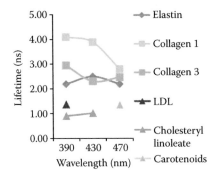

Figure 19.2 Literature values of average lifetimes of common plaque fluorophores at different wavelengths. (Adapted from Marcu, L. et al., Time-resolved laser-induced fluorescence spectroscopy for staging atherosclerotic lesions. In Mycek, M. A. and Pogue, B. W., eds., *Handbook of Biomedical Fluorescence*. New York: Marcel Dekker, 397–430, 2003.)

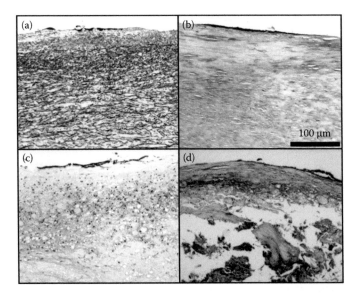

Figure 19.3 Trichrome stained carotid artery tissue sections. (a) Normal artery: black wavy lines indicate elastin; red cells are smooth muscle cells (SMCs). (b) Collagen-rich plaque: blue indicates collagen; red cells are SMCs. (c) "Inflamed" plaque: inflammatory cell infiltration into the intima. (d) Necrotic core plaque: blue fibrous cap; necrotic core is depicted by empty spaces from lipid washout and inflammatory cells.

Tissue autofluorescence lifetime spectroscopy and imaging: Applications

19.1.3 HISTORY OF FLUORESCENCE STUDIES OF ATHEROSCLEROSIS

Fluorescence spectroscopy techniques detect elastin, collagen, lipids, and other sources of autofluorescence in normal and diseased arterial walls and have been used to characterize the biochemical composition of atherosclerotic plaques both *ex vivo* and *in vivo*. Edholm and Jacobson (1965) first suggested that quantitative optical techniques could be used *in vivo* to improve the diagnosis of atherosclerosis. They found that reflectance of human aorta at 500 and 550 nm differed between healthy vs. atherosclerotic tissue from 15 postmortem samples. The next significant study using fluorescence to study atherosclerosis came in 1985 when Kittrell et al. (1985) demonstrated that normal aorta and fibrous plaque could be distinguished using a single-fiber optic device.

Following this work, the majority of fluorescence studies in atherosclerosis focused on developing techniques for assisting in laser ablation procedures. In the early 1990s, fluorescence analysis of arterial wall was studied in terms of how it could be helpful for improving plaque laser ablation methods. Bartorelli et al. (1991) recorded fluorescence spectra from 268 locations *in vivo* in 48 patients undergoing open-heart surgery or percutaneous catheterization using a 325 nm low power helium–cadmium laser excitation through a flexible 200 μm fiber. They found that spectra from normal arterial wall, noncalcified plaque, and calcified plaque differed in peak intensity, peak intensity wavelength, and shape. When compared with normal artery, calcified and noncalcified plaques showed decreased intensity (more so for calcified), higher shape index, and a shift of peak wavelength location to longer values for noncalcified plaques. Using these factors, normal artery was identified with 100% specificity, and atherosclerotic sites were identified with 73% sensitivity (Bartorelli et al. 1991). In 1994, another application of laser-induced fluorescence to laser ablation guidance was reported. Morguet et al. (1994) used a single XeCl excimer laser (308 nm excitation through 600 μm silica fiber) for simultaneous tissue ablation and fluorescence excitation. Arterial media, lipid-rich plaques, and calcified plaques were identified by fluorescence spectra in ten human cadaver aortas. These two laser ablation studies sought to find a way to improve laser ablation by using fluorescence to identify when and where the technique should be applied. This strategy has not been pursued following these initial studies in the early 1990s.

Also in the 1990s, fluorescence first began to be used for characterizing atherosclerotic plaque composition. Fitzmaurice et al. (1989) reported *en face* fluorescence measurements from frozen sections of human coronary arteries and aorta with an Olympus BH-2 metallurgic bright-field microscope and an argon ion laser (476 nm excitation). Normal artery was distinguished from atherosclerotic artery based on differences in fluorescence properties of elastin, collagen, lipid, and calcium. Following this study, Verbunt et al. (1992) specifically studied the fluorescence properties of ceroid deposits in lipid and calcium-rich plaques. They found that the autofluorescence spectrum of ceroid (a type of lipopigment) was wider and right-shifted compared to collagen and elastin. In 1994, Papazoglou et al. (1994) measured the autofluorescence of abdominal aorta from cadavers and femoral artery from bypass surgeries using a He–Cd laser (442 nm excitation). They found that the autofluorescence spectra could be used to discriminate between fibrous, aneurysmal, and normal samples but not between lipid and calcium-rich samples. Then with the fluorescence tagging of hypocrellin by incubation with the tissue prior to imaging, they were able to distinguish between lipid and calcium-rich plaques. This was one of the first studies to use molecular imaging to characterize atherosclerotic plaque. Christov et al. (2000) demonstrated *in vivo* fluorescence emission analysis of plaque disruption in a rabbit model.

Concurrently in the 1990s, the characteristics of plaque that indicate higher risk of rupture were rigorously studied (Virmani et al. 2000). Hence, fluorescence analysis started to focus on applications to assess features of plaque vulnerability. Arakawa et al. (2002) studied fluorescence spectra from human femoral and coronary arteries obtained from cadavers to assess how fluorescence analysis could be used to identify plaque with a thin fibrous cap. A He–Cd laser (325 nm excitation) was used to induce autofluorescence, and arterial specimen autofluorescence was compared to autofluorescence of collagen, elastin, and various lipid components. It was found that lipid core fluorescence spectra were red-shifted and broader compared to normal tissue and collagenous plaques. Spectra from lipid cores were similar to oxidized low-density lipoprotein (oxLDL), indicating the strong presence of this molecule in the lipid core. Fibrous plaques exhibited similar fluorescence as collagen, and normal artery exhibited similar fluorescence as elastin. Based on the fluorescence spectra of these three molecules (collagen, elastin, oxLDL), classification of arterial segments into normal artery, lipid core, atheroma, or preatheroma was performed with 86% accuracy. Additionally, fibrous cap thickness was correlated with spectral collagen content ($R = 0.65$, $P < 0.0001$) (Arakawa et al. 2002). More recently, Angheloiu et al. (2006) reported using intrinsic fluorescence and diffuse reflectance spectroscopy (IFS and DRS, respectively) to identify superficial foam cells in coronary plaques. They studied IFS and DRS signal from human coronary segments and were able to determine percentages of plaque area covered by superficial foam cells.

19.1.4 HISTORY OF TIME-RESOLVED FLUORESCENCE STUDIES OF ATHEROSCLEROSIS

19.1.4.1 Early point spectroscopy findings

In the early 1990s, time-resolved fluorescence spectroscopy (TRFS) began to be used to investigate atherosclerosis. The first of these studies were performed by Baraga et al. (1989, 1990) and Andersson-Engels et al. (1990) in 1989 and 1990. Baraga et al. (1989) found that fluorescence emission intensity from normal and atherosclerotic aorta at 340 and 380 nm wavelengths differed with 308 nm UV laser excitation. These wavelength bands characterized the tryptophan content (340 nm) and the elastin-to-collagen ratio (380 nm) that allowed for a binary classification scheme to distinguish normal from atherosclerotic aorta in 56 of 60 total cases (Baraga et al. 1990). Also, part of this study consisted of time-resolved fluorescence measurements acquired by a time-correlated single-photon counting (TCSPC) technique. It was found that the lifetime was shorter at the tryptophan region (340 nm) and longer at the collagen/elastin region (380 nm) (Baraga et al. 1989). Andersson-Engels et al. (1990) used a continuous wave dye laser for picosecond pulse generation at 320 nm and photon-counting techniques to acquire time-resolved signal from human aorta samples *in vitro*. Also plaque (both collagenous and calcified) exhibited longer fluorescence lifetimes than normal artery at both 400 and 480 nm (Andersson-Engels et al. 1990). In 1997, Maarek et al. (1997) reported the spectrally resolved TRFS characterization of arterial wall constituents (collagen, elastin), and in 1998, two additional studies were published from this group (Maarek et al. 1998; Marcu et al. 1998), one delineating the contribution of lipid fluorescence to time-resolved fluorescence measurements of atherosclerosis and the other characterizing specific lesion types with time-resolved fluorescence.

19.1.4.2 Early imaging findings

Also in the late 1990s, time-resolved microscopy techniques were being developed for microscope-based analysis of atherosclerotic samples. In 1998, Dowling et al. (1998) presented a fluorescence lifetime imaging microscopy (FLIM) system with ~10 ps temporal resolution to be used for biomedical applications. It used a commercial Ti:sapphire laser (excitation at 415 nm), a time-gated image intensifier, and an intensified charge-coupled device (CCD) camera that allowed images of fluorescence lifetimes to be acquired (Dowling et al. 1998). This publication demonstrated differences in lifetime between elastin, collagen, and aorta. In a later publication, this same system was used to perform optical sectioning of rat tissue. In particular, fluorescence lifetime was used as contrast to distinguish elastic cartilage, hair, artery, vessel wall, and vein (Cole et al. 2001). More recent advances in FLIM technology revolutionized its potential for use in clinically relevant biomedical applications (Talbot et al. 2010).

19.1.4.3 Contemporary work

In the 2000s systematic TRFS investigations were performed in human aortic, coronary, and carotid plaques. These studies particularly assessed TRFS and FLIM variables associated with specific atherosclerotic plaque functional (macrophages and MMPs) and biochemical (collagen, elastin, and lipid content) features. TRFS has recently been shown capable of assessing macrophage infiltration into plaques (Marcu et al. 2005) as well as the presence of MMPs (Phipps et al. 2011a). In addition, a TRFS analysis of atherosclerosis was conducted *in vivo* in a rabbit model by Marcu et al. in 2005. This work demonstrated for the first time that TRFS could discriminate between macrophage and collagen content in atherosclerotic plaques and is described in more detail below (Marcu et al. 2005).

19.2 TIME-RESOLVED FLUORESCENCE OF HUMAN ARTERIES

The more extensive evaluation of fluorescence lifetime techniques in atherosclerotic plaques employed customized TRFS and FLIM instrumentation based on point-spectroscopy pulse sampling and wide-field fiber-bundle imaging methods, respectively. The principles of these techniques are described in Chapters 4 and 8 and earlier publications (Elson et al. 2007; Marcu et al. 2009). The Laguerre expansion method to deconvolve the sample fluorescence from the measured fluorescence was used in most cases. This method is discussed in detail in Chapter 11. Briefly, the Laguerre method of deconvolution expands the sample fluorescence on an nth-order Laguerre basis, where n is determined by the number of Laguerre functions and corresponding coefficients required to accurately model the measured fluorescence. The order increases with the complexity of the fluorescence decay, and for arterial samples, Laguerre order 4 was most often used. There were two main ways of describing the fluorescence decays following Laguerre deconvolution, both of which define average lifetime in different ways. First, the fluorescence decay can be fitted with a multiexponential model to obtain time-decay and amplitude constants that are used to calculate an average lifetime. Second, the overall decay intensity dynamics can be described by the Laguerre coefficients (LECs) with average lifetime defined as the time that the normalized fluorescence decay falls to $1/e$ of its original intensity. Thus, each fluorescence decay measurement can be described by multiple parameters that are all a function of emission wavelength: intensity (I), decay constants (τ_M) and amplitudes (A_M, M = order of multiexponential fitting), LECs (LEC-N_λ, N = order of Laguerre deconvolution), and average lifetime (τ_λ, λ = emission wavelength or wavelength band). These parameters were often represented as their acronym or symbol followed by the emission wavelength in subscript (i.e., τ_{380} would represent average lifetime at 380 nm emission wavelength and LEC-2_{450} would represent LEC 2 at 450 nm emission wavelength). Both methods of describing the fluorescence decays are used to characterize the time-resolved properties of atherosclerotic tissue in the following paragraphs.

19.2.1 TRFS AND FLIM OF ATHEROSCLEROTIC AORTA

19.2.1.1 Point spectroscopy

The first extensive characterization of the time-resolved fluorescence at multiple emission wavelengths of normal artery and various stages of atherosclerosis was conducted in human aortic specimens (94 samples, postmortem) (Maarek et al. 2000; Marcu et al. 2003). Fluorescence excitation (nitrogen laser, 337 nm) and emission collection occurred through silica fibers combined in a single probe. A scanning monochromator was used for wavelength selection, and the emission waveform was measured with a gated microchannel plate photomultiplier and sampled by a digitizing oscilloscope (Figure 19.4). The Laguerre deconvolution method for TRFS data analysis was applied to estimate the intrinsic fluorescence decays, from which the time-integrated fluorescence emission spectrum and fluorescence lifetime were computed using a multiexponential approximation. Time-resolved fluorescence emission spectra of aortic samples were found to vary with the progression of atherosclerosis (fluorescence intensity as well as average lifetime derived from multiexponential parameters). For example, average lifetime at 390 nm gradually increased from 2.4 ± 0.1 ns (normal aorta) to 3.9 ± 0.1 ns (advanced lesions), while fluorescence intensity was markedly decreased above 430 nm in intermediate and advanced lesions. Additionally, specific changes in time-resolved parameters were associated with different lesion types as characterized by histological examination.

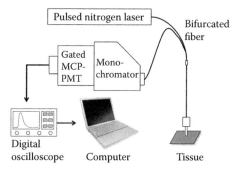

Figure 19.4 Typical TRFS instrumentation.

Tissue autofluorescence lifetime spectroscopy and imaging: Applications

Lesions ranging from early type I to advanced type V demonstrated unique combinations of spectral and temporal fluorescence emission features that were derived from the makeup of intimal fluorophore content of each lesion type. A set of five predictor variables (fluorescence time-decay parameters at 390 and 460 nm and fluorescence intensity values at 490 nm) allowed for discrimination of the lipid-rich lesions (rupture-prone) from collagenous/fibrous lesions (stable) with a sensitivity and specificity higher than 95% (Marcu et al. 2003). These results demonstrate substantial differences in time-resolved properties between clinically relevant plaque compositions, suggesting that such differences provide contrast mechanisms and could translate to diagnostic markers in the analysis of the arterial wall.

19.2.1.2 Fluorescence lifetime imaging

A FLIM investigation of atherosclerotic aortas was also published (Phipps et al. 2011b). A total of 11 human aorta samples were imaged with a FLIM system (48 locations). The system was composed of a pulsed nitrogen laser (337 nm excitation, 700 ps width), a fast-gated (up to 400 ps) intensified CCD camera, and an imaging bundle with 10,000 optical fibers within an ~0.6 mm outer diameter probe (Figure 19.5). Two filters were used: F377: 377/50 nm and F460: 460/60 nm (center wavelength/bandwidth), because this combination of wavelengths can provide discrimination between intrinsic fluorophores related to plaque vulnerability. Average lifetime derived from a Laguerre deconvolution of the fluorescence impulse response function as well as LECs at each wavelength were used to discriminate between tissue types (Jo et al. 2005). Based on histopathology, 81 regions of interest (ROIs) within the FLIM images were divided into four distinct ROIs: lipid-rich (LR), collagen-rich (CR), elastin-rich (ER), and elastin + macrophage-rich (E + M). Certain FLIM-derived characteristics (τ and LECs) identified each group based on these averaged values. For example, ER regions had similar values for τ_{F377} and τ_{F460} (1.8 and 2.0 ns, respectively); E + M and ER samples had similar τ_{F377} values, but E + M samples had significantly longer τ_{F460} values (2.3 ns). LR samples had slightly longer τ_{F377} values than ER and E + M samples (2.0 ns) but slightly shorter τ_{F460} values (1.9 ns). CR samples had the longest τ_{F377} values (2.7 ns) and relatively shorter τ_{F460} values (Figure 19.6).

These results showed for the first time that FLIM images of human atherosclerotic plaque can be used to derive parameters that discriminate between luminal areas that are elastin-rich, elastin- and macrophage-rich, collagen-rich, and lipid-rich by only using intrinsic fluorescence decay dynamics without the need for fluorescence intensity information (Phipps et al. 2011b).

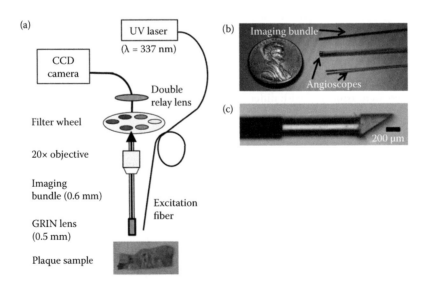

Figure 19.5 (a) Typical endoscopic FLIM instrumentation. (b) Demonstration of the small size of the FLIM imaging bundle. (c) A prism at the tip of the imaging bundle demonstrates how the probe could be adjusted to direct and receive light from the arterial wall for intravascular application. (Adapted from Phipps, J. E. et al., *Journal of Biomedical Optics*, 16(9), 096018, 2011.)

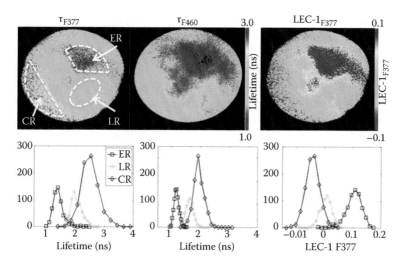

Figure 19.6 FLIM images from one location in an aortic plaque that demonstrates elastin-rich (ER), collagen-rich (CR), and lipid-rich (LR) compositions. The top panel displays three images: average lifetime (τ) at F377, τ at F460, and LEC-1 at F377. Below each image is a histogram corresponding to the three regions outlined in white dashes from each FLIM image. (Adapted from Phipps, J. E. et al., *Journal of Biomedical Optics*, 16(9), 096018, 2011.)

19.2.1.3 Summary of lifetime spectroscopy and imaging of aorta

Both TRFS and FLIM studies in aorta show that collagen, elastin, and lipid-rich tissue could be distinguished with time-resolved fluorescence variables. The TRFS study used a Laguerre deconvolution to retrieve time-resolved spectra, but did not incorporate LECs into the tissue characterization, as the FLIM study did. Additionally, both studies assessed time-resolved parameters at similar wavelength bands, one near 350–400 nm and the second around 460 nm. This is because these are important wavelength bands for discriminating between elastin, collagen, and lipid components of plaque. It is important to note that the FLIM analysis did not require the use of fluorescence intensity information; this increases the potential of a fluorescence lifetime technique to be translated clinically because the presence of blood in arteries affects fluorescence intensity but not lifetime (Stephens et al. 2009; Sun et al. 2011b).

19.2.2 TRFS AND FLIM OF ATHEROSCLEROTIC CORONARY ARTERY

Time-resolved fluorescence emission of normal and atherosclerotic human coronary arteries (58 coronary segments from 11 subjects, postmortem) were analyzed using the Laguerre deconvolution method with a similar system as described in Section 19.2.1.1 (Marcu et al. 2001, 2003). Comparable to the study in aorta, these results showed that analysis of the time-resolved fluorescence spectra could be used to enhance discrimination between normal and different grades (I, II, III, IV, V_a, and V_b) of atherosclerotic lesions as defined by the American Heart Association (AHA). Lipid-rich lesions (e.g., average lifetime at 390 nm: 2.6 ± 0.1 ns) were differentiated from the other lesion types—in particular, fibrous lesions (3.2 ± 0.15 ns) and normal arterial wall (2.1 ± 0.1 ns). Spectroscopic features derived from lipid components were reflected in the emission of lipid-rich lesions, whereas characteristics of type I collagen were identified in the emission of fibrous lesions. In addition, the results suggest that six spectroscopic parameters that combine spectral features at longer wavelengths (>430 nm) and time-resolved characteristics from the peak arterial emission region (390 nm) are optimal for coronary artery lesion discrimination. For example, parameters derived from time-resolved spectra, such as the average lifetime and the fast time decay constants determined using a multiexponential approximation of the intensity decay, are most likely to differentiate between lipid-rich (more unstable) and fibrous lesions (more stable). These findings further validated the potential of TRFS for discriminating lipid-rich plaques from fibrotic lesions.

A FLIM investigation of *ex vivo* human coronary arteries that showed discrimination in FLIM images between collagen-rich plaques and normal (elastin-rich) artery was reported (Thomas et al. 2010). However, no lipid-rich plaques were included in this study; thus, extensive comparison with TRFS results cannot be made. This study used similar instrumentation as previous work (Phipps et al. 2009; Sun et al. 2009b).

19.2.3 TRFS AND FLIM OF ATHEROSCLEROTIC CAROTID ARTERY

Carotid arteries are a useful model of atherosclerosis for imaging purposes because they are exposed and removed during carotid endarterectomy procedures, which allows for the possibility of *in vivo* imaging prior to excision as well as immediate removal and analysis of tissue, important for preservation of tissue integrity. Additionally, it is the only human atherosclerotic plaque that can be measured *in vivo* in live patients that can also be analyzed histopathologically. TRFS and FLIM studies have been performed in carotid artery both implementing the Laguerre deconvolution technique and using LECs for tissue characterization. The following investigations include new variables compared to previous work by correlating TRFS signal (including LECs) with functional composition including inflammatory components of the arterial wall (macrophage infiltration) and MMP expression, as well as plaque biochemical content including collagen, elastin, and lipids.

19.2.3.1 Point spectroscopy

The first TRFS study of carotid plaques was conducted in freshly excised tissue from 65 endarterectomy patients (831 distinct areas) with a TRFS apparatus (Marcu et al. 2009) that allows for higher temporal resolution and faster acquisition speed (Fang et al. 2004) than instrumentation from previous work. Spectral- and time-resolved fluorescence measurements were collected. Similar to other experiments in this section, autofluorescence was induced with a 337 nm wavelength excitation pulsed nitrogen laser and a custom-made bifurcated fiber-optic probe. A monochromator was used to disperse the fluorescence that was then detected with a gated microchannel-plate photomultiplier tube and temporally resolved with a digital oscilloscope. The measurements scanned a spectral range from 360 to 550 nm in 10 nm increments, and the temporal resolution was 0.2 ns. Results indicated that time-resolved parameters derived from the normalized time-integrated fluorescence emission spectra of carotid atherosclerotic plaques could provide means for discriminating normal and intimal thickening (IT) samples from more advanced stable lesions such as fibrotic (FP) and fibrocalcified (FC) and unstable lesions such as those with macrophages infiltrated into the fibrous cap (inflamed: INF) and with a necrotic core (NEC). Figure 19.7a and b demonstrates the average spectra and lifetimes for all plaque types identified (IT: thin, fibrotic; FC: calc; NEC: Necr; INF: infl; and low-inflamed: LowInfl).

Also demonstrated was the important role that the LECs play in atherosclerotic plaque discrimination. In particular, one of these expansion coefficients (LEC-2) enabled discrimination of both INF and NEC lesions from FP and FC lesions. Figure 19.7c demonstrates that INF and NEC lesions show markedly higher LEC-2 values at 550 nm compared to the other lesion types. Figure 19.7d shows LEC-2 values for LDL, elastin, and collagen I correspond to the main plaque fluorophores in the plots in Figure 19.7c. Moreover, a combination of spectral and time-resolved fluorescence parameters, including LECs, were used as features in a linear discriminant analysis (LDA)-based classification, which discerned INF and NEC lesions from IT and FP lesions with high sensitivity (SE > 80%) and specificity (SP > 90%). The overall cross-validation classification performance was 74.3%. IT was discriminated with SE and SP larger than 80%. Also, the INF and NEC lesions were detected with high SE (larger than 80%) and high SP (larger than 90%). These results provide strong indication that the LECs may offer a novel domain for representing time-resolved characteristics of tissue fluorescence emission in a condensed, detailed, thorough, and computationally efficient way. It was clearly demonstrated that the LECs derived from the TRFS data analysis could characterize arterial tissue composition and, most importantly, detect features of vulnerable plaques, such as superficial necrosis and inflammation (Marcu et al. 2009).

This TRFS analysis of carotid plaques also demonstrates that a linear correlation between plaque biochemical content and spectroscopic parameters can be determined. Plaque elastin content, ranging from near 0% (in the advanced FP, FC, INF, and NEC) to 50% (in IT), showed a positive correlation with the intensity values at 440 nm emission (Figure 19.8a). Necrotic area, ranging from 0% (in stable plaques) to 70% (in NEC) presented a negative correlation with LEC-0 at 550 nm emission (Figure 19.8b). A similar correlation was determined for the average lifetime values. The inflammatory cell (lymphocyte/

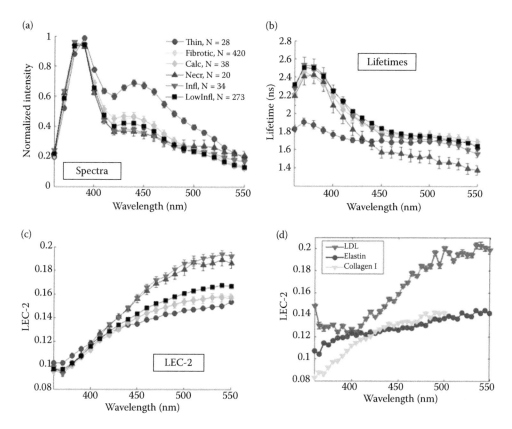

Figure 19.7 TRFS-derived variables from human carotid plaque. (a) Normalized intensity spectra; (b) average lifetimes; (c) LEC-2; (d) average LEC-2 measurements from pure plaque components LDL, elastin, and collagen I demonstrate the source of signal in (c). (Adapted from Marcu, L. et al., *Atherosclerosis*, 204(1), 156–164, 2009.)

macrophage) content, ranging from 0% (in stable plaques) to 60% (in INF and NEC), was positively correlated with LEC-2 at 550 nm (Figure 19.8c). In contrast, LEC-2 at 550 nm was negatively correlated to the collagen/SMC content (Figure 19.8d). Thus, it was demonstrated that a TRFS technique could detect clinically relevant biochemical features of human coronary atherosclerotic plaques with high specificity.

Importantly, results were reported from *in vivo* TRFS measurements of carotid plaque (Marcu et al. 2009). Sample size was too small to perform classification, but parameters acquired *in vivo* were found highly correlated with those acquired *ex vivo*. This demonstrates that information determined *ex vivo* is still translatable to a clinical implementation of this technique.

19.2.3.2 Matrix metalloproteinases

An additional application of the data described in Section 19.2.3.1 established the relationship between TRFS signal and MMP content. Structural proteins (e.g., elastin, collagens), lipids, and lipoproteins are the main autofluorescent constituents of atherosclerotic plaques (Arakawa et al. 2002), and MMP activity directly affects the biomolecular structure of these proteins and the presence of lipid components in plaques. MMPs do not fluoresce themselves, but it was determined that TRFS could be used to detect their footprint, that is, the effect of the enzymes on the plaque composition.

TRFS data from 29 carotid plaques out of the total dataset in Section 19.2.3.1 were compared to tissue sections stained with an antibody for MMP-2 and -9. It was found that both MMP-2 and -9 expression levels were elevated in plaques with higher risk of rupture and correlated with TRFS spectroscopic parameters. TRFS spectroscopic parameters correlated highly with MMP-2 and -9 expression levels 0, 1, and 2 ($R > 0.5$ and $P < 0.05$) and thus could be used to predict the levels of MMP-2 and -9 (Figure 19.9). Each level was defined by a pathologist based on the number of macrophages and smooth muscle cells

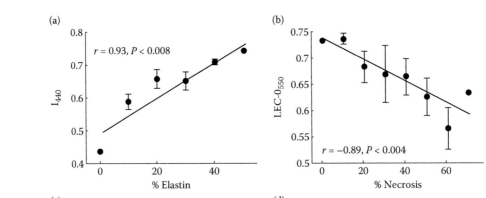

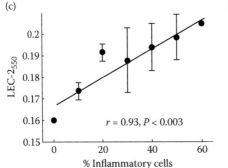

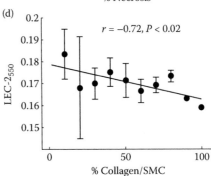

Figure 19.8 Pearson correlations between plaque components and TRFS-derived variables from human carotid plaque measurements. (a) Intensity at 440 nm is positively correlated to elastin content. (b) LEC-0 at 550 nm is negatively correlated with necrosis. LEC-2 at 550 nm is positively correlated with (c) inflammatory cell infiltration and (d) combined collagen/SMC content. (Reproduced from Marcu, L. et al., *Atherosclerosis*, 204(1), 156–164, 2009. With permission.)

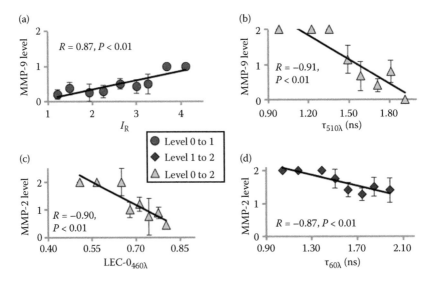

Figure 19.9 Pearson correlations between varying MMP levels (level 0-to-1, level 1-to-2, and level 0-to-2) and TRFS-derived variables in TRFS measurements of carotid plaque. (a) An MMP-9 level change from 0-to-1 is positively correlated with an intensity ratio in the range 380–390/430–450 nm (I_R); an MMP-9 level change from 0-to-2 is negatively correlated with (b) lifetime at 510 nm and (c) LEC-0 at 460 nm. (d) An MMP-2 level change from 1-to-2 is negatively correlated with lifetime at 460 nm. (Reproduced from Phipps, J. E. et al., *Journal of Biophotonics*, 4(9), 650–658, 2011a. With permission.)

that stained positive for MMP per high-power field of view. This work represents the first successful examination of MMP content in human plaques without labeling or destroying the measured tissue. While the relationships between active MMPs and TRFS parameters were not determined, an initial framework was established that will allow for the future analysis of such relationships (Phipps et al. 2011a).

19.2.3.3 Fluorescence lifetime imaging

Recent work presented a method of determining plaque biochemical composition and risk of rupture based on parameters derived from FLIM images of human carotid plaque (Phipps et al. 2012). Autofluorescence of atherosclerotic plaque (18 carotid endarterectomy patients) was measured at 43 locations through three filters—F377: 377/50 nm, F460: 460/66 nm, and F510: 510/84 nm (center wavelength/bandwidth). The fluorescence decay dynamic was described by τ and four LECs retrieved from a Laguerre deconvolution technique. LDA classified the images in 10×10 pixel regions based on biochemical composition (collagen-rich vs. lipid-rich) with sensitivity and specificity >80%. Additionally, a risk level was assigned based on the surface area composed of each constituent. Results from this investigation showed that an LDA method of classifying regions of FLIM images of carotid plaque into collagen- or lipid-rich groups is capable of being automated and used to rate the risk of plaque rupture based on autofluorescence decay dynamics and without the need for fluorescence intensity or contrast agents.

Both FLIM and TRFS studies of carotid plaque were able to implement LDA classification to identify collagen-rich (more stable) from lipid-rich (less stable) atherosclerotic compositions. Additionally, TRFS was able to identify functional components of the arterial wall, inflammatory cells, and MMPs. The FLIM study showed for the first time that an automated method using FLIM images of atherosclerotic plaque could assess risk of plaque rupture by determining lipid vs. collagen content. Automated classification methods will be important for clinical translation of new imaging devices.

19.2.4 TRFS- AND FLIM-BASED DETECTION OF BIOCHEMICAL AND FUNCTIONAL FEATURES OF ATHEROSCLEROSIS

In the past 10 years, much progress has been made toward the development of a clinically viable TRFS or FLIM method for characterizing plaque composition. It has been shown that both techniques are capable of identifying biochemical features: collagen-rich (more stable) and lipid-rich (more unstable) plaques as well as normal artery wall (elastin-rich). But it has also been shown that TRFS is capable of investigating the functional features in terms of macrophage infiltration and MMP expression. The ability of time-resolved parameter average lifetime and LECs to characterize the arterial wall without the need for contrast agents or fluorescence intensity information further increases the potential of this technique for clinical translation. In the following section, advances toward this goal are discussed.

19.3 *IN VIVO* POINT SPECTROSCOPY (TRFS) FOR MACROPHAGE DETECTION

A TRFS system was used to investigate *in vivo* the aortas of 12 male New Zealand white rabbits (6 control, 6 high-cholesterol fed) (Marcu et al. 2005). The system consisted of a pulsed nitrogen laser (337 nm excitation, 700 ps pulse width) and a custom-made bifurcated fiber-optic probe for delivery of excitation light and collection of emitted fluorescence. The collected fluorescence was dispersed by an imaging spectrometer/monochromator and detected with a gated multichannel plate photomultiplier tube. Temporal resolution was achieved with a digital phosphor oscilloscope. Spectral resolution was 5 nm and time resolution was 300 ps at a scanning speed of 0.8 s per wavelength.

The goals of this study were to determine fluorescence-derived parameters that allow discrimination between atherosclerotic intima rich in macrophages, from intima rich in collagen (fibrous tissue), and to evaluate the accuracy of this technique for predicting macrophage accumulation in the fibrous cap. Using parameters derived from the time-resolved fluorescence emission of plaques measured *in vivo* in this rabbit atherosclerotic model (including LEC-0_{500}, LEC-1_{450}, LEC-3_{390}, and a ratio of LEC-2_{500}/LEC-2_{390}), it was determined that intima rich in macrophages can be distinguished from intima rich in collagen with high sensitivity (>85%) and specificity (>95%).

These results demonstrate that TRFS may be useful in a clinical setting for the diagnosis of inflammatory activity in the fibrous cap (Marcu et al. 2005), well known to be important in the progression of atherosclerosis (Libby 2002). Additionally, the TRFS parameters and wavelengths shown to be related to macrophage content will be focused on in future experiments.

19.4 CHALLENGES FOR INTRAVASCULAR TRFS/FLIM TECHNIQUES

Despite the demonstrated potential of fluorescence lifetime techniques for diagnosis of critical arterial pathologies, these techniques were virtually untested in an intravascular setting. In part, this is due to two major challenges that hamper their practical use in detection and characterization of atherosclerotic plaques. First, the technique is limited by light penetration depth (<300 μm; Marcu et al. 2001); thus, tissue diagnosis can only focus on variation of the distribution of endogenous fluorophores at the tissue surface. Moreover, no features describing structural or anatomical changes of vessel wall are attainable with this method. Therefore, an imaging technique with deeper penetration depth would need to be integrated with TRFS to solve these issues and to provide complementary information to improve tissue diagnosis. Second, although the TRFS techniques are reported as having the ability of enabling discrimination of different types of plaques and determination of risk of plaque rupture (Marcu et al. 2009), practical instrumentation is presently unavailable for clinical applications. The development of a catheter-based intravascular system will be critical for applying this technique clinically. Recent efforts addressing these challenges have been made in order to advance the TRFS technique to intravascular diagnosis. In the following, we outline some of these studies.

19.4.1 SOLUTIONS FOR INTRAVASCULAR APPLICATIONS OF THE TRFS/FLIM TECHNIQUE

For efficient imaging of the arterial wall in an intravascular setting, an imaging technique is beneficial over a point spectroscopy technique because it allows for the visualization of extent of areas associated with features of plaque vulnerability. Additionally, image acquisition via a scanning mode, for example, a rotational single fiber, would be advantageous over an imaging fiber bundle used in wide-field FLIM techniques. The former configuration is more compact and robust when compared with the latter that would face the challenge of remaining still in a beating heart through blood flow during image acquisition. A recently reported fast data acquisition TRFS system allowing for simultaneous wavelength- and time-resolved fluorescence emission recording in multiple spectra provides a viable solution for future scanning TRFS measurements (Sun et al. 2011a). This new approach was initially reported in 2008 (Sun et al. 2008a; Figure 19.10) and allowed simultaneous acquisition and storage of fluorescence data from three wavelength bands in a few microseconds. When coupled with a compact x–y–z 3D positioning stage, this system also allowed collection of point spectroscopic data in a fast-scanning mode that can then be used for the formation of FLIM images (Sun et al. 2011a). This scanning TRFS system was validated in *ex vivo* atherosclerotic tissue (Sun et al. 2011b). Results of this study indicated that distinct compositions of human aortic plaques were clearly discriminated over different scanning positions using fluorescence lifetime and the intensity ratio at 390 and 452 nm wavelength bands.

19.4.2 SOLUTIONS FOR MULTIMODAL ANALYSIS OF ATHEROSCLEROTIC PLAQUES

19.4.2.1 Combining optical and ultrasonic technologies

Ultrasound has been used in clinical practice as a tool for evaluation and diagnosis of cardiovascular disease (Konig et al. 2008). Because ultrasound has much greater depth penetration than fluorescence techniques, one technique to address the limited penetration depth of optical techniques is to combine ultrasound imaging with TRFS or FLIM for plaque localization and for the creation of a combined diagnostic tool. The use of high-frequency transducers can provide high-resolution images with sufficient penetration depth. For example, the ultrasonic backscatter microscopy (UBM) with a 40 MHz transducer can achieve a penetration depth of up to 5 mm with a resolution of 50 μm (Sun et al. 2008b). The surface scanning UBM system can be adapted to intravascular ultrasound (IVUS) via a rotational catheter with

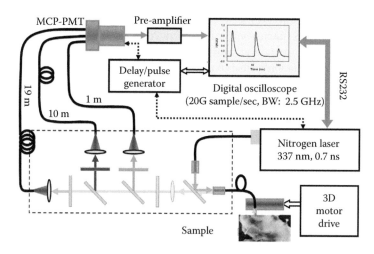

Figure 19.10 Schematic diagram of the scanning TRFS system with a single optical fiber as probe for excitation and fluorescence collection. (Adapted from Sun, Y. et al., *Optics Express*, 19(5), 3890–3901, 2011b.)

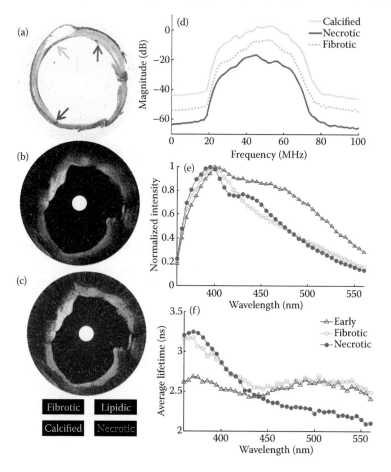

Figure 19.11 Bimodal data from an *ex vivo* section of human carotid artery. (a) Trichrome stained section of the carotid sample with arrows labeling the locations of the TRFS measurements. (b) UBM image warped into a circular shape compared with (c) color-coded image characterizing different tissue types. The plaque in (c) was colored by the automated tissue characterization algorithm. The accuracy of this characterization is confirmed by the histological analysis in (a). (d) UBM spectrum analysis for different tissue types. (e) Fluorescence emission spectrum and (f) average lifetime measured at the three locations indicated in (a).

a single element ultrasound transducer. IVUS is used in clinic as a minimally invasive diagnostic tool for atherosclerotic plaques. IVUS systems operating at frequencies greater than 40 MHz have previously been reported for the analysis of arterial wall morphology in patients (Yuan et al. 2008) and can be used to provide structural information related to plaque vulnerability (Low et al. 2009). Moreover, by using the spectral analysis of the ultrasound radiofrequency (RF) data, tissue characterization has been achieved by analysis of such features as integrated backscatter and spectral similarity (Okubo et al. 2008; Sathyanarayana et al. 2009). Therefore, the high-resolution ultrasound (UBM or IVUS) will effectively compensate the penetration limitation of the TRFS technique and provide complementary anatomical information for plaque characterization, such as the remodeling of the lumen, the size of the lipid pool, and the thickness of the fibrous cap. Moreover, since IVUS is an established technique used in clinical practice, a device that integrates TRFS within an IVUS catheter will take advantage of established clinical protocols for intravascular assessment of arterial vessels.

An additional advantage of combining optical and ultrasonic techniques provides also the opportunity to integrate photoacoustic imaging (PAI) within a TRFS-UBM catheter. PAI has been suggested as a potential modality for detection of vulnerable plaques (B. Wang et al. 2010), showing potential to resolve the lipid-rich from collagen-rich regions due to the different optical properties of these constitutes (Sethuraman et al. 2007; B. Wang et al. 2010).

19.4.2.2 Examples of multimodal scanning instrumentation

A multimodal imaging prototype was reported that included a scanning TRFS subsystem capable of creating FLIM images, a UBM subsystem with a high-frequency single-element focused transducer, and a PAI subsystem that shared electronics with the UBM subsystem and the illumination fiber with the TRFS subsystem (Sun et al. 2011a). Application of the bimodal (TRFS+UBM) or multimodal

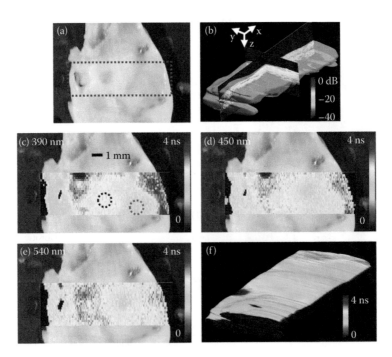

Figure 19.12 Representative coregistered scanning TRFS-based FLIM and UBM measurements of an *ex vivo* human carotid plaque sample. (a) Photo of the carotid plaque sample with the imaging area outlined by the red dashed line. (b) 3D UBM volume data from the investigated region. Surface renderings of the UBM volume with orthogonal planes are shown with 45 dB displaying dynamics range and jet color map. (c–e) FLIM images overlaid on top of the plaque photo at three different wavelengths, where the color represented average lifetime values from 0 to 4 ns for each pixel. Three regions were labeled with circles in (c) corresponding to three tissue types (pink: intimal thickening; black: fibro-lipidic plaque; and cyan: fibrotic plaque). (f) FLIM map at 390 nm coregistered with and overlaid onto a 3D UBM volume rendering. (Adapted from Sun, Y. et al., *Optics Express*, 19(5), 3890–3901, 2011b.)

(TRFS+UBM+PAI) system was achieved using a hybrid imaging probe that allowed acquisition of coregistered multimodal images. This combined probe integrated coaxially a 0.6 mm silica optical fiber (for fluorescence excitation collection and delivery of light for PAI) and a press-focused ultrasonic transducer (for UBM imaging and PAI signal detection; Sun et al. 2008b, 2009a, 2011a).

This pilot multimodal system was used to investigate whether compositional, structural, and functional characteristics of atherosclerotic plaques could be evaluated concurrently. FLIM has demonstrated potential for characterization and mapping of the biochemical composition within the superficial layer of plaques (Phipps et al. 2009). UBM enables imaging of plaque microanatomy and morphology with high resolution (~50 μm) and penetration depth (~5 mm) (S.Z. Wang et al. 2010). PAI can map the differences in optical absorption of various molecular constituents of plaques. In summary, the multimodal tissue diagnostic technique combining FLIM, UBM, and PAI provides more complete information regarding features potentially predicative of plaque rupture.

Another multimodal system was recently reported that combines OCT with FLIM for applications in atherosclerosis as well as epithelial cancer (Park et al. 2010).

19.4.2.3 Validation of the multimodal technique in *ex vivo* human carotid samples

A representative result demonstrating the complementary information acquired with a combined UBM and point TRFS system, from a study recently reported (Sun et al. 2011a), is depicted in Figure 19.11.

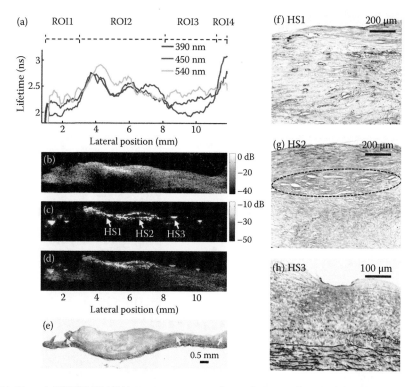

Figure 19.13 Multimodal TRFS/UBM/PAI measurements and correlation analysis with histology from an *ex vivo* human carotid plaque sample. (a) Plots of the average lifetime as a function of scanning position. Four ROIs were analyzed to show the changes in lumen-surface composition. (b) Transverse UBM images from the same position as (a) showed the reconstruction of the plaque structure, where an acoustic shadowing area indicated calcific deposits. (c) PAI image provided optical absorption contrast and structural information. HS: hot spot. (d) Coregistered UBM and PAI images were fused as one combined image. (e) The corresponding trichrome stained tissue section. (f–h) Zoomed-in histological images of the regions labeled with HS1 to HS3 in (c): (f) HS1, trichrome staining showing fibro-lipidic plaque and calcification; (g) HS2, trichrome staining demonstrating the fibrotic collagen-rich region (blue lines) and lipid-rich area underneath (white clefts); and (h) HS3, dense fibrotic knot and intimal thickening with elastin fibers (black lines) and smooth muscle cells (red). (Adapted from Sun, Y. et al., *Optics Express*, 19(5), 3890–3901, 2011b.)

The preliminary results indicated that ultrasound provided important plaque structural information that complements the compositional information obtained by TRFS, resulting in a more robust characterization of vulnerable plaques. Coregistered 3D UBM and FLIM images of *ex vivo* human carotid plaque samples were also demonstrated by Sun et al. (2011a) (Figure 19.12). This study also showed solutions for simultaneous recordings of the scanning-TRFS, UBM, and photoacoustic data that presented a wealth of complementary information concerning the plaque composition, structure, and function indicating plaque vulnerability (Figure 19.13). Specifically, FLIM images at 390, 450, and 540 nm mapped areas with distinct biochemical compositions (such as fibrotic, fibrolipidic, and intimal thickening) and enhanced detection of critical plaque features. Two high-resolution imaging techniques (UBM and PAI) were able to provide specific anatomical and functional images of the plaques (cross sections) and overcome the limited penetration depth of the TRFS/FLIM approach. This multimodal system potentially can be translated into a catheter-based system for intravascular studies.

19.4.2.4 Intravascular validation of scanning TRFS and IVUS

Recent studies demonstrate progress toward incorporating time-resolved fluorescence detection into clinically available IVUS catheters. This was achieved by advancing the surface scanning system described above (Stephens et al. 2009) to intravascular use (Sun et al. 2011b). Although in this first TRFS-UBM catheter configuration, the fiber optics required a lateral deployment to reach the ROI (Figure 19.14), this

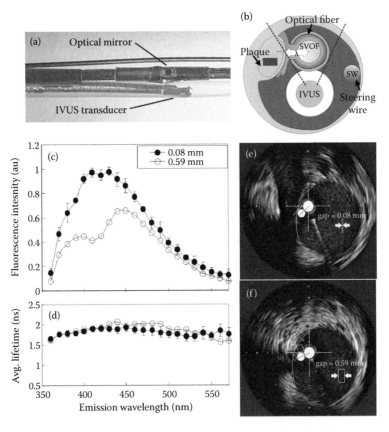

Figure 19.14 (a) The 1.8 mm multimodal catheter used in coregistration experiments. The catheter tip detail showing the orientation of the IVUS transducer aligned with the fiber optics. (b) Cross-sectional view of the catheter in operation within a vessel with saline flushing to remove the blood in the optical path. The steering wire is shown in its position against the vessel wall to deploy the catheter to the opposite vessel wall. The comparison of (c) fluorescence intensity and (d) lifetime at two pull-back positions for demonstration of blood effects. The black dot curves show the TRFS spectroscopic results with the catheter in contact (0.08 mm) with the arterial wall, and the white dot curves show the results far away (0.59 mm) from the arterial wall. The positions of the catheter are labeled in the two IVUS images shown in (e) and (f).

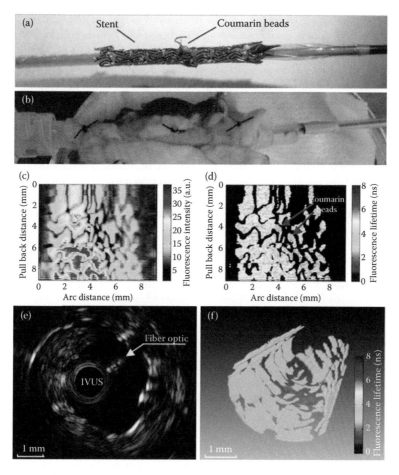

Figure 19.15 Coregistered IVUS and FLIM: evaluation in a coronary artery phantom. (a) Stent before deployment; (b) stent inserted in porcine coronary artery; (c) fluorescence intensity image; (d) fluorescence lifetime image; (e) IVUS cross-sectional image; and (f) reconstructed 3D FLIM image of the coronary lumen. (From Bec, J. et al., *Journal of Biomedical Optics*, 17(10), 106012, 2012.)

work demonstrated the ability of a single-fiber probe integrated with an ultrasound imaging catheter to operate *in vivo* in blood flow in pig femoral arteries. The results demonstrated the potential of using time-resolved parameters as a tool for real-time optical characterization of arterial tissue composition, which was found more robust than using fluorescence intensity in a blood absorption environment.

Moreover, very recent reports demonstrate successful FLIM "pull-back" image scans in a helical motion through a luminal surface with a single side-viewing optical fiber (Xie et al. 2012). This technology was also incorporated into a combined FLIM/IVUS catheter (8 French outer diameter) for simultaneous FLIM and IVUS intravascular imaging (Figure 19.15; Bec et al. 2012). Both of these studies were performed in *ex vivo* porcine arteries and establish the potential of this technology for clinical translation.

19.5 DISCUSSION

19.5.1 LIMITATIONS

There are several limitations to a time-resolved fluorescence technique for the characterization and diagnosis of atherosclerotic plaque: (1) Time-resolved fluorescence measurements acquire signal from less than 300 μm of intimal tissue and thus are only capable of determining biochemical information regarding the fibrous cap region of plaques and not overall plaque burden; (2) Time-resolved measurements provide an average signal from multiple fluorophores within the investigated volume of tissue, and thus, it is not completely specific to any single fluorophore; and (3) In an intravascular implementation, blood will need to be flushed from the field of view since it is a strong absorber of UV fluorescence.

19.5.2 FUTURE DIRECTION OF THE FIELD

Time-resolved fluorescence measurements provide information that is vital to determining risk of plaque rupture, which is presently unavailable clinically. Currently time-resolved fluorescence measurement capabilities are being incorporated into clinically available IVUS and OCT catheters, adding biochemical and functional imaging capabilities to otherwise structural imaging techniques. Progress has been made toward intravascular implementation of time-resolved fluorescence techniques, and this combined with the fact that no contrast agent is required for imaging and that the technique is being combined with FDA-approved imaging modalities increases the potential of time-resolved techniques for clinical translation. Clinical implementation of time-resolved techniques not only will allow for improved atherosclerotic plaque characterization but could also be used to study drug efficacy and stent endothelialization, both important factors in the treatment of atherosclerosis.

REFERENCES

Andersson-Engels, S., Johansson, J., Stenram, U., Svanberg, K. and Svanberg, S. (1990) Time-resolved laser-induced fluorescence spectroscopy for enhanced demarcation of human atherosclerotic plaques. *Journal of Photochemistry and Photobiology. B, Biology,* 4(4), 363–369.

Angheloiu, G. O., Arendt, J. T., Muller, M. G., Haka, A. S., Georgakoudi, I., Motz, J. T., Scepanovic, O. R., Kuban, B. D., Myles, J., Miller, F., Podrez, E. A., Fitzmaurice, M., Kramer, J. R. and Feld, M. S. (2006) Intrinsic fluorescence and diffuse reflectance spectroscopy identify superficial foam cells in coronary plaques prone to erosion. *Arteriosclerosis, Thrombosis, and Vascular Biology,* 26(7), 1594–1600.

Arakawa, K., Isoda, K., Ito, T., Nakajima, K., Shibuya, T. and Ohsuzu, F. (2002) Fluorescence analysis of biochemical constituents identifies atherosclerotic plaque with a thin fibrous cap. *Arteriosclerosis, Thrombosis, and Vascular Biology,* 22(6), 1002–1007.

Baraga, J. J., Rava, R. P., Taroni, P., Kittrell, C., Fitzmaurice, M. and Feld, M. S. (1990) Laser-induced fluorescence spectroscopy of normal and atherosclerotic human aorta using 306–310 nm excitation. *Lasers in Surgery and Medicine,* 10(3), 245–261.

Baraga, J. J., Taroni, P., Park, Y. D., An, K., Maestri, A., Tong, L. L., Rava, R. P., Kittrell, C., Dasari, R. R. and Feld, M. S. (1989) Ultraviolet-laser induced fluorescence of human aorta. *Spectrochimica Acta Part A-Molecular and Biomolecular Spectroscopy,* 45(1), 95–99.

Bartorelli, A. L., Leon, M. B., Almagor, Y., Prevosti, L. G., Swain, J. A., McIntosh, C. L., Neville, R. F., House, M. D. and Bonner, R. F. (1991) In vivo human atherosclerotic plaque recognition by laser-excited fluorescence spectroscopy. *Journal of the American College of Cardiology,* 17(6 Suppl B), 160B–168B.

Bec, J., Xie, H., Yankelevich, D. R., Zhou, F., Sun, Y., Ghata, N., Aldredge, R. and Marcu, L. (2012) Design, construction, and validation of a rotary multifunctional intravascular diagnostic catheter combining multispectral fluorescence lifetime imaging and intravascular ultrasound. *Journal of Biomedical Optics,* 17(10), 106012.

Christov, A., Dai, E., Drangova, M., Liu, L., Abela, G. S., Nash, P., McFadden, G. and Lucas, A. (2000) Optical detection of triggered atherosclerotic plaque disruption by fluorescence emission analysis. *Photochemistry and Photobiology,* 72(2), 242–252.

Cole, M. J., Siegel, J., Webb, S. E., Jones, R., Dowling, K., Dayel, M. J., Parsons-Karavassilis, D., French, P. M., Lever, M. J., Sucharov, L. O., Neil, M. A., Juskaitis, R. and Wilson, T. (2001) Time-domain whole-field fluorescence lifetime imaging with optical sectioning. *Journal of Microscopy,* 203(Pt 3), 246–257.

Dowling, K., Dayel, M. J., Lever, M. J., French, P. M., Hares, J. D. and Dymoke-Bradshaw, A. K. (1998) Fluorescence lifetime imaging with picosecond resolution for biomedical applications. *Optics Letters,* 23(10), 810–812.

Edholm, P. and Jacobson, B. (1965) Detection of aortic atheromatosis in vivo by reflection spectrophotometry. *Journal of Atherosclerosis Research,* 5(6), 592–595.

Elson, D. S., Jo, J. A. and Marcu, L. (2007) Miniaturized side-viewing imaging probe for fluorescence lifetime imaging (FLIM): Validation with fluorescence dyes, tissue structural proteins and tissue specimens. *New Journal of Physics,* 9, 127.

Fang, Q. Y., Papaioannou, T., Jo, J. A., Vaitha, R., Shastry, K. and Marcu, L. (2004) Time-domain laser-induced fluorescence spectroscopy apparatus for clinical diagnostics. *Review of Scientific Instruments,* 75(1), 151–162.

Fitzmaurice, M., Bordagaray, J. O., Engelmann, G. L., Richards-Kortum, R., Kolubayev, T., Feld, M. S., Ratliff, N. B. and Kramer, J. R. (1989) Argon ion laser-excited autofluorescence in normal and atherosclerotic aorta and coronary arteries: Morphologic studies. *American Heart Journal,* 118(5 Pt 1), 1028–1038.

Jo, J. A., Fang, Q. and Marcu, L. (2005) Ultrafast method for the analysis of fluorescence lifetime imaging microscopy data based on the Laguerre expansion technique. *IEEE Journal of Quantum Electronics*, 11(4), 835–845.

Kittrell, C., Willett, R. L., de los Santos-Pacheo, C., Ratliff, N. B., Kramer, J. R., Malk, E. G. and Feld, M. S. (1985) Diagnosis of fibrous arterial atherosclerosis using fluorescence. *Applied Optics*, 24(15), 2280–2281.

Konig, A., Margolis, M. P., Virmani, R., Holmes, D. and Klauss, V. (2008) Technology insight: In vivo coronary plaque classification by intravascular ultrasonography radiofrequency analysis. *Nature Clinical Practice Cardiovascular Medicine*, 5(4), 219–229.

Libby, P. (2002) Inflammation in atherosclerosis. *Nature*, 420(6917), 868–874.

Libby, P., DiCarli, M. and Weissleder, R. (2010) The vascular biology of atherosclerosis and imaging targets. *Journal of Nuclear Medicine*, 51(Suppl 1), 33S–37S.

Low, A. F., Kawase, Y., Chan, Y. H., Tearney, G. J., Bouma, B. E. and Jang, I. K. (2009) In vivo characterisation of coronary plaques with conventional grey-scale intravascular ultrasound: Correlation with optical coherence tomography. *EuroIntervention*, 4(5), 626–632.

Maarek, J. M. I., Marcu, L., Fishbein, M. C. and Grundfest, W. S. (2000) Time-resolved fluorescence of human aortic wall: Use for improved identification of atherosclerotic lesions. *Lasers in Surgery and Medicine*, 27(3), 241–254.

Maarek, J. M. I., Marcu, L. and Grundfest, W. S. (1998) Characterization of atherosclerotic lesions with laser-induced time-resolved fluorescence spectroscopy. *Proceedings of SPIE*, 3250, 181–187.

Maarek, J. M. I., Snyder, W. J. and Grundfest, W. S. (1997) Time-resolved laser-induced fluorescence of arterial wall constituents: Deconvolution algorithm and spectrotemporal characteristics. *Proceedings of SPIE*, 2980, 278–285.

Marcu, L., Fang, Q. Y., Jo, J. A., Papaioannou, T., Dorafshar, A., Reil, T., Qiao, J. H., Baker, J. D., Freischlag, J. A. and Fishbein, M. C. (2005) In vivo detection of macrophages in a rabbit atherosclerotic model by time-resolved laser-induced fluorescence spectroscopy. *Atherosclerosis*, 181(2), 295–303.

Marcu, L., Fishbein, M. C., Maarek, J. M. and Grundfest, W. S. (2001) Discrimination of human coronary artery atherosclerotic lipid-rich lesions by time-resolved laser-induced fluorescence spectroscopy. *Arteriosclerosis, Thrombosis, and Vascular Biology*, 21(7), 1244–1250.

Marcu, L., Grundfest, W. S. and Fishbein, M. (2003) Time-resolved laser-induced fluorescence spectroscopy for staging atherosclerotic lesions. In Mycek, M. A. and Pogue, B. W., eds., *Handbook of Biomedical Fluorescence*. New York: Marcel Dekker, 397–430.

Marcu, L., Jo, J. A., Fang, Q., Papaioannou, T., Reil, T., Qiao, J. H., Baker, J. D., Freischlag, J. A. and Fishbein, M. C. (2009) Detection of rupture-prone atherosclerotic plaques by time-resolved laser-induced fluorescence spectroscopy. *Atherosclerosis*, 204(1), 156–164.

Marcu, L., Maarek, J. M. I. and Grundfest, W. S. (1998) Time-resolved laser-induced fluorescence of lipids involved in development of atherosclerotic lesion lipid-rich core. *Proceedings of SPIE*, 3250, 158–167.

Morguet, A. J., Korber, B., Abel, B., Hippler, H., Wiegand, V. and Kreuzer, H. (1994) Autofluorescence spectroscopy using a XeCl excimer-laser system for simultaneous plaque ablation and fluorescence excitation. *Lasers in Surgery and Medicine*, 14(3), 238–248.

Okubo, M., Kawasaki, M., Ishihara, Y., Takeyama, U., Yasuda, S., Kubota, T., Tanaka, S., Yamaki, T., Ojio, S., Nishigaki, K., Takemura, G., Saio, M., Takami, T., Fujiwara, H. and Minatoguchi, S. (2008) Tissue characterization of coronary plaques—Comparison of integrated backscatter intravascular ultrasound with virtual histology intravascular ultrasound. *Circulation Journal*, 72(10), 1631–1639.

Packard, R. R. S. and Libby, P. (2008) Inflammation in atherosclerosis: From vascular biology to biomarker discovery and risk prediction. *Clinical Chemistry*, 54(1), 24–38.

Papazoglou, T. G., Liu, W. Q., Katsamouris, A. and Fotakis, C. (1994) Laser-induced fluorescence detection of cardiovascular atherosclerotic deposits via their natural emission and hypocrellin (Ha) probing. *Journal of Photochemistry and Photobiology B-Biology*, 22(2), 139–144.

Park, J., Jo, J. A., Shrestha, S., Pande, P., Wan, Q. and Applegate, B. E. (2010) A dual-modality optical coherence tomography and fluorescence lifetime imaging microscopy system for simultaneous morphological and biochemical tissue characterization. *Biomedical Optics Express*, 1(1), 186–200.

Phipps, J. E., Hatami, N., Galis, Z. S., Baker, J. D., Fishbein, M. C. and Marcu, L. (2011a) A fluorescence lifetime spectroscopy study of matrix metalloproteinases-2 and -9 in human atherosclerotic plaque. *Journal of Biophotonics*, 4(9), 650–658.

Phipps, J. E., Sun, Y., Fishbein, M. C. and Marcu, L. (2012) A fluorescence lifetime imaging classification method to investigate the collagen to lipid ratio in fibrous caps of atherosclerotic plaque. *Lasers in Surgery and Medicine*, 44(7), 564–571.

Phipps, J. E., Sun, Y., Saroufeem, R., Hatami, N., Fishbein, M. C. and Marcu, L. (2011b) Fluorescence lifetime imaging for the characterization of the biochemical composition of atherosclerotic plaques. *Journal of Biomedical Optics*, 16(9), 096018.

Tissue autofluorescence lifetime spectroscopy and imaging: Applications

Phipps, J., Sun, Y., Saroufeem, R., Hatami, N. and Marcu, L. (2009) Fluorescence lifetime imaging microscopy for the characterization of atherosclerotic plaques. *Proceedings—Society of Photo-Optical Instrumentation Engineers,* 7161, 71612G.

Roger, V. L., Go, A. S., Lloyd-Jones, D. M., Adams, R. J., Berry, J. D., Brown, T. M., Carnethon, M. R., Dai, S., de Simone, G., Ford, E. S., Fox, C. S., Fullerton, H. J., Gillespie, C., Greenlund, K. J., Hailpern, S. M., Heit, J. A., Ho, P. M., Howard, V. J., Kissela, B. M., Kittner, S. J., Lackland, D. T., Lichtman, J. H., Lisabeth, L. D., Makuc, D. M., Marcus, G. M., Marelli, A., Matchar, D. B., McDermott, M. M., Meigs, J. B., Moy, C. S., Mozaffarian, D., Mussolino, M. E., Nichol, G., Paynter, N. P., Rosamond, W. D., Sorlie, P. D., Stafford, R. S., Turan, T. N., Turner, M. B., Wong, N. D. and Wylie-Rosett, J. (2010) Heart disease and stroke statistics—2011 update: A report from the American Heart Association. *Circulation,* 123(4), e18–e209.

Ross, R. (1993) The pathogenesis of atherosclerosis: A perspective for the 1990s. *Nature,* 362(6423), 801–809.

Sathyanarayana, S., Carlier, S., Li, W. G. and Thomas, L. (2009) Characterisation of atherosclerotic plaque by spectral similarity of radiofrequency intravascular ultrasound signals. *EuroIntervention,* 5(1), 133–139.

Sethuraman, S., Amirian, J. H., Litovsky, S. H., Smalling, R. W. and Emelianov, S. Y. (2007) Ex vivo characterization of atherosclerosis using intravascular photoacoustic imaging. *Optics Express,* 15(25), 16657–16666.

Sharif, F. and Murphy, R. T. (2010) Current status of vulnerable plaque detection. *Catheterization and Cardiovascular Interventions,* 75(1), 135–144.

Stephens, D. N., Park, J., Sun, Y., Papaioannou, T. and Marcu, L. (2009) Intraluminal fluorescence spectroscopy catheter with ultrasound guidance. *Journal of Biomedical Optics,* 14(3), 030505.

Sun, Y., Chaudhari, A. J., Lam, M., Xie, H., Yankelevich, D. R., Phipps, J., Liu, J., Fishbein, M. C., Cannata, J. M., Shung, K. K. and Marcu, L. (2011a) Multimodal characterization of compositional, structural and functional features of human atherosclerotic plaques. *Biomedical Optics Express,* 2(8), 2288–2298.

Sun, Y., Liu, R., Elson, D. S., Hollars, C. W., Jo, J. A., Park, J. and Marcu, L. (2008a) Simultaneous time- and wavelength-resolved fluorescence spectroscopy for near real-time tissue diagnosis. *Optics Letters,* 33(6), 630–632.

Sun, Y., Park, J., Stephens, D. N., Jo, J. A., Sun, L., Cannata, J. M., Saroufeem, R. M., Shung, K. K. and Marcu, L. (2009a) Development of a dual-modal tissue diagnostic system combining time-resolved fluorescence spectroscopy and ultrasonic backscatter microscopy. *Review of Scientific Instruments,* 80(6), 065104.

Sun, Y., Phipps, J., Elson, D. S., Stoy, H., Tinling, S., Meier, J., Poirier, B., Chuang, F. S., Farwell, D. G. and Marcu, L. (2009b) Fluorescence lifetime imaging microscopy: In vivo application to diagnosis of oral carcinoma. *Optics Letters,* 34(13), 2081–2083.

Sun, Y., Stephens, D. N., Park, J., Marcu, L., Cannata, J. M. and Shung, K. K. (2008b) Development of a multi-modal tissue diagnostic system combining high frequency ultrasound and photoacoustic imaging with lifetime fluorescence spectroscopy. *Proceedings IEEE Ultrasonics Symposium,* 570–573.

Sun, Y., Stephens, D., Xie, H., Phipps, J., Saroufeem, R., Southard, J., Elson, D. S. and Marcu, L. (2011b) Dynamic tissue analysis using time- and wavelength-resolved fluorescence spectroscopy for atherosclerosis diagnosis. *Optics Express,* 19(5), 3890–3901.

Talbot, C., McGinty, J., McGhee, E., Owen, D., Grant, D., Kumar, S., Beule, P., Auksorius, E., Manning, H., Galletly, N., Treanor, B., Kennedy, G., Lanigan, P., Munro, I., Elson, D., Magee, A., Davis, D., Neil, M., Stamp, G., Dunsby, C. and French, P. (2010) Fluorescence lifetime imaging and metrology for biomedicine. In Tuchin, V. V., ed., *Handbook of Photonics for Biomedical Science.* Boca Raton, FL: CRC Press, 159–196.

Thomas, P., Pande, P., Clubb, F., Adame, J. and Jo, J. A. (2010) Biochemical imaging of human atherosclerotic plaques with fluorescence lifetime angioscopy. *Photochemistry and Photobiology,* 86(3), 727–731.

Verbunt, R. J., Fitzmaurice, M. A., Kramer, J. R., Ratliff, N. B., Kittrell, C., Taroni, P., Cothren, R. M., Baraga, J. and Feld, M. (1992) Characterization of ultraviolet laser-induced autofluorescence of ceroid deposits and other structures in atherosclerotic plaques as a potential diagnostic for laser angiosurgery. *American Heart Journal,* 123(1), 208–216.

Virmani, R., Burke, A. P., Farb, A. and Kolodgie, F. D. (2006) Pathology of the vulnerable plaque. *Journal of the American College of Cardiology,* 47(8 Suppl), C13–C18.

Virmani, R., Kolodgie, F. D., Burke, A. P., Farb, A. and Schwartz, S. M. (2000) Lessons from sudden coronary death—A comprehensive morphological classification scheme for atherosclerotic lesions. *Arteriosclerosis, Thrombosis, and Vascular Biology,* 20(5), 1262–1275.

Wang, B., Su, J. L., Amirian, J., Litovsky, S. H., Smalling, R. and Emelianov, S. (2010) Detection of lipid in atherosclerotic vessels using ultrasound-guided spectroscopic intravascular photoacoustic imaging. *Optics Express,* 18(5), 4889–4897.

Wang, S. Z., Huang, Y. P., Saarakkala, S. and Zheng, Y. P. (2010) Quantitative assessment of articular cartilage with morphologic, acoustic and mechanical properties obtained using high-frequency ultrasound. *Ultrasound in Medicine and Biology,* 36(3), 512–527.

Xie, H., Bec, J., Liu, J., Sun, Y., Lam, M., Yankelevich, D. and Marcu, L. (2012) Multispectral scanning time-resolved fluorescence spectroscopy (TRFS) technique for intravascular diagnosis. *Biomedical Optics Express,* 3(7), 1521–1533.

Yuan, J., Rhee, S. and Jiang, X. N. (2008) 60 MHz PMN-PT based 1-3 Composite Transducer for IVUS Imaging. *Proceedings IEEE Ultrasonics Symposium*, 682–685.

20

Ophthalmic applications of FLIM

Dietrich Schweitzer

Contents

20.1 MOTIVATION

The first signs of the onset of pathological processes are frequently changes in metabolism. If such changes can be detected and treated in time, the pathological process can often be reversed. Metabolism can be studied *in vivo* in the human eye (particularly the fundus), where indicators of external and internal aspects of metabolism are presented. External aspects include blood flow and oxygen saturation in retinal vessels that characterize the microcirculation. By determining the product of blood flow and the arterial oxygen saturation, the supply of oxygen can be determined for specific regions of the retina (Schweitzer et al. 1995). Internal aspects of metabolism can be observed by studying the fluorescence of endogenous fluorophores that accumulate in anatomical structures of the eye or act as coenzymes in basic processes of energy production, for example, respiration or glycolysis, for which fluorescence lifetime imaging microscopy (FLIM) can provide useful quantitative information. While measurements of microcirculation are helpful for the early detection of various diseases, this approach has its limitations. For example, in a study of diabetes mellitus (Hammer et al. 2009; Schweitzer et al. 2007b), the blood flow was found to be higher and the arteriovenous difference in oxygen saturation was found to be lower compared to age-matched healthy subjects. These results point to an oversupply of oxygen to the retinal tissue, but in fact, the diffusion of oxygen from vessels in the tissue is hindered because of thickening of membranes in diabetes. Furthermore, reduced oxygen consumption was found for bovine retinal pericytes when exposed with high glucose (Trudeau 2011). The internal aspects of metabolism can be studied by measuring the fluorescence of endogenous fluorophores, as shown in pioneering works of Chance (1976). The redox pairs of coenzymes NAD-NADH and FAD-FADH$_2$ are of special interest, because they are electron transporters in fundamental mechanisms of energy production (glycolysis, respiratory chain). The presence of additional endogenous fluorophores in biological tissue, however, complicates their measurement, and it is generally necessary to take account of the specific excitation and emission spectra, as well as fluorescence decay profiles, in order to discriminate between the signals from specific fluorophores.

20.2 CONSTRAINTS FOR FLIM MEASUREMENTS IN THE EYE

In comparison with measurements of time-resolved autofluorescence of cell or organ cultures, or in isolated tissues or small animals, *in vivo* FLIM measurements at the human eye are limited by several conditions. Most important is the maximal permissible exposure according to the ANSI Standard Z 136.1 2000 since the eye is highly photosensitive and damage by the excitation radiation must be avoided under all circumstances. Studies have shown that parallel excitation of the whole ocular fundus for FLIM requires excitation radiation levels above the permitted exposure, and so wide-field frequency domain FLIM or time-gated FLIM is precluded (Schweitzer et al. 2000). The optimum approach for *in vivo* FLIM of the eye is therefore to implement FLIM in a confocal laser scanning system with low power excitation of the fundus and sensitive detection of the extremely weak fluorescence emission by time-correlated single-photon counting (TCSPC). As a consequence of the low fluorescence signal, long measurement times are required to achieve a useful signal-to-noise ratio. Unfortunately, the eye is a stochastically moving object, and so image registration is required to permit the correct accumulation of the fluorescence photons corresponding to each pixel of the fluorescence image.

FLIM measurements at the fundus are also constrained by the transmission of the ocular media. Whereas the cornea blocks all light having wavelengths shorter than 350 nm, the absorption edge of the lens is at 400 nm (Boettner and Wolter 1962; Geeraets and Berry 1968), and in a young eye, there is only 50% transmission at 450 nm. From 550 nm up to the absorption lines of water at about 900 nm, the transmission of the ocular media is about 96%. This transmission decreases at all wavelengths with increasing age. As the maxima of excitation spectra are shorter than 400 nm for several fluorophores (Berezin and Achilefu 2010; Schweitzer et al. 2007c), an efficient discrimination of fundus fluorophores is not possible by selective excitation. Figure 20.1 shows effective excitation spectra of FAD, advanced glycation end-products, and A2E. For these spectral profiles, the original excitation spectra are multiplied by the transmission of the ocular media.

A further constraint is imposed by the anatomy of the eye. In contrast to typical fluorescence microscopy measurements, where samples are approximately uniform throughout the depth of focus, eye has a strongly layered structure. The anterior part (cornea, lens) should be distinguished from the posterior part (fundus) of the eye, and the ocular fundus itself consists of several functional layers (of nerve fibers

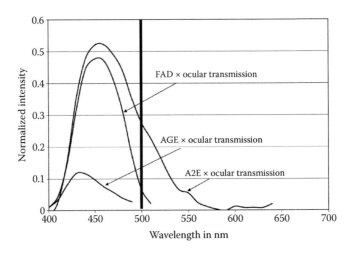

Figure 20.1 Effective excitation spectra of N-retinylidene-N-retinylethanolamine (A2E), flavin adenine dinucleotide (FAD), and advanced glycation end-products (AGE). The excitation spectra of the pure substances are multiplied by the transmission of the ocular media (OT). (Schweitzer, D. et al.: Towards metabolic mapping of the human retina. *Microscopy Research and Technique*. 2007c. Volume 70. 410–419. Copyright Wiley-VCH Verlag GmbH & Co. KGaA. Reproduced with permission.)

and of ganglion cells, of bipolares and of receptors, retinal pigment epithelium [RPE], and choroid). Although laser scanning confocal imaging provides optical sectioning, the fluorescence of the ocular lens still influences the minimum detectable fundus fluorescence. The suppression of the unwanted fluorescence background from the lens depends on the diameter of the aperture diaphragm (i.e., the pupil of the eye) and on the diameter of the confocal pinhole at the fundus (Schweitzer et al. 2005). The depth of focus in a confocal laser scanning ophthalmoscope is limited by the numerical aperture of the eye to ~300 µm, which extends over several fundus layers. As several fluorophores, located in different layers, are typically excited simultaneously, it is hard to determine the signal from specific fluorophores. A better axial resolution could, in principle, be provided by two-photon excitation, but, as shown by Wang (2006), there is only a factor 2 to 3 between the excitation intensity required to obtain a useful fluorescence signal and the threshold to damage the absorbing RPE. This technique is therefore excluded from *in vivo* application for functional diagnostic imaging of the fundus.

20.3 EXAMPLE OF A CORRECTED *IN VIVO* AUTOFLUORESCENCE SPECTRUM OF THE EYE GROUND

To understand which fluorophores dominate the emission from the fundus, the fluorescence of the whole eye and the fluorescence of the lens were measured using the experimental setup shown in Figure 20.2. The system consisted of a confocal laser scanner ophthalmoscope and an imaging spectrograph adapted with a control unit. The fundus was excited by a 446 nm pulse laser (LDH-440, Picoquant, Berlin, Germany) in a 20° field. The spectrally resolved fluorescence was accumulated for all pixels over the acquisition time to measure the whole eye. To measure the fluorescence of the lens alone, a nonexcited field was imaged onto the entrance slit of the spectrograph. Figure 20.3 shows the fluorescence spectrum of the lens of a 63 year old man measured *in vivo*. This "background" fluorescence spectrum was subtracted from the fluorescence spectrum of the whole eye. As the shape of the fundus fluorescence is modified by absorption in the ocular lens, the fluorescence spectrum from the whole eye was corrected for the ocular transmission, and it was also corrected to account for absorption by the macular pigment xanthophyll.

The resulting corrected spectrum for the fundus fluorescence is shown in Figure 20.4. This spectrum is dominated by the fluorescence of A2E, for which the spectrum is also presented. A2E is the component VIII of the fluorescent fraction from the chloroform extract of RPE from human donors (Eldred and Katz 1988). Local maxima and shoulders indicate the contribution of other fluorophores. The local maxima attributed to FAD and AGE are quite narrow because of the low number of notes for spectra construction. Unknown fluorophores until now, with emission maxima longer than 600 nm also contribute to the fundus fluorescence.

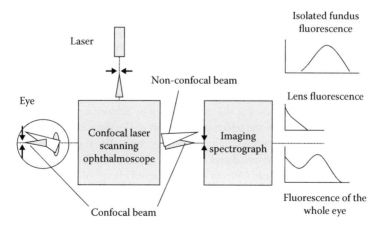

Figure 20.2 Setup for measurement of the fluorescence of the whole eye and separate lens fluorescence. The fluorescence of the fundus is the difference of both measurements.

Tissue autofluorescence lifetime spectroscopy and imaging: Applications

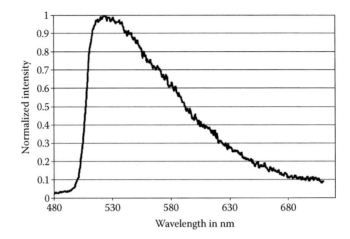

Figure 20.3 Fluorescence spectrum of the crystalline lens, measured *in vivo* of a 63-year-old man.

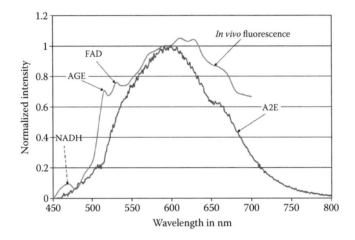

Figure 20.4 Comparison of *in vivo* measured fundus fluorescence spectrum with emission spectrum of A2E, component VIII of lipofuscin. The fundus fluorescence spectrum was corrected for ocular transmission and absorption in the macular pigment xanthophyll.

20.4 OVERVIEW OF LIFETIMES OF ENDOGENOUS FLUOROPHORES AND ISOLATED OCULAR STRUCTURES

For discrimination between endogenous fluorophores, specific knowledge of excitation and emission spectra, as well as of fluorescence lifetimes, is required. The excitation and emission spectra of the endogenous fluorophores that are expected at the eye are given in Schweitzer et al. (2007c). The transmission of the ocular media limits the selective excitation of the fluorophores such that the effective excitation spectra are the pure excitation spectra multiplied by the ocular transmission. As shown in Figure 20.1, the endogenous fluorophores have broad and overlapping excitation spectra, and most of these have effective excitation peak wavelengths near 450 nm. A2E is the only fluorophore that can be excited by wavelengths longer than 500 nm and discriminated by selective excitation.

The endogenous fluorophores can be better discriminated by resolving the emission spectra. As demonstrated in Figure 20.5, the emission of AGE, FAD, and other fluorophores is detectable in the spectral range below 560 nm, while the spectral range longer than 560 nm is dominated by A2E. In addition to these fluorescence components, Figure 20.5 also shows the emission spectrum of the

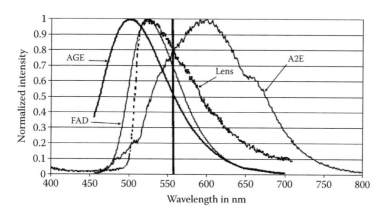

Figure 20.5 Emission spectra of AGE, FAD, and A2E together with the fluorescence of the crystalline lens. AGE, FAD, lens fluorescence, and to a lesser degree A2E are detectable in the short wavelength range <560 nm. In the range >560 nm, A2E dominates. (Schweitzer, D. et al.: Towards metabolic mapping of the human retina. *Microscopy Research and Technique.* 2007c. Volume 70. 410–419. Copyright Wiley-VCH Verlag GmbH & Co. KGaA. Reproduced with permission.)

crystalline lens, which overlaps the fundus fluorescence in the spectral range shorter than 560 nm. The fluorescence of the crystalline lens was measured *in vivo* by excitation at 446 nm and detected above 500 nm.

Whereas the spectral properties of endogenous fluorophores are generally consistent across the literature, there is considerable variation in the published lifetime data. One reason for this is the variation in the number of exponential components in the fitting models. An overview of the spectroscopic parameters of endogenous fluorophores is given by Berezin and Archilefu (2010), where maxima of excitation and emission spectra are presented as well as mean fluorescence lifetimes. Lifetimes and amplitudes for more specific ophthalmic endogenous fluorophores are given in Table 20.1.

New results for fundus autofluorescence lifetime components and their amplitudes of *in toto* fundus structures of porcine eyes, determined using two-photon excitation, are given in Table 20.2 (Peters et al. 2011).

Because of the improved optical sectioning in two-photon applications, lifetime measurements are possible within single layers of the retina. The fundus layers were excited by 150 fs pulses at 760 nm, corresponding to 380 nm at single-photon excitation, and the fluorescence was detected in a short (500–560 nm) and in a long (560–700 nm) wavelength spectral channel, as used in the *in vivo* lifetime mapper (Schweitzer et al. 2007c). The autofluorescence lifetime components and their amplitudes of ocular structures excluding fundus layers are given in Table 20.3.

The values presented in Tables 20.1 and 20.3 were measured using single-photon excitation at 446 nm with picosecond pulses (~100 ps full width at half-maximum [FWHM]), and the fluorescence was detected in the spectral range 500–700 nm. The presented data were obtained from 10 dissected samples (Schweitzer et al. 2007a) for which all the anatomical structures were carefully separated from the porcine eyes. Figure 20.6 shows fluorescence lifetime images of representative porcine eyes where it is apparent that the lifetime in RPE is equal in both spectral channels and much shorter than in the receptor layer which is different in the spectral channels 500–550 and 560–700 nm.

Considering these data from porcine eyes, the shortest lifetime components in fundus measurements (about 40–170 ps) originate from the RPE, the median lifetime components (about 350–450 ps) are related to the receptor layer and the ganglion cell layer, and the longest lifetime components (about 1.5–4 ns) originate from the lens and connective tissue. It is difficult to attribute these lifetimes to single fluorophores since the lifetimes associated with particular anatomical structures relate to mixtures of fluorophores found in each layer. The effect of changes in single fluorophores can be determined by provocation tests of tissue cultures or in case of some diseases by using prior knowledge of pathomechanisms. The fluorescence spectra of isolated anatomical structures of porcine eyes are given in Schweitzer (2010).

Table 20.1 **Lifetimes and amplitudes of substances, expected in ocular tissue**

SUBSTANCE	τ_1 (PS)	a_1 (%)	τ_2 (PS)	a_2 (%)	REFERENCE
Collagen 1	670	68	4040	32	Own measurement
Collagen 2	470	64	3150	36	Own measurement
Collagen 3	345	69	2800	31	Own measurement
Collagen 4	740	70	3670	30	Own measurement
Elastin	380	72	3590	28	Own measurement
AGE	865	62	4170	28	Own measurement
A2E	170	98	1120	2	Own measurement
Lipofuscin	390	48	2240	52	Own measurement
DOPA melanin	40		1200		Ehlers et al. 2007
Pheomelanin	340		2300		Ehlers et al. 2007
Eumelanin	30		900		Ehlers et al. 2007
Melanocyten Naevus	136 ± 33	94 ± 3	1061 ± 376	6 ± 3	Dimitrow et al. 2008
Melanin	280	70	2400	30	Own measurement
FAD free	330	18	2810	82	Own measurement
FAD bound	110–130		Monomer		Skala et al. 2007
Bound	40		Dimer		Nakashima et al. 1980
NADH bound	6040				Vishwasrao et al. 2005
Bound	2154				
Bound	599				
NADH free folded	600–700				
Free extended	155				
NADH free	387	73	3650	27	Own measurement
NADH protein bound	>1 ns				Wu et al. 2006

Source: Schweitzer, D. et al.: Towards metabolic mapping of the human retina. *Microscopy Research and Technique.* 2007c. Volume 70. 410–419. Copyright Wiley-VCH Verlag GmbH & Co. KGaA.

20.5 REQUIRED PHOTON NUMBER FOR DISCRIMINATION OF FLUOROPHORES

It is first useful to estimate the number of photons excited during a typical measurement of the fundus. According to the ANSI standard (ANSI Z-136.1 2000), the mean radiant power is more critical in the pulse regime than in continuous-wave exposure. In Schweitzer et al. (2000), a mean radiant power of 390 µW with a repetition rate of 77.3 MHz was assumed in the cornea plane. The number of photons excited by each pulse of 5 pJ can be estimated according to the following equation:

$$N_{Det} = \frac{E_{Ex} \cdot \tau_{Ex} \cdot \Phi_{Fl} \cdot \Delta\lambda \cdot \Omega \cdot \tau_{Em} \cdot \lambda_{Device} \cdot \tau_F \cdot \Phi_{Detector}}{h \cdot \nu_{Em}} \tag{20.1}$$

where

N_{Det} = detectable fluorescence photons
E_{Ex} = energy of excitation pulse in cornea plane = 5 pJ

Table 20.2 Lifetimes and amplitudes of fundus structures, determined by two-photon excitation at 790 nm

		NFL	GCL	GCL-C	GCL-M	INL	ONL	PRIS	RPE
500–550 nm	a_1 (%)	72.5 ± 5.7	74.1 ± 6.9	80.0 ± 3.4	66.6 ± 3.7	73.1 ± 4.1	68.6 ± 3.4	76.2 ± 3.3	98.4 ± 0.9
	τ_1 (ns)	0.39 ± 0.05	0.39 ± 0.06	0.36 ± 0.04	0.43 ± 0.05	0.40 ± 0.06	0.39 ± 0.07	0.43 ± 0.06	0.07 ± 0.01
	a_2 (%)	27.5 ± 5.7	25.9 ± 6.9	20.0 ± 3.4	33.4 ± 3.7	26.9 ± 4.1	31.4 ± 3.4	23.8 ± 3.3	1.6 ± 0.9
	τ_2 (ns)	2.35 ± 0.20	2.36 ± 0.25	2.22 ± 0.24	2.48 ± 0.20	2.36 ± 0.24	2.64 ± 0.26	2.33 ± 0.27	0.61 ± 0.23
550–700 nm	a_1 (%)	83.4 ± 2.4	85.3 ± 3.0	86.7 ± 2.2	81.9 ± 2.2	85.7 ± 2.1	88.6 ± 2.1	91.6 ± 2.5	98.3 ± 1.2
	τ_1 (ns)	0.23 ± 0.03	0.21 ± 0.04	0.20 ± 0.04	0.23 ± 0.04	0.20 ± 0.04	0.17 ± 0.03	0.17 ± 0.03	0.07 ± 0.01
	a_2 (%)	16.6 ± 2.4	14.7 ± 3.0	13.3 ± 2.2	18.1 ± 2.2	14.3 ± 2.1	11.4 ± 2.1	8.4 ± 2.5	1.7 ± 1.2
	τ_2 (ns)	1.98 ± 0.20	1.93 ± 0.25	1.78 ± 0.19	2.13 ± 0.18	1.98 ± 0.22	2.07 ± 0.24	1.81 ± 0.19	0.54 ± 0.22

Source: Peters, S. et al., SPIE 8086: 808604-1–808604-10, 2011. With permission of SPIE.

Note: GCL, ganglion cell layer (c: cell body, m: Müller cell); INL, inner nuclear layer; NFL, nerve fiber layer; ONL, outer nuclear layer; PRIS, photoreceptor inner segments; RPE, retinal pigment epithelium.

Tissue autofluorescence lifetime spectroscopy and imaging: Applications

T_{Ex} = ocular transmission for excitation light = 0.9 (at about 488 nm)
Φ_{Fl} = quantum yield of fluorophore = 100 nJ nm^{-1} sr^{-1}/J (Delori 1994)
$\Delta\lambda$ = wavelength range of emission = 100 nm
Ω = solid angle of detection = 0.02 sr
τ_{Em} = ocular transmission for emitted light = 0.9
λ_{Device} = transmission of the ophthalmoscope = 0.9
τ_F = transmission of the cutoff filter = 0.8
$\Phi_{Detector}$ = quantum yield of photo-detector = 0.1
ν_{Em} = frequency of the emitted light $\equiv$ 600 nm
h = Planck constant

This indicates that there is 0.06 potentially detectable fluorescence photon from the fundus per 5 pJ excitation pulse. Thus, approximately 10 excitation pulses are required for each detectable fluorescence photon. These conditions fit optimally the requirements for the application of TCSPC. Equation 20.1 can be used to optimize the detection system and indicates that a larger detection solid angle and higher photodetector quantum yield will increase the number of detectable fluorescence photons. Unfortunately, the given transmission of the ophthalmoscope (λ_{Device} = 0.9) is over-optimistic. As the excitation of

Table 20.3 **Lifetimes and amplitudes of ocular structures exclusive of fundus tissue**

STRUCTURE	τ_1 (PS)	a_1 (%)	τ_2 (PS)	a_2 (%)
Cornea	570	70	3760	30
Sclera	450	65	3110	35
Choroid	500	70	3400	30
Lens	490	69	3600	31

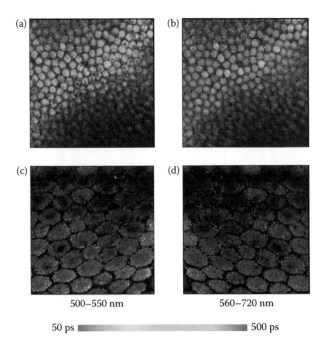

500–550 nm   560–720 nm

50 ps   500 ps

Figure 20.6 Mean lifetime image of fundus structures after two-exponential fit, excited by fs two-photon pulses at 800 nm. Receptor layer: (a) emission 500–550 nm, (b) emission 560–700 nm; RPE: (c) emission 500–550 nm, (d) emission 560–700 nm.

endogenous fluorophores should be between 450 and 470 nm, the transmission of the ocular media is reduced to about 0.5. It is the crystalline lens that limits the short wavelength range of the incoming light in the eye, and, in the case of artificial lenses implanted after cataract surgery, the transmission of the ocular media can approach 1 both for excitation and emission light.

To investigate how many photons are required for the determination of component lifetimes within acceptable error boundaries, noisy two-exponential decay profiles were simulated with different lifetime components and amplitudes and fitted to a two-exponential decay model. Independent of the actual value of lifetimes, the relative difference of lifetimes, for example, $(\tau_2 - \tau_1)/\tau_2$, can be considered to determine the number of photons required to distinguish lifetimes. The lifetime τ_1 was set to be 100 ps corresponding to that found in RPE, while the lifetime τ_2 was set to 400 ps corresponding to the lifetime in the neuronal retina. Thus, the relative difference of lifetimes is 0.75. Besides the lifetimes, the relative contribution Q_i is important:

$$Q_i = \frac{\tau_i \cdot a_i}{\sum_j \tau_j \cdot a_j} \tag{20.2}$$

Thus, both the lifetime values, τ_i, and the relative amplitudes, a_i, determine the required number of detected photons. Figure 20.7 shows how the sum of relative errors, ES, varies as a function of the relative contribution, Q_1, and with the number of detected photons where ES was calculated according to the following equation:

$$ES = \left| \frac{\tau_{1set} - \tau_{1actual}}{\tau_{1set}} \right| \cdot 100 + \left| \frac{\tau_{2set} - \tau_{2actual}}{\tau_{2set}} \right| \cdot 100 \tag{20.3}$$

where, τ_{iset} are the lifetime components of the synthesized fluorescence decay profiles, and $\tau_{iactual}$ are the lifetime values determined by fitting the noisy decay profiles to a two-exponential decay model.

Figure 20.7 shows that the minimum number of required photons is reached when the relative contributions of the two lifetime components are equal, that is, an equal number of photons originate from each component. For an error sum of 20%, 10,000 photons are sufficient if the relative contribution of components $Q_1 = Q_2$ and if $\tau_1 = 100$ ps and $\tau_2 = 400$ ps. To achieve an error sum of about 10%, 40,000 photons are required. This result is in good agreement with the data presented in Köllner (1992). From

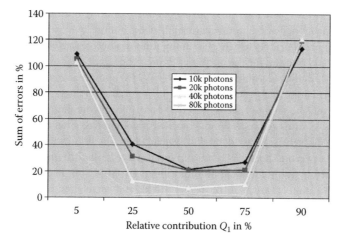

Figure 20.7 Sum of errors of lifetimes in simulation of two-exponential decay as a function of relative contribution Q_1, parameter photon number.

these data, the required data acquisition times can be estimated. For excitation by pulses at 80 MHz repetition rate, a detection probability of 0.1, and a field of 150 × 150 pixels, 4.69 min is required to accumulate 10,000 photons and 18.75 min is required for 40,000 photons. Improvements in the quantum efficiency of the detectors could reduce the measuring time, and this would permit the excitation power to be further reduced below the safety threshold for irradiating the eye.

20.6 LIFETIME LASER SCANNER OPHTHALMOSCOPE

Based on the theoretical and experimental considerations outlined above, a laser scanning lifetime ophthalmoscope was developed with TCSPC detection. The sequential pixel acquisition of the confocal laser scanning system permits higher exposure levels per pixel compared to simultaneous excitation of the whole fundus. It has the disadvantage, however, of increased data acquisition times, during which eye movements can result in distortion of the acquired images. A further advantage of the confocal scanning approach is that the volume of the crystalline lens that is also excited during fundus fluorescence measurements is much smaller than when using fundus cameras where the background lens fluorescence significantly impairs the contrast of fundus fluorescence images. According to Schweitzer et al. (2005), the suppression U_r of the reflection at the crystalline lens is determined by the following equation:

$$U_r = \frac{d_{field}^2}{d_{aperture}^2} \qquad (20.4)$$

where d_{field} = diameter of the pinhole or of the confocal spot at the fundus and $d_{aperture}$ = diameter of the aperture diaphragm or of the pupil of the eye.

Taking into consideration the fluorescent layers of the eye, the thicknesses of the crystalline lens z_{lens} and of the fundus z_{fundus} should also be taken into account for calculation of the suppression U_f of fluorescence light:

$$U_f = \frac{d_{field}^2 \cdot z_{lens}}{d_{aperture}^2 \cdot z_{fundus}} \qquad (20.5)$$

Assuming d_{field} = 0.1 mm, $d_{aperture}$ = 7 mm, z_{lens} = 5 mm, and z_{fundus} = 0.3 mm, then the fluorescence of the lens is suppressed by a factor U_f = 0.0034. In practice, a contribution from the crystalline lens fluorescence is detectable in fundus fluorescence measurements made using laser scanning ophthalmoscopes.

Figure 20.8 shows a block schematic of the developed laser scanning ophthalmoscope for mapping of time-resolved fluorescence. The basic optomechanical device is a modified laser scanning ophthalmoscope HRA2 (Heidelberg Engineering, Heidelberg, Germany) that has been adapted to provide inputs for the pulsed lasers at 448 or 468 nm and for the output of fluorescence light. The excitation lasers BLD 440 or BLD 470 (Becker/Hickl, Berlin, Germany) emit 70 ps (FWHM) pulses at a repetition rate of 80 MHz. Most clinical studies were done with 446 nm excitation. The excitation lasers are coupled via a 3 μm core diameter optical fiber, and the fluorescence is detected via a 100 μm core diameter optical fiber, which acts as a detection pinhole. The excitation light is blocked by a long pass filter at 488 nm (LP02-488 RS, Semrock, Rochester, NY) in the separate detection system where the fluorescence light is divided by a dichroic beam splitter between a short (490–560 nm) and a long (560–700 nm) wavelength channel and detected in both channels by multichannel photomultipliers (R3809U-50, Hamamatsu, Herrsching, Germany). After amplification (HFAC-26dB, Becker/Hickl) and correlation by a router (HRT-41, Becker/Hickl), the photon signals are counted using a TCSPC board SPC 150 (Becker/Hickl).

Images of the fluorescence intensity decay profiles are recorded in both spectral channels, while, simultaneously with the pulsed excitation, the fundus is also illuminated by a continuous (CW) near-infrared (NIR) laser (817 nm) to acquire NIR reflection fundus images that are recorded using a frame grabber. These NIR contrast-rich fundus images are recordable also in case of mild cataract lenses. The

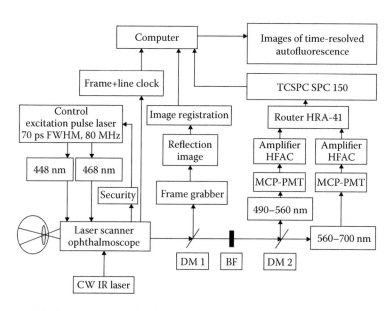

Figure 20.8 Scheme of the laser scanner ophthalmoscope for time-resolved autofluorescence. For better discrimination of fluorophores, the emission is detected in two spectral channels. An IR laser delivers contrast-rich images permitting image registration also in the case of cataract lens. The dichroic mirror DM 1 separates the IR light. The excitation light and additionally the remaining IR light are blocked by a bandpass filter (BF). The fluorescence light is separated by DM 2 in a short and a long wavelength channel. There is a security system for switching off the lasers if the scanners are not moving.

HRA2 laser scanning ophthalmoscope has an algorithm for online registration of the NIR images. The image registration data are combined with the TCSPC data (i.e., frame and line clock, as well as information about the channel where the photon was detected) such that each detected photon is assigned to the correct TCSPC time channel, spectral channel, and image pixel. Thus, the correct fluorescence decay profiles are recorded for each spatial pixel in each spectral channel from which spectrally resolved fluorescence lifetime and intensity images can be obtained. A field subtending 30° is excited at the fundus, and the image registration can accommodate eye movements for deviations of the actual image up to half of the reference image. This field is divided in 150 × 150 pixels, corresponding to a pixel size of 40 × 40 μm², which is approximately one third of the diameter of a large retinal vessel. The interval between consecutive pulses (12.5 ns) is divided in 1024 time channels, resulting in a time resolution of 12.2 ps. The excitation power is about 120 μW in the cornea plane, corresponding to about 1% of the maximal permissible exposure according to the ANSI standard (ANSI Z 136.1 2000). The instrument incorporates a safety system that interrupts the laser immediately if one of the scanners is not working. During measurement, the accumulation of photons at each pixel in both spectral channels, the decay curve of the whole image, and the actual color-coded images of fluorescence intensity can be observed.

The measurement time depends on the number of accumulated photons. The lowest count rate of fluorescence photons is obtained from the macula. The macular pigment xanthophyll has an absorption spectrum that peaks at 460 nm, and, depending on the optical density of xanthophyll, some of the excitation light is absorbed. A practical criterion for the measuring time is given by the requirement to accumulate at least 1000 photons in the macula. After binning by $B = 2$, the decay curve is determined by 25,000 photons. In regions outside the macula and the optic disc, the accumulated photon number is then ~30,000–40,000, corresponding to measurement time of 10 to 15 min. During this time, the patient can blink, close their eyes, or move their head without losing the image registration. The measurement is interrupted if no vessel structure is detectable in IR fundus images and continues automatically if contrast-rich fundus images of the reference region are detectable again.

For evaluation of the fluorescence intensity decay profiles, the data are fitted to a three-exponential model function according to the following equation:

Tissue autofluorescence lifetime spectroscopy and imaging: Applications

$$\frac{I(t)}{I_o} = \sum_{i=1}^{3} a_i \cdot e^{-\frac{t}{\tau_i}} + b \tag{20.6}$$

where I_o is the number of photons at $t = 0$; $I(t)$ is the number of photons at time t; a_i is the amplitude or pre-exponential factor; τ_i is the lifetime of exponent i; and b is the background.

The detected fluorescence intensity decay profile is a convolution of the actual sample decay profile with the excitation pulse and the temporal instrument response of the detectors. In the fitting process, the model function in Equation 20.6 is convolved with the experimentally determined instrumental response function (IRF). The criterion for an optimal approximation is the minimization of χ_r^2:

$$\chi_r^2 = \frac{1}{n-q} \cdot \sum_{j=1}^{n} \frac{[N(t_j) - N_c(t_j)]^2}{N(t_j)} \tag{20.7}$$

where n are the time channels (here applied 1024); q are the free parameters (a_i, τ_i, b); $N(t_j)$ are the detected photons in time channel j; and $N_c(t_j)$ are the photons of convolution of model function and IRF.

Assuming the detection of photons as a Poisson process, the limiting value in Equation 20.7 is 1.

The fitting process yields the lifetime components, τ_i, the amplitudes, a_i, and the relative contributions, Q_i according to Equation 20.2, from which the mean lifetime, τ_m, can be calculated:

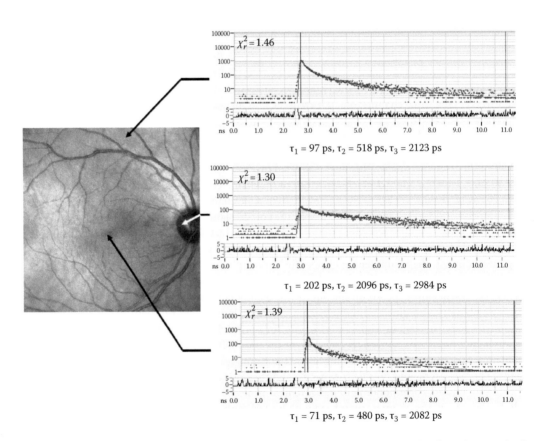

Figure 20.9 Fluorescence decay at selected fundus sites. The shortest decay is detectable from the macula, the location of the highest visual acuity. At the optic disc, the blind spot, the longest decay of connective tissue is detectable. (From Schweitzer, D. et al., *SPIE* 8087: 808750-1–808750-9, 2011. With permission of SPIE.)

$$\tau_m = \frac{\sum\limits_{i=1}^{3} a_i \cdot \tau_i}{\sum\limits_{i=1}^{3} a_i} \tag{20.8}$$

Typical lifetime values are presented in Figure 20.9 for characteristic locations at the fundus. The shortest lifetime, τ_1, is observed for the macula and the longest for the optic disc. The lifetimes τ_2 and τ_3 are comparable for macula and the range outside macula and optic disc. These generally longer lifetimes in the optic disc indicate contributions from the fluorescence of connective tissue in this structure.

20.7 INTERPRETATION OF FLIM MEASUREMENTS OF THE EYE

The program SPCImage 3080 (Becker/Hickl) was used to fit the time-resolved fluorescence data. The fitting algorithm in this program typically assumes a steep slope of the fluorescence decay after excitation. In fluorescence measurements of the eye, however, there is a delay between the onset of the fluorescence of the lens and the fundus fluorescence. This delay results in a stepped fluorescence decay profile that cannot be correctly handled by the "standard" fitting algorithms, and so the data were fitted using "tail fits" with the left border of the fitting interval set to the maximum of the fluorescence curve. An algorithm that includes the stepped slope of fluorescence in a layered structure will be explained in Section 20.9.

The three-exponential decay model is an approximation to the complex fluorescence decay profiles resulting from the multiple fluorophores in the eye. Nevertheless, this provides good correspondence between anatomical ocular structures and the fitted amplitudes and lifetime components, particularly in the short wavelength channel. Figure 20.10 illustrates how images obtained using optical coherence

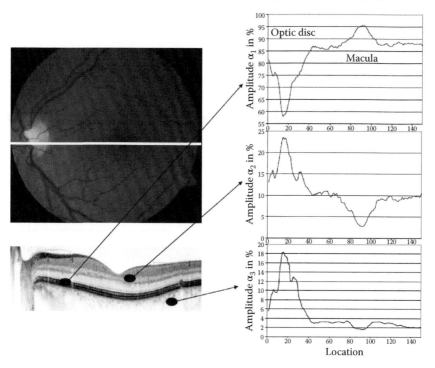

Figure 20.10 Correspondence between amplitudes in three-exponential fit of fluorescence decay and anatomical structures of the fundus, determined by OCT. In the macula, where the neuronal retina is thin, the fluorescence of the RPE (α_1) dominates up to 95%. The amplitude of α_2, which corresponds to the neuronal retina, reduces down to 3% in the macula. The fluorescence of connective tissue dominates in the optic disc. (From Schweitzer, D. et al., *SPIE* 8087: 808750-1–808750-9, 2011. With permission of SPIE.)

Tissue autofluorescence lifetime spectroscopy and imaging: Applications

tomography (OCT) correlate with line scans of amplitudes of specific lifetime components across a color fundus image.

The relative amplitude, a_1, is about 85% and increases up to 95% in the macula, where there is no nerve fiber layer but only receptor axons. That means that most of the macular fluorescence originates from the RPE. In the optic disc, where there is no RPE, a_1 decreases considerably down to less than 60%. The relative amplitude a_2 is ~10% and lower in the macula according to the reduced thickness of the neuronal retina, decreasing down to ~3%. In the optic disc, where retinal nerves leave the eye, a_2 increases to more than 20%. The relative amplitude a_3 is generally weak and decreases in the macula to ~2%. It increases up to 18% in the optic disc, corresponding to the dominating neuronal and connective tissue in this region. The variation in the lifetimes of these exponential components is shown in Figure 20.11.

The lifetime τ_1, corresponding to RPE, is in the order of 60 ps and is reduced to 40 ps in the macula. In the optic disc, τ_1 increases to more than 120 ps. The lifetime τ_2 is about 400 ps at the fundus. There is no change in the macular region despite the neuronal retina being very thin in the macula. This behavior confirms that the lifetime is independent of the concentration of fluorophores. In the optic disc, τ_2 is elongated up to 1200 ps. The lifetime τ_3 is in the order of 3000 ps and is elongated slightly in the macula but is increased up to 5000 ps in the optic disc. These lifetimes correspond to connective tissue. τ_3 in the spectral range 490–560 nm is influenced by the fluorescence of the lens because it is shortened both in the long wavelength spectral channel and in eyes with artificial intraocular lenses.

The correspondence between RPE and the exponential component with the shortest lifetime, τ_1, as well as between neuronal retina and the exponent with the lifetime τ_2, was confirmed by two-photon

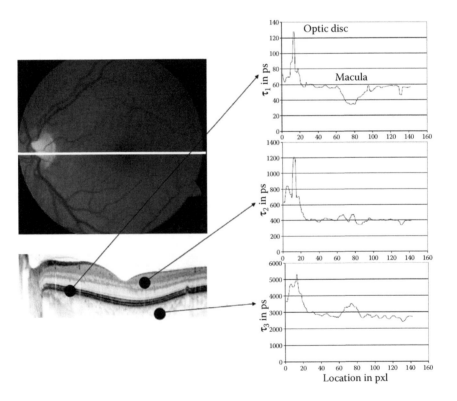

Figure 20.11 Correspondence between lifetimes in three-exponential fit of fluorescence decay in the short wavelength channel and anatomical structures of the fundus, determined by OCT. The short component, which corresponds to RPE, is about 60 ps and goes down to 40 ps in the macula. The middle lifetime, corresponding to the neuronal retina, is constant at the fundus at about 400 ps. The long lifetime is in the order of 3000 ps and originates from connective tissue but is also influenced by the long fluorescence decay of the crystalline lens. (From Schweitzer, D. et al., *SPIE* 8087: 808750-1–808750-9, 2011. With permission of SPIE.)

measurements on porcine fundus layers. Images of mean lifetime τ_m of RPE and of receptor layer after excitation at 760 nm and biexponential approximation were already shown in Figure 20.6. Whereas τ_m of RPE is relatively short (~50 ps) and equal in both spectral channels, the mean lifetime τ_m of the receptor layer is ~500 ps in the short wavelength channel but ~200 ps in the long wavelength channel.

FLIM measurements in the eye can facilitate discrimination between healthy subjects and patients presenting disease. There are many different methods to interpret such FLIM data, for example, detecting local alterations in images of lifetimes, amplitudes, or relative contribution Q_i. For determination of global metabolic changes, histograms of the values of FLIM parameters, depicting their frequency of occurrence in a region of interest, are practically useful. Such histograms should be determined for regions of interest with the same size and relative position in the fundus in all images. Because the optic disc can appear in different relative positions in the 30° images of different subjects and its relatively long fluorescence decay time could lead to artifacts misrepresenting variations in pathology, the optic disc was not included in the evaluated region of interests in the studies described below. The lifetime in the optic disc is, however, of particular interest for glaucoma studies.

Histograms of FLIM parameters can be evaluated in different ways, for example, by comparing global parameters like the median, mean, or modal values of histograms of each subject. More detailed information can be obtained if the frequencies of FLIM parameter values are compared in intervals across the individual histograms of healthy subjects and patients, for example, according to the Holm–Bonferroni algorithm (Holm 1979). Here, the range of the FLIM parameter is divided in n intervals, and the frequency of the FLIM parameter value in each interval is compared between healthy subjects and patients by the Wilcoxon rank sum test. The level of significance, for example, $p = 0.05$, is then divided by the number n of intervals. Calculation of all intervals of the considered FLIM parameter provides a useful discrimination of the considered groups. The values of FLIM parameters in significantly different intervals between healthy subjects and patients can be used for calculation of sensitivity and specificity, which are important to evaluate FLIM as a diagnostic tool.

Histograms of lifetimes can be used for discovering substance-specific pathologic alterations. For this, the difference between the mean histograms of all patients suffering from a specific disease, and, of all healthy subjects, is calculated. Negative values in a particular lifetime interval indicate that fluorophores having this lifetime are absent at certain pixels in the region of interest in the pathologic case. Positive values at certain lifetime intervals correspond to the appearance of additional fluorophores at some pixels in the ROI of patients.

20.8 CLINICAL EXAMPLES

20.8.1 AGE-RELATED MACULAR DEGENERATION

A key potential application of FLIM is the detection of early pathological alterations of metabolic parameters, particularly when no changes are detectable at the eye by conventional diagnostic methods. In a first example, FLIM parameters of patients suffering from early stages of age-related macular degeneration (AMD), according to the AREDS classification (AREDS Research Group 2001), were compared with age-matched healthy subjects (Schweitzer et al. 2009b). As demonstrated in Figure 20.12, significant differences were found in the short wavelength channel, with τ_2 increasing in patients with AMD, while no differences in τ_2 were found in the long wavelength channel. According to the correspondence between the lifetimes of exponential components and anatomical structures, alterations in τ_2 in the short wavelength channel represent changes in the neuronal retina. This illustrates how a subjectively detected impairment of vision can be objectively detected using FLIM.

Local variations in lifetimes τ_i (in ps), amplitudes a_i (in %), and relative contributions Q_i (in %) are demonstrated for advanced AMD in Figure 20.13. Besides lifetime values, images of lifetime component amplitudes and relative contribution are helpful for detection and interpretation of pathological changes. Interestingly, the variation in FLIM parameters appears more clearly in the short wavelength channel images, where endogenous fluorophores emit to a greater extent than in the long wavelength channel.

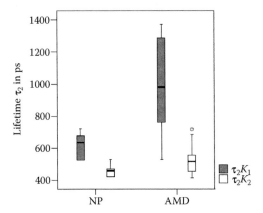

Figure 20.12 Changes of fluorescence intensity of τ_2 in early age-related macular degeneration. The alteration is significant only in the short wavelength channel. (With kind permission from Springer Science+Business Media: *Der Ophthalmologe*, Vergleich von Parametern der zeitaufgelösten Autofluoreszenz bei Gesunden und Patienten mit früher AMD, 106, 2009, Schweitzer et al., Abb. 2.)

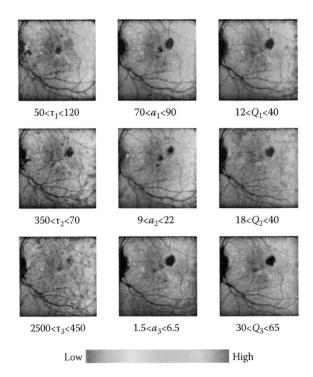

Figure 20.13 Images of lifetimes, amplitudes, and relative contribution in advanced age-related macular degeneration after three-exponential fit of the fluorescence in the short wavelength channel.

20.8.2 BRANCH ARTERIAL OCCLUSION

In branch arterial occlusion, the supply by oxygen is interrupted in defined regions at the fundus. The FLIM parameters in these regions of changed metabolism can be compared with data obtained from normally supplied regions of the same size. In Schweitzer et al. (2010), a shift to longer τ_1 lifetimes was found in the undersupplied regions for both spectral channels. The strongest difference between both regions was detected for τ_2 in the short wavelength channel: the distribution of the τ_2 histogram varied from 350 up to 650 ps with a modal value at 480 ps in the supplied regions, while in the undersupplied regions, the histogram of τ_2 was extended up to 1500 ps. In the long wavelength channel, almost identical τ_2 histograms were found ranging from 350 up to 650 ps with a modal value at about 490 ps in both regions. The extension of τ_2 values from about 480 up to 1500 ps only in the emission range 490–560 nm is probably influenced by changes from free NADH (380–700 ps) to protein-bound form of NADH (>1 ns). This change is interpretable as an increased contribution of glycolysis to the energy production in the retina.

20.8.3 DIABETIC RETINOPATHY

Diabetes mellitus is a systemic metabolic disease. As long-term consequence of diabetes, diabetic retinopathy can lead to blindness, when untreated. Early detection of metabolic alterations at the fundus can avoid the onset of diabetic retinopathy or at least lead to a patient-optimized therapy. First results of a clinical study are published in Schweitzer et al. (2011) for which well-controlled diabetic patients, having no signs of diabetic retinopathy (RD 0), were compared with age-matched healthy subjects (NP). A study of the change in FLIM parameters with increasing severity of diabetes was also undertaken with patients presenting first signs of diabetic retinopathy (RD 1). An overview of the data from both groups is given in Table 20.4.

To achieve valid comparisons, the FLIM parameters were determined in a region of 100×70 pixels, which was located at the same relative position in all acquired images of time-resolved autofluorescence. The macula was included in the region, but the optic disc was excluded because it was not completely contained in all measurements. A tail fit to a three-exponential decay model was undertaken for the fluorescence decay profiles at each pixel in the region, and the histograms of amplitudes, lifetimes, and relative contributions for each lifetime component were compared between healthy subjects and diabetic patients using the Holm–Bonferroni method. The interval sizes and ranges for lifetime values, amplitudes, and relative contribution are given in Table 20.5.

Statistical comparisons of the FLIM parameters (lifetime component values, amplitudes, and relative contributions) in both spectral channels between age-matched healthy subjects and diabetic patients in the stage RD 0 of diabetic retinopathy can potentially provide different contrast between the different patient groups. In this data set, changes in amplitudes are significantly different only for a_3 at 4% and 4.5% in the shorter wavelength detection channel (K1: 490–560 nm). The relative contributions in K1 also provide a means to differentiate NP and RD 0 (Q_1: 19%, 21%, 37%, and Q_2: 21%, 33%, error probability 5%).

Table 20.4 **Characterization of healthy subjects and diabetic patients, included in the FLIM study**

PROBANDS	NUMBERS	AGE IN YEARS	DURATION IN YEARS	HBA1C IN %	GLUCOSE IN MMOL/L	RR DIAS IN MMHG	RR SYS. IN MMHG
Healthy subjects (crystalline lens)	47	53 ± 18.2				131 ± 11	80 ± 9.3
Diabetes mellitus RD 0 (no diabetic retinopathy)	57	56.1 ± 14.3	10.6 ± 7.2	7.9 ± 1.3	8.4 ± 2.2	139 ± 15	85 ± 11
Diabetes mellitus RD 1 (mild diabetic retinopathy)	18	60.2 ± 16.5	20 ± 10	7.7 ± 0.93	7.7 ± 2.3	141 ± 16.3	85 ± 7.7
Healthy subjects (artificial lens)	8	75.3 ± 6.5				6.5	
Diabetes mellitus RD 0 (artificial lens)	9	74.6 ± 5.4				5.4	

Source: Schweitzer, D. et al., *SPIE* 8087: 808750-1–808750-9, 2011. With permission of SPIE.

Tissue autofluorescence lifetime spectroscopy and imaging: Applications

Table 20.5 **Ranges and interval size of amplitudes, lifetimes, and relative contributions for statistical comparison of pixel frequency in images of time-resolved autofluorescence of healthy subjects and diabetic patients**

PARAMETER	INTERVAL SIZE	RANGE
Amplitude a_1	2%	60%–100%
Amplitude a_2	1%	5%–30%
Amplitude a_3	0.5%	0.5%–8%
Lifetime τ_1	5 ps	40–120 ps
Lifetime τ_2	20 ps	300–800 ps
Lifetime τ_3	100 ps	K1: 2000–6200 ps K2: 1000–4000 ps
Rel. contribution Q_1	2%	5%–50%
Rel. contribution Q_2	2%	20%–50%
Rel. contribution Q_3	2%	20%–70%

Source: Schweitzer, D. et al., *SPIE* 8087: 808750-1–808750-9, 2011. With permission of SPIE.

Table 20.6 **Lifetime intervals of significant differences in frequency of pixels between healthy subjects and diabetic patients, suffering not yet from diabetic retinopathy**

LIFETIME	LIFETIME INTERVALS FOR SIGNIFICANT DIFFERENCES IN FREQUENCY (IN PS)	ERROR (IN %)	CHANNEL
τ_1	45, 50	5	490–560 nm
	340, 360, 380, 460, 480, 500, 520	5	
τ_2	340, 360, 460, 430	0.5	
	460	0.1	
τ_3	2700, 2800 2900, 3000, 3700, 3800, 3900	5	
	2800	1	
τ_1	60, 65, 75, 80	5	560–700 nm
	60, 65, 80	1	
	60, 80	0.5	
	60	0.1	
τ_2	360	5	
	2000, 2100, 2500, 2600, 2700	5	
τ_3	2000, 2500, 2600	1	
	2000, 2600	0.5	

Source: Schweitzer, D. et al., *SPIE* 8087: 808750-1–808750-9, 2011. With permission of SPIE.

The best discrimination between healthy subjects and diabetic patients, having no signs of diabetic retinopathy, was observed in the lifetime histograms. This indicates that it is not so much the relative contributions from different layers that is changed in diabetes, but more the fluorophore composition in the layers. Table 20.6 shows the lifetime intervals, in which the frequency is different between NP and RD 0. The most sensitive lifetime components in the lower wavelength channel (K1: 490–560 nm) are τ_2 and τ_3, while the lifetime τ_1 provides the least statistically significant comparison of NP and RD 0 subjects. In the longer wavelength channel (K2: 560–700 nm), no discrimination is possible between NP and RD 0 both for amplitudes and relative contributions. In contrast to K1, the lifetime τ_1 and τ_3 provide the best discrimination between NP and RD 0 with the lifetime τ_2 providing only weak differentiation.

As presented in histograms of lifetimes in Figure 20.14, there is a shift to longer lifetimes in both spectral channels with increasing severity of diabetic retinopathy. For τ_2 and τ_3, this shift is more prominent in K1 than in K2. The influence of the long fluorescence decay of the crystalline lens is clearly observable in the histogram of τ_3 in K1 already for healthy subjects. Whereas the modal value of τ_3 for NP is at 3050 ps in K1, this lifetime is most frequent at 2150 ps in K2. In case of RD 0, the modal value of τ_3

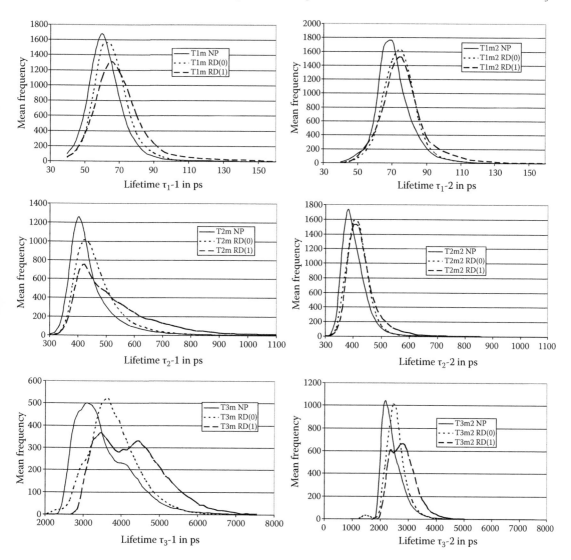

Figure 20.14 Histograms of lifetimes in the short wavelength channel (τ_i-1) and in the long wavelength channel (τ_i-2) for healthy subjects and diabetic patients, having no (RD 0) or first signs (RD 1) of diabetic retinopathy. In all lifetimes, there is a shift to longer values with increasing severity of diabetic retinopathy.

Tissue autofluorescence lifetime spectroscopy and imaging: Applications

Tissue autofluorescence lifetime spectroscopy and imaging: Applications

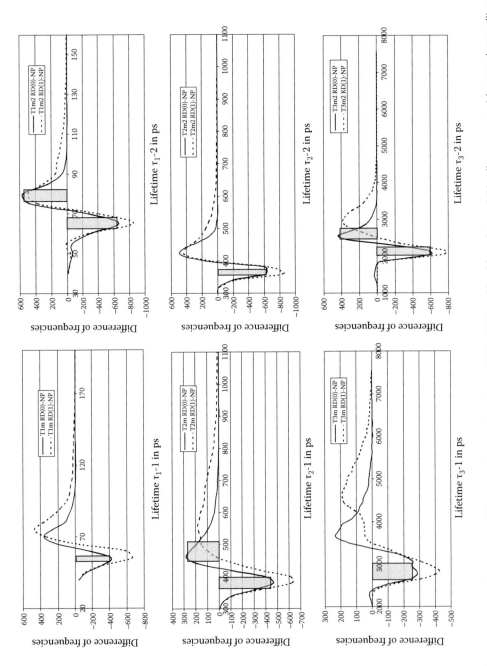

Figure 20.15 Difference of mean histograms of lifetimes of diabetic patients (RD 0, RD 1) and healthy subjects (NP). All eyes were with natural crystalline lens. Negative value mean pixels are missing in the region of interest, having corresponding lifetimes. Pixels with additionally appearing lifetimes in diabetes are positive. Pixels with elongated lifetimes in diabetes are more frequent in lifetimes of the short wavelength channel.

is enlarged to 3650 ps in K1 and to 2450 ps in K2. The larger difference of lifetime τ_3 in RD 0 and NP in the short wavelength channel than in the long wavelength channel points to changes in lifetimes in both the crystalline lens and in the fundus.

Calculating the differences between the distribution of lifetimes can indicate where fluorophores are missing or additionally present in diabetes patients. Figure 20.15 shows the differences in frequency distribution of lifetimes for RD 0-NP and RD 1-NP. The lifetime intervals for significant different frequent pixels in the region of interest are added. These intervals are located at the absolute values of maximal differences.

The modal values of pixels with lifetimes indicating missing and of additionally appearing fluorophores are given in Table 20.7. The distribution of pixels with additional fluorophores is asymmetric with considerably more pixels having longer lifetimes than the modal value in comparison to pixels having shorter lifetimes. Interestingly, the modal value of pixels with missing lifetime τ_2 is exactly at 380 ps for both spectral channels. This lifetime matches the lifetime of free NADH. The appearance of pixels with longer τ_2 lifetime can be interpreted as a change from free to protein-bound NADH, despite the fact that NADH is only weakly excitable at 448 nm. This modification probably originates in the neuronal retina.

The considerably increased lifetimes in the short wavelength channel 490–560 nm, where the crystalline lens fluoresces, point to an accumulation of advanced glycation end-products in the lens in diabetes. To clarify this assumption, the difference in lifetime histograms of diabetic patients (RD 0) and healthy subjects was determined for subjects wearing artificial intraocular lenses (IOLs). Despite having only four healthy subjects and seven diabetic patients with RD 0 in the study, clear trends are apparent. Figure 20.16 shows the difference histograms of τ_2 and τ_3 for both groups. Pixels with the long decay time τ_2 up to 700 ps and up to 5000 ps for τ_3 are missing in difference histograms of both channels. As shown in Figure 20.5, the fluorescence spectrum of AGEs is detectable in both spectral channels but is dominant in the short wavelength channel. Thus, there is an accumulation of AGEs in the crystalline lens of diabetic patients, in addition to changes in the fundus fluorescence. It is important to note that the fluorescence of the lens, which has a mean lifetime of ~3 ns, also influences the lifetime component τ_2, when the fluorescence decay profile of the whole eye is approximated to a three-exponential model function.

The regions with a reduced number of pixels for lifetime $340 < \tau_2 < 380$ ps or increased number of pixels for $450 < \tau_2 < 480$ ps in K1: 490–560 nm are color-coded as light blue and yellow in Figure 20.17. Both regions appear randomly distributed and not correlated with specific fundus locations.

In advanced diabetic retinopathy, regions of elongated lifetime τ_2 are detectable. In branch arterial occlusion, the lifetime τ_2 in the short wavelength channel was considerably increased up to 1500 ps in the undersupplied region. Thus, the spot-like fields of increased lifetime in diabetic retinopathy point to locally limited regions of reduced metabolism. The demarcation of such regions can be used to optimize therapy. For example, these regions can be targeted for coagulation in a personalized treatment mode.

The histograms of fluorescence lifetime τ_2 in the short wavelength channel can be used for calculation of sensitivity (correct detection of early diabetic retinopathy, RD 0) and specificity (correct detection of healthy subject). For this, the difference in frequency of occurrence of pixels in the lifetime intervals $\tau_2 = 460$ ps and $\tau_2 = 340$ ps was calculated. All subjects with a pixel difference above 294 were classified as diabetic patients, suffering from diabetic retinopathy in stage RD 0. This resulted in both sensitivity and

Table 20.7 **Lifetimes at modal values of missing and additionally appearing pixels in diabetic patients in comparison with age-matched healthy subjects**

LIFETIME (IN PS)	MISSING		ADDITIONAL	
	K1	K2	K1	K2
τ_1	55	65	70	80
τ_2	380	380	460	420
τ_3	2850	2050	3650	2550

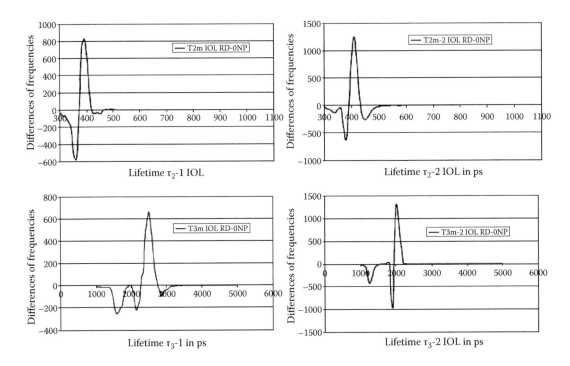

Figure 20.16 Difference of mean histograms of fluorescence lifetimes τ_2 and τ_3 for diabetic patients not yet suffering from diabetic retinopathy and healthy subjects. Only a few subjects are included, wearing intraocular lenses.

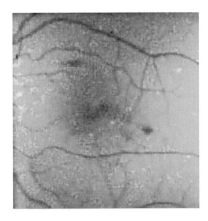

Figure 20.17 Image of lifetime t_2 in the short wavelength channel with missing (blue) and additional (yellow) appearing lifetimes according to the modal values in difference histograms of Figure 20.15 (blue: 340–380 ps; red: 450–480 ps; green: 300–600 ps).

specificity of 75%. Thus, FLIM measurements could be used to optimize the time between inspections of diabetic patients and treatment by a clinician.

20.9 OUTLOOK

Improvements in the sensitivity of detectors will increase the number of detected photons. In test measurements, the hybrid detector HPM-100-40 (Hamamatsu) was 4 to 10 times more sensitive, depending on the wavelength, than the MCP-PMT (R3809U-50, Hamamatsu) used for the work discussed here. Thus, the measurement time or the excitation power could be reduced. Measurements with increased temporal resolution down to 3 ps are also possible.

The diagnostic benefit of FLIM measurements in the eye will be evaluated for detection of early metabolic alterations in further diseases. It may be that individual lifetime components or global parameters, such as the median lifetime of each patient group suffering from a particular disease, form specific clusters in diagrams plotting τ_3, τ_2 vs. τ_1, etc. A preclassification of disease state might be possible according to such clusters. Further, by comparing individual lifetime (τ_3, τ_2 vs. τ_1) clusters with clusters of lifetime measurements of isolated substances, it may be possible to determine the composition of substances measured *in vivo* (Schweitzer et al. 2004).

The calculated lifetimes have been based on three-exponential tail fit to the measured fluorescence decay profiles, but this approach does not take into account the structure of fluorescent layers in the eye. Figure 20.18 shows a typical fluorescence intensity decay profile measured after pulsed excitation that exhibits a stepped behavior. This stepped slope is the result of multiple fluorescence decay profiles detected at different arrival times from different layers in the eye, as represented schematically in Figure 20.19.

As demonstrated in Schweitzer (2009b), the time-resolved fluorescence of a layered structure can be approximated according to the following equation:

$$\frac{I(t)}{I_o} = \sum_{i,j=1}^{i=p,j=r} a_{i,j} \cdot e^{\frac{t-t_{ci}}{\tau_{i,j}}} + b \tag{20.9}$$

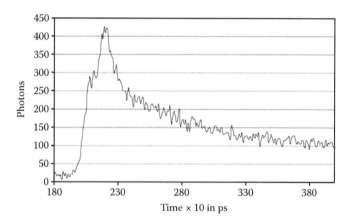

Figure 20.18 Stepped slope of fluorescence intensity in the short wavelengths channel after pulse excitation in an eye with crystalline lens. The fluorescence of the lens appears earlier than the fundus fluorescence.

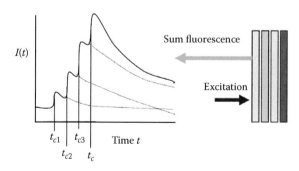

Figure 20.19 Stepped slope of fluorescence intensity after pulse excitation originating symbolically from a layered structure. Topographic information can be calculated from differences in appearing times t_{ci} and the lifetimes characterize functional properties. (From Schweitzer, D. et al., *SPIE* 7368: 736804-1–736804-9, 2009. With permission of SPIE.)

Tissue autofluorescence lifetime spectroscopy and imaging: Applications

where i is the parameter of the layer; j is the parameter of the lifetime in layer i; $a_{i,j}$ is the amplitude of lifetime j in layer i; t is time; t_{ci} is the appearance time of fluorescence of layer i; $\tau_{i,j}$ is lifetime j in layer i; and b is the background.

The distance d between fluorescent layers can be calculated from t_{ci} values according to the following equation:

$$d = \frac{(t_{cl} - t_{ck}) \cdot c}{2 \cdot n} \tag{20.10}$$

where c is the velocity of light, and n is the refractive index.

By applying this algorithm, both geometric (distance of layers) and functional information (lifetimes) can be obtained from the same measurement. In this way, FLIM measurements have the potential for functional tomography in the eye. For example, the difference in the appearance time between light from the lens and fundus fluorescence was measured to be ~174 ps using the laser scanning system discussed above, which has a time resolution of 12.2 ps. Assuming the refractive index of the vitreous to be $n = 1.3668$, the distance between lens and fundus can be calculated as 18.8 mm, which is in good agreement with 19 mm, the value expected for the normal eye. To achieve a resolution of less than 10 μm in z-direction, a time resolution of <30 fs is necessary.

REFERENCES

ANSI. 2000. "American National Standard for the safe use of laser. ANSI Z 136.1-2000." Laser Institute of America, Orlando, FL.

AREDS Research Group. 2001. "A randomized, placebo-controlled, clinical trial of high-dose supplementation with vitamins C and E, beta carotene, and zinc for age-related macular degeneration and vision loss: AREDS Report No 8." *Arch Ophthalmol* 119(10): 1417–1436.

Berezin, M.Y., and S. Archilefu. 2010. "Fluorescence lifetime measurements and biological imaging." *Chem Rev* 110: 2641–2684.

Boettner, E.A., and J.R. Wolter. 1962. "Transmission of the ocular media." *Invest Ophthalmol* 1: 776–783.

Chance, B. 1976. "Pyridine nucleotide as an indicator of the oxygen requirements for energy-linked functions of mitochondria." *Circ Res* 38(5 Suppl 1): 31–38.

Delori, F.C. 1994. "Spectrometer for non-invasive measurement of intrinsic fluorescence and reflectance of ocular fundus." *Appl Opt* 33(31): 7439–7452.

Dimitrow, E., I. Riemann, A. Ehlers, M.J. Köhler, J. Norgauer, P. Elsner, K. König, and M. Kaatz. 2009. "Spectral fluorescence lifetime detection and selective melanin imaging by multiphoton laser tomography for melanoma diagnosis." *Exp Dermatol* 18: 509–515.

Ehlers, A., I. Riemann, M. Stark, and K. König. 2007. "Multiphoton fluorescence lifetime imaging of human hair." *Microsc Res Tech* 70: 154–161.

Eldred, G.E., and M.L. Katz. 1988. "Fluorophores of the human retinal pigment epithelium: Separation and spectral characterization." *Exp Eye Res* 47: 71–86.

Geeraets, W.J., and E.R. Berry. 1968. "Ocular spectral characteristics as related to hazards from lasers and other light sources." *Am J Ophthalmol* 66(1): 15–20.

Hammer, M., W. Vilser, T. Riemer, A. Mandecka, D. Schweitzer, U. Kühn, J. Dawczynski, F. Liemt, and J. Strobel. 2009. "Diabetic patients with retinopathy show increased retinal venous oxygen saturation." *Graefes Arch Clin Exp Ophthalmol* 247(8): 1025–1030.

Holm, S. 1979. "A simple sequentially rejective multiple test procedure." *Scand J Stat* 6(2): 65–70.

Köllner, M., and J. Wolfrum. 1992. "How many photons are necessary for fluorescence-lifetime measurements?" *Chem Phys Lett* 200(1–2): 199–204.

Nakashima, N., K. Yoshihara, F. Tanaka, and K. Yagi. 1980. "Picosecond fluorescence lifetime of the coenzyme of D-amino acid oxidase." *J Biol Chem* 255(11): 5261–5263.

Peters, S., M. Hammer, and D. Schweitzer. 2011. "Two-photon excited fluorescence microscopy application for ex vivo investigation of ocular fundus samples." *SPIE* 8086: 808604-1–808604-10.

Schweitzer, D. 2010. "Metabolic mapping." In *Medical Retina—Focus on Retinal Imaging*, edited by F.G. Holz, and R.F. Spaide. Springer, Berlin, Heidelberg, 107–123.

Schweitzer, D., L. Leistritz, M. Hammer, M. Scibor, U. Bartsch, and J. Strobel. 1995. "Calibration-free measurement of the oxygen saturation in retinal vessels of men." *SPIE* 2393: 210–218.

Schweitzer, D., A. Kolb, M. Hammer, and E. Thamm. 2000. "Tau-mapping of the autofluorescence of the human ocular fundus." *SPIE* 4164: 79–89.

Schweitzer, D., M. Hammer, F. Schweitzer, R. Anders, T. Doebbecke, S. Schenke, and E.R. Gaillard. 2004. "*In vivo* measurement of time-resolved autofluorescence at the human fundus." *J Biomed Opt* 9(6): 1214–1222.

Schweitzer, D., M. Hammer, and F. Schweitzer. 2005. "Limits of confocal laser scanning technique in measurements of time-resolved autofluorescence of the ocular fundus." *Biomed Technik* 50(9): 263–267 (in German).

Schweitzer, D., S. Jentsch, S. Schenke, M. Hammer, C. Biskup, and E.R. Gaillard. 2007a. "Spectral and time-resolved studies on ocular structures." *SPIE* 6628: 662807-1–662807-12.

Schweitzer, D., A. Lasch, S. van der Vorst, K. Wildner, M. Hammer, U. Voigt, M. Jütte, and U.A. Müller. 2007b. "Alteration of retinal oxygen saturation in healthy subjects and early stages of diabetic retinopathy after breathing 100% oxygen." *Klin Monatsbl Augenheilkd* 224: 402–410 (in German).

Schweitzer, D., S. Schenke, M. Hammer, F. Schweitzer, S. Jentsch, E. Birckner, W. Becker, and A. Bergmann. 2007c. "Towards metabolic mapping of the human retina." *Microsc Res Tech* 70: 410–419.

Schweitzer, D., M. Klemm, M. Hammer, S. Jentsch, and F. Schweitzer. 2009a. "Method for simultaneous detection of functionality and tomography." *SPIE* 7368: 736804-1–736804-9.

Schweitzer, D., S. Quick, S. Schenke, M. Klemm, S. Gehlert, M. Hammer, S. Jentsch, and J. Fischer. 2009b. "Comparison of parameters of time-resolved autofluorescence of healthy subjects and patients, suffering from early age-related macular degeneration." *Ophthalmologe* 106: 714–722 (in German).

Schweitzer, D., S. Quick, M. Klemm, M. Hammer, S. Jentsch, and J. Dawczynski. 2010. "Time-resolved autofluorescence in retinal vascular occlusions." *Ophthalmologe* 107(12): 1145–1152 (in German).

Schweitzer, D., M. Klemm, L. Deutsch, S. Jentsch, M. Hammer, J. Dawczynski, C. Kloos, and U.A. Müller. 2011. "Detection of early metabolic alterations in the ocular fundus of diabetic patients by time-resolved autofluorescence of endogenous fluorophores." *SPIE* 8087: 808750-1–808750-9.

Skala, M.C., K.M. Riching, A. Gendron-Fitzpatrick, J. Eickhoff, K.W. Eliceiri, J.G. White, and N. Ramanujam. 2007. "*In vivo* multiphoton microscopy of NADH and FAD redox states, fluorescence lifetimes, and cellular morphology in precancerous epithelia." *PNAS* 104(49): 19494–19499.

Trudeau, K., A.J.A. Molina, and S. Roy. 2011. "High glucose affects mitochondrial metabolic capacity and extracellular acidification in retinal pericytes." *Invest Ophthalmol Vis Sci* 52(12): 8657–8664.

Vishwasrao, H.D., A.A. Heikal, K.A. Kasischke, and W.W. Webb. 2005. "Conformational dependence of intracellular NADH on metabolic state revealed by associated fluorescence anisotropy." *J Biol Chem* 280(26): 25119–25126.

Wang, B.-G. 2006. "Corneal laser surgery and multiphoton microscopy by means of NIR—Femtosecond laser at rabbits." MD diss., University of Jena (in German).

Wu, Y., W. Zheng, and J.Y. Qu. 2006. "Sensing cell metabolism by time-resolved autofluorescence." *Opt Lett* 31(21): 3122–3124.

21 Fluorescence lifetime imaging applications in tissue engineering

Bernard Y. Binder, J. Kent Leach, and Laura Marcu

Contents

21.1 INTRODUCTION

The field of tissue engineering has emerged from the related areas of cellular/molecular biology and engineering as a unique discipline focused on the development and implementation of functional biological replacements for tissues and organs (Langer and Vacanti 1993). The successful creation of tissues that can effectively serve as physiological replacements or accurate biological models of development and pathology requires the appropriate combination of numerous variables including cellular populations, an underlying matrix, and soluble signals. Due to the complexity and dynamic nature of biological systems, establishing a firm understanding of the intracellular and intercellular processes in target tissues is paramount for successful advances in tissue engineering and regenerative medicine (Georgakoudi et al. 2008).

The functional goals of tissue engineering and regenerative medicine necessitate the characterization of intracellular biochemical regulation, as well as cellular and molecular interactions with surrounding substrates and the local milieu. Upon association with an underlying matrix, cells engage the surface material through integrin-mediated interactions and begin to differentiate and secrete a tissue-associated extracellular matrix (ECM) that is characteristic of the cell phenotype (Decaris and Leach 2011; Singelyn et al. 2009; Zhang et al. 2009). In many cases, differentiated cells lose their unique phenotypes and functionality when interactions with ECM are restricted, and tissue proliferation and development are severely hampered in the absence of the appropriate substrate (Zhang et al. 2009).

Currently employed methods to assess differentiation state, quantity, and quality of requisite ECM constituents, construct maturity, and resulting functional mechanical properties require destructive biopsies. These biopsies reduce the amount of available tissue, expose the construct to potential contamination, and may compromise the integrity of the remaining neotissue. Furthermore, such techniques are limited to single time point measurements and may not completely capture the nature of native cellular states and interactions. After biopsy collection, the progression and activity of the cells of interest are evaluated using histology, biochemical assays, or determination of mechanical properties through compression, tension, or other deformative procedures. Therefore, it is imperative to develop new solutions for assessing the quality of engineered tissue to address these challenges (Ashjian et al. 2004).

Effective tissue engineering strategies often involve a progression from 2D cellular and molecular experiments, commonly performed on tissue culture plastic (TCP), to the design of 3D delivery systems and eventual *in vivo* implementation. The behavior of cells in 2D is commonly characterized by various microscopy methods including brightfield and fluorescence microscopy, and various cytochemical stains are often performed to assess the presence of cellular constituents associated with a certain phenotype (e.g., von Kossa stain for calcium in bone, Oil Red O for lipids in fat, keratin for epithelium, etc.). Biochemical assays for matrix content or protein activity and secretion yield quantitative values and enable objective comparisons over time in culture. However, each of these techniques is destructive, eliminates the ability to conduct longitudinal studies of cell populations, and does not lend itself to *in situ* applications.

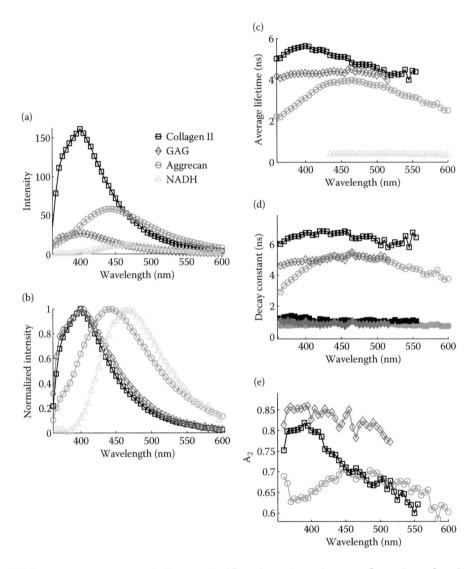

Figure 21.1 Fluorescence spectroscopic data acquired from the major endogenous fluorophores found in the cartilage samples including collagen type II, GAG powder, aggrecan, and NADH in cells. The time-resolved fluorescence parameters were retrieved with biexponential deconvolution. (a) Plots of fluorescence spectrum, (b) normalized spectrum, (c) average lifetime, (d) fast decay constant τ_1 (filled symbols) and slow decay constant τ_2 (open symbols) for collagen, GAG, and aggrecan (since NADH has only one exponential component, NADH was not included here), and (e) the fractional contribution of the slow decay component A_2. GAG, glycosaminoglycans; NADH, the reduced form of nicotinamide adenine dinucleotide. (From Sun, Y. et al., *Tissue Eng Part C Methods*, 18, 215–26, 2012.)

In contrast to the destructive effects of traditional screening methods, a number of imaging modalities enable the noninvasive monitoring of cellular processes in many stages of tissue engineering development, such as x-ray computed tomography (CT), positron emission tomography (PET), magnetic resonance (MR), and combinations of these approaches. Since the 1990s, multimodal PET/CT and PET/MRI systems have been used to characterize perfusion and metabolic function in *in vivo* models (Wehrl et al. 2009). Similarly, micro-CT has been used to examine angiogenesis in engineered scaffolds for bone repair (Young et al. 2008). However, the ionizing radiation inherent in these techniques reduces their versatility and can add unnecessary complications to biological studies.

Optical imaging and spectroscopy represent an exciting solution to these challenges and can yield high-quality, real-time information on cellular processes, physiology, and tissue development (Ashjian et al. 2004; Benesch et al. 2007; Berezin and Achilefu 2010; Borst and Visser 2010; Fite et al. 2011; Georgakoudi et al. 2008; Guo et al. 2008; Levitt et al. 2009; Wessels et al. 2010). Fluorescence lifetime spectroscopy and imaging techniques hold promise for many areas of tissue engineering due to their versatility, high specificity, and lack of dependence on measurement conditions (Borst and Visser 2010). As described in Chapter 3, several endogenous fluorescent biomolecules play significant structural and functional roles in engineered tissues. For example, distinct types of collagen and their cross-links are major ECM components that are responsible for the toughness of connective tissues, while proteoglycans such as aggrecan and their constituent glycosaminoglycans (GAGs) confer compressive resistance in cartilage. Similarly, cellular metabolism depends on NADH-mediated reactions, while flavoproteins and lipofuscin are involved in cellular responses to oxidative stress (Rice et al. 2010). The autofluorescent properties of these biomolecules can play an important role in distinguishing between different stages or conditions in engineered tissues. Representative fluorescence emission characteristics of biomolecules present in cartilage are given in Figure 21.1. This chapter describes key principles of cell biology and tissue engineering that lend themselves to investigation by fluorescence lifetime techniques. Additionally, this chapter presents a variety of noninvasive lifetime fluorescence applications for the characterization of tissue-engineered approaches in multiple physiological systems.

21.2 TISSUE ENGINEERING APPLICATIONS OF FLUORESCENCE LIFETIME IMAGING

21.2.1 CELLULAR DIFFERENTIATION AND FLUORESCENCE LIFETIME TECHNIQUES

Many adult and terminally differentiated tissue types are restricted in their regenerative capacity and proliferative rates, thereby limiting their potential and promise for numerous applications in tissue engineering. Consequently, there is a strong focus on strategies involving multipotent progenitor cells from a variety of tissue compartments including bone marrow and adipose tissue, as well as embryonic and induced pluripotent cells, that can be manipulated and grown into a number of different cell types and tissues (Ashjian et al. 2004; Guo et al. 2008). Recent efforts to understand and direct the potency and differentiation of multipotent stem cells hold promise for tissue engineering strategies involving a wide range of tissue types and biological systems (Kim et al. 2009; Nishiyama et al. 2009).

In order to take advantage of the proliferative potential of stem and progenitor cells for tissue engineering and regenerative medicine, a common approach is to expand a small pool of cells using efficient protocols to generate a sufficiently large population and then differentiate them to the desired lineage for subsequent implantation or use in fabricating a functional implant (He et al. 2010; Singelyn et al. 2009; Tanaka et al. 2004; Zhang et al. 2009). Stem and progenitor cells are distinguishable from more committed, differentiated cells based on morphology, metabolic and proliferative characteristics, receptor presentation, and ECM composition (Decaris and Leach 2011; Singelyn et al. 2009; Zhang et al. 2009). By tracking these biological properties, the success of particular engineering techniques can be effectively assessed (Ashjian et al. 2004; Fite et al. 2011; Guo et al. 2008). The application of fluorescence lifetime techniques can facilitate a noninvasive determination of the degree of tissue differentiation and development (Berezin and Achilefu 2010; Sun et al. 2012).

Tissue autofluorescence lifetime spectroscopy and imaging: Applications

Mesenchymal stem cells (MSC) can be induced to differentiate into bone, cartilage, and fat cells, and this cellular population is often used in studies focused on bone formation. While immunohistological targets may be used to analyze development, undifferentiated MSC are known to have a lower metabolic profile than osteogenic cells (Guo et al. 2008; Lonergan et al. 2007). Stem cells rely on glycolysis for generating cellular energy rather than producing adenosine triphosphate (ATP) by oxidative phosphorylation, resulting in reduced amounts of intracellular NADH (Guo et al. 2008). The autofluorescent properties of NADH provide a distinction that enables its use as a contrast mechanism for fluorescence lifetime imaging microscopy (FLIM) characterization. For example, Figure 21.2 provides FLIM images of NADH in MSC differentiated for up to 3 weeks. Each pixel in the pictures corresponds to an average fluorescence lifetime, with shorter lifetimes on the red end of the spectrum and longer lifetimes shifted toward blue. Histograms of each image's spectra reflect a progressive increase in mean NADH lifetime, allowing for osteogenically induced MSC to be clearly distinguished from undifferentiated populations (Guo et al. 2008). In this study, the cell fluorescence was induced using two-photon excitation, detected using a single-photon-counting photomultiplier tube and board (Beckel & Hickl GmbG, Germany) and analyzed using a biexponential decay model. As an added benefit, the utilization of cellular autofluorescence obviates the need for exogenous probes that could affect cellular development and metabolism. Fluorescence lifetime techniques may also find applicability in the evaluation of cellular viability in tissue scaffolds. A recent study employing two-photon excitation and detection of autofluorescence intensity has demonstrated the potential of this technique for assessment of cell viability based solely on autofluorescence without the need of tissue staining or enzymatic processing, thus preserving the scaffold integrity (Rice et al. 2010). By examining NADH, flavoproteins, and lipofuscin as markers of metabolism and oxidative stress, the differentiation and hypoxic response of MSC can be

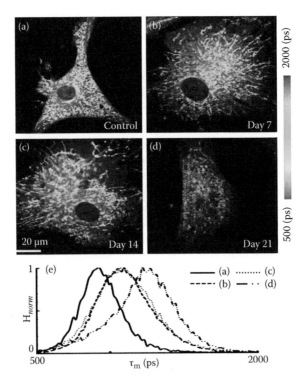

Figure 21.2 FLIM results of NADH-based characterization of osteogenically differentiated MSC. Each pixel in images (a–d) is color-coded to represent the average fluorescence lifetime, with a red-to-blue spectrum corresponding to a range of 0.5 to 2.0 ns, respectively. Increased MSC differentiation corresponds to a shift toward longer NADH fluorescent lifetimes, as shown in the histogram in (e). (From Guo, H. W. et al., *J Biomed Opt*, 13, 050505, 2008.)

recorded in real time. The use of a fluorescence lifetime rather than intensity-based technique is likely to improve the specificity of the fluorescence measurement.

MSC development can also be assessed using the autofluorescence of ECM components such as collagen. Since Type I collagen is the primary organic component in fully developed bone matrix (Martin et al. 1998), it is commonly targeted using histology (e.g., methyl blue or Masson's trichrome stain) or immunohistochemistry as an indicator of osteogenic differentiation. However, a few reports demonstrate that the relative expression of collagens (Types I, III, IV, V) in newly forming osteogenic ECM within the engineered tissue can be evaluated using fluorescence spectroscopy, including time-resolved measurements (Ashjian et al. 2004). Pentosidine and pyridinoline cross-links within the collagen molecules represent the main fluorescent constituents of this matrix protein (Ashjian et al. 2004; Berezin and Achilefu 2010; Georgakoudi et al. 2008), and changes in time-resolved fluorescence characteristics mirror histological differences in osteogenically differentiating adipose-derived stem cells over time. In this study, the formation of osteogenic ECM produced by putative stem cells (PLA cells) derived from human adipose tissue was monitored using a point spectroscopy time-resolved technique (Ashjian et al. 2004). A customized instrument based on pulse sampling and single-photon excitation at 337 nm and employing a fast-rise MCP-PMT and digitizer and a biexponential fluorescence decay model was used to analyze the characteristics of the fluorescence emission of the PLA samples (cells and matrices). This study showed that changes in fluorescence emission characteristics are associated with expression of Type 1 collagen in the ECM. The results suggested that this time-resolved fluorescence method could provide a viable alternative to traditional destructive approaches.

While collagen cross-linking levels offer insight into the stage of differentiation of the target, these additional properties can be used not only to investigate the progressive formation of collagen structure during osteogenic differentiation (Ashjian et al. 2004) but also to predict physical and mechanical properties based on matrix cross-linking without the need for mechanical testing. By noninvasively assessing fluorescence lifetime changes over time, the biochemical and physical maturation of engineered tissue can be monitored in a rapid, consistent manner. However, the use of fluorescent lifetime analysis to anticipate physical properties of tissue must be rigorously tested and corroborated by conventional approaches, especially in samples containing large variations in ECM types or where significant structural integrity is conferred by nonfluorescent components such as mineralization.

A two-photon excitation implementation of time-resolved fluorescence spectroscopy (TRFS) or FLIM may offer additional advantages in studying cellular differentiation as it limits cellular phototoxicity. Two-photon excitation of fluorescence was used to monitor the differentiation of human MSC toward the osteogenic and adipogenic lineage, as well as to assess the role of oxidative stress. Changes in the redox ratio of NADH and flavoprotein, determined by fluorescence intensity of each endogenous fluorophore, were distinct for MSC cultured in lineage-specific medium. In addition, second harmonic generation (SHG) images enabled detection of collagen fibril deposition by cells undergoing osteogenic differentiation, providing yet another method to identify the progress of cellular differentiation in near real time without destructive methods (Rice et al. 2010).

21.2.2 FLUORESCENCE LIFETIME ANALYSIS OF SCAFFOLDS

The examination of cellular differentiation and molecular processes in cells in monolayer culture is a critical first step in tissue engineering strategies, yet culture and delivery systems such as cellular scaffolds or gels are essential for the development of functional 3D tissues and *in vivo* implementation (Benesch et al. 2007). Scaffolds are often constructed from porous, biocompatible, and degradable polymers intended to facilitate cellular attachment and proliferation and enable *in vivo* implantation or serve as biologically relevant model systems. Many groups have demonstrated the importance of dimensionality in such studies. For example, adipose-derived stem cells exhibited reduced osteogenic potential when cultured on 3D polymeric scaffolds compared to maintenance on TCP (He et al. 2010). Other groups have shown that proliferation of rat bone marrow cells is significantly enhanced by culture on 3D alginate hydrogel tubes (Barralet et al. 2005), and separate studies have demonstrated increased chondrogenesis of mouse embryonic stem cells grown as 3D embryoid bodies, compared to 2D culture (Tanaka et al. 2004). Likewise, malignant breast tumor cells exhibit distinct regulation of signaling pathways for β-integrin and epidermal growth factor receptor in 3D culture compared to monolayer culture (Wang et al. 1998).

Consequently, it is of significant importance to characterize interactions between scaffold materials and cellular proteins and matrix (Benesch et al. 2007; Fite et al. 2011).

Protein adsorption is an important property of biocompatible scaffolds that affects the response of both seeded cells and the *in vivo* host. In particular, a host's immune response is critical in determining the success of an implanted polymer or gel, and inflammatory reactions to foreign materials and proteins can result in fibrosis and rejection. Similarly, the recruitment of autologous cells to promote neovascularization and effective incorporation of a construct into surrounding tissue can be enhanced by presenting growth factors and chemokines on the material surface (Thevenot et al. 2010). While it can be difficult to examine protein adsorption using biochemical assays alone, FLIM can be applied to measure protein conformational changes and deduce molecular interactions (Benesch et al. 2007). On a basic level, the intrinsic fluorescence of amino acids such as tryptophan can be utilized for fluorescent spectroscopy. Tryptophan's fluorescent characteristics are affected by local polarity, resulting in emission spectra that shift depending on environmental conditions and can then be used to deduce structural changes. However, some scaffold polymers and other biological components (such as collagen) can emit background fluorescence that leads to a poor signal-to-noise ratio (SNR). In such cases, more specific protein adsorption information can be obtained by analyzing the fluorescence lifetimes of exogenous probes such as Nile red and fluorescein isothiocyanate (FITC). For example, FITC may be conjugated at several ratios to bovine serum albumin (BSA) and adsorbed to polymer scaffolds, and the corresponding variations in lifetime have been used to effectively quantify aggregation properties and distinguish between adsorbed and free protein (Benesch et al. 2007). In this study by Benesch et al., a customized inverted confocal microscope including excitation from a pulsed diode laser and a time-correlated single-photon counting module was used to time-resolve the fluorescence emission. This information can, in turn, aid in the improvement or tuning of biomaterial characteristics for tissue growth by revealing the release and adsorption profiles of chemotactic and morphogenic proteins.

21.2.3 CHARACTERIZATION OF TISSUE DEVELOPMENT IN *EX VIVO* 3D CULTURE

After applying results from 2D studies to scaffolds or other 3D constructs, further analysis is needed to evaluate the mechanical, structural, and biological properties of engineered tissues. Intrinsic fluorescence of cell components can be utilized in much the same way as in 2D culture, although the interactions are significantly more complex (Fite et al. 2011).

Standard techniques to assess the stage of differentiation, cell behavior, and quality of cartilaginous tissues require destructive biochemical and histological processing. Recently, fluorescence-based imaging has been applied to assess the differentiation of progenitor cells and maturation of resulting cartilage tissue (Fite et al. 2011). Adipose-derived stem cells were seeded on biocompatible polymer scaffolds and induced toward the chondrogenic lineage. Adult-derived stem and progenitor cells, as well as chondrocytes, secrete and deposit significant levels of collagen type II when differentiating toward cartilage, while cells undergoing dedifferentiation, a common phenomenon for these cells in 2D culture over time (Fite et al. 2011; Tanaka et al. 2004; Zhang et al. 2009), produce less collagen type II and more collagen type I. Both of these molecules are highly fluorescent with unique signatures due to the presence of the protein's cross-links. Fite et al. applied fluorescence lifetime imaging to determine the presence and distribution of collagen type II *in vitro*. Since these properties are closely tied to the mechanical properties and function of cartilage and chondrocytes, it was proposed that results from FLIM could be used to deduce the physical characteristics and functionality of engineered tissue without damage or conventional mechanical testing (Fite et al. 2011). The experimental setup involved the use of pulse sampling TRFS and wide-field FLIM techniques. The fluorescence emission of tissue scaffolds was excited at 337 nm. Indeed, a combination of FLIM and TRFS allowed for the discrete identification of collagen II and GAGs in engineered constructs, yielding results consistent with histological staining (Figure 21.3). Furthermore, the collagen content identified by fluorescent intensity was associated with a significant correlation with compressive moduli, suggesting that similar optical methods may be used to supplant mechanical testing experiments for characterizing 3D constructs (Fite et al. 2011).

The validity of lifetime imaging for determining structural and biochemical properties of engineered tissue has been further demonstrated in self-assembled, scaffold-less cartilage constructs (Sun et al. 2012). In contrast to the previously described study, which followed the deposition of ECM over the course

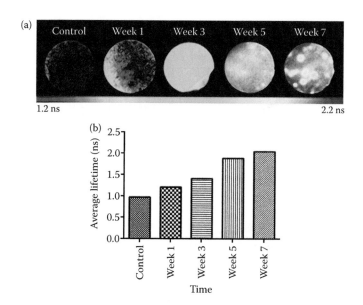

Figure 21.3 (a) Representative FLIM images from adipose-derived stem cells cultured on highly porous, biodegradable polymeric scaffolds. Samples are arranged from left to right in order of culture duration. Samples exhibit longer lifetimes over increased growth periods, reflecting an increase in the formation of collagen II and resultant chondrogenic properties. Part (b) shows the average fluorescent lifetime of samples over 7 weeks. Increased culture duration resulted in longer average lifetimes. The blue-shifted increased fluorescence intensity and longer lifetime values observed after week 3 suggest that collagen and its cross-links began to contribute to cartilage construct fluorescence, hence providing a minimally invasive imaging modality to assess the maturation of engineered cartilage. (Adapted from Fite, B. Z. et al., *Tissue Eng Part C Methods*, 17, 495–504, 2011.)

of progenitor cell differentiation, using similar experimental setups, Sun et al. investigated changes in fluorescence lifetime in well-developed matrices as a result of controlled treatment with several chemicals. Ribose, collagenase, and chondroitinase-ABC were applied to separate constructs in order to increase collagen cross-linking, degrade collagen, and decrease GAG content, respectively. TRFS assessment of the post-treatment ECM structure and composition of each of these groups was able to distinguish changes in abundance of collagen that correlated strongly with histological and biochemical characterization (Sun et al. 2012). Results from these studies are depicted in Figure 21.4. Weak emission from GAG components prevented the detection of significant correlation between optical parameters and total GAG content, but indentation testing revealed significant relationships between collagen fluorescence lifetime and Young's modulus, further supporting the ability of optical techniques to replace destructive testing for interrogating the mechanical properties of engineered tissue (Sun et al. 2012).

21.2.4 FLUORESCENCE LIFETIME IMAGING OF TISSUE IMPLANTS

Since the ultimate goal of tissue engineering is to create constructs and implants that function *in vivo*, fluorescence lifetime imaging could have many uses for examining the integration and interaction of engineered tissue in a host organism. FLIM and lifetime spectroscopy could be implemented to track properties such as gene expression and angiogenesis using exogenous dyes. Furthermore, tissue development and its physical and biochemical interactions with surrounding cellular systems could also be easily monitored without the need to sacrifice model animals (Berezin and Achilefu 2010; Bloch et al. 2005; Kumar et al. 2008; Wessels et al. 2010). Unfortunately, conventional optical imaging modalities that rely on excitatory light in the ultraviolet spectrum are limited by poor penetration depth in tissue and engineered constructs (Berezin and Achilefu 2010; Fite et al. 2011; Hwang et al. 2011). Generally speaking, UV light can only travel through samples less than 250 μm thick, which restricts imaging applications to subcutaneous levels (Berezin and Achilefu 2010; Fite et al. 2011). As such, a number of techniques are currently under investigation to overcome this significant disadvantage.

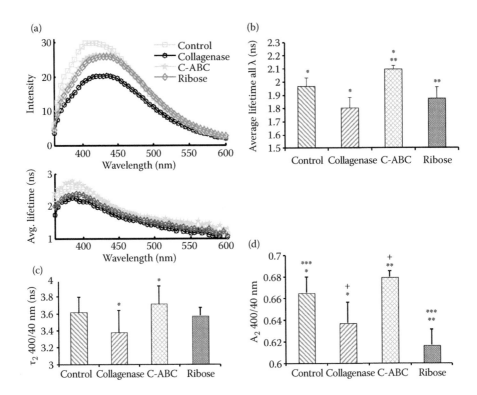

Figure 21.4 (a) Fluorescence emission spectrum and average lifetime from four groups of cartilage samples that underwent distinct chemical treatments (control, collagenase, C-ABC, and ribose). The time-resolved fluorescence parameters were retrieved with biexponential deconvolution. Quantification of the TRFS parameters for the four groups showing (b) the mean value of the average lifetime from 360 to 600 nm, (c) slow decay constant τ_2 at 400/40 nm spectral bandwidth, and (d) fractional contribution of the slow decay component A_2. Bars carrying the same marker (*,**,***,+) are statistically significantly different ($p < 0.05$). These findings demonstrate that changes in fluorescence spectra correlate with destruction or formation of fluorescent moieties attributed to the deposition of ECM and resulting tissue formation. (Adapted from Sun, Y. et al., *Tissue Eng Part C Methods*, 18, 215–26, 2012.)

For example, two-photon excitation, which involves the application of highly focused, extremely short pulses of low-energy photons, allows for increases in penetration depth up to approximately 350 μm without damaging local tissue (Berezin and Achilefu 2010; Borst and Visser 2010; Durr et al. 2011; Fite et al. 2011). Although this represents an improvement over single-photon microscopy and allows for the use of longer wavelengths, significant advances need to be made to noninvasively characterize tissue-engineered solutions implanted below the surface of the skin. Multimodal imaging techniques are also being developed, including the integration of ultrasound backscatter microscopy (UBM) with FLIM or TRFS. However, while the addition of ultrasound allows for improved penetration depth and morphological resolution, sample thicknesses are still limited to only 5–6 mm (Fite et al. 2011). Such challenges will likely restrict the use of whole-body fluorescence lifetime imaging to small animal models. Investigation in humans may require development of endoscopic-type probes and multimodal studies involving conventional techniques such as CT, MR, or PET imaging.

21.3 CONCLUSION AND SUMMARY

Tissue engineering is an important area of research focused on developing functional replacements for tissues and organs (Langer and Vacanti 1993). Although many techniques currently exist to assess levels of growth, development, and molecular interactions, they are frequently invasive or destructive and reveal only discrete snapshots of cellular function. Fluorescence lifetime spectroscopy and imaging overcome these obstacles by providing a sensitive, noninvasive method to probe tissue development and interactions.

As such, it is an attractive option for evaluating success and implementation of results in nearly every stage of the tissue engineering process and is likely to see broader acceptance and application in the future.

ACKNOWLEDGMENTS

BYB was supported by Award Number T32EB003827 from the National Institute of Biomedical Imaging and Bioengineering and the California Institute for Regenerative Medicine UC Davis Stem Cell Training Program (CIRM T1-00006, CIRM TG2-01163).

REFERENCES

Ashjian, P., Elbarbary, A., Zuk, P., Deugarte, D. A., Benhaim, P., Marcu, L. & Hedrick, M. H. 2004. Noninvasive in situ evaluation of osteogenic differentiation by time-resolved laser-induced fluorescence spectroscopy. *Tissue Eng,* 10, 411–20.

Barralet, J. E., Wang, L., Lawson, M., Triffitt, J. T., Cooper, P. R. & Shelton, R. M. 2005. Comparison of bone marrow cell growth on 2D and 3D alginate hydrogels. *J Mater Sci Mater Med,* 16, 515–9.

Benesch, J., Hungerford, G., Suhling, K., Tregidgo, C., Mano, J. F. & Reis, R. L. 2007. Fluorescence probe techniques to monitor protein adsorption-induced conformation changes on biodegradable polymers. *J Colloid Interface Sci,* 312, 193–200.

Berezin, M. Y. & Achilefu, S. 2010. Fluorescence lifetime measurements and biological imaging. *Chem Rev,* 110, 2641–84.

Bloch, S., Lesage, F., McIntosh, L., Gandjbakhche, A., Liang, K. & Achilefu, S. 2005. Whole-body fluorescence lifetime imaging of a tumor-targeted near-infrared molecular probe in mice. *J Biomed Opt,* 10, 054003.

Borst, J. W. & Visser, A. J. W. G. 2010. Fluorescence lifetime imaging microscopy in life sciences. *Meas Sc Technol,* 21, 1–21.

Decaris, M. L. & Leach, J. K. 2011. Design of experiments approach to engineer cell-secreted matrices for directing osteogenic differentiation. *Ann Biomed Eng,* 39, 1174–85.

Durr, N. J., Weisspfennig, C. T., Holfeld, B. A. & Ben-Yakar, A. 2011. Maximum imaging depth of two-photon autofluorescence microscopy in epithelial tissues. *J Biomed Opt,* 16, 026008.

Fite, B. Z., Decaris, M., Sun, Y., Lam, A., Ho, C. K., Leach, J. K. & Marcu, L. 2011. Noninvasive multimodal evaluation of bioengineered cartilage constructs combining time-resolved fluorescence and ultrasound imaging. *Tissue Eng Part C Methods,* 17, 495–504.

Georgakoudi, I., Rice, W. L., Hronik-Tupaj, M. & Kaplan, D. L. 2008. Optical spectroscopy and imaging for the noninvasive evaluation of engineered tissues. *Tissue Eng Part B Rev,* 14, 321–40.

Guo, H. W., Chen, C. T., Wei, Y. H., Lee, O. K., Gukassyan, V., Kao, F. J. & Wang, H. W. 2008. Reduced nicotinamide adenine dinucleotide fluorescence lifetime separates human mesenchymal stem cells from differentiated progenies. *J Biomed Opt,* 13, 050505.

He, J., Genetos, D. C., Yellowley, C. E. & Leach, J. K. 2010. Oxygen tension differentially influences osteogenic differentiation of human adipose stem cells in 2D and 3D cultures. *J Cell Biochem,* 110, 87–96.

Hwang, J. Y., Wachsmann-Hogiu, S., Ramanujan, V. K., Nowatzyk, A. G., Koronyo, Y., Medina-Kauwe, L. K., Gross, Z., Gray, H. B. & Farkas, D. L. 2011. Multimodal wide-field two-photon excitation imaging: Characterization of the technique for in vivo applications. *Biomed Opt Express,* 2, 356–64.

Kim, D., Kim, C. H., Moon, J. I., Chung, Y. G., Chang, M. Y., Han, B. S., Ko, S., Yang, E., Cha, K. Y., Lanza, R. & Kim, K. S. 2009. Generation of human induced pluripotent stem cells by direct delivery of reprogramming proteins. *Cell Stem Cell,* 4, 472–6.

Kumar, A. T., Raymond, S. B., Dunn, A. K., Bacskai, B. J. & Boas, D. A. 2008. A time domain fluorescence tomography system for small animal imaging. *IEEE Trans Med Imaging,* 27, 1152–63.

Langer, R. & Vacanti, J. P. 1993. Tissue engineering. *Science,* 260, 920–6.

Levitt, J. A., Matthews, D. R., Ameer-Beg, S. M. & Suhling, K. 2009. Fluorescence lifetime and polarization-resolved imaging in cell biology. *Curr Opin Biotechnol,* 20, 28–36.

Lonergan, T., Bavister, B. & Brenner, C. 2007. Mitochondria in stem cells. *Mitochondrion,* 7, 289–96.

Martin, R. B., Burr, D. B. & Sharkey, N. A. 1998. *Skeletal Tissue Mechanics.* New York, Springer.

Nishiyama, A., Xin, L., Sharov, A. A., Thomas, M., Mowrer, G., Meyers, E., Piao, Y., Mehta, S., Yee, S., Nakatake, Y., Stagg, C., Sharova, L., Correa-Cerro, L. S., Bassey, U., Hoang, H., Kim, E., Tapnio, R., Qian, Y., Dudekula, D., Zalzman, M., Li, M., Falco, G., Yang, H. T., Lee, S. L., Monti, M., Stanghellini, I., Islam, M. N., Nagaraja, R., Goldberg, I., Wang, W., Longo, D. L., Schlessinger, D. & Ko, M. S. 2009. Uncovering early response of gene regulatory networks in ESCs by systematic induction of transcription factors. *Cell Stem Cell,* 5, 420–33.

Rice, W. L., Kaplan, D. L. & Georgakoudi, I. 2010. Two-photon microscopy for non-invasive, quantitative monitoring of stem cell differentiation. *PLoS One,* 5, e10075.

Singelyn, J. M., Dequach, J. A., Seif-Naraghi, S. B., Littlefield, R. B., Schup-Magoffin, P. J. & Christman, K. L. 2009. Naturally derived myocardial matrix as an injectable scaffold for cardiac tissue engineering. *Biomaterials,* 30, 5409–16.

Sun, Y., Responte, D., Xie, H., Liu, J., Fatakdawala, H., Hu, J., Athanasiou, K. & Marcu, L. 2012. Nondestructive evaluation of tissue engineered articular cartilage using time-resolved fluorescence spectroscopy and ultrasound backscatter microscopy. *Tissue Eng Part C Methods,* 18, 215–26.

Tanaka, H., Murphy, C. L., Murphy, C., Kimura, M., Kawai, S. & Polak, J. M. 2004. Chondrogenic differentiation of murine embryonic stem cells: Effects of culture conditions and dexamethasone. *J Cell Biochem,* 93, 454–62.

Thevenot, P. T., Nair, A. M., Shen, J., Lotfi, P., Ko, C. Y. & Tang, L. 2010. The effect of incorporation of SDF-1alpha into PLGA scaffolds on stem cell recruitment and the inflammatory response. *Biomaterials,* 31, 3997–4008.

Wang, F., Weaver, V. M., Petersen, O. W., Larabell, C. A., Dedhar, S., Briand, P., Lupu, R. & Bissell, M. J. 1998. Reciprocal interactions between beta1-integrin and epidermal growth factor receptor in three-dimensional basement membrane breast cultures: A different perspective in epithelial biology. *Proc Natl Acad Sci U S A,* 95, 14821–6.

Wehrl, H. F., Judenhofer, M. S., Wiehr, S. & Pichler, B. J. 2009. Pre-clinical PET/MR: Technological advances and new perspectives in biomedical research. *Eur J Nucl Med Mol Imaging,* 36 Suppl 1, S56–68.

Wessels, J. T., Yamauchi, K., Hoffman, R. M. & Wouters, F. S. 2010. Advances in cellular, subcellular, and nanoscale imaging in vitro and in vivo. *Cytometry A,* 77, 667–76.

Young, S., Kretlow, J. D., Nguyen, C., Bashoura, A. G., Baggett, L. S., Jansen, J. A., Wong, M. & Mikos, A. G. 2008. Microcomputed tomography characterization of neovascularization in bone tissue engineering applications. *Tissue Eng Part B Rev,* 14, 295–306.

Zhang, Y., He, Y., Bharadwaj, S., Hammam, N., Carnagey, K., Myers, R., Atala, A. & Van Dyke, M. 2009. Tissue-specific extracellular matrix coatings for the promotion of cell proliferation and maintenance of cell phenotype. *Biomaterials,* 30, 4021–8.

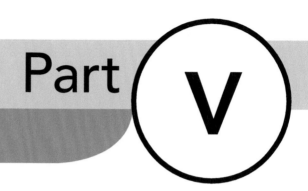

Part V

Fluorescence lifetime imaging based on exogenous probes

22 Tomographic fluorescence lifetime imaging

Anand T. N. Kumar

Contents

22.1 INTRODUCTION

Fluorescence lifetime imaging microscopy (FLIM) is a well-established technique (Bastiaens and Squire 1999; Berezovska et al. 2003; Selvin 2000; Vogel et al. 2006) that combines microscopic techniques with time-resolved detection to provide high-resolution lifetime images of thin tissue sections. This chapter concerns the *in vivo* time domain imaging of fluorescent contrast agents embedded in deep tissue. Optical molecular imaging is a rapidly growing field of interest (Bremer et al. 2003; Bugaj et al. 2001; Massoud and Gambhir 2003), with several contrast agents reported to date that cover the visible to the near infrared (NIR) spectral range. Although lifetime microscopy techniques are well established, whole body molecular imaging of lifetime contrast is relatively recent (Berezin et al. 2011; Bloch et al. 2005; Goiffon et al. 2009; Nothdurft et al. 2009; Raymond et al. 2010). One possible explanation for this discrepancy is the general concern among researchers about the viability of fluorescence lifetime as a useful marker for *in vivo* imaging, given that the fluorophore can undergo complex interactions with the biochemical environment *in vivo*, altering its photophysical properties in indeterminable ways. In fact, two distinct phenomena can alter the fluorescence lifetimes as measured on the surface of a living subject. The first is the interaction of the fluorophore with the biological environment (e.g., pH, viscosity, and protein binding). The effect of tissue environment on the lifetime can be characterized in advance using careful control measurements (Raymond et al. 2010). The second phenomenon is the interaction of light with tissue, which can also indirectly affect the temporal response of the fluorophore as measured on the surface. This chapter is mainly concerned with the second phenomenon, namely, the influence of tissue scattering and absorption on the lifetime of fluorophores embedded in biological tissue. The details of this influence are incorporated through differential equations that describe light propagation in scattering media. We will, in particular, derive a tomographic FLIM model, which is valid under a widely applicable condition, namely, that the fluorescence lifetime is longer than the intrinsic diffusive timescales in the medium. Under this "FLIM condition," the temporal decay of fluorescence from deep tissue can be directly used to recover both the *in vivo* lifetimes and their corresponding yield distributions. Further, this model naturally leads to an elegant algorithm for tomographic FLIM, which allows the complete 3D separation of multiple lifetimes present within biological tissue. We will also discuss experimental aspects of performing tomographic FLIM in turbid media and present *in vivo* results using organ-specific contrast agents.

22.2 THEORY

22.2.1 GENERAL FORWARD PROBLEM STATEMENT

A typical tomography measurement involves optical sources and detectors placed on the boundary of the imaging specimen. The detected fluorescence can be described as a sequential propagation of the excitation light from the source(s) to the fluorophore, fluorophore emission, and propagation of the emission field from the fluorophore to the detector. This is described using coupled equations for light transport at the excitation and emission wavelengths. Let the source and detector locations be $\mathbf{r}_s$ and $\mathbf{r}_d$. Let $\eta(\mathbf{r})$ be the yield distribution (product of the quantum yield Q, concentration, and extinction coefficient) of the fluorophore, with $\mathbf{r}$ denoting the location of a point within the medium (voxel). The expression for the detected fluorescence in the time domain (TD) can then be written as a double convolution of the excitation ($G^x(\mathbf{r},\mathbf{r}_s,t)$) and emission ($G^m(\mathbf{r}_d,\mathbf{r},t)$) Green's functions (GFs) with the fluorescence decay term ($e^{-t/\tau(\mathbf{r})}$) (additional scaling factors are necessary when considering experimental data, including the source and detector coefficients, geometrical factors, and fluorescence filter attenuation):

$$U(\mathbf{r}_s,\mathbf{r}_d,t) = \int_\Omega d^3 r\, W(\mathbf{r}_s,\mathbf{r}_d,\mathbf{r},t)\eta(\mathbf{r}), \tag{22.1}$$

where the weight function (also called the "sensitivity" function) is given by

$$W(\mathbf{r}_s,\mathbf{r}_d,\mathbf{r},t) = \int_0^t dt' \int_0^{t'} dt''\, G^m(\mathbf{r}_d,\mathbf{r},t-t')e^{-\Gamma(\mathbf{r})(t'-t'')}G^x(\mathbf{r},\mathbf{r}_s,t''), \tag{22.2}$$

where $\tau(\mathbf{r}) = 1/\Gamma(\mathbf{r})$ is the fluorescence lifetime distribution. The above equation ignores re-emission of the fluorescence by the fluorophore, an assumption used widely in applications of tomographic fluorescence imaging and also termed the "Born approximation." Besides this approximation, the accuracy of Equation 22.1 depends on the level of approximation used for estimating the GFs, which depend on the intrinsic tissue optical properties, namely, absorption $\left(\mu_a^x(\mathbf{r}),\mu_a^m(\mathbf{r})\right)$ and scattering $\left(\mu_s^x(\mathbf{r}),\mu_s^m(\mathbf{r})\right)$ distributions at the excitation (λ_x) and emission (λ_m) wavelengths, in addition to the tissue anisotropy factor g. In general, the absorption and scattering are heterogeneous and include tissue components (such as water, melanin, and blood) and the absorption of the fluorophore (at both λ_x and λ_m). Generally, all the parameters, $\eta(\mathbf{r})$, $\tau(\mathbf{r})$, $\mu_a^{(x,m)}(\mathbf{r}),\mu_s^{(x,m)}(\mathbf{r})$ are unknown. A common starting point is the homogeneous approximation where the optical properties are assumed uniform throughout and fluorophore absorption is ignored for evaluating the GFs. In this case, the GFs in Equation 22.2 are the solutions to the homogeneous diffusion or transport equations. Note that $\mu_a^{x,m}$ and $\mu_s^{x,m}$ can, in practice, be determined independently using two separate "excitation" measurements at wavelengths λ_x and λ_m. In this case, Equation 22.2 can provide a highly accurate description of time-resolved fluorescence in turbid media. With $\mu_a^{x,m}$ and $\mu_s^{x,m}$ known, the GFs can be calculated using either the diffusion approximation or the radiative transport equation.

Evaluation of the sensitivity (Equation 22.2) can become computationally intractable for multiple source–detector (S–D) pairs (~10^2), medium voxels (~10^4), and time points (~10^3). One simplifying approach is to solve Equation 22.2 in the frequency (Fourier) domain (FD), since the double convolution then simplifies to a product for each frequency. While convenient in some scenarios (Nothdurft et al. 2009; Soloviev et al. 2007), this approach has the limitation that the Fourier transform is nontrivial for multiexponential decays, and multiple frequencies are required to reliably extract multiple lifetimes or complex decay profiles. Existing TD approaches to solve Equation 22.2 are overly simplistic (such as assuming point fluorophores in an infinite homogeneous medium; Hall et al. 2004), while the general formalism is quite intractable (Arridge and Schotland 2009). In what follows, we will take an approach to solve Equation 22.1 that is motivated by a practical observation, namely, that typical fluorescence lifetimes (τ) of fluorophores from the visible to NIR wavelengths are longer than the timescales for intrinsic diffusive relaxation in small volumes, τ_D (which is shorter than the absorption timescale $\tau_a =$ $(v\mu_a)^{-1}$; Haselgrove et al. 1992). We will make more precise definition of τ_D in Section 22.2.4. Under the approximation that $\tau > \tau_D$, it will be shown that Equations 22.1 and 22.2 can be cast into an elegant and rigorous formalism for inverting multiple lifetimes within turbid media (Kumar et al. 2005) that is analogous to the multiexponential model in FLIM. We will derive this "tomographic-FLIM" approach both in the frequency domain, using complex integration, and in time domain using Equation 22.2. Before proceeding, it is convenient to recast Equations 22.1 and 22.2 in the following way using the commutativity of the convolution:

$$U(\mathbf{r}_s,\mathbf{r}_d,t) = \int_\Omega d^3r \int_0^t dt' W^B(\mathbf{r}_s,\mathbf{r}_d,\mathbf{r},t')\left[\sum_n e^{-\Gamma_n(t-t')}\eta_n(\mathbf{r})\right], \tag{22.3}$$

where we have defined a "background" weight function as

$$W^B(\mathbf{r}_s,\mathbf{r}_d,\mathbf{r},t') = \int_0^{t'} dt'' G^x(\mathbf{r}_s,\mathbf{r},t'-t'')G^m(\mathbf{r},\mathbf{r}_d,t''), \tag{22.4}$$

and $\Gamma_n = 1/\tau_n$ are discretized values of the *in vivo* lifetime distribution with corresponding yield distributions $\eta_n(\mathbf{r})$. In other words, each lifetime has a distinct yield distribution. We will see below that the $\eta_n(\mathbf{r})$'s can be separately reconstructed from analyzing the TD data.

22.2.2 FREQUENCY DOMAIN

One way to analyze the TD forward problem is to Fourier-transform the data. In the frequency domain, the double convolution in the integrand of Equation 22.4 reduces to a simple product:

$$U(\mathbf{r}_s,\mathbf{r}_d,t) = \int_\Omega d^3r \int_{-\infty}^{\infty} d\omega e^{-i\omega t}\tilde{W}^B(\mathbf{r}_s,\mathbf{r}_d,\mathbf{r},\omega)\left[\underbrace{\sum_n \frac{\tau_n\eta_n(\mathbf{r})}{(1-i\omega\tau_n)}}_{F(\mathbf{r},\omega)}\right]. \tag{22.5}$$

where

$$\tilde{W}^B(\mathbf{r}_s,\mathbf{r}_d,\mathbf{r},\omega) = \tilde{G}^x(\mathbf{r}_s,\mathbf{r},\omega)\tilde{G}^m(\mathbf{r}_d,\mathbf{r},\omega) \tag{22.6}$$

is the FD sensitivity obtained as a Fourier transform of the TD sensitivity (Equation 22.4), with ω as the modulation frequency. The standard FD approach (see, for example, Chapter 13) is to reconstruct the spatial distribution of $F(\mathbf{r},\omega)$ (quantity within the square brackets in Equation 22.5) from the FD measurements at a given frequency ω. The lifetime and yield distributions are obtained as the phase and the real part of $F(\mathbf{r},\omega)$, respectively. As is clear from Equation 22.5, the TD data provide multiple frequencies. Handling multiple frequencies with the above forward problem becomes complicated since $F(\mathbf{r},\omega)$ inseparably involves both a measurement parameter ω and the unknown lifetime $\tau(\mathbf{r})$. This necessitates a nonlinear approach as described, for example, in Milstein et al. (2003). However, it has also been shown that multiple frequencies do not necessarily improve the quality of the reconstruction (Milstein et al. 2004), with the number of useful frequencies restricted to the first three or four frequency components from zero.

22.2.3 TOMOGRAPHIC FLIM MODEL

The TD forward problem in Equations 22.3 and 22.4 can be reduced to a sum of exponential decays, analogous to the signal in microscopic FLIM (Bastiaens and Squire 1999), the difference being that the amplitude coefficients of the lifetimes will now correspond to a measurement set for tomographic reconstructions, while in microscopy, they directly relate to the amplitude of the lifetime component in a particular tissue location. We will prove this result using two different methods, one using the FD form in Equation 22.5 and the other directly from the general TD forward problem, viz., Equations 22.3 and 22.4. The result derived using the direct TD approach is applicable to the entire TD fluorescence signal, while the FD approach is applicable only to the asymptotic or long-time portion of the TD fluorescence. The TD approach results in a tomographic FLIM model that clearly elucidates the individual contributions of the diffusive background and fluorescence decay to the measured TD fluorescence signal. We will also discuss the conditions when the tomographic FLIM model is valid.

22.2.3.1 Frequency domain derivation of tomographic FLIM

We apply contour integration (from complex analysis; Matthews and Walker 1970) to solve the FD integral in Equation 22.5. Without loss of generality, we consider an infinite homogeneous medium (the results will, however, be shown later to be valid for bounded heterogeneous media) and examine the analytic nature, in the complex variable sense, of the integrand of Equation 22.5. The homogeneous GFs are $\tilde{G}^{(x,m)}(\mathbf{r}_1,\mathbf{r}_2,\omega) = \exp(ik^{(x,m)}\rho_{12})/4\pi D^{(x,m)}$, where $\rho_{12} = |\mathbf{r}_1 - \mathbf{r}_2|$, $k^{(x,m)} = \left[\left(-\upsilon\mu_a^{(x,m)} + i\omega\right)/D^{(x,m)}\right]^{1/2}$, $D^{(x,m)} = \upsilon/3\mu_s^{(x,m)}$ is the diffusion coefficient, and υ is the velocity of light in the medium. It is evident that the integrand in Equation 22.1 possesses simple pole singularities distributed along the negative imaginary axis at $\omega_n = -i\Gamma_n$ (see Figure 22.1) due to the fluorescence decay term, $F(\mathbf{r},\omega)$. In addition, the homogeneous GF and its spatial derivatives are bi-valued owing to the square root in k, implying branch points (Matthews and Walker 1970) in the lower half plane at $\omega = -i\upsilon\mu_a^{(x,m)}$. To evaluate U (Equation 22.5), we choose a contour (C) shown in the complex "ω-plane" in Figure 22.1 and apply Cauchy's integral theorem

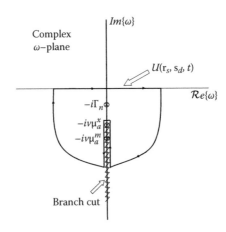

Figure 22.1 Complex ω-plane structure of the integrand of Equation 22.5 for a homogeneous diffuse medium, showing the simple pole singularities at $\omega = -i\Gamma_n$ and the branch points at $\omega = -i\upsilon\mu_a^{(x,m)}$. It is assumed that $\Gamma_n < \upsilon\mu_a^{(x,m)}$. Also shown is the contour for calculating Equation 22.5 using Cauchy's theorem. The integral of Equation 22.5 corresponds to the contribution of the real-ω axis extended from $-\infty$ to ∞.

(Matthews and Walker 1970), whence U separates into a sum of two terms, the first term arising from the residue at the simple poles at $-i\Gamma_n$ (which will lead to a sum of pure fluorescence decays) and the second term arising from the integration on either side of the branch cut (Matthews and Walker 1970) along the branch points in the lower imaginary axis (which will lead to a fluorescent diffuse photon density wave):

$$U(\mathbf{r}_s,\mathbf{r}_d,t) = \sum_n a_{Fn}(\mathbf{r}_s,\mathbf{r}_d)e^{-\Gamma_n t} + a_D(\mathbf{r}_s,\mathbf{r}_d,t)e^{-v\mu_a t} \tag{22.7}$$

where a_{Fn} denote the decay amplitudes of the fluorophore with lifetime $\tau_n = 1/\Gamma_n$ and are readily obtained as the residue (Matthews & Walker 1970) of the integrand in Equation 22.5 at the simple poles at $-i\Gamma_n$. This leads to the following linear inverse problem for the yield distribution $\eta_n(\mathbf{r})$ of the nth fluorophore:

$$a_{Fn}(\mathbf{r}_s,\mathbf{r}_d) = \int_V d^3 r \tilde{W}_n^B(\mathbf{r}_s,\mathbf{r}_d,\mathbf{r},-i\Gamma_n)\eta_n(\mathbf{r}). \tag{22.8}$$

with a weight matrix $\tilde{W}_n^B$ for inverting the nth lifetime component τ_n simply given by the FD weight matrix (Equation 22.6) but evaluated at an imaginary frequency of $\omega = -i\Gamma_n$. For this imaginary frequency, we have

$$(k^{(x,m)})^2 = \frac{\left(-v\mu_a^{(x,m)} + \Gamma_n\right)}{D^{(x,m)}} = \frac{-v\left(\mu_a^{(x,m)} - \Gamma_n/v\right)}{D^{(x,m)}} = \frac{-v\mu_{an}'^{(x,m)}}{D^{(x,m)}} \tag{22.9}$$

where we have defined a "reduced" absorption, $\mu_{an}' = \mu_a - \Gamma_n/v$. But Equation 22.9 is identical to the definition of k for the CW diffusion (Helmholtz) equation, showing that $\tilde{W}_n^B$ are sensitivity functions for the CW diffusion equation with a reduced absorption. If we assume that $\Gamma_n < v\mu_a^{(x,m)}$, that is, $\tau_n > 1/v\mu_a^{(x,m)}$, it is clear that the first term in Equation 22.7 dominates the long-time behavior of the net fluorescence signal. We call this region the asymptotic region, where $t > \tau_a$. We thus see that the TD fluorescence forward problem reduces "asymptotically" to multiple CW forward problems with a separate effective absorption of μ_{an}' for each lifetime. As can be checked easily, $\tau_a < 0.5$ ns for $\mu_a > 0.1/$cm (typical value for biological tissue), which means that the asymptotic TD decay is dominated by the fluorescence lifetimes that are longer than a few subnanoseconds. We will, in fact, see in the next section that the diffusive timescale from a finite volume is even shorter than τ_a. It is also possible to evaluate the second term of Equation 22.7 as the contribution of the integration on either side of the branch cuts. But this signal is negligible from the point of view of fluorescence lifetime, and we therefore do not derive it here. We will show in the next section that the multiexponential form of the first term of Equation 22.7 also holds for the entire temporal response but with a time-dependent decay amplitude.

Although the above results were derived assuming a homogeneous infinite medium, a general solution for the inhomogeneous diffusion equation in a bounded volume may be written in terms of the homogeneous GF and its normal derivatives at the boundary (Barton 1989). It is thus plausible to assert that the complex plane structure in Figure 22.1 also applies to arbitrary heterogeneous media. This implies that Equation 22.8 can be generalized to arbitrary media by simply substituting the GF solutions of the heterogeneous diffusion equation with finite boundary models. In the next section, we will indeed derive a general time-dependent, reduced-absorption decay amplitude that smoothly goes over to Equation 22.8 in the asymptotic limit and is valid for bounded heterogeneous media. The definition of the asymptotic region, when the fluorescence decay is dominant, will also be made more precise.

22.2.3.2 Time domain derivation of tomographic FLIM

Consider the full TD forward problem given in Equations 22.3 and 22.4. We start with the following general result (Kumar et al. 2006) for the GF for the heterogeneous transport equation (Arridge 1999; Chandrasekhar 1960):

$$G^{(x,m)}(\mathbf{r},t) = G_\nabla^{(x,m)}(\mathbf{r},t)e^{-v\mu_a^{(x,m)}(\mathbf{r})t}, \tag{22.10}$$

where the functions $G_\nabla^{(x,m)}$ are dependent only on the *gradient* of the absorption coefficient, $\nabla\mu_a^{(x,m)}(\mathbf{r})$, and independent of $\mu_a^{(x,m)}(\mathbf{r})$ itself (note that this result is a generalization of Equation 2 in Durduran et al. 1997, which is for a homogeneous medium). Thus, the functions $G_\nabla^{(x,m)}$ are invariant to constant shifts in the absorption. This means that under the long lifetime condition, $\Gamma_n < v\mu_a^{(x,m)}(\mathbf{r})$, $\forall \mathbf{r} \in \Omega$, the exponential factor in the integrand of Equation 22.3 can be absorbed into the GFs as follows: since $e^{\Gamma_n t'} = e^{\Gamma_n(t'-t'')}e^{\Gamma_n t''}$, we can write, using Equation 22.4,

$$W^B(\mathbf{r}_s,\mathbf{r}_d,\mathbf{r},t')e^{\Gamma_n t'} = \int_0^{t'} dt'' \left[G^x(\mathbf{r}_s,\mathbf{r},t'-t'')e^{\Gamma_n(t'-t'')} \right]\left[G^m(\mathbf{r},\mathbf{r}_d,t'')e^{\Gamma_n t''} \right].$$

Using Equation 22.10 and the property that $G_\nabla^{(x,m)}$ are unaffected by a constant shift in the absorption, and defining a heterogeneous reduced absorption $\mu_a'^{(x,m)}(\mathbf{r}) = \mu_a^{(x,m)}(\mathbf{r}) - \Gamma_n/v$, the right-hand side of the above equation becomes

$$\int_0^{t'} dt'' \left[G_\nabla^x(\mathbf{r}_s,\mathbf{r},t'-t'')e^{-v\mu_a'^x(\mathbf{r})(t'-t'')} \right]\left[G_\nabla^m(\mathbf{r},\mathbf{r}_d,t'')e^{-v\mu_a'^m(\mathbf{r})t''} \right]$$

$$= \int_0^{t'} dt'' G_n^x(\mathbf{r}_s,\mathbf{r},t'-t'')G_n^m(\mathbf{r},\mathbf{r}_d,t'') \tag{22.11}$$

$$= W_n^B(\mathbf{r}_s,\mathbf{r}_d,\mathbf{r},t').$$

Here we have defined a TD sensitivity W_n^B (counterpart of the FD sensitivity $\tilde{W}_n^B$ in Equation 22.8), which is evaluated as in Equation 22.4 but using transport GFs, $G_n^{(x,m)}$, that are evaluated with a reduced absorption, $\mu_a'^{(x,m)}$. We finally substitute Equation 22.11 into Equation 22.3 and get the following form for the TD fluorescence signal when $\Gamma_n < v\mu_a^{(x,m)}(\mathbf{r})$, $\forall \mathbf{r} \in \Omega$:

$$U(\mathbf{r}_s,\mathbf{r}_d,t) = \sum_n e^{-\Gamma_n t}\int d^3r \left[\int_0^t dt' W_n^B(\mathbf{r}_s,\mathbf{r}_d,\mathbf{r},t') \right]\eta_n(\mathbf{r}). \tag{22.12}$$

The above equation reduces to an elegant multiexponential form:

$$U(\mathbf{r}_s,\mathbf{r}_d,t) = \sum_n A_n(\mathbf{r}_s,\mathbf{r}_d,t)e^{-\Gamma_n t} \tag{22.13}$$

but with time-dependent decay amplitudes A_n, given by

$$A_n(\mathbf{r}_s,\mathbf{r}_d,t) = \int d^3r \left[\int_0^t dt' W_n^B(\mathbf{r}_s,\mathbf{r}_d,\mathbf{r},t') \right]\eta_n(\mathbf{r}). \tag{22.14}$$

The time dependence of the decay amplitudes A_n reflects the evolution of the background diffusive response. In Figure 22.2, the simulated temporal evolution of $A(t)$ is shown for diffusive slabs of thicknesses 2 and 10 cm, with a 2 mm³ fluorophore inclusion of 1 ns lifetime embedded at the center of the slab. The net fluorescence signal calculated using tomo-FLIM model in Equations 22.13 and 22.14 is compared with the fluorescence signal computed directly using Equations 22.3 and 22.4 and confirms the accuracy of tomo-FLIM model over a wide range of medium thicknesses that includes small animal to human imaging applications. Note that $A(t)$ rapidly reaches a constant value, beyond which the temporal evolution of $U(\mathbf{r}_s,\mathbf{r}_d,t)$ is purely exponential. Indeed, it is clear that Equation 22.14 approaches Equation 22.8 in

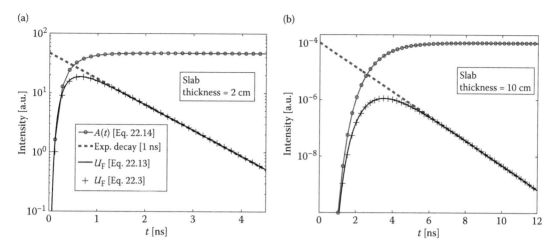

Figure 22.2 Simulations to elucidate the diffuse and pure fluorescent decay components as revealed by the time domain fluorescence model presented in Equation 22.13. The medium was an infinite slab of thickness 2 cm (a) and 10 cm (b), with optical properties $\mu_s^x = \mu_s^m = 10/\text{cm}$, $\mu_a^x = \mu_a^m = 0.1/\text{cm}$. The fluorescence signal was calculated for a single source detector pair, with a small fluorescent inclusion at the center. The signal calculated using the conventional approach in Equation 22.3 (+ symbol) is compared with that calculated using an effective-absorption-based model, viz., Equation 22.13 (solid black line). The decay amplitude, $A(t)$ (dotted blue line) and pure fluorescence decay (dashed red line) are also delineated for both cases.

the asymptotic limit, since $\tilde{W}_n^B$ is just the CW (or time integrated) version of W_n^B, so that $A_n(\mathbf{r}_s, \mathbf{r}_d, t) \rightarrow a_n(\mathbf{r}_s, \mathbf{r}_d)$ in the asymptotic limit. The time constant for the rise of $A(t)$ toward a_n (see the next section) will depend on the intrinsic tissue absorption, scattering, and size of the imaging volume. We define the "asymptotic regime" for times when the time average of W_n^B (which is the integrand of $A(t)$) in Equation 22.14 will become nearly time independent and approach the CW sensitivity function, which we denote by $\bar{W}_n^B \left(= \tilde{W}_n^B(-i\Gamma_n) \right)$. We thus have

$$U(\mathbf{r}_s, \mathbf{r}_d, t) \xrightarrow{t > \tau_D} \sum_n e^{-\Gamma_n t} \underbrace{\int d^3 r \, \bar{W}_n^B(\mathbf{r}_s, \mathbf{r}_d, \mathbf{r}) \eta_n(\mathbf{r})}_{a_{Fn}(\mathbf{r}_s, \mathbf{r}_d)} . \tag{22.15}$$

In other words, Equation 22.14 approaches Equation 22.8 in the asymptotic limit. Equation 22.13 therefore constitutes a generalized multiexponential forward problem for tomographic FLIM that includes both early and late arriving photons and is rigorous within the radiative transport model of photon propagation in turbid media.

22.2.4 CONDITIONS FOR RECOVERY OF *IN VIVO* FLUORESCENCE LIFETIMES

The natural question in the mind of any researcher interested in whole-body FLIM is, under what conditions can intrinsic fluorescence lifetimes reliably be measured on the surface of a living subject? As mentioned in Section 22.1, the intrinsic fluorescence lifetime of a fluorophore can be altered *in vivo* either because of (a) the tissue environment or (b) light scattering and absorption in tissue. This chapter concerns the latter phenomenon, with the understanding that the "intrinsic" lifetime refers to the lifetime of the fluorophore in the biological environment, which can be separately characterized in advance (Raymond et al. 2010). There are two timescales involved. Firstly, $\tau_a = 1/v\mu_a$ is the asymptotic decay time of the intrinsic diffuse temporal response in the limit of homogenous semi-infinite media (Patterson et al. 1989). The second is the actual decay time, τ_D, for the intrinsic photon diffusion in a finite-sized object (which is also related to the rise-time constant of the TD decay amplitude, $A(t)$; Figure 22.2). The time constant

τ_D is always shorter than τ_a due to the presence of boundaries (Haselgrove et al. 1992; Patterson et al. 1989), that is, $\tau_D < \tau_a$ (a numerical evaluation of τ_D for a range of tissue optical properties can be found in Kumar et al. 2005). Since the net fluorescence signal is a convolution of the pure fluorescence decay with intrinsic diffusive response, the condition $\tau_n > \tau_a$ guarantees that the intrinsic lifetimes within tissue can be measured asymptotically from surface fluorescence decays,* irrespective of scattering and medium size. Furthermore, the tomo-FLIM model presented in Equation 22.13 is valid.

The condition $\tau_n > \tau_a$ is easily satisfied for approximately nanosecond lifetime fluorophores in biomedical applications ($\mu_a = 0.1$/cm corresponds to $\tau_a = 0.5$ ns). The simple rule that $\tau_n > \tau_a$ with τ_a denoting the average absorption for a heterogeneous medium dictates the condition for measuring intrinsic lifetimes from whole animals and for the applicability of the tomographic FLIM model. For heterogeneous media, care should be exercised in using the reduced absorption model in Equation 22.13 in avoiding regions with a negative value for μ_a'. The exact influence of low absorption regions on the reduced absorption model will be studied in future work.

22.2.5 INVERSE PROBLEM

The previous sections were concerned with the forward problem for time domain fluorescence tomography. The inverse problem consists of determining the fluorescence yield and lifetime distributions (and the background tissue optical properties, if they are not known) from measured fluorescence and excitation data. In the tomo-FLIM approach, the lifetimes are first obtained as discrete values by fitting the decay (asymptotic) portion of the TD data (which constitutes an inversion problem itself, via the inverse Laplace transform; Kumar et al. 2001). We only need to determine the fluorescence yield distributions, $\eta(\mathbf{r})$. We refer the reader to standard texts for a general treatment of linear inverse problems (Bertero and Boccacci 1998). Here, we present a simple approach that illustrates what is involved. Equation 22.8 is in the form of an underdetermined problem of the form $y = Wx$, where W is a general sensitivity matrix. The pseudo-inverse W_s^{-1} defined by $x = W_s^{-1} y$ is given as

$$W_s^{-1} = L^{-1}\tilde{W}^T (\tilde{W}\tilde{W}^T + \alpha\lambda I)^{-1}, \tag{22.16}$$

where $\tilde{W} = WL^{-1}$, with L as a diagonal matrix whose diagonal elements are $\sqrt{(diag(W^T W))}$, λ is the regularization parameter, and $\alpha = (max(diag(\tilde{W}\tilde{W}^T)))$. Note that since $(W^T W)_{ij} = \sum_l W_{li}W_{lj}$, we have

$$L_{ii} = \sqrt{(W^T W)_{ii}} = \sqrt{\sum_l W_{li}^2}. \tag{22.17}$$

Thus, the diagonal elements of L are the column norms of the sensitivity function W or the net sensitivity at each voxel summed over all measurement pairs. The normalization by L therefore has the effect of annulling the strong spatial variations in the sensitivity function. In order to reduce the computational time for evaluating the inverse in Equation 22.16 for multiple regularization parameters, an SVD analysis can be employed. Writing $\tilde{W} = USV^T$, it can be readily shown that

$$x = L^{-1}VS(S^2 + \alpha\lambda I)^{-1}U^T y. \tag{22.18}$$

---

* Note that for heterogeneous media, it is known that the intrinsic diffusive decay time τ_a is relatively constant on the measurement surface (Haselgrove et al. 1992), so that we can use the average or "baseline" absorption of the tissue medium to estimate τ_a.

22.3 EXPERIMENTAL METHODS

22.3.1 IMAGING SYSTEM

The imaging system for tomographic FLIM essentially consists of a light source, a fiber to deliver the light to the surface of the animal or subject, and a detection system that can be either wide-field-based using a CCD camera or point-based using fibers. In order to provide nanosecond time resolution, either a gated camera (Nothdurft et al. 2009; Raymond et al. 2010) or time-correlated single-photon counting (TCSPC) (Bloch et al. 2005) is used. These experimental methods are discussed in detail in Chapters 6 and 9. A detailed description of a small animal TD fluorescence tomography system can be found in Kumar et al. (2008b). Below we briefly discuss some experimental techniques specific to wide-field detected tomographic FLIM.

22.3.2 IMPULSE RESPONSE AND TIME ORIGIN

An important parameter for tomographic FLIM is the system impulse response function (IRF) as well as an accurate estimate t_0, which is the time when the excitation pulse is incident on the surface of the imaging medium (we have set $t_0 = 0$ in the derivations above). A correct estimate of the time origin t_0 ensures that the relative amplitudes of the multiple lifetime components are correct (Kumar et al. 2006), and is crucial for estimating the optical properties using TD data, as well as for fluorescence reconstructions using the early time gates (Niedre et al. 2008). An advantage of noncontact detection is that t_0 can be estimated from the IRF, which can, in turn, be measured directly from the source illumination. This is possible since the position of the CCD camera (i.e., the detectors) is unchanged before and after the sample is placed on the imaging plate, unlike in contact geometries with fiber-based detection. Note that the time t_0 measured from the IRF includes the time for free space propagation of light from the source to the camera. However, the propagation time for the fluorescence emitted from the imaging surface to the detector will be slightly shorter than t_0, due to the finite thickness of the mouse. Before analyzing the fluorescence data, t_0 should therefore be offset by the sample thickness, for example, ≈50 ps for a mouse of thickness 2 cm. The measured IRF can be directly forward-convolved into the model before tomographic inversion. This procedure is superior to a deconvolution of the IRF from the raw fluorescence data, which is a highly ill-posed problem. For the asymptotic reconstructions, the effect of the IRF on the forward model can be calculated analytically for a square IRF. Using Equation 22.15, the influence of a square IRF of width T is given as

$$
\lim_{t > \tau_D} U_T(\mathbf{r}_s, \mathbf{r}_d, t) = \int_t^{t+T} \lim_{t > \tau_D} U(\mathbf{r}_s, \mathbf{r}_d, t)
$$
$$
= \sum_n \tau_n \left(1 - e^{-T/\tau_n}\right) a_n(\mathbf{r}_s, \mathbf{r}_d) e^{-\Gamma_n t}.
$$

(22.19)

Thus, the effect of the system IRF on the measured decay amplitudes is a scaling factor that depends on the lifetime and the width of the IRF. The decay amplitudes a_{F_n} need to be scaled by the coefficient $\tau_n \left(1 - e^{-T/\tau_n}\right)$ before being inverted using Equation 22.18. This scaling factor is essentially the result of a partial averaging of the exponential decay in Equation 22.19 and approaches the CW signal in the limit $T \to \infty$.

22.3.3 MULTIEXPONENTIAL FITS USING GLOBAL LIFETIME ANALYSIS

An important step in the asymptotic approach is the extraction of the decay amplitudes for all the lifetimes present in the medium. The recovery of lifetimes and decay amplitudes from multiexponential fits to decay data is a nonlinear problem, which is further complicated by the presence of noise. However, the tomographic-FLIM model (Equation 22.8) allows for a significant simplification of the fitting process. Since the set of discretized lifetime components, τ_n, are independent of the measurement locations (i.e., the S–D coordinates $(\mathbf{r}_s, \mathbf{r}_d)$), the multiexponential analysis can be performed in two stages. First, the total fluorescence

Fluorescence lifetime imaging based on exogenous probes

decay is calculated as a sum over all S–D pairs to obtain a high signal-to-noise ratio temporal data set. This "global" signal is composed of all the lifetime components present in the system and allows a more robust determination of the lifetimes through a nonlinear analysis. In the second step, the lifetimes determined from the S–D-integrated decays are used in a *linear* fit of the decays for each individual S–D pair. Besides improving the robustness of the fitting procedure for the lifetimes and decay amplitudes, the global analysis is computationally much less cumbersome than performing a nonlinear fit for every S–D measurement. Chapters 10 through 12 present a detailed review of existing methods for FLIM data analysis, which can be directly incorporated for tomographic FLIM in turbid media using Equation 22.8.

22.3.4 SINGLE VS. MULTIEXPONENTIAL ANALYSIS

When two or more fluorophores are present simultaneously in an animal, the measured fluorescence is a sum of photons from the respective fluorophores. Monoexponential analysis of a mixed signal provides an average lifetime for each detector or pixel, but can be misleading given the contribution of the fluorophores at different concentrations and quantum efficiencies. In contrast, a multiexponential analysis based on expected lifetime components can produce a quantitative measure of multiple fluorophores. Moreover, a monoexponential analysis does not exploit the full power of lifetime-based tomographic separability as afforded by Equation 22.8. As an example, consider multiplexing Osteosense 750 (PerkinElmer), a bone-targeted NIR probe, with X-Sight 761 (Kodak), a NIR probe that remains in the blood stream and accumulates slowly in the liver. When just Osteosense 750 is present in the animal, the monoexponential lifetime map exhibits a narrow lifetime distribution with mean $\tau = 835$ ps (Figure 22.3b). Administration

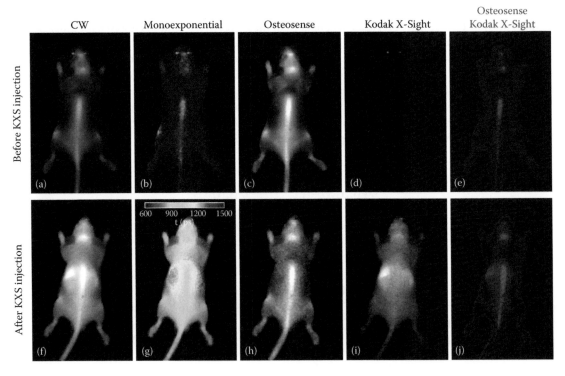

Figure 22.3 Planar FLIM of Osteosense 750 and Kodak X-Sight 761. A nude mouse received 2 nmol Osteosense 24 h prior to imaging; planar fluorescence time-resolved images were collected immediately before (a–e) and after (f–j) administration of Kodak X-Sight 761 (3 nmol). Image pixels were fit for the amplitude and lifetime of a monoexponential function or were fit with a linear biexponential function. (a, f) Continuous wave (CW) images. (b, g) Lifetime maps from monoexponential fit; color bar indicates the lifetime in picoseconds. The decay amplitude distributions from a biexponential fit to the composite probe data are shown for Osteosense ($\tau_1 = 835$ ps) in (c, h) and for Kodak X-Sight 761 ($\tau_2 = 1343$ ps) in (d, i). (e, j) Merged amplitude distribution map shown as RGB images with amplitudes of Osteosense (blue) and X-Sight (red). (Reproduced from Raymond, S. B. et al., *Journal of Biomedical Optics* 15(4), 046011, 2010. With permission.)

of Kodak X-Sight results in a lifetime map with a large distribution of lifetimes, ranging from 1000 to 1500 ps (Figure 22.3g). Average lifetime at bony structures is significantly higher than with Osteosense alone, due to the contribution of vascular Kodak X-Sight. Biexponential fitting using average lifetimes for Osteosense (τ = 835 ps) and Kodak X-Sight (τ = 1343 ps) components results in clear anatomical separation of the two probes (Figure 22.3c–e, h–j). The Osteosense components before and after Kodak X-Sight administration are qualitatively similar (compare Figure 22.3c, h), and no Kodak X-Sight is detected by the biexponential fit before administration (Figure 22.3d).

22.3.5 NOISE CONSIDERATIONS FOR LIFETIME MULTIPLEXING

In the global analysis approach, we employ a linear fit with *a priori* lifetimes (estimated from separate measurements) to determine the amplitudes. The uncertainty (variance) in the decay amplitudes recovered from the linear fit, σ_a, depends on both the measurement noise and the separation ($\Delta\tau$) of the lifetimes involved. To estimate σ_a, we analyze the propagation of noise from the measurement (time domain image) to the recovered amplitudes using linear regression theory (Press et al. 1992). The time-dependent measurement, **y**, is linearly related to the amplitudes of the individual decay components as **y** = X**a**, where the columns of X are normalized single-exponential decays, and **a** are the component amplitudes. The uncertainty in the recovered amplitudes, σ_a, is dependent upon the measurement noise at each time point, expressed as a diagonal matrix Σ, where $\sum_{ii} = 1/\sigma_{yi}^2$, and the respective basis functions X:

$$\sigma_{aj}^2 = \left[\left(X^T \sum X \right)^{-1} \right]_{jj}. \tag{22.20}$$

Equation 22.20 allows calculation of recovered amplitude uncertainty given known system noise parameters (which determine Σ) and fluorophore lifetimes (which dictate the basis functions, X). Figure 22.4 shows the relative amplitude uncertainty (σ_a/a) simulated assuming a specific noise model for Σ relevant to a time-gated imaging system. It is seen that σ_a/a increases as the lifetime separation between probes, $\Delta\tau$, decreases (Figure 22.4a). For a given lifetime separation, the relative amplitude uncertainty for one component increases as the amplitude of the other component increases (Figure 22.4b). This means that if one component is much weaker than the other, it will have increased relative uncertainty. Fluorophore separability, defined arbitrarily as the conditions under which relative uncertainty is

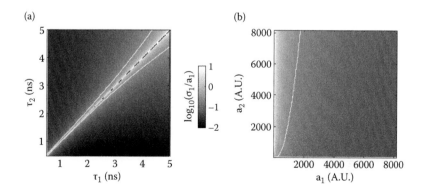

Figure 22.4 Influence of noise on the amplitude uncertainty for biexponential fits. Simulated biexponential fluorescence decays were generated for a noise model, $\sigma_y^2 = \beta y + \sigma_r^2$, with $\beta = 6.53$ and a dynamic range of 2^{14} (4 × 4 hardware and 2 × 2 software binning) for the time-gated intensifier camera (Kumar et al. 2008a). (a) Relative amplitude uncertainty for equal amplitude probes of varying lifetimes τ_1 and τ_2. (b) Relative amplitude uncertainty of the first amplitude component for varying amplitudes, a_1 and a_2, over the dynamic range of the instrument, with fixed lifetimes, τ_1 = 1 ns, τ_2 = 1.2 ns. The white lines in (a, b) indicate the 30% relative uncertainty contours. (Reproduced from Raymond, S. B. et al., *Journal of Biomedical Optics* 15(4), 046011, 2010. With permission.)

<30%, can be estimated directly as described above. For example, assuming one fluorophore with lifetime τ_1 = 1000 ps, quantitative unmixing is possible for $\Delta\tau \geq 200$ ps at relative concentrations as low as 1:5 (Figure 22.4b). Measurement noise also affects the determination of the lifetime, τ, from a monoexponential fit, for example, when measuring the *in vivo* lifetime characteristics of a fluorophore. We estimated the propagation of measurement noise by simulating a fluorophore that has a fixed τ and realistic amplitude distribution (chosen from the mean lifetime and amplitude distribution of a similar *in vivo* measurement) and noise according to a conservative empirical model; the simulated measurement was then fit at each pixel for τ. As shown in Figure 22.5, the uncertainty in τ due to noise is ≈20%, which shows that a majority of the apparent *in vivo* lifetime heterogeneity is due to noise. This sets a limit on the number of fluorophores that may be multiplexed. Figure 22.6 shows the amplitude uncertainty with a simulation of lifetime multiplexing using up to five lifetimes using the shot noise model. This indicates that under the noise statistics used here, the maximum number of fluorophores that can be multiplexed reasonably (less than 30% relative amplitude uncertainty) is three.

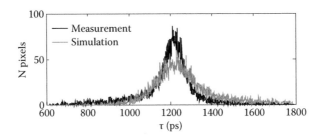

Figure 22.5 Influence of noise on lifetime uncertainty for monoexponential fits. The effects of measurement noise on estimation of τ for a single fluorophore were determined by simulating a measurement of Kodak X-Sight *in vivo*, using the mean lifetime and amplitude distribution of an actual Kodak X-Sight measurement. Noise was added according to an empirical noise model, $\sigma_y^2 = \beta y + \sigma_r^2$, with $\beta = 6.53$; each pixel was fit for τ and then plotted as a histogram. The measurement is shown above in black and the simulation in gray. The standard deviation for the measurement was 157 and the simulation was 193, which reflects the conservative noise model (adds slightly more noise than needed). (Reproduced from Raymond, S. B. et al., *Journal of Biomedical Optics* 15(4), 046011, 2010. With permission.)

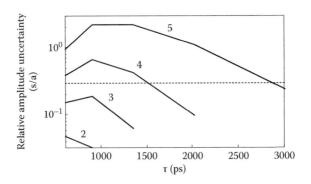

Figure 22.6 Amplitude uncertainty for multiple lifetime components was tested by simulating n = 2–5 lifetimes, with the initial lifetime τ_1 = 600 ps and each additional lifetime 1.5× the previous; amplitudes were split evenly between the components as 4096 × 4/n. Noise was added according to the conservative noise model for 4 × 4 hardware binning and 2 × 2 software binning. The relative uncertainty was calculated for each lifetime component as additional components were added. The 30% uncertainty cutoff is shown as a dotted line. (Reproduced from Raymond, S. B. et al., *Journal of Biomedical Optics* 15(4), 046011, 2010. With permission.)

22.4 *IN VIVO* TOMOGRAPHIC FLIM

As a demonstration of tomographic FLIM in living mice, we show lifetime multiplexing of Kodak X-Sight and Osteosense simultaneously injected in living nude mice ($n = 3$), using a time-resolved free-space tomographic imaging system (Kumar et al. 2008b; Raymond et al. 2010). Excitation and fluorescence measurements were collected 1 h after Kodak X-Sight administration from a grid (~3 × 3 mm separation) of 44 sources and ~100 detectors (Figure 22.7b). The asymptotic region was approximated as times t such that the quantity $\int_{0}^{t} e^{t'/\tau} U_{exc}\, dt'$, where U_{exc} is the excitation measurement, approached 99% of its maximum value (Figure 22.4a; this is an experimental approximation for the quantity τ_D discussed in Section 22.2.4). The decays for all S–D pairs were fit in the asymptotic region for *a priori* lifetimes, $\tau_1 = 844$ ps and $\tau_2 = 1242$ ps, which were the mean *in vivo* lifetimes of X-sight and Osteosense injected separately in mice. Note that the scale for the Kodak X-Sight amplitude component is approximately fivefold greater than the Osteosense amplitude scale. The decay amplitudes for the two lifetimes at all S–D pairs (Figure 22.7f) were employed in Equation 22.8 to recover the full 3D yield distributions for Osteosense and X-sight. The 3D tomographic images shown in Figure 22.7g through j indicate that

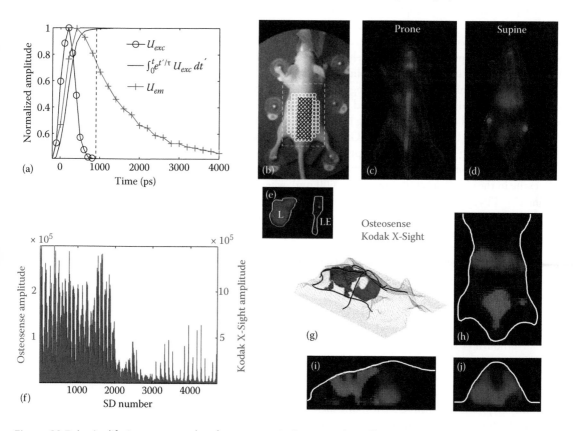

Figure 22.7 *In vivo* lifetime tomography of two anatomically targeted NIR fluorophores. Anesthetized nude mice were administered Osteosense (24 h prior) and Kodak X-Sight (1 h prior), and tomographic data were acquired for 44 sources and 107 detectors. Reconstructions used excitation (750/40 nm BP filter), emission (800 nm LP filter), and 3D-surface measurements using a 3D camera (Kumar et al. 2008b). (a) Representative time-resolved data for tomographic reconstructions. (b) Bright field image of with source (black "x") and detector locations (white "o"). The reconstructed ROI is indicated with the dashed white rectangle. (c, d) Dual color, lifetime-unmixed, planar fluorescence images showing the amplitudes of Osteosense (blue) and Kodak X-Sight (red). (e) Postmortem organs (liver, L; amputated lower extremity, LE). (f) Recovered amplitude components for 44 × 107 SD pairs. (g) 3D rendering of surface (grid) and the Osteosense and Kodak X-Sight distributions. The location of slices in g–i are shown in bold black lines. (h–j) Slices from lifetime-separated reconstruction. Surface boundaries are indicated as solid white lines. (Reproduced from Raymond, S. B. et al., *Journal of Biomedical Optics* 15(4), 046011, 2010. With permission.)

Osteosense is confined to bony structures, including the spinal column, pelvis, skull, and long bones, whereas Kodak X-Sight has localized to the liver, with residual Kodak X-Sight in the vascular compartment (Figure 22.7c, d). These findings are consistent with the expected behavior of these two fluorophores and were confirmed by postmortem imaging of the organs and bones (Figure 22.7e). Thus, the tomographic-FLIM model separates and correctly localizes the spatial distribution of two lifetime components simultaneously present in a living animal.

22.5 CONCLUSIONS AND FUTURE OUTLOOK

In this chapter, we have presented theoretical and experimental methods for tomographic FLIM in turbid media. A forward model for TD fluorescence was presented that directly translates the multiexponential approach used in microscopic FLIM to 3D tomographic reconstructions in the whole body. The basis of this algorithm is the experimental observation that typical fluorophore lifetimes are longer than the intrinsic timescales for diffusive propagation of light through biological tissue and can therefore be directly measured from surface fluorescence decays. Further, we presented a rigorous formalism for isolating the individual 3D fluorescence yield distribution of each lifetime from a mixture of fluorophores using their tomographic surface decay amplitudes. Several commercial NIR fluorescent dyes are already available with capability for targeting specific disease pathologies in small animals. By a careful selection of dyes with lifetime contrast, the TD lifetime technology presented here will thus create new avenues to visualize multiple biological processes noninvasively, accelerating the drug discovery process in preclinical imaging, and can potentially offer interesting applications in the clinical settings as well.

The application areas for *in vivo* tomographic FLIM can be broadly classified into three categories based on contrast mechanism. The first is based on intrinsic contrast, for instance, the tracking of fluorescent protein (FP) labeled cancer cells or using intrinsic tissue autofluorescence (AF) to separate tissue types. FPs have revolutionized biological research (Giepmans et al. 2006) by allowing the visualization of cellular level processes using intravital microscopy. The ability to image FPs in whole animals will open several new avenues for monitoring biological processes in intact environments. However, a majority of existing FPs excite in the visible spectral range (400–600 nm), where tissue (AF) is significant. Previous attempts for imaging FPs in whole animals used CW techniques (Deliolanis et al. 2008; Hoffman & Yang 2006), where the AF signal is indistinguishable from the FP fluorescence since it is intensity based. Recently, it has been shown that fluorescence lifetime allows improved detection of FPs against background tissue AF (Kumar et al. 2009; McCormack et al. 2007; Soloviev et al. 2007). Specially, the distinct nonexponential temporal response of AF decays as compared with the pure exponential decay of the fluorescence of several FPs *in vivo* has been used to enhance the imaging of GFP tumors in live mice (Kumar et al. 2009). The second application of preclinical whole-body lifetime imaging uses extrinsic fluorophores that are designed to undergo lifetime shifts *in vivo* due to a change in biochemical environment, such as pH (Berezin et al. 2011), protein binding (Goiffon et al. 2009), and binding to disease targets such as tumors (Bloch et al. 2005; Nothdurft et al. 2009).

Perhaps the most interesting and underexplored application of tomographic FLIM concerns activatable probes. For instance, Forster resonance energy transfer (FRET) probes have been widely used in FLIM (Vogel et al. 2006) and are based on nonradiative quenching of a pair of fluorophores located in close proximity (typically attached to a backbone such as a polymer) with overlapping absorption and emission spectra. Binding to a target enzyme or protein results in a conformational change or cleavage of the original pair, resulting in a dramatic increase in both fluorescence intensity and lifetime. While this mechanism has been explored for whole-body imaging using intensity-based CW techniques (Nahrendorf et al. 2007; Weissleder et al. 1999), lifetime-based applications of activatable probes are relatively recent (Goergen et al. 2012; McGinty et al. 2011; Solomon et al. 2011). Coupled with the fact that lifetime is a concentration- and intensity-independent functional marker of nonradiative quenching, this suggests that the future is bright for *in vivo* fluorescence lifetime tomography of activatable probes.

The central challenge for applying tomographic FLIM in the clinical setting (or whole-body human imaging) is the same as that for diffuse optical NIR spectroscopy, namely, the larger imaging volumes and the high absorption and scattering suffered by light photons in tissue. Although sophisticated algorithms exist for modeling light transport through complex tissue, the limited penetration of diffuse light implies poor depth penetration and spatial resolution. However, recent advances in multimodality imaging have demonstrated the potential of NIR-diffuse optical tomography (DOT) for providing functional contrast in clinical applications (Fang et al. 2011). Tomographic FLIM can capitalize on these multimodal techniques since it relies on the same mathematical formalism as intrinsic contrast DOT. A second challenge for the clinical use of tomographic FLIM concerns the availability of disease-specific contrast agents and the subsequent long approval process for human use. With concurrent advances in lifetime-based targeted probe development (Berezin and Achilefu 2010), it is hoped that an increasing number of disease-specific fluorochromes will be available for human use, at which point tomographic FLIM is likely to play an important role in the clinical setting as well.

REFERENCES

Arridge, S. (1999), 'Optical tomography in medical imaging,' *Inverse Problems* 15(2), R41–R93.

Arridge, S. R. & Schotland, J. C. (2009), 'Optical tomography: Forward and inverse problems,' *Inverse Problems* 25(12), 123010.

Barton, G. (1989), *Elements of Greens Functions and Propagation.* Oxford University Press, Oxford.

Bastiaens, P. I. H. & Squire, A. (1999), 'Fluorescence lifetime imaging microscopy: Spatial resolution of biochemical processes in the cell,' *Trends in Cell Biology* 9(2), 48–52.

Berezin, M. Y. & Achilefu, S. (2010), 'Fluorescence lifetime measurements and biological imaging', *Chemical Reviews* 110(5), 2641–2684.

Berezin, M. Y., Guo, K., Akers, W., Northdurft, R. E., Culver, J. P., Teng, B., Vasalatiy, O., Barbacow, K., Gandjbakhche, A., Griffiths, G. L. & Achilefu, S. (2011), 'Near-infrared fluorescence lifetime pH-sensitive probes', *Biophysical Journal* 100(8), 2063–2072.

Berezovska, O., Ramdya, P., Skoch, J., Wolfe, M. S., Bacskai, B. J. & Hyman, B. T. (2003), 'Amyloid precursor protein associates with a nicastrin-dependent docking site on the presenilin 1-gamma-secretase complex in cells demonstrated by fluorescence lifetime imaging', *Journal of Neuroscience*, 23(11), 4560–4566.

Bertero, M. & Boccacci, P. (1998), *Introduction to Inverse Problems in Imaging.* Institute of Physics Publishing, London.

Bloch, S., Lesage, F., McIntosh, L., Gandjbakhche, A., Liang, K. X. & Achilefu, S. (2005), 'Whole-body fluorescence lifetime imaging of a tumor-targeted near-infrared molecular probe in mice', *Journal of Biomedical Optics* 10(5), 054003.

Bremer, C., Ntziachristos, V. & Weissleder, R. (2003), 'Optical-based molecular imaging: Contrast agents and potential medical applications', *European Radiology* 13(2), 231–243.

Bugaj, J. E., Achilefu, S., Dorshow, R. B. & Rajagopalan, R. (2001), 'Novel fluorescent contrast agents for optical imaging of in vivo tumors based on a receptor-targeted dye-peptide conjugate platform', *Journal of Biomedical Optics* 6(2), 122–133.

Chandrasekhar, S. (1960), *Radiative Transfer.* Dover, New York.

Deliolanis, N. C., Kasmieh, R., Wurdinger, T., Tannous, B. A., Shah, K. & Ntziachristos, V. (2008), 'Performance of the red-shifted fluorescent proteins in deep-tissue molecular imaging applications', *Journal of Biomedical Optics* 13(4), 044008.

Durduran, T., Yodh, A. G., Chance, B. & Boas, D. A. (1997), 'Does the photon-diffusion coefficient depend on absorption?', *Journal of the Optical Society of America A* 14, 3358–3365.

Fang, Q., Selb, J., Carp, S. A., Boverman, G., Miller, E. L., Brooks, D. H., Moore, R. H., Kopans, D. B. & Boas, D. A. (2011), 'Combined optical and x-ray tomosynthesis breast imaging', *Radiology* 258(1), 89–97.

Giepmans, B. N. G., Adams, S. R., Ellisman, M. H. & Tsien, R. (2006), 'Review—The fluorescent toolbox for assessing protein location and function', *Science* 312, 217–224.

Goergen, C. J., Chen, H. H., Bogdanov, Jr., A., Sosnovik, D. E. & Kumar, A. T. N. (2012), 'In vivo fluorescence lifetime detection of an activatable probe in infarcted myocardium', *Journal of Biomedical Optics* 17(5), 056001.

Goiffon, R. J., Akers, W. J., Berezin, M. Y., Lee, H. & Achilefu, S. (2009), 'Dynamic noninvasive monitoring of renal function in vivo by fluorescence lifetime imaging', *Journal of Biomedical Optics* 14, 020501.

Hall, D., Ma, G., Lesage, F. & Yong, W. (2004), 'Simple time-domain optical method for estimating the depth and concentration of a fluorescent inclusion in a turbid medium', *Optics Letters* 29(19), 2258–2260.

Haselgrove, J. C., Schotland, J. C. & Leigh, J. S. (1992), 'Long-time behavior of photon diffusion in an absorbing medium: Application to time-resolved spectroscopy', *Applied Optics* 31, 2678–2683.

Hoffman, R. M. & Yang, M. (2006), 'Whole-body imaging with fluorescent proteins', *Nature Protocols* 1(3), 1429–1438.

Kumar, A. T. N., Chung, E., Raymond, S. B., van de Water, J., Shah, K., Fukumura, D., Jain, R. K., Bacskai, B. J. & Boas, D. A. (2009), 'Feasibility of in vivo imaging of fluorescent proteins using lifetime contrast', *Optics Letters* 34(13), 2066–2068.

Kumar, A. T. N., Raymond, S. B., Bacskai, B. J. & Boas, D. A. (2008a), 'Comparison of frequency-domain and time-domain fluorescence lifetime tomography', *Optics Letters* 33(5), 470–472.

Kumar, A. T. N., Raymond, S. B., Bacskai, B. J. & Boas, D. A. (2008b), 'A time domain fluorescence tomography system for small animal imaging', *IEEE Transactions on Medical Imaging* 27(8), 1152–1163.

Kumar, A. T. N., Raymond, S. B., Boverman, G., Boas, D. A. & Bacskai, B. J. (2006), 'Time resolved fluorescence tomography of turbid media based on lifetime contrast', *Optics Express* 14(25), 12255–12270.

Kumar, A. T. N., Skoch, J., Bacskai, B. J., Boas, D. A. & Dunn, A. K. (2005), 'Fluorescence-lifetime-based tomography for turbid media', *Optics Letters* 30(24), 3347–3349.

Kumar, A., Zhu, L., Christian, J., Demidov, A. & Champion, P. M. (2001), 'On the rate distribution analysis of kinetic data using the maximum entropy method: Applications to myoglobin relaxation on the nanosecond and femtosecond timescales', *Journal of Physical Chemistry B* 105, 7847–7856.

Massoud, T. & Gambhir, S. (2003), 'Molecular imaging in living subjects: Seeing fundamental biological processes in a new light', *Genes & Development* 17(5), 545–580.

Matthews, J. & Walker, R. (1970), *Mathematical Methods of Physics*, 2nd ed. Addison-Wesley, Reading, MA.

McCormack, E., Micklem, D. R., Pindard, L. E., Silden, E., Gallant, P., Belenkov, A., Lorens, J. B. & Gjertsen, B. T. (2007), 'In vivo optical imaging of acute myeloid leukemia by green fluorescent protein: Time-domain autofluorescence decoupling, fluorophore quantification, and localization', *Molecular Imaging* 6(3), 193–204.

McGinty, J., Stuckey, D. W., Soloviev, V. Y., Laine, R., Wylezinska-Arridge, M., Wells, D. J., Arridge, S. R., French, P. M. W., Hajnal, J. V. & Sardini, A. (2011), 'In vivo fluorescence lifetime tomography of a FRET probe expressed in mouse', *Biomedical Optics Express* 2(7), 1907–1917.

Milstein, A. B., Oh, S., Webb, K. J., Bouman, C. A., Zhang, Q., Boas, D. A. & Millane, R. P. (2003), 'Fluorescence optical diffusion tomography', *Applied Optics* 42(16), 3081–3094.

Milstein, A. B., Stott, J. J., Oh, S., Boas, D. A., Millane, R. P., Bouman, C. A. & Webb, K. J. (2004), 'Fluorescence optical diffusion tomography using multiple-frequency data', *Journal of the Optical Society of America A-Optics Image Science and Vision* 21(6), 1035–1049.

Nahrendorf, M., Sosnovik, D. E., Waterman, P., Swirski, F. K., Pande, A. N., Aikawa, E., Figueiredo, J. L., Pittet, M. J. & Weissleder, R. (2007), 'Dual channel optical tomographic imaging of leukocyte recruitment and protease activity in the healing myocardial infarct', *Circulation Research* 100(8), 1218–1225.

Niedre, M. J., de Kleine, R. H., Aikawa, E., Kirsch, D. G., Weissleder, R. & Ntziachristos, V. (2008), 'Early photon tomography allows fluorescence detection of lung carcinomas and disease progression in mice in vivo', *Proceedings of the National Academy of Sciences of the United States of America* 105(49), 19126–19131.

Nothdurft, R. E., Patwardhan, S. V., Akers, W., Ye, Y. P., Achilefu, S. & Culver, J. P. (2009), 'In vivo fluorescence lifetime tomography', *Journal of Biomedical Optics* 14(2), 024004.

Patterson, M. S., Chance, B. & Wilson, B. C. (1989), 'Time-resolved reflectance and transmittance for the non-invasive measurement of tissue optical properties', *Applied Optics* 28, 2331–2336.

Press, W. H., Flannery, B. P., Teukolsky, S. A. & Vetterling, W. T. (1992), *Numerical Recipes in C: The Art of Scientific Computing*, 2nd ed. Cambridge University Press, New York.

Raymond, S. B., Boas, D. A., Bacskai, B. J. & Kumar, A. T. N. (2010), 'Lifetime-based tomographic multiplexing', *Journal of Biomedical Optics* 15(4), 046011.

Selvin, P. R. (2000), 'The renaissance of fluorescence resonance energy transfer', *Nature Structural Biology* 7(9), 730–734.

Solomon, M., Guo, K., Sudlow, G. P., Berezin, M. Y., Edwards, W. B., Achilefu, S. & Akers, W. J. (2011), 'Detection of enzyme activity in orthotopic murine breast cancer by fluorescence lifetime imaging using a fluorescence resonance energy transfer-based molecular probe', *Journal of Biomedical Optics* 16(6), 066019.

Soloviev, V. Y., McGinty, J., Tahir, K. B., Neil, M. A. A., Sardini, A., Hajnal, J. V., Arridge, S. R. & French, P. M. W. (2007), 'Fluorescence lifetime tomography of live cells expressing enhanced green fluorescent protein embedded in a scattering medium exhibiting background autofluorescence', *Optics Letters* 32(14), 2034–2036.

Vogel, S., Thaler, C. & Koushik, S. V. (2006), 'Fanciful fret', *Science's STKE: Signal Transduction Knowledge Environment*, 2006(331), re2.

Weissleder, R., Tung, C. H., Mahmood, U. & Bogdanov, A. (1999), 'In vivo imaging of tumors with protease-activated near-infrared fluorescent probes', *Nature Biotechnology* 17(4), 375–378.

Photosensitizers and PDT

Rinaldo Cubeddu, Paola Taroni, and Gianluca Valentini

Contents

23.1 PHOTOSENSITIZATION AND PHOTOSENSITIZERS

23.1.1 WHAT IS PHOTOSENSITIZATION AND HOW IT PROCEEDS

Photosensitization is a response to light that is mediated by a light-absorbing molecule, called a "photosensitizer," which is not the final target and is not destroyed during the process. When it absorbs a photon, the photosensitizer alters another molecule, called "substrate" or "acceptor," that can be part of a chemical system or a living organism. Both photosensitizer and light are required for photosensitization to occur, while generally light levels and sensitizer nature and concentration are such that individually they have no main effect on the system or organism.

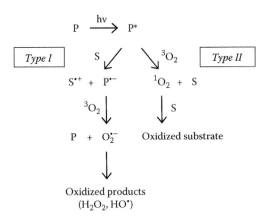

Figure 23.1 Type I vs. Type II photosensitization reactions. P: Photosensitizer; S: substrate.

The first step of the process is absorption of light by a photosensitizer P to produce an excited state (P*). In the presence of oxygen, P* undergoes internal reactions that ultimately result in the chemical alteration of the substrate S. As shown in Figure 23.1, P* can react directly either with the substrate (*Type I reaction*) or with oxygen (*Type II reaction*).

Type I reactions result in either hydrogen atom or electron transfer, yielding radicals or radical ions, with the excited sensitizer generally acting as an oxidant. The radical species that are formed are often highly reactive, such as superoxide and hydroxyl radicals.

Type II reactions mainly lead to excited state singlet oxygen ($^1O_2$). Singlet oxygen then reacts with the substrate to generate oxidized products. For most photosensitizers, the excited singlet state has very short lifetime and rapidly undergoes intersystem crossing, that is, a transition to a slightly lower energy level involving an electron spin flip. The spin flip leads to a triplet energy state. Transitions between triplet and singlet energy states are forbidden. Consequently, the excited triplet state generally has a longer lifetime than an excited singlet state and hence is more likely to enter into an energy transfer reaction. Singlet oxygen is a highly reactive oxygen species that has an excited state lifetime of a few microseconds in most biological environments.

Type II (singlet oxygen) processes are favored over Type I (radical) processes by lower substrate concentration and higher oxygen concentration. Most important photosensitization reactions, either spontaneous or induced ones, occur via Type II processes.

It is also worth noting that oxygen is an important reactant in most photosensitization reactions, whether it is part of the initial reaction with the excited photosensitizer (Type II) or gets involved later, transforming the initial radicals into oxidized products (Type I).

Photodynamic therapy (PDT) is a form of treatment that requires oxygen for its therapeutic effect. This implies that in PDT, a photosensitizer, light, and oxygen are all required. The Type II reaction pathway is expected to be the main one, but Type I reactions involving oxygen may also be implied. PDT has been investigated and developed mostly for the treatment of cancer and precancerous lesions (actinic keratosis). However, various other applications are under investigation (e.g., the treatment of age-related macular degeneration [AMD]). More details can be found in Section 23.1.5.

23.1.2 BASIC PROPERTIES OF PHOTOSENSITIZERS

Let us now examine in more detail excitation and de-excitation processes of a generic molecule to highlight properties that are key for an efficient photosensitizer.

Molecules that absorb light and are excited to the first singlet state subsequently lose the acquired energy through either radiative or radiation-less decay (Figure 23.2). When the higher vibrational levels of the lower electronic state overlap the lower vibrational levels of the higher electronic state (i.e., the nuclear configurations and energies of the two electronic states are identical), vibrational coupling occurs, and crossover from a higher to a lower excited singlet state is possible. This is a very fast process (10–12 s), known as internal conversion.

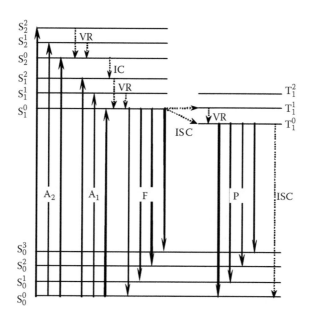

Figure 23.2 Jablonsky diagram showing excitation and de-excitation transitions. A1 (A2): absorption to the first (second) excited singlet state; F: fluorescence; IC: internal conversion; VR: vibrational relaxation; ISC: intersystem crossing; P: phosphorescence.

If the vibrational levels of the two electronic states are separated, a radiative transition (*fluorescence*) may also occur from the lowest vibrational level of the higher electronic state to any vibrational levels of the lower electronic state. Emission takes place from thermally equilibrated excited states, namely, from the lowest vibrational level of the lowest excited singlet state, because relaxation from the excited vibrational levels (through emission of infrared quanta or as kinetic energy lost during collisions) is much faster than emission (10^{-14}–10^{-12} s vs. $\approx 10^{-9}$–10^{-8} s). The transition may lead to any vibrational levels of the ground state because, even though the lifetime of the fluorescent molecule is of the order of 10^{-9}–10^{-8} s, the actual electronic transition occurs much faster (10^{-15} s), so that vibrational relaxation cannot take place during the transition. Thus, the emission spectrum generally consists of a broadband, not a narrow line. Moreover, for biomolecules, the emitted photons have most often frequency in the ultraviolet (UV) or visible range. Following radiative transition, the molecule undergoes vibrational relaxation to the lowest vibrational level (10^{-14}–10^{-12} s). Which one of the two deactivation mechanisms is favored depends on the number of vibrational levels and the difference in energy between the vibrational levels of the two electronic states. The closer they are, the more likely internal conversion will occur. This is the reason why molecules with no rigid skeleton, such as aliphatic molecules, rarely exhibit fluorescence, while aromatic molecules with their rigid ring structures are usually characterized by strong fluorescent emission.

For the same reason, fluorescence occurs almost always only as a deactivation mechanism between the first excited singlet state and the ground state. Higher excited states are generally much closer to each other. Hence, internal conversion is favored and precludes the possibility of fluorescence emission. Consequently, different from the absorption spectrum, the emission line shape is made of just one band. The presence of more than one band in the spectrum of isolated molecules indicates that more than one species is fluorescing. This does not necessarily mean that the emitting molecules have different chemical structure. Actually, when the emitting molecules interact with a complex environment or aggregate, distinct fluorescence bands can be observed, corresponding to different binding sites, states of aggregation, etc.

A feature that is commonly used to characterize each deactivation process is its *quantum yield*. Let us, for example, consider deactivation by fluorescent emission. The *quantum yield of fluorescence* ϕ_f is the fraction of excited molecules that fluoresce. It can be expressed as

Fluorescence lifetime imaging based on exogenous probes

$$\phi_f = \frac{k_f}{k_f + \sum k_d} = \frac{\tau_f}{\tau_f^o} \qquad (23.1)$$

where k_f is the *rate constant for fluorescence* (i.e., the probability that the excited molecule will fluoresce); $\sum k_d$ is the sum of the rate constants for all radiation-less deactivation mechanisms of the first excited singlet state; $\tau_f = \left(k_f + \sum k_d\right)^{-1}$ is the *lifetime* of the first excited singlet state, that is, the average time the molecule spends in the excited state; and $\tau_f^o = (k_f)^{-1}$ is the *radiative lifetime*, that is, the average time the molecule would spend in the excited state if fluorescence were the only deactivation mechanism.

As described previously, when molecules are in the lowest vibrational state of the first excited singlet state, de-excitation may also occur via *intersystem crossing*, which is a transition to a slightly lower energy level involving an electron spin flip. The spin flip leads to a triplet energy state. Transitions between triplet and singlet energy states are forbidden, and consequently, the excited triplet state generally has a longer lifetime than an excited singlet state and hence is more likely to enter into an energy transfer reaction. However, de-excitation of the triplet state may also occur through long-living ($\geq 10^{-3}$ s) photon emission (*phosphorescence*).

Photosensitizers are molecules that efficiently transfer an electron to or from another molecule, or transfer their excitation energy to other molecules (typically molecular oxygen). In most cases, the reason why some molecules are so effective at electron transfer and/or energy transfer is that they very efficiently populate their excited triplet states. The relatively longer lifetime of the triplet state allows more time for energy and/or electron transfer to occur. So, most highly effective photosensitizers are characterized by high quantum yields for the production of their excited triplet state, which is a competing mechanism against radiative decay (fluorescence).

The *action spectrum* is often used to quantify the efficiency of a photosensitizer to produce a specific effect (e.g., cell killing). An action spectrum shows the dependence of the measured effect on the excitation wavelength. As an example, the action spectrum for the photochemical inactivation of cells can be found by measuring the dose of light necessary for inactivating the same cell fraction at different wavelengths and plotting the results as required dose vs. excitation wavelength.

23.1.3 PHOTOSENSITIZER INTERACTIONS WITH BIOLOGICAL TISSUES

Ground-state molecular oxygen has low reactivity and diffuses rapidly through most biological environments, including cell membranes, while singlet oxygen reacts with several biological substrates. Efficient interaction occurs with certain amino acids in proteins (e.g., tryptophan, tyrosine, histidine), DNA and RNA (guanine bases), as well as a variety of unsaturated lipids (including cholesterol and unsaturated fatty acids). On the contrary, no significant effect is produced on carbohydrates. This reactivity limits the ability of singlet oxygen to diffuse great distances. Thus, Type II photosensitization can affect biological substrates only at a moderate distance, up to fractions of micrometers, from the photosensitizer itself.

In turn, hydroxyl radicals generated by Type I reactions are so reactive that they typically interact with the first molecule they get close to, limiting their effects to the site where they are generated.

Therefore, for both types of reactions, the photosensitizer localization is of key importance to determine and control biological effects. Some selectivity toward the environment can inherently be provided by the sensitizer itself. For example, if it is highly soluble in lipids, it will likely tend to localize in the cell membrane rather than in its cytoplasm. However, stronger and more selective targeting can be achieved, attaching the sensitizer to a target-specific molecule.

The heat released during photosensitization processes may also be used to cause selective damage in a process that is named photothermal sensitization.

23.1.4 TYPES OF PHOTOSENSITIZERS

Photosensitizers can either be naturally present in living systems (i.e., endogenous) or generated by an external source and administered to the living system (i.e., exogenous).

Endogenous photosensitizers include molecules such as porphyrins or chlorophyll. Under native conditions, their potential photosensitizing effects are not apparent, either because their concentrations are too low or because they form complexes that inhibit photosensitization reactions. *Exogenous photosensitizers* include many varieties of dyes and biomolecules. Some are natural products (e.g., from plants), while others are synthesized for different applications (e.g., medical or agricultural ones).

A third interesting type of photosensitizers also exists, namely, *exogenously induced endogenous photosensitizers*. These photosensitizers are intermediate between endogenous and exogenous ones. A precursor is administered, which is not active. The metabolic activity of the receiving organism then transforms it into a fully functional photosensitizer.

23.1.4.1 Porphyrin-based photosensitizers

Hundreds of different compounds act as photosensitizers for biological systems. They are often classified based on their chemical structure.

Porphyrin-based photosensitizers deserve particular attention as they include important endogenous sensitizers, as well as exogenously induced endogenous sensitizers and exogenous sensitizers that are approved for clinical PDT and/or for fluorescence-based cancer detection (Berg et al. 2005; Sternberg and Dolphin 1998).

Porphyrins contain four pyrrole subunits linked by methine bridges (Figure 23.3). Tetrapyrroles are naturally occurring pigments, involved in many biological processes. They include the metallopigments heme (the prosthetic group of proteins like hemoglobin), vitamin B12, and chlorophyll. All these compounds allow coordination of different metals at the ring center. While the presence of the metal prevents photosensitization, its removal yields efficient photosensitizers. Most efficient porphyrin-based photosensitizers generally lack coordinated metal ions. On the other hand, several metallophotosensitizers for clinical purposes have been developed along the years, balancing opposing needs. In most cases, they are less effective than they would be in the absence of metal ions; however, they have other properties of basic importance for clinical applications, like improved solubility and stability.

Except for methylene blue, all clinically approved photosensitizers used in PDT are porphyrin-based sensitizers, with substituents in the peripheral positions of the pyrrole rings, on the methine bridges that link the pyrroles, and/or with coordinated metals. These derivates are synthesized to influence the water/lipid solubility, amphiphilicity, and stability of the compounds, which, in turn, determine their biodistribution and pharmacokinetics.

Hematoporphyrin was the first photosensitizer applied in humans. In the early 1960s, attempts to purify it led to hematoporphyrin derivative (HpD), a mixture of monomers and oligomers. A more purified version (Photofrin®) was then produced and commercialized. The idea was to enrich the solution in oligomers that were recognized as the tumor-localizing fraction of the porphyrin mixture. HpD and Photofrin are far from being ideal photosensitizers for use in PDT. Their composition is not accurately reproducible. They have long half-life in the body, leading to undesired skin and eye photosensitivity for 4–6 weeks after injection. Even more important limitations are due to their nonideal spectral properties. Their absorption spectrum consists of a strong absorption Soret band in the UV (350–400 nm) and four much weaker Q-bands in the visible range. Tissue attenuation is strong at short wavelengths (UV and blue-green), mostly due to melanin and hemoglobin. Thus, the longest wavelength (red) Q-band is exploited for PDT when deep tissues need to be reached with enough light for an effective treatment, even though this

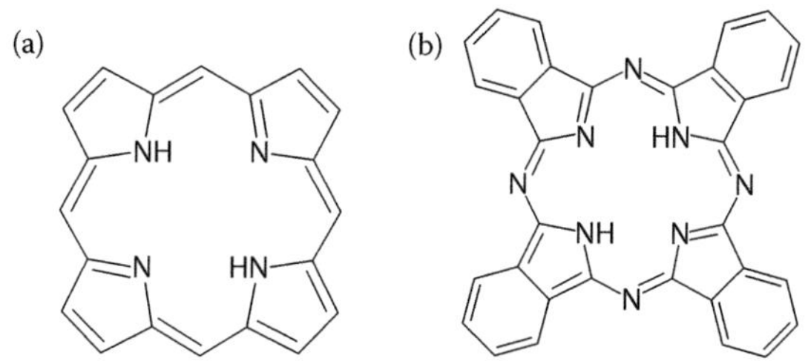

Figure 23.3 Porphyrin (a) and phthalocyanine (b) structure.

provides weak photosensitizer absorption. Notwithstanding all these limitations, porphyrin derivatives are approved for a wide range of clinical applications, and a large number of patients have been treated up to now.

It is interesting to note that, for diagnostic purposes, shorter wavelengths (within the Soret band) have often been applied, favoring efficient excitation at the expenses of deep tissue penetration. This choice is particularly suitable when surface targets are investigated, as in the case of skin pathologies or mucosal lesions that can be reached endoscopically.

Increased absorption in the red–infrared range can be achieved by reducing one (chlorines) or two (bacteriochlorins) double bonds in the conjugated ring structure. Specifically, meso-tetra(hydroxyphenyl) chlorin (m-THPC, Temoporfin, Foscan®) is a chlorine that has been approved for the clinical treatment of head and neck cancer.

The possibility to develop new photosensitizers with higher absorption at long wavelengths is actively being explored. However, it has to be taken into account that an upper wavelength limit exists around 850–900 nm, due to the minimum energy photons needed to absorb to induce singlet oxygen formation. Moreover, due to water absorption, tissue attenuation above 900 nm raises considerably. Thus, the therapeutic window for effective *in vivo* treatments extends approximately from 600 to 800 nm.

23.1.4.2 Nonporphyrin-based photosensitizers

A variety of endogenous and exogenous sensitizers with nonporphyrin structure exist (Wainwright 1996). Just a few examples are mentioned here, as they are much less investigated and used for PDT and fluorescence detection of pathologic lesions than porphyrins and related compounds.

Cyanines are synthetic dyes, originally developed to extend the sensitivity range of photographic emulsions. *Phthalocyanines*, in particular, are macrocyclic compounds, featuring four pyrrole-like subunits, similar to porphyrins (Figure 23.3). They have intense blue-green color (corresponding to strong absorption in the red) and are widely used in dyeing. Phthalocyanines form coordination complexes with most elements. These complexes are also intensely colored and used as dyes or pigments (e.g., in paints). Substitution can increase solubility and shift the absorption to longer (near-infrared) wavelengths, with advantage for *in vivo* applications.

Hypericin is isolated from the plant known as St. John's wort (*Hypericum perforatum*) and is possibly the most powerful photosensitizer present in nature, with quantum yields of singlet oxygen formation up to 0.8. It acts through both Type I and Type II reactions, and its spectral properties are not too unfavorable, with peak absorption in the red (around 595 nm). It has been tested clinically for several cancer indications and also for the detection of bladder cancer. Furthermore, Hypericin has antibiotic and antiviral properties.

23.1.4.3 Precursor-induced photosensitizers

The most common couple of precursor/sensitizer is δ-aminolevulinic acid/Protoporphyrin IX (Berg 2005; Collaud et al. 2004).

Protoporphyrin IX (PpIX) is an intermediate in biosynthetic pathways to produce cytochromes, hemoglobin (the oxygen-binding protein in red blood cells), and myoglobin (which binds oxygen in muscles). Under normal conditions, free PpIX is present at too low concentration to produce photosensitization reactions. However, if excess δ-aminolevulinic acid (ALA) is provided (either systemically or topically, depending on application), PpIX may accumulate, causing photosensitization reactions. The administration of ALA-hexyl ester is approved for the therapy of bladder cancer, while ALA-methyl ester is approved for the treatment of Bowen's disease, actinic keratosis, and basal cell carcinoma.

PpIX is also characterized by intense red fluorescence, when excited by violet light. Thus, it is also effectively applied for the detection of lesions to be treated with PDT and for the identification of surgical resection margins.

23.1.5 WHY IS PHOTOSENSITIZATION IMPORTANT?

Photosensitization reactions affect our lives in many different ways. For example, they are used in synthetic chemistry to produce products that would be much more difficult or expensive to produce by other means. However, in the following, we will focus mostly on therapeutic applications.

23.1.5.1 PDT and detection of tumors

In the last decades, photosensitizers and light have extensively been tested and applied to treat malignant tumors (e.g., in the bladder, prostate, lung, brain) and to remove through PDT other unwanted tissue (e.g., in the precancerous condition represented by Barrett's esophagus; Hamblin and Mroz 2008; MacDonald and Dougherty 1999; O'Connor et al. 2009).

The sensitizer is most often administered systemically and reaches both healthy and diseased tissues. The possibility to perform an efficient PDT relies on the fact that the sensitizer tends to accumulate more and is retained longer in the tumor than in the surrounding healthy tissue, due to concurrent factors that differentiate normal from pathologic tissue (metabolism, angiogenesis, etc.). This makes the latter more prone to the photodynamic action, when irradiated with light of suitable wavelength. Obviously, also selective irradiation of the area of interest can contribute to limit undesired photodynamic actions in healthy tissue.

When ALA/PpIX PDT is performed, therapeutic benefit comes also from the fact that the production of PpIX occurs more efficiently in lesions than in surrounding normal tissue.

One of the most interesting phenomena observed in the PDT of cancer is also an advantage over other therapeutic modalities: PDT has the potential to induce strong and long-lasting antitumor immune response (Canti et al. 1994; Korbelik and Dougherty 1999).

Besides therapeutic purposes, the fluorescence properties of photosensitizers may be effectively exploited for the detection of cancers. This use is generally referred to as photodynamic diagnosis (PDD) or fluorescence diagnosis. For example, intravesical instillation of ALA-hexyl ester (Hexvix®) is approved for the detection of bladder cancer, in particular, for carcinoma *in situ*, which is generally difficult to detect. PPIX fluorescence-guided resection of bladder cancer has also shown promise for treatment purposes. Similarly, systemic administration of ALA (Gliolan®) is approved for intraoperative fluorescence-guided detection and resection of malignant glioma.

23.1.5.2 PDT in ophthalmology

PDT was originally introduced as a modality for cancer treatment, but one of its most successful applications is certainly the treatment of choroidal neovasculature associated with AMD. PDT with Verteporfin (Visudyne®), a benzoporphyrin derivative, was the first therapy approved for subfoveal lesions. Since 2000, it spared hundreds of thousands of eyes from blindness and is now a standard treatment for AMD (Bessler 2004; Lim 2002). The drug is injected intravenously, and 15 min after the start of the infusion, the retina is irradiated with 690 nm light ($50 \, J/cm^2$). At the time of irradiation, the drug is still mostly localized in the blood vessels of the pathologic choroidal neovasculature and causes their occlusion.

23.1.5.3 PDT in dermatology

Dermatologic pathologies are especially suitable for PDT, as they are exposed. So local administration of the photosensitizer is often applied and light can easily be conveyed to the treatment site.

One of the main applications of PDT in dermatology is the treatment of nonmelanoma skin cancer and its precursors, such as actinic keratosis. However, several other pathologic conditions have been successfully treated with PDT, including inflammatory and immune diseases (e.g., psoriasis) as well as infections (e.g., human papilloma virus) (Choudhary et al. 2009; Nestor et al. 2006). ALA-PDT seems to be a safe and suitable alternative for a variety of conditions encountered in dermatology. The technique is effective in patients of all ages and typically results in better clinical and cosmetic outcomes than conventional surgery, which is an important benefit, when exposed lesions are treated, especially on the head and neck.

23.1.5.4 PDT of cardiovascular diseases

Atherosclerosis and its complications are a leading cause of morbidity and mortality in industrialized countries. In cardiovascular medicine, PDT has been applied to treat atherosclerosis and to inhibit the restenosis due to intimal hyperplasia that often occurs after vascular interventions.

Phase I trials were performed recently, suggesting that good therapeutic response can be achieved and that the therapy is well tolerated (Kereiakes et al. 2003; Rockson et al. 2000).

Fluorescence lifetime imaging based on exogenous probes

23.1.5.5 Antimicrobial PDT

From the beginning of the 20th century onward, many reports have been published on the photodynamic inactivation of various species of bacteria, fungi, and viruses.

More recently, fundamental differences in susceptibility to PDT were observed between Gram (+) and Gram (–) bacteria. In general, neutral or anionic photosensitizers are efficiently bound to and allow the photodynamic inactivation of Gram (+) bacteria (Malik et al. 1992; Merchat et al. 1996), but they are unable to photoinactivate Gram (–) bacteria. The latter result can be achieved in various ways. The permeability of the cell outer membrane can be increased administering substances, such as EDTA, together with the photosensitizer. Alternatively, one can use a cationic photosensitizer molecule with an intrinsic positive charge or polycationic sensitizer conjugates formed from polymers such as polylysine.

The demonstration of efficient photoinactivation of several classes of microorganisms, the fact that antibiotic resistant bacteria are as susceptible to photodynamic inactivation as their naive counterparts (Wainwright et al. 1998), together with the increasing appearance of antibiotic resistance among pathogenic bacteria, suggest that PDT may be a useful tool to treat infectious diseases (Hamblin and Hasan 2004) and will likely be a growing application of PDT.

23.1.5.6 Photoactivated pesticides

Photoresponsive systems are widespread in nature, and life processes such as photosynthesis and vision are linked with structural changes of molecules caused by sunlight. Similar photochemical transformations can be employed to make synthetic agrochemicals and drugs active or to control their bioavailability at the site of action. Photoactivated pesticides have been used as insecticides, fungicides, and herbicides. Their activity relies on preferential accumulation of photosensitizers (e.g., porphyrins and xanthenes) in target organisms and on phototoxicity elicited by sunlight illumination (Ben Amor and Jori 2000; Heitz and Downum 1987).

Photoactivated pesticides are known as green pesticides as they are environment friendly: they exert their toxic action selectively on the target where they preferentially localize, degrade themselves, and cause no contamination of the environment.

23.2 FLUORESCENCE IMAGING OF PHOTOSENSITIZERS IN SMALL ANIMALS

In Section 23.1, it was pointed out that the activation of photosensitizers requires radiation of proper wavelength. Hence, the design of a PDT treatment should consider the absorption spectrum of the drug and the optical properties of biological tissues at the activation wavelength. A wealth of papers deals with PDT dosimetry, that is, mathematical models that take into account the scattering (μ_s') and the absorption (μ_a) properties of the tissue in order to predict the amount of light absorbed by the photosensitizer. The scattering μ_s', in fact, modulates the power density distribution in the tissue, while the absorption μ_a, beyond contributing to determine the light pattern, indicates unspecific absorption of photons, which does not induce any photochemical effect. It is worth noting that most of the dosimetry studies assume that the absorption spectrum of the photosensitizer is the one measured from injectable solutions or in simple biological models, like micelles or similar systems. Nevertheless, it has been found that the interaction of photosensitizers with the biological substrate can lead to an *in vivo* spectrum that cannot be predicted on the basis of *in vitro* studies. The following section deals with a time-resolved technique that can be used to measure the actual absorption spectrum of sensitizers upon intravenous injection in mice.

23.2.1 ABSORPTION SPECTRA OF PHOTOSENSITIZERS MEASURED *IN VIVO*

The optical properties of tissues and exogenous substances administered for diagnostic or therapeutic purposes may be influenced by functional state, metabolism, blood perfusion, and so on. Therefore, *in vitro* measurements could lead to an inaccurate spectrum. This problem can be avoided by means of time-resolved reflectance, which allows one to determine the optical properties (absorption and scattering coefficients) of tissues *in vivo* (Cubeddu et al. 1994a; Patterson et al. 1989). This can be easily done in anesthetized mice, but could be theoretically performed in humans as well, since the procedure is completely noninvasive. Hereafter, only measurements made on mice are reported.

To this purpose, short (picoseconds) light pulses generated by a laser tunable over a wide spectral range (e.g., 600–700 nm) are delivered to an anesthetized mouse through an optical fiber. The light traveling inside the animal, and partially remitted by its surface, is collected by a second fiber placed a few millimeters apart (e.g., 1 cm). The distal end of the fiber is coupled to a fast photomultiplier tube (PMT) whose output is processed by a time-correlated single photon counting (TCSPC) apparatus.

As discussed in detail in Chapter 13, photons undergo multiple scattering and absorption inside the tissue. This results in broadening and shaping of the laser pulses in such a way that the remitted light carries information about the absorption (μ_a) and scattering (μ_s) coefficients of the part of the tissue that is traversed by photons. Assuming an ideally short input pulse (Dirac delta function), the output pulse, measured by the TCSPC apparatus, gives the temporal point spread function (TPSF) of the overall system, including the response of the experimental equipment and the properties of the analyzed sample. It can be demonstrated that, by fitting the TPSF with a proper mathematical model for photon propagation in turbid media, convoluted with the response of the experimental system, the dependencies of the equipment can be removed and the optical parameters of the sample (μ_a and μ_s) can be recovered. In most cases, the mathematical model used to predict the photon migration inside the tissue relies on the diffusion equation, with proper boundary conditions. This assumes that photons lose information about their original direction after traveling a distance (about 1 mm or less) that corresponds to a few mean free paths. Thereafter, they behave like heat diffusing in a homogeneous medium.

Coming back to the problem of estimating the absorption properties of photosensitizers *in vivo*, a differential measurement is performed (Cubeddu et al. 1994b). Actually, by scanning the laser wavelength all over the spectral range of interest, the absorption coefficient μ_a of anesthetized animals is measured twice, as a function of the wavelength, before and after the injection of a therapeutic dose of the sensitizer. The true absorption spectrum of the sensitizer is then recovered through a simple difference between the acquired spectra, in such a way that the absorption properties of tissue are removed.

Experiments performed on mice after the injection of 25 mg/kg body weight (b.w.) of HpD showed that the absorption spectrum of HpD measured in the range 600–640 nm (Figure 23.4) has a peak close to 620 nm. This is in good agreement with the spectrum measured in a solution of HpD bounded to low-density lipoproteins (LDLs), which simulated the biological environment. This result demonstrates that the absorption properties of HpD are not remarkably modified, at least in the 600–640 nm range (which is of interest for PDT), by interactions with the biological substrate that may occur only *in vivo*. It is worth noting that the described method, besides providing the true spectrum of the sensitizer *in vivo*, also estimates the increase in the absorption caused by the accumulation of the drug in the tissue. This resulted to be close to 20% for the specific case considered in this study. The simultaneous measurement

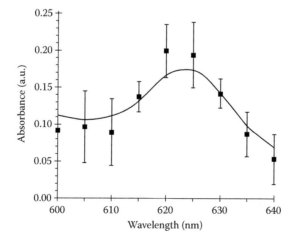

Figure 23.4 Absorption line shape of HpD measured *in vivo* in tumor-bearing mice (25 mg/kg b.w.) (■) and in LDL (10 µM HpD in 0.3 mg/mL LDL) (solid line). (Reproduced from Cubeddu R. et al., *Photochem. Photobiol.* 60: 582–585, 1994.)

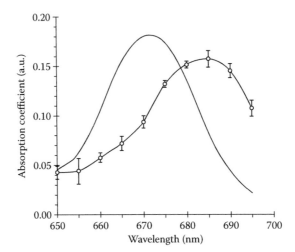

Figure 23.5 Absorption line shape of 2.5 mg/kg b.w. AlS$_2$Pc measured *in vivo* in tumor-bearing mice (o) and 10 µg/ml AlS$_2$Pc in aqueous solution (solid line). (Reproduced from Cubeddu R. et al., *J. Photochem. Photobiol. B* 34: 229–235, 1996.)

of scattering and absorption properties leads to the knowledge of the penetration depth of light in tissue. This allows one to determine the optimal light dose for therapeutic purposes, while the measurement of the absorption lineshape yields the optimal irradiation wavelength.

Different from what reported for HpD, measurements performed following the same protocol on the absorption spectrum of disulfonated aluminum phthalocyanine (AlS$_2$Pc) showed a marked difference with respect to the absorption spectrum measured in aqueous solution.

Phthalocyanines have been considered as possible second-generation photosensitizers for PDT (MacDonald and Dougherty 1999). AlS$_2$Pc shows an absorption spectrum peaked at 672 nm or red-shifted of no more than 4 nm in various solution environments. Nevertheless, experiments on the action spectrum for PDT in tumor-bearing mice resulted in good therapeutic efficacy for wavelengths greater than 670 and up to 710 nm, in a range where the absorption measured in solution decreases rapidly (Canti et al. 1992).

Actually, the absorption spectrum of AlS$_2$Pc measured by time-resolved reflectance in the 650–695 nm range in tumor-bearing mice (dose = 2.5 mg/ml b.w.) peaks at 685 nm instead of 672 nm (Figure 23.5; Cubeddu et al. 1996). This red shift with respect to the absorption maximum in solution is consistent with the therapeutic efficacy in the treatment of tumor models that proved to be significantly better at 685 nm than at 672 nm.

The spectral change seems to indicate a strong interaction between the sensitizer and the biological substrate. A modification of the chemical structure can perhaps be speculated, since, as above-mentioned, various solvents and environments (e.g., micelles) could never lead to such a remarkable red shift of the AlS$_2$Pc absorption spectrum.

This result highlights the importance of the *in vivo* measurement of the optical properties of drugs whenever the interaction with light is a key element for activation, as it is the case of PDT.

23.2.2 TIME-GATED FLUORESCENCE IMAGING OF PHOTOSENSITIZERS IN TUMOR-BEARING MICE

23.2.2.1 *In vivo* studies

Fluorescence spectroscopy and imaging offer effective opportunities for cancer diagnosis, which are being carefully investigated since long ago (Wagnières et al. 1998). Fluorescence techniques are minimally invasive, are relatively inexpensive with respect to other diagnostic methodologies, and can be easily applied to any part of the human body that can be reached by light, either directly or by means of an endoscope. In diagnostic procedures, fluorescence can, in principle, provide the clues for the detection of several disorders and, in particular, tumors. Yet, unfortunately, a broadband emission is a general characteristic of both cancerous and noncancerous tissues. Therefore, the mere presence of a fluorescence signal often provides only a limited diagnostic aid. In order to increase the specificity for tumor detection, the exogenous emission of suitable markers can be considered. To this purpose, great attention has been devoted to photosensitizers originally developed to treat tumors with PDT. Some photosensitizers are also promising for diagnosis, since they accumulate in cancerous tissues with a good selectivity, are fluorescent, and, last but not least, have already been approved for human administration (as described in Section 23.1).

To exploit exogenous fluorescence for tumor diagnosis, a selectivity criterion is needed to discriminate the emission of the fluorophore of interest from the background signal, which is mainly due to tissue autofluorescence. The discrimination can be obtained in the spectral domain by selecting two suitable excitation or observation wavelengths (Andersson-Engels et al. 1991). The subtraction between the images acquired at the two wavelengths leads to an image that mainly contains the contribution of the fluorophore of interest. However, this technique requires a nontrivial normalization; moreover, the fluorescence spectra of organic compounds often overlap, reducing the selectivity of the spectral approach. Using a pulsed excitation, various events characterized by a different timescale take place. Such events, like scattered light and emission of different fluorophores, may be discriminated with a properly gated acquisition.

Most intensified charge-coupled device (CCD) cameras allow very fast electronic gating, almost equivalent to a fast shutter with an exposure time of a few nanoseconds or even less. Using such a device, it is possible to acquire only the fluorescence light that falls within a definite time window, properly delayed with respect to the excitation pulses. Taking into account the (multi-)exponential behavior of the fluorescence decays, a suitable combination of gate width and delay provides an effective discrimination among fluorophores having lifetimes that differ at least by a few nanoseconds. To apply this time-gated fluorescence imaging technique, no narrowband filters are required. Moreover, the images show the fluorophore localization without any image processing, being the selectivity intrinsic in a single acquisition process. This feature led to the development of a tumor detection technique capable of providing good selectivity and real-time operation at a time when the computer performances were several orders of magnitude worse than now.

Time-gated fluorescence imaging was devoted, since the beginning, to the detection of cancer through the labeling of tissues with photosensitizing drugs. Hereafter, a synthetic review is reported of time-gated imaging performed on tumor-bearing mice sensitized with different drugs.

It has been already observed (Section 23.1) that all clinically approved photosensitizers used in PDT are porphyrin-based drugs. The detection of tumors on the basis of the HpD fluorescence dates back to the 1960s (Lipson et al. 1961). In fact, HpD, when excited at 405 nm (Soret band), emits a strong fluorescence with a multicomponent relaxation dynamics, resulting in an average lifetime of about 15 ns, whereas tissue autofluorescence mainly extinguishes within 3–6 ns. This large difference in the decay time of the two emission components is well suited for the application of the time-gated approach. Assuming a preferred localization of HpD in the tumor, the neoplasia can be easily singled out by acquiring the fluorescence of the suspected portion of tissue after a delay greater than 15 ns with respect to the excitation pulses. In fact, such an image will record almost exclusively the long-living HpD emission, with a negligible contribution from autofluorescence.

This paradigm was tested on mice bearing experimental tumors implanted intraderma (Cubeddu et al. 1993). Mice received an intravenous injection with different doses of HpD (5–25 mg/kg b.w.) and underwent a fluorescent measurement after an uptake time of 12 h, required for the drug to preferentially localize in the tumor. Figure 23.6a shows a time-gated image acquired after a delay of 20 ns with respect to 1 ns long excitation pulses at 405 nm. The gate width was 50 ns, long enough to collect most of the HpD fluorescent emission. For comparison, Figure 23.6b shows the same field of view observed synchronously with the excitation pulses. This arrangement simulates a continuous wave (CW) acquisition. The excitation light was removed in both cases with a spectral filter. The tumor, which corresponds to the bright spot, can be distinguished more easily in the delayed image, as expected due to the selective acquisition of the exogenous signal and the effective rejection of the background fluorescence with the delayed gate. It is worth noting that the tumor was clearly detected in spite of the screen action of the skin against light penetration.

A second set of experiments was performed on mice injected with a second-generation photosensitizer, that is, AlS_2Pc (Cubeddu et al. 1997a). In that case, the excitation light was set to 650 nm, while the

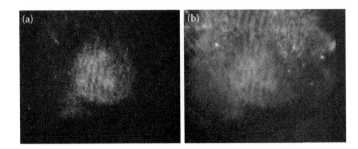

Figure 23.6 Fibrosarcoma on the back of a mouse treated with 10 mg/kg b.w. of HpD 12 h before the experiment: (a) image acquired using a 20 ns delay after the excitation pulses; (b) image acquired synchronously with the excitation.

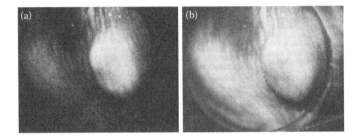

Figure 23.7 Fibrosarcoma on the back of a mouse treated with 5 mg/kg b.w. of AlS$_2$Pc 3 h before the experiment: (a) image acquired using a 2 ns delay after the excitation pulses; (b) image acquired synchronously with the excitation. (Reproduced from Cubeddu R. et al., *Photochem. Photobiol.* 57: 480–485, 1993.)

detection was at 670 nm, through a high-pass spectral filter. The red excitation light gave two advantages with respect to the blue/violet light used to excite HpD: a deeper penetration in tissues and a strong reduction of the autofluorescence. However, being the excitation wavelength not far from the detection one, a simple high-pass filter revealed to be insufficient to eliminate the strong laser light scattered by the sample, which is orders of magnitude greater than the fluorescence signal. In this case, the acquisition delay was set to ≈1–2 ns in order to remove the scattered laser light, which was by far the greatest source of noise, since the endogenous fluorescence excited in the red was almost negligible. Figure 23.7a shows the tumor area of a mouse treated with phthalocyanine acquired with delay of 2 ns. For comparison, Figure 23.7b was acquired at zero delay. The improvement in contrast is evident, thus demonstrating the superior performance of the gated acquisition over the CW one.

23.2.2.2 *Ex vivo* studies

The application of the time-gated imaging to the detection of tumors in locations other than the dermis represents a natural extension of this technique. A prerequisite for reliable tumor detection is the knowledge of the fluorescence properties, after porphyrin sensitization, of healthy organs, which may represent the peritumoral tissue.

Measurements of the fluorescence of selected tissues (skin, bone, bowel, brain, muscle, lymph node) in tumor-bearing mice were performed after the administration of HpD or Photofrin II (Cubeddu et al. 1995a). The emissions from different tissue types were compared to that of the tumor using the time-gated approach (30 ns delay) for image acquisition. The animals were sacrificed and dissected; the tumor with the surrounding skin was removed, and its fluorescence was measured from the dermal side. Tissue samples of muscle, fat, brain, lymph nodes, bowel, and bone were excised and their fluorescence was measured as well.

The average intensity measured in different organs, as a function of the drug dose, is shown in Figure 23.8. The strongest fluorescence was observed from the tumor through the inner mucosa (dermal side). The fluorescence of lymph node, fat, and muscle only differs in intensity, while presenting similar dependence on the drug dose. The most fluorescent tissue in this group is lymph node, whose intensity is 0.4 times that of the tumor. The fluorescence signal of the fat is intermediate, whereas the signal recorded

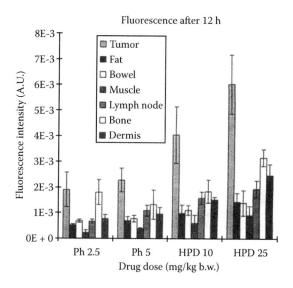

Figure 23.8 Average fluorescence intensity of tumor, fat, bowel, muscle, lymph node, bone, and dermis. (Reproduced from Cubeddu R. et al., *J. Photochem. Photobiol. B* 29: 171–178, 1995.)

from the muscle is appreciably lower (<0.2 times that of the tumor). The fluorescence of the bowel partially depends on the mouse feed and deceases with uptake time (data not shown). Hence, the contrast with respect to the tumor is rather high after 12 h.

The fluorescence of the bone in control animals is by far higher than that of other tissues, and it is more than half the fluorescence of bone in sensitized animals. Finally, the fluorescence of the brain is negligible for all the examined animals due to the blood-brain barrier that prevents drug uptake.

The signal detected in the tumor is very high, yet also the fluorescence of the derma, which in this model is the peritumoral tissue, is remarkable. Better contrast could be expected with peritumoral tissues other than skin.

Time-resolved spectra taken in all the examined tissues confirmed that the only fluorophore with a long lifetime is the sensitizer.

From the diagnostic standpoint, the rather high fluorescence observed in adipose tissues and in bones could be misinterpreted and cause false positives. This is especially true for bones due to the long lifetime emission that is naturally present in the tissue. As a consequence of this endogenous emission, the fluorescence intensity of bones can compete with that of the tumor, mainly at a low dose of a sensitizer. The awareness of this possible mistake should suggest particular care to avoid any wrong or uncertain diagnosis.

In summary, it can be affirmed that, except for bones, the endogenous fluorescence does not impair the tumor detection by the time-gated imaging technique, whose reliability is strictly related to the selectivity of the drug.

In conclusion, it is worth noting that the time-gated imaging technique provides several interesting features for cancer detection after sensitization with exogenous compounds, mainly porphyrin-based ones, among them: real-time operation, complete removal of laser excitation light, and operation under ambient light, since the synchronous acquisition effectively removes any CW illumination. On the other hand, it requires a pulsed laser and a rather expensive gated camera.

23.2.3 FLUORESCENCE LIFETIME IMAGING OF PHOTOSENSITIZERS IN TUMOR-BEARING MICE

In the previous section, we observed that the concentration of HpD in the peritumoral tissue is often not negligible. The ratio of fluorescence intensity in the tumor to the intensity in the nearby tissues is dose-dependent and sets a lower limit to the drug dose that is required for reliable tumor detection using porphyrins.

It is well known that the time-resolved emission of HpD presents three main components, having different lifetimes and amplitudes. Moreover, the biological environment influences the fluorescence properties of the drug. This observation suggests one to study the spatial distribution of the HpD fluorescence decay time in sensitized mice, with the aim to discriminate tumor from healthy tissue.

Fluorescence lifetime imaging microscopy (FLIM), which is the natural evolution of time-gated imaging, is the preferred option for cancer detection through exogenous sensitization. To fully understand its effectiveness, the fluorescence lifetime maps of the HpD fluorescence were measured in tumor-bearing mice as a function of the drug dose (Cubeddu et al. 1997b). Mice bearing a fibrosarcoma (MS-2, diameter of ~6 mm) on the back were injected intraperitoneally with the following HpD doses: 0.1, 0.25, 1, 2.5, and 10 mg/kg b.w. The highest dose was considered for comparison with time-gated imaging (see Section 23.2.2.1). A very simple approach, consisting of the acquisition of only two images delayed by 10 and 30 ns with respect to the excitation pulses, allows one to measure the map of the effective lifetime in the field of view, in a very short time. In fact, the acquisition of the images takes a fraction of a second, and the image processing is also very fast since only algebraic functions are required to recover the lifetime. Moreover, since the first delay is longer than the typical decay time of the endogenous fluorescence, the lifetime map is minimally affected by the endogenous emission of the tissues.

Figure 23.9a shows the fluorescence lifetime map of the back of a mouse sensitized with 0.25 mg/kg body weight of HpD. The tumor boundaries are clearly outlined in the time domain map, since the neoplasia corresponds to a fluorescence lifetime longer than that of the healthy tissue. On the contrary, an intensity-based fluorescence image, which is shown in Figure 23.9b for comparison, is much less effective for tumor detection. It is worth noting that Figure 23.9b displays only the exogenous fluorescence since it has been acquired 20 ns after the excitation. Fluorescence images taken at zero delay are even worse, since the tissue natural fluorescence completely masks the drug signal, leading to useless images. A similar result was achieved with all the drug doses, but the highest one. In fact, even though the FLIM technique still works at high dose, delayed intensity images also allow a reliable detection of the neoplasia, as already observed in Section 23.2.2.1. As an example, Figure 23.10a and b shows a decay time image and an intensity image, respectively, of the back of a mouse sensitized with 10 mg/kg b.w. of HpD.

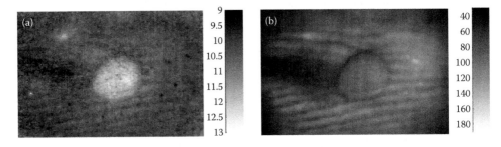

Figure 23.9 Decay time image (a) and time-gated image (b) of a tumor on the back of a mouse sensitized with 0.25 mg/kg b.w. of HpD. (Reproduced from Cubeddu R. et al., *Opt. Lett.* 20: 2553–2555, 1995.)

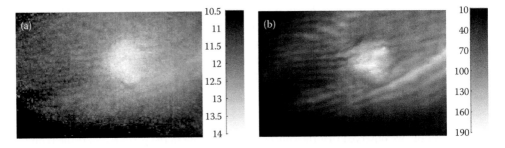

Figure 23.10 Decay time image (a) and time-gated image (b) of a tumor on the back of a mouse sensitized with 10 mg/kg b.w. of HpD. (Reproduced from Cubeddu R. et al., *Opt. Lett.* 20: 2553–2555, 1995.)

Fluorescence lifetime imaging based on exogenous probes

It is worth noting that the lifetime measured by means of this technique does not represent any true physical parameter. However, it gives a diagnostic index resulting from a weighted average of the lifetimes of the different HpD fractions that localize in tumor and healthy tissues.

As a matter of fact, the imaging study indicates that the HpD fluorescence emission is characterized by a longer lifetime when incorporated into neoplastic tissues. This is in agreement with the dependence on the environment typically observed for various photophysical properties of HpD. In particular, as described in Section 23.1.4.1, HpD and its commercial version Photofrin are mixtures of porphyrin monomers and different types of oligomers. Monomers and oligomers have different fluorescence properties for what concerns both the emission spectrum and the corresponding lifetimes. The tumor-localizing fraction consists mostly of oligomers; thus, different fluorescence properties are expected from tumors as compared to healthy tissues.

23.3 FLUORESCENCE IMAGING OF PHOTOSENSITIZERS IN HUMANS

23.3.1 DETECTION OF SKIN LESIONS BY FLUORESCENCE IMAGING

As described in Section 23.1.4.3, a very special photosensitizer is δ-aminolevulinic acid (ALA), which is a naturally occurring precursor in the cycle of heme biosynthesis (Kennedy et al. 1990). The exogenous administration of this substance bypasses a physiologic regulation mechanism and gives rise to an excess of the intermediate molecule protoporphyrin IX (PpIX), which is photoactive and strongly fluorescent. Even though this process takes place also in healthy tissue, in cancerous tissue, it is more effective due to enhanced uptake of ALA and enzymatic deficiencies occurring in neoplastic cells. This imbalance in the content of PpIX provides a selectivity criterion that can be profitably exploited not only for PDT but also to detect tumors by means of fluorescence techniques. In fact, PpIX, when excited with light at 400 nm, emits a characteristic red fluorescence (peaking around 635 nm) with lifetime of about 18 ns, that is, much longer than that of the endogenous tissue fluorescence (3–6 ns or shorter).

Topical administration of ALA allows the patient to avoid skin sensitization. It is especially suitable for the detection of exposed pathologies, like skin tumors, even though pigmented lesions, like the very aggressive melanoma, cannot be detected by fluorescence imaging because no appreciable emission can be revealed from these strongly absorbing lesions. This notwithstanding, a great interest exists in dermatology for the classification of lesions that present a different degree of malignancy. In addition, a technique capable of showing the actual extent of the pathologic area would allow surgeons to perform conservative excisions, while reducing the recurrence rate. In the following, we will report on the use of FLIM to distinguish basal cell carcinomas and squamous cell carcinomas from benign lesions (Cubeddu et al. 1999).

The excitation light is provided by a dye laser pumped by a subnanosecond nitrogen laser. The laser emission peaks around 400 nm with average power of 800 μW. Such a power level does not raise any safety concern and allows one to perform the diagnostic procedure even on the patient's head, close to the eyes. This is very important since most dermatological lesions are on parts of the body exposed to sunlight and, in particular, on the face. An optical fiber is used to deliver the light to the patient, while the fluorescence signal is collected using a high aperture photographic lens. The system was designed in such a way that even nontechnical personnel can operate it for routine clinical examinations.

The diagnostic procedure involves the preparation of an ointment made of 2% ALA powder in a lipid emulsion. The ALA ointment is applied onto the lesion with a margin of at least 10 mm in the visibly normal skin, and the lesion is covered with an occlusive dressing. The patient is kept at rest for about 1 h to allow the metabolism to transform ALA into PpIX. Then, the excess cream is gently removed from the lesion, and the fluorescence test is carried out. A preliminary examination is performed using only two gated images. In this case, the map of the fluorescence lifetime is calculated in real time using an algebraic equation and displayed in pseudocolors at a few frames per second. Then, for the regions that present a clinical relevance, a more precise measurement is performed by acquiring more images with delays ranging from 0 to 20 ns, relative to the excitation pulses. The images are processed offline immediately after the acquisition to calculate the maps of the average fluorescence lifetime and of its amplitude.

The classification of the lesions from fluorescence images is made according to the following considerations. It is proved that the malignant character of a skin lesion is associated with an excess of exogenous fluorescence and a reduction in natural emission, as compared to healthy tissues. As a consequence, the average lifetime in tumors is longer than in healthy tissues since the amplitude of the long-living component (PpIX fluorescence) is higher than that of the short-living emission (endogenous fluorescence). The use of the fluorescence lifetime as a diagnostic index, instead of intensity or spectral features, gives additional advantages. In fact, the detection index is based on the relaxation dynamics of the fluorescence emission. Hence, it is almost insensitive to artifacts due to the spatial variation of the fluorescence intensity. In particular, the time domain approach gets rid of an uneven distribution of the excitation light and—most important—of local differences in skin absorption. These may be due to anomalies in pigmentation, vascularization, or blood perfusion, due to an inflammatory status, which might lead to severe artifacts if one looks at the fluorescence intensity. Finally, the diagnostic procedure can be carried out under normal illumination.

Thirty-four patients affected by 48 lesions, either malignant (mainly basal cell carcinomas) or benign ones, were included in the study considered hereafter. For all the patients, the classification of the lesions on the basis of fluorescence lifetime maps was compared to histology, which was routinely performed. A typical fluorescence lifetime image of a malignant lesion is shown in Figure 23.11a, while the map of the CW fluorescence intensity is displayed in Figure 23.11b. For comparison, the color photo of the lesion is also reported in Figure 23.11c. The tumor, which was classified as a basal cell carcinoma, is characterized by a significantly longer lifetime (18 ns) than that of the surrounding healthy tissue (≈10 ns). It is worth noting that the region where the ALA ointment was applied is larger than the lesion itself and can be easily distinguished in Figure 23.11a. The fluorescence amplitude is lower in the lesion, as expected. However, the clinical experience demonstrates that, while the fluorescence amplitude presents a strong variability from patient to patient, the lifetime is much more stable and thus reliable for the classification of the lesions. In the present trial, 27 out of 35 malignant lesions were correctly identified, while only 1 out of 13 benign lesions was misinterpreted. On the basis of these outcomes, the diagnostic procedure showed a sensitivity of 88% and a specificity of 74%. However, it is very important to observe that the test was not conducted according to a strictly blind protocol. Hence, the results are only indicative. Actually, it has been reported

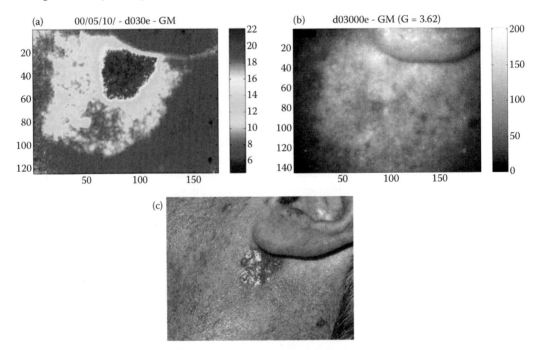

Figure 23.11 Fluorescence lifetime map (a) and intensity image (b) of a basal cell carcinoma on the cheek, 1 h after the topical administration of 1% ALA cream. The color image of the lesion is also shown (c).

that the typical PpIX red fluorescence can be directly observed on the skin lesions of ALA-sensitized patients under CW excitation. This is certainly true only when the marker dose is rather high. When the ALA dose is reduced to a few percents in the ointment, as it is required in a minimally invasive diagnostic protocol, the interference of the natural skin fluorescence and the uneven pigmentation prevent reliable lesion identification by means of an intensity-based approach. In such a condition, the lifetime technique exhibits its maximum effectiveness as it has been previously shown.

23.3.2 FLUORESCENCE IMAGING OF BRAIN TUMORS

Brain tumors, either primary or resulting from metastasis, are among the most aggressive neoplastic diseases. In particular, gliomas account for a large fraction of all primary brain tumors and are characterized by high morbidity. In fact, in almost all but very few patients, such a tumor leads to inevitable death within a relatively short time after diagnosis. Moreover, treatment of gliomas poses a challenge due to their tendency to infiltrate the surrounding normal brain and to a high rate of recurrence.

The most common treatment of gliomas is surgery followed by chemotherapy and radiation therapy. It has been observed that the extent of tumor resection plays a significant role in patient prognosis and disease progression, resulting in the most important factor for longer survival. Nevertheless, the extent of resection compares with the need to minimize the removal of normal brain, which might result in neurological impairment and affect the patient's quality of life. Although of paramount importance, complete tumor resection is limited by the difficulty in visually detecting differences between normal brain and malignant tissue during surgery. Thus, patients with malignant gliomas often have a subtotal resection.

As a consequence, there is a pressing need to develop new strategies to improve the intraoperative imaging of malignant glioma. To this purpose, fluorescence imaging of brain tumor during surgery has been extensively studied since long ago (Moore et al. 1948). Yet, it was not until the last decades when the technological advancement and the advent of ALA–PpIX sensitization fostered this research field toward a widespread use. This led to multicenter clinical trials, mainly performed in Germany (Stummer et al. 2006), while other clinical efforts were carried out in Japan focusing on a different class of markers (i.e., fluorescein labeling; Kuroiwa et al. 1998; Shinoda et al. 2003).

More generally, in a wealth of studies, either experimental or clinical ones, three major labeling strategies exploiting fluorescence emerged for brain tumor imaging, with a different degree of maturity (Pogue et al. 2010): endogenous fluorophores (autofluorescence); exogenous agents that are routinely used in humans (e.g., ALA–PpIX, fluorescein, indocyanine green); and exogenous agents developed for first-time use in humans, with molecular targeting potential.

Autofluorescence does not require the administration of any drug. The signal is largely attributed to collagen, nicotinamide adenine dinucleotide (NADH), flavin adenine dinucleotide (FAD), and endogenous porphyrins. Yet, it is usually very dim and imposes a rather long acquisition time, which might hamper surgical procedures. Moreover, fluorescence images can be affected by artifacts caused by blood and scattered light. Different strategies, mainly based on spectral ratios, have been devised to gain specificity for tumor tissue versus normal one. However, the reliability of the autofluorescence approach for tumor demarcation is still critical and site dependent.

The ALA–PpIX system has been extensively studied for PDT and photodiagnosis, including fluorescence-guided resection of brain tumors (Stummer et al. 2000). In fact, a significant accumulation of PpIX has been found in brain tumors, mainly high-grade gliomas, even if a still detectable PpIX emission characterizes also low-grade gliomas, meningiomas, and brain metastases (Valdés et al. 2011).

Other low molecular weight markers, like sodium fluorescein or indocyanine green, have been studied for tumor labeling, because they are retained within tumor tissue as a result of slow clearance. In particular, fluorescein labeling of brain tumors dates back to the 1940s (Moore et al. 1948) and has been applied even recently in Japan (Kuroiwa et al. 1998).

Finally, functionalized nanoparticles and other molecular probes providing active tumor targeting are intensively studied for tumor detection but are still not used in clinical applications due to safety concerns.

From the technological point of view, several devices were proposed for brain tumor demarcation, ranging from point-like spectroscopes to modified surgical microscopes and commercial systems with built-in fluorescence channels. Laser-induced fluorescence spectroscopy has been used since long time ago for diagnosis of brain cancer. Most of the measurements were performed under steady-state excitation. Yet, in the last years, time-resolved spectroscopy of several types of brain tumors demonstrated the potential of this strategy on patients. In particular, a recent study (Butte et al. 2011) showed that time-resolved fluorescence holds the potential to diagnose brain tumors intraoperatively and to provide a valuable tool for aiding the neurosurgeon to rapidly distinguish between tumor and normal brain tissue.

Point-like spectroscopy proved very effective to detect biochemical differences in tissues leading to diagnostic outcomes; yet the choice of the measurement site requires *a priori* information or clear clues of the disease in the operation field. This is not always the case, mainly close to the tumor margins. Time is a precious resource during surgery, and imaging devices are certainly preferred over point-like instruments, since they can provide an immediate view of critical areas. Moreover, the margins between tumor and healthy tissue could be displayed in real time, thus giving the surgeon an effective guidance. Among imaging devices, the ones based on the time-resolved approach, like FLIM, provide extra valuable features, like good immunity versus unspecific absorbers (e.g., blood) and uneven excitation patterns.

The first attempt to develop a FLIM system for fluorescence-guided brain resection was recently made by Sun et al. (2010), exciting with a nitrogen laser and detecting the endogenous fluorescence with a gated CCD coupled to a fiber optic endoscope. Even if the results of the study were still preliminary, they suggest that a FLIM-assisted surgical microscope could be very beneficial for conservative brain surgery with a high degree of selectivity for tumor. Such a device would certainly result in an increase in patient survival time.

As the reader has likely noted from what reported here above, none of the attempts to the fluorescence imaging of the brain made up to now has combined exogenous markers with time domain detection. The advantages of either of the two approaches were demonstrated on patients separately, but they have not been combined yet.

23.4 CONCLUSIONS

As outlined in this chapter, one of the major demands on photosensitizers developed for therapeutic applications is their ability to localize in the target tissue. On the other hand, photosensitizers are also often characterized by intense fluorescence. The combination of these two features makes them eligible, at least in principle, as contrast media for diagnostic purposes based on fluorescence detection. Moreover, in several cases, the fluorescence lifetimes of photosensitizers are markedly different from the typical lifetimes of endogenous tissues. This is especially true for porphyrin-based sensitizers (Section 23.1.4.1) and precursor-induced sensitizers (Section 23.1.4.3), which again rely on a porphyrin compound. Actually, the average lifetime of such compounds is much longer than typically observed in untreated tissue. Consequently, they look particularly appealing to perform selective diagnosis through FLIM and time-gated fluorescence imaging.

REFERENCES

Andersson-Engels S., Johansson J., Svanberg K., Svanberg S. 1991. Fluorescence imaging and point measurements of tissue: Applications to the demarcation of malignant tumors and atherosclerotic lesions from normal tissue. *Photochem. Photobiol.* 53: 807–814.

Ben Amor T., Jori G. 2000. Sunlight-activated insecticides: Historical background and mechanisms of phototoxic activity. *Insect Biochem. Mol. Biol.* 30: 915–925.

Berg K., Selbo P. K., Weyergang A. et al. 2005. Porphyrin-related photosensitizers for cancer imaging and therapeutic applications. *J. Microsc.* 218: 133–147.

Bessler N. M. 2004. Verteporfin therapy in age-related macular degeneration (VAM): An open-label multicenter photodynamic therapy study of 4,435 patients. *Retina* 24: 512–520.

Butte P. V., Mamelak A. N., Nuno M., Bannykh S. I., Black K. L., Marcu L. 2011. Fluorescence lifetime spectroscopy for guided therapy of brain tumors. *NeuroImage* 54: S125–S135.

Canti G., Lattuada D., Leroy E., Cubeddu R., Taroni P., Valentini G. 1992. Action spectrum of photoactivated phthalocyanine AlS$_2$Pc in tumor bearing mice. *Anti-Cancer Drug* 3: 139–142.

Canti G., Lattuada D., Nicolin A., Taroni P., Valentini G., Cubeddu R. 1994. Antitumor immunity induced by photodynamic therapy with aluminum disulfonated phthalocyanines and laser light. *Anti-Cancer Drug* 5: 443–447.

Choudhary S., Nouri K., Elsaie M. L. 2009. Photodynamic therapy in dermatology: A review. *Lasers Med. Sci.* 24: 971–980.

Collaud S., Juzeniene A., Moan J., Lange N. 2004. On the selectivity of 5-aminolevulinic acid-induced protoporphyrin IX formation. *Curr. Med. Chem. Anticancer Agents* 4: 301–316.

Cubeddu R., Canti G., Taroni P., Valentini G. 1993. Time-gated fluorescence imaging for the diagnosis of tumors in a murine model. *Photochem. Photobiol.* 57: 480–485.

Cubeddu R., Musolino M., Pifferi A., Taroni P., Valentini G. 1994a. Time-resolved reflectance: A systematic study for the application to the optical characterization of tissue. *IEEE J. Quantum Electron.* 30: 2421–2430.

Cubeddu R., Canti G., Musolino M., Pifferi A., Taroni P., Valentini G. 1994b. Absorption spectrum of hematoporphyrin derivative in vivo in a murine tumor model. *Photochem. Photobiol.* 60: 582–585.

Cubeddu R., Canti G., Taroni P., Valentini G. 1995a. Study of porphyrin fluorescence in tissue samples of tumor-bearing mice. *J. Photochem. Photobiol. B* 29: 171–178.

Cubeddu R., Pifferi A., Taroni P., Valentini G., Canti G. 1995b. Tumor detection in mice by measurements of fluorescence decay time matrices. *Opt. Lett.* 20: 2553–2555.

Cubeddu R., Canti G., Musolino M., Pifferi A., Taroni P., Valentini G. 1996. In vivo absorption spectrum of disulphonated aluminum phthalocyanine in a murine tumor model. *J. Photochem. Photobiol. B* 34: 229–235.

Cubeddu R., Canti G., Taroni P., Valentini G. 1997a. Tumor visualisation in a murine model by time-delayed fluorescence of sulfonated aluminium phthalocyanine. *Lasers Med. Sci.* 12: 200–208.

Cubeddu R., Canti G., Pifferi A., Taroni P., Valentini G. 1997b. Fluorescence lifetime imaging of experimental tumors in HpD-sensitised mice. *Photochem. Photobiol.* 66: 229–236.

Cubeddu R., Pifferi A., Taroni P. et al. 1999. Fluorescence lifetime imaging: An application to the detection of skin tumors. *IEEE J. Sel. Top. Quantum Electron.* 5: 923–929.

Hamblin M. R., Hasan T. 2004. Photodynamic therapy: A new antimicrobial approach to infectious disease? *Photochem. Photobiol. Sci.* 3: 436–450.

Hamblin M. R., Mroz P. 2008. *Advances in Photodynamic Therapy: Basic, Translational and Clinical.* Norwood, MA: Artech House.

Heitz J. R., Downum K. R. 1987. *Light-Activated Pesticides (ACS Symposium Series 339).* Washington, DC: American Chemical Society.

Kennedy J. C., Pottier R., Pross D. C. 1990. Photodynamic therapy with endogenous protoporphyrin IX: Basic principles and present clinical experience. *J. Photochem. Photobiol. B* 6: 143–148.

Kereiakes D. J., Szyniszewski A. M., Wahr D. et al. 2003. Phase I drug and light dose-escalation trial of motexafin lutetium and far red light activation (phototherapy) in subjects with coronary artery disease undergoing percutaneous coronary intervention and stent deployment: Procedural and long-term results. *Circulation* 108: 1310–1315.

Korbelik M., Dougherty G. J. 1999. Photodynamic therapy-mediated immune response against subcutaneous mouse tumors. *Cancer Res.* 59: 1941–1946.

Kuroiwa T., Kajimoto Y., Ohta T. 1998. Development of a fluorescein operative microscope for use during malignant glioma surgery: A technical note and preliminary report. *Surg. Neurol.* 50: 41–49.

Lim J. I. 2002. Photodynamic therapy for choroidal neovascular disease: Photosensitizers and clinical trials. *Ophthalmol. Clin. North Am.* 15: 473–478.

Lipson R. L., Baldes E. J., Olsen A. M. 1961. The use of a derivative of hematoporphyrin in tumor detection. *J. Natl. Cancer Inst.* 26: 1–11.

MacDonald I. J., Dougherty T. J. 1999. Basic principles of photodynamic therapy. *J. Porphyrins Phthalocyanines* 5: 105–129.

Malik Z., Ladan H., Nitzan Y. 1992. Photodynamic inactivation of Gram-negative bacteria: Problems and possible solutions. *J. Photochem. Photobiol. B* 14: 262–266.

Merchat M., Bertolini G., Giacomini P., Villanueva A., Jori G. 1996. Meso-substituted cationic porphyrins as efficient photosensitizers of Gram-positive and Gram-negative bacteria. *J. Photochem. Photobiol. B* 32: 153–157.

Moore G. E., Peyton W. T., French L. A., Walker W. W. 1948. The clinical use of fluorescein in neurosurgery; the localization of brain tumors. *J. Neurosurg.* 5: 392–398.

Nestor M. S., Gold M. H., Kauvar A. N. B. et al. 2006. The use of photodynamic therapy in dermatology: Results of a consensus conference. *J. Drugs Dermatol.* 5: 140–154.

O'Connor A. E., Gallagher W. M., Byrne A. T. 2009. Porphyrin and nonporphyrin photosensitizers in oncology: Preclinical and clinical advances in photodynamic therapy. *Photochem. Photobiol.* 85: 1053–1074.

Patterson M. S., Chance B., Wilson B. C. 1989. Time-resolved reflectance and transmittance for the non-invasive measurement of tissue optical properties. *Appl. Opt.* 28: 2331–2336.

Pogue B. W., Gibbs-Strauss S., Valdés P. A., Samkoe K., Roberts D. W., Paulsen K. D. 2010. Review of neurosurgical fluorescence imaging methodologies. *IEEE J. Sel. Top. Quantum Electron.* 16: 493–505.

Rockson S. G., Kramer P., Razavi M. et al. 2000. Photoangioplasty for human peripheral atherosclerosis: Results of a phase I trial of photodynamic therapy with motexafin lutetium (Antrin). *Circulation* 102: 2322–2324.

Shinoda J., Yano H., Yoshimura S. et al. 2003. Fluorescence-guided resection of glioblastoma multiforme, by using high-dose fluorescein sodium—Technical note. *J. Neurosurg.* 99: 597–603.

Sternberg E. D., Dolphin D. 1998. Porphyrin-based photosensitizers for use in photodynamic therapy. *Tetrahedron* 54: 4151–4202.

Sun Y., Hatami N., Yee M., Phipps J., Elson D. S., Gorin F., Schrot R. J., Marcu L. 2010. Fluorescence lifetime imaging microscopy for brain tumor image-guided surgery. *J. Biomed. Opt.* 15: 056022.

Stummer W., Novotny A., Stepp H., Goetz C., Bise K., Reulen H. J. 2000. Fluorescence-guided resection of glioblastoma multiforme by using 5-aminolevulinic acid-induced porphyrins: A prospective study in 52 consecutive patients. *J. Neurosurg.* 93: 1003–1013.

Stummer W., Pichlmeier U., Meinel T. et al. 2006. Fluorescence-guided surgery with 5-aminolevulinic acid for resection of malignant glioma: A randomised controlled multicentre phase III trial. *Lancet Oncol.* 7: 392–401.

Valdés P. A., Leblond F., Kim A. et al. 2011. Quantitative fluorescence in intracranial tumor: Implications for ALA-induced PpIX as an intraoperative biomarker. *J. Neurosurg.* 115: 11–17.

Wagnières G. A., Star W. M., Wilson B. C. 1998. In vivo fluorescence spectroscopy and imaging for oncological applications. *Photochem. Photobiol.* 68: 603–632.

Wainwright M. 1996. Non-porphyrin photosensitizers in biomedicine. *Chem. Soc. Rev.* 25: 351–359.

Wainwright M., Phoenix D. A., Laycock S. L., Wareing D. R., Wright P. A. 1998. Photobactericidal activity of phenothiazinium dyes against methicillin-resistant strains of Staphylococcus aureus. *FEMS Microbiol. Lett.* 160: 177–181.

Fluorescence lifetime imaging of ions in biological tissues

Christoph Biskup and Thomas Gensch

Contents

24.1 INTRODUCTION

24.1.1 ION COMPOSITION OF BIOLOGICAL TISSUES

In all living cells, ions play fundamental roles. The intracellular ion composition of a cell is tightly controlled, and also the cells of an organism are situated in a carefully regulated environment, i.e. the tissue. However, the fluid compartments of the human body differ considerably in their composition. Figure 24.1 displays the typical ion concentrations in the intracellular and extracellular space and illustrates the tools at the disposal of a prototypic cell for managing its intracellular composition.

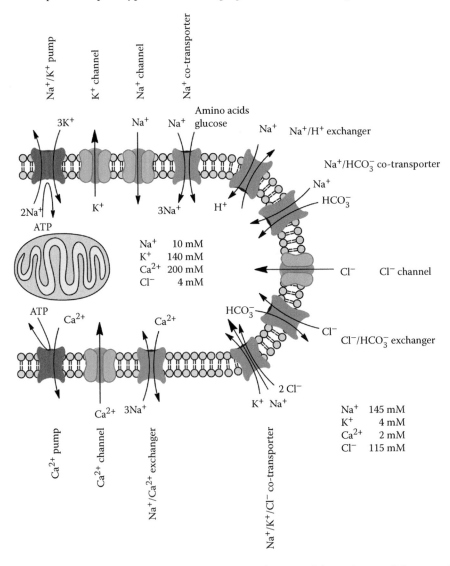

Figure 24.1 Ion transport mechanisms and ion concentrations in the intracellular and extracellular space. The intracellular ion composition is tightly controlled by ion pumps and transporters. The Na$^+$/K$^+$ pump generates a gradient for Na$^+$ and K$^+$ ions. The Na$^+$ gradient is used to fuel a variety of other secondary active transport processes such as the Na$^+$/H$^+$ exchange, the Na$^+$/Ca^{2+} exchange, or the co-transport of glucose, amino acids, HCO$_3^-$, or Cl$^-$. (Concentrations according to Lodish, H. et al., *Molecular Cell Biology*, 7th ed., 485. New York: W.H. Freeman & Co., 2013.)

The most important gradients across the cell membranes are those for sodium, potassium, calcium, and chloride. In the extracellular space, sodium, calcium, and chloride are high compared to those in the intracellular space, whereas the potassium concentration is low in the extracellular space and high in the intracellular space. The cell maintains these concentrations by regulating the membrane permeability to these ions. It uses a Na^+/K^+ pump to actively extrude Na^+ from the cell and transport K^+ into the cell.

The *sodium* gradient, which is created in this way, provides the basis for rapid electrical signaling in many excitable cells (Hille 2001). It also supplies the energy for important secondary active transport processes, in which the sodium influx along the gradient is used to power the co- or counter-transport of other ions or nutrients (Hediger et al. 2004). For instance, the Na^+/H^+ exchanger and the $Na^+–HCO_3^-$ co-transporter help to maintain intracellular pH (Orlowski and Grinstein 2004; Soleimani and Burnham 2001). The Na^+/Ca^{2+} exchanger contributes to the fast restoration of diastolic calcium concentrations in heart cells and also constitutes the dominant calcium efflux pathway in the outer segments of photoreceptor cells (Philipson et al. 2002). In epithelial cells of the intestine and the kidney, nutrient uptake (i.e., glucose and amino acids) is powered by a co-transport with sodium (Wright and Turk 2004).

As discussed above, the Na^+/K^+ pump also creates a *potassium* gradient across the cell membrane. In the cytosol, the potassium concentration is high whereas the extracellular potassium concentration is low. Due to this gradient, the equilibrium potential of K^+ is close to the resting membrane potential. K^+ efflux via potassium channels helps to stabilize the resting membrane potential and to repolarize the neuronal cell membrane after an action potential has been elicited.

Whereas the concentration of *calcium* in the extracellular space is about 2 mM, that in the intracellular fluid is about 200 nM. Thus, the concentration gradient is 10^4 fold, which is the steepest of all ions. Because of the negative membrane potential, the inward directed gradient for Ca^{2+} is enormous, so that Ca^{2+} can enter rapidly via voltage or ligand-gated ion channels. This Ca^{2+} influx creates a transient rise in the intracellular $[Ca^{2+}]$, which can be further magnified by Ca^{2+}-induced Ca^{2+} release from the endoplasmic/sarcoplasmic reticulum. The increase in cytosolic $[Ca^{2+}]$ can trigger other events such as muscle contraction or activate Ca^{2+}-dependent enzymes. The intracellular $[Ca^{2+}]$ is reduced to resting levels by Ca^{2+} pumps in the endoplasmic reticulum (SERCA) and the plasma membrane as well as by action of the Na^+/Ca^{2+} exchanger (NCX) of the plasma membrane. The NCX is especially important in restoring the Ca^{2+} concentrations when large influxes of Ca^{2+} occur.

The intracellular *chloride* concentration (4 mM) is below the chloride concentration of the extracellular space (e.g., blood, 115 mM) with some exceptions like excitatory neurons or certain epithelial cells. All cells have anion-selective channels, through which chloride can permeate passively. Active chloride uptake can be mediated by a variety of co-transporters such as the Cl^-/HCO_3^- exchanger, the K^+/Cl^- co-transporter or the $Na^+/K^+/Cl^-$ co-transporter.

For an understanding of the molecular action of these ion-transporting membrane proteins, and their role under physiological and pathological processes, it is desirable to know the extracellular and intracellular ion concentrations (Pieske et al. 2003). Moreover, many ion channel and ion-dependent carriers are the target of specific drugs (Hediger et al. 2004). Knowledge of the actual intracellular and extracellular ion concentrations is a key to quantify the action of these drugs.

24.1.2 MEASUREMENTS OF ION CONCENTRATIONS IN BIOLOGICAL TISSUES

The accurate determination of intracellular ion concentrations in living cells and tissues is not an easy task. Methods employed to measure intracellular and extracellular ion concentrations include analytical chemical methods, such as atomic absorption or atomic emission spectroscopy (Jackson and Mahmood 1994) or nuclear magnetic resonance (NMR) measurements (Gupta et al. 1984), which provide only an averaged measurement of the ion concentration in the tissue and lack spatial resolution. To measure exclusively the intracellular concentration, the extracellular solution needs to be removed carefully. Quantitative x-ray microanalysis (= electron probe microanalysis) in the electron microscope can be used to measure ion concentrations in subcellular compartments (Hall 1979; Schumann et al. 1976; Wendt-Gallitelli and Isenberg 1989), but this method is only applicable to frozen specimens and does not allow tracing changes in the concentration of ions in living cells or tissues. Moreover, all mentioned methods measure the total ion concentration including the fraction bound to proteins or stored in certain cell compartments.

Fluorescence lifetime imaging based on exogenous probes

However, in most cases, it is desirable to measure the free, biologically active fraction, since this is the concentration that is relevant for all transport processes. This can be achieved with ion-sensitive electrochemical or optical sensors (Janata et al. 1994, 1998; Shortreed et al. 1997; Wolfbeis 2004). These methods, however, allow only for measurements in a few individual cells. The electrode has to be impaled into the cell or tissue and facilitates only the measurement at one single spot.

All the above-mentioned problems can be circumvented by the use of fluorescent ion indicators. They can be used to measure ion concentrations with a high spatial and temporal resolution. Fluorescent dyes are incorporated into cells by using their acetoxymethylester derivatives (Tsien 1981). However, to fulfill its task, an ion sensor needs to meet a number of requirements, which will be discussed below. But let us first have a look on the general principles, on which ion sensing is based.

24.2 TRANSDUCTION PRINCIPLES

24.2.1 OVERVIEW

Fluorescent sensors usually consist of three moieties: a receptor or ionophore (which is responsible for the selective binding of the analyte), a fluorophore (whose properties change upon ion binding to the receptor),

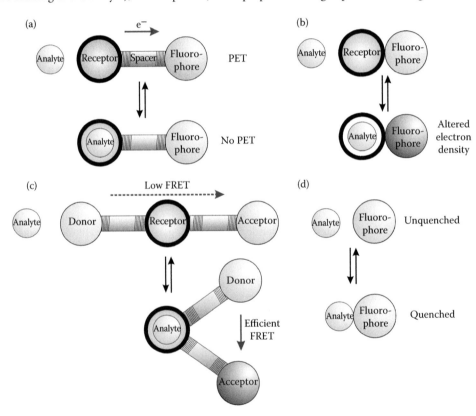

Figure 24.2 Signal transduction mechanisms of fluorescent sensors. (a) Signal transduction via PET. When the fluorophore is excited, electrons can be transferred from lone electron pairs of the receptor to lower-energy orbitals of the fluorophore, thereby quenching the fluorescence effectively. Upon binding of a cation, the energy of the free electron pair is lowered. Electron transfer is no longer possible and the fluorophore is not quenched. (b) Signal transduction via PCT. Receptor and fluorophore are directly connected so that coordination of the analyte affects directly the electron density distribution in the fluorophore, resulting in a change of the intensity and wavelength of fluorescence emission. (c) Signal transduction by conformational changes of the sensor structure. Those can be detected by fusing donor and acceptor fluorophores to a molecule that undergoes conformational changes upon binding of the analyte. Changes in the distance and the orientations of the fluorophores will result in a change of the Förster resonance energy transfer (FRET) efficiency, which can be detected. (d) Signal transduction via collisional quenching. Collisional (dynamic) quenching of the analyte (especially anions) can decrease the fluorescence intensity and lifetime of fluorophores.

and a spacer (which links receptor and fluorophore). Binding of the analyte can be transduced by several mechanisms into a change of fluorescence. The most common approaches are based on photoinduced electron transfer (PET) or photoinduced charge transfer (PCT). In sensors based on these mechanisms, analyte binding modifies the electronic properties of the receptor and fluorophore (Figure 24.2a and b).

Complexation of an ion can also change the conformation of the receptor moiety. In sensors based on this principle, the conformational change is detected by fluorophores attached to both ends of the receptor. Binding of the analyte changes the distance and orientation between the fluorophores and thereby the resonance energy transfer between them (Figure 24.2c). Fluorescence can also be quenched by a wide variety of substances including anions such as chloride, bromide, or iodide and *vice versa*, the change in fluorescence intensity and fluorescence lifetime can be exploited to measure the concentration of the respective quenchers. Here, the indicator does not need to have a binding site to which the analyte binds (Figure 24.2d).

These transduction principles can be exploited to build fluorescent sensors for ions. Depending on the mechanism, binding of the ion might change the absorption properties and/or its fluorescence intensity, its fluorescence spectrum, and its fluorescence lifetime. How these changes can be exploited for quantitative measurements is discussed in Sections 24.2.2–24.2.5.

24.2.2 SENSORS BASED ON PET

As discussed above, a typical PET sensor consists of three components: the fluorophore, the receptor, and a spacer. Figure 24.3a demonstrates the general principle. Examples for PET sensors are shown in Figure 24.3b and c.

When no ion is bound to the ionophore, the fluorescence of the fluorophore is quenched by electron transfer from the ionophore (donor) to the fluorophore (acceptor). In most cases, the donor atom is an aminic nitrogen atom of the ionophore structure and usually plays also an active part in coordinating the analyte. Upon ion binding, the free nitrogen electron pair is coordinated to the metal cation, whereby its

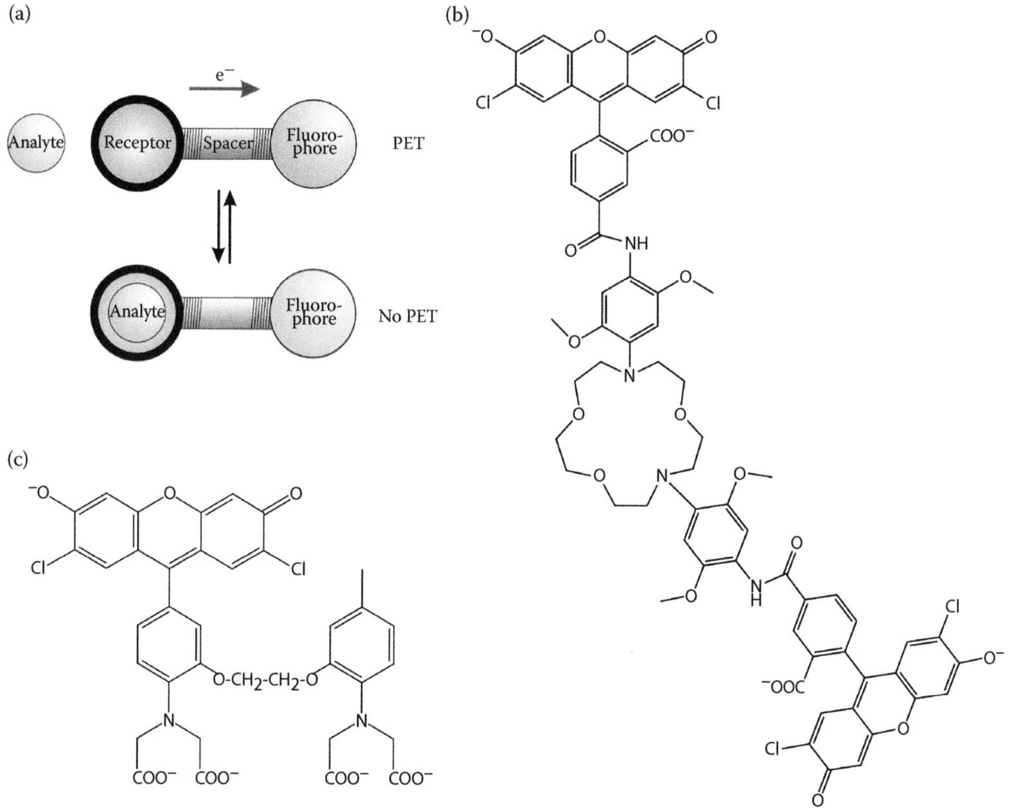

Figure 24.3 Examples for PET sensors. (a) Signal transduction principle; (b) Sodium Green; (c) Fluo-3.

energy is decreased so that electron transfer from the ionophore to the fluorophore is no longer favored, and as a consequence, the fluorescence is no longer quenched (De Silva 1997; Valeur and Leray 2000). Along with an increase in quantum yield as a result of ion binding in PET sensors, there is also a change in the excited state lifetime, as would be expected when a fluorophore is dequenched (Valeur and Leray 2000). This means that PET sensors can be ideally used in fluorescence lifetime imaging microscopy (FLIM) applications. PET fluorescent ion sensors used in biomedical research include the sodium ion sensor Sodium Green (Kuhn and Haugland 1995) and the calcium ion chelating sensors Fluo-3, Fluo-4, Oregon Green BAPTA 1, Calcium Green 1, Calcium Orange, and Calcium Crimson.

Many other PET-based fluorescent ion probes have been developed. However, most of them are not soluble in aqueous solutions and are not suited for biomedical applications.

24.2.3 SENSORS BASED ON PCT

PCT fluorescent ion sensors undergo a change in their internal charge distribution upon excitation. Such a charge transfer occurs when the fluorophore has an electron donor group (such as an amino group) and an electron accepting group (such as a carbonyl group). The interaction of an ion with one of these groups will affect the charge separation within the molecule, which is directly linked to dipole moment changes. This leads to shifts of the absorption and emission spectra.

Typical ion sensors that operate in this way include the Na^+ sensor SBFI (Figure 24.4b; Minta and Tsien 1989), the K^+ sensor PBFI (Jezek et al. 1990; Kasner and Ganz 1992), and the Ca^{2+} sensors Indo-1 and Fura-2

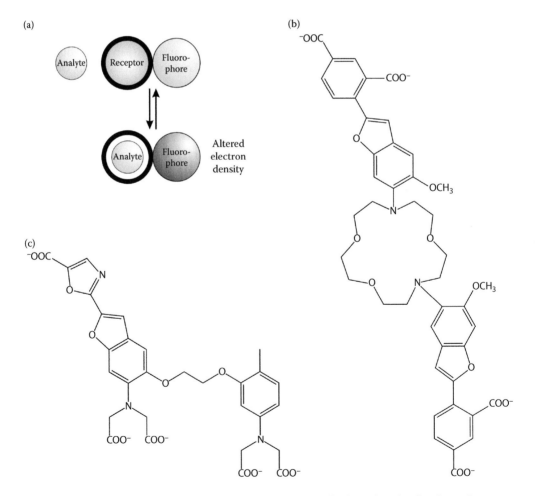

Figure 24.4 Examples for PCT sensors. (a) Signal transduction principle; (b) sodium-binding benzofuran isophthalate (SBFI); (c) Fura-2.

(Figure 24.4c; Grynkiewicz et al. 1985). These indicators are chelators for the respective ions. The spectral shift that occurs upon ion binding allows for ratiometric measurements. The spectral shifts are often accompanied by fluorescence lifetime changes that—in principle—can also be used for determining absolute ion concentrations.

24.2.4 SENSORS BASED ON CONFORMATIONAL CHANGES OF THE RECEPTOR (FRET PROBES)

Genetically engineered probes typically take advantage of a third transduction mechanism (Figure 24.2c). Here fluorescent proteins (FPs) are genetically fused to both ends of the ion binding moiety of an enzyme or receptor. The fluorophores are chosen such that Förster resonance energy transfer (Förster 1946; Jares-Erijman and Jovin 2003) can occur between them. Ideally, the emission spectrum of the fluorophore absorbing at shorter wavelengths (donor) and the absorption spectrum of the acceptor fluorophore should strongly overlap, while the absorption and emission spectra of both fluorophores should be well separated. As the ion of interest binds to the receptor moiety, the molecule undergoes a conformational change, and the distance and orientation between the fluorophores change, which result in a change of energy transfer efficiency. Sensors based on this principle are discussed in more detail in Section 24.6.

24.2.5 SENSORS BASED ON COLLISIONAL QUENCHING

Collisional quenching is another mechanism that can be exploited for ion sensing. During a collisional encounters quencher molecules and fluorophore molecules in its excited singlet state form a transient complex in which the rate constant for non-radiative relaxation is increased by orders of magnitude in this way diminishing the probability for fluorescence. It results in a decrease in fluorescence intensity and lifetime of the fluorophore that is proportional to the quencher concentration. It is characterized by the Stern–Volmer equation

$$\frac{I_0}{I} = \frac{\tau_0}{\tau} = 1 + K_{SV}[Q] \tag{24.1}$$

where I_0 and τ_0 are the fluorescence intensity and lifetime of the fluorophore in the absence of the quencher, respectively; I and τ are the fluorescence intensity and lifetime in the presence of the quencher, respectively; $[Q]$ is the concentration of the quencher; and K_{SV} is the so-called Stern–Volmer constant (Lakowicz 2006).

Typical collisional quenchers are heavy anions like bromine or iodine. Consequently, the change in fluorescence intensity and lifetime can be exploited to measure the concentration of these quenchers. For biomedical application, the need to measure chloride is more crucial. Since chloride is a good quencher of quinolinium derivatives (Figure 24.5), these fluorophores can be used as chloride sensors (Biwersi et al. 1992, 1994; Geddes 2001; Verkman et al. 1989, 1990).

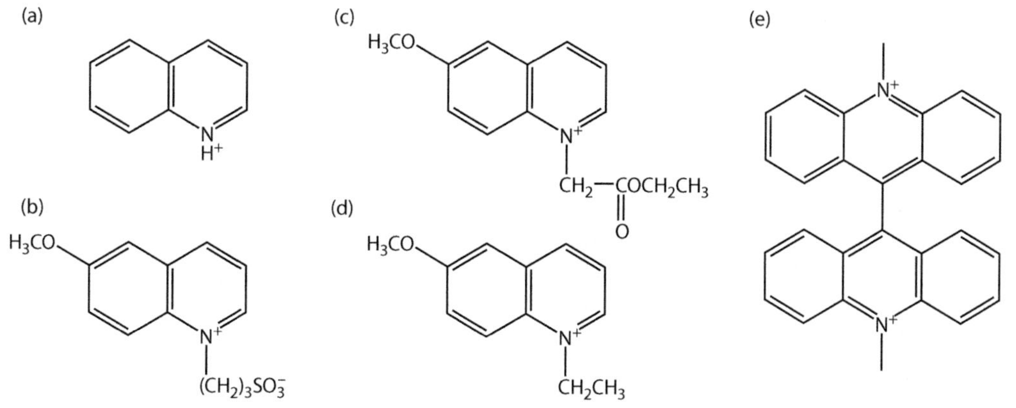

Figure 24.5 Examples for sensors based on collisional quenching. (a) Quinolinium; (b) 6-Methoxy-*N*-(3-sulfopropyl)quinolinium (SPQ); (c) *N*-(Ethoxycarbonylmethyl)-6-methoxyquinolinium (MQAE); (d) 6-Methoxy-*N*-ethylquinolinium (MEQ); (e) Lucigenin.

24.3 WISH LIST FOR AN IDEAL ION SENSOR

Sensors that are built according to one of the principles discussed above emit fluorescent light whose intensity or lifetime is dependent on the analyte concentration. Then—with proper calibration—the measured fluorescence intensity or fluorescence lifetime can be, in principle, attributed to the concentration of the analyte. But, except for being able to transduce the analyte concentration into a measurable optical signal, a good sensor molecule needs to meet some more criteria:

1. Criteria concerning the receptor/ionophore:
   a. The sensitivity of the sensor must be as such that fluctuations of the analyte occurring near physiological concentrations do lead to significant changes of the fluorescence signal of the sensor. This implies for indicators whose transduction mechanism is based on ion binding that the dissociation constant K_d of the receptor/ionophore for its target ion should be in the physiological concentration range. (If K_d is too high, the analyte does not bind. If K_d is too low, the indicator will be saturated at physiological concentrations and will not be able to respond to changes in the analyte concentration.) For indicators based on collisional quenching, the same applies. Here the reciprocal value of the Stern–Volmer constant K_{SV} should be in the physiological range.
   b. The indicator should discriminate against other ions so that variations in those ions have little effect. For indicators whose transduction mechanism is based on ion binding, the analyte always competes with ions having similar chemical properties. Especially when the competing ion is more abundant, the competing ion might outmatch the analyte at the binding site. Since $[K^+]_i$ is 10- to 20-fold higher than $[Na^+]_i$, the affinity of a sodium indicator should be at least 20-fold higher to sodium, which means that the K_d for potassium should be greater than 150 mM. Likewise, indicators for Ca^{2+} must be able to discriminate against Mg^{2+}. For indicators based on collisional quenching, one requirement is that its fluorescence is not quenched by other intracellular components.
2. Criteria concerning the fluorophore:
   a. Excitation wavelengths should exceed 350 nm, preferably 400 nm, because shorter wavelengths demand expensive quartz optics rather than fused silica optics. Besides, lasers emitting at wavelengths below 400 nm are expensive and not available in common scanning microscopes (Gratton and Vande Ven 2006). This applies particularly for pulsed lasers (with a pulse width in the picosecond range), which are necessary for fluorescence lifetime measurements in the time domain. Multiphoton excitation with fs- or ps-pulsed NIR lasers constitute an elegant but expensive alternative preventing those limitations.
   b. Emission wavelengths should be longer than ~550 nm to avoid overlap with cellular autofluorescence.
   c. To yield an intense fluorescence signal, the brightness of the fluorophore should be high. This means that it should have a large extinction coefficient ($>5 \times 10^4$ M^{-1} cm^{-1}) and a quantum yield close to unity.
   d. For fluorescence lifetime measurements, the fluorophore should exhibit a monoexponential fluorescence decay larger than 1 ns in the absence of the analyte. This simplifies the analysis and interpretation of fluorescence lifetime measurements considerably (see the next section for a detailed explanation).
3. Criteria concerning the entire sensor molecule:
   a. The response of the sensor to its analyte should be rapid enough to accurately reflect the true dynamics of the analyte.
   b. The sensor should be water soluble, which means that it should have enough polar groups such as carboxylates or sulfates.
   c. To load the cells with the indicator, the polar groups should be maskable by nonpolar protecting groups that can be cleaved by intracellular enzymes. One protecting group that has found wide application is the acetoxymethyl ester (Tsien 1981). The nonpolar ester can easily pass the lipophilic plasma membrane of cells. In the cytosol, the acetomethoxyester groups are cleaved by endogenous esterases. The "unmasked" polar fluorophore is trapped inside the cell.
   d. The sensor should be distributed evenly within the compartment of interest and not bind to any endogenous macromolecules or structures.
   e. The sensor (in ground and excited states) should not have any side effects that would alter the normal function and behavior of the cell.

24.4 DETECTION PRINCIPLES

24.4.1 INTENSITY MEASUREMENTS

When a fluorescent ion sensor complexes with its analyte (as distinct from ion sensors based on quenching), fluorescence intensities can be used to calculate the analyte concentration $[A]$ according to the formula

$$[A] = K_d \, (I - I_{min})/(I_{max} - I) \tag{24.2}$$

where I is the fluorescence intensity, I_{min} is the fluorescence intensity of the fluorophore in the absence of the analyte, I_{max} is the intensity of the analyte saturated fluorophore, and K_d is the dissociation constant (Grynkiewicz et al. 1985).

Accurate intensity measurements require a homogeneous distribution of the indicator inside the compartment of interest. Otherwise, a high fluorescence signal, which is caused by an accumulation of the indicator dye, might be misinterpreted as a local increase in the analyte concentration. Likewise, the sample should be homogenously illuminated, and absorption along the optical path from each point of the specimen to the detector should be the same. In reality, all these conditions are not so easy to satisfy: in the case of epifluorescence microscopes, where arc or filament lamps are used as the excitation source, there is no uniformity of light delivered from the light emitting area (i.e., from the filament or arc themselves; Nolte et al. 2006). Even when assuming that all illumination inhomogeneities and optical path length differences can be accounted for, there still remains the problem of changes in sensor concentration within the cell due to interaction with cellular proteins or sequestration into organelles. Another problem affecting an accurate determination of the ion concentration is photobleaching of the fluorophore, which will inevitably occur over the time it takes to collect the images.

Yet, fluorescence intensities can be used to measure intracellular ion concentrations accurately, regardless of the aforementioned complications, if there is an excitation or emission wavelength at which the indicator fluorescence is independent of the analyte concentration. Fluorescence emitted at this wavelength is only proportional to the indicator concentration, the illumination intensity, and the transmission efficiencies along the optical path. Thus, this value can be used to correct the indicator signal for these factors. Sensors that possess such an analyte-independent standard are referred to as *ratiometric* sensors (Tsien 1989a,b). Most ratiometric sensors are PCT probes. The downside of most ratiometric indicators is that they are typically excited in the blue region of the spectrum (Valeur 2002).

Another way to circumvent the problems associated with fluorescence intensity measurements is to measure fluorescence lifetimes. As opposed to fluorescence intensities, fluorescence lifetimes have the advantage of being independent of the fluorophore concentration, illumination strength, or inhomogeneities along the optical path.

24.4.2 FLUORESCENCE LIFETIME MEASUREMENTS

Fluorescence lifetimes are only dependent on the intrinsic properties of the fluorophore and its interaction with its immediate environment. Because of this, fluorescence lifetime measurements can circumvent the need for ratiometric probes.

The fluorescence intensity decay function of a fluorophore that follows a monoexponential decay is

$$I(t) = I_0 \, \exp(t/\tau) \tag{24.3}$$

where I_0 is the intensity right after excitation with a light pulse whose length is short compared to the fluorescence lifetime τ. $I(t)$ is the intensity at time t after the excitation pulse.

As discussed above, collisional quenching can shorten the lifetime. But also electron transfer from a donor can decrease the lifetime. If we assume that a short lifetime, τ_1, can be attributed to the ion-free form of the indicator and a long component, τ_2, can be attributed to the ion-bound form, then in an analyte-containing solution, in which the indicator is not saturated, both forms should contribute to the overall fluorescence decay, and accordingly, one should observe a biexponential fluorescence decay.

$$I(t) = A_1 \exp(t/\tau_1) + A_2 \exp(t/\tau_2) \tag{24.4}$$

By fitting a biexponential function to the fluorescence decay, the lifetime components τ_1 and τ_2, as well as the amplitudes A_1 and A_2, can be recovered. Provided that this simple model holds and that the fit is appropriate, then A_1 and A_2 would be equal to the fraction of the ion-free and the ion-bound form of the indicator.

In order to then calculate the analyte concentration $[A]$, using only the bound amplitude data, a modified form of Equation 24.2 can be used:

$$[A] = K_d (A_2 - A_{2,min})/(A_{2,max} - A_2) \tag{24.5}$$

$A_{2,min}$ is amplitude A_2 determined in the absence of the analyte and should ideally be equal to 0. $A_{2,max}$ is the intensity of the analyte saturated fluorophore and should ideally be equal to 1. Equation 24.5 holds only in case of a biexponential fluorescence decay. However, in practice, this is not always the case as the example shown in the following section demonstrates. Often, the fluorescence decay needs to be fitted with a triexponential or multiexponential decay (Despa et al. 2000; Lakowicz et al. 1992). Although it is unclear to which form the third lifetime component can be attributed, the mean or average fluorescence lifetime can be used to estimate the ion concentration on the basis of a calibration curve (Szmacinski and Lakowicz 1995).

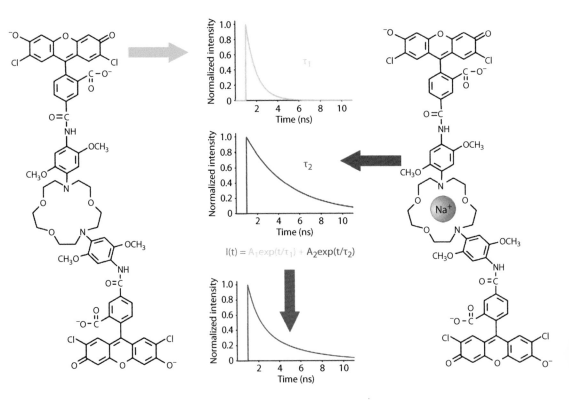

Figure 24.6 The concept behind fluorescence lifetime measurements. In the ideal case, a fluorescence lifetime component, τ_1, can be attributed to the ion-free form of the indicator dye, and another component, τ_2, can be attributed to the ion-bound form, so that the overall fluorescence decay exhibits a biexponential time course. By fitting a biexponential function to the fluorescence decay, the lifetime components τ_1 and τ_2, as well as the amplitudes A_1 and A_2, can be recovered. Provided that this simple model holds and that the fit is appropriate, then A_1 and A_2 would be equal to the fraction of the ion-free and the ion-bound form of the indicator. (Reproduced from Dietrich, S. et al., *SPIE Proc.* 7569:14, 2010.)

For fluorescent ion sensors that are based on collisional quenching, shorter fluorescence lifetimes are measured for higher analyte concentrations according to Equation 24.1. If K_{SV} and τ_0 are known, the ion concentration $[Q]$ can be calculated from τ. Instead of knowing τ_0, one can also determine τ_i at several $[Q_i]$ in an intracellular calibration experiment. From the plot τ_i^{-1} vs. $[Q_i]$ that will show a linear dependence (see Equation 24.1), an unknown ion concentration can be determined from the estimated τ.

24.4.3 QUANTITATIVE ANALYSIS AND INTERPRETATION OF FLUORESCENCE LIFETIME IMAGES

The quantitative analysis and interpretation of fluorescence lifetimes of ion sensing fluorescent dyes or proteins is often difficult. This applies to cuvette measurements and, in particular, to measurements in cells or biological tissues. Fluorescence decays of fluorophores are rarely single exponential (see for fluorescence lifetime standards, Boens et al. 2007). This is especially true for organic fluorescent ion-sensing dyes, which have been optimized for maximal fluorescence intensity change upon binding, cell penetration, or low phototoxicity, but not for having monoexponential decay kinetics. An example is Oregon Green BAPTA-1 (OGB1), one of the best calcium dyes for FLIM (Wilms et al. 2006). In addition, the dyes are often based on a hydrophobic molecule core. Therefore, they tend to be associated to hydrophobic surface patches of proteins or to localize in or near lipid membranes of the cell, which both can lead to different photophysics with additional or different fluorescence decay components as well as changed ion sensitivities (see Wilms and Eilers [2007] for OGB-1 and comments on Sodium Green in Section 24.5.1). The same is true for FPs in general (see, for instance, Cotlet et al. [2001] for EGFP-1) and genetically encoded sensors in particular (see, for instance, Geiger et al. [2012] for TN-XXL and Laptenok et al. [2012] for YC3.60), which have in more than 95% multiexponential fluorescence decays. Environmental parameters—like refractive index or pH—that differ substantially inside cells also contribute to the complicated fluorescence behavior of FP-based ion sensors. As a consequence of these nonideal fluorescence decays of fluorescent ion sensors, an intracellular calibration of the fluorescence decay parameters—determining not only ion affinity but also zero and maximal ion interaction decay characteristics—is unavoidable for a full quantitative analysis of FLIM data. In some studies, the researchers have not performed intracellular calibrations themselves but use values from other studies determined in other (similar or very different) cell types. This is not at all a valid procedure. For illustration, see the Stern–Volmer constant K_{SV} and the fluorescence lifetime τ_0 at 0 mM Cl⁻ of MQAE, an often used fluorescence lifetime sensor for Cl⁻, determined in buffer and in different cells (Table 24.1).

Depending on the model function used for the analysis of the fluorescent ion sensor signal, different fit parameters or derived values (average lifetime, relative amplitudes, amplitude ratio, etc.) can be calculated resulting in an intensity-independent readout of the ion concentration. Surprisingly—and in favor of the application of FLIM in biological tissue—the fluorescence lifetime (or other parameter) maps often reveal that, despite of the above-mentioned problems, individual cells in a tissue show very distinct values (see, e.g., Figures 24.7, 24.10, 24.11, and 24.12) making them distinguishable from the neighboring cells.

Table 24.1 Stern–Volmer constant K_{SV} and fluorescence lifetime τ_0 at 0 mM [Cl⁻] from intracellular calibrations of the Cl⁻ sensor dye MQAE (see also Section 24.5.4 on Cl⁻ sensing)

CELL TYPE	K_{SV} (mM⁻¹)	τ_0 (ns)	REFERENCE
Phosphate buffer	185	25.1	Kaneko et al. 2002
Rat dorsal root ganglion (DRG) neurons	9.1	6.8	Kaneko et al. 2002
Perineuronal satellite cells from rat DRG	29.1	9.1	Kaneko et al. 2002
Rat olfactory sensory (OSN) neurons	18	6.5	Kaneko et al. 2004
Mouse olfactory sensory (OSN) neurons	13	4.9	Kaneko et al. 2004
Mice DRG neurons	3.05	3.9	Gilbert et al. 2007
Cockroach salivary duct cells	9.6	5.8	Hille et al. 2009
Cockroach salivary acinar peripheral cells	5.8	5.5	Lahn et al. 2011

Fluorescence lifetime imaging based on exogenous probes

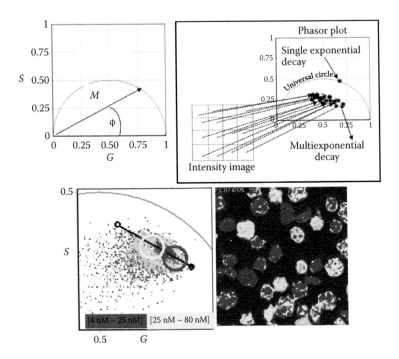

Figure 24.7 Upper left panel: representation of a fluorescence decay as a phasor in the phasor plot with the two projection vectors G and S and the universal circle, where the phasors of all monoexponential decays end. Upper right panel: every pixel of a fluorescence image (with time-resolved fluorescence detection) is represented by its own phasor in the phasor plot. Lower panel: intracellular calibration of Calcium Green 1 in erythrocytes (black line) and the phasors from the fluorescence image on the right. The red and yellow circle selects pixels with low and higher Ca^{2+} concentrations. (Adapted from Sanchez, S. et al., *PLoS One* 6:e21127, 2011.)

In the past few years, the phasor plot—an old concept, which has its origin in fluorescence lifetime measurements based on frequency modulation of continuous wave excitation light (frequency domain measurements; Weber 1981; Jameson et al. 1984)—has been used successfully for the analysis and interpretation of a number of different FLIM data (Behne et al. 2011; Celli et al. 2010a; Digman et al. 2008; James et al. 2011; Sanchez et al. 2011; Štefl et al. 2011). The charm (and power) of the method consists of the fact that no fitting is required. Instead, the fluorescence intensity time trace $I(t)$ (measured in the time domain) of *every* pixel is represented by a vector (the phasor) with length M and phase angle φ (Figure 24.7). The phasor is the sum of the two vector projections $g(\omega)$ and $s(\omega)$ obtained by a Fourier transform of $I(t)$:

$$g(\omega) = \int_0^\infty I(t)\cos(\omega t)dt \bigg/ \int_0^\infty I(t)dt = M\cos(\varphi) \tag{24.6}$$

$$s(\omega) = \int_0^\infty I(t)\sin(\omega t)dt \bigg/ \int_0^\infty I(t)dt = M\sin(\varphi) \tag{24.7}$$

where ω is the laser repetition angular frequency or the angular frequency of the light modulation.

If $I(t)$ can be described as a single exponential decay ($I(t) = I_0 \exp(-t/\tau)$), $g(\omega)$ and $s(\omega)$ take the following form:

$$g(\omega) = 1/(1 + (\omega\tau)^2) \tag{24.8}$$

$$s(\omega) = \omega\tau/(1 + (\omega\tau)^2) \tag{24.9}$$

It can be mathematically shown that every monoexponential fluorescence decay lies on the universal circle (origin (0.5,0) and radius 0.5; see Figure 24.7). The vectors $g(\omega)$ and $s(\omega)$ of a fluorescence intensity trace, which is described as a sum of exponential functions, can be written as

$$g(\omega) = \sum_k \frac{f_k}{1 + (\omega\tau_k)^2} \tag{24.10}$$

$$s(\omega) = \sum_k \frac{f_k\omega\tau_k}{1 + (\omega\tau_k)^2} \tag{24.11}$$

For a biexponential decay function, for example, signals of an ion sensing dye with lifetimes τ_{free} and τ_{bound}, all phasors of fluorescence intensity traces from pixels inside cells lie on the straight line that connects the points of τ_{free} and τ_{bound} on the universal circle. For many sensors, the fluorescence decays at the extremes—free and bound—are not monoexponential; still all phasors from intracellular volume elements with concentrations in between lie on a straight line between these extremes.

As an example, the intracellular calibration curve of a Ca^{2+} sensing dye (Calcium Green 1 [CG1]) is shown in Figure 24.7 and the phasors (black line and points in lower left image) of all the pixels from the image of erythrocytes loaded with CG1 (lower right image). The authors measure in this study the Ca^{2+} influx into erythrocytes upon exposure to the toxin α-hemolysin (Sanchez et al. 2011). Highlighted in red and yellow are pixels that belong to a certain Ca^{2+} concentration range. The blue pixels deviate too much from the model to be interpreted by the simplified two-state model (one reason could be a too high contribution of autofluorescence). While the fluorescence decays of Ca^{2+} free and Ca^{2+} bound indicators can be determined by applying a Ca^{2+} ionophore and different extracellular Ca^{2+} concentrations, the Ca^{2+} affinity can only be measured accurately in buffer solutions with added chemical compounds assuring an environment for CG1 as alike as possible to the intracellular situation. It is therefore only an approximation, and the determined Ca^{2+} concentrations are still not completely absolute values but depend on how "alike" the buffer conditions were chosen compared to the intracellular situation. Nevertheless, in Figure 24.7, cells with low and medium Ca^{2+} concentrations can be easily identified.

The fluorescence change of other fluorescent ion sensors is based on collisional quenching or FRET. In such cases, the trajectories of the phasors of concentrations between no and full interaction of the ion sensor and the respective ion are more complicated but possible to be described mathematically (see for FRET, e.g., Albertazzi et al. 2009; Hinde et al. 2012). There are only two constraints when applying the phasor plot: the choice of the correct model and a signal that is large enough compared to the autofluorescence, so that autofluorescence cannot distort the phasors. The conventional procedure of fitting multiexponential decay functions to all fluorescence intensity time traces of an image often requires simplified fitting functions (less exponential functions) or the help of weighted global fitting. In combination with a thorough intracellular calibration of the ion sensing dye, it allows an analysis of FLIM data with similar results, interpretation, and visualization as the phasor plot approach. However, it is less intuitive and more error prone compared to the phasor plot.

A last word is devoted to the determination of relative vs. absolute ion concentrations. As discussed here and in the sections of the different ions, there are many difficulties in obtaining absolute ion concentrations especially in biological tissue, where intracellular calibrations often fail. Nevertheless, relative ion concentration differences among different cells from the same cell type in a biological tissue as well as changes of ion concentrations in living cells in its natural tissue reacting on physiological stimuli are, for many research projects, very valuable information and can be perfectly determined in the conventional way (fitting of multiexponential functions) or using a phasor plot.

In principle, several methods are suitable for FLIM measurements, including wide field, scanning confocal, and scanning two-photon (or multiphoton) fluorescence microscopy. Also, frequency-modulated constant as well as pulsed excitation sources (nowadays usually lasers) can be used. With respect to studies in alive biological tissues, two-photon fluorescence microscopy using femtosecond (or picosecond) pulsed near-infrared lasers with high repetition rates is the method of choice due to the lower overall excitation load on the tissue thanks to the confined excitation limiting the phototoxic effects to the focal plane.

24.5 OVERVIEW OF FLUORESCENT ION INDICATOR DYES AND THEIR USE IN FLIM APPLICATIONS

In this review, we set out to describe the best strategies and indicators that are available to measure intracellular ion concentrations and fluctuations in living cells using fluorescent sensors. We will restrict ourselves to the investigation of intracellular Na$^+$, K$^+$, Ca^{2+}, Cl$^-$, and H$^+$, which are the ions of most interest to cell biologists and physiologists. Since sensing of these ions has been reviewed quite extensively (Lakowicz 2006; Valeur 2002), we only give a short overview of the currently available sensors in the field and then focus on their use for fluorescence lifetime measurements in biological tissues.

24.5.1 SODIUM INDICATORS

24.5.1.1 Overview of currently used indicators

During recent decades, fluorescent indicators have been developed that selectively bind sodium ions and change their absorption and/or emission properties upon binding. However, despite considerable efforts in synthesizing new compounds, only a few indicators have been proven to be suitable for measurements in aqueous solutions and biological specimens. Among them, only sodium-binding benzofuran phthalate (SBFP), SBFI, and Sodium Green have a K_d value in aqueous solutions that is close to typical intracellular sodium concentrations (Johnson and Spence 2010; Minta and Tsien 1989). These ionophores are characterized by a cyclic crown ether structure with an electron-rich cavity to bind the cation (see Figures 24.3b and 24.4b). The cavity has just the right size to bind the sodium cation.

One important advantage the PCT-based fluorescent sensors SBFP and SBFI exhibit over Sodium Green is that changes in [Na$^+$] do not just cause a change in the fluorescence intensity but also induce a wavelength shift of the excitation maximum (Minta and Tsien 1989). Such a shift permits "ratioing" between signals obtained at two excitation wavelengths (Tsien 1989a,b). By calculating the ratio of fluorescence signals elicited by excitation at wavelengths that are shorter (e.g., 340 nm) or longer (e.g., 380 nm) than the isosbestic point, artifacts that are caused by an inhomogeneous dye distribution inside the cell can be cancelled out (Harootunian et al. 1989). Because of this property, especially SBFI has been widely used. However, it has a couple of disadvantages due to its short excitation wavelength in the UV region. First, excitation with UV light can result in considerable photodamage, and second, NADH autofluorescence is also excited by UV light. Therefore, the SBFI signal must be separated from cellular autofluorescence, which is not trivial under experimental conditions such as hypoxia and metabolic impairment, where a reliable measurement of the intracellular sodium concentration is of great interest. SBFI cannot be used in most laser scanning microscopes because only a few of them are equipped with expensive UV lasers for one-photon or Titanium:Sapphire lasers for multiphoton excitation. These disadvantages preclude a wider use of SBFI.

In contrast to SBFI, the PET-based fluorescent sensor Sodium Green has superior optical properties. First, it can be excited with the 488 nm line of argon ion lasers, with which most confocal laser scanning microscopes are equipped, and second, its emission wavelengths are well-separated from cellular autofluorescence. Hence, it can also be used under conditions where cellular autofluorescence is likely to change. Furthermore, Sodium Green has a higher fluorescence quantum yield than SBFI (0.2 vs. 0.08) and shows greater selectivity for sodium over potassium (41-fold vs. 18-fold) (Johnson and Spence 2010).

Figure 24.8 summarizes some of the chemical and physical properties of the Sodium Green indicator. In aqueous solution, its excitation maximum exhibits a peak at 505 nm, and its emission spectrum has a maximum at 533 nm (Figure 24.8a). Upon binding to sodium ions, Sodium Green exhibits an increase in fluorescence intensity with small shift toward longer wavelengths (Figure 24.8b). Figure 24.8c shows the relative fluorescence (F/F_{min}) as a function of [Na$^+$]. The data can be fitted to Equation 24.1 yielding a value of 8.6 mM for Sodium Green's dissociation constant (K_d). In solutions containing both sodium and potassium (with a total cation concentration of 145 mM), K_d is elevated to a value of 22.4 mM

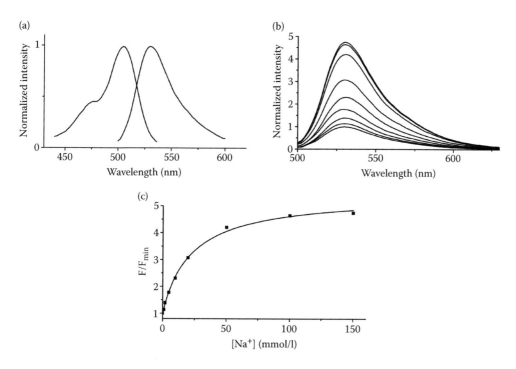

Figure 24.8 Optical and chemical properties of Sodium Green. (a) Fluorescence excitation (λ_{em} = 530 nm) and fluorescence emission spectrum (λ_{ex} = 490 nm) of Sodium Green in buffered solution (pH 7.4) containing 0 mM sodium. (b) Fluorescence emission spectra of Sodium Green (λ_{ex} = 488 nm) at sodium chloride concentrations of 0 (bottom), 1, 2, 5, 10, 20, 50, 100, and 150 mM (top). For all solutions containing less than 145 mM sodium chloride, potassium chloride was added so that the total cation concentration was 145 mM. The emission spectrum at a sodium concentration of 0 mM was normalized to unity. All other spectra were scaled by the same factor. (c) Fluorescence intensity as a function of free sodium concentration. The data were derived from the emission spectra shown in panel b and fitted to Equation 24.2. The fit yielded a K_d of 22.4 mM. (Reproduced from Dietrich, S. et al., *SPIE Proc.* 7569:14, 2010.)

(Figure 24.8c). This finding is consistent with a model in which the bigger potassium ion blocks the sodium binding site of the crown ether. Because both K_d values match the range of intracellular sodium concentrations, physiological changes in $[Na^+]_i$ are able to exert an effect on the ratio of free to sodium-bound indicator molecules and the overall fluorescence intensity.

By knowing these data, it is, in principle, possible to determine $[Na^+]$ in a biological sample. But, since Sodium Green is not a ratiometric dye, there is no way to correct for spatial inhomogeneities of the indicator distribution. High fluorescence signals may not only be due to a high sodium concentration but could also have been caused by dye accumulation in cellular compartments. Likewise, inhomogeneities in the illumination path can bias the fluorescence signal. One way to circumvent these problems is to measure fluorescence lifetimes, which are independent of these factors.

24.5.1.2 Indicators used for lifetime measurements

Fluorescence lifetimes are independent of the above-mentioned factors. They are a property of the fluorophore and its environment. If one fluorescence lifetime component can be attributed to the sodium-free form of the dye, and one component to the sodium-bound form, then the overall fluorescence decay should exhibit a biexponential time course (see Figure 24.6).

Figure 24.9 shows a series of fluorescence decay measurements obtained with a streak camera system. The fit of the data with a biexponential model, however, did not yield acceptable results. The fitted curves deviated considerably from the measured intensity decays (Figure 24.9a), and the residuals were not randomly distributed (Figure 24.9b). The parameters recovered by the fit are shown in Table 24.2. A triexponential fit could describe the fluorescence decay satisfactorily (Figure 24.9c and Table 24.2). Similar results were already obtained by measurements in the frequency domain by Szmacinski and Lakowicz (1997) and Despa et al.

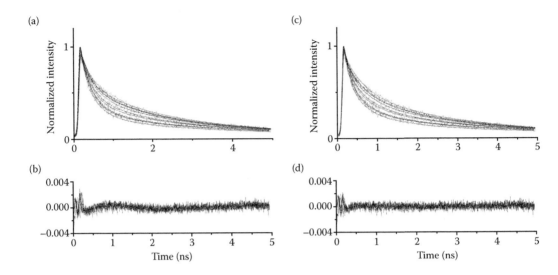

Figure 24.9 Global fit of fluorescence intensity decays of Sodium Green recorded at varying sodium concentrations ranging from 0, 1, 2, 5, 10, 20, 50, to 100 mM. Sodium chloride was complemented with tetramethylammonium chloride to yield a constant ionic strength of 290 mM (pH = 7.4). (a) Fluorescence intensity decays (black) and results of the biexponential fit (red). (b) Residuals. (c) Fluorescence intensity decays (black) and results of the triexponential fit (red). (d) Residuals. The parameters recovered by the fit are shown in Table 24.2. (Reproduced from Dietrich, S. et al., *SPIE Proc.* 7569:14, 2010.)

(2000). Apparently, the fluorescence decay cannot be explained by such a simple model as the one presented in Figure 24.6. Most likely, the sodium free and the sodium bound exist in several conformational states with different fluorescence lifetimes contributing to the overall fluorescence decay.

In the presence of potassium ions, even more complicated decay models are necessary to describe the fluorescence decays of Sodium Green (not shown). In principle, the average lifetime (τ_m) or the amplitudes (A_1, A_3) of the fast and slow lifetime component (τ_1, τ_3) could still be used to estimate the sodium concentration, despite the lack of a good model. But, in practice, these measurements are biased by interactions of the dye with intracellular components, which make an analysis of the fluorescence decay impossible. No successful use of other Na$^+$ indicator dyes like SBFI or Asante Green in FLIM studies has been reported until now (see also Lahn et al. 2011).

Table 24.2 Global fit of the fluorescence intensity decays shown in Figure 24.9

	BIEXPONENTIAL FIT				TRIEXPONENTIAL FIT				
c (mmol/l)	A_1 (%)	A_2 (%)	τ_m (ns)	χ^2	A_1 (%)	A_2 (%)	A_3 (%)	τ_m (ns)	χ^2
0	75.4	24.6	0.83	1.81	46.8	36.5	16.7	0.73	1.42
1	72.5	27.5	0.89	1.84	47.1	33.6	19.4	0.78	1.54
2	69.9	30.1	0.94	2.06	43.2	35.0	21.7	0.85	1.79
5	64.2	35.8	1.07	1.92	38.0	35.2	26.9	0.99	1.63
10	59.4	40.6	1.17	1.42	31.2	37.2	31.5	1.12	1.08
20	54.4	45.6	1.28	1.32	26.6	37.0	36.4	1.25	1.03
50	48.3	51.7	1.41	1.31	18.9	38.4	42.7	1.43	1.06
100	47.8	52.2	1.42	1.37	16.1	40.2	43.8	1.46	1.07

Source: Reproduced from Dietrich, S. et al., *SPIE Proc.* 7569:14, 2010.
Note: The biexponential fit yielded the global lifetimes of τ_1 = 0.30 ns and τ_2 = 2.44 ns. The triexponential fit yielded the global lifetimes of τ_1 = 0.13 ns, τ_2 = 0.54 ns, and τ_3 = 2.80 ns.

24.5.2 POTASSIUM INDICATORS

24.5.2.1 Overview of currently used indicators

The range of potassium indicators that are useful for biological measurements is as limited as for sodium indicators. The reasons for this are the same. There are a few structures that are able to bind effectively the monovalent potassium ion in aqueous solutions. Moreover, sodium and potassium ions compete for the binding site. Each cation has to be measured in the presence of the other cation. This becomes a challenge for measurements of sodium in the cytosol and for measurements of potassium in the extracellular space, where the concentration of the analyzed ion is exceeded by the competing ion by factors of 15 and 35, respectively (see Figure 24.1).

The structures employed to bind potassium are similar and are based on diaza-crown ethers. To match the size of the bigger potassium cation, the crown-ring structure has to be extended by one [$-CH_2-CH_2-$ O$-$] unit. In this way, potassium-binding benzofuran phthalate (PBFP) and potassium-binding benzofuran isophthalate (PBFI) are obtained as the homologous compounds to SBFP and SBFI, respectively (Minta and Tsien 1989). PBFP and PBFI are also spectroscopically similar to the sodium indicator. However, the spectral shift in absorption is much lower than observed for SBFP and SBFI (Minta and Tsien 1989; Szmacinski and Lakowicz 1999). The dissociation constants of PBFP and PBFI are 83 and 8 mM (measured in 0.1 M tetramethylammonium), respectively (Minta and Tsien 1989). Thus, PBFB has the right affinity for intracellular measurements, whereas PBFI appears (at the first sight) to be suited for extracellular measurement. Unfortunately, PBFI is of little practical use for extracellular K$^+$ measurements, since the affinity for sodium is only slightly lower (K_d = 21 mM).

For extracellular measurements, the affinity to potassium must be increased. This can be achieved by using ligands that better enclose the potassium ion, such as in a class of compounds, that are derived from the crown ether structure, but in which the nitrogen atoms are bridged by three instead of two [$-CH_2-CH_2-O-CH_2-CH_2-O-CH_2-CH_2-$] units. Because potassium is not only complexed but also completely encapsulated by the ether bridges, this class of compounds is also called cryptands. The name suggests that the cation is completely hidden from the bulk solvent. Jean-Marie Lehn, who coined this name, suggested also a system of nomenclature, in which the number of oxygen atoms in the ether chains linking the nitrogen atoms is indicated assuming the oxygen atoms are separated by ethyl residues. Thus, a cryptand with three [$-CH_2-CH_2-O-CH_2-CH_2-O-CH_2-CH_2-$] bridges is called a [2.2.2] cryptand (Gokel 1991).

The first potassium indicator of this class was obtained by fusing a [2.2.2] cryptand with a coumarin dye (Masilamani et al. 1992). This structure, however, suffered from a strong pH and Na$^+$ interference. The pH sensitivity could be overcome by aromatizing the cryptand leading to the coumarin diacid cryptand [2.2.2] (CD222) (Crossley et al. 1994). The K_D of this compound was found to be 1 mM in the absence of sodium but was highly dependent on the sodium concentration (Crossley et al. 1994).

24.5.2.2 Indicators used for lifetime measurements

Like the fluorescence decay of SBFI, the fluorescence decay of PBFI is complex. At least three decay times of 0.005, 0.30, and 0.83 ns have to be used to describe the fluorescence decay adequately (Szmacinski and Lakowicz 1999; see also Meuwis et al. 1995). The mean lifetime increases only from 0.40 to 0.72 ns upon K$^+$ binding and is less sensitive to Na$^+$ than the changes in intensity.

Likewise, analysis of the fluorescence decay of CD222 yields three lifetimes of 0.04, 0.15, and 0.82 ns (Szmacinski and Lakowicz 1999). The mean lifetime increases only from 0.17 ns for the free form to 0.71 ns upon K$^+$ binding. In the presence of 100 mM Na$^+$, the lifetime is 0.26 ns. Hence, the increase in the mean lifetime upon binding of Na$^+$ is much smaller than the increase in the mean lifetime upon binding of K$^+$, opening the possibility to measure K$^+$ with a better selectivity in the presence of high concentrations of Na$^+$ (Szmacinski and Lakowicz 1999). By using modulated UV light-emitting diode, this feature of CD222 can be exploited to measure the K$^+$ concentration in blood (Szmacinski and Chang 2000). However, due to the small changes in the average fluorescence lifetime introduced by binding of K$^+$ (small dynamic range), the accuracy of the method is limited.

24.5.3 CALCIUM INDICATORS

24.5.3.1 Overview of currently used indicators

Due to the important physiological role of calcium ions as second messenger, the development of fluorescent calcium sensors was a very active area of research to improve imaging methods in physiology for about three decades. All dyes in use today are based on modified versions of the calcium chelating group 1,2-bis(o-aminophenoxy)ethane-N,N,-N′,N′-tetraacetic acid (BAPTA). BAPTA shows large absorbance changes (in the region 240–320 nm) and a threefold fluorescence increase (at 360 nm) upon chelating of one Ca^{2+} with an affinity of about 100 nM. Among the first dyes useful for physiological measurements was Quin-2 (also named Quin2; Tsien 1980; Tsien et al. 1982a,b), where one side of the symmetric BAPTA moiety was extended to create a larger conjugated system of alternating single and double bonds. Its spectra show substantial red shifts of absorption (320–380 nm) and fluorescence (510–525 nm), and increased fluorescence quantum yields of calcium free and calcium bound form compared to BAPTA. Some of the early studies were devoted to lymphocyte mitogenesis (Tsien et al. 1982a), T-cell stimulation (Tsien et al. 1982b), and the change in intracellular free Ca^{2+} concentrations upon different extracellular stimuli (Ashley et al. 1984; Rogers et al. 1983). Those early studies built the basis for the nearly countless physiological studies that use the observation of transient calcium changes—mostly elevations—by fluorescence microscopy as a readout for cellular responses upon different external stimuli.

Already the absorption spectra of the calcium-free and calcium-bound forms of Quin-2 are spectrally so different that a ratiometric readout is possible (consecutive excitations at the two absorption maxima with emission detection at one wavelength), but not practicable due to the low extinction coefficients and fluorescence quantum yields. In 1985, the group of Roger Tsien published some new calcium sensing dyes, where the methylquinoline group in Quin-2 was replaced by stilbene derivatives with aromatic rings stabilizing the stilbene double bond and preventing photoisomerization (Grynkiewicz et al. 1985). Two of those, Fura-2 and Indo-1, are still very much in use due to their ratiometric readout that is possible because of the large shifts in absorption (Fura-2) or emission (Indo-1) upon Ca^{2+} binding. In addition, the extinction coefficients as well as the fluorescence quantum yields are larger by a factor of five or more compared to Quin-2 allowing to work with much lower intracellular dye concentrations (well below 100 μM compared to 1–2 mM in the case of Quin-2). Fura-2 and Indo-1—despite their excitation in the near UV—are still popular especially for studies where absolute Ca^{2+} concentrations are reported. Yet another property of the newly developed fluorescent indicator dyes was missing, to say it with the title words of the original publication, "a non-disruptive technique for loading" into living cells and tissue (Tsien 1981). It was again Roger Tsien who suggested a chemical modification that made fluorescent dyes "temporarily membrane permeable" by attaching acetoxymethylester groups to the polar groups of the fluorescent sensor (the four carboxylates of the BAPTA moiety in case of calcium-sensitive dyes), which often prevent membrane passage of the dyes. Unspecific esterases active in nearly all cell types cleave the acetoxymethylester groups leaving the bare fluorescent indicator dye in the cell, which is now hardly capable of passing the membrane and therefore trapped and enriched in the cell. Starting with Quin-2 (Tsien 1981), only with this invention the newly created sensing fluorescent indicator dyes could develop their full power.

The next step of Tsien and coworkers was to produce similar ratiometric calcium sensing dyes with absorption maxima in the visible spectrum rather than in the near-UV as for Indo-1 and Fura-2. They replaced the stilbene chromophores by fluoresceins and rhodamines, keeping the concept of integrating one half of the Ca^{2+} chelating BAPTA group into the chromophore, resulting in calcium indicators with excitation in the range of 500–550 nm and emissions from 500 to 600 nm (Fluo- and Rhod- series; Minta et al. 1989). The most important one was Fluo-3, which for one decade was the calcium sensor of choice, allowing for the first time physiological studies with time resolutions down to seconds and intracellular dye concentrations of a couple of micrometers only (Kao et al. 1989). The excitation with light of the visible spectrum also opened the way for the application of confocal fluorescence microscopy in physiological studies involving the observation of intracellular Ca^{2+} concentrations. Despite the great improvement achieved with these new calcium indicator dyes, the authors were disappointed that one important property was missing, that is, the possibility for ratiometric imaging. Neither absorption nor emission spectra of the fluorescein- and rhodamine-based calcium sensors show pronounced shifts in spectral position. In fact, this behavior is a consequence of a change in the photophysics. In Fura-2 and Indo-1, PCT is the basis of the spectral changes caused by Ca^{2+} binding. Instead, the Fluo- and Rhod-sensors show PET after excitation. Binding of Ca^{2+} prevents PET and increases the fluorescence intensity.

In the following years, a couple of fluorophores with improved photophysical parameters (e.g., spectral position, reduced photobleaching) were synthesized and introduced for physiological studies, among them Fluo-4 (Chambers et al. 1999; Gee et al. 2000; Miriel et al. 1999), Calcium Green 1 (CG1, Eberhard and Erne 1991; Miriel et al. 1999), Oregon Green 1 Bapta (OGB1, Brain and Bennett 1997), Calcium Orange (Duffy and MacVicar 1995; Eberhard and Erne 1991), and Calcium Crimson (Eberhard and Erne 1991), none of them owing properties necessary for ratiometric imaging. They all have properties similar to Fluo-3 (large increase in fluorescence quantum yield upon calcium binding, negligible changes of absorption and fluorescence spectra). The photophysical explanation for this behavior is the occurrence of efficient intramolecular PET (see Section 24.2) in the calcium-free form. PET competes with the radiative deactivation channel (= fluorescence) and in this way reduces the fluorescence quantum yield considerably (3 to 50 times). Upon binding of a Ca^{2+}, the PET process becomes less efficient. Consequently, the fluorescence quantum yield is increased. Calcium indicator dyes with affinities for Ca^{2+} binding from 40 nM to 90 mM have been established. There exist numerous reviews on the properties of calcium sensing dyes and important issues for application; two recent ones are Paredes et al. (2008) and Russell (2011). In the recent past, PET-based calcium-sensitive dyes with absorption and emission in the red and near-infrared spectral region have been generated (Egawa et al. 2011; Matsui et al. 2011; Ozmen and Akkaya 2000). These dyes will be of great help in multimodal fluorescent imaging of physiological processes leaving the blue/green and yellow/orange spectral region for other fluorescent sensors, for example, genetically encoded FRET-based sensors for other second messengers, the cells redox properties, or kinase activities that emit in these spectral regions. A very important parameter of the calcium indicators is the affinity. It has been shown in several studies that the affinities are lower in cells (two to three times; Thomas et al. 2000; Zhao et al. 1996). In addition, the affinities are usually measured at room temperature, but they are significantly higher at physiological temperature (37°C; see, e.g., Larsson et al. 1999; Lattanzio 1990; Woodruff et al. 2002). For absolute measurements, a calibration of the calcium sensor fluorescence in the cell type studied applying calcium ionophores is unavoidable.

24.5.3.2 Indicators used for lifetime measurements

A good number of the many well working fluorescent calcium indicator dyes (see above) also show significant fluorescence lifetime changes upon Ca^{2+} binding. Many of them have been characterized already in the 1990s. But several experimental problems, especially the interaction of the calcium indicator dyes with proteins, lipids, and other components of the cell, have prevented a general use of FLIM for the determination of intracellular Ca^{2+} concentrations. Quin-2 was the first Ca^{2+}-sensitive dye for which a large fluorescence decay change was found (Lakowicz et al. 1992a). The fluorescence decay at low calcium is almost ten times faster than at saturating calcium concentration (1.3 ns vs. 11.6 ns). This is the largest fluorescence lifetime change reported ever for calcium-sensitive dyes, yet Quin-2 has not become a commonly used indicator for FLIM studies. Apart from the unfavorable excitation wavelengths (around 340 nm) in the near UV, a reason might be that the group of Lakowicz reported soon after their first publication that Quin-2 forms a photoproduct with changed calcium affinity (Lakowicz et al. 1994). They introduced a correction procedure for this effect that changed the obtained intracellular calcium concentrations by a factor of ten. This complicated procedure puts it into question, whether trustable absolute values for the intracellular calcium concentration can be obtained from FLIM measurements with Quin-2. Parallel or soon after Quin-2, most of the other prominent calcium-sensitive dyes have been tested for their suitability in FLIM measurements. Fluo-3 (Sanders et al. 1994) and Indo-1 (Szmacinski et al. 1993) show only small changes of the average fluorescence lifetime. Indo-1, in addition, also has complicated photophysics with excitation and emission wavelengths-dependent behavior (Scheenen et al. 1996; Szmacinski et al. 1993). The PET-probes Calcium Crimson (Herman et al. 1997; Lakowicz et al. 1992b), Calcium Orange (Lakowicz et al. 1992b), and especially Calcium Green 1 (CG1; Lakowicz et al. 1992b; Sanders et al. 1994) show a decent increase in the average fluorescence lifetime upon Ca^{2+} binding. Their fluorescence intensity decays can be well described by a global fit to a biexponential or three-exponential function (Gensch and Wirth 2011; Lakowiczs et al. 1992b; Oliver et al. 2000; Sanders et al. 1994; Wilms et al. 2006; Wilms and Eilers 2007). For CG1, the dynamic range is largest, the two with lifetimes of about 0.45 and 3.5 ns are well separated while the ratio of their amplitudes is changing by more than one order of magnitude. (Gensch and Wirth 2011; Lakowicz et al. 1992b; Oliver et al. 2000; Sanders et al. 1994; Wilms et al. 2006). The group of Pansu described the fluorescence intensity decays of CG1 as a linear combination of those in Ca^{2+}-free and Ca^{2+}-bound state (Schoutteten et al. 1999b). They also performed the only intracellular calibration of a calcium-sensitive dye for FLIM—with only limited success (Schoutteten et al. 1999a). Besides adsorption to proteins and pH changes, also the temperature has an influence on the fluorescence lifetimes and the dynamic range of their change (Oliver et al. 2000). In later studies, Calcium Orange (Wilms and Eilers 2007), Oregon Green Bapta 1 (Wilms and Eilers 2007), Oregon Green Bapta-6F (Yoshiki et al. 2005), Fura-2 (Gensch and Wirth 2011; Oliver et al. 2000), Fura-FF (Gensch and Wirth 2011), and Calcium Green 5N (Celli et al. 2010a; CG5N; Yoshiki et al. 2005) were identified as suitable FLIM probes for intracellular calcium sensing.

A number of FLIM measurements have been performed in different biological tissues. Calcium concentrations have been determined in rodent neuronal tissues (Gensch and Wirth 2011; Kuchibhotla et al. 2009; Wilms et al. 2006), rodent and human epidermis (Behne et al. 2011; Celli et al. 2010a,b), and roots of plants (Guo et al. 2009). Wilms et al. (2006) quantified calcium elevations upon electrical stimulation via fluorescence lifetime changes in distal dendrites of Purkinje cells in 200 μm thick brain slices that were filled with OGB1 through a patch pipette. For both fast signals (about 100 ms) and prolonged calcium elevations, basal calcium concentrations as well as the maximum transient calcium concentrations were obtained. Kuchibhotla et al. (2009) investigated in a study dedicated to the functional response of astrocytes in Alzheimer's disease the calcium concentration of astrocytes and neuropils by FLIM through a cranial window 100–150 μm deep in the brain of living mice. The authors demonstrated that the resting Ca^{2+} concentration in astrocytes and the neuropil is elevated in Alzheimer model compared to wild-type mice. Gensch and Wirth (2011) proved that in neurons of the auditory brainstem in chicken embryo brain slices—more specifically in neurons of nucleus magnocellularis and nucleus laminaris— different intracellular calcium concentrations can be found. The neurons were filled with the dye— Calcium Green 1 (CG1) coupled to dextran—via retrograde tracing. Guo et al. (2009) used FLIM (with CG1) for the observation of proton and calcium fluxes as well as pH and cytosolic Ca^{2+} concentrations in root meristem cells of Arabidopsis plants upon salt stress and blocking of Na^+–H^+ antiporters on the timescale of 5 min. Finally, three studies (Behne et al. 2011; Celli et al. 2010a,b) investigating the Ca^{2+} concentrations and gradients in different layers of rodent and human epidermis using CG5N all benefit from applying the phasor approach to visualize Ca^{2+} concentrations in many layers as deep as 120 μm. It was found that most of the Ca^{2+} resists in internal stores rather than in the extracellular space. In addition, the redistribution of Ca^{2+} in the skin after a barrier insult and disruption was documented.

In none of these studies, an intracellular calibration of the fluorescence lifetime of the used calcium-sensitive dye is performed, although absolute values are given with the exception of the study of Gensch and Wirth (2011). While the latter discuss only relative concentrations all other authors rely on calibrations in calcium buffers of known calcium concentration. As a consequence some of the given absolute Ca^{2+} concentrations may be systematically offset. Only the group of Gratton determines the dynamic range (the change of the fluorescence decay upon Ca^{2+} binding) *in vivo* (the epidermis) via the phasor method. In addition they investigate the general, non-specific influence of the presence of proteins on the affinity of the Ca^{2+} dye. Nevertheless, no intra-tissue calibration (intracellular calibration in the natural and alive tissue) with calcium ionophores is presented and may also not be possible due to the role of Ca^{2+} as intra- and extracellular messenger.

24.5.4 CHLORIDE INDICATORS

24.5.4.1 Overview of currently used indicators

Chloride is an important ion with a great impact on the properties of biological fluids. This is true for both intracellular (cytosol as well as subcellular compartments) as well as extracellular (gastric and pancreatic juices, sweat, urine) fluids. Chloride transport across cell membranes is an essential part of many physiological processes like synaptic transmission, pH and cell volume regulation, or transport processes in epithelial tissues. Besides patch clamp and microelectrodes, fluorescence is a very useful method—especially in those physiological questions, where not the performance of a single cell but of a whole tissue is of interest—since it is noninvasive and in combination with multiphoton fluorescence well applicable to living tissue.

The development of useful chloride-sensitive fluorescent probes for biological and physiological investigations was for more than two decades almost entirely carried out in the laboratory of Alan Verkman (University of California, San Francisco; see Verkman 1990, 2009). The same is true for most of the breakthrough applications. Caused by the nonexistence of effective and selective Cl⁻ chelators, his group decided to search among quinoline-based fluorophores, whose fluorescence was known to be quenched by halides. Already in the late 1980s, a group of heterocyclic fluorescent compounds containing a quaternary nitrogen was identified (for a review, see Verkman 1990), synthesized, and applied in cellular studies; among them, the most often used are SPQ (Illsley and Verkman 1987) and MQAE, which have the highest Cl⁻ sensitivity (Verkman et al. 1989). The Cl⁻ sensitivity is caused by efficient collisional quenching and is named physical or dynamic quenching. In this, fluorescent molecules form short-lived (picoseconds) encounter complexes with the Cl⁻ ions of the solvent. In the encounter complex, the photophysical properties of the fluorescent dye are different. In particular, the rate constant for nonradiative deactivation in this encounter complex is orders of magnitudes higher than for the noncomplexed fluorophore causing a reduction of the fluorescence proportional to the Cl⁻ concentration. It was shown that a charge-transfer process is playing a central role for the quenching (Jayaraman and Verkman 2000). Collisional quenching can be quantitatively described by the Stern–Volmer equation that is valid for both the fluorescence intensity and the fluorescence lifetime (see Section 24.2.5). More complicated situations may occur due to occurrence of static quenching or the existence of several forms of the fluorophore of interest in biological measurements (e.g., fluorophore–protein association, dimers or higher oligomers, [partial] chemical transformation by endogenous cellular enzymes, different isomers; see, e.g., Geddes 2001). In the course of time, the Verkman group developed a couple of Cl⁻-sensitive dyes with improved properties for their application like membrane permeability (di-MHQ, Biwersi and Verkman 1991), excitation at longer wavelength (Biwersi et al. 1994; Jayaraman et al. 1999a), or ratiometric fluorescent readout possibility (Jayaraman et al. 1999b). But in most studies, the quinolinium dyes with absorption in the near-UV (310–380 nm) are used. Those are well excitable in two-photon mode with femtosecond- or picosecond-pulsed light from 700 to 780 nm from the widely used Titan:Sapphire lasers. This excitation mode allows deep-tissue studies as well as FLIM based on time-correlated single photon counting. As in all intracellular studies, a critical point for quantitative measurements is the intracellular calibration of the Cl⁻-sensitive fluorescent dyes. Two methods are suggested using either tributiltin (a Cl⁻/OH⁻ antiporter, together with a K⁺/H⁺ antiporter to stabilize the pH; Kaneko et al. 2002; Krapf et al. 1988) or a triterpenoid (β-escin, Waseem et al. 2010).

24.5.4.2 Indicators used for lifetime measurements

For most of the quinolinium dyes like SPQ, MEQ, or MQAE, the fluorescence lifetime has been determined and is very long (up to 25 ns, see, e.g., Verkman et al. 1989) compared to the classical fluorophore families (fluoresceins, rhodamines, oxazines, porphyrines, flavins, or cyanines). However, only one of these dyes (MQAE) has been used in FLIM studies to determine the intracellular Cl⁻ concentration. While the extinction coefficient is rather low, both for one- and two-photon excitation (2800 $cm^{-1}\,M^{-1}$ [350 nm] and 1.05 GM [750 nm]; hence 30 times smaller than those of fluorescein), the fluorescence quantum yield is high and the dynamic range is large (Hille et al. 2009; Mansoura et al. 1999). The fluorescence lifetime in the absence of Cl⁻ is extremely long (25 ns; Kaneko et al. 2002; Verkman et al. 1989). Noteworthy, the intracellular Stern–Volmer constant of MQAE determined in different cell types differs by almost one order of magnitude (see Table 24.1). Altogether, MQAE is a well-suited fluorescent dye for FLIM studies.

Only two research groups have been investigating Cl⁻ concentrations by FLIM in biological tissues so far—in neuronal tissue of rodents (Frings/Gensch) and in insect salivary glands (Dosche/Hille). In all studies, the Cl⁻-sensitive dye MQAE was used, and its largely reduced intracellular Cl⁻ affinity—compared to plain buffer solution—has been quantified (Stern–Volmer constant) allowing the calculation of absolute Cl⁻ concentrations. The neuronal tissues (olfactory epithelium and dorsal root ganglion) contain neuron types (olfactory sensory neurons [OSNs] and nociceptors (DRG neurons), respectively) for which elevated Cl⁻ concentrations have been proposed (based, e.g., on results of electrophysiological experiments). For the dendritic knobs in rat OSNs, values of 54 mM (50 mM extracellular Cl⁻) and 69 mM (150 mM extracellular Cl⁻) have been determined as well as a Cl⁻ gradient from the knob down the dendrite for the high extracellular chloride concentration (Kaneko et al. 2004). The knobs of OSN in mice olfactory epithelium showed slightly lower resting Cl⁻ concentrations (37 mM at 50 mM extracellular Cl⁻). The Cl⁻ concentration in DRG neurons (Gilbert et al. 2007) was very high in dorsal root ganglia from newborn mice (P1–P4, 77 mM) and lower in DRG neurons in DRGs from mice in the third postnatal week (62 mM). These results represent a nice visualization of a physiological phenomenon in neuron maturation named the "chloride switch" (see Figure 24.10). A closer inspection revealed three pools of neurons with high, medium, and low intracellular Cl⁻ concentrations in the third postnatal week. The pool with the low Cl⁻ concentration cannot be found in dorsal root ganglia of mice from P1 to P4. It has to be noted that an intratissue calibration was not possible in this study. Instead, the Stern–Volmer constant was determined in primary cultures of DRG neurons. In a follow-up study, the influence of inflammatory mediators on the Cl⁻ concentration of DRG neurons in intact, vital dorsal root ganglia was investigated (Funk et al. 2008). An increase in a subset of DRG neurons was proven, but no absolute Cl⁻ concentrations are given due to the lack of intratissue calibration of MQAE.

The salivary gland system of insects—here from cockroach—serves as a model system for the study of epithelial ion transport in higher organisms like mammals. Hille et al. (2009) determined a high intracellular Cl⁻ concentration of 59 mM in duct cells of the cockroach salivary glands. In three later reports, the same group observes an increase in intracellular Cl⁻ concentrations when the extracellular Cl⁻ concentrations are raised (Koberling et al. 2010) and determines the intracellular Cl⁻ concentration of a second cell type (acinar peripheral cells, 49 mM) of the salivary gland lobes (Lahn et al. 2011). The authors unwantedly highlighted the need for an intracellular calibration of the indicator dye sensitivity in the cell of interest. They first used the Stern–Volmer constant of MQAE and the fluorescence lifetime at 0 mM Cl⁻ determined in duct cells (reasoning that those cells are from the same tissue and with a very similar function) and calculated a too high intracellular Cl⁻ concentration (Lahn et al. 2010) that was corrected in a later publication (Lahn et al. 2011). Based on the intracellular Cl⁻ concentration determined by FLIM, the Cl⁻ transport in acinar peripheral cells is characterized with respect to blocking of anion transporters and dopamine stimulation and compared to Na⁺ transport phenomena. For the latter, no absolute intracellular Na⁺ concentration could be determined by FLIM (see Section 24.5.1), but only relative concentration changes were observed.

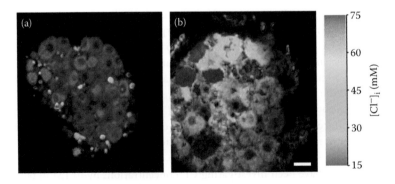

Figure 24.10 Fluorescence lifetime images of MQAE in vital dorsal root ganglia of (a) newborn (day 1–day 4) and (b) (at least) 3 week old mice with two-photon excitation at 750 nm (scale bar: 30 µm).

24.5.5 pH INDICATORS

24.5.5.1 Overview of currently used indicators

The proton is one of the most important ions in the biological context. The properties of all major building blocks of cells (e.g., nucleic acids, proteins, enzymes, lipids) are highly influenced when interacting with protons. Therefore, the pH-value, which is defined as the negative decadic logarithm of the concentration of free protons, is an important parameter for biochemical and cell biological studies. The intracellular pH in most mammalian cells is near to 7.2; in cell organelles, it can vary from 8 in mitochondria to 4–6 in lysosomes and endocytic or synaptic vesicles (Demaurex 2002; Matsuyama et al. 2000; Roos and Boron 1981; Santo-Domingo and Demaurex 2012). Many physiological processes, for example, apoptosis, are accompanied by a transient or permanent change of the intracellular pH. In addition, abnormal intracellular pH values are found in cells when involved in diseases like ischemic stroke, traumatic brain injury, epilepsy, Parkinson's disease, and Alzheimer's disease (Fang et al. 2010) as well as cancer (Webb et al. 2011).

To investigate these effects, it is important to have noninvasive ways to observe the intracellular pH in living cells and biological cell tissues. Since the use of fluorescence in cell biology, a large number of organic fluorescent dyes with pH-dependent fluorescence properties have been suggested, and a couple of them have been established for regular use in cells and tissues (for a comprehensive overview, see Han and Burgess 2010). Although many fluorescent dyes alter their fluorescence properties upon pH change, in today's scientific applications, only two fluorophore families are of importance: the fluoresceins and the benzoxanthenes with the most widely used pH sensors—BCECF and SNARF—belonging to the two families, respectively.

BCECF—introduced in 1982 (Rink et al. 1982)—is almost perfect for measurements in the cytosol of cells. It is sensitive near neutral pH (pK$_a$ 7.0), has a high fluorescence quantum yield, and can be used in ratiometric fluorescence measurements using two excitation and one emission wavelengths. Among the many fluorescein derivatives with similar spectral and pH-dependent properties, one fluorescein class—the halogenated fluoresceins (Oregon Green dyes)—needs to be mentioned. The halogen atoms have an electron-withdrawing effect and lower the pK$_a$ by more than two pH units to below pH 5.

The SNARF (and SNAFL) dyes belong to the benzoxanthenes. Their major advantage is their longer absorption and emission maxima compared to fluorescein. They can be used in the more popular ratiometric fluorescence microscopy method with one excitation and two emission wavelengths. Their pK$_a$'s are also in the neutral pH range; their fluorescence quantum yields, however, are distinctively lower compared to the fluoresceins.

24.5.5.2 Indicators used for lifetime measurements

A number of the available pH-sensitive fluorescent dyes show—apart from spectral and brightness differences—a distinct fluorescence lifetime dependence upon pH change. Most of them have been investigated by time-resolved fluorescence measurements already in the 1990s in the laboratory of J.R. Lakowicz (e.g., Lakowicz and Szmacinski 1993; Szmacinski and Lakowicz 1993; summarized in Lakowicz 2006). Some have large (various SNAFL and SNARF dyes) and others decent (BCECF, Oregon green) changes of the average fluorescence lifetime at different pH in buffer solutions that, in principle, could be exploited for FLIM-based determinations of intracellular pH in cells and cell tissues.

However, the few FLIM studies that exist exclusively use BCECF. This might be due to the fact that the SNARF and SNAFL dyes, in general, have a much lower fluorescence quantum yield compared to BCECF. In addition, SNARF and SNAFL dyes show a threefold to tenfold higher fluorescence quantum yield of acidic vs. basic form or *vice versa*, which is a disadvantage for FLIM measurements since the signal contributions of the two forms are so different. For BCECF, the fluorescence quantum yield is nearly equal for both forms.

Only three FLIM studies determining the pH in living biological tissue are known, one in skin (Hanson et al. 2002), one in artificial skin model (Niesner et al. 2005), and one in salivary glands of cockroach (Hille et al. 2008; Lahn et al. 2010). Here, the intracellular pH of duct (pH 7.1) and peripheral cells (pH 6.9) was estimated (Lahn et al. 2010). The authors themselves question the low value determined for peripheral cells and report problems with the intracellular calibration of BCECF with a highly reduced dynamic range, the absence of a sigmoidal shape of the curve, and a clear pK$_a$. Hanson et al. determined the pH in mice skin as deep as 17 μm, while Niesner et al. investigated cells and intercellular space in an artificial skin construct as deep as 30 μm. In both samples, a low pH (≤pH 6) in the surface layer and near neutral pH values (pH 7–7.2) in the stratum corneum were determined. Both studies fail to determine the fluorescence lifetime dependence of BCECF in the cells but apply corrections for the difference in refractive index in the skin that affects the fluorescence lifetime according to the Strickler–Berg relationship (Strickler and Berg 1962). However, the corrections are not convincing in the sense that a doubtless determination for every cell would be possible.

At the end, another problem has to be pointed out. The fluorescence lifetime change of BCECF in pure buffer is well defined and an easy measurement task. Nevertheless, we find six different titration curves that show, in part, dramatic differences in the dynamic range, absolute lifetime values, and apparent pK$_a$: 3.05–3.9 ns (7.3, Hille et al. 2008/Lahn et al. 2010); 3.2–4.5 ns (7.0, Lakowiczs 2006); 2.8–3.9 ns (6.0, Hanson et al. 2002); 2.8–3.55 ns (7.2, Ohta et al. 2010); 2.5–3.6 ns (6.5, Niesner et al. 2005); and 3.0–3.8 ns (n.d., Ryder et al. 2001). These large differences may have different origins but highlight again the absolute necessity to perform calibrations of the sensors (not only pH sensors but all; both synthetic organic fluorophores as well as fluorescent proteins) in appropriate buffers and in the cells of interest.

24.6 GENETICALLY ENCODED FLUORESCENT ION SENSORS

24.6.1 OVERVIEW

In the 1970s (e.g., Chiu and Haynes 1976; Clements et al. 1971)—somehow paralleled to the development of new and better fluorescence microscopy methods—the research community started to synthesize organic molecules with high fluorescence brightness that change their fluorescence properties in an ion-dependent manner (pH and Ca^{2+} being the major targets) and can be transferred into and survive nondeteriorated in living cells for a decent time (hours). In the early 1990s, despite impressive developments and successful applications in living cells and tissue and after more than two decades of research, important issues were far from being solved or even tackled: The power to control temporally the presence or absence of the fluorescent sensors in living cells and biological tissue as well as the possibility to spatially confine the fluorescent sensors to certain cell types or compartments of the cells. Both problems can be addressed thanks to the introduction of the green fluorescent protein (GFP) from *Aequora victoria* and its many relatives (the family of [auto]fluorescent proteins [FPs]) as a tool in cell biology since 1992–1994 (Chalfie et al. 1993, 1994; Prasher et al. 1992). From that moment, it was possible to produce (under aerobic conditions) fluorescent molecules—the FPs—in basically every cell and to fuse the FP to every protein of interest. In addition, the expression can be limited to specific cell types and started or stopped enabling spatial and temporal control. Fusion of the FPs to proteins or small peptides that are directed to certain cell organelles and cell compartments helps to target the FPs inside the cells. The prospects of using FPs for the construction of sensors were known early on, and the first sensors (pH, Ca^{2+}) were introduced in 1997/1998. Meanwhile, there exist more than 150 FP-based biosensors (Newman et al. 2011); among them are many for pH and Ca^{2+}, some decent ones for Cl$^-$, and only a few or none for the other relevant ions (Na$^+$, K$^+$). Many sensors are made of a single FP and often rely on a change of the protonation equilibrium of the FP. But the majority of the sensors is based on Förster Resonance Energy Transfer (FRET). Here, two FPs with spectra that allow FRET between them are linked together by a sensing element. Often, the FP pairs used in these sensors are mutants of GFP such as cyan FP (CFP) and yellow FP (YFP). Upon the sensing event (e.g., binding of a ligand, phosphorylation, or proteolytic cleavage), the FRET efficiency is modulated due to changes in distance and orientation of the donor and the acceptor. All such biosensors can, in principle, be targeted to certain cell types in a tissue or cell compartments, the latter by fusion to peptides or full proteins. Of course, properties of the sensor (e.g., affinity, dynamic range) may be different in the fusion protein. A thorough characterization of the sensor *in vitro* and *in vivo* is necessary to obtain quantitative results. Sections 24.6.2–24.6.5 give an overview about the genetically encoded sensors for the physiologically most important ions. Special emphasis is put on the suitability of the sensors for use in FLIM applications.

24.6.2 CALCIUM

Miyawaki et al. (1997) reported fluorescent calcium indicators that had been genetically encoded for the measurement of calcium ions in cells. These fluorescent genetically encoded calcium indicators (fGECIs) were collectively called "Cameleons" by the authors. Cameleons are multidomain proteins consisting of two FPs separated by calmodulin (a calcium-binding protein) and a calmodulin-binding peptide called M13. Binding of (four) Ca^{2+} ions to the calmodulin domain causes it to wrap around the M13 domain. This movement, in turn, induces a conformational change in the multidomain cameleon protein. The latter either brings the two GFPs closer together or alters the orientation of the GFPs with respect to each other such that FRET is more likely to occur; thus, there is reduction in the donor fluorescence intensity and an increase in the acceptor fluorescence intensity upon calcium binding (Mank and Griesbeck 2008). By inducing slight changes in the calmodulin motif within the plasmid that encodes for the sensor, Miyawaki et al. were also able to alter the binding efficiencies of their Cameleons to produce a series of Ca^{2+} sensors with a range of $K_{Ca^{2+}}$'s (100 nM to 100 μM). Further alterations in mutations in Cameleons have enabled them to be localized in different cell compartments, and even transgenic animals have been produced that have Cameleon sensors expressed as part of their genome (Hara et al. 2004; Higashijima et al. 2003).

When it was discovered that there are several sites within the GFP that can host alternative protein fragments, without hindering the overall fluorescence of the protein, new single FP-based genetically encoded calcium sensors were created (Baird et al. 1999). This was done by inserting a Ca^{2+}-binding protein, such as calmodulin, directly into the GFP analogue. These forms of fGECIs do not rely on FRET but on a protonation state change of the FP chromophore caused by the binding of Ca^{2+}. The Camgaroos are another example using Citrine (Griesbeck et al. 2001) instead of EYFP (Baird et al. 1999). GCaMPs (Nakai et al. 2001) and Pericams (Nagai et al. 2001) also use these specific sites, where the GFP amino acid chain can be interrupted but the fluorescence is preserved. They consist of calmodulin, the M13 peptide, and a circular permuted GFP, where the original GFP amino acid sequence is interrupted and the two pieces are interchanged. The circular permuted EGFP (GCaMP; Nakai et al. 2001) or EYFP (Pericam; Nagai et al. 2001) is flanked by the M13 binding peptide and calmodulin. Ca^{2+}-dependent binding of calmodulin to the M13 peptide induces conformational changes of the circular permuted GFPs and shifts protonation/deprotonation equilibrium of their chromophores and in this way detects Ca^{2+} concentration changes. As some of these single-FP genetically encoded sensors also exhibit spectral shifts upon Ca^{2+} binding, they can be used as ratiometric sensors (Pericams). Single-FP-based genetically encoded Ca^{2+} sensors have a better signal-to-noise ratio than their FRET-based counterparts, but they do have a downside in that they are often more pH dependent (Demaurex 2005; Mank and Griesbeck 2008). The GCaMP family has been further improved in the past 10 years addressing different aspects including the maturation and cell toxicity as well as the dynamic range. The cell toxicity is a general problem for all fGECIs based on calmodulin and stems, in part, from binding to the many endogenous calmodulin-binding sites.

By replacing YFP with other GFPs, such as circularly permuted Venus, improved Cameleons were created that are less pH dependent and have a better dynamic range (Evanko et al. 2005). The amino acid sequence of calmodulin and the M13 peptide has been optimized to minimize the cell toxicity of the Cameleons (Palmer et al. 2006). Last, there are also FRET-based fGECIs that use the skeletal and cardiac muscle calcium-binding protein troponin C instead and are referred to as TnC sensors (Heim and Griesbeck 2004). They do not show any binding to the endogenous calmodulin-binding sites.

Despite the long development history, still every year, new or improved fGECIs are published since for certain cellular processes the fGECI signals are not yet sufficient (e.g., the reliable detection of single action potentials). Recently very successful were the advanced GCaMP versions from the Looger lab (GCaMP3 [Tian et al. 2009] and GCaMP5 [Akerboom et al. 2012]). For many sensors (in particular, yellow cameleon 3.6 [YC3.6], GCaMP2, and CerTnL15), transgenic animals (*Caenorhabditis elegans*, mice, drosophila) exist. In the past 5 years, viral transfection—especially of the brain—greatly widened the possibilities of using fGECIs in tissue and even living organisms. For example, the observation of neuron activity in living animals during stimulation, chronic imaging of neurons (over periods of weeks), or localized Ca^{2+} elevations in cells are new exciting developments (see, e.g., Gensch and Kaschuba 2012). The combination of fGECIs with optogenetic tools like proteins that change the membrane potential of

Fluorescence lifetime imaging based on exogenous probes

cells or cell signaling in a light-dependent way (Miesenböck 2009; Pastrana 2011) will further expand the application fields of fGECIs.

Despite these developments, FLIM of intracellular Ca^{2+} with fGECIs is rarely applied in tissue, and no prominent publication can be found. Interestingly, the group of Bacskai applied ratiometric imaging of a Cameleon (YC3.6) for the *in vivo* characterization of mouse models of Alzheimer's disease (Kuchibhotla et al. 2008). In a closely related publication, they used indeed FLIM for the estimation of intracellular Ca^{2+} concentrations in astrocytes in the brain; however, they did not use YC3.6 but an organic fluorophore (OGB1, Kuchibhotla et al. 2009) (see also Section 24.5.3.2).

The possible reasons, which will be outlined below, are not limited to fGECIs but also valid for the rest of the genetically encoded sensors.

1. The improvement of fGECIs until now never aimed to improve the dynamic range of the fluorescence lifetime change but other parameters (like intensity change, ratio change, maturation, brightness, affinity, or pH stability). It is a fact that there exists no fGECI with a change of the average fluorescence lifetime larger than 50%; instead most of them are below 10%–20%. Since the physiologically relevant Ca^{2+} changes often cause only a fraction of the full dynamic range, the FLIM signal changes are very low (often only a few percent).

2. For the FRET-based fGECIs, incomplete maturation of donor or acceptor FP is leading to a lower dynamic range but also wrong absolute fluorescence lifetime values.

3. Only a few of the more than 100 FPs show (nearly) monoexponential fluorescence decay with a lifetime of a few nanoseconds. The others have very complicated decays that have to be described as the sum of two, three, or more exponential decay functions.

In particular, ECFP, the donor in many genetically encoded FRET-based biosensors (also in yellow cameleons and the TnC sensors), has a four-exponential decay behavior if the time resolution is good enough (better than 30 ps; Villoing et al. 2008) but definitely has three-exponential decay behavior (Geiger et al. 2012; Villoing et al. 2008). This often prevents an accurate fluorescence decay determination, and often simplifying fitting models (like a two-exponential decay) have to be used. For two yellow cameleons, the time course of donor and acceptor fluorescence was analyzed in detail and could only be described with complicated models (Borst et al. 2008; Habuchi et al. 2002; Laptenok et al. 2012). Visser and coworkers suggest to use the rise term(s) in the acceptor fluorescence (or—even more complicated—the acceptor fluorescence anisotropy) time course (Borst et al. 2008), but this is not practicable in FLIM measurements of fGECIs in cells or biological tissue due to the much lower count rates compared to cuvette measurements.

The so far best-suited sensor for FLIM studies comes from the TnC sensor family (TN-XXL, Mank et al. 2008). In the Ca^{2+}-free form, TN-XXL shows very low FRET and therefore an almost unperturbed donor fluorescence lifetime, while at saturating Ca^{2+} concentrations, the average fluorescence lifetime is halved (Geiger et al. 2012). With biochemical and structural methods, it could be shown that TN-XXL

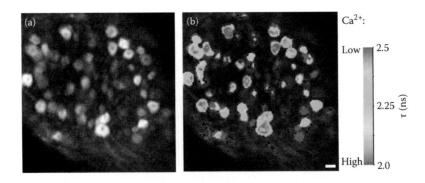

Figure 24.11 Fluorescence intensity (a) and lifetime image (b) of nociceptors in an isolated, vital dorsal root ganglion of a transgenic mice (Thy1-CerTN-L15; Heim et al. 2007) with two-photon excitation (865 nm) and emission detected at wavelengths smaller than 500 nm (scale bar: 40 μm). (From Gensch, T., O. Griesbeck, and K. Funk., Resting concentrations of Ca^{2+} in vital dorsal root ganglion neurons of transgenic Thy1-CerTN-L15 mice measured within two-photon fluorescence lifetime imaging, *unpublished*, 2008.)

undergoes a large conformational change from a flexible elongated, almost linear to a rigid globular shape upon binding of Ca^{2+}. The 50% change in fluorescence lifetime (detectable with a time resolution of about 300 ps, which is also found in most FLIM microscopes) offers the greatest signal-to-noise ratio available, although no *in vivo* data are available. However, with one of the earlier TnC variants (CerTN-L15, i.e., TN-L15 with Cerulean replacing ECFP), it was possible to see differences in the resting concentration of Ca^{2+} in neurons of the dorsal root ganglion—the nociceptors (see Figure 24.11; Gensch et al. 2008). The nociceptors were 0 to 150 µm deep inside isolated, still vital dorsal root ganglia of a transgenic animal expressing CerTN-L15 in many areas of the central nervous system (Heim et al. 2007).

24.6.3 CHLORIDE

The detection of intracellular Cl^- concentrations is not as popular and important compared to Ca^{2+}, but Cl^- efflux or influx as well as Cl^- transport is a key element in many physiological processes. Therefore, FP-based Cl^- sensors have a long history. A recent review gives a nice overview about the different sensors and their applications (Bregestovski and Arosio 2012). Early on in the generation of spectral variants of wt-GFP, a mutant with red-shifted absorption and emission spectra was generated (YFP [S65G/V68L/S72A]; Wachter et al. 1998) that got very important as a partner for FRET with CFP as a donor. It was found that YFP was strongly pH and halide sensitive, properties that affected its performance as a FRET partner. Soon, pH- and halide-insensitive variants of YFP were produced. At the same time, the development of Cl^- sensors had been started (Galietta et al. 2001; Jayaraman et al. 2000; Wachter and Remington 1999). In all those variants, the binding of the Cl^- anion leads to a shift of the protonation equilibrium of the YFP chromophore toward the nonemissive, neutral form. The concentration of the anionic form with absorption around 515 nm is decreased upon Cl^- binding; the sensor gets dimmer. These sensors found great applications, for instance, in screening mutants of Cl^- transporters and channels. The Cl^- binding, however, can be described like a static quenching, and therefore, no influence on the fluorescence lifetime and no FLIM was possible with these sensors.

A FRET-based genetically encoded ratiometric fluorescent sensor for Cl^- has been described by Kuner and Augustine (2001), which they refer to as a "Clomeleon." To make their sensor ratiometric, they fused YFP with another, Cl^--insensitive GFP analogue *CFP* in order to create a FRET pair, whereby a ratio between the FRET-dependent emissions can be analyzed in the presence and absence of Cl^- ions. Some impressive examples of imaging physiological processes in vital brain or retina slices have been reported in neuronal tissue of transgenic "Clomeleon" mice (Berglund et al. 2008; Duebel et al. 2006; Dzhala et al. 2010). The group of Kuner produced many more variants with different linkers or YFPs. While in cultured neurons some fluorescence lifetime contrast was reported, no application in neuronal tissue has been published. Besides the relatively small dynamic range of Clomeleon near physiological pH, the very complicated donor (CFP) fluorescence decay and its pH dependence (Jose et al. 2007; Nair et al. 2006) may be responsible. In 2008, an improved FRET-based Cl^- sensor named "Cl-sensor" has been introduced with a higher Cl^- affinity (Markova et al. 2008). It is much more sensitive at physiological pH and shows good performance in neuronal tissue (Mukhtarov et al. 2008; Waseem et al. 2010). While the dynamic range looks good enough, it still uses CFP as a donor and no FLIM application has been performed.

A problem of all GFP-based Cl^- sensors is caused by the linkage of Cl^- and pH dependence. In fact, Cl^- is not influencing the fluorescence properties of the FP directly but rather its protonation state, which is also affected by the environmental pH. As a consequence, all sensors have high dependence of their Cl^- affinity on pH. For quantitative measurements, the intracellular pH has to be determined independently in every cell to calculate the Cl^- concentration. Arosio and coworkers generated an elegant sensor that may circumvent this problem. They fused a dual-color fluorescent GFP (E^2GFP) that shows a pH-dependent fluorescence (Arosio et al. 2007), where neutral and anionic forms are both fluorescent, to an orange emitting FP whose fluorescence is neither pH nor Cl^- dependent (Arosio et al. 2010). In addition to its pH dependence, the fluorescence of E^2GFP (neutral and anionic) is also quenched by Cl^- by static quenching without changing the protonation state of the E^2GFP chromophore (Arosio et al. 2007). In this way, intracellular pH and intracellular Cl^- concentration can be determined simultaneously. Unfortunately, the static quenching mechanism leads to no fluorescence lifetime contrast.

It has to be concluded that the currently available Cl⁻ sensors are not suitable for Cl⁻ concentration measurements (relative or absolute) based on FLIM in cells or tissue. The FRET-based sensors may be improved to get a larger dynamic range and a less complicated fluorescence decay behavior of the donor allowing FLIM in tissue samples. The pH dependence, however, will remain a problem there.

24.6.4 pH

From a look at the chemical structure of the chromophore of wtGFP from *A. victoria* (Tsien 2009)—many of its mutants and FPs from different species—it seems straightforward to develop pH sensors. The protonation state of the chromophore's phenolic group, which originates from Tyr66, has a great influence on the position of the visible absorption spectrum very similar to the situation in tyrosine. This is also true for many of its mutants and FPs from other species. The absorption maximum of the neutral form is blue-shifted by at least 70 nm compared to that of the anionic form. Therefore, if the local pH around the FP changes, a change in the protonation state can be caused and used as a readout parameter. Interestingly, the first two investigated GFPs from the jellyfish *A. victoria* and the sea pansy *Renilla reniformis* show a large pH stability of their absorption and fluorescence properties near physiological pH (Ward 1981). Many of the GFP mutants, however, indeed show pH transitions suitable for the development of pH sensors that rely on simple intensity (Abad et al. 2004; Kneen et al. 1998; Llopis et al. 1998; Miesenboeck et al. 1998) or ratiometric changes of excitation or emission wavelengths (Arosio et al. 2007; Bizzarri et al. 2006; Hanson et al. 2002; Mahon 2011; Miesenboeck et al. 1998). These sensors have pK_a's from pH 6 to 8 and were targeted early on to cell compartments like organelles or synaptic vesicles demonstrating the advantages of genetically encoded sensors. They have a widespread use in cell biological studies. The fluorescence lifetime of some of them is also significantly pH dependent (Arosio et al. 2007; Bizzarri et al. 2006; Tantama et al. 2011), but only recently, FLIM measurements in cultured mammalian cell lines have been reported (Battisti et al. 2012; Tantama et al. 2011).

Despite these pH sensors based on a single FP, a number of FRET-based pH sensors have been generated (Awaji et al. 2001; Esposito et al. 2008), but their use is limited. The FRET-based sensors have intrinsic disadvantages (larger size that may affect targeting or protein fusions, need for complete maturation of donor and acceptor FP), but the fluorescence decay of the donor will reflect the FRET change. As for Ca²⁺ and Cl⁻, no FLIM studies on intracellular pH in biological tissue using genetically encoded pH sensors are known. Nevertheless, the large dynamic range that can be achieved in FLIM measurements is shown in Figure 24.12.

24.6.5 OTHER IONS AND GENERAL CONSIDERATIONS

For the other ions highly relevant for cells (K⁺, Na⁺), no genetically encoded sensors exist up to now despite their importance. Convincing ideas for such sensors are missing. For Zn²⁺ (Dittmer et al. 2009; Vinkenborg et al. 2009) and Cu²⁺ (Hötzer et al. 2011), relevant ions in some cell types or organs of higher

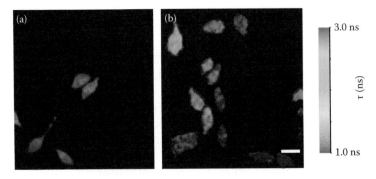

Figure 24.12 Fluorescence lifetime images of HEK293 cells expressing E¹GFP (Serresi et al. 2009) at (a) low pH (average fluorescence lifetime of a biexponential fit) and (b) high pH (monoexponential fit) (10 μM nigericin and buffers of pH 5.0 and pH 8.6) (scale bar: 10 μm). (From Gensch, T., A. Baumann, R. Bizzarri, and A. Franzen, E¹GFP as a probe for intracellular pH measurements with fluorescence lifetime imaging, *unpublished*, 2009.)

organisms, FP-based sensors have been reported that have a fluorescence lifetime contrast suitable for FLIM. But again, no FLIM study in a biological tissue has been reported.

In many ways, genetically encoded fluorescent sensors are better than their synthetic counterparts. Firstly, they can be expressed in cells by a commonplace transfection procedure that requires only a miniscule volume of DNA to be transferred into a cell. Genetically encoded fluorosensors can also be targeted to specific regions within cells, and if stably expressed, new cell lines can be produced that permanently express the fluorescent sensors. As mentioned above, transgenic animals expressing these fluorescent sensors organ- or cell-type specific *in vivo* can also be bred enabling systemic research. However, they do have their downsides too in that many can be more pH dependent than their synthetic counterparts (Evanko et al. 2005; Tsien 1998), their response time is slower, and they exhibit smaller variations in fluorescence upon ion concentration changes compared to their organic fluorophore-based counter parts (Mank and Griesbeck 2008).

In summary, up to the year 2011, there has been no use of genetically encoded ion sensors in biological tissue using FLIM as the readout mode. It seems that although there exist many well-performing genetically encoded sensors and the fact that the noninvasive determination of resting or slowly (minutes) changing ion concentrations of cells in a biological tissue is important in physiological studies, the imperfections of most sensors for FLIM prevents the realization of such studies. The reasons for the imperfections are manifold:

1. Only very few FP-based biosensors (Lee et al. 2009) have been developed for their detection with FLIM; therefore, most of the existing sensors have suboptimal properties.
2. Most sensors use FPs with multiexponential fluorescence decays. Especially in FRET-based sensors, this complicates (often prevents) a quantitative analysis tremendously.
3. The dynamic range of the sensors is often too small for FLIM, since it is less sensitive compared to simple intensity or ratiometric fluorescence readout, because FLIM needs more photons to determine the readout parameter (fluorescence lifetime) compared to the other two readout modes.

Since the genetically encoded sensors are steadily improving and fluorescence imaging in living tissue and animals is getting more and more important—also in the context of optogenetic experiments where the cell-specific noninvasive observation of pH, Cl^-, or Ca^{2+} (transient changes and resting concentrations) is required—it is to be expected that, in the future, genetically encoded ion sensors optimized for FLIM will be generated and find their use in the context of modern physiology investigations.

24.7 CONCLUSIONS

The use of FLIM for imaging ion concentrations in biological tissue is rather limited. As explained in the previous section, the dynamic range and the fluorescence decay properties of genetically encoded ion sensors are simply not good enough yet, with a few exceptions like TN-XXL (Ca^{2+}, Geiger et al. 2012; Mank et al. 2008) and E^{1}GFP (pH, Bizzarri et al. 2006). This explains why no FLIM study in biological tissue is known to the authors that uses genetically encoded ion sensors despite their advantages in tissue measurements (e.g., targeting to the specific cells or cell compartments).

Among the many fluorescent organic molecules with ion sensing capability, there are a number of dyes that show a robust ion-dependent lifetime change with a large dynamic range. Still, not even 20 studies can be found where ion concentrations are determined by FLIM with fluorescent dye molecules in biological tissue. The reasons for that will be discussed below. The rare use of FLIM for ion imaging in biological tissue is surprising since the dye incorporation into cells in biological tissue is simple and well established. One reason might be that fluorescent dyes have a high tendency of interactions with cellular components like proteins, nucleic acids, and small cytosolic molecules affecting their sensor and fluorescence properties. For some fluorescent ion-sensing dyes, large differences of dynamic range (minimal and maximal lifetime) and affinity have been found when comparing measurements in aqueous solution and in cells (e.g., CGN5 [Ca^{2+}], Celli et al. 2010a; MQAE [Cl^-], Kaneko et al. 2002; and BCECF [pH], Lahn et al. 2010). In addition, the time for experiments is limited to a couple of hours after loading at most when working with organic fluorescent ion-sensing dyes. The dyes are prone to degradation in cells leading to the production of nonfluorescent or fluorescent molecules with different (and unknown) fluorescence and ion-sensing properties. Finally, the dye molecules themselves and, in particular, their photoproducts are cell toxic. Therefore, the dye concentration as well as the excitation power and time have to be limited to levels that do not induce physiological reactions (like apoptosis) of the cell or changes of the intracellular properties. Different cells and tissues have specific tolerance levels and no exact quantitative numbers exist. Consequently, only great care of the investigators and through control experiments can prevent an erroneous determination of the ion concentration.

At the end, one has to conclude that ion imaging with FLIM in biological tissue is much less used than one would expect from examining the prospects, possibilities, and ease of technical implementation, especially on two-photon fluorescence microscopes.

ACKNOWLEDGMENTS

T.G. thanks S. Balfanz, A. Franzen, J. Schmidt, P. Thelen, and A. Baumann (ICS-4, Forschungszentrum Jülich, Germany) for their help with molecular biology and cell culture work in the institute. Collaborations with the groups of S. Frings (University of Heidelberg, Germany), M. Wirth (RWTH Aachen, Germany), O. Griesbeck (Max-Planck-Institute of Neurobiology, Munich, Germany), and R. Bizzarri (University of Pisa, Italy) are highly appreciated and have helped in writing this chapter. C.B. greatly appreciates the work of Sascha Dietrich, who did the Sodium Green fluorescence intensity and lifetime measurements shown in Figures 24.8 and 24.9.

REFERENCES

Abad, M.F., G. Di Benedetto, P.J. Magalhaes, L. Filippin, and T. Pozzan. 2004. Mitochondrial pH monitored by a new engineered green fluorescent protein mutant. *J. Biol. Chem.* 279:11521–11529.

Akerboom, J., T.-W. Chen, T.J. Wardill, L. Tian, J.S. Marvin, S. Mutlu et al. 2012. Optimization of a GCaMP calcium indicator for neural activity imaging. *J. Neurosc.*, 32:13819–13840.

Albertazzi, L., D. Arosio, L. Marchetti, F. Ricci, and F. Beltram. 2009. Quantitative FRET analysis with the E0GFP-mCherry fluorescent protein pair. *Photochem. Photobiol.* 85:287–297.

Arosio, D., G. Garau, F. Ricci, L. Marchetti, R. Bizzarri, R. Nifosı, and F. Beltram. 2007. Spectroscopic and structural study of proton and halide ion cooperative binding to GFP. *Biophys. J.* 93:232–244.

Arosio, D., F. Ricci, L. Marchetti, R. Gualdani, L. Albertazzi, and F. Beltram. 2010. Simultaneous intracellular chloride and pH measurements using a GFP-based sensor. *Nat. Methods* 7:516–518.

Ashley, R.H., M.J. Brammer, and R. Marchbanks. 1984. Measurement of intrasynaptosomal free calcium by using the fluorescent indicator Quin-2. *Biochem. J.* 219:149–158.

Awaji, T., A. Hirasawa, H. Shirakawa, G. Tsujimoto, and S. Miyazaki. 2001. Novel green fluorescent protein-based ratiometric indicators for monitoring pH in defined intracellular microdomains. *Biochem. Biophys. Res. Commun.* 289:457–462.

Baird, G.S., D.A. Zacharias, and R.Y. Tsien. 1999. Circular permutation and receptor insertion within green fluorescent proteins. *Proc. Natl. Acad. Sci. USA* 96:11241–11246.

Battisti, A., M.A. Digman, E. Gratton, B. Storti, F. Beltram, and R. Bizzarri. 2012. Intracellular pH measurements made simple by fluorescent protein probe and phasor approach to fluorescence lifetime imaging. *Chem. Comm.* 48:5127–5129.

Behne, M.J., S. Sanchez, N.P. Barry, N. Kirschner, W. Meyer, T.M. Mauro, I. Moll, and E. Gratton. 2011. Major translocation of calcium upon epidermal barrier insult: Imaging and quantification via FLIM/Fourier vector analysis. *Arch. Dermatol. Res.* 303:103–115.

Berglund, K., W. Schleich, H. Wang, G. Feng, W.C. Hall, T. Kuner, and G.J. Augustine. 2008. Imaging synaptic inhibition throughout the brain via genetically targeted Clomeleon. *Brain Cell Biol.* 36:101–118.

Biwersi, J., and A.S. Verkman. 1991. Cell-permeable fluorescent indicator for cytosolic chloride. *Biochemistry* 30:7879–7883.

Biwersi, J., N. Farah, Y.X. Wang, R. Ketcham, and A.S. Verkman. 1992. Synthesis of cell-impermeable Cl⁻-sensitive fluorescent indicators with improved sensitivity and optical properties. *Am. J. Physiol. Cell Physiol.* 262:C243–C250.

Biwersi, J., B. Tulk, and A.S. Verkman. 1994. Long-wavelength chloride-sensitive fluorescent indicators. *Anal. Biochem.* 219:139–143.

Bizzarri, R., C. Arcangeli, D. Arosio, F. Ricci, P. Faraci, F. Cardarelli, and F. Beltram. 2006. Development of a novel GFP-based ratiometric excitation and emission pH indicator for intracellular studies. *Biophys. J.* 90:3300–3014.

Boens, N., W. Qin, N. Basaric, J. Hofkens, M. Ameloot, J. Pouget, J.P. Lefevre, B. Valeur, E. Gratton, M. van de Ven, N.D. Silva Jr., Y. Engelborghs, K., Willaert, A. Sillen, G. Rumbles, D. Phillips, A.J.W.G. Visser, A. van Hoek, J.R. Lakowicz, H. Malak, I. Gryczynski, A.G. Szabo, D.T. Krajcarski, N. Tamai, and A. Miura. 2007. Fluorescence lifetime standards for time and frequency domain fluorescence spectroscopy. *Anal. Chem.* 79:2137–2149.

Borst, J.W., S.P. Laptenok, A.H. Westphal, R. Kuhnemuth, H. Hornen, N.V. Visser, S. Kalinin, J. Aker, A. van Hoek, C.A.M. Seidel, and A.J.W.G. Visser. 2008. Structural changes of yellow cameleon domains observed by quantitative FRET analysis and polarized fluorescence correlation spectroscopy. *Biophys. J.* 95:5399–5411.

Brain, K.L., and M.R. Bennett. 1997. Calcium in sympathetic varicosities of mouse vas deferens during facilitation, augmentation and autoinhibition. *J. Physiol.* 502:521–536.

Bregestovski, P., and D. Arosio. 2012. Green fluorescent protein-based chloride ion sensors for in vivo imaging. In G. Jung (ed.), *Fluorescent Proteins II, Springer Ser. Fluoresc.*, 12:99–124. Berlin, Heidelberg: Springer-Verlag.

Celli, A., S. Sanchez, M. Behne, T. Hazlett, E. Gratton, and T. Mauro. 2010a. The epidermal Ca^{2+} gradient: Measurement using the phasor representation of fluorescent lifetime imaging. *Biophys. J.* 98:911–921.

Celli, A., D.S. Mackenzie, D.S. Crumrine, C.L. Tu, M. Hupe, D.D. Bikle, P.M. Elias, and T.M. Mauro. 2010b. Endoplasmic reticulum Ca^{2+} depletion activates XBP1 and controls terminal differentiation in keratinocytes and epidermis. *Brit. J. Dermatol.* 164:16–25.

Chalfie, M., Y. Tu, and D.C. Prasher. 1993. Glow worms—A new method of looking at C. elegans gene expression. *Worm Breed Gaz* 13:19.

Chalfie, M., Y. Tu, G. Euskirchen, W.W. Ward, and D.C. Prasher. 1994. Green fluorescent protein as a marker for gene expression. *Science* 263:802–805.

Chambers, R.S., D. Ames, A. Bergsma, L.R. Muir, G. Fitzgerald, G.M. Hervieu, J.J. Dytko, J. Foley, W.S. Martin, J. Liu, C. Park, S. Ellis, S. Ganguly, R. Cluderay, S. Leslie, S. Wilson, and H.M. Sarau. 1999. Melanin-concentrating hormone is the cognate ligand for the orphan G-protein coupled receptor SLC-1. *Nature* 400:261–265.

Chiu, V.C.K., and D.H. Haynes. 1976. High and low affinity Ca^{2+} binding to the sarcoplasmatic reticulum use of a high-affinity fluorescent calcium indicator. *Biophys. J.* 18:3–22.

Clements, R.L., J.I. Read, and G.A. Sergeant. 1971. 4-[Bis(carboxymethyl)aminomethyl]-3-hydroxy-2-naphthoic acid as a fluorescent indicator for the complexometric titration of calcium plus magnesium. *Analyst* 96:656–658.

Cotlet, M., J. Hofkens, M. Maus, T. Gensch, M. van der Auweraer, J. Michiels, G. Dirix, M. van Guyse, J. Vanderleyden, A.J.W.G. Visser, and F.C. de Schryver. 2001. Excited-state dynamics in the enhanced green fluorescent protein mutant probed by picosecond time-resolved single photon counting spectroscopy. *J. Phys. Chem. B* 105:4999–5006.

Crossley, R., Z. Goolamali, and P.G. Sammes. 1994. Synthesis and properties of a potential extracellular fluorescent probe for potassium. *J. Chem. Soc. Perkin Trans.* 2:1615–1623.

De Silva, A., H. Gunaratne, T. Gunnlaugsson, A. Huxley, C. McCoy, J. Rademacher, and T. Rice. 1997. Signaling recognition events with fluorescent sensors and switches. *Chem. Rev.* 97:1515–1566.

Demaurex, N. 2002. pH homeostasis of cellular organelles. *News Physiol. Rev.* 17:1–5.

Demaurex, N. 2005. Calcium measurements in organelles with Ca^{2+}-sensitive fluorescent proteins. *Cell Calcium* 38:213–222.

Despa, S., J. Vecer, P. Steels, and M. Ameloot. 2000. Fluorescence lifetime microscopy of the Na^+ indicator Sodium Green in HeLa cells. *Anal. Biochem.* 281:159–175.

Dietrich, S., S.E. Stanca, C.G. Cranfield, B. Hoffmann, K. Benndorf, and C. Biskup. 2010. New strategies to measure intracellular sodium concentrations. *SPIE Proc.* 7569:14.

Digman, M.A., V.R. Caiolfa, M. Zamai, and E. Gratton. 2008. The phasor approach to fluorescence lifetime imaging analysis. *Biophys. J.* 93:L14–L16.

Dittmer, P.J., J.G. Miranda, J.A. Gorski, and A.E. Palmer. 2009. Genetically encoded sensors to elucidate spatial distribution of cellular zinc. *J. Biol. Chem.* 284:16289–16297.

Duebel, J., S. Haverkamp, W. Schleich, G. Feng, G.J. Augustine, T. Kuner, and T. Euler. 2006. Two-photon imaging reveals somatodendritic chloride gradient in retinal ON-type bipolar cells expressing the biosensor Clomeleon. *Neuron* 49:81–94.

Duffy, S., and B.A. MacVicar. 1995. Adrenergic calcium signaling in astrocyte networks within the hippocampal slice. *J. Neurosci.* 15:5535–5550.

Dzhala, V.I., K.V. Kuchibhotla, J.C. Glykys, K.T. Kahle, W.B. Swiercz, G. Feng, T. Kuner, G.J. Augustine, B.J. Bacskai, and K.J. Staley. 2010. Progressive NKCC1-dependent neuronal chloride accumulation during neonatal seizures. *J. Neurosci.* 30:11745–11761.

Eberhard, M., and P. Erne. 1991. Calcium binding to fluorescent calcium indicators: Calcium green, calcium orange and calcium crimson. *Biochem. Biophys. Res. Comm.* 180:209–215.

Egawa, T., K. Hanaoka, Y. Koide, S. Ujita, N. Takahashi, Y. Ikegaya, N. Matsuki, T. Terai, T. Ueno, T. Komatsu, and T. Nagano. 2011. Development of a far-red to near-infrared fluorescence probe for calcium ion and its application to multicolor neuronal imaging. *J. Am. Chem. Soc.* 133:14157–14159.

Esposito, A., M. Gralle, M.A.C. Dani, D. Lange, and F.S. Wouters. 2008. pHlameleons: A family of FRET-based protein sensors for quantitative pH imaging. *Biochemistry* 47:13115–13126.

Evanko, D.S., and P.G. Haydon. 2005. Elimination of environmental sensitivity in a cameleon FRET-based calcium sensor via replacement of the acceptor with Venus. *Cell Calcium* 37:341–348.

Fang, B., D. Wang, M. Huang, G. Yu, and H. Li. 2010. Hypothesis on the relationship between the change in intracellular pH and incidence of sporadic Alzheimer's disease or vascular dementia. *Int. J. Neurosc.* 120:591–595.

Förster, T. 1946. Energiewanderung und Fluoreszenz. *Naturwissenschaften* 33:166–175.

Funk, K., A. Woitecki, C. Franjic-Würtz, T. Gensch, F. Möhrlen, and S. Frings. 2008. Modulation of chloride homeostasis by inflammatory mediators in dorsal root ganglion neurons. *Mol. Pain* 4:32.

Galietta, L.J., P.M. Haggie, and A.S. Verkman. 2001. Green fluorescent protein-based halide indicators with improved chloride and iodide affinities. *FEBS Lett.* 499:220–224.

Geddes, C.D. 2001. Optical halide sensing using fluorescence quenching: Theory, simulations and applications—A review. *Meas. Sci. Technol.* 12:R53–R88.

Gee, K.R., K.A. Brown, W.N. Chen, J. Bishop-Stewart, D. Gray, and I. Johnson. 2000. Chemical and physiological characterization of fluo-4 Ca^{2+}-indicator dyes. *Cell Calcium* 27:97–106.

Geiger, A., L. Russo, T. Gensch, T. Thestrup, S. Becker, K.-P. Hopfner, C. Griesinger, G. Witte, and O. Griesbeck. 2012. Correlating calcium binding, FRET and conformational change in the biosensor TN-XXL. *Biophys. J.* 102:2401–2410.

Gensch, T., O. Griesbeck, and K. Funk. 2008. Resting concentrations of Ca^{2+} in vital dorsal root ganglion neurons of transgenic Thy1-CerTN-L15 mice measured with in two-photon fluorescence lifetime imaging. *unpublished*.

Gensch, T., A. Baumann, R. Bizzarri, and A. Franzen. 2009. E^{1}GFP as a probe for intracellular pH measurements with fluorescence lifetime imaging. *unpublished*.

Gensch, T., and M. Wirth. 2011. Determination of calcium concentrations in cells and tissue with fluorescence lifetime imaging (FLIM). In *Multiphoton Microscopy in the Biomedical Sciences XI, Proceedings of SPIE* 7903:790322.

Gensch, T., and D. Kaschuba. 2012. Fluorescent genetically encoded calcium indicators and their in vivo application. In G. Jung (ed.), *Fluorescent Proteins II, Springer Series on Fluorescence* 12:125–162.

Gilbert, D., C. Franjic-Wuertz, K. Funk, T. Gensch, S. Frings, and F. Moehrlen. 2007. Differential maturation of chloride homeostasis in primary afferent neurons of the somatosensory system. *Int. J. Dev. Neurosci.* 25:479–489.

Gokel, G. 1991. *Crown Ethers and Cryptands*. Cambridge: Royal Society of Chemistry.

Gratton, E., and M.J. vande Ven. 2006. Laser sources for confocal microscopy. In J.B. Pawley (ed.), *Handbook of Biological Confocal Microscopy*, 80–118. New York: Springer.

Griesbeck, O., G. Baird, R. Campbell, D. Zacharias, and R.Y. Tsien. 2001. Reducing the environmental sensitivity of yellow fluorescent protein. *J. Biol. Chem.* 276:29188–29194.

Grynkiewicz, G., M. Poenie, and R.Y. Tsien. 1985. A new generation of Ca^{2+} indicators with greatly improved fluorescence properties. *J. Biol. Chem.* 260:3440–3450.

Guo, K.-M., O. Babourinaa, and Z. Rengel. 2009. Na$^+$/H$^+$ antiporter activity of the *SOS1* gene: Lifetime imaging analysis and electrophysiological studies on Arabidopsis seedlings. *Physiol. Plant.* 137:155–165.

Gupta, R.K., P. Gupta, and R.D. Moore. 1984. NMR studies of intracellular metal ions in intact cells and tissues. *Annu. Rev. Biophys. Bioeng.* 13:221–246.

Habuchi, S., M. Cotlet, J. Hofkens, G. Dirix, J. Michiels, J. Vanderleyden, V. Subramaniam, and F.C. De Schryver. 2002. Resonance energy transfer in a calcium concentration-dependent cameleon protein. *Biophys. J.* 83:3499–3506.

Hall, T.A. 1979. Biological X-ray microanalysis. *J. Microsc.* 117:145–163.

Han, J., and K. Burgess. 2010. Fluorescent indicators for intracellular pH. *Chem. Rev.* 110:2709–2728.

Hanson, G.T., T.B. McAnaney, E.S. Park, M.E. Rendell, D.K. Yarbrough, S. Chu, L. Xi, S.G. Boxer, M.H. Montrose, and S.J. Remington. 2002. Green fluorescent protein variants as ratiometric dual emission pH sensors. 1. Structural characterization and preliminary application. *Biochemistry* 41:15477–15488.

Hanson, K.M., M.J. Behne, N.P. Barry, T.M. Mauro, E. Gratton, and R.M. Clegg. 2002. Two-photon fluorescence lifetime imaging of the skin stratum corneum pH gradient. *Biophys. J.* 83:1682–1690.

Hara, M., V. Bindokas, J.P. Lopez, K. Kaihara, L.R. Landa Jr., M. Harbeck, and M.W. Roe. 2004. Imaging endoplasmic reticulum calcium with a fluorescent biosensor in transgenic mice. *Am. J. Physiol. Cell Physiol.* 287:C932–C938.

Harootunian, A.T., J.P.Y. Kao, B.K. Eckert, and R.Y. Tsien. 1989. Fluorescence ratio imaging of cytosolic free Na$^+$ in individual fibroblasts and lymphocytes. *J. Biol. Chem.* 264:19458–19467.

Hediger, M.A., M.F. Romero, J.B. Peng, A. Rolfs, H. Takanaga, and E.A. Bruford. 2004. The ABCs of solute carriers: Physiological, pathological and therapeutic implications of human membrane transport proteins. *Pflug. Arch. Eur. J. Phys.* 447:465–468.

Heim, N., and O. Griesbeck. 2004. Genetically encoded indicators of cellular calcium dynamics based on troponin C and green fluorescent protein. *J. Biol. Chem.* 279:14280–14286.

Heim, N., O. Garaschuk, M.W. Friedrich, M. Mank, R.I. Milos, Y. Kovalchuk, A. Konnerth, and O. Griesbeck. 2007. Improved calcium imaging in transgenic mice expressing a troponin C-based biosensor. *Nat. Methods* 4:127–129.

Herman, B., P. Wodnicki, S. Kwon, A. Periasamy, G.W. Gordon, N. Mahajan, and W.X. Feng. 1997. Recent developments in monitoring calcium and protein interactions in cells using fluorescence lifetime microscopy. *J. Fluoresc.* 7:85–92.

Higashijima, S.-I., M.A. Masino, G. Mandel, and J.R. Fetcho. 2003. Imaging neuronal activity during zebrafish behavior with a genetically encoded calcium indicator. *J. Neurophysiol.* 90:3986–3997.

Hille, B. 2001. *Ion Channels of Excitable Membranes.* Sunderland, MA: Sinauer Associates.

Hille, C., M. Berg, L. Bressel, D. Munzke, P. Primus, H.-G. Löhmannsröben, and C. Dosche. 2008. Time-domain fluorescence lifetime imaging for intracellular pH sensing in living tissues. *Anal. Bioanal. Chem.* 391:1871–1879.

Hille, C., M. Lahn, G.-H. Loehmannsroeben, and C. Dosche. 2009. Two-photon fluorescence lifetime imaging of intracellular chloride in cockroach salivary glands. *Photochem. Photobiol. Sci.* 8:319–327.

Hinde, E., M.A. Digman, C. Welch, K.M. Hahn, and E. Gratton. 2012. Biosensor Förster resonance energy transfer detection by the phasor approach to fluorescence lifetime imaging microscopy. *Microsc. Res. Technol.* 75:271–281.

Hötzer, B., R. Ivanov, S. Altmeier, R. Kappl, and G. Jung. 2011. Determination of copper(II) ion concentration by lifetime measurements of green fluorescent protein. *J. Fluoresc.* 2143–2153.

Illsley, N.P., and A.S. Verkman. 1987. Membrane chloride transport measured using a chloride-sensitive fluorescent probe. *Biochemistry* 26:1215–1219.

Jackson, K.W., and T.M. Mahmood. 1994. Atomic absorption, atomic emission and flame emission spectrometry. *Anal. Chem.* 66:252R–279R.

James, N.G., J.A. Ross, M. Štefl, and D.M. Jameson. 2011. Applications of phasor plots to in vitro protein studies. *Anal. Biochem.* 410:62–69.

Jameson, D.M., E. Gratton, and R. Hall. 1984. The measurement and analysis of heterogeneous emissions by multifrequency phase and modulation fluorometry. *Appl. Spectrosc. Rev.* 20:55–106.

Janata, J., M. Josowicz, and D.M. DeVaney. 1994. Chemical sensors. *Anal. Chem.* 66:207R–228R.

Janata, J., M. Josowicz, P. Vanysek, and D.M. DeVaney. 1998. Chemical sensors. *Anal. Chem.* 70:179R–208R.

Jares-Erijman, E.A., and T.M. Jovin. 2003. FRET imaging. *Nat. Biotechnol.* 21:1387–1395.

Jayaraman, S., L. Teitler, B. Skalski, and A.S. Verkman. 1999a. Long-wavelength iodide-sensitive fluorescent indicators for measurement of functional CFTR expression in cells *Am. J. Phys. Cell Phys.* 277:C1008–C1018.

Jayaraman, S., J. Biwersi, and A.S. Verkman. 1999b. Synthesis and characterization of dual-wavelength Cl⁻-sensitive fluorescent indicators for ratio imaging. *Am. J. Physiol. Cell Physiol.* 276:C747–C757.

Jayaraman, S., and A.S. Verkman. 2000. Quenching mechanism of quinolinium type chloride-sensitive fluorescent indicators. *Biophys. Chem.* 85:49–57.

Jayaraman, S., P. Haggie, R.M. Wachter, S.J. Remington, and A.S. Verkman. 2000. Mechanism and cellular applications of a green fluorescent protein-based halide sensor. *J. Biol. Chem.* 275:6047–6050.

Jezek, P., F. Mahdi, and K.D. Garlid. 1990. Reconstitution of the beef heart and rat liver mitochondrial K⁺/H⁺ (Na⁺/H⁺) Antiporter. *J. Biol. Chem.* 265:10522–10526.

Johnson, I., and M.T.Z. Spence. 2010. *The Molecular Probes Handbook. A Guide to Fluorescent Probes and Labeling Technologies,* 11th ed. Eugene, OR: Molecular Probes.

Jose, M., D.K. Nair, C. Reissner, R. Hartig, and W. Zuschratter. 2007. Photophysics of clomeleon by FLIM: Discriminating excited state reactions along neuronal development. *Biophys. J.* 92:2237–2254.

Kaneko, H., I. Putzier, S. Frings, and T. Gensch. 2002. Determination of intracellular chloride concentration in dorsal root ganglion neurons by fluorescence lifetime imaging. In Fuller C.M. (Ed.), *Calcium-Activated Chloride Channels, Book Series: Current Topics in Membranes,* 53:167–194.

Kaneko, H., I. Putzier, S. Frings, U.B. Kaupp, and T. Gensch. 2004. Chloride accumulation in mammalian olfactory sensory neurons. *J. Neurosci.* 24:7931–7938.

Kao, J.P., A.T. Harootunian, and R.Y. Tsien. 1989. Photochemically generated cytosolic calcium pulses and their detection by fluo-3. *J. Biol. Chem.* 264:8179–8184.

Kasner, S.E., and M.B. Ganz. 1992. Regulation of intracellular potassium in mesangial cells: A fluorescence analysis using the dye, PBFI. *Am. J. Physiol. Ren. Physiol.* 262:F462–F467.

Kneen, M., J. Farinas, Y. Li, and A.S. Verkman. 1998. Green fluorescent protein as a noninvasive intracellular pH indicator. *Biophys. J.* 74:1591–1599.

Koberling, F., V. Buschmann, C. Hille, M. Patting, C. Dosche, A. Sandberg, A. Wheelock, and R. Erdmann. 2010. Fast raster scanning enables FLIM in macroscopic samples up to several centimeters. In *Multiphoton Microscopy in the Biomedical Sciences X, Proceedings of SPIE* 7569:756931.

Krapf, R., C.A. Berry, and A.S. Verkman. 1988. Estimation of intracellular chloride activity in isolated perfused rabbit proximal convoluted tubules using a fluorescent indicator. *Biophys. J.* 53:955–962.

Kuchibhotla, K.V., S.T. Goldman, C.R. Lattarulo, H.-Y. Wu, B.T. Hyman, and B.J. Bacskai. 2008. Aβ plaques lead to aberrant regulation of calcium homeostasis in vivo resulting in structural and functional disruption of neuronal networks. *Neuron* 59:214–225.

Kuchibhotla, K.V., C.R. Lattarulo, B.T. Hyman, and B.J. Bacskai. 2009. Synchronous hyperactivity and intercellular calcium waves in astrocytes in Alzheimer mice. *Science* 323:1211–1215.

Kuhn, M., and R.P. Haugland. 1995. U.S. Patent 5 405 975.

Kuner, T., and G.J. Augustine. 2000. A genetically encoded ratiometric indicator for chloride: Capturing chloride transients in cultured hippocampal neurons. *Neuron* 27:447–459.

Lahn, M., C. Hille, F. Koberling, P. Kapusta, and C. Dosche. 2010. pH and chloride recordings in living cells using two-photon fluorescence lifetime imaging microscopy. In *Multiphoton Microscopy in the Biomedical Sciences X, Proceedings of SPIE* 7569:75690U.

Lahn, M., C. Dosche, and C. Hille. 2011. Two-photon microscopy and fluorescence lifetime imaging reveal stimulus induced intracellular Na^+ and Cl^- changes in cockroach salivary acinar cells. *Am. J. Physiol. Cell Physiol.* 300:C1323–C1336.

Lakowicz, J.R., H. Szmacinsky, K. Nowaczyk, and M.L. Johnson. 1992a. Fluorescence lifetime imaging of calcium using Quin-2. *Cell Calcium* 13:131–147.

Lakowicz, J.R., H. Szmacinsky, and M.L. Johnson. 1992b. Calcium imaging of using fluorescence lifetimes and long-wavelength probes. *J. Fluoresc.* 2:47–62.

Lakowicz, J.R., and H. Szmacinski. 1993. Fluorescence lifetime-based sensing of pH, Ca^{2+}, K^+ and glucose. *Sens. Actuators B* 11:133–143.

Lakowicz, J.R., H. Szmacinsky, K. Nowaczyk, W.J. Lederer, and M.L. Johnson. 1994. Fluorescence lifetime imaging of intracellular calcium in COS cells using Quin-2. *Cell Calcium* 15:7–27.

Lakowicz, J.R. 2006. *Principles of Fluorescence Spectroscopy.* New York: Springer.

Laptenok, S.P., I.H.M. van Stokkum, J.W. Borst, B. van Oort, A.J.W.G. Visser, and H. van Amerongen. 2012. Disentangling picosecond events that complicate the quantitative use of the calcium sensor YC3.60. *J. Phys. Chem. B* 3013–3020.

Larsson, D., B. Larsson, T. Lundgren, and K. Sundell. 1999. The effect of pH and temperature on the dissociation constant for fura-2 and their effects on $[Ca2^+](i)$ in enterocytes from a poikilothermic animal, Atlantic cod (*Gadus morhua*). *Anal. Biochem.* 273:60–65.

Lattanzio, F.A. Jr. 1990. The effects of pH and temperature on fluorescent calcium indicators as determined with Chelex-100 and EDTA buffer systems. *Biochem. Biophys. Res. Commun.* 171:102–108.

Lee, S.-J.R., Y. Escobedo-Lozoya, E.M. Szatmari, and R. Yasuda. 2009. Activation of CaMKII in single dendritic spines during long-term potentiation. *Nature* 458:299–304.

Llopis, J., J.M. McCaffery, A. Miyawaki, M.G. Farquhar, and R.Y. Tsien. 1998. Measurement of cytosolic, mitochondrial, and Golgi pH in single living cells with green fluorescent proteins. *Proc. Natl. Acad. Sci. USA* 95:6803–6808.

Lodish, H., A. Berk, C.A. Kaiser, M. Krieger, A. Bretscher, H. Ploegh, A. Amon, and M.P. Scott. 2013. *Molecular Cell Biology*, 7th ed. New York: W.H. Freeman & Co.

Mahon, M.J. 2011. PHluorin2: An enhanced, ratiometric, pH-sensitive green fluorescent protein. *Adv. Biosci. Biotechnol.* 2:132–137.

Mank, M., and O. Griesbeck. 2008. Genetically encoded calcium indicators. *Chem. Rev.* 108:1550–1564.

Mank, M., A.F. Santos, S. Direnberger, T.D. Mrsic-Flogel, S.B. Hofer, V. Stein, T. Hendel, D.F. Reiff, C. Levelt, A. Borst, T. Bonhoeffer, M. Hubener, and O. Griesbeck. 2008. A genetically encoded calcium indicator for chronic in vivo two-photon imaging. *Nat. Methods* 5:805–811.

Mansoura, M.K., J. Biwersi, M.A. Ashlock, and A.S. Verkman. 1999. Fluorescent chloride indicators to assess the efficacy of CFTR cDNA delivery. *Hum. Gene Ther.* 10:861–875.

Markova, O., M. Mukhtarov, E. Real, Y. Jacob, and P. Bregestovski. 2008. Genetically encoded chloride indicator with improved sensitivity. *J. Neurosci. Methods* 170:67–76.

Masilamani, D., M.E. Lucas, and G.S. Hammond. 1992. U.S. Patent 5 162 525.

Matsui, A., K. Umezawa, Y. Shindo, T. Fujii, D. Citterio, K. Oka, and K. Suzuki. 2011. A near-infrared fluorescent calcium probe: A new tool for intracellular multicolour Ca^{2+} imaging. *Chem. Commun.* 47:10407–10409.

Matsuyama, S., J. Llopis, Q.L. Deveraux, R.Y. Tsien, and J.C. Reed. 2000. Changes in intramitochondrial and cytosolic pH: Early events that modulate caspase activation during apoptosis. *Nat. Cell Biol.* 2:318–325.

Meuwis, K., N. Boens, F.C. De Schryver, J. Gallay, and M. Vincent. 1995. Photophysics of the fluorescent K^+ indicator PBFI. *Biophys J.* 68:2469–2473.

Miesenböck, G., D.A. De Angelis, and J.E. Rothman. 1998. Visualizing secretion and synaptic transmission with pH-sensitive green fluorescent proteins. *Nature* 394:192–196.

Miesenböck, G. 2009. The optogenetic catechism. *Science* 326:395–399.

Minta, A., and R.Y. Tsien. 1989. Fluorescent indicators for cytosolic sodium. *J. Biol. Chem.* 264:19449–19457.

Minta, A., J.P. Kao, and R.Y. Tsien. 1989. Fluorescent indicators for cytosolic calcium based on rhodamine and fluorescein chromophores. *J. Biol. Chem.* 264:8171–8178.

Miriel, V.A., J.R. Mauban, M.P. Blaustein, and W.G. Wier. 1999. Local and cellular Ca^{2+} transients in smooth muscle of pressurized rat resistance arteries during myogenic and agonist stimulation. *J. Physiol.* 518:815–824.

Miyawaki, A., J. Llopis, R. Heim, J.M. McCaffery, J.A. Adams, M. Ikura, and R.Y. Tsien. 1997. Fluorescent indicators for Ca^{2+} based on green fluorescent proteins and calmodulin. *Nature* 388:882–887.

Mukhtarov, M., O. Markova, E. Real, Y. Jacob, S. Buldakova, and P. Bregestovski. 2008. Monitoring of chloride and activity of glycine receptor channels using genetically encoded fluorescent sensors. *Phil. Trans. A Math. Phys. Eng. Sci.* 366:3445–3462.

Nagai, T., A. Sawano, E.S. Park, and A. Miyawaki. 2001. Circularly permuted green fluorescent proteins engineered to sense Ca^{2+}. *Proc. Natl. Acad. Sci. USA* 98:3197–3202.

Nair, D.K., M. Jose, T. Kuner, W. Zuschratter, and R. Hartig. 2006. FRET-FLIM at nanometer spectral resolution from living cells. *Opt. Express* 14:12217–12229.

Nakai, J., M. Ohkura, and K. Imoto. 2001. A high signal-to-noise Ca^{2+} probe composed of a single green fluorescent protein. *Nat. Biotechnol.* 19:137–141.

Newman, R.H., M.D. Fosbrink, and J. Zhang. 2011. Genetically encodable fluorescent biosensors for tracking signaling dynamics in living cells. *Chem. Rev.* 111:3614–3666.

Niesner, R., B. Peker, P. Schlüsche, K.H. Gericke, C. Hoffmann, D. Hahne, and C. Müller-Goymann. 2005. 3D-resolved investigation of the pH gradient in artificial skin constructs by means of fluorescence lifetime imaging. *Pharm. Res.* 22:1079–1087.

Nolte, A., J.B. Pawley, and L. Horing. 2006. Non-laser light sources for three-dimensional microscopy. In J.B. Pawley (ed.), *Handbook of Biological Confocal Microscopy*, 126–144. New York: Springer.

Ohta, N., T. Nakabayashi, S. Oshita, and M. Kinjo. 2010. Fluorescence lifetime imaging spectroscopy in living cells with particular regards to pH dependence and electric field effect. In *Reporters, Markers, Dyes, Nanoparticles, and Molecular Probes, Proceedings of SPIE* 7576:0G1–0G13.

Oliver, A.E., G.A. Baker, R.D. Fugate, F. Tablin, and J.H. Crowe. 2000. Effects of temperature on calcium-sensitive fluorescent probes. *Biophys. J.* 78:2166–2176.

Orlowski, J., and S. Grinstein. 2004. Diversity of the mammalian sodium/proton exchanger SLC9 gene family. *Pflug. Arch. Eur. J. Phys.* 447:549–565.

Ozmen, B., and E.U. Akkaya. 2000. Infrared fluorescence sensing of submicromolar calcium: Pushing the limits of photoinduced electron transfer. *Tetrahedron Lett.* 41:9185–9188.

Palmer, A.E., M. Giacomello, T. Kortemme, S.A. Hires, V. Lev-Ram, D. Baker, and R.Y. Tsien. 2006. Ca^{2+} indicators based on computationally redesigned calmodulin-peptide pairs. *Chem. Biol.* 13:521–530.

Paredes, R.M., J.C. Etzler, L.T. Watts, W. Zheng, and J.D. Lechleiter. 2008. Chemical calcium indicators. *Methods* 46:143–151.

Pastrana, E. 2011. Optogenetics: Controlling cell function with light. *Nat. Methods* 8:24–25.

Philipson, K.D., D.A. Nicoll, M. Ottolia, B.D. Quednau, H. Reuter, S. John, and Z. Qiu. 2002. The Na^+/Ca^{2+} exchange molecule. *Ann. N.Y. Acad. Sci.* 976:1–10.

Pieske, B., S.R. Houser, G. Hasenfuss, and D.M. Bers. 2003. Sodium and the heart: A hidden key factor in cardiac regulation. *Cardiovasc. Res.* 57:871–872.

Prasher, D.C., V.E. Eckenrode, W.W. Ward, F.G. Prendergast, and M.J. Cormier. 1992. Primary structure of the Aequorea victoria green-fluorescent protein. *Gene* 111:229–233.

Rink, T.J., R.Y. Tsien, and T. Pozzan. 1982. Cytoplasmic pH and free Mg2+ in lymphocytes. *J. Cell Biol.* 95:189–196.

Rogers, J., T.R. Hesketh, G.A. Smith, M.A. Beavent, J.C. Metcalfe, P. Johnson, and P.B. Garland. 1983. Intracellular pH and free calcium changes in single cells using Quene-1 and Quin-2 probes and fluorescence microscopy. *FEBS Lett.* 161:21–27.

Roos, A., and W.F. Boron. 1981. Intracellular pH. *Physiol. Rev.* 61:296–434.

Russell, J.T. 2011. Imaging calcium signals *in vivo*: A powerful tool in physiology and pharmacology. *Br. J. Pharmacol.* 163:1605–1625.

Ryder, A.G., S. Power, T.J. Glynn, and J.J. Morrison. 2001. Time-domain measurements of fluorescence lifetime variation with pH. Biomarkers and biological spectral imaging. *Proc. SPIE* 4259:102–109.

Sanchez, S., L. Baka, E. Gratton, and V. Herlax. 2011. Alpha hemolysin induces an increase of erythrocytes calcium: A FLIM 2-photon phasor analysis approach. *PLoS One* 6:e21127.

Sanders, R., H.C. Gerritsen, A. Draaijer, P.M. Houpt, and Y.K. Levine. 1994. Fluorescence lifetime imaging of free calcium in single cells. *Bioimaging* 2:131–138.

Santo-Domingo, J., and N. Demaurex. 2012. The renaissance of mitochondrial pH. *J. Gen. Physiol.* 139:415–423.

Scheenen, W.J., L.R. Makings, L.R. Gross, T. Pozzan, and R.Y. Tsien. 1996. Photodegradation of indo-l and its effect on apparent Ca^{2+} concentrations. *Chem. Biol.* 3:765–774.

Schoutteten, L., P. Denjean, G. Joliff-Botrel, C. Bernard, D. Pansu, and R.B. Pansu. 1999a. Development of intracellular calcium measurement by time-resolved photon-counting fluorescence. *Photochem. Photobiol.* 70:701–709.

Schoutteten, L., P. Denjean, J. Faure, and R.B. Pansu. 1999b. Photophysics of calcium green 1 in vitro and in live cells. *Phys. Chem. Chem. Phys.* 1:2463–2469.

Schumann, H., A.V. Somlyo, and A.P. Somyo. 1976. Quantitative electron probe microanalysis of biological thin sections: Methods and validity. *Ultramicroscopy* 1:317–339.

Serresi, M., R. Bizzarri, F. Cardarelli, and F. Beltram 2009. Real-time measurement of endosomal acidification by a novel genetically encoded biosensor. *Anal. Bioanal. Chem.* 393:1123–1133.

Shortreed, M.R., S. Dourado, and R. Kopelman. 1997. Development of a fluorescent optical potassium-selective ion sensor with ratiometric response for intracellular applications. *Sens. Actuators B-Chem.* 38:8–12.

Soleimani, M., and C.E. Burnham. 2001. The Na^+/HCO_3^- co-transporters. *J. Membr. Biol.* 183:71–84.

Štefl, M., N.G. James, J.A. Ross, and D.M. Jameson. 2011. Applications of phasors to in vitro time-resolved fluorescence measurements. *Anal. Biochem.* 410:70–76.

Strickler, S.J., and R.A. Berg. 1962. Relationship between absorption intensity and fluorescence lifetime of molecules. *J. Chem. Phys.* 37:814–822.

Szmacinski, H., I. Gryczynski, and J.R. Lakowicz. 1993. Calcium-dependent fluorescence lifetimes of Indo-1 for one- and two-photon excitation of fluorescence. *Photochem. Photobiol.* 58:341–345.

Szmacinski, H., and J.R. Lakowicz. 1993. Optical measurements of pH using fluorescence lifetimes and phase-modulation fluorometry. *Anal. Chem.* 65:1668–1674.

Szmacinski, H., and J.R. Lakowicz. 1995. Possibility of simultaneously measuring low and high calcium concentrations using Fura-2 and lifetime-based sensing. *Cell Calcium* 18:64–75.

Szmacinski, H., and J.R. Lakowicz. 1997. Sodium Green as a potential probe for intracellular sodium imaging based on fluorescence lifetime. *Anal. Biochem.* 250:131–138.

Szmacinski, H., and J.R. Lakowicz. 1999. Potassium and sodium measurements at clinical concentrations using phase-modulation fluorometry. *Sens. Actuators B-Chem.* 60:8–18.

Szmacinski, H., and Q. Chang. 2000. Micro- and sub–nanosecond lifetime measurements using a UV light-emitting diode. *Appl. Spectrosc.* 54:106–109.

Tantama, M., Y.P. Hung, and G. Yellen. 2011. Imaging intracellular pH in live cells with a genetically encoded red fluorescent protein sensor. *J. Am. Chem. Soc.* 133:10034–10037.

Thomas, D., S.C. Tovey, T.J. Collins, M.D. Bootman, M.J. Berridge, and P. Lipp. 2000. A comparison of fluorescent Ca^{2+} indicator properties and their use in measuring elementary and global Ca^{2+} signals. *Cell Calcium* 28:213–223.

Tian, L., S.A. Hires, T. Mao, D. Huber, M.E. Chiappe, S.H. Chalasani et al. 2009. Imaging neural activity in worms, flies and mice with improved GCaMP calcium indicators. *Nat. Meth.* 6:875–881.

Tsien, R.Y. 1980. New calcium indicators and buffers with high selectivity against magnesium and protons: Design, synthesis, and properties of prototype structures. *Biochemistry* 19:2396–2404.

Tsien, R.Y. 1981. A non-disruptive technique for loading calcium buffers and indicators into cells. *Nature* 290:527–528.

Tsien, R.Y., T. Pozzan, and T.J. Rink. 1982a. T-cell mitogens cause early changes in cytoplasmic free Ca^{2+} and membrane potential in lymphocytes. *Nature* 295:68–71.

Tsien, R.Y., T. Pozzan, and T.J. Rink. 1982b. Calcium homeostasis in intact lymphocytes: Cytoplasmic free calcium monitored with a intracellularly trapped fluorescent indicator. *J. Cell Biol.* 94:325–334.

Tsien, R.Y. 1983. Intracellular measurements of ion activities. *Annu. Rev. Biophys. Biol.* 12:91–116.

Tsien, R.Y. 1988. Fluorescence measurement and photochemical manipulation of cytosolic free calcium. *Trends Neurosci.* 11:419–424.

Tsien, R.Y. 1989a. Fluorescent probes of cell signaling. *Annu. Rev. Neurosci.* 12:227–253.

Tsien, R.Y. 1989b. Fluorescent indicators of ion concentrations. *Method Cell Biol.* 30:127–156.

Tsien, R.Y. 1998. The green fluorescent protein. *Annu. Rev. Biochem.* 67:509–544.

Tsien, R.Y. 2009. Constructing and exploiting the fluorescent protein paintbox (Nobel Lecture). *Angew. Chem. Int. Ed.* 48:5612–5626.

Valeur, B. 2002. Principles of fluorescent probe design for ion recognition. In J.R. Lakowicz (ed.), *Topics in Fluorescence Spectroscopy, Volume 4: Probe Design and Chemical Sensing*, pp. 21–46. New York: Plenum.

Valeur, B., and I. Leray. 2000. Design principles of fluorescent molecular sensors for cation recognition. *Coord. Chem. Rev.* 205:3–40.

Verkman, A.S., M.C. Sellers, A.C. Chao, T. Leung, and R. Ketcham. 1989. Synthesis and characterization of improved chloride-sensitive fluorescent indicators for biological applications. *Anal. Biochem.* 178:355–361.

Verkman, A.S. 1990. Development and biological applications of chloride-sensitive fluorescent indicators for biological applications. *Am. J. Physiol.* 259(3 Pt 1):C375–C388.

Verkman, A.S. 2009. Chemical and GFP-based fluorescent chloride indicators. In F.J. Alvarez-Leefmans, and E. Delpire (eds), *Physiology and Pathology of Chloride Transporters and Channels in the Nervous.* San Diego, CA: Academic Press, pp. 111–123.

Villoing, A., M. Ridhoir, B. Cinquin, M. Erard, L. Alvarez, G. Vallverdu, P. Pernot, R. Grailhe, F. Merola, and H. Pasquier. 2008. Complex fluorescence of the cyan fluorescent protein: Comparisons with the H148D variant and consequences for quantitative cell imaging. *Biochemistry* 47:12483–12492.

Vinkenborg, J.L., T.J. Nicolson, E.A. Bellomo, M.S. Koay, G.A. Rutter, and M. Merkx. 2009. Genetically encoded FRET sensors to monitor intracellular Zn^{2+} homeostasis. *Nat. Methods* 6:737–740.

Wachter, R.M., M.A. Elsliger, K. Kallio, G.T. Hanson, and S.J. Remington. 1998. Structural basis of spectral shifts in the yellow-emission variants of green fluorescent protein. *Structure* 6:1267–1277.

Wachter, R.M., and S.J. Remington. 1999. Sensitivity of the yellow variant of green fluorescent protein to halides and nitrate. *Curr. Biol.* 9:R628–R629.

Ward, W.W. 1981. Properties of coelenterate green-fluorescent proteins. In M. DeLuca, and D.W. McElroy (eds) *Bioluminescence and Chemiluminescence: Basic Chemistry and Analytical Applications.* New York: Academic Press, pp. 235–242.

Waseem, T., M. Mukhtarov, S. Buldakova, I. Medina, and P. Bregestovski. 2010. Genetically encoded Cl-Sensor as a tool for monitoring of Cl^- dependent processes in small neuronal compartments. *J. Neurosci. Methods* 193:14–23.

Webb, B.A., M. Chimenti, M.P. Jacobson, and D.L. Barber. 2011. Dysregulated pH: A perfect storm for cancer progression. *Nat. Rev. Cancer* 11:671–677.

Weber, G. 1981. Resolution of the fluorescence lifetimes in a heterogeneous system by phase and modulation measurements. *J. Phys. Chem.* 85:949–953.

Wendt-Gallitelli, M.F., and G. Isenberg. 1989. X-ray microanalysis of single cardiac myocytes frozen under voltage-clamp conditions. *Am J. Physiol.* 256: H574–H583.

Wilms, C.D., H. Schmidt, and J. Eilers. 2006. Quantitative two-photon Ca^{2+} imaging via fluorescence lifetime analysis. *Cell Calcium* 40:73–79.

Wilms, C.D., and J. Eilers. 2007. Photophysical properties of Ca^{2+}-indicator dyes suitable for two-photon fluorescence-lifetime recording. *J. Microsc.* 225:209–213.

Wolfbeis, O.S. 2004. Fiber-optic chemical sensors and biosensors. *Anal. Chem.* 76:3269–3284.

Woodruff, M.L., A.P. Sampath, H.R. Matthews, N.V. Krasnoperova, J. Lem, and G.L. Fain. 2002. Measurement of cytoplasmic calcium concentration in the rods of wild-type and transducin knock-out mice. *J. Physiol.* 542:843–854.

Wright, E.M., and E. Turk. 2004. The sodium/glucose co-transport family SLC5. *Pflug. Arch. Eur. J. Phys.* 447:510–518.

Yoshiki, K., H. Azuma, K. Yoshioka, M. Hashimoto, and T. Araki. 2005. Finding of optimal calcium ion probes for fluorescence lifetime measurement. *Opt. Rev.* 12:415–419.

Zhao, M., S. Hollingworth, and S.M. Baylor. 1996. Properties of tri- and tetracarboxylate Ca^{2+} indicators in frog skeletal muscle fibers. *Biophys. J.* 70:896–916.

Index

Page numbers followed by f and t indicate figures and tables, respectively.

Printed and bound by CPI Group (UK) Ltd, Croydon, CR0 4YY

24/10/2024

01778285-0008